Sustainability in Plant and Crop Protection

Volume 20

Series Editor

Aurelio Ciancio, Sostenibile delle Piante, CNR IPSP, Istituto per la Protezione, Bari, Italy

The series describes new perspectives recently developed in pest and disease management, based on innovative tools emerging from basic and applied research. The volumes will aim at interested readers involved in plant protection and crop management, for whom soil biodiversity, crop sustainability and, in general, organic approaches are fundamental. Different cropping systems will be treated by researchers involved in cutting edge studies worldwide. A number of basic issues including sustainability, life-cycle assessment, evolution, plant nutrition and organic cropping technologies will provide a common framework, within which different components of the crop production cycle will be focused on. These will range from roots and endophytes to pest and disease control, through the management of soil microbiome and fertility. These issues will be examined at the field and crop levels, including the effects of invasive species and climate changes on agroecosystems. Recent advancements in massive sequencing will represent the basis of dedicated volumes, dealing with transcriptomics and related approaches. They will illustrate the potentials and benefits of extensive DNA and RNA data analyses and studies, for practical purposes in crop protection, disease management and food production.

Contributions on any of the above cited topics are welcome. Potential Editors proposing a new volume are requested to contact the Series Responsible Editor and provide a short CV (400 words max) listing a selection of their most significant publications. In order to broaden the base of contributors and avoid redundancies, only one volume per Editor is allowed. Exceptionally, in case of many contributed chapters, a two-issues volume can eventually be considered.

Piyush Mathur • Swarnendu Roy
Editors

Plant Microbiome and Biological Control

Emerging trends and applications

Springer

Editors
Piyush Mathur
Department of Botany, Institute of Science
Banaras Hindu University
Varanasi, Uttar Pradesh, India

Swarnendu Roy
Department of Botany
University of North Bengal
Siliguri, West Bengal, India

ISSN 2567-9805 ISSN 2567-9821 (electronic)
Sustainability in Plant and Crop Protection
ISBN 978-3-031-75847-8 ISBN 978-3-031-75845-4 (eBook)
https://doi.org/10.1007/978-3-031-75845-4

This Springer imprint is published by the registered company Springer Nature Switzerland AG
The registered company address is: Gewerbestrasse 11, 6330 Cham, Switzerland

Preface

Plant microbiome involves the varied microbial diversity associated with plants in their different organs, such as stems, leaves, roots, fruits, etc. This plant-associated microbiome plays a pivotal role in plant growth promotion, stress alleviation, and crop improvement. Therefore, we emphasize and delineate in this volume the intricate relationships established between plants and microorganisms, essential for eco-friendly approaches to protect plants from several biotic stress factors. The first chapters of the book stress the role of beneficial microbes and the molecular mechanisms underlying plant immunity. Additionally, other chapters highlight the importance of rhizomicrobiome and phyllomicrobiome in the enhancement of sustainable crop protection approaches. The volume also provides detailed insights into the role of arbuscular mycorrhizal fungi, plant growth promoting rhizobacteria, and several other plant growth promoting fungi in safeguarding plants from pathogens. Further chapters focus on post-harvest disease management strategies in different crops using bio-priming with microbial consortia. Authors also discuss cutting-edge, multi-omic approaches to unravel the yet poorly explored potential of the plant microbiome, such as metagenomics for rhizosphere and phyllosphere study as well as the applications in disease management.

The volume caters useful information to students, researchers, and teachers working on topics in microbiology, mycology, plant-microbe interactions, plant pathology, disease control, sustainable agriculture, etc. Overall, this book represents an ultimate guide to understand the potential of plant microbiomes exploitation in sustainable agriculture, in the current world scenario.

Varanasi, Uttar Pradesh, India Piyush Mathur

Siliguri, West Bengal, India Swarnendu Roy

Contents

About the Editors

Piyush Mathur is presently working as an Assistant Professor in the Department of Botany, Institute of Science, Banaras Hindu University, Varanasi, Uttar Pradesh, India. He has been teaching for the past 6 years and has taught both at the undergraduate and postgraduate level. He completed his Ph.D. in Botany with a specialization in Mycology and Plant Pathology from Delhi University. Presently, he is supervising four Ph.D. research scholars who are working on microbial diversity, bioprospecting, and plant disease management. He has published quite a number of articles in well-known international journals of good repute. He has also contributed articles as book chapters in International and National books. He is serving as a member of editorial board of BMC Plant Biology. He has previously edited three books with Springer and is working on other books too. Dr. Mathur is an active member of the Society of Rapeseed Mustard Research (SRMR), Bharatpur, and Delhi University Botanical Society (DUBS), Delhi, India, and Phytopathological Society, New Delhi, India.

Swarnendu Roy works as an Assistant Professor at the University of North Bengal, Siliguri, India. He has more than 14 years of teaching experience. He has received the University Gold Medal for securing the first position at B.Sc. and the University Silver medal for securing the second position at M.Sc. level. He received the Post-Graduate Merit Scholarship for University Rank Holder—awarded by the Ministry of Human Resource Development (HRD) and University Grants Commission (UGC) (2006–2008), Government of India. He also received the Council of Scientific & Industrial Research (CSIR) Junior Research Fellowship, which was awarded by CSIR-UGC (2008). He is actively engaged in research topics encompassing plant stress management by engineered nanoparticles and the synthesis of starch-based biofilms and edible coatings. He has published more than 40 articles with high-impact factors in reputed journals, such as *Water Research* and *Scientific Reports*, among many others. He has recently edited two books published by Springer and has also published several book chapters in edited volumes to date. He is a life member of prestigious societies like "The Indian Science Congress Association," "Association of Food Scientists and Technologists" (India), and "Society of Plant Research" (India).

Contributors

Samarpita Adhikary Microbiology and Computational Biology Laboratory, Department of Botany, Raiganj University, Raiganj, West Bengal, India

Sk Naim Aktar Department of Agronomy, Palli Siksha Bhavana (Institute of Agriculture), Visva-Bharati, Birbhum, India

Jasima Ali Department of Plant Pathology, Punjab Agricultural University, Ludhiana, India

Elif Özlem Arslan Aydoğdu Department of Biology, Faculty of Science, Istanbul University, Istanbul, Turkey

Kakan Ball Plant Molecular Biology Laboratory, Department of Botany, Raiganj University, Raiganj, Uttar Dinajpur, West Bengal, India

Prakash Balu Department of Biotechnology, School of Life Sciences, Vels Institute of Science, Technology and Advanced Studies (VISTAS), Chennai, Tamil Nadu, India

Partha Pratim Baruah Plant Ecology Laboratory, Department of Botany, Gauhati University, Guwahati, Assam, India

Kuntal Bera Plant Molecular Biology Laboratory, Department of Botany, Raiganj University, Raiganj, West Bengal, India
Department of Seed Science and Technology, Uttar Banga Krishi Viswavidyalaya, Cooch Behar, West Bengal, India

Renu Bhardwaj Department of Botanical and Environmental Sciences, Guru Nanak Dev University, Amritsar, India

Tamanna Bhardwaj Department of Botanical and Environmental Sciences, Guru Nanak Dev University, Amritsar, India

Shaswatee Bhattacharjee Plant Ecology Laboratory, Department of Botany, Gauhati University, Guwahati, Assam, India

V. Bhuvaneswari Department of Plant Pathology, Assam Agricultural University, Jorhat, Assam, India

N. Boopathi Department of Plant Pathology, Centre for Plant Protection Studies, Tamil Nadu Agricultural University, Coimbatore, Tamil Nadu, India

Munmi Borkataky Department of Life Sciences, Dibrugarh University, Assam, India

M. Chaithra ICAR-Central Plantation Crop Research Institute, Research Center, Kahikuchi, Guwahati, Assam, India

Madhumita G. Chaki Microbiology and Computational Biology Laboratory, Department of Botany, Raiganj University, Raiganj, West Bengal, India

Trideep Chetia Plant Ecology Laboratory, Department of Botany, Gauhati University, Guwahati, Assam, India

Debjani Choudhury Department of Plant Pathology, School of Agriculture, Lovely Professional University, Phagwara, Punjab, India

Shouvik Chowdhury Department of Biotechnology, School of Life Sciences, Vels Institute of Science, Technology and Advanced Studies (VISTAS), Chennai, Tamil Nadu, India

Ankan Das Department of Horticulture, Institute of Agricultural Science, University of Calcutta, Kolkata, West Bengal, India

Pratisha Das Plant Ecology Laboratory, Department of Botany, Gauhati University, Guwahati, Assam, India

Sukanya Das Department of Botany, Gauhati University, Guwahati, Assam, India

Keshab Das Department of Botany, Balurghat Mahila Mahavidyalaya Balurghat, Dakshin Dinajpur, West Bengal, India

Rashadin Md Bodeul Ddoza Department of Agronomy, Palli Siksha Bhavana (Institute of Agriculture), Visva-Bharati, Birbhum, India

Priya Dubey Department of Bioscience, Integral University, Lucknow, Uttar Pradesh, India

Puspendu Dutta Department of Seed Science and Technology, Uttar Banga Krishi Viswavidyalaya, Cooch Behar, West Bengal, India

R. Gangaraj ICAR-Indian Institute of Vegetable Research, Regional Research Station, Sargatia, Kushinagar, Uttar Pradesh, India

Maunata Ghorui Department of Biotechnology, School of Life Sciences, Vels Institute of Science, Technology and Advanced Studies (VISTAS), Chennai, Tamil Nadu, India

Nihal Doğruöz Güngör Department of Biology, Faculty of Science, Istanbul University, Istanbul, Turkey

Md Saddam Hosen Department of Agronomy, Palli Siksha Bhavana (Institute of Agriculture), Visva-Bharati, Birbhum, India

V. Jayanth Division of Entomology, ICAR-Indian Agricultural Research Institute, New Delhi, India

Subashini Jegadeesan Department of Plant Sciences, School of Biological Sciences, Madurai Kamaraj University, Madurai, Tamil Nadu, India

Chaitanya Kumar Jha Microbiology Department, Gujarat Arts and Science College, Ahmedabad, India

Nilakhi Kakoti Department of Life Sciences, Dibrugarh University, Assam, India

Sonika Kalia Department of Biotechnology, School of Applied and Life Sciences, Uttaranchal University, Dehradun, India

Nilamjyoti Kalita Plant Ecology Laboratory, Department of Botany, Gauhati University, Guwahati, Assam, India

Nitika Kapoor Department of Botanical and Environmental Sciences, Guru Nanak Dev University, Amritsar, India

Rupinder Kaur Academy of Scientific and Innovative Research (AcSIR), Ghaziabad, India
Entomology Laboratory, Agrotechnology Division, CSIR-Institute of Himalayan Bioresource Technology (IHBT), Palampur, India

Amrit Kumar Department of Botany, Nowgong College (Autonomous), Nagaon, Assam, India

Deepak Kumar Department of Zoology, Guru Nanak Dev University, Amritsar, India

Prasann Kumar Department of Agronomy, School of Agriculture, Lovely Professional University, Phagwara, Punjab, India

Rahul Kumar Academy of Scientific and Innovative Research (AcSIR), Ghaziabad, India
Entomology Laboratory, Agrotechnology Division, CSIR-Institute of Himalayan Bioresource Technology (IHBT), Palampur, India

Shailendra Kumar Department of Microbiology, Dr. Ram Manohar Lohia Avadh University, Ayodhya, Uttar Pradesh, India

Arun Kumar Lovely Professional University, Phagwara, Punjab, India

Sagar Mal Kumawat College of Agriculture, Mandawa, Jhunjhunu (Rajasthan) under S. K. Rajasthan Agricultural University, Bikaner, India

Rishabh Maheshwari Department of Plant Pathology, Punjab Agricultural University, Ludhiana, India

Smriti Mall Molecular Plant Pathology Laboratory, Department of Botany, DDU Gorakhpur University, Gorakhpur, UP, India

Prahlad Masurkar Department of Plant Pathology, School of Agriculture, Lovely Professional University, Phagwara, Punjab, India

Debasis Mitra Department of Microbiology, Graphic Era (Deemed to be University), Dehradun, Uttarakhand, India

R. K. Mohana Pradeep Department of Plant Pathology, Centre for Plant Protection Studies, Tamil Nadu Agricultural University, Coimbatore, Tamil Nadu, India

Banabasini Mohanty Department of Agronomy, Palli Siksha Bhavana (Institute of Agriculture), Visva-Bharati, Birbhum, India

Kalaivani K. Nadarajah Department of Biological Sciences and Biotechnology, Faculty of Science and Technology, Universiti Kebangsaan Malaysia, Bangi, Selangor, Malaysia

Gireesh Nadda Academy of Scientific and Innovative Research (AcSIR), Ghaziabad, India
Entomology Laboratory, Agrotechnology Division, CSIR-Institute of Himalayan Bioresource Technology (IHBT), Palampur, India

Ratul Nath Department of Life Sciences, Dibrugarh University, Dibrugarh, Assam, India

Shanvi Nautiyal Department of Biotechnology, School of Applied and Life Sciences, Uttaranchal University, Dehradun, India

K. Nishmitha Division of Plant Pathology, ICAR-Indian Agricultural Research Institute, New Delhi, India

Puja Ohri Department of Zoology, Guru Nanak Dev University, Amritsar, India

Ayon Pal Microbiology and Computational Biology Laboratory, Department of Botany, Raiganj University, Raiganj, West Bengal, India

Parkirti Department of Zoology, Guru Nanak Dev University, Amritsar, India

Pooja S. Patel Department of Plant Pathology, College of Agriculture, V. C. Farm, Mandya, Karnataka, India

Kalipada Pramanik Department of Agronomy, Palli Siksha Bhavana (Institute of Agriculture), Visva-Bharati, Birbhum, India

Aniruddh Rabari Department of Microbiology and Biotechnology, University School of Science, Gujarat University, Ahmedabad, India

Jintu Rabha Department of Botany, Gauhati University, Guwahati, Assam, India

Nahdhoit Ahamada Rachid Institute of Graduate Studies in Sciences, Istanbul University, Istanbul, Turkey

Aditya Singh Ranout Academy of Scientific and Innovative Research (AcSIR), Ghaziabad, India
Entomology Laboratory, Agrotechnology Division, CSIR-Institute of Himalayan Bioresource Technology (IHBT), Palampur, India

S. M. Reddy Department of Botany, Kakatiya University, Warangal, Telangana, India

Vinay Rojaria Department of Genetics and Plant Breeding, BRD PG College, Deoria, India

Janki Ruparelia Department of Microbiology and Biotechnology, University School of Science, Gujarat University, Ahmedabad, India
Microbiology Department, Gujarat Arts and Science College, Ahmedabad, India

Sanjoy Sadhukhan Plant Molecular Biology Laboratory, Department of Botany, Raiganj University, Raiganj, Uttar Dinajpur, West Bengal, India

Debajyoti Saha Department of Botany, Siksha Bhavana, Visva-Bharati, Santiniketan, Birbhum, West Bengal, India

Nabanita Sarkar Department of Botany, Ballygunge Science College, University of Calcutta, Kolkata, West Bengal, India

R. Z. Sayyed Department of Microbiology, PSGVP Mandal's S. I. Patil Arts, G. B. Patel Science and STKV Sangh Commerce College, Shahada, India

Maksud Hasan Shah Department of Agronomy, Bidhan Chandra Krishi Viswavidyalaya (BCKV), Nadia, India

N. K. Sharma College of Agriculture, Chandgothi, Jhunjhunu (Rajasthan) under S. K. Rajasthan Agricultural University, Bikaner, India

Supriya Sharma Department of Plant Pathology, Assam Agricultural University, Jorhat, Assam, India

Amandeep Singh Department of Botanical and Environmental Sciences, Guru Nanak Dev University, Amritsar, India

Balwant Singh Department of Botany, K. S. Saket PG College Ayodhya, Dr. Ram Manohar Lohia Avadh University, Ayodhya, Uttar Pradesh, India

Hausila Prasad Singh Department of Genetics and Plant Breeding, BRD PG College, Deoria, India

Laishram Shantikumar Singh Department of Microbiology, Assam Down Town University, Guwahati, Assam, India

Swati Singh Department of Phytochemistry, CSIR- Central Institute of Medicinal and Aromatic Plants, Lucknow, India

Jatinder Singh Lovely Professional University, Phagwara, Punjab, India

Kiran Sunar Department of Botany, Balurghat Mahila Mahavidyalaya Balurghat, Dakshin Dinajpur, West Bengal, India

Department of Botany, Satavahana University, Karimnagar, Telangana, India

M. Surekha Department of Botany, Kakatiya University, Warangal, Telangana, India

Jebasingh Tennyson Department of Plant Sciences, School of Biological Sciences, Madurai Kamaraj University, Madurai, Tamil Nadu, India

Abhishek Tripathi Department of Biotechnology, Institute of Engineering and Technology, Abdul Kalam Technical University, Lucknow, Uttar Pradesh, India

Shivangi Tripathi Department of Bioscience, Integral University, Lucknow, Uttar Pradesh, India

Santvana Tyagi Department of Molecular Biology and Genetic Engineering, G. B. Pant University of Agriculture and Technology, Pantnagar, Uttarakhand, India

Shanmugam Velmurugan Division of Plant Pathology, ICAR- Indian Agricultural Research Institute, New Delhi, India

Vikanksha Lovely Professional University, Phagwara, Punjab, India

Vikram Department of Zoology, Guru Nanak Dev University, Amritsar, India

Pratima Vishwakarma Department of Botany, St. Joseph's College for Women, Gorakhpur, UP, India

Ragini Vishwakarma Molecular Plant Pathology Laboratory, Department of Botany, DDU Gorakhpur University, Gorakhpur, UP, India

List of Figures

List of Tables

Chapter 1
Beneficial Microbes in Plant Health, Immunity and Resistance

Sk Naim Aktar, Maksud Hasan Shah, Banabasini Mohanty, Kalipada Pramanik, Md Saddam Hosen, Rashadin Md Bodeul Ddoza, Sagar Mal Kumawat, N. K. Sharma, and R. Z. Sayyed

Abstract Beneficial microorganisms play an important role in plant health and growth. They induce the host to improve resistance against a wide range of pathogens, a process known as induced systemic resistance (ISR). A recent advancement in plant immunity engineering technology is the setup of immunity inducers. In this review, we discuss about the role of beneficial microbes in plant health immunity, application and discuss actual prospects of immunity inducers. We also examine the role of beneficial microbes in plant disease resistance through early ISR events, reactive oxygen species, callose deposition, Ca^{2+} influx, defense-related genes and resistant mechanisms.

Keywords Beneficial microbes · Ca^{2+} influx · Callose deposition · Defense response · Induced systemic resistance · ROS

S. N. Aktar · B. Mohanty · K. Pramanik · Md. S. Hosen · R. Md. B. Ddoza
Department of Agronomy, Palli Siksha Bhavana (Institute of Agriculture), Visva-Bharati, Birbhum, India
e-mail: kalipada.pramanik@visva-bharati.ac.in

M. H. Shah (✉)
Department of Agronomy, Bidhan Chandra Krishi Viswavidyalaya (BCKV), Nadia, India

S. M. Kumawat
College of Agriculture, Mandawa, Jhunjhunu (Rajasthan) under S. K. Rajasthan Agricultural University, Bikaner, India

N. K. Sharma
College of Agriculture, Chandgothi, Jhunjhunu (Rajasthan) under S. K. Rajasthan Agricultural University, Bikaner, India

R. Z. Sayyed
Department of Microbiology, PSGVP Mandal's S. I. Patil Arts, G. B. Patel Science and STKV Sangh Commerce College, Shahada, India

P. Mathur, S. Roy (eds.), *Plant Microbiome and Biological Control*, Sustainability in Plant and Crop Protection 20,
https://doi.org/10.1007/978-3-031-75845-4_1

Abbreviations

ETI	Effector triggered immunity
HR	Hypersensitive reaction.
PAMPs/MAMPs	Pathogen- or microbial-associated molecular patterns.
PGPRs	Plant growth promoting rhizobacteria.
PRRs	Pattern recognition receptors.
PTI	Pattern-triggered immunity.
ISR	Induced systemic resistance.
PR	Pathogenesis Related Proteins.
SAR	Systemic acquired resistance.

1.1 Introduction

The world growing population requires a significant improvement of agricultural productions. The quest for productivity often resulted in the misuse of pesticides and fertilizers, seriously harming the ecosystem. Among crop production factors, various microorganisms including bacteria, fungi, protozoa, actinomycetes, play a key role by forming complex food webs in soil. Fertility is hence not only dependent upon chemical and physical traits of soil, but also on the biology and activity of its inhabiting microorganisms (Saranraj et al., 2022).

Over time, the agricultural sector top concerns switched from productivity to food quality and environmental effects. Since then, there has been a rise in the search and application of environmentally friendly agricultural inputs. Various types of microbial interactions involve plants, and are an essential segment of the agro-ecosystem. Beneficial organisms are used in biological control to stifle undesirable species and encourage plant growth. Species of well-known genera of bacteria and fungi such as *Bacillus*, *Pseudomonas,* or *Trichoderma* are just a few of the beneficial microbes that are presently employed as biocontrol agents, to address plant diseases and pests in cultivated fields. Plants have the intrinsic capacity to detect and identify potentially invasive microbes and to launch defense mechanisms (Pandey et al., 2017). On the contrary, they evolved similar but separate cell surface receptors to recognize the advantageous microbes and establish a symbiotic connection with them (Zipfel & Oldroyd, 2017). Transmembrane pattern recognition receptors (PRRs), such as bacterial flagellin and fungal chitin, enable plants to identify pathogen- or microbial-associated molecular patterns (PAMPs or MAMPs). This recognition triggers pattern-triggered immunity (PTI), the first step of immune defense (Bigeard et al., 2015). Pathogens can, however, bypass the initial line of defense by subduing PTI signals or dodging the diagnosis of PRRs by secreting virulence effectors. Pathogens infiltrate plant cells or the apoplast through effectors, a group of virulence-associated chemicals, and locally weaken host defense (Guo et al., 2009). In order to identify pathogen effectors using polymorphisms in proteins containing nucleotide-binding and leucine-rich repeat domains (NB-LRR), the level-II of plant immunity, plants established an alias effector triggered immunity (ETI). This results in a hypersensitive reaction (HR), which is a second layer of plant immunity.

Increasing crop resistance through plant immunity inducers appears as a sound strategy for disease management. Rhizosphere immunity may be the foremost goal for the integrated management of plant immunity. Curiously, subsequent research revealed that PRRs are necessary for ETI as well (Yuan et al., 2021). Targeting the essential components of PTI and ETI by manipulating host short RNAs enables beneficial bacteria to induce plant immunity through the explicit and complex immune system developed through the host-pathogen conflict. Induced systemic resistance (ISR) and Systemic acquired resistance (SAR), which are produced by pathogenic and non-pathogenic bacteria, respectively, can be used to categorize plant resistance (Desai et al., 2022). Beneficial microbial colonisation causes the plant host to enter a physiological state known as "priming." As a common characteristic of systemic resistance established by helpful microbes, plants exhibit stronger and faster defense responses upon activation of "priming" against the ensuing invasion of pathogens (Conrath et al., 2006). Plant growth promoting rhizobacteria (PGPRs) also produce elicitors that get the plant immune system ready by encouraging the activation its disease-related genes (Enebe & Babalola, 2019). The arbuscular mycorrhizal fungi (AMF) also release a large number of stress- and defense-related gene products (Saboor et al., 2021). We herein review the recognition of advantageous microorganisms and the early occurrences of ISR, with an emphasis on callose deposition, reactive oxygen species burst, that can prevent pathogen infection and spread, Ca^{++} signaling, and transcription factors that are crucial for regulating the expression of downstream genes involved in defense and disease prevention.

1.2 Role of Beneficial Microbes in Plant Health Immunity

Due to higher and considerable pace of their evolutionary processes, pathogens can readily overcome host resistance. Alternative methods to restore the lost plant immune activities through genetic engineering or by applying large amounts of pesticides are problematic, because they lack a legal framework, public support, and may have direct harmful effects on genetic loads and environmental health. As a result, it became necessary to improve output without substantially relying on the usage of such approaches, and mostly of pesticides. Therefore, new crop protection measures are required to ensure the long-term sustainability of agricultural outputs. One strategy to enhance plant disease management is to abandon the reductionist perspective. As opposed to a more comprehensive framework where plant immunity is thought to arise from the interactions with plant-associated microbes and environmental factors, plant health is here examined by focusing on specific components in isolation. The enhancement of crop resistance to diseases using plant immunity

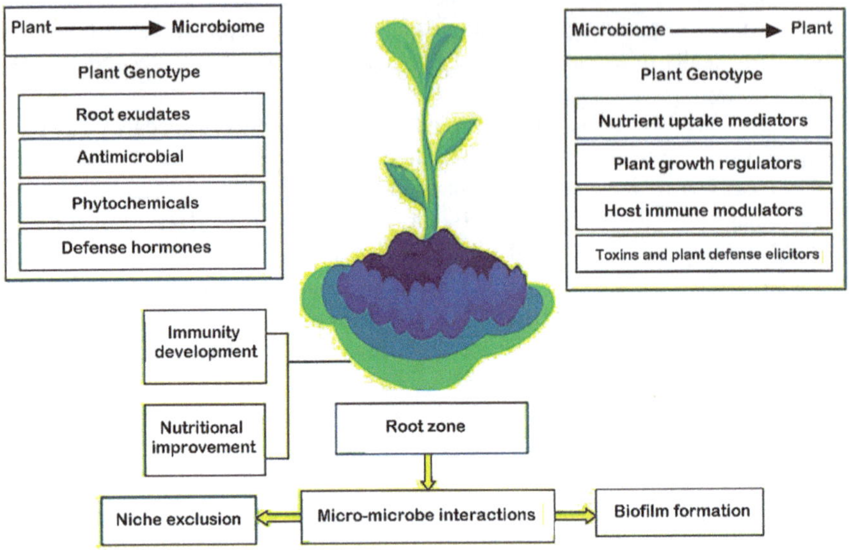

Fig. 1.1 Beneficial features of microbes-plant interactions for development of plant immunity

inducers is an environmentally sound strategy for management. Rhizosphere immunity build up is the main goal in the search for the conditions leading to an integrated management of plant immunity (Khan et al., 2021) (Fig. 1.1).

1.2.1 Applications of Plant Immunity Inducers

In recent years, research has been carried out to obtain plant immunity inducers based on recognized elicitors. One famous instance is the harpin protein that Cornell University and Eden Biotechnology identified from the plant pathogen *Erwinia amylovora*. The harpin protein, which has been converted into the messenger protein biological pesticide, has a special immune-activating role. This messenger was developed into a well-known immunity-inducing agent with protective action against a variety of plant diseases after years of use (Swarnalatha et al., 2022).

The study of oligosaccharides that induce plant immunity has seen significant advancements by several research teams, worldwide. For instance, researchers of the Dalian Institute of Chemical Physics of the Chinese Academy of Sciences made significant strides in discovering oligosaccharide uses. Particularly, chitosan oligosaccharides have been employed as starting materials to create oligosaccharides that promote plant immunity. These oligosaccharides have obtained several permits for pesticide application and have significantly improved agricultural production and quality. Additionally, scientists of Guizhou University have produced a number of synthetic chemicals that trigger plant immunity, such dufulin. In Japan, isotianil, a synthetic plant immunity inducer based on isothiazole, has been extensively utilized to manage rice diseases.

1.2.2 Promising Strategies Utilizing Plant Immunity Inducers

The discovery of plant immunity inducers from microbes or plants poses a significant hurdle for their widespread use. A further successful tactic is the ectopic expression of the genes in plants that code for elicitors. For instance, numerous transgenic plants (such as soybean, rice, tobacco, and sugar beet) achieve increased resistance to diseases by overexpressing the genes that code for harpin. Additionally to transgenic plants, modified bacteria are frequently employed to create plant immunity inducers. For instance, harpin protein was commercially generated in a strain of *Escherichia coli* that was genetically modified. Moreover, the costly protein purification process requires the target proteins to be freed from bacterial cells using sonication or high pressure. Ineffective gene expression, development of protein inclusion bodies, and the generation of bacterial toxin are additional issues that arise at the time of synthesis and purification of proteins in *E. coli*. Therefore, considering the downstream output process, microorganisms that secrete proteins directly into the fermentation broth offer significant benefits.

Excellent bacteria for use in the generation of secreted proteins include a number of *Bacillus* species. One of the most well-known Gram-positive bacteria, *Bacillus subtilis*, has been employed extensively due to its high product yields, total lack of hazardous byproducts and superior fermentation capabilities (Nithyapriya et al., 2021). For instance, the harpin protein from *X. oryzae* pv. *oryzicola*, which is a component of the HpaGXooc protein, is expressed by *B. subtilis* and induces the hypersensitive response while promoting tobacco plant development. The US Food and Drug Administration has also designated *B. subtilis* as "generally regarded as safe," and a number of *B. subtilis* strains encourage plant development and boost plant resistance. Application of the plant Pep elicitor to increase crop resistance to plant-parasitic nematodes and the use of *B. subtilis* for effective delivery of these elicitors have recently been presented as a new technique to manage plant diseases. This tactic will open up new opportunities for the simultaneous use of immunity-inducing agents and advantageous microorganisms to safeguard plant health.

1.2.3 Prospects for the Development of Plant Immunity Inducers

A recent advancement in plant immunity engineering technology is the production of inducers of immunity. This rapidly growing discipline, which supports the production of bio-pesticides and allows scientists in theory to rely on plant immunity to manage pests and diseases, is anticipated to emerge as a new key sector with a broad growth potential. Plant immunity inducers have a number of benefits over chemical pesticides, including their exceptionally low dosage, lack of harmful effects on people, animals, or the environment, capacity to significantly increase disease resistance, induction of long-lasting and wide-ranging plant resistance, low risk of microbial resistance, and capacity to decrease the use of chemical pesticides — a reduction that can help with environmental protection attempts. The discovery, production, and use of plant immunity inducers can contribute significantly to the development of sustainable agriculture and the strategic needs of national development. They have significant research and development potential and value, and can be used to support agricultural productivity.

1.3 Immunity in Roots

1.3.1 Immune Signaling

The roots are home to a wide range of microorganisms. Known as the second genome of the plant, the complex microbial communities connected to roots are thought to be essential to their health and growth. The rhizosphere, an underground region, may be considered as an immune system nerve center for plants. Here, a complex network made of soil, plants, and microbes interact together to balance the spread of pathogens and pests. Our knowledge of plant immunological signaling has been improved as a result of research on the interactions between microbes and aboveground plant tissues. However, upon PRR-mediated MAMP identification, plant roots may generate considerable immune responses through callose deposition, camalexin synthesis, and activation of defense-related genes. It is intriguing to observe how closely pathogen immunogenic MAMPs mimic non-pathogenic, beneficial bacteria (Stringlis et al., 2018). When helpful bacteria first come into touch with plant roots, plant PRRs identify them and start immunological signaling. It was shown that the root immune system is activated in a number of root-microbe relationships. For instance, early in the course of infection, *Bradyrhizobium japonicum* greatly increases the expression of genes connected to defense in soybean root hair cells (Libault et al., 2010). The AMF *Glomus versiforme* also releases a large number of stress- and defense-related genes when first coming into contact with *Medicago truncatula* (Hamidian et al., 2023). Parallel to this, the cellular components of *Pseudomonas simiae* WCS417 and *Pseudomonas capeferrum* WCS358 cause immunological reactions in tobacco cells and *Arabidopsis* roots, resulting in the production of ROS, the expression of MAMP-responsive genes, and the formation of callus (Stringlis et al., 2018).

Another indication that these PGPRs may be recognized by plant PRRs was the significantly decreased colonization of the roots of an *Arabidopsis* mutant pub22/23/24 sensitive to MAMP (Jacobs et al., 2011). These studies demonstrate that beneficial microorganisms can cause responses as well. This stimulation appears to be restricted to the first stages of these beneficial relationships, which raises the possibility that beneficial bacteria are actively compromising their own immunity. Pathogens can now produce and release effector proteins that prevent the immune system from responding to MAMPs or their detection. As a result, they and their host plants are now competing for resources (Anderson et al., 2010). Despite the fact that plants may step up their immune defense in a response to illness, they do not possess immunological memory, which would enable their immune systems to specifically recognize pathogens they have already met.

1.4 Role of Microorganism in Phytohormone Production

Microorganisms produce the plant hormone auxin, also known as indole-3-acetic acid or indole acetic acid (IAA). Auxins may be produced and secreted by 80% of the bacteria that are present around various crops, according to various reports (Table 1.1).

1.5 Role of Beneficial Microbes in Plant Disease Resistance

-Beneficial bacteria promote the growth and health of nearby plants, causing the host to develop improved resistance, known as induced systemic resistance (ISR), *vs* a variety of diseases. Plants also use long-distance systemic communication when ISR is activated to protect distal tissue against pathogen incursions, by triggering quick and potent immune responses (Fig. 1.2). It was widely believed that the transmission of ISR signaling was dependent on ethylene and jasmonic acid but not salicylic acid. Systemic induction of resistance favors a microbial defense response (Karimian et al., 2023; Rafique et al., 2023).

1.5.1 Early ISR Events Triggered by Beneficial Microbes

Host plant defense mechanisms may be activated by helpful microbes providing plants with a range of disease resistance. ISR has shown that some bacteria, including *Bacillus amyloliquefaciens*, *B. atrophaeus*, *B. cereus*, and *Pseudomonas fluorescens*, are efficient against fungal, bacterial, and viral invasion (Table 1.1). Early plant ISR events are thought to be caused by beneficial microbes, such as elevated expression of PR genes linked to pathogenesis, elevated activity of defense-related molecules such as 1, 3-glucanase, polyphenol oxidase, peroxidase, and chitinase, and accumulation of reactive oxygen species (ROS) (Table 1.2) (Guo et al., 2019; Wang et al., 2020).

Table 1.1 Production of plant growth regulators by various microorganisms and their effect on plant growth and development

Microorganism	Observed effects on plants
Auxin production	
Azospirillum, Gluconacetobacter, Herbaspirillum	The amount of free active IAA and IBA increased in infected corn seedlings.
Azospirillum, Rhizobium, Bradyrhizobium	Reduction in root length and growth of root hair
Pseudomonas syringae pv. *savastanoi*	Gall and tumor development induction
P. denitrificans, P. rathonis	In pot trials, it was discovered that every bacterial strain accelerated the development of wheat and maize plants.
Klebsiella sp.	Observed increase in root surface and branching
Agrobacterium tumefaciens	Increased vulnerability of the host plant
Erwinia herbicola pv. *gypsophilae*	Promoted growth of galls on *Gypsophila paniculata*
Rhizobacterial isolates	Wheat and rice plants significantly benefited from the growth-promoting effects of rhizobacterial isolate injection.
Indole-3-acetic acid	
Rhizobacteria	In modified jar tests, a significant association was found between auxin production by PGPR *in vitro* and growth enhancement of infected rapeseed seedlings
Pseudomonas sp.	Wheat grain nutritional value increased with the bio-inoculation of diacetylphloroglucinol-producing PGPR and AMF together
Pisolithus tinctorius	Plant growth promotion
Pseudomonas tolaasii ACC23	PGPR strains protected canola plants from antagonistic effects of cadmium
Azotobacter sp., *Pseudomonas* sp.	Sesbania and mung bean crops grew less when tryptophane levels increased from 1 mg to 5 mg ml^{-1}
Rhizobium leguminosarum	Significant growth-promoting impacts of *R. Leguminosarum* priming in rice seedlings
Bacillus sp.	Isolate SVPR 30 was effective in rising the root and shoot sections of rice plants
Mesorhizobium loti MP6	Compared to control, seeds coated with *M. Loti* MP6 improved germination, early vegetative development, and grain production.
Bacillus cereus RC 18, *Bacillus licheniformis* RC08	All strains of bacteria produced indole acetic acid (IAA) and effectively and considerably accelerated the development of wheat and spinach
Kluyvera ascorbata SUD 165	Decreased ability of some heavy metals (nickel, lead, zinc) to impede plant growth
Azospirillum brasilense	Compared to the uninoculated control, all bacterial strains enhanced rice grain yields
Azotobacter sp.	Inoculating maize seedlings with a strain that produces IAA showed a significant growth-promoting impact

(continued)

Table 1.1 (continued)

Microorganism	Observed effects on plants
Gibberellins production	
Gibberella fujikuroi	"Bakanae" effect in maize, rice, and other plants
Azospirillum brasilense	Promotion of shoot elongation, growth, and root hair density
A. brasilense, A. lipoferum	Reversion of dwarfism in maize and rice
Cytokinins production	
Erwinia herbicola	Induced gall formation
Rhizobium leguminosarum, Azospirillum	Plant growth promotion
P. syringae pv. *savastanoi*	Induction of gall and tumor formation

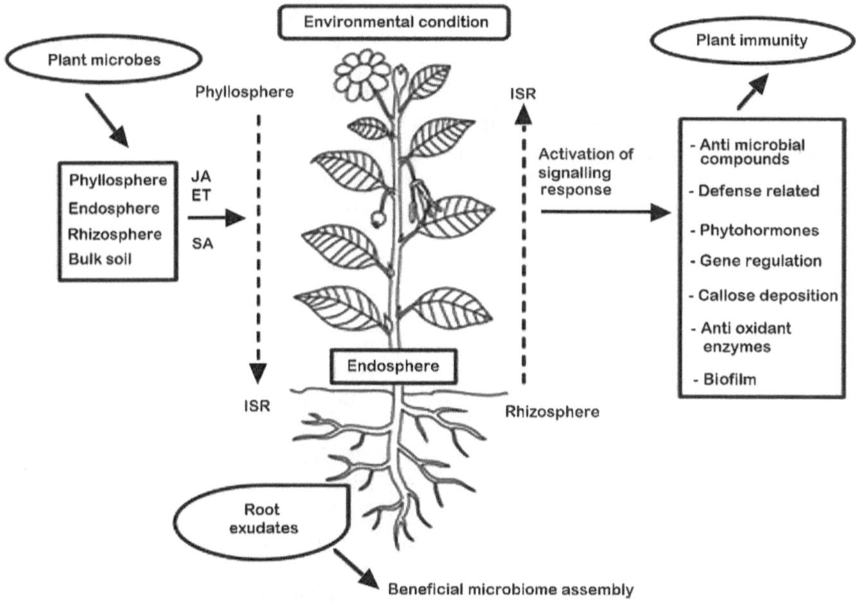

Fig. 1.2 Plant microbiome mechanisms conferring disease resistance in plants

1.5.2 Reactive Oxygen Species

Plants generate several ROS under biotic or abiotic stress, including superoxide anion (O_2), hydroxyl radical (OH), hydrogen peroxide (H_2O_2), and others. Numerous activities, including stomatal closure, programmed cell death, and pathogen defense are regulated by the generation of ROS. Respiratory burst oxidase homologs D (RBOHD)-mediated ROS are produced instantaneously, accurately, and strongly when *Arabidopsis* recognizes MAMPs. The primary mechanisms by which Ca^{2+} regulates RBOHD, a plant NADPH oxidase, are direct binding to EF hand motifs and phosphorylation by Ca^{2+}-dependent protein kinases (Wang et al., 2018). However, the buildup of ROS also causes tissue cells to deteriorate. Therefore,

Table 1.2 Resistance mechanism induced by beneficial microbes

Strains	Pathogens/pests	Diseases	Main resistance mechanisms
Bacillus amyloliquefaciens Ba13	Tomato yellow leaf curl virus	Tomato yellow leaf curl virus disease	Improved beta-1,3glucanase (inhibition of mycelial development and spore germination), phenylalanine ammonia lyase (plant growth and stress tolerance), and chitinase activity all result from the PR1, PR2, and PR3 genes (inhibition of mycelial growth)
Bacillus amyloliquefaciens FZB42	*Phytophthora nicotianae*, *Rhizoctonia solani*	Leaf disease, bottom rot	ABA/SA (FZB42-induced stomatal closure); stomatal closure (reduced pathogen invasion); secondary metabolites—surfactin, fengycin, and bacillomycin D; toxin-related genes PR-la, LOX, and ERF1 (defence effects); direct antagonistic effect and induction of defense-related genes (Paliwal et al., 2023).
Bacillus atrophaeus GBSC56	*Meloidogyne incognita*	Root-knot nematode	Volatiles include 2-undecanone, methyl isovalerate, and dimethyl disulfide (regulation of antioxidant enzymes, protection from oxidative stress and *M. incognita*)
Bacillus cereus AR156	*Botrytis cinerea*, *Cochliobolus heterostrophus*	Foliar and soil diseases	Dimethyl disulfide is a volatile metabolite (induction of ISR).
Bacillus megaterium DE BABY TRS-4	*Fomes lamaoensis*	Brown root rot	Peroxidase, chitinase, beta-1,3-glucanase, and phenyl alanine ammonia lyase are active enzymes that control plant growth and stress tolerance. Phosphate solubilization is improved, and IAA is produced more readily, which promotes plant growth.
Bacillus subtilis OTPB1	*Alternaria solani*, *Phytophthora infestans.*	Early and late blight	Enzymes involved in defense include peroxidase, polyphenol oxidase, and superoxide dismutase (inhibition of the mycelial growth and spore germination, and protection from oxidative stress, Ilyas et al., 2022)
Bacillus subtilis UMAF6639	*Podosphaera fusca*	Cucurbit powdery mildew	ROS (inhibiting mycelial growth and spore germination), surfactin lipopeptide (metabolite), and cell wall reinforcement reducing pathogen penetration (immune response stimulation).

(continued)

Table 1.2 (continued)

Strains	Pathogens/pests	Diseases	Main resistance mechanisms
Paenibacillus alvei K165	*Verticillium dahliae*		PR-1, PR2, and PR-5 genes (antimicrobial effects, beta-1,3 glucanase, and chitinase activities, markers for SA-mediated activation of SAR).
Pseudomonas aeruginosa 7NSK2	*Magnaporthe grisea, Rhizoctonia solani, B. cinerea*	Rice blast and sheath blight	Mycelial development and spore germination are inhibited by metabolites phenazine pyocyanin and pyochelin, ROS, and SA (expression of acquired resistance).
Pseudomonas fluorescens PTA-CT2	*Plasmopara viticola, B. cinerea*	Downy mildew and gray mold diseases	Defense pathways SA, JA, and ABA are activated, with HR (reduction in pathogen invasion).
Streptomyces lydicus M01	*Alternaria alternata*	Foliar disease	ROS-inhibition of mycelium development and spore germination (Omar et al., 2022)
Streptomyces pactum	Tomato yellow leaf curl virus	Tomato yellow leaf curl, virus disease	ROS (preventing mycelium growth and spore germination); enzyme activity, such as peroxidase, chitinase, beta-1,3-glucanase, and phenylalanine ammonia lyase (regulating plant growth and stress tolerance); defense-related genes PR-1, PR2, and PR-5 (antimicrobial effects), beta-1,3 glucanase, and chitin (induction of immune response and reduction in pathogen invasion).
Acrophialophora jodhpurensis	*Rhizoctonia solani* AG4-HG II	Tomato root and crown rot	Iron limitation, direct antagonistic effect, ROS (inhibiting mycelium development and spore germination), enzyme activity, such as that of peroxidase, chitinase, beta-1,3-glucanase, and phenylalanine ammonia lyase regulating plant growth and stress tolerance (inhibition of pathogen growth and promotion of plant growth).
Serendipita vermifera	*Bipolaris sorokiniana*		ROS; hydrolytic enzyme activity; suppression of mycelial development and spore germination (activation of defense).
Trichoderma atroviride	*B. cinerea*		Glutamate: WRKY transcription factors; glyoxylate aminotransferase (GGAT1) (stimulates plant development and induces systemic resistance in plants) (active defense response to biotics and abiotic stresses).

(continued)

Table 1.2 (continued)

Strains	Pathogens/pests	Diseases	Main resistance mechanisms
Trichoderma harzianum	*Bipolaris sorokiniana, R. solani*	Spot blotch, wilt	Phenylpropanoid activity (increased suberization and lignification of plant cells, decreased breakdown of cell walls and tissue disintegration), secondary metabolite Harzianic acid (inducing the expression of several genes involved in defence response).
Trichoderma longibrachiatum MK1	*B.s cinerea, A. alternata, Pythium ultimum,* and *R. solani*		Type II hydrophobins have direct antifungal, molecular patterns related with microbes, and plant growth promotion (PGP) properties (Rai et al., 2023)
Trichoderma harzianum	*A. solani, P. infestans*	Early and late blight	Peroxidase, polyphenol oxidase, and superoxide dismutase are defense-related enzymes that restrict mycelial development, spore germination, and provide protection from oxidative stress.

effective ROS scavenging requires both enzymatic and nonenzymatic techniques. Peroxidase (POX), polyphenol oxidase (PPO), superoxide dismutase (SOD), ascorbate peroxidase (APX), glutathione peroxidase (GPX), and catalase (CAT) are crucial to the defense against ROS by converting superoxide to H_2O during the enzymatic ROS scavenging activities in plants.

By encouraging the generation of hydrogen peroxide and callose in plants as well as the activation of POD and SOD enzymes, mostly through SA and MAPK signaling pathways, *B. cereus* AR156 promotes plant defense response. By raising H_2O_2 in both local and distant leaves, the pyocyanin produced by *Pseudomonas aeruginosa* 7NSK2 confers resistance to blast disease (*Magnaporthe grisea*), but not sheath blight (*Rhizoctonia solani*). Reactive oxygen species can affect disease resistance in both negative and positive ways, as demonstrated by the ability of the antioxidant sodium ascorbate to counteract the opposite impact when combined with the pyocyanin pigment (Kusale et al., 2021).

1.5.3 Callose Deposition

The 1, 3-glucan polymer callose occurs in plant cell walls that are porous or vulnerable to pathogen invasion and is necessary for the sieve tube metabolism. Its growth and metabolism are closely related to how plants typically develop and function (Fig. 1.2). The callose response involves specifically induced resistance-related genes active in callose depositionsuch as PEN2 and PEN3 genes acting in disease resistance, which are necessary for the buildup of callose. In the process, SA levels rise as a result of ROS being produced and MAMPs being released by PGPR. By controlling the callose synthase gene (*CALS10*) PDLP5-dependent expression, high levels of SA cause callose deposition. By accelerating the synthesis of callose and chitinases, endophytic *P. fluorescens* strain 63–28 improved tomato resistance to *Fusarium oxysporum*. Moreover, *Trichoderma harzianum* T-203 was shown to increased peroxidase and chitinase activity, forming callose barriers, and eliciting systemic defense responses in plants (Afrouz et al., 2023).

1.5.4 Ca^{2+} Influx

Plant immune rsponse, cell growth, and signal transmission depend on elicitors that swiftly cause ion fluxes, such as K^+/H^+ exchange, Cl^- effluxes, and Ca^{2+} influx. One of these ion fluxes, the Ca^{2+} influx, is considered as one of the most important processes as it serves as a second messenger for a variety of physiological changes and cellular functions. Elicitor-induced Ca^{2+} influx not only regulates following events but also strengthens Ca^{2+} signaling via Ca^{2+}-dependent H_2O_2 generation, which can augment Ca^{2+} influx from external sources (Lecourieux et al., 2002). Forskolin, dibutyryl cAMP, or the Ca^{2+} ionophore A23187 were used as pretreatments on bean (*Phaseolus vulgaris*) to encourage the production of ROS that were harmful to the pathogenic fungus *Colletotrichum lindemuthianum*. The Ca^{2+} channel blocker, on the other hand, reduced the oxidative burst, demonstrating that Ca^{2+} influx is necessary for ROS. Common Ca^{2+} sensors like calmodulin can be turned on by Ca^{2+} binding. Protein phosphatase, transcription factors, membrane-bound enzymes, and Ca^{2+}/calmodulin-dependent protein kinase are all further activated by Ca^{2+} and activated calmodulin.

Ca^{2+} binding also regulates the large kinase family members known as Ca^{2+} dependent protein kinases (CDPK), essential for controlling plant defense responses. By increasing the expression of CPK32, PGPR, *Pseudomonas putida* MTCC 5279 increased Ca-dependent signaling (Srivastava et al., 2012). Upon binding of Ca^{2+}, Ca^{2+} sensor relay proteins, calmodulin-binding transcription activators, and controlled transcription in plants, the Ca^{2+} signal may be nonlinearly enhanced (Iqbal et al., 2020; Tanvere et al., 2023).

1.5.5 Defense-Related Genes

For ISR defense mechanisms to work, gene expression must be precisely and specifically controlled. A regulatory network is created by the interactions of genes with the complex circuits that receive a feedback from their own products. Using gene expression and time series data, Timmermann et al. (2020) investigated the control mechanism of the ISR defensive response induced by the bacterium *Paraburkholderia phytofirmans* PsJN. A marker of the JA signaling pathway, Plant Defensin 1.2 (encoded by PDF1.2; AT5G44420) has been demonstrated to rise systemically in *Arabidopsis* leaves in response to fungal pathogen challenge in previous studies (Manners et al., 1998).

Several SA-dependent PR genes express antimicrobial proteins, as previously described. PDF1.2, PR3, and PR4 are activated by JA-dependent and SA-independent pathways, respectively, whereas PR1, PR2, and PR5 genes depend on SA signaling. *Arabidopsis thaliana* pretreatment with non-pathogenic *B. cereus* AR156 induced the expression of PR1, PR2, PR5, and PDF1.2, which indicated the activation of the SA and JA/ET signaling pathways, respectively (Nie et al., 2017). ISR or SAR cannot be produced by the loss-function mutant of NPR1, a crucial regulatory component in the SA-dependent pathway. According to findings from earlier studies, NPR1 modulates downstream defense response genes and coordinates the SA and JA signaling pathways (Spoel et al., 2003).

1.5.6 Stomatal Regulation

Plant respiration, transpiration, and photosynthesis are all significantly influenced by stomata. The reduction in gas exchange caused by stomatal closure lowers photosynthetic output, although this response is really just a natural plant defense mechanism attempting to thwart bacterial invasion. Controlling stomatal opening requires the presence of abscisic acid (ABA) which is produced as a result of stress. When cellular ABA receptors connect with a collection of type 2C protein phosphatases (PP2C), SnRK2 protein kinase OST1 is activated (Umezawaa et al., 2009), whilst the PP2C inhibitory regulatory function is abolished (Ma et al., 2009). The direct association and phosphorylation of the slow anion channel associated 1 (SLAC1) by activated OST1 facilitates anion release from guard cells and stomatal closure (Brandt et al., 2012). Guard cells plasma membrane must have ABA-regulated, hyperpolarization-activated Ca^{2+} channels. OST1 has the ability to catalyze the formation of H_2O_2 (Raghavendra et al., 2010). Indications suggest that the JA signaling pathway is involved in the control of stomatal defense, including the production of the lipoxygenase-encoding gene LOX1 in guard cells in response to PAMPs, needed for stomatal defense activation (Montill et al., 2013). In response to biotic stress, PGPR of *B. amyloliquefaciens* FZB42 was found to promote ABA and JA pathways, as well as producing acetoin and 2,3-butanediol to induce stomatal closure (Wu et al., 2018; Xie et al., 2018). These findings demonstrate how different signaling components cooperate to regulate stomatal function.

References

Afrouz, M., Sayyed, R. Z., Fazeli-Nasab, B., Piri, R., Almalki, W. H., & Fitriatin, B. N. (2023). Seed bio-priming with beneficial *Trichoderma harzianum* alleviates cold stress in maize. *PeerJ, 11*, e15644. https://doi.org/10.7717/peerj.15644

Anderson, J. P., Gleason, C. A., Foley, R. C., Thrall, P. H., Burdon, J. B., & Singh, K. B. (2010). Plants versus pathogens: An evolutionary arms race. *Functional Plant Biology, 37*, 499–512.

Bigeard, J., Colcombet, J., & Hirt, H. (2015). Signaling mechanisms in pattern-triggered immunity (PTI). *Molecular Plant, 8*, 521–539.

Brandt, B., Brodsky, D. E., Xue, S., Negi, J., Iba, K., Kangasjärvi, J., et al. (2012). Reconstitution of abscisic acid activation of SLAC1 anion channel by CPK6 and OST1 kinases and branched ABI1 PP2C phosphatase action. *PNAS, 109*, 10593–10598.

Conrath, U., Beckers, G. J. M., Flors, V., García-Agustín, P., Jakab, G., Mauch, F., et al. (2006). Priming: Getting ready for battle. *Molecular Plant Microbe Interaction, 19*, 1062–1071.

Desai, A., Ruparelia, J., Jha, C. K., Sayyed, R. Z., Mitra, D., Priyadarshini, A., et al. (2022). Articulating beneficial rhizobacteria mediated plant defenses through induced systemic resistances. *Pedosphere, 33*, 556–566. https://doi.org/10.1016/j.pedsph.2022.10.003

Enebe, M. C., & Babalola, O. O. (2019). The impact of microbes in the orchestration of plants' resistance to biotic stress: A disease management approach. *Applied Microbiology and Biotechnology, 103*, 9–25.

Guo, M., Tian, F., Wamboldt, Y., & Alfano, J. R. (2009). The majority of the type III effector inventory of *Pseudomonas syringae* pv. *tomato* DC3000 can suppress plant immunity. *Molecular Plant Microbe Interaction, 22*, 1069–1080.

Guo, Q., Li, Y., Lou, Y., Shi, M., Jiang, Y., Zhou, J., et al. (2019). *Bacillus amyloliquefaciens* Ba13 induces plant systemic resistance and improves rhizosphere microecology against tomato yellow leaf curl virus disease. *Applied Soil Ecology, 137*, 154–166.

Hamidian, M., Movahhedi-Dehnavi, M., Sayyed, R. Z., Almalki, W. H., Gafur, A., & Fazeli-Nasab, B. (2023). Co-inoculation of mycorrhiza and methyl jasmonate regulates morpho-physiological and antioxidant responses of *Crocus sativus* (saffron) under salinity stress conditions. *Scientific Reports, 13*, 7378. https://doi.org/10.1038/s41598-023-34359-6

Ilyas, N., Akhtar, N., Naseem, A., Qureshi, R., Majeed, A., & Sayyed, R. Z. (2022). The potential of *Bacillus subtilis* and phosphorus in improving the growth of wheat under chromium stress. *Journal of Applied Microbiology, 133*, 3307–3321. https://sfamjournals.onlinelibrary.wiley.com/doi/10.1111/jam.15676

Iqbal, Z., Shariq Iqbal, M., Singh, S. P., & Buaboocha, T. (2020). Ca^{2+}/Calmodulin complex triggers CAMTA transcriptional machinery under stress in plants: Signaling cascade and molecular regulation. *Frontiers in Plant Science, 11*, 598327.

Jacobs, S., Zechmann, B., Molitor, A., Trujillo, M., Petutschnig, E., et al. (2011). Broad-spectrum suppression of innate immunity is required for colonization of *Arabidopsis* roots by the fungus *Piriformospora indica. Plant Physiology, 156*, 726–740.

Karimian, M. A., Nasab, B. F., Sayyed, R. Z., Ilyas, N., Almalki, W. H., Vats, S., et al. (2023). Salicylic acid foliar spray promotes yield, yield components, and physiological characteristics in foxtail millet under drought stress. *Pakistan Journal of Botany, 55*(SI). https://doi.org/10.30848/PJB2023-SI(11)

Khan, N., Ali, A., Shahi, M. A., Mustafa, A., Sayyed, R. Z., & Curaá, J. A. (2021). Insights into the interactions among roots, rhizosphere and rhizobacteria for improving plant growth and tolerance to abiotic stresses: A review. *Cells, 10*, 1551. https://doi.org/10.3390/cells10061551

Kusale, S. P., Attar, Y. C., Sayyed, R. Z., Malek, R. A., Ilyas, N., Suriani, N. L., et al. (2021). Production of plant beneficial and antioxidants metabolites by *Klebsiella variicola* under salinity stress. *Molecules, 26*, 1894.

Lecourieux, D., Mazars, C., Pauly, N., Ranjeva, R., & Pugin, A. (2002). Analysis and effects of cytosolic free calcium increases in response to elicitors in *Nicotiana plumbaginifolia* cells. *Plant Cell, 14*, 2627–2641.

Libault, M., Farmer, A., Brechenmacher, L., Drnevich, J., Langley, R. J., et al. (2010). Complete transcriptome of the soybean root hair cell, a single-cell model, and its alteration in response to *Bradyrhizobium japonicum* infection. *Plant Physiology, 152*, 541–552.

Ma, Y., Szostkiewicz, I., Korte, A., Moes, D., Yang, Y., Christmann, A., & Grill, E. (2009). Regulators of PP2C phosphatase activity function as abscisic acid sensors. *Science, 324*, 1266–1068.

Manners, J. M., Penninckx, I. A., Vermaere, K., Kazan, K., Brown, R. L., Morgan, A., et al. (1998). The promoter of the plant defensin gene PDF1.2 from *Arabidopsis* is systemically activated by fungal pathogens and responds to methyl jasmonate but not to salicylic acid. *Plant Molecular Biology, 38*, 1071–1080.

Montill, J. L., Leonhardt, N., Mondy, S., Tranchimand, S., Rumeau, D., et al. (2013). An abscisic acid-independent oxylipin pathway controls stomatal closure and immune defense in *Arabidopsis*. *PLoS Biology, 11*, e1001513.

Nie, P., Li, X., Wang, S., Guo, J., Zhao, H., & Niu, D. (2017). Induced systemic resistance against *Botrytis cinerea* by *Bacillus cereus* AR156 through a JA/ET- and NPR1-dependent signaling pathway and activates PAMP-triggered immunity in *Arabidopsis*. *Frontiers in Plant Science, 8*, 238.

Nithyapriya, S., Lalitha, S., Sayyed, R. Z., Reddy, M. S., Dailin, D. J., Enshasy, H. E., et al. (2021). Production, purification, and characterization of bacillibactin siderophore of *Bacillus subtilis* and its application for improvement in plant growth and oil content in sesame. *Sustainability, 13*, 5394. https://doi.org/10.3390/su13105394

Omar, A. F., Abdelmageed, A. H. A., Al-Turki, A., Abdelhameid, N. M., Sayyed, R. Z., & Rehan, M. (2022). Exploring the plant growth-promotion of four *Streptomyces* strains from rhizosphere soil to enhance cucumber growth and yield. *Plants, 11*, 3316. https://doi.org/10.3390/plants11233316

Paliwal, K., Jajoo, A., Tomar, R. S., Anil, P., Syed, S., Bright, J. P., & Sayyed, R. Z. (2023). Enhancing biotic stress tolerance in soybean affected by *Rhizoctonia solani* root rot through an integrated approach of biocontrol agent and fungicide. *Current Microbiology, 80*, 304. https://doi.org/10.1007/s00284-023-03404-y

Pandey, P., Irulappan, V., Bagavathiannan, M. V., & Senthil-Kumar, M. (2017). Impact of combined abiotic and biotic stresses on plantgrowth and avenues for crop improvement by exploiting physio-morphological traits. *Frontiers in Plant Science, 8*, 537.

Rafique, N., Aqeel, M., Raja, N. I., Shabbir, G., Ajaib, M., et al. (2023). Interactive effects of melatonin and salicylic acid on *Brassica napus* under drought condition. *Plant and Soil*. https://doi.org/10.1007/s11104-023-05942-7

Raghavendra, A. S., Gonugunta, V. K., Christmann, A., & Grill, E. (2010). ABA perception and signaling. *Trends in Plant Sciences, 15*, 395–401.

Rai, S., Omar, A. F., Rehan, M., Al-Turki, A., Sagar, A., et al. (2023). Crop microbiome: Their role and advances in molecular and omic techniques for the sustenance of agriculture. *Planta, 257*, 27. https://doi.org/10.1007/s00425-022-04052-5

Saboor, A., Muhammad, A. A., Hussain, S., El Enshasy, H. E., Hussain, S., Ahmed, N., et al. (2021). Zinc nutrition and arbuscular mycorrhizal symbiosis effects on maize (*Zea mays* L.) growth and productivity. *Saudi Journal of Biological Sciences, 28*, 6339–6351. https://doi.org/10.1016/j.sjbs.2021.06.096

Saranraj, P., Sayyed, R. Z., Kokila, M., Sudha, A., Sivasakthivelan, P., Devi, D. M., et al. (2022). Plant growth-promoting and biocontrol metabolites produced by *Pseudomonas fluorescence*. In R. Z. Sayyed & V. Gavicho Uarrota (Eds.), *Secondary metabolites and volatiles of PGPR in plant-growth promotion* (pp. 349–381). Springer. https://doi.org/10.1007/978-3-031-07559-9_18

Spoel, S., Koornneef, A., Claessens, S. M. C., Korzelius, J. P., Van Pelt, J. A., Mueller, M. J., et al. (2003). NPR1 modulates cross-talk between salicylate- and jasmonate-dependent defense pathways through a novel function in the cytosol. *Plant Cell, 15*, 760–770.

Srivastava, S., Chaudhry, V., Mishra, A., Chauhan, P. S., Rehman, A., et al. (2012). Gene expression profiling through microarray analysis in *Arabidopsis thaliana* colonized by *Pseudomonas putida* MTCC5279, a plant growth promoting rhizobacterium. *Plant Signaling & Behavior, 7*, 235–245.

Stringlis, I. A., Proietti, S., Hickman, R., Van Verk, M. C., Zamioudis, C., & Pieterse, C. M. (2018). Root transcriptional dynamics induced by beneficial rhizobacteria and microbial immune elicitors reveal signatures of adaptation to mutualists. *The Plant Journal, 93*, 166–180.

Swarnalatha, G. V., Goudar, V., Reddy, E. C., Reddy, G. S., Tawaha, A. R. M. A., & Sayyed, R. Z. (2022). Siderophores and their applications in sustainable management of plant diseases. In R. Z. Sayyed & V. Gavicho Uarrota (Eds.), *Secondary metabolites and volatiles of PGPR in plant-growth promotion* (Vol. 99, pp. 289–302). Springer.

Tanvere, S., Akhtar, N., Ilyas, N., Sayyed, R. Z., Fitriatin, B. N., Parveen, K., & Bukhari, N. A. (2023). Interactive effects of *Pseudomonas putida* and salicylic acid for mitigating drought tolerance in canola (*Brassica napus* L.). *Heliyon, 9*, e14193. https://doi.org/10.1016/j.heliyon.2023.e14193

Timmermann, T., Gonzalez, B., & Ruz, G. A. (2020). Reconstruction of a gene regulatory network of the induced systemic resistance defense response in *Arabidopsis* using boolean networks. *BMC Bioinformatics, 21*, 142.

Umezawaa, T., Sugiyamab, N., Mizoguchia, M., Hayashid, S., Myougae, F., et al. (2009). Type 2C protein phosphatases directly regulate abscisic acid-activated protein kinases in Arabidopsis. *PNAS, 106*, 17588–17593.

Wang, W., Chen, D. D., Zhang, X. P., Liu, D., Cheng, Y. Y., & Shen, F. F. (2018). Role of plant respiratory burst oxidase homologs in stress responses. *Free Radical Research, 52*, 826–839.

Wang, M., Xue, J., Ma, J., Feng, X., Ying, H., & Xu, H. (2020). *Streptomyces lydicus* M01 regulates soil microbial community and alleviates foliar disease caused by *Alternaria alternata* on cucumbers. *Frontiers in Microbiology, 11*, 942.

Wu, L., Huang, Z., Li, X., Ma, L., Gu, Q., Wu, H., Liu, J., Borriss, R., Wu, Z., & Gao, X. (2018). Stomatal closure and SA-, JA/ET signaling pathways are essential for *bacillus amyloliquefaciens* FZB42 to restrict leaf disease caused by *Phytophthora nicotianae* in *Nicotiana benthamiana*. *Frontiers in Microbiology, 9*, 847.

Xie, S., Jiang, H., Ding, T., Xu, Q., Chai, W., & Cheng, B. (2018). *Bacillus amyloliquefaciens* FZB42 represses plant miR846 to induce systemic resistance via a jasmonic acid-dependent signaling pathway. *Molecular Plant Pathology, 19*, 1612–1623.

Yuan, M., Jiang, Z., Bi, G., Nomura, K., Liu, M., Wang, Y., Cai, B., Zhou, J. M., He, S. Y., & Xin, X. F. (2021). Pattern-recognition receptors are required for NLR-mediated plant immunity. *Nature, 592*, 105–109.

Zipfel, C., & Oldroyd, G. E. D. (2017). Plant signaling in symbiosis and immunity. *Nature, 543*, 328–336.

Chapter 2
Understanding Plant Immunity Through Plant-Microbe Interaction

Nilakhi Kakoti, Munmi Borkataky , and Ratul Nath

Abstract The plant immune system is crucial in the context of sustainable agriculture and soil health, in order to understand the dynamics of plant-microbe interaction. Aspart of their immunity, plants recognize the microbial invasion and respond to invaders through a series of immune responses, including MTI (MAMP Triggered Immunity/PAMP Triggered Immunity), ETI (Effector Triggered Immunity), and SAR (Systemic Acquired Resistance). This chapter highlights the multifaceted relationships between plants and rhizosphere microbes and the response mechanisms deployed by the host immune system. Additionally, we also discuss the advantages of utilizing -omics tools to understand the molecular and genetic aspects that will be helpful for crop management, sustainable agriculture, and ecological conservation.

Keywords ETI · MTI · Omics · Plant-microbe interactions · Plant immune system · Plant health · SAR

Abbreviations

MAMPS	Microbe-Associated Molecular Pattern
PAMP	Pathogen Associated Molecular Pattern
MTI	MAMP Triggered Immunity
PTI	Pattern Triggered Immunity
ETI	Effector Triggered Immunity
SAR	Systemic Acquired Resistance
ISR	Induced Systemic Resistance
PRR	Pattern Recognition Receptors
NAMPs	Nematode-Associated Molecular Patterns

N. Kakoti · M. Borkataky (✉) · R. Nath
Department of Life Sciences, Dibrugarh University, Assam, India
e-mail: munmiborkataky@dibru.ac.in; ratulnath@dibru.ac.in

P. Mathur, S. Roy (eds.), *Plant Microbiome and Biological Control*, Sustainability in Plant and Crop Protection 20,
https://doi.org/10.1007/978-3-031-75845-4_2

DAMP	Damage-Associated Molecular Pattern (DAMP)
MIR	Mycorrhiza-Induced Resistance
FLS	Flagellin Sensitive
CERK	Chitin Elicitor-Binding Protein
BIK	Botrytis-Induced Kinase
GAP	GTPase Accelerating Protein
CNGC	Cyclic Nucleotide-Gated Channel
NLR	Nucleotide-Binding Domain Leucine-Rich Repeat Containing Receptor
SA	Salicylic Acid
JA	Jasmonic acid
ET	Ethylene
ABA	abscisic acid
CPK	Ca^{2+} dependent protein kinase
EDS	Enhanced Disease Susceptibility
NPR	Non-Expressor of Pathogenesis Related Genes
BTH	benzothiodiazole
DIR	Defective in induced resistance
FMO	Flavin Dependent Monooxygenase
NHP	N-Hydroxypipecolic Acid
DAPG	2, 4- diacetylphloroglucinol
EPS	Exopolysaccharide
SYMRK	Symbiosis Receptor-like Kinase
CCaMK	Calcium and Calmodulin Dependent Protein Kinase
NSP	Nodulation Signalling Pathway
ACC	Amino Cyclopropane-1-Carboxylate
MAP	Mitogen-Activated Protein
PGPR	Plant growth promoting rhizobacteria
AMF	Arbuscular mycorrhizal fungi

2.1 Introduction

The rhizosphere provides a home for beneficial microbes with a symbiotic relationship between plants and emphasize nutrient uptake and plant growth (Gaskins et al., 1995). Plant microbial interactions signify the promotion of nitrogen fixation, denitrification, growth hormones, and phytoalexins that employ insurance of the plant health (Gupta et al., 2000). Microbes are the architecture surrounding the possible niche of the soil. The biological function of the rhizomicrobes includes biodegradation, decomposition, and remediation and masks the destruction of environmental stress, which directly promotes plant growth and sustainability (Bhatt et al., 2022). These interactions mediate plant dynamics and help to understand the effects of invasive species by altering pH, temperature, and nutrient availability to diminish the ecological and economic impacts and provide a safeguard to the native flora and fauna (Sanon et al., 2009). Positive plant-microbial interactions include symbiosis, nitrogen fixation, and others beneficial factors for plant growth, whereas negative interactions mostly involve pest and pathogen attacks (Haldar & Sengupta, 2015).

Root exudates provide organic carbon to the soil through rhizodeposition, with an impact shaping the resident hizosphere microbial populations. These carbon source compounds include sugars, organic acids, amino acids, phenolics, vitamins, secondary metabolites (low-molecular-weight), mucilage, and proteins (high molecular weight) (Chaparro et al., 2012; Khare et al., 2020). Root exudates organic compounds mediate chemotaxis and electrotaxis for root colonization (Bais et al., 2004).

During abiotic and biotic stress conditions, root-microbe interactions may enhance/induce the secretion of plant growth hormones (auxin, cytokinin, gibberellin, abscissic acid, and strigolactones) and secondary metabolites (phenols, flavonoids, essential oil) as a response to attacks. Moreover, quorum sensing facilitates defensive mechanisms against competitors in plants. Biocontrol agents i.e., *Pseudomonas* and *Bacillus* spp., may also prime plants for Induced Systemic Resistance (ISR) against pathogens (Badri et al., 2009; Doornbos et al., 2012). Host tissues destroyed by pathogens also release a broad range of effector molecules and proteins. A general scheme of plant-mirobe interactions is given in Fig. 2.1.

Other strategies active in plant-microbe interaction for defense include phenol production via the phenylpropanoid pathway, lignin deposition, ACC deaminase production, nutrient mobilization, and transgenerational defense (Mishra et al., 2015; Etesami & Beattie, 2017). Beneficial plant-microbial interactions may increase the crop yield, also by improving soil structure and regime, with a potential

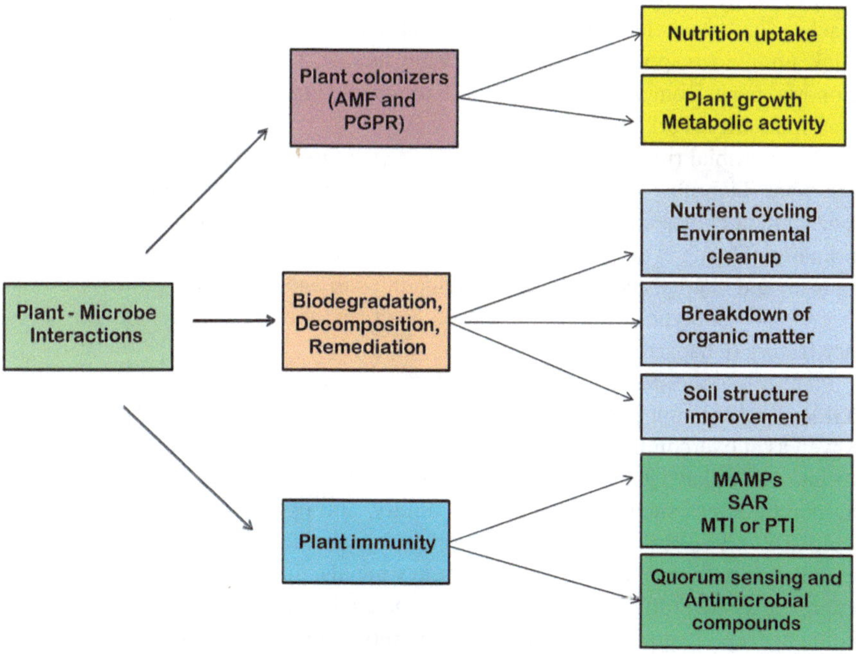

Fig. 2.1 Initiation of several plant responses during plant-microbe interactions

to alleviate the adverse effects of pollution (Singh et al., 2011). Plant growth-promoting rhizobacteria (PGPR), Arbuscular Mycorrhizal Fungi (AMF), diazotrophs, and endophytes promote plant growth by boosting nutrition through phosphorus provision and nitrogen fixation (Morel & Sowinski, 2013; Bhatt et al., 2022).

Plants recognize root microbes via MAMPS (Microbe-Associated Molecular Patterns) molecules eliciting a defense mechanism that helps generate phytohormone signaling (Finkel et al., 2017). Most useful immunity mechanisms triggered by plant-microbe interactions are: Systemic acquired resistance (SAR), pathogen-associated molecular pattern (PAMPs)-triggered immunity (MTI or PTI) using pattern recognition receptors (PRR) and effector-triggered immunity (ETI), Nematode-Associated Molecular Patterns (NAMPs), and Damage-Associated Molecular Pattern (DAMP).

The antibiotics produced by rhizobacteria are also beneficial for plants against pathogens. 2.4-diacetyl phloroglucinol, neomycin A, amphibian, phenazine, pyoluteorin, pyrrolnitrin, hydrogen cyanide, cyclic lipopeptides, zwitterion A, xanthobaccin, kanosamine are well-known antibiotics produced by rhizobacteria. Further beneficial compounds i.e., bacteriocins (proteinaceous peptides) and lytic enzymes that are produced by bacteria, are of great importance against fungal pathogens (Mabood et al., 2014).

Plant-microbe interactions also involve cellular organelles that play a vital role in plant defense. Rapid initiation of signaling cascades occurs from the cell membrane to the trans-Golgi network and multivesicular bodies to respond to pathogenic attacks. The cytoskeleton (microtubules and microfilaments) may also modulate a quick immune response during pathogenic threats. Chloroplasts, nucleus, mitochondria, peroxisomes, and vacuoles are the major organelle-based contributors to plant immunity (Park et al., 2017).

The microbial rhizosphere communities depend in a species-specific way on the host plant (Doornbos et al., 2012). Plant-microbial interactions sustain various biogeochemical cycles and influence the mechanisms for metals and metalloid transformation (Huang et al., 2005; Singh et al., 2011). However, the interactions are not constant and change according to the plant and soil age, structure, growth patterns, species and genotype, climate, cultivation practices, and availability of other necessary factors (Chaparro et al., 2012).

Plant-microbe interactions are crucial in sustainable agriculture. Inoculation of PGPR and AMF aim at sustaining fundamental biological process for plant growth as well as at overcoming stress factors (Parasuraman et al., 2019; Kour et al., 2020). Recent studies have gained significant attention on the function of PGPR in mitigating abiotic stress such as drought and salinity, and optimizing crop yield (Gupta et al., 2022). PGPR lessen the use of chemical pesticides and contribute to the bioremediation and decontamination from heavy metals, through the production of siderophores (pyoverdine and pseudobactin-c), chelating activities, degradation of pollutants (i.e., hydrocarbons, herbicides, explosives), decomposition of chemicals and mineralization of organic contaminants (Bhatt et al., 2022).

Climatic variations alsp control the microbial communities. Some climatic conditions favour growth while can be hazardous for other species. Such changes affect not only plants but also their microbiome, thus retarding or diversifying their interactions (Aqeel et al., 2023).

Many -omic tool, including metatranscriptomics, proteomics, genomics, and amplicon sequencing techniques, have been able to identify the profiles of microbial communities and understand the relationship of specific rhizosphere taxa with plants. Microarrays have been used to analyze the detrimental effect of plant-pathogen interactions. Multispecies transcriptomics can monitor the plant and microbial genes specifically expressed in the regulation of disease resistance (Schenk et al., 2012; Finkel et al., 2017). Metabolomic studies help to recognize the presence of carbohydrates, amino acids, polyamines, polyols, and quaternary ammonium compounds (such as b-alanine betaine, glycine betaine, and proline betaine), which play crucial functions in environmental stress (Gupta et al., 2022).

2.2 Induction of Plant Immune Response

Plants develop various physical barriers (cytoskeleton, plasma membrane, cuticles, waxes, trichomes) to overcome pathogenic stress. However, if pathogens penetrate these barriers, they have to face a competent plant immune system regulated by the microbial community. Plants do not have any adaptive immune system. Instead, they depend on the signaling cascade of the infected tissues (Nishad et al., 2020). Jones and Dangl (2006) proposed two branched innate immune systems in plants. The first branch, called Pattern-Triggered Immunity (PTI), induces Microbe-Associated or Pathogen-Associated Molecular Patterns (PAMPs, or MAMPs). The second is called Effector-Triggered immunity (ETI) (Jones & Dangl, 2006; Pritchard & Birch, 2014). Plants communicate with MAMP (Microbe-Associated Molecular Pattern) against pathogenic microbes via PRR (Pattern Recognition Receptors, PRRS). These microbes overcome MTI by producing some effector molecules which are further recognized by resistance (R) proteins that are intracellular receptors, activating ETI (Cook et al., 2015). Plant receptor molecules triggering MAMPs or PAMPs include lipopolysaccharides, peptidoglycans, flagellin, sterols, fatty acids, proteins, double-stranded RNA and methylated DNA, elongation factor EF-Tu, harpin (HrpZ), NEP1-like proteins (NLPs), b-glucan and Pep13.

Many R proteins (PRRs) contain leucine-rich repeat domains responsible for the recognization of specific pathogens (Iriti & Varoni, 2015). Plants have specific receptors for disease-resistant genes and interact with the molecules of microbial receptors to activate a defense response (Cohn et al., 2001). PRRs recognize bacterial flagellin FLS2 (FLAGELLIN SENSITIVE 2) located in the plasma membrane. Mutation of FLS2 left plants susceptible to bacterial pathogens (Boller & He, 2009).

2.2.1 MTI (MAMP Triggered Immunity)

Beneficial plant microbe interactions involve the sensing of MAMPs by PRRs triggering immune response (Hacquard et al., 2017). Oligogalacturonides (OGs) are produced as DAMPs (DAMAGE-ASSOCIATED MOLECULAR PATTERNS) during pathogenicity. OGs mediate the production of endogenous PGs (polygalacturonases) during wound infection and mechanical stress (Lorenzo et al., 2018). Plants respond to microbial pathogens via PAMPs. The immune response is induced by bacterial flagellin and elongation factor-Tu (EFR), observed when restricting the *Arabidopsis* transformation by *Agrobacterium tumefaciens* (Zipfel et al., 2006). Miya et al. (2007) demonstrated that CERK1 (CHITIN ELICITOR-BINDING PROTEIN, CEBiP), consisting Ser/Thr kinase domain and LysM motifs with kinase activity or autophosphorylation, acts as a MAMPs in *Arabidopsis*.

The G proteins (Gα XLG2, Gβ AGB1, Gγ AGG1, AGG2) are mandatory for FLS2-facilated immune responses (Liang et al., 2016). The plant G proteins connect cytosol signalling molecules to the plasma membrane receptors (Trusov & Botella, 2012).. The G proteins are self-activating and act independently. Their regulation occurs either by slowing nucleotide exchange or speeding nucleotide hydrolysis. However, mutations of any subunits of G proteins can deteriorate the development process and make them less resistant to pathogens in plants (Urano & Jones, 2013). The FLS2 receptor interacts with G proteins to regulate the immune response in *Arabidopsis* via a central cytoplasmic kinase BIK1 (Botrytis-Induced Kinase 1) (Liang et al., 2016). According to Liang et al. (2018) *Arabidopsis* G protein is maintained by a GAP (GTPase accelerating protein) RGS1. As FLS2 is activated by flg22, a bacterial flagellin epitope, RGS1 is phosphorylated by BIK1/PBL triggering the immune response.

FLS2 regulates RbohD, a component of the FLS2 immune receptor, that is phosphorylated by BIK1 at Ser39, Ser343, and Ser347 and Ser343. Phosphorylation as a result induces calcium influx that mediates stomatal defense and ROS production (Li et al., 2014). BIK1 initiates, on activation of FLS2 and EF-Tu, the early defense response by opening the Ca-associated plasma membrane anion channel. The rise in intracellular calcium activates the Cl^-, K^+, $NO^-{}_3$, H^+, K^+ ion fluxes thus depolarizing the plasma membrane (Jeworutzki et al., 2010). Under elevated Ca concentrations, cyclic nucleotide-gated channel (CNGC) proteins, CNGC2 and CNGC4, trigger PAMP-induced calcium signaling. During resting phase both CNGC2 and CNGC4 together act as a calcium channel and are blocked by calmodulin. Upon pathogenicity, BIK1 phosphorylates the channel and increase the calcium influx (Tian et al., 2019).

2.2.2 ETI (Effector-Triggered Immunity)

Plants activate a stronger immune response called ETI to counteract a pathogen attacks. ETI is initiated upon recognition of pathogen receptors by nucleotide-binding domain leucine-rich repeat-containing receptors (NLRs). In plants, NLRs contain three domains: a N-terminal variable domain, a nucleotide-binding domain, and a C-terminal LRR domain. NLRs are classified into three groups on the basis of the N terminal: i) coiled-coil (CC)-type NLRs (CNLs), ii) Toll/interleukin-1 receptor/Resistance protein (TIR)-type NLRs (TNLs) and iii) Resistance to powdery mildew 8-like domain (RPW8)-type NLR (RNLs). They play a significant role in the recognizion of pathogen receptors (Yuan et al., 2021). ETI leads to the biosynthesis of salicylic acid (SA), jasmonic acid (JA), ethylene (ET), antimicrobial products found in endoplasmic reticulum and vacuoles of plant cells. This further initiate a hypersensitive response which is a feature of the programmed cell death.

The events that occur during such responses are: cytoplasmic shrinkage, mitochondrial swelling, vacuolization, chloroplast disruption and chromatin condensation (Iwai et al., 2006; Schäfer & Eichmann, 2012; Pennell & Lamb, 1997; Hayward et al., 2009; Coll et al., 2011; Muthamilarasan & Prasad, 2013). The hypersensitive response leads to an increase in cytosolic Ca and accumulation of hydrogen peroxide (Grant et al., 2000). In *Arabidopsis*, NLRs protein activates six Ca^{2+}-dependent protein kinases (CPKs) for immune response. CPK 4/5/6/11 phosphorylates transcription factors such as WRKY 8/28/48 that help to curb the pathogen growth. On the other hand, CPK 1/2/4/11 phosphorylates the NADPH oxidases present on the plasma membrane, resulting in ROS production (Gao et al., 2013).

Autoinhibition of NLRs depends on ADP (off), and its activation depends on ATP (on) (Hu et al., 2013). ETI is activated by NLR signaling pathways such as lipase-like protein EDS1 (enhanced disease susceptibility 1) and CC-containing NLRs requiring NDR1 (non-race specific disease resistance 1) (Zhang et al., 2017). NLR induces an immune response either by recognizing effectors or host proteins, which involves the activation of an NBS–LRR protein in *Arabidopsis* called RPS5 (Ade et al., 2007).

Another NLR protein coiled-coil (CC)–NLR Hopz-Activated Resistance 1 (ZAR1) complex with resistance-related kinase 1 (RKS1) acts *vs* microbial pathogens (Wang et al., 2019). The pathogen effector protein AvrAC (uridylyl transferase) is recognized by ZAR1 via resistance-related kinase 1 (RKS1) forming a ZAR1- RKS1 complex. AvrAC uridylylates a receptor-like cytoplasmic kinases (RLCK), PBS1-like protein 2 (PBL2 or PBL2UMP). Uridylylated PBL2 binds to the ZAR1-RKS1 complex leading to a conformational change in NB domain. The ensuing ADP release from inactive ZAR1 leads to a nucleotide depleted ZAR1, i.e. the ZAR1- RKS1-PBL2UMP complex. As ATP activates ZAR-RKS1-PBL2UMP, oligomerization occurs in NLR proteins. A second conformational change may be observed in reconstruction of NBD and CC domains, forming a complex called resistosome, activating an immune response (Wang et al., 2019).

2.2.3 SAR (Systemic Acquired Resistance)

SAR is an effectual immune response of plants to react to pathogens. The role of SA in SAR was first advocated in the study of transgenic tobacco and *Arabidopsis* plants (Gaffney et al., 1993; Delaney et al., 1994). This immune response occurs in distal plant regions due to the massive function of MTI and ETI. It is regulated by Non-Expressor of Pathogenesis-Related Genes 1 (NPR).

To manage pathogens, SAR can also be induced by increasing the concentration of salicylic acid (SA) or by a treatment with its biologically active analogs benzothiadiazole (BTH) and 2,6-dichloro isonicotinic acid (INA). Elevation of SA occurs upon pathogenicity due to the accumulation of the metabolite azelaic acid, a nine-carbon dicarboxylic acid, in the vascular bundle of plants inducing SAR, as observed in *Arabidopsis* when infected by *Pseudomonas syringae* (Jung et al., 2009).

In case of a second attack (after primary infection), an immediate and effective immune response elicits the SAR state. SAR depends on the past experience of pathogenicity, in the form of a sort of "plant memory". Hydrogen peroxide burst, gaseous methyl salicylate, and ethylene are signals inducing SAR (Conrath, 2006). Methyl salicylate (MeSA) is defective in induced resistance1 (DIR1), and the Glycerol-3-phosphate dependent factor (G3P) activates SAR in darkness. FMO1 (Flavin-Dependent-Monooxygenase 1), Lysine catabolite pipecolic acid (Pip), and N-hydroxy pipecolic acid (NHP) are essential signaling compounds that induce SAR in presence of light (Bernsdorff et al., 2016; Chen et al., 2018; Hartmann et al., 2018). SAR acts as an alarmone in plants against pathogen infection priming, via accumulation of PIP (Lysine catabolite pipecolic acid). One such benefit of SAR is that it passes its memory to the next generation in plants so that the progeny activates a rapid immune response on new pathogens attack (Shah & Zeier, 2013).

2.3 Plants-AMF Interactions and Induction of Immunity

Plants interacting with Arbuscular mycorrhizal fungi (AMF) show higher levels of resistance against soil-borne pathogens. Upon infection, a number of metabolic changes take place in roots, including the accumulation of indolic derivatives and phenolic compounds such as ferulic acid and chlorogenic acid, folic acid and riboflavin vitamins. AMF stimulates mycorrhiza-induced resistance (MIR), a characteristic of systemic acquired resistance (SAR) following root colonization (Cameron et al., 2013). Initially, plants identify the beneficial microbes in the rhizosphere as attackers, inducing an immune response. Meanwhile, the beneficial microbes establish a successful interaction and colonize the root (Pieterse et al., 2012). Interaction of AMF with plants involves a temporal activation of defense response. This symbiosis induces plant mechanisms, including elicitor degradation, secondary messenger modulation, regulation of plant hormones and nutrition, triggering regulatory genes for symbiosis (Garcia-Garrido & Ocampo, 2002). SA shows an effect in colonization of AMF, and inhibition of SA responsive genes is hence essential for a successful AMF symbiotic establishment (Dumas-Gaudot et al., 2000).

AMF express effector proteins to suppress pathogenicity during colonization (Kloppholz et al., 2011). Jasmonic acid (JA) is known to play a major role in influencing the establishment of AMF symbiosis for regulating MIR (Hause et al., 2007; Yogendra et al., 2015). AMF activates JA regulation of immune responses that acts against necrotrophs, whereas SA matches up the defense mechanisms against biotrophs (Pozo et al., 2009).

AMF colonization causes a phenomenon called the mycorrhizosphere effect, which directly attracts some beneficial bacterial populations such as PGPR (Plant growth-promoting Rhizobacteria). PGPR promote plant growth and protect the plant from biotic and abiotic stress. Cameron et al. (2013) proposed four phases in the AMF-plant interaction:

Phase I: Recruitment of Food Source by AMF

Recruitment of root exudation chemicals by AMF, i.e., sugars, amino acids, carboxylic acids, phenolics, and other secondary metabolites that are rich in energy, as nutrients for AMF (Akiyama et al., 2005). Strigolactone is known to be the most important signal for AMF infection. Terpenoid lactones stimulate seed germination and hyphal branching in AMF, required to establish themselves in roots and facilitate infection. This further promotes the production of isoflavonoids and creates a center of attraction for nitrogen-fixing bacteria (Cameron et al., 2013).

Phase II: Immune Response of Plants to AMF Infection

The initial infection of AMF causes the plants to deploy an immune response similar to that observable during a pathogen attack. At a later stage, the immune response is modulated via MAMP that activates the production of plant hormones for defense (Zhang & Zhou, 2010; De Jonge et al., 2011).

Phase III: Suppression of Immune Response by AMF and Attraction of PGPR

Certain specific effector molecules of AMF lower the immune response of plants for successful colonization, such as Ca/calmodulin kinases DMI3, SP7 (Siciliano et al., 2007). On the other hand AMF induce the production of abscisic acid (ABA) and benzoxazinoid for infection and promote nutrients uptake. These changes and progression in root exudation shape the colonization of beneficial rhizosphere bacteria (Jung et al., 2012).

Phase IV: Establishment of Mycorrhizosphere Bacteria Triggering Systemic Resistance

Rhizobacteria induce a MAMP immune response similar to that of AMF. The possible defense elicitors shown by bacteria are the elongation factor Tu, rhamnolipids, flagellin, lipopolysaccharides and quorum sensing. PGPR synthesizes autoinducers - phenazine and 2,4-DAPG - that are crucial for JA-dependent ISR (Cameron et al., 2013).

2.4 Plants-Rhizobacterial Interactions and Induction of Immunity

Beneficial microbes elicit immune responses against pathogenic diseases by achieving shelter within and around the plant tissues. Beneficial microbes also produce toxins and antimicrobial moieties active against harmful species. This type of interaction is referred to as symbiotic association (Nishad et al., 2020). Beneficial rhizosphere microbes produce antimicrobial compounds such as oligomycin A, phenazine-1-carboxylic acid (PCA), ammonia, butyrolactones, cyfluthrin, and pyrrolnitrin to overcome biotic stress. Bacteria or fungi such as *Pseudomonas fluorescens*, *Agrobacterium radiobacter*, *Bacillus subtilis*, *Aspergillus niger*, *Penicillium citrinum*, *Trichoderma harzianum* produce antibiotics and hydrolytic enzymes such as chitinase and β-1,3 glucanase, agrocin 84, 2,4- diacetylphloroglucinol (DAPG), siderophore to hold back the growth of pests or pathogens such as *Meloidogyne incognita*, *Fusarium oxysporum*, *Botrytis cinerea*, *Colletotrichum lindemuthianum*, *C. orbiculare* or cucumber mosaic virus (Bukhata et al., 2020).

Many studies have shown evidence of PGPR having a potent role in inducing immune response for stress tolerance in plants, by promoting a central mechanism called defense priming (Balmer et al., 2015). Phytohormones are known as key regulators of plant defense. Upon a pathogen attack, PGPR trigger the systemic response initiated by phytohormones including SA, ET and JA (Dempsey & Klessig, 2012). ISR triggered by PGPR in plants is regulated by phytohormones. ISR provides the priming, i.e. an immune response against attackers, from the root to the above ground part of the plant. An established feature of plant-PGPR interactions is the production of antibiotics during pathogenicity. DAPG, lipopeptides (LP), surfactin, iturin, and fengycin promote plant growth, auxin production, and develop the root structure (Rosier et al., 2018). Rhizobacteria-plant interactions stimulate MTI which is suppressed in later stages of nodule formation. Rhizobacteria produce the exopolysaccharide (EPS) succinoglycan, which helps to establish a symbiotic relationship with plants, providing a signal for successful infection. However, in absence of this signal, plants cease the infection process by stimulating an immune response (Zamioudis & Pieterse, 2012; Jones et al., 2008). Exopolysaccharides suppress MTI due to their polyanionic nature and chelate divalent Ca ions (Aslam et al., 2008).

Rhizobium spp. infection, yielding root nodules, is the most prominent and significant symbiotic relationship in agriculture, known for fixing atmospheric nitrogen. Root nodules formation is induced by Nod Factors (NFs), chitin (N -acetylglucosamine oligomers) derivatives that determine the specificity of *Rhizobium* for its host plant, required for a successful infection (Kouchi et al., 2010). The recognition of the Nod factors involves several components, including: extracellular LysM domains, a receptor-like kinase that binds with chitooligosaccharides activating symbiosis receptor-like kinase (SYMRK); components of the nuclear pore and two cation channels; the Ca oscillation, i.e., a Ca and calmodulin-dependent protein kinase (CCaMK) phosphorylating and activating CYCLOPS. The activation of CCaMK and CYCLOPS induces the transcription factors of the nodulation signalling pathway (NSP) NSP1, NSP2, nodule inception (NIN) and ERF, required for nodulation (ERN)1 a complex that regulates NF-induced gene expression (Oldroyd et al., 2011).

Rhizobial Nod factors suppress MTI of the host plant (Liang et al., 2013). *Rhizobium* spp. developed several pathways to minimize the recognition of their MAMPs, including secretion of lipopolysaccharide (LPS) and exopolysaccharide (EPS) and type 3 secretion system (T3SS). The mechanism of rhizobacteria is similar to that of the pathogenic bacterial infection. The Nod signaling pathway reduces SA-dependent defense response (Zamioudis & Pieterse, 2012). Roots interact with rhizobial transcription factor Nod D to synthesize flavonoids that activate a type III secretion gene cluster which suppresses the host plant immunity (Okazaki et al., 2013; Gourion et al., 2015).

Etylene signalling is also important in plant defense. However, the rhizobacteria utilize 1-aminocyclopropane-1-carboxylate (ACC), the ET precursor, to suppress MAMP (Millet et al., 2010). Rhizomicrobes show some negative effects against SA and ET, while a positive impact on JA is observed in plant immunity.

2.5 Understanding Plant-Microbe Interactions Through Omics

Plant–microbe interactions mediate vital fundamental processes such as the management of environmental stress (biotic and abiotic), phytohormones, nutrient uptake and many more. The interactions between host plants and microbes (beneficial/detrimental) need to be understood in more detail at the molecular level to enhance crops productivity. Omics technologies (genomics, proteomics, transcriptomics, and metabolomics) unravel the plant microbe interactions and give insights that help to understand most associations, at a molecular level. During the interaction, a regulation of protein expression takes place for recognization of the pathogen, initiation of immune response and protection of plant integrity (Jain et al., 2021).

Genomic analyses are essential to understand the overall mechanism of pathogen invasion and infection process. For example, *Myrothecium roridum* and *Rhizoctonia solani* are pathogenic fungi that infect many food crops, including members of Solanaceae, Fabaceae, Asteraceae, Poaceae, and Brassicaceae. Transcriptomics analyses disclosed that the mitogen-activated protein (MAP) kinase pathway encompasses a vital role in fungal infection (Wibberg et al., 2014).

Metatranscriptomics helps to analyze the functional genes expressed during exposure of plants to various environmental stress factors, as well as the related specificities in the rhizosphere microbial profiles. A comparative wheat rhizosphere metatranscriptomic study carried out in suppressive and non-suppressive soils by Hayden et al. (2018) on *Rhizoctonia solani* showed that *Arthrobacter* sp. and *Pseudomonas* sp. were abundant in the non-suppressive soil, whereas *Stenotrophomonas* sp. and *Buttiauxella* sp. were dominant in the suppressive soil. In addition, the suppressive soil comprised cold shock proteins and polyketide cyclase, while non-suppressive soil showed antibiotic expression genes engaged in defense with phenazine and pyrrolnitrin synthesis family protein F (phzF) (Hayden et al., 2018).

Proteomics analysis first introduced in *Pinus* sp. showed protein expression genes that were different in both resistant and susceptible seedlings (Ekramoddoullah & Hunt, 1993). A proteomic study conducted by Mathesius et al. (2001) reported that plants can differentiate the signal of beneficial and detrimental microbes during their interaction. Proteomics analyses have been used to identify plant proteins that are stimulated to suppress PTI (Whitham et al., 2006; Lee et al., 2006). Proteomics and metabolomics analyses revealed the phenotypic expression of the induced immune response (Feussner & Polle, 2015). Proteomics analysis of inoculated PGPR *Kosakonia radicincitans* and uninoculated samples in *Arabidopsis* showed that 12 proteins were related with the cellular stress response (Witzel et al., 2017).

Proteomics studies also showed that ethylene perception plays a significant role in the plant-PGPR interaction (Ibort et al., 2018). Various comparative analyses revealed the initiation of PR proteins during an immune response in plant-microbe interaction. Twenty-five different proteins were found to be involved in several metabolic processes, including photosynthesis, respiration, protein synthesis, stress regulation, phenylpropanoid metabolism, nitrogen metabolism, and defense-related pathways (Jain et al., 2021).

Omics are, therefore, a reliable ensemble of techniques useful to decode the plant-microbe interaction and an employable tool in sustainable agriculture. Such approaches unravel the sets of genes and proteins associated with different environmental setups, focusing on the type of interaction, whether it is beneficial or detrimental for plants.

2.6 Conclusions

Plants overcome damage due to pests and pathogens by enhancing complex infection-induced defense and producing protective secondary compounds. Beneficial microbes, via symbiosis, promote plant growth and induce strong immunity and survival strategies active against pathogens. In conventional farming, chemical-based fertilizers, pesticides, and fungicides are used for agricultural productivity and to combat pathogens but are considered environmentally unfriendly. Therefore, biofertilizers (PGPR and AMF) are actively investigated and applied to enhance the quality and quantity of yields.

The colonization of valuable, beneficial microbes within and around roots triggers the plant innate immunity without activating an immune response (Nishad et al., 2020). Beneficial plant–microbe interactions may hence assist in the management of pathogenic species by inducing a diverse set of immune responses, i.e., via MAMPs/PAMPs (MTI). If the pathogenic infection worsens even after these first lines of defense, the plant may activate a second response profile, i.e., ETI or SAR, induced systemically from belowground to the upper tissues.

Omics approaches represent a further, advanced point of interest unraveling the insights of resistant genes, proteins and molecular mechanisms induced during plant innate immune response. Understanding protein expression is necessary in plant - microbial interactions to hold information on the cellular mechanisms and pathways activated by biotic and abiotic stress factors.

References

Ade, J., DeYoung, B. J., Golstein, C., & Innes, R. W. (2007). Indirect activation of a plant nucleotide binding site-leucine-rich repeat protein by a bacterial protease. *Proceedings of the National Academy of Sciences of the United States of America, 104*, 253–256.

Akiyama, K., Matsuzaki, K., & Hayashi, H. (2005). Plant sesquiterpenes induce hyphal branching in arbuscular mycorrhizal fungi. *Nature, 435*, 824–827.

Aqeel, M., Ran, J., Hu, W., Irshad, M. K., Dong, L., et al. (2023). Plant-soil-microbe interactions in maintaining ecosystem stability and coordinated turnover under changing environmental conditions. *Chemosphere, 318*.

Aslam, S. N., Newman, M. A., Erbs, G., Morrissey, K. L., Chinchilla, D., Boller, T., Jensen, T. T., Castro, C. D., Ierano, T., Molinaro, A., Jackson, R. W., Knight, M. R., & Cooper, R. M. (2008). Bacterial polysaccharides suppress induced innate immunity by calcium chelation. *Current Biology, 18*, 078–1083.

Badri, D. V., Weir, T. L., Lelie, D. V., & Vivanco, J. M. (2009). Rhizosphere chemical dialogues: Plant–microbe interactions. *Current Opinion in Biotechnology, 20*, 642–650.

Bais, H. P., Park, S. W., Weir, T. L., Callaway, R. M., & Vivanco, J. M. (2004). How plants communicate using the underground information superhighway. *Trends in Plant Sciences, 9*, 1360–1385.

Balmer, A., Pastor, V., Gamir, J., Flors, V., & Mani, B. M. (2015). The 'prime-ome': Towards a holistic approach to priming. *Trends in Plant Science, 20*, 443–452.

Bernsdorff, F., Doering, A., Gruner, K., Schucka, S., Bräutigamb, A., & Zeier, J. (2016). Pipecolic acid orchestrates plant systemic acquired resistance and defense priming via salicylic acid-dependent and independent pathways. *Plant Cell, 28*, 102–129.

Bhatt, K., Suyal, D. C., Kumar, S., Singh, K., & Goswami, P. (2022). New insights into engineered plant-microbe interactions for pesticide removal. *Chemosphere, 309*.

Boller, T., & He, S. Y. (2009). Innate immunity in plants: An arms race between pattern recognition receptors in plants and effectors in microbial pathogens. *Plant-Microbe Interactions, 324*, 742–744.

Bukhata, S., Imranb, A., Javaidc, S., Shahid, M., Majeed, A., & Naqqash, T. (2020). Communication of plants with microbial world: Exploring the regulatory networks for PGPR mediated defense signaling. *Microbiological Research, 126486*.

Cameron, D. D., Neal, A. L., van Wees, S. C. M., & Ton, J. (2013). Mycorrhiza induced resistance: More than the sum of its parts? *Trends in Plant Science, 18*, 539–545.

Chaparro, J. M., Sheflin, A. M., Manter, D. K., & Vivanco, J. M. (2012). Manipulating the soil microbiome to increase soil health and plant fertility. *Biology and Fertility of Soils, 48*, 489–499.

Chen, Y. C., Holmes, E. C., Rajniak, J., Kima, J. G., Tang, S., Fischer, C. R., Mudgett, M. B., & Sattely, E. S. (2018). N-hydroxy-pipecolic acid is a mobile metabolite that induces systemic disease resistance in Arabidopsis. *Proceedings of the National Academy of Sciences of the United States of America, 115*, E4920–E4929.

Cohn, J., Sessa, G., & Martin, G. B. (2001). Innate immunity in plants. *Current Opinion in Immunology, 13*, 55–62.
Coll, N. S., Epple, P., & Dangl, J. L. (2011). Programmed cell death in the plant immune system. *Cell Death and Differentiation, 18*, 1247–1256.
Conrath, U. (2006). Systemic acquired resistance. *Plant Signaling & Behavior, 1*, 179–184.
Cook, D. E., Mesarich, C. H., & Thomma, B. P. H. J. (2015). Understanding plant immunity as a surveillance system to detect invasion. *Annual Review of Phytopathology, 53*, 541–563.
de Jonge, R., Bolton, M. D., & Thomma, B. P. H. J. (2011). How filamentous pathogens co-opt plants: The ins and outs of fungal effectors. *Current Opinion in Plant Biology, 14*, 400–406.
Delaney, T. P., Uknes, S., Vernooij, B., Friedrich, L., Weymann, K., et al. (1994). Central role of salicylic acid in plant disease resistance. *Science, 266*, 1247.
Dempsey, D. M. A., & Klessig, D. F. (2012). SOS – Too many signals for systemic acquired resistance? *Trends in Microbiology, 17*, 538–545.
Doornbos, R. F., Loon, L. C. V., & Bakker, P. A. H. M. (2012). Impact of root exudates and plant defense signaling on bacterial communities in the rhizosphere. *A Review. Agronomy for Sustainable Development, 32*, 227–243.
Dumas-Gaudot, E., Gollotte, A., Cordier, C., Gianinazzi, S., & Pearson, V. (2000). Modulation of host defence systems. In Y. Kapulnik & D. Douds (Eds.), *Arbuscular mycorrhizas: Physiology and function* (pp. 173–200). Kluwer Academic Publishers.
Ekramoddoullah, A. K. M., & Hunt, R. S. (1993). Changes in protein profile of susceptible and resistant sugar pine foliage infected with the white pine blister rust fungus *Cronartium ribicola. Canadian Journal of Plant Pathology, 15*, 259–264.
Etesami, H., & Beattie, G. A. (2017). Plant-microbe interactions in adaptation of agricultural crops to abiotic stress conditions. *Probiotics and Plant Health*, 163–200.
Feussner, I., & Polle, A. (2015). What the transcriptome does not tell – Proteomics and metabolomics are closer to the plants' patho-phenotype. *Current Opinion in Plant Biology, 26*, 26–31.
Finkel, O. M., Castrillo, G., Paredes, S. H., González, I. S., & Dangl, J. L. (2017). Understanding and exploiting plant beneficial microbes. *Current Opinion in Plant Biology, 38*, 155–163.
Gaffney, T., Friedrich, L., Vernooij, B., Negrotto, D., Nye, G., et al. (1993). Requirement of salicylic acid for the induction of systemic acquired resistance. *Science, 261*, 754–754.
Gao, X., Chen, X., Lin, W., Chen, S., Lu, D., Niu, Y., Li, L., Cheng, C., McCormack, M., Sheen, J., Shan, L., & He, P. (2013). Bifurcation of Arabidopsis NLR immune signaling via Ca(2+)-dependent protein kinases. *PLoS Pathogens, 9*, 1003127.
Garrido, J. M. G., & Ocampo, J. A. (2002). Regulation of the plant defense response in arbuscular mycorrhizal symbiosis. *Journal of Experimental Botany, 53*, 1377–1386.
Gaskins, M. H., Albrecht, S. L., & Hubbell, D. H. (1995). Rhizosphere bacteria and their use to increase plant productivity: A review. *Agriculture, Ecosystems and Environment, 12*, 99–116.
Gourion, B., Berrabah, F., Ratet, P., & Stacey, G. (2015). Rhizobium–legume symbioses: The crucial role of plant immunity. *Trends in Plant Science, 20*, 186–194.
Grant, M., Brown, I., Adams, S., Knight, M., Ainslie, A., & Mansfield, J. (2000). The RPM1 plant disease resistance gene facilitates a rapid and sustained increase in cytosolic calcium that is necessary for the oxidative burst and hypersensitive cell death. *The Plant Journal, 23*, 441–450.
Gupta, A., Gopal, M., & Tilak, K. V. (2000). Mechanism of plant growth promotion by rhizobacteria. *Indian Journal of Experimental Biology, 38*, 856–862.
Gupta, A., Bano, A., Rai, S., Mishra, R., Singh, M., et al. (2022). Mechanistic insights of plant-microbe interaction towards drought and salinity stress in plants for enhancing the agriculture productivity. Plant. *Stress, 4*.
Hacquard, S., Spaepen, S., Oter, R. G., & Lefert, P. S. (2017). Interplay between innate immunity and the plant microbiota. *Annual Review of Phytopathology, 55*, 565–589.
Haldar, S., & Sengupta, S. (2015). Plant-microbe cross-talk in the rhizosphere: Insight and biotechnological potential. *The Open Microbiology Journal, 9*, 1–7.

Hartmann, M., Zeier, T., Bernsdorff, F., Holzel, T., Ganter, C., & Zeier, J. (2018). Flavin monooxygenase-generated N-hydroxypipecolic acid is a critical element of plant systemic immunity. *Cell, 173*(456), e16–469.e16.

Hause, B., Mrosk, C., Isayenkov, S., & Strack, D. (2007). Jasmonates in arbuscular mycorrhizal interactions. *Phytochemistry, 68*, 101–110.

Hayden, H. L., Savin, K. W., Wadeson, J., Gupta, V. V. S. R., & Mele, P. M. (2018). Comparative metatranscriptomics of wheat rhizosphere microbiomes in disease suppressive and non-suppressive soils for Rhizoctonia solani AG8. *Frontiers in Microbiology, 9*.

Hayward, A. P., Tsao, J., & Kumar, D. S. P. (2009). Autophagy and plant innate immunity: Defense through degradation. *Seminars in Cell & Developmental Biology, 20*, 1041–1047.

Hu, Z., Yan, C., Liu, P., Huang, Z., Ma, R., et al. (2013). Crystal structure of NLRC4 reveals its autoinhibition mechanism. *Science, 341*, 172–175.

Huang, P. M., Wang, M. K., & Chui, C. Y. (2005). Soil mineral-organic matter-microbe interactions: Impacts on biogeochemical process and biodiversity in soils. *Pedobiologia, 4*, 609–635.

Ibort, P., Imai, H., Uemura, M., & Aroca, R. (2018). Proteomic analysis reveals that tomato interaction with plant growth promoting bacteria is highly determined by ethylene perception. *Journal of Plant Physiology, 220*, 43–59.

Iriti, M., & Varoni, E. M. (2015). Chitosan-induced antiviral activity and innate immunity in plants. *Environmental Science and Pollution Research, 22*, 2935–2944.

Iwai, T., Miyasaka, A., Seo, S., & Ohashi, Y. (2006). Contribution of ethylene biosynthesis for resistance to blast fungus infection in young rice plants. *Plant Physiology, 142*, 1202–1215.

Jain, A., Singh, H. B., & Das, S. (2021). Deciphering plant-microbe crosstalk through proteomics studies. *Microbiological Research, 242*.

Jeworutzki, E., Roelfsema, M. R., Anschutz, U., Krol, E., Elzenga, J. T. M., Felix, G., Boller, T., Hedrich, R., & Becker, D. (2010). Early signaling through the arabidopsis pattern recognition receptors fls2 and efr involves ca-associated opening of plasma membrane anion channels. *The Plant Journal, 62*, 367–378.

Jones, J. D. G., & Dangl, J. L. (2006). The plant immune system. *Nature, 444*, 323–329.

Jones, K. M., Sharopova, N., Lohar, D. P., & Walker, G. C. (2008). Differential response of the plant medicago truncatula to its symbiont *Sinorhizobium meliloti* or an exopolysaccharide-deficient mutant. *PNAS, USA, 105*, 704–709.

Jung, H. W., Tschaplinski, T. J., Wang, L., Glazebrook, J., & Greenberg, J. T. (2009). Priming in systemic plant immunity. *Science, 324*, 89–91.

Jung, S. C., Medina, A. M., Raez, J. A. L., & Pozo, M. J. (2012). Mycorrhiza-induced resistance and priming of plant defenses. *Journal of Chemical Ecology, 38*, 651–664.

Khare, E., Tyagi, S., & Patil, K. S. (2020). Language of plant-microbe-microbe interactions in rhizospheric ecosystems. *Molecular Aspects of Plant Beneficial Microbes in Agriculture*, 59–76.

Kloppholz, S., Kuhn, H., & Requena, N. (2011). A secreted fungal effector of glomus intraradices promotes symbiotic biotrophy. *Current Biology, 21*, 1204–1209.

Kouchi, H., Anraku, H. I., Hayashi, M., Hakoyama, T., Nakagawa, T., Umehara, Y., Suganuma, N., & Kawaguchi, M. (2010). How many peas in a pod? Legume genes responsible for mutualistic symbioses underground. *Plant & Cell Physiology, 51*, 1381–1397.

Kour, D., Rana, K. L., Yadav, A. N., Yadav, N., Kumar, M., Kumar, V., Vyas, P., Dhaliwal, H. S., & Saxena, A. K. (2020). Microbial biofertilizers: Bioresources and eco-friendly technologies for agricultural and environmental sustainability. *Biocatalysis and Agricultural Biotechnology, 23*.

Lee, B. J., Kwon, S. J., Kim, S. K., Kim, K. J., Park, C. J., et al. (2006). Functional study of hot pepper 26S proteasome subunit RPN7 induced by tobacco mosaic virus from nuclear proteome analysis. *Biochemical and Biophysical Research Communications, 351*, 405–411.

Li, L., Li, M., Yu, L., Zhou, Z., Liang, X., Liu, Z., Cai, G., Gao, L., Zhang, X., Wang, Y., Chen, S., & Zhou, J. M. (2014). The FLS2-associated kinase bik1 directly phosphorylates the nadph oxidase rbohd to control plant immunity. *Cell Host & Microbe, 15*, 329–338.

Liang, Y., Cao, Y., Tanaka, K., Thibivilliers, S., Wan, J., Choi, J., Kang, C. H., Qiu, J., & Stacey, G. (2013). Nonlegumes respond to rhizobial nod factors by suppressing the innate immune response. *Science, 341*, 6152.

Liang, X., Ding, P., Lian, K., Wang, J., Ma, M., Li, L., Li, L., Li, M., Zhang, X., & Zhou, J. M. (2016). Arabidopsis heterotrimeric G proteins regulate immunity by directly coupling to the FLS2 receptor. *eLife, 5*, e13568.

Liang, X., Ma, M., Zhou, Z., Wang, J., Yang, X., et al. (2018). Ligand triggered de-repression of Arabidopsis heterotrimeric G proteins coupled to immune receptor kinases. *Cell Research, 28*, 529–543.

Lorenzo, G. D., Ferrari, S., Cervone, F., & Okun, E. (2018). Extracellular DAMPs in plants and mammals: Immunity, tissue damage and repair. *Trends in Immunology, 39*, 937–350.

Mabood, F., Zhou, X., & Smith, D. L. (2014). Microbial signaling and plant growth promotion. *Canadian Journal of Plant Science, 94*, 1051–1063.

Mathesius, U., Keijzers, G., Natera, S. H. A., Weinman, J. J., Djordjevic, M. A. M., & Rolfe, B. G. (2001). Establishment of a root proteome reference map for the model legume Medicago truncatula using the expressed sequence tag database for peptide mass fingerprinting. *Proteomics, 1*.

Millet, Y. A., Danna, C. H., Clay, N. K., Songnuan, W., Simon, M. D., et al. (2010). Innate immune responses activated in *Arabidopsis* roots by microbe-associated molecular patterns. *Plant Cell, 22*, 973–990.

Mishra, S., Singh, A., Keswani, C., Saxena, A., Sarma, B. K., & Singh, H. B. (2015). Harnessing plant-microbe interactions for enhanced protection against phytopathogens. *Plant Microbes Symbiosis*, 111–125.

Miya, A., Albert, P., Shinya, T., & Shibuya, N. (2007). CERK1, a LysM receptor kinase, is essential for chitin elicitor signaling in *Arabidopsis*. *PNAS, USA, 104*, 19613–19618.

Morel, M. A., & Sowinski, S. C. (2013). The complex molecular signaling network in microbe–plant interaction. *Plant Microbe Symbiosis: Fundamentals and Advances*, 169–199.

Muthamilarasan, M., & Prasad, M. (2013). Plant innate immunity: An updated insight into defense mechanism. *Journal of Biosciences, 38*, 433–449.

Nishad, R., Ahmed, T., Rahman, V. J., & Kareem, A. (2020). Modulation of plant defense system in response to microbial interactions. *Frontiers in Microbiology, 11*, 1298.

Okazaki, S., Kaneko, T., Sato, S., & Saeki, K. (2013). Hijacking of leguminous nodulation signaling by the rhizobial type III secretion system. *Proceedings of the National Academy of Sciences of the United States of America, 110*, 17131–17136.

Oldroyd, G. E., Murray, J. D., Poole, P. S., & Downie, J. A. (2011). The rules of engagement in the legume-rhizobial symbiosis. *The Annual Review of Genomics, 45*, 119–144.

Parasuraman, P., Pattnaik, S., & Busi, S. (2019). Plant-microbe interactions in ecosystems functioning and sustainability. *New and Future Developments in Microbial Biotechnology and Bioengineering*, 255–266.

Park, E., Nedo, A., Caplan, J. L., & Kumar, S. P. D. (2017). Plant–microbe interactions: organelles and the cytoskeleton in action. *New Phytologist, 217*, 1012–1028.

Pennell, R. I., & Lamb, C. (1997). Programmed cell death in plants. *Plant Cell, 9*, 1157–1168.

Pieterse, C. M. J., Does, D. V., Zamioudis, C., Reyes, A. L., & Wees, S. C. M. V. (2012). Hormonal modulation of plant immunity. *Annual Review of Cell and Developmental Biology, 28*, 489–521.

Pozo, M. J., Verhage, A., Andrade, G. J., Garcia, J. M., & Aguilar, C. A. (2009). Priming plant defence against pathogens by arbuscular mycorrhizal fungi. *Mycorrhizas-Functional Processes and Ecological Impact*, 123–135.

Pritchard, L., & Birch, P. R. J. (2014). The zigzag model of plant–microbe interactions: Is it time to move on? *Molecular Plant Pathology, 15*(9), 865–870.

Rosier, A., Medeiros, F. H. V., & Bais, H. P. (2018). Defining plant growth promoting rhizobacteria molecular and biochemical networks in beneficial plant-microbe interactions. *Plant and Soil, 428*, 35–55.

Sanon, A., Andrianjaka, Z. N., Prin, Y., Bally, R., Comte, T. G., & Duponnois, R. (2009). Rhizopshere microbiota interfers with plant-plant interactions. *Plant and Soil, 321*, 259–278.

Schäfer, P., & Eichmann, R. (2012). The endoplasmic reticulum in plant immunity and cell death. *Frontiers in Plant Science, 3*.

Schenk, P. M., Carvalhais, L. C., & Kazan, K. (2012). Unraveling plant–microbe interactions: Can multi-species transcriptomics help? *Trends in Biotechnology, 30*, 177–184.

Shah, J., & Zeier, J. (2013). Long-distance communication and signal amplification in systemic acquired resistance. *Frontiers in Plant Science, 4*.

Siciliano, V., Genre, A., Balestrini, R., Cappellazzo, G., deWit, P. J. G. M., & Bonfante, P. (2007). Transcriptome analysis of arbuscular mycorrhizal roots during development of the prepenetration apparatus. *Plant Physiology, 144*, 1455–1466.

Singh, J. S., Pandey, V. C., & Singh, D. P. (2011). Efficient soil microorganisms: A new dimension for sustainable agriculture and environmental development. *Agriculture, Ecosystems and Environment, 140*, 339–353.

Tian, W., Hou, C. C., Ren, Z. J., Wang, C., Zhao, F., et al. (2019). A calmodulin-gated calcium channel links pathogen patterns to plant immunity. *Nature, 572*, 131–135.

Trusov, Y., & Botella, J. R. (2012). New faces in plant innate immunity: Heterotrimeric G proteins. *Journal of Plant Biochemistry and Biotechnology, 21*, 40–47.

Urano, D., & Jones, A. M. (2013). Heterotrimeric G protein–coupled signaling in plants. *Annual Review of Plant Biology, 65*, 365–384.

Wang, J., Hu, M., Wang, J., Qi, J., Han, Z., et al. (2019). Reconstitution and structure of a plant NLR resistosome conferring immunity. *Science, 364*, eaav5870.

Whitham, S. A., Yang, C., & Goodin, M. M. (2006). Global impact: Elucidating plant responses to viral infection. *Molecular Plant-Microbe Interactions, 19*, 1207–1215.

Wibberg, D., Jelonek, L., Rupp, O., Krober, M., Goesmann, A., et al. (2014). Transcriptome analysis of the phytopathogenic fungus Rhizoctonia solani AG1-IB 7/3/14 applying high-throughput sequencing of expressed sequence tags (ESTs). *Fungal Biology, 118*, 800–813.

Witzel, K., Üstün, S., Schreiner, M., Grosch, R., Bornke, F., & Ruppel, S. (2017). A proteomic approach suggests unbalanced proteasome functioning induced by the growthpromoting bacterium Kosakonia radicincitans in *Arabidopsis*. *Frontiers in Plant Science, 8*.

Yogendra, K. N., Kumar, A., Sarkar, K., Li, Y., Pushpa, D., Mosa, K. A., Duggavathi, R., & Kushalappa, A. C. (2015). Transcription factor StWRKY1regulates phenylpropanoid metabolites conferring late blight resistance in potato. *Journal of Experimental Botany, 66*, 7377–7389.

Yuan, M., Ngou, B. P. M., Ding, P., & Xin, F. X. (2021). PTI-ETI crosstalk: An integrative view of plant immunity. *Current Opinion in Plant Biology, 62*.

Zamioudis, C., & Pieterse, C. M. J. (2012). Modulation of host immunity by beneficial microbes. *Molecular Plant-Microbe Interactions, 25*, 139–150.

Zhang, J., & Zhou, J. M. (2010). Plant immunity triggered by microbial molecular signatures. *Molecular Plant, 3*, 783–793.

Zhang, X., Dodds, P. N., & Bernoux, M. (2017). What do we know about NOD like receptors in plant immunity? *Annual Review of Phytopathology, 55*, 205–229.

Zipfel, C., Kunze, G., Chinchilla, D., Jones, J. D. G., Boller, T., & Felix, G. (2006). Perception of the bacterial PAMP EF-Tu by the receptor EFR restricts *agrobacterium*-mediated transformation. *Cell, 125*, 749–760.

Chapter 3
Plant-Microbe Interactions, Signalling Events, and Cellular Networks Against Harmful Pathogens

Tamanna Bhardwaj, Deepak Kumar, Parkirti, Vikram, Amandeep Singh, Nitika Kapoor, Puja Ohri, and Renu Bhardwaj

Abstract Plants have to face several pathogenic agents that pose a threat to growth and development. Their immune response strategy terminates progression of pathogenic microbial agents. Failure in the plant immune system permits entrance of pathogens, whosr reproduction harnesses primary production. Plants recognise pathogens through membrane receptors that trigger immunity (MAMP-triggered Immunity, MTI). This involves rapid calcium influx and oxidative burst. In response to MTI pathogens release effector proteins in cells. In response to MTI plants activate a second type of immune response, called Effector Triggered Immunity (ETI), which involves a cascade of complex signalling events leading to the pathogen suppression. Plants exhibit a further, effective immune response following a pathogen encounter, via a process known as Systemic Acquired Resistance (SAR). In addition, plants also evolved an induced immune response due to their intricate association with unharmful, beneficial microbes. Association with mycorrhizae, rhizobia and endophytespositively induce such plant immune response. The present study reviews how plant-microbe interactions boost the plant immune system and how they can be exploited in plant protection.

Keywords Immunity · Pathogenic agents · Oxidative burst

T. Bhardwaj · A. Singh · N. Kapoor · R. Bhardwaj (✉)
Department of Botanical and Environmental Sciences, Guru Nanak Dev University, Amritsar, India
e-mail: renu.botenv@gndu.ac.in

D. Kumar · Parkirti · Vikram · P. Ohri
Department of Zoology, Guru Nanak Dev University, Amritsar, India

P. Mathur, S. Roy (eds.), *Plant Microbiome and Biological Control*, Sustainability in Plant and Crop Protection 20,
https://doi.org/10.1007/978-3-031-75845-4_3

Abbreviations

ETI	Effector Triggered Immunity
MAMP	triggered Immunity
SAR	Systemic Acquired Resistance

3.1 Introduction

Plants encounter a variety of microorganisms throughout their lifetimes, with interactions that can either be advantageous or detrimental, leading to mutualistic or pathogenic connections, respectively (Rodriguez et al., 2019). They also have ability to modify their immune response towards beneficial or pathogenic microorganisms (Pieterse et al., 2014). Plants possess a sophisticated immune system capable to deploy resistance towards a majority of microbes. It involves physical barriers such as cuticle or cell wall, and an arsenal of antimicrobial secondary metabolites. Once a microorganism succeeds in overcoming these barriers, it establishes a connection with the plant system that can be either beneficial or a detrimental (Nishad et al., 2020).

Microorganisms have the ability to induce immune responses. Mutualistic associations can induce an immune response *vs* other microorganisms. The role of beneficial non-pathogenic microbes in defence priming of host plants indicates that plant immunity is triggered after an effective interaction is achieved (Singh et al., 2016). Examples of immune responses induced by beneficial microbes include both mycorrhizal and rhizobial associations. Nevertheless, the detailed signalling mechanism working behind this induced immunity still needs to be completely understood, to develop sustainable control measures *vs* various plant pests and diseases. The use of beneficial microbes to activate the plant immune system against pathogenic microbes will serve as a green technology in disease management, alternative to environment-damaging pesticides.

This chapter discusses plant defence system involving both innate and systemic acquired immunity. Further, we focussed on the plant interactions with diverse microbial communities, both mycorrhizal and rhizobia, which can help in designing new synthetic microbial communities (SynComs) with an increased plant protection potential (Nishad et al., 2020).

3.2 Plant Defence Responses

Although plants are attacked by a broad variety of pathogens, only a few of them can cause a disease (Kaur et al., 2022; Kumar & Ohri, 2023). To protect themselves from microbial infections, plants rely on a complex immune system. However, a constitutive development of defence measures is resource-intensive and typically accompanied with slower growth (Karasov et al., 2017). Because inducible immune responses enable a timely constrained investment in defence, when the microbes/

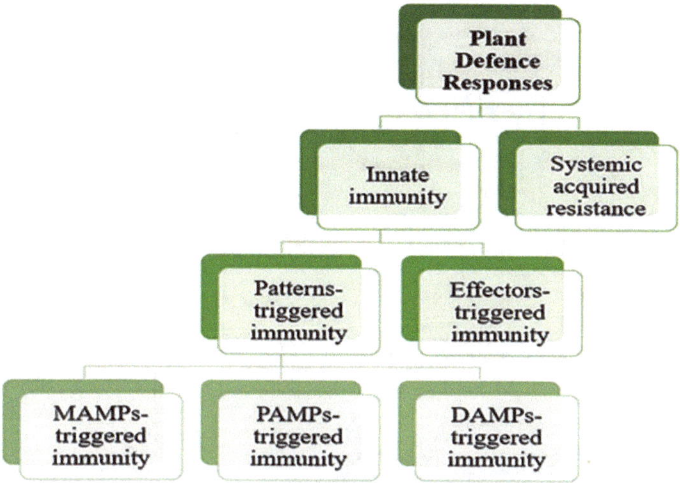

Fig. 3.1 General scheme of plant defence responses. (*MAMPs* Microbe-associated molecular pattern, *PAMPs* Pathogen-associated molecular patter and *DAMPs* Damage-associated molecular pattern)

pathogens attack takes place, an acceptable compromise between defence and growth must be reached (Nobori & Tsuda, 2019; Zeier, 2021). Through its innate immune system the plant can defend itself fron the onslaught of these pathogens.

The two main components of the innate immune system are pattern-triggered immunity (PTI) and effector-triggered immunity (ETI) (Pruitt et al., 2021). In contrast to ETI, which frequently causes a hypersensitive response, PTI confers a lasting resistance (Pruitt et al., 2021; Huang et al., 2023). Further, to protect healthy tissues from further pathogen invasions, the defence mechanism that aroused at the infection site activates a systemic defence response in distal plant parts. This defence mechanism is known as systemic acquired resistance (SAR), which offers long-term protection against a wide variety of pests and diseases (Kachroo & Kachroo, 2020; Kamle et al., 2020). A general illustration of plant defence responses is given in Fig. 3.1.

3.2.1 *Innate Immunity*

During their coevolution, plants and microorganisms have been constantly subjected to a survival race. Many plants in nature appear immune to most infections; however, several pathogenic microbes evolved necrotrophic and pathogenic behaviors that cause severe illnesses (Lyu et al., 2021). To protect themselves from a variety of biotic assailants, plants depend on their innate immune systems (Fig. 3.2). Their pre-fabricated defence layer includes the cell wall as well as pre-produced metabolites, that serve as primary response against pathogenic invasions (Yuan et al., 2021). Plants evolved multilevel protection and tracking networks to effectively respond to and combat harmful microorganisms (Kumar et al., 2023; Sharma et al., 2023).

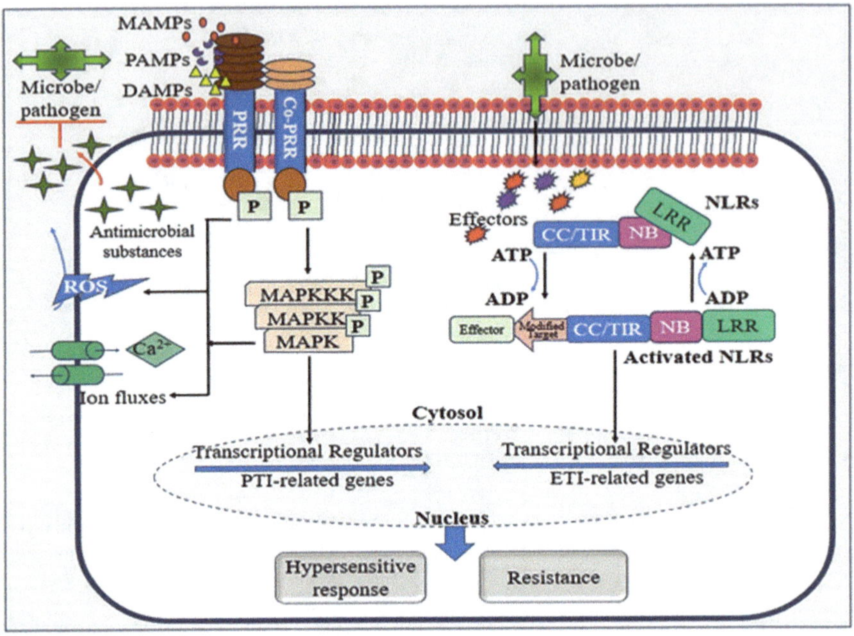

Fig. 3.2 Schematic illustration of plants innate immune response composed of patter-triggered immunity (PTI) and effector triggered immunity (ETI). PRRs identification of MAMPs is necessary for PTI activation. Effectors modify the structure of the R protein to activate it. PTI and ETI both use their downstream signaling pathways to regulate the expression of genes relevant to such processes. (*MAMPs* Microbe-associated molecular patterns, *PAMPs* Pathogen-associated molecular patterns, *DAMPs* Damage-associated molecular patterns, *PRR* Pattern recognition receptors, *MAPK* Mitogen-activated protein kinase, *NLRs* nucleotide binding (NB) and leucin rich repeat (LRR) containing receptor, *CC* Coiled-coil domain, *TIR* Toll-interleukin receptor domain)

PTI, the initial level of the immune response, is triggered by microbe/pathogen-associated molecular patterns (MAMPs/PAMPs), which are conserved pathogen molecular structures such as fungal chitin or bacterial flagellin Damage-associated molecular patterns (DAMPs) are molecules produced because of the interactions with pathogens/microbes i.e., peptides and oligosaccharides (DeFalco & Zipfel, 2021). Pattern recognition receptors (PRRs)are plasma membrane-localized plant immune receptors, that are mostly found in the form of receptor-like proteins and receptor-like protein kinases, and that recognise these inducers (Garcia & Steinbrenner, 2023). These receptors are activated upon infection, which results in a variety of plant defence reactions that prevent the pathogen colonisation and dissemination. PTI activates numerous signaling pathways in the host cells. However, many microbes use a range of effector proteins to thwart MAMPs-triggered immunity (MTI) (Bernoux et al., 2022).

The second level of plant immune response, known as ETI, is activated when effector proteins are detected by special plant receptors (Nguyen et al., 2021; Remick et al., 2023). There are two conserved traits in cuch compounds, namelythe nucleotide-binding (NB) and leucine-rich repeat (LRR) domains, collectively known as NLRs, found during investigations of several R (resistance) proteins (Ray et al., 2019; Zhai et al., 2022). Indirectly or directly, NLRs preferentially detect the effectors and eventually trigger a hypersensitive reaction, a type of rapid and localised programmed cell death. Although the duration and intensity of the ETI responses are frequently significantly higher, immunological responses induced by NLRs and PRRs are similar (Han & Jung, 2013; Chang et al., 2022).

3.2.1.1 MAMPs-Triggered Immunity (MTI)

The plant membrane receptors or PRRs, detect MAMPs to trigger MTI, which leads to a swift influx of calcium, the accumulation of reactive oxygen species (ROS, commonly referred to as the oxidative burst), phosphorylation cascades of mitogen-activated protein kinase (MAPK), associated to cell wall modifications and to the expression of defence-related genes (Moeder et al., 2019; Gonçalves Dias et al., 2022) (Fig. 3.3). Among the conserved MAMPs found in pathogens are bacterial flagellin, elongation factor, lipopolysaccharides, and the fungal chitin. Given that MTI is a crucial immunological response in plants, focus has recently been given to understanding how PRRs detect MAMPs, and to the intricate network that underlies signaling events. The RLK family member Flagellin-Insensitive 2 (fls2) is a well-researched PRR in *Arabidopsis* (Lü et al., 2022; Nishad et al., 2020). For innate immune signaling, fls2 needs the G protein. Since the discovery of the G protein role in plant defence, researchers have developed a strong interest in this process. Plant G proteins are connected to receptor kinases, such as fls2, but unlike animal G proteins, they are self-activating and therefore do not depend on the G protein-coupled receptors (Xu et al., 2019; Maruta et al., 2021).

By enlisting the coreceptor BRI1-associated receptor kinase (BAK1), the receptor-like cytoplasmic kinase *Botrytis*-induced kinase 1 (BIK1) and the G protein — that trigger defence responses —, fls2 creates a dynamic complex in response to the detection of the flagellin epitope (flg22) (Offor et al., 2020; Zhang & Kong, 2022). The regulator of G protein signaling 1 (RGS1) is phosphorylated by BIK1 in response to the activation of FLS2 by Flg22. This phosphorylation causes RGS1 to separate from the FLS2-G protein complex (Wang et al., 2020; Zhou & Zhang, 2020). Immune signallig is positively regulated after the RGS1-mediated suppression of G protein is lifted.

The rapid ROS burst is triggered by MAMP sensing and is carried out by the NADPH oxidase RBOHD (Kadota et al., 2015; Pfeilmeier et al., 2021). BIK1 effectively phosphorylates RBOHD to activate flg22-induced ROS. Superoxide dismutase (SOD) transforms the membrane-impermeable superoxide (O_2^-) produced by RBOHD into H_2O_2 in apoplasts, which reaches the cytosol and causes an increase in cytosolic Ca^{2+} (Marcec & Tanaka, 2021; Blekemolen et al., 2023).

One of the early reactions to MAMP sensing is an extracellular Ca^{2+} inflow in the cytosol, sometimes referred to as Ca^{2+} burst and positively controlled by BIK1 (de Azevedo et al., 2021; Bai et al., 2023). Further, Ca^{2+} stimulates phospholipase C, a

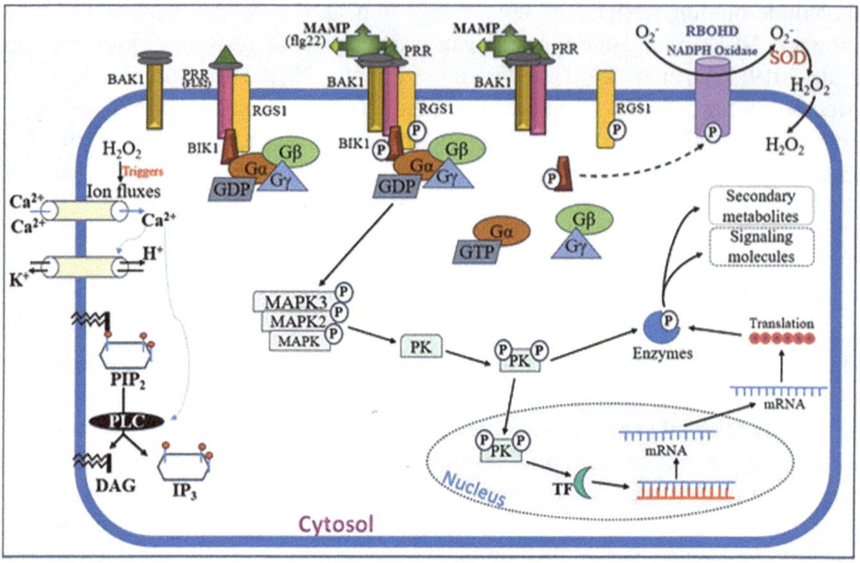

Fig. 3.3 Schematic representation of microbe-associated molecular patterns-triggered immunity (MTI), based on the *Arabidopsis* fls2 model. An immune response complex develops immediately after fls2 (PRR) detects flg22 (MAMP). When the latter is detected by the receptor complex (FLS2, BIK1, BAK1 and G protein), an active FLS2-BAK1 receptor complex is assembled with the phosphorylation of RGS1 by BIK1, separating RGS1 from the G proteins and fls2. The initial response to MAMP recognition is an increased Ca^{2+} influx that activates the H^+/K^+ exchange. PLC is also activated by Ca^{2+}, which results in the escape of a secondary messenger at a later stage. When the G protein or Ca^{2+} activate PLC, PIP_2 is broken down into DAG and IP_3, which ultimately leads to a structural cell instability. The NADPH oxidase (RBOHD) is phosphorylated by BIK1, which causes a generation of H_2O_2. The protein kinase is phosphorylated by the active sensor molecule, which either causes the expression of the defence gene or the activation of the enzyme responsible for metabolite formation. TF is phosphorylated as a result of the MAPK cascade, which also activates phytohormone signaling. (*BAK1* BRI1-associated receptor kinase, *PRR* Pattern recognition receptors, *fls2* flagellin-insensitive 2, *BIK1 Botrytis*-induced kinase 1, *RGS1* Regulator of G protein signaling 1, *MAMP* Microbe-associated molecular patters, *flg22* flagellin epitope, *PIP_2* Phosphatidyl-4,5-bisphosphate, *PLS* Phospholipase C, *DAG* diglyceride, *IP_3* Inositol 3-phosphate, H_2O_2 Hydrogen peroxide, *SOD* Superoxide dismutase, *MAPK* mitogen-activated protein kinase, *PK* Protein kinase, *RBOHD* Respiratory burst oxidase homologue D, *TF* Transcription factor)

secondary messenger released downstream, that causes more Ca^{2+} to be released. A sensor molecule (such as calmodulin or Ca^{2+}-dependent protein kinases) binds to Ca^{2+} (Dey & Ghosh, 2022). The protein kinase is phosphorylated by the active sensor molecule, which either causes the expression of the defence gene or the activation of the enzyme responsible for metabolite formation. The efflux Cl^-, K^+ and NO_3^- is also activated by the Ca^{2+} influx, which results in external alkanization and depolarization of the cell membrane (Nishad et al., 2020). Almost at the same time of Ca^{2+} burst, few minutes following the perception of flg22, MAPK stimulation takes place (Chi et al., 2021; Kozyulina et al., 2023). The phosphorylation of the ethylene-dependent transcription factor, the activation of the ethylene responsive genes and the production of ethylene are all caused by MAPK activation (Vanthana et al., 2019). Multiple transcription factors that regulate numerous genes, including those that play a role in phytohormone signaling and the synthesis of antimicrobial substances, are phosphorylated because of the MAPK activity (Chakraborty, 2023; Chandra & Sharma, 2023). Finally, this intricate signaling network results in a plant-induced defence.

3.2.1.2 Effector Triggered Immunity (ETI)

ETI frequently offers resistance when pattern-triggered immunity (PTI), the initial immune response layer, is rendered ineffective due to pathogen effectors. In ETI, host plants detect pathogen effectors directly or indirectly through R proteins and mount a more effective and quickening defensive response. The striking characteristic of ETI is the hypersensitive response, which exhibits a rapid induction of programmed cell death at the local infection site (localized PCD). The primary purpose of cell death is to kill biotrophic pathogens, which obtain nutrients from living cells (Dangl et al., 2013). LRRs and NB proteins are the two types of resistance proteins that are most frequently discovered in plants. NLR may also change structurally and become associated with other NLRs, after effector recognition. The ETI response is the result of "helper" NLRs being activated by the dimerization or oligomerization of NLRs, which also sends signals to downstream components. PTI was initially believed to have a minor impact on ETI (Richard et al., 2018).

In ETI, the pathogen virulence-promoting effector protein(s) are recognized by a R gene product, which is typically found inside the plant cell. Most R genes produce NB-LRR proteins. ETI happens when an effector is recognized by a corresponding NB-LRR protein (Van Wersch & Li, 2019). Several NB-LRR proteins detect the outcomes of the effector actions to recognize them indirectly. For example, the *Arabidopsis* RPM1-interacting protein 4 (RIN4) is targeted for degradation by the bacterial effector AvrRpt2. The matching R gene, involvedin resistance to *P. syringae* expressing AvrRpt2 (RPS2), detects RIN4 degradation to initiate ETI. The equivalent R gene product, eliciting resistance to *P. syringae* expressing AvrRpm1 (RPM1), detects the phosphorylation of RIN4 to initiate ETI. Two additional, unrelated effectors from *P. syringae*, AvrRpm1 and AvrB, target RIN4 for phosphorylation (Mackey et al., 2003). As a result, three unrelated effectors have the same host protein as a target, which raises the possibility that RIN4 manipulation is crucial for pathogenesis. RIN4, with the H^+-ATPases AHA1 and AHA2, have recently been found to govern stomatal openings, regulating the entry of bacterial pathogens into the leaf. However, NB-LRR recognition events are not always indirect. For instance, the RRS1-R protein from *Arabidopsis* binds directly to PopP2, an effector protein from *Ralstonia solanacearum*, causing ETI.

3.2.2 Systemic Acquired Response

Many plants become more resistant to subsequent pathogen attacks after contracting necrotizing infections, even in organs that have not been inoculated or affected by the disease. Systemic acquired resistance (SAR) refers to this kind of increased resistance (Ryals et al., 1996). When in the SAR state plants are sensitive and primed to defend themselves against a second pathogen attack (Sticher et al., 1997). A model supporting the presence of a sort of "memory" is the acquired disease resistance, which is dependent on the capacity to access a prior attack. Although the long-distance signal for SAR identification is unknown, it is claimed that salicylic acid (SA) is necessary to establish SAR in a distant tissue (Durrant & Dong, 2004). The expression of a group of genes known as SAR genes, some of which encode pathogenesis-related (PR) proteins, is linked to the insurgence of SAR (Ward et al., 1991). It has been discovered that some PR proteins are acidic −1,3-glucanases (BGL2) and chitinases (PR-3), which may be able to hydrolyze microbial cell wall components (Van Loon & Van Strien, 1999). As a result, the buildup of PR proteins has frequently been proposed as the molecular cause of SAR. Production of SA is triggered by recognition of bacterial MAMPs like flg22. Since flg22-triggered immunity was only partially restored in SA-deficient mutants, SA signaling plays a role in flg22-triggered immunity. Additionally, the expression of some groups of MAMP-regulated genes is impacted by the breakdown of SA signaling components.

3.3 Host Immunity Failure and Plant Disease

Once a pathogen has defeated the host plant defences, it can use the plant primary production to reproduce, which in most cases results in noticeable disease signs (Jones & Takemoto, 2004). Pathogenic microorganisms are drawn to plants by a number of signal molecules, creating an ideal habitat to cause disease. Some pathogens manipulate host cellular machinery by producing toxic substances, such as HC toxin, a histone deacetylase inhibitor produced by the phytopathogenic fungus *Cochliobolus carbonum*. In maize the fungus reprograms the host transcriptional response to its challenge, resulting in an ineffective immune defence response (Walley et al., 2018). Pathogens evolved to utilize effector proteins and other small molecules to suppress MAMP activated responses (Zheng et al., 2012), allowing them to establish effector-triggered vulnerability by decreasing MTI (Pel & Pieterse, 2013) (Fig. 3.4).

Effectors can also modify host target proteins to deactivate the plant immune system. An effector from phytoplasma binds to host Teosinte branched 1/Cincinnata/proliferating cell factor (TCP)/TFs to block JA production (Sugio et al., 2011). Oomycete effectors inhibit host extracellular defence enzymes and suppress its defences by interfering with processes within the host cell. The AVR3a effector from *Phytophthora infestans* affects ubiquitin ligase to prevent hypersensitive-like

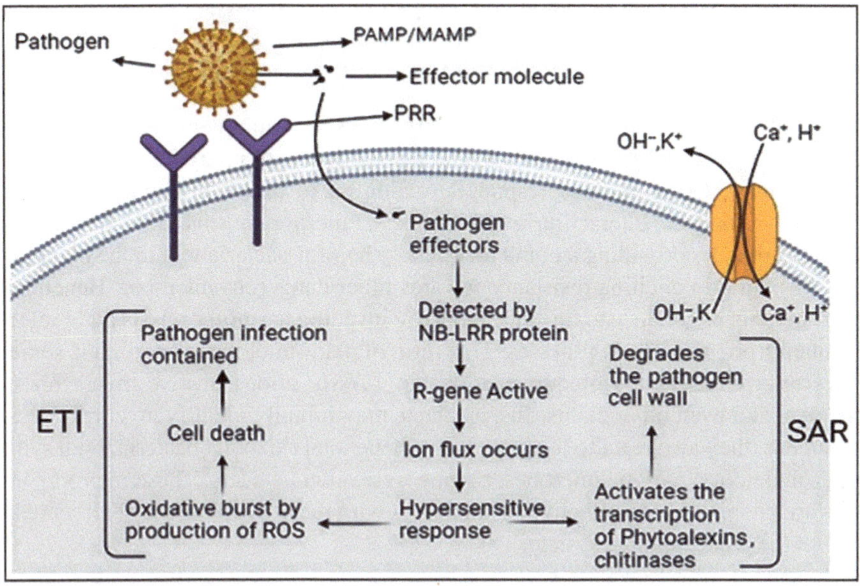

Fig. 3.4 A diagrammatic representation of Effector Triggered Immunity (ETI) and Systemic Acquired Response (SAR) in plants. (*PAMP*: Pathogen Associated Molecular Patterns, *MAMP*: Microbe Associated Molecular Patterns, *ROS*: Reactive Oxygen Species, *PRR*: Pattern Recognition receptor)

host cell death, thus allowing its infiltration into further cells (Stassen & Van den Ackerveken, 2011). In addition to host immunity failure, processes such as connection to the host plant, prviding room for infection structures in the cells of the plant, and pathogen food provision may increase plant susceptibility (Lapin & Van den Ackerveken, 2013).

Plant signals may also enhance the pathogen development, which has been extensively researched in several host species. Gene expression experiments on germinating spores of the hemibiotrophic fungus *Colletotrichum higginsianum* demonstrated the influence of host plant signals on the early pathogen growth. When *C. higginsianum* spores were attached to the host plant surface, transcription of more than 1700 genes was accelerated compared to adherence to artificial polystyrene (O'Connell et al., 2012). Some plant-derived compounds, such as alcohols, aldehydes, fatty acids, and flavonoids, attract pathogens, encourage their attachment, and eventually cause the production of appressoria, which are the first host penetration structures (Lapin & Van den Ackerveken, 2013).

Plants have the innate ability to recognise and respond to the pathogen attacks, as well as to resist most possible infections. However, before practicing the transfer of beneficial microorganisms for their exploitation into the field, extensive studies must be conducted to ensure that they do not acquire deleterious genes from pathogens, through a process known as horizontal gene transfer (León & Montesano, 2013).

3.4 Mutualistic Associations and Induced Plant Immune Response

A kind of mutualistic relationship that is crucial in boosting plant defences *vs* soil-borne pathogens and pests is the specific association with microorganisms. Plants produce an induced immune response in addition to innate immunity and SAR through mutualistic interaction with beneficial microorganisms. This reaction is accomplished by providing a home for friendly helpful bacteria within the plant tissues, resulting in eliciting resistance towards other dangerous microbes. Beneficial microorganisms produce toxins in response to invading harmful bacteria and protect the plant from pathogenic illnesses. This form of plant-microbe interaction is sometimes referred to as “symbiotic relationship”. Associations involve mycorrhizae, rhizobia, and even other plants. Just as plants may initially identify mycorrhizae as pathogens, they also react to the presence of beneficial rhizobial bacteria, both symbiotic and non-symbiotic microbes, as they were alien invaders. These microorganisms are unable to form effective relationships with the host unless they deliberately disrupt the plant immune system.

3.4.1 Mycorrhizal Association and Its Role in Plant Immune Response

Plant-mycorrhizae interaction is an example of mutualistic association that is vital in increasing host immune response *vs* soil-borne pathogens and pests (Sanchez-Bel et al., 2016). Plants develop an increased defensive strategy known as mycorrhiza-induced resistance (MIR) in response to an arbuscular mycorrhizal fungus (AMF) colonisation. AMF recruit root-associated bacteria, such as plant growth-promoting rhizobacteria, after establishing a successful relationship with a plant, to upregulate immune responses against pathogens MIR results from a direct plant response to arbuscular mycorrhizal colonisation and an indirect response to rhizobacteria that live in the mycorrhizosphere (Cameron et al., 2013). Through induced systemic resistance, AMF contributes to inhibit plant diseases and pests (Song et al., 2015). Successful mycorrhizal connection improves photosynthate delivery to the roots and modulates sugar-dependent signalling pathways (Smith & Smith, 2011). Plants react to helpful rhizobial bacteria, including symbiotic and non-symbiotic taxa, as they were invaders, just as they may perceive pathogenic mycorrhizae. These microorganisms cannot form successful relationships with the host unless they strongly interfere with their immune system.

Cameron et al. (2013) developed a four-phase spatio-temporal model to explain the overall mechanism of mycorrhizal attachment and MIR. The release of root exudates such as sugars, amino acids, phenolic compounds, and other secondary metabolites, which attract soil microbes to the roots, is the first step. More critically, strigolactone-containing root exudates can recruit AMF to roots (Akiyama et al.,

2005). The plant mounts an immune response against AMF in the second phase. This immunity induction depends on the PRRs recognising the AMF MAMPs, hence increasing SA accumulation and activating the immune response (Zhang & Zhou, 2010). The next stage is mycorrhizal reduction of host immunity and mycorrhizosphere recruitment. The mechanism of plant–pathogenic fungal association, in which pathogenic fungi decrease plant immunity by secreting particular effectors and subsequently establishing a successful infection, is comparable to the immunity suppression caused by AMF (de Jonge et al., 2011). According to Ton et al. (2009), AMF suppresses the SA-dependent defensive reaction, and in turn enhances its own colonisation of the root system by inducing the synthesis of the hormone abscisic acid (ABA). The basis of SA-based defence may lie in the early SA response ability to provide long-range SAR signals, even when SA induction is inhibited during successful AMF colonisation (Conrath, 2011). The formation of the mycorrhizosphere and ISR constitutes the final phase of the four-phase spatio-temporal model.

A fruitful mycorrhizal relationship impacts the sugar-dependent signalling pathways and improves photosynthate transfer to the roots (Smith & Smith, 2011). The mycorrhizal fungus is however an alien, "non-self" organism for the plant. As such, it can recognise it as a disease and mount an immunological defence, which the microbe must then overcome to forge a fruitful partnership. The early defence responses triggered against these microorganisms are comparable because AMF resides within the plant, making the colonisation procedure similar to infection by a biotrophic fungus (Pieterse et al., 2012) (Fig. 3.5).

3.4.2 Rhizobial Association and Role in Disease Suppression

Several *Rhizobia* act as beneficial soil bacteria that live in mutualistic symbiotic relationship with leguminous plants and provide benefit to their host by fixing atmospheric nitrogen. These beneficial bacteria are first recognised by plants as invaders during the initial step of interaction. As an outcome, plants activate their defence system. However, for a successful symbiotic interaction, the modulation of the plant immune system is an imperative process (Tonelli et al., 2020).

During the symbiotic interaction the host secretes flavonoid derivatives that attract rhizobia and accumulate auxin for infection and nodulation. It also activates nodulation genes such as *nol*, *noe* and *nod*. This results in the synthesis of nodulation (Nod) factors (Jamil et al., 2022). Nodulation is a precise process in which plants develop certain mechanism to restrict symbiosis with rhizobia. At the time of nodulation, Nod factors (NF) and receptors such as Nod factor perception (NFPs) and Nod factor receptors (NFRs), are involved in root hair infection and nodule organogenesis. NFs partially supress MTI (MAMP-triggered immunity) and play significant role in immune suppression. NFs increase auxin and cytokinin level in cortical cells and via hormonal crosstalk, and activate the SA signalling pathway (Mbengue et al., 2020). The host plants differentiate pathogens from rhizobial bacteria with the help of MTI. Rhizobia contain bacterial structure such as flagellin, a

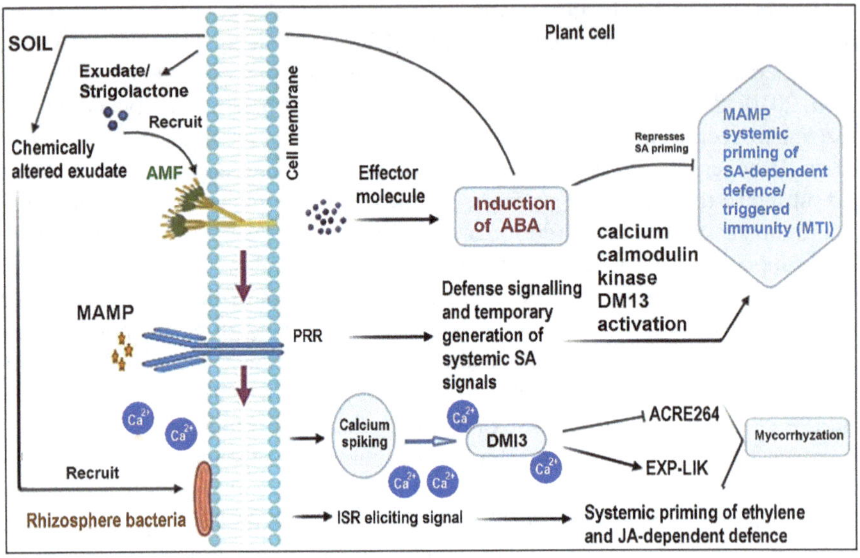

Fig. 3.5 A diagram illustrating mycorrhizal association and role in disease suppression/resistance caused by mycorrhiza. Arbuscular mycorrhizal fungi (AMF) are attracted to roots by strigolactone present in exudates. Pattern recognition receptors (PRR) detect mycorrhizal microbe-associated molecular patterns (MAMPs) and initiate MAMP-triggered immunity (MTI) and systemic priming of salicylic acid (SA)-dependent defences. SA priming is repressed and MTI is suppressed by ABA induced by mycorrhizal effectors. When Ca^{2+} levels rise due to MAMP, the Ca^{2+}/calmodulin kinase DMI3 is activated. This suppresses the production of early defence genes and increases mycorrhization. Rhizobacterial recruitment, chemical modification of exudates, and mycorrhizosphere development are the outcomes of a successful mycorrhizal connection. Mycorrhizosphere bacteria and defence against ethylene (ET) and jasmonic acid (JA) results in mycorrhiza-induced systemic resistance.

MAMP in many plant-pathogen interactions, that activates MTI defensive responses. To perceive bacterial MAMPs, host plants rely on PRR (pattern recognition receptor) membrane proteins. Other rhizobial MAMPs such as exopolysaccharides (EPSs), that participate in nodule formation, suppress host defence during invasion by inhibiting Ca^{2+} influx. Thus, EPSs induces calcium chelation that may inhibit MTI defence before diversion of rhizobial flagellin (Grundy et al., 2023).

Flavonoids along with regulating Nod genes are responsible for the induction of TtsI (a regulatory protein) transcriptional activator. Therefore Nod dependent flavonoids induce type III secretion system (T3SS) transcription by activating *ttsl*. T3SS forms a system that helps bacterial cells to inject proteins into the host cells. T3SS also induces the production of Nops (nodulation outer proteins). Therefore, host-induced flavonoids eventually secrete Nops, some of which interfere with hosts signalling machinery and promote symbiosis more directly. These proteins hence support symbiosis and suppress the host defence response to pathogens. These Nops are recognised by nucleotide binding site-leucine-rich repeat receptor (NLR) in resistant plants and inhibit effector-triggered immunity (ETI) (Dong & Song, 2020).

Studies shown that the inoculation with endophytic bacteria exerts a direct action by alleviating ethylene (ET), whose levels increase generally during pathogenic stress conditions. Endophytic bacteria release an enzyme ACC (1-aminocyclopropa ne-1-carboxylate) whose main function is to cleave ET into ammonia and α-ketobutyrate (Oukala et al., 2021). Besides ET other plant growth regulators, such as JA and SA, also participate in the regulation of plant resistance *vs* diverse pathogens. In plants JA/ET pathway mediates the ISR response that is independent of SAR mediated by SA (Salwan et al., 2023). Several rhizobacteria through ISR induce defence in plants. Some genera include *Bacillus*, *Xenorhabdus*, *Agrobacterium*, *Pseudomonas*, *Burkholderia*, *Photorhabdus*, *Streptomyces*, etc. (Disi et al., 2019) (Fig. 3.6).

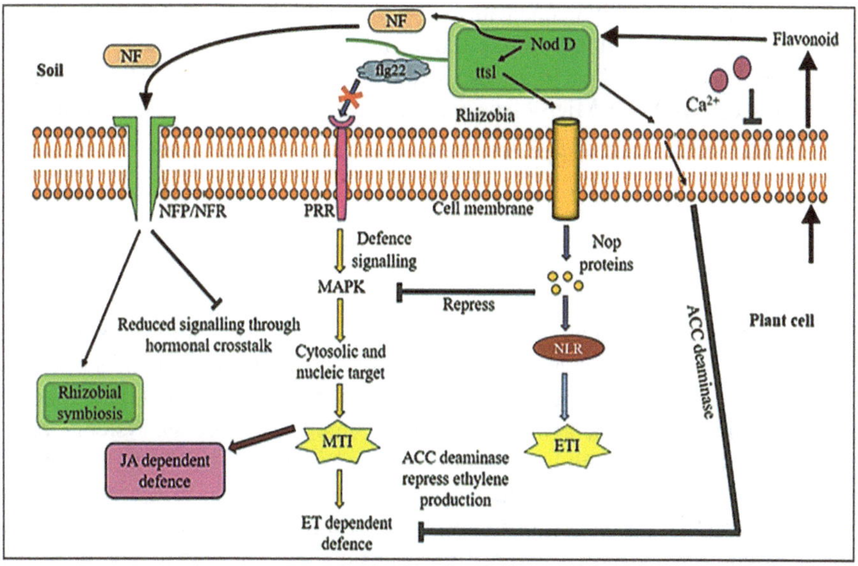

Fig. 3.6 Diagrammatic representation of symbiotic rhizobial interaction and plant immunity response. Plant secretes flavonoids that activate Nod genes, transcript nod factor (NF) and *ttsl*, whose expression induces the production of effector molecules that inhibit defence signalling in plants. Plant NLR receptors recognize effectors and this results in effector triggered immunity (ETI) activation. While NF interacts with their host receptors (Nod factor perception, NFPs and Nod factor receptors, NFRs) it initiates root nodule formation, supporting the *Rhizobium*-plant symbiosis. The pattern recognition receptors (PRR) recognise flagellin [microbe-associated molecular patterns (MAMPs)] and induce MAMPs pattern-triggered immunity (MTI). Initially, plants recognise *Rhizobium* as a pathogenic bacterium and induce MTI response that results in JA and ET-dependent defence response. To suppress ET response, the bacterium secretes enzyme such as 1-aminocyclopropane-1-carboxylic acid (ACC) deaminase. JA-dependent pathway remains unaffected during rhizobial association. This indicates that it plays a significant role in induced systemic resistance (ISR) and provides immunity to plant

3.5 Synthetic Microbial Communities (SynComs) Promising Increased Plant Protection

The functions of complex microbial communities in the root or phyllosphere microbiome are empirically tested using SynComs of well-characterized bacterial or fungal taxa (Liu et al., 2022). However, single—mode and mini SynCom studies showed that the SynCom enhancement of disease resistance can be modest, as only one or a few strains can be enough to boost resistance *vs* certain pathogens (Prigigallo et al., 2022). According to SynCom research, microbial communities have developed avoidance mechanisms to overcome immune systems that plants evolved to deal with pathogens. Another study compared SynComs made up of bacteria *vs* multi-kingdom members, to avoid growth-defense trade—offs when the plant was growing under distinct wavelengths Hao et al. (2023) studied that a SynCom composed of bacteria, fungi, and oomycetes prevented the growth-defense trade-offs observed when plants were grown in low—light conditions. However, an interaction between the plant defence mechanisms and the microbial communities in its various compartments is required to avoid dysbiosis and to maximise the benefit from plant-growth promoting microbes, in its associated microbiome. Table 3.1 shows promising effects of Synthetic Microbial Communities (SynComs), to increase disease resistance.

3.6 Conclusions

The main synergistic mechanism acting between plants and microbes is mediated by signalling pathways that affect the main mechanism of invasion and survival of beneficial microbes. Plants deploy MAMP-triggered immunity (MTI) and effector-triggered immunity (ETI), a two-tiered innate immunity or induced systemic resistance (ISR). Plants also exhibit systemic acquired resistance (SAR) in response to biotic stress factors. These provide wide-spectrum resistance at the whole plant level. Various phytohormones also play a significant role in plant-microbe associations, via hormonal crosstalks. When microbes interact with plants, the microbial elicitors or MAMPs trigger ISR or SAR. However, rhizobia and mycorrhizae induce Nod and Myc factors that help to induce root symbiosis. They release proteins that interfere with host defence signalling pathways, and make hosts more suitable for symbiosis. These beneficial microbes are able to suppress plant immunity responses, including the production of proteins that prevent the release of hormones, interfering with signalling pathways and suppressing the host immune system. As a feedback effect, the immune system failure may allow the entry of pathogens into the host plants, and increase food availability that makes them more susceptible to pathogens. By studying the mechanistic pathways of microbes and plants interaction, it is possible to decipher the role of selected microorganisms in enhancing plant immunity *vs* various pathogens. With advancement in technology, it is now

Table 3.1 Plant protection in *in vivo* experiments with synthetic microbial communities (SynComs)

Plant	Pathogen	Disease	Microorganisms	References
Musa spp.	*Fusarium oxysporum* f. sp. *cubense* (TR4)	Fusarium wilt	*Pseudomonas, Bacillus, Trichoderma* spp.	Prigigallo et al. (2022)
Nicotiana attenuata	*Fusarium Alternaria*	Sudden wilt disease	*Bacillus mojavensis, Anthrobacter nitroguajacolicusPriestia megaterium, Pseudomonas azotoformans, P. frederiksbergensis.*	Santhanam et al. (2015)
Solanum lycopersicum	*Ralstonia solanacearum*	Bacterial wilt disease	*Brevibacterium frigoritolerans, Bacillus niacin, Solibacillus silvestris, B. luciferensis*	Lee et al. (2021)
Triticum spp.	*Fusarium pseudograminear um*	Crown rot	*Priestia megaterium, Bacillus* sp., *Acinetobacter pittii, Enterobacter asburiae, Bacillus pumilus, Xanthamonas, Burkholderia* spp.	Liu et al. (2022)
Zea mays	*Rhizoctonia solani*	Root rot	*Pseudomonas* sp., *Rhodococcus erythropolis, Chryseobacterium soldanellicola, Pedobacter* sp.	Yin et al. (2022)
Capsicum annuum	*Phytophthora capsici*	Root rot	*Trichoderma harzianum, Streptomyces rochei, Bacillus cereus, B. subtilis, Serratia* sp.	Ezziyyani et al. (2007)
Capsicum annuum	*Phytophthora capsici*	Phytophthora blight	*Bacillus cereus, B. subtilis, Serratia* sp.	Zhang et al. (2019)
Solanum lycopersicum	*Fusarium oxysporum*	Vascular wilt	*B. subtilis, B. atrophaeus, Burkholderia cepacia*	Shanmugam and Kanoujia (2011)
Musa spp.	*Fusarium oxysporum*	Fusarium wilt	*Pseudomonas aeruginosa, T. harzianum*	Wong et al. (2019)
Aconitum	*Sclerotium rolfsii*	Southern blight	*Bacillus aryabhattai, Streptomyces fungicidicus, S. rochei, S. enissocaesilis*	Li et al. (2022)
Solanum lycopersicum	*Rhizoctonia solani*	Root rot	*Streptomyces atrovirens, T. lixii*	Solanki et al. (2019)

(continued)

Table 3.1 (continued)

Plant	Pathogen	Disease	Microorganisms	References
Beta vulgaris	*Rhizoctonia solani*	Fungal root	*Flavobacterium* sp., *Chitinophaga* spp.	Carrión et al. (2019)
Zea mays	*Fusarium verticillioides*	Seedling blight	*Stenotrophomonas maltophilia, Ochrobactrum pituitosum, Curtobacterium pusillum, Enterobacter cloacae, Chryseobacterium indologenes, Herbaspirillum frisingense, Pseudomonas putida*	Niu et al. (2017)
Solanum lycopersicum	*Ralstonia solanacearum*	Bacterial wilt	*Pseudomonas brassicacearum, P. protegens, P. fluorescens*	Hu et al. (2016)
Solanum tuberosum	*Phytophthora infestans*	Late blight	*Pseudomonas putida, P. chloraphis, P. fluorescens, P. frederiksbergensis.*	De Vrieze et al. (2018)

possible to engineer such beneficial species for plant health by using synthetic microbial communities (SynComs) that contain species with desired microbiome functions. SynComs are giving insights into the rhizosphere structure and dynamics, including how these dynamics influence plant behaviour and fitness.

References

Akiyama, K., Matsuzaki, K. I., & Hayashi, H. (2005). Plant sesquiterpenes induce hyphal branching in arbuscular mycorrhizal fungi. *Nature, 435*(7043), 824–827.

Bai, J., Zhou, Y., Sun, J., Chen, K., Han, Y., Wang, R., et al. (2023). BIK1 protein homeostasis is maintained by the interplay of different ubiquitin ligases in immune signaling. *Nature Communications, 14*(1), 4624.

Bernoux, M., Zetzsche, H., & Stuttmann, J. (2022). Connecting the dots between cell surface- and intracellular-triggered immune pathways in plants. *Current Opinion in Plant Biology, 69*, 102276.

Blekemolen, M. C., Liu, Z., Stegman, M., Zipfel, C., Shan, L., & Takken, F. L. (2023). The PTI-suppressing Avr2 effector from *Fusarium oxysporum* suppresses mono-ubiquitination and plasma membrane dissociation of BIK1. *Molecular Plant Pathology, 24*(10), 1273–1286.

Cameron, D. D., Neal, A. L., van Wees, S. C., & Ton, J. (2013). Mycorrhiza-induced resistance, more than the sum of its parts? *Trends in Plant Science, 18*, 539–545.

Carrión, V. J., Perez-Jaramillo, J., Cordovez, V., Tracanna, V., De Hollander, M., Ruiz-Buck, D., et al. (2019). Pathogen-induced activation of disease-suppressive functions in the endophytic root microbiome. *Science, 366*(6465), 606–612.

Chakraborty, J. (2023). Microbiota and the plant immune system work together to defend against pathogens. *Archives of Microbiology, 205*(10), 347.

Chandra, S., & Sharma, I. P. (2023). Elicitins as microbe-associated molecular patterns and their role in plant defense. In D. Chandra & P. Bhatt (Eds.), *Developments in applied microbiology and biotechnology, unravelling plant-microbe synergy* (pp. 77–86). Academic.

Chang, M., Chen, H., Liu, F., & Fu, Z. Q. (2022). PTI and ETI, convergent pathways with diverse elicitors. *Trends in Plant Science, 27*(2), 113–115.

Chi, Y., Wang, C., Wang, M., Wan, D., Huang, F., Jiang, Z., et al. (2021). Flg22-induced Ca^{2+} increases undergo desensitization and resensitization. *Plant, Cell and Environment, 44*(12), 3793–3805.

Conrath, U. (2011). Molecular aspects of defence priming. *Trends in Plant Science, 16*, 524–531.

Dangl, J. L., Horvath, D. M., & Staskawicz, B. J. (2013). Pivoting the plant immune system from dissection to deployment. *Science, 341*, 746–751.

de Azevedo, E., Manhães, A. M., Ortiz-Morea, F. A., He, P., & Shan, L. (2021). Plant plasma membrane-resident receptors, surveillance for infections and coordination for growth and development. *Journal of Integrative Plant Biology, 63*, 79–101.

De Jonge, R., Bolton, M. D., & Thomma, B. P. (2011). How filamentous pathogens co-opt plants, the ins and outs of fungal effectors. *Current Opinion in Plant Biology, 14*, 400–406.

De Vrieze, M., Germanier, F., Vuille, N., & Weisskopf, L. (2018). Combining different potato-associated *Pseudomonas* strains for improved biocontrol of *Phytophthora infestans*. *Frontiers in Microbiology, 9*, 2573.

DeFalco, T. A., & Zipfel, C. (2021). Molecular mechanisms of early plant pattern-triggered immune signaling. *Molecular Cell, 81*, 3449–3467.

Dey, M., & Ghosh, S. (2022). Arbuscular mycorrhizae in plant immunity and crop pathogen control. *Rhizosphere, 22*, 100524.

Disi, J., Simmons, J., & Zebelo, S. (2019). Plant growth-promoting rhizobacteria-induced defense against insect herbivores. In D. K. Maheshwari & S. Dheeman (Eds.), *Field crops, sustainable management by PGPR* (pp. 385–410). Springer Verlag.

Dong, W., & Song, Y. (2020). The significance of flavonoids in the process of biological nitrogen fixation. *International Journal of Molecular Sciences, 21*, 5926.

Durrant, W. E., & Dong, X. (2004). Systemic acquired resistance. *Annual Review of Phytopathology, 42*, 185–209.

Ezziyyani, M., Requena, M. E., Egea-Gilabert, C., & Candela, M. E. (2007). Biological control of *Phytophthora* root rot of pepper using *Trichoderma harzianum* and *Streptomyces rochei* in combination. *Journal of Phytopathology, 155*, 342–349.

Garcia, A. G., & Steinbrenner, A. D. (2023). Bringing plant immunity to light, a genetically encoded, bioluminescent reporter of pattern-triggered immunity in *Nicotiana benthamiana*. *MPMI, 36*, 139–149.

Gonçalves Dias, M., Soleimani, F., & Monaghan, J. (2022). Activation and turnover of the plant immune signaling kinase BIK1, a fine balance. *Essays in Biochemistry, 66*, 207–218.

Grundy, E. B., Gresshoff, P. M., Su, H., & Ferguson, B. J. (2023). Legumes regulate symbiosis with rhizobia via their innate immune system. *International Journal of Molecular Sciences, 24*, 2800.

Han, S. W., & Jung, H. W. (2013). Molecular sensors for plant immunity; pattern recognition receptors and race-specific resistance proteins. *Journal of Plant Biology, 56*, 357–366.

Hao, X., Wang, X., Chen, C., Liu, R., Yin, Y., Yao, J., ... & Liu, X. (2023). Synthetic bacterial communities reshape microbial communities and enhance nutrient supply in desertified land of Northwest China. *Applied Soil Ecology, 189*, 104972.

Hu, J., Wei, Z., Friman, V. P., Gu, S. H., Wang, X. F., Eisenhauer, N., et al. (2016). Probiotic diversity enhances rhizosphere microbiome function and plant disease suppression. *MBio, 7*, e01790–e01716. https://doi.org/10.1128/mBio.01790-16

Huang, S., Jia, A., Ma, S., Sun, Y., Chang, X., Han, Z., & Chai, J. (2023). NLR signaling in plants, from resistosomes to second messengers. *Trends in Biochemical Sciences, 48*, 776–787.

Jamil, F., Mukhtar, H., Fouillaud, M., & Dufossé, L. (2022). Rhizosphere signaling, insights into plant-rhizomicrobiome interactions for sustainable agronomy. *Microorganisms, 10*, 899.

Jones, D. A., & Takemoto, D. (2004). Plant innate immunity-direct and indirect recognition of general and specific pathogen-associated molecules. *Current Opinion in Immunology, 16*, 48–62.

Kachroo, A., & Kachroo, P. (2020). Mobile signals in systemic acquired resistance. *Current Opinion in Plant Biology, 58*, 41–47.

Kadota, Y., Shirasu, K., & Zipfel, C. (2015). Regulation of the NADPH oxidase RBOHD during plant immunity. *Plant & Cell Physiology, 56*, 1472–1480.

Kamle, M., Borah, R., Bora, H., Jaiswal, A. K., Singh, R. K., & Kumar, P. (2020). Systemic acquired resistance (SAR) and induced systemic resistance (ISR), role and mechanism of action against phytopathogens. In A. E. L. Hesham et al. (Eds.), *Fungal biotechnology and bioengineering. Fungal Biology* (pp. 457–470). Springer Nature.

Karasov, T. L., Chae, E., Herman, J. J., & Bergelson, J. (2017). Mechanisms to mitigate the trade-off between growth and defense. *Plant Cell, 29*, 666–680.

Kaur, S., Samota, M. K., Choudhary, M., Choudhary, M., Pandey, A. K., Sharma, A., & Thakur, J. (2022). How do plants defend themselves against pathogens-biochemical mechanisms and genetic interventions. *Physiology and Molecular Biology of Plants, 28*, 485–504.

Kozyulina, P. Y., Pavlova, O. A., Bovin, A. D., Shirobokova, S. A., Dolgikh, A. V., Dymo, A. M., & Dolgikh, E. A. (2023). Transcriptomic analysis of pea plant responses to chitooligosaccharides' treatment revealed stimulation of mitogen-activated protein kinase cascade. *Frontiers in Plant Science, 14*, 1092013.

Kumar, D., & Ohri, P. (2023). Say "NO" to plant stresses, unravelling the role of nitric oxide under abiotic and biotic stress. *Nitric Oxide, 130*, 36–57.

Kumar, D., Manhas, R. K., & Ohri, P. (2023). Deciphering the growth promoting and stress curtailing role of nitric oxide in *Meloidogyne incognita* infested *Solanum lycopersicum* seedlings. *Scientia Horticulturae, 319*, 112147.

Lapin, D., & Van den Ackerveken, G. (2013). Susceptibility to plant disease, more than a failure of host immunity. *Trends in Plant Science, 18*, 546–554.

Lee, S. M., Kong, H. G., Song, G. C., & Ryu, C. M. (2021). Disruption of Firmicutes and Actinobacteria abundance in tomato rhizosphere causes the incidence of bacterial wilt disease. *The ISME Journal, 15*, 330–347.

León, I. P., & Montesano, M. (2013). Activation of defense mechanisms against pathogens in mosses and flowering plants. *International Journal of Molecular Sciences, 14*, 3178–3200. https://doi.org/10.3390/ijms14023178

Li, Z., Bai, X., Jiao, S., Li, Y., Li, P., Yang, Y., et al. (2021). A simplified synthetic community rescues *Astragalus mongholicus* from root rot disease by activating plant-induced systemic resistance. *Microbiome, 9*, 1–20.

Li, Y., He, F., Guo, Q., Feng, Z., Zhang, M., Ji, C., et al. (2022). Compositional and functional comparison on the rhizosphere microbial community between healthy and *Sclerotium rolfsii*-infected monkshood (*Aconitum carmichaelii*) revealed the biocontrol potential of healthy monkshood rhizosphere microorganisms. *Biological Control, 165*, 104790.

Liu, H., Qiu, Z., Ye, J., Verma, J. P., Li, J., & Singh, B. K. (2022). Effective colonisation by a bacterial synthetic community promotes plant growth and alters soil microbial community. *Journal of Sustainable Agriculture and Environment, 1*, 30–42.

Lü, P., Liu, Y., Yu, X., Shi, C. L., & Liu, X. (2022). The right microbe-associated molecular patterns for effective recognition by plants. *Frontiers in Microbiology, 13*, 1019069.

Lyu, D., Msimbira, L. A., Nazari, M., Antar, M., Pagé, A., Shah, A., et al. (2021). The coevolution of plants and microbes underpins sustainable agriculture. *Microorganisms, 9*, 1036.

Mackey, D., Belkhadir, Y., Alonso, J. M., Ecker, J. R., & Dangl, J. L. (2003). *Arabidopsis* RIN4 is a target of the type III virulence effector AvrRpt2 and modulates RPS2-mediated resistance. *Cell, 112*, 379–389.

Marcec, M. J., & Tanaka, K. (2021). Crosstalk between calcium and ROS signaling during flg22-triggered immune response in *Arabidopsis* leaves. *Plants, 11*, 14.

Maruta, N., Trusov, Y., Jones, A. M., & Botella, J. R. (2021). Heterotrimeric G proteins in plants, canonical and atypical G-subunits. *International Journal of Molecular Sciences, 22*, 11841.

Mbengue, M. D., Hervé, C., & Debellé, F. (2020). Nod factor signaling in symbiotic nodulation. *Advances in Botanical Research, 94*, 1–39.

Moeder, W., Phan, V., & Yoshioka, K. (2019). Ca^{2+} to the rescue-Ca^{2+} channels and signaling in plant immunity. *Plant Science, 279*, 19–26.

Nguyen, Q. M., Iswanto, A. B. B., Son, G. H., & Kim, S. H. (2021). Recent advances in effector-triggered immunity in plants, new pieces in the puzzle create a different paradigm. *International Journal of Molecular Sciences, 22*, 4709.

Nishad, R., Ahmed, T., Rahman, V. J., & Kareem, A. (2020). Modulation of plant defense system in response to microbial interactions. *Frontiers in Microbiology, 11*, 1298.

Niu, B., Paulson, J. N., Zheng, X., & Kolter, R. (2017). Simplified and representative bacterial community of maize roots. *PNAS, 114*, E2450–E2459.

Nobori, T., & Tsuda, K. (2019). The plant immune system in heterogeneous environments. *Current Opinion in Plant Biology, 50*, 58–66.

O'Connell, R. J., Thon, M. R., Hacquard, S., Amyotte, S. G., Kleemann, J., Torres, M. F., et al. (2012). Lifestyle transitions in plant pathogenic *Colletotrichum* fungi deciphered by genome and transcriptome analyses. *The Nature Generators, 44*, 1060–1065.

Offor, B. C., Dubery, I. A., & Piater, L. A. (2020). Prospects of gene knockouts in the functional study of mamp-triggered immunity, a review. *International Journal of Molecular Sciences, 21*, 2540.

Oukala, N., Aissat, K., & Pastor, V. (2021). Bacterial endophytes, the hidden actor in plant immune responses against biotic stress. *Plants, 10*, 1012.

Pantigoso, H. A., Newberger, D., & Vivanco, J. M. (2022). The rhizosphere microbiome, plant-microbial interactions for resource acquisition. *Journal of Applied Microbiology, 133*, 2864–2876.

Pel, M. J., & Pieterse, C. M. (2013). Microbial recognition and evasion of host immunity. *Journal of Experimental Botany, 64*, 1237–1248.

Pfeilmeier, S., Petti, G. C., Bortfeld-Miller, M., Daniel, B., Field, C. M., Sunagawa, S., & Vorholt, J. A. (2021). The plant NADPH oxidase RBOHD is required for microbiota homeostasis in leaves. *Nature Microbiology, 6*, 852–864.

Pieterse, C. M., Van der Does, D., Zamioudis, C., Leon-Reyes, A., & Van Wees, S. C. (2012). Hormonal modulation of plant immunity. *Annual Review of Cell and Developmental Biology, 28*, 489–521.

Pieterse, C. M., Zamioudis, C., Berendsen, R. L., Weller, D. M., Van Wees, S. C., & Bakker, P. A. (2014). Induced systemic resistance by beneficial microbes. *Annual Review of Phytopathology, 52*, 347–375.

Prigigallo, M. I., Gómez-Lama Cabanás, C., Mercado-Blanco, J., & Bubici, G. (2022). Designing a synthetic microbial community devoted to biological control, the case study of Fusarium wilt of banana. *Frontiers in Microbiology, 13*, 967885.

Pruitt, R. N., Gust, A. A., & Nürnberger, T. (2021). Plant immunity unified. *Nature Plants, 7*, 382–383.

Ray, S. K., Macoy, D. M., Kim, W. Y., Lee, S. Y., & Kim, M. G. (2019). Role of RIN4 in regulating PAMP-triggered immunity and effector-triggered immunity, current status and future perspectives. *Molecules and Cells, 42*, 503.

Remick, B. C., Gaidt, M. M., & Vance, R. E. (2023). Effector-triggered immunity. *Annual Review of Immunology, 41*, 453–481.

Richard, M. M. S., Gratias, A., Meyers, B. C., & Geffroy, V. (2018). Molecular mechanisms that limit the costs of NLR-mediated resistance in plants. *Molecular Plant Pathology, 19*, 2516–2523.

Rodriguez, P. A., Rothballer, M., Chowdhury, S. P., Nussbaumer, T., Gutjahr, C., & Falter-Braun, P. (2019). Systems biology of plant-microbiome interactions. *Molecular Plant, 12*, 804–821.

Ryals, J. A., Neuenschwander, U. H., Willits, M. G., Molina, A., Steiner, H.-Y., & Hunt, M. D. (1996). Systemic acquired resistance. *Plant Cell, 8*, 1809–1819.

Salwan, R., Sharma, M., Sharma, A., & Sharma, V. (2023). Insights into plant beneficial microorganism-triggered induced systemic resistance. *Plant Stress, 7*, 100140.

Sanchez-Bel, P., Troncho, P., Gamir, J., Pozo, M. J., Camañes, G., Cerezo, M., et al. (2016). The nitrogen availability interferes with mycorrhiza-induced resistance against *Botrytis cinerea* in tomato. *Frontiers in Microbiology, 7*, 1598. https://doi.org/10.3389/fmicb.2016.01598

Santhanam, R., Luu, V. T., Weinhold, A., Goldberg, J., Oh, Y., & Baldwin, I. T. (2015). Native root-associated bacteria rescue a plant from a sudden-wilt disease that emerged during continuous cropping. *PNAS, 112*, E5013–E5020.

Shanmugam, V., & Kanoujia, N. (2011). Biological management of vascular wilt of tomato caused by *Fusarium oxysporum* f. sp. *lycospersici* by plant growth-promoting rhizobacterial mixture. *Biological Control, 57*, 85–93.

Sharma, N., Khanna, K., Jasrotia, S., Kumar, D., Bhardwaj, R., & Ohri, P. (2023). Metabolites and chemical agents in the plant roots, an overview of their use in plant-parasitic nematode management. *Nematology, 25*, 243–257.

Singh, U. B., Malviya, D., Singh, S., Pradhan, J. K., Singh, B. P., Roy, M., et al. (2016). Bio-protective microbial agents from rhizosphere eco-systems trigger plant defense responses provide protection against sheath blight disease in rice (*Oryza sativa* L.). *Microbiological Research, 192*, 300–312.

Smith, S. E., & Smith, F. A. (2011). Roles of arbuscular mycorrhizas in plant nutrition and growth, new paradigms from cellular to ecosystem scales. *Annual Review of Plant Biology, 62*, 227–250.

Solanki, M. K., Yandigeri, M. S., Kumar, S., Singh, R. K., & Srivastava, A. K. (2019). Co-inoculation of different antagonists can enhance the biocontrol activity against *Rhizoctonia solani* in tomato. *Antonie van Leeuwenhoek, 112*, 1633–1644.

Song, Y., Chen, D., Lu, K., Sun, Z., & Zeng, R. (2015). Enhanced tomato disease resistance primed by arbuscular mycorrhizal fungus. *Frontiers in Plant Science, 6*, 786.

Stassen, J. H., & Van den Ackerveken, G. (2011). How do oomycete effectors interfere with plant life? *Current Opinion in Plant Biology, 14*, 407–414.

Sticher, L., Mauch-Mani, B., & Métraux, J. P. (1997). Systemic acquired resistance. *Annual Review of Phytopathology, 35*, 235–270.

Sugio, A., Kingdom, H. N., MacLean, A. M., Grieve, V. M., & Hogenhout, S. A. (2011). Phytoplasma protein effector SAP11 enhances insect vector reproduction by manipulating plant development and defense hormone biosynthesis. *PNAS, 108*, E1254–E1263.

Tena, G. (2021). PTI and ETI are one. *Nature Plants, 7*, 1527–1527.

Ton, J., Flors, V., & Mauch-Mani, B. (2009). The multifaceted role of ABA in disease resistance. *Trends in Plant Science, 14*, 310–317.

Tonelli, M. L., Figueredo, M. S., Rodríguez, J., Fabra, A., & Ibañez, F. (2020). Induced systemic resistance-like responses elicited by rhizobia. *Plant and Soil, 448*, 1–14.

Van Loon, L. C., & Van Strien, E. A. (1999). The families of pathogenesis-related proteins, their activities, and comparative analysis of PR-1 type proteins. *Physiological and Molecular Plant Pathology, 55*, 85–97.

Van Wersch, S., & Li, X. (2019). Stronger when together, clustering of plant NLR disease resistance genes. *Trends in Plant Science, 24*, 688–699.

Vanthana, M., Nakkeeran, S., Malathi, V. G., Renukadevi, P., & Vinodkumar, S. (2019). Induction of in planta resistance by flagellin (Flg) and elongation factor-TU (EF-Tu) of *Bacillus amyloliquefaciens* (VB7) against groundnut bud necrosis virus in tomato. *Microbial Pathogenesis, 137*, 103757.

Walley, J. W., Shen, Z., McReynolds, M. R., Schmelz, E. A., & Briggs, S. P. (2018). Fungal-induced protein hyperacetylation in maize identified by acetylome profiling. *PNAS, 115*, 210–215.

Wang, W., Feng, B., Zhou, J. M., & Tang, D. (2020). Plant immune signaling, advancing on two frontiers. *Journal of Integrative Plant Biology, 62*, 2–24.

Ward, E. R., Uknes, S. J., Williams, S. C., Dincher, S. S., Wiederhold, D. L., Alexander, D. C., et al. (1991). Coordinate gene activity in response to agents that induce systemic acquired resistance. *Plant Cell, 3*, 1085–1094.

Wong, C. K. F., Saidi, N. B., Vadamalai, G., Teh, C. Y., & Zulperi, D. (2019). Effect of bioformulations on the biocontrol efficacy, microbial viability and storage stability of a consortium of biocontrol agents against Fusarium wilt of banana. *Journal of Applied Microbiology, 127*, 544–555.

Xu, L., Yao, X., Zhang, N., Gong, B. Q., & Li, J. F. (2019). Dynamic G protein alpha signaling in *Arabidopsis* innate immunity. *Biochemical and Biophysical Research Communications, 516*, 1039–1045.

Yin, C., Hagerty, C. H., & Paulitz, T. C. (2022). Synthetic microbial consortia derived from rhizosphere soil protect wheat against a soilborne fungal pathogen. *Frontiers in Microbiology, 13*, 908981.

Yuan, M., Ngou, B. P. M., Ding, P., & Xin, X. F. (2021). PTI-ETI crosstalk, an integrative view of plant immunity. *Current Opinion in Plant Biology, 62*, 102030.

Zeier, J. (2021). Metabolic regulation of systemic acquired resistance. *Current Opinion in Plant Biology, 62*, 102050.

Zhai, K., Liang, D. I., Li, H., Jiao, F., Yan, B., Liu, J., et al. (2022). NLRs guard metabolism to coordinate pattern-and effector-triggered immunity. *Nature, 601*(7892), 245–251.

Zhang, M., & Kong, X. (2022). How plants discern friends from foes. *Trends in Plant Science, 27*, 107–109.

Zhang, J., & Zhou, J. M. (2010). Plant immunity triggered by microbial molecular signatures. *Molecular Plant, 3*, 783–793.

Zhang, L. N., Wang, D. C., Hu, Q., Dai, X. Q., Xie, Y. S., Li, Q., Liu, H. M., Guo, J. H., et al. (2019). Consortium of plant growth-promoting rhizobacteria strains suppresses sweet pepper disease by altering the rhizosphere microbiota. *Frontiers in Microbiology, 10*, 1668.

Zhao, H., Liu, Y. P., & Zhang, L. Q. (2019). *In silico* and genetic analyses of cyclic lipopeptide synthetic gene clusters in *Pseudomonas* sp. 11K1. *Front Microbiology, 10*, 544.

Zheng, X. Y., Spivey, N. W., Zeng, W., Liu, P. P., Fu, Z. Q., Klessig, D. F., et al. (2012). Coronatine promotes *pseudomonas syringae* virulence in plants by activating a signaling cascade that inhibits salicylic acid accumulation. *Cell Host & Microbe, 11*, 587–596.

Zhou, J. M., & Zhang, Y. (2020). Plant immunity, danger perception and signaling. *Cell, 181*, 978–989.

Chapter 4
Building Resilience: Engineering the Plant Microbiome for Biotic Stress Management

Kalaivani K. Nadarajah

Abstract Crop diseases result in significant losses and affect farmers' livelihoods, making it crucial for the agricultural sector to remain resilient against the onset and spread of these diseases. Recent studies have emphasized building this resilience through the engineering of plant microbiomes. This chapter explores the role of plant-microbe interactions in shaping microbiome communities and examines how root exudates influence the recruitment and suppression of microorganisms in the soil. Microbe-microbe interactions also plays a crucial role in regulating the growth and decline of specific microbial populations within the soil environment. Importantly, the chapter discusses how engineering the soil microbiome can enhance disease suppression and minimize disease occurrence and transmission.

Keywords Biotic stress · Disease suppressiveness · Microbiome engineering · Plant immunity

Abbreviations

ISR	induced systemic resistance
SA	salicylic acid
JA	jasmonic acid
ETH	ethylene
VOC	volatile organic compounds
CAT	cationic amino acid transporter
MATE	multidrug and toxic compound extrusion

K. K. Nadarajah (✉)
Department of Biological Sciences and Biotechnology, Faculty of Science and Technology, Universiti Kebangsaan Malaysia, Bangi, Selangor, Malaysia
e-mail: vani@ukm.edu.my

P. Mathur, S. Roy (eds.), *Plant Microbiome and Biological Control*, Sustainability in Plant and Crop Protection 20,
https://doi.org/10.1007/978-3-031-75845-4_4

RSA	root system architecture
PKS	polyketide synthases
NRPS	non-ribosomal peptide synthases
BGC	biosynthetic gene clusters
GDU	glutamine dumper
ALMT	Aluminum-activated Malate Transporters
ETI	effector trigger immunity
PTI	PAMP triggered immunity
PAMP	pathogen-associated molecular patterns
PRR	pathogen recognition receptors
QQ	quorum quenching
QS	quorum sensing

4.1 Introduction

The environment and hosts both contain complex and varied microbiomes that interact with their corresponding micro-environments. According to Berg et al. (2014) and Lebeis (2014), the dynamic interaction of the host and its associated microbiome with the environmental microbiota benefits host growth and fitness. Different niche habitats within the host and its surroundings provide varying circumstances for the development of particular microbiomes. Natural processes such as root exudation, essential for plant growth and development, provide plants the opportunity to modify their rhizospheric microorganisms. Organic acids, amino acids, fatty acids, flavonols, glucosinolates, indole compounds, polysaccharides, and proteins are just a few of the chemicals that are secreted from microorganisms and roots into the rhizosphere. Once recipient organisms recognize these signals, communication and interaction can begin (Upadhyay et al., 2022; Jamil et al., 2022). The fact that plants exude a wide variety of chemicals into the rhizosphere, accounting for up to 40% of their photosynthetically absorbed carbon, is astonishing (Canarini et al., 2019). Root-exudates inhibit growth of harmful microbes, and alter the microbial diversity and dynamics in soil, favoring the growth of microorganisms that can contribute positively to crop productivity and health (Santoyo et al., 2021). These interactions linking host, environment, microbiome, and pathogens are affected by climate, land use, management techniques, and other environmental factors (soil characteristics, nutritional status, etc.). Choosing a certain combination of these environmental elements can have an impact on the microbiome, with an impact on how a pathogen infection turns out. Similarly, altering host-associated microbiomes can result in the creation of novel methods in managing plant diseases (Cheng et al., 2019).

Currently, there is a consistent interest in the crop root exudation patterns as they have the capacity to alter the rhizosphere microbiota and potentially replace the usage of herbicides in agricultural fields (Olanrewaju et al., 2019). Entropy-based engineering of the rhizosphere is a concept recently developed by scientists with the goal of increasing production while requiring less input (Sharma et al., 2023). With increasing knowledge connected to function of root exudates in regulating the rhizospheric microbial communities, it is now possible to predict the environmental factors that change the root associated microbiome (Canarini et al., 2019). The nature and quantity of root exudates are strongly influenced by specific plant species or host genotypes, existing environmental circumstances, and the plant stress level in addition to the edaphic factors mentioned above. The root exudates function as signaling molecules, assisting in the recruitment of advantageous root-associated mutualists such as *Bacillus, Pseudomonas, Trichoderma*, and arbuscular mychorrizal species. Furthermore, they may assist in the eradication of detrimental microorganisms in the rhizosphere by creating shifts the microbial community (Dastogeer et al., 2020).

In order to prime their host against invasion, plant-associated rhizospheric microbes activate induced systemic resistance (ISR), where phytohormones such as salicylic acid (SA), jasmonic acid (JA), ethylene (ETH), volatile organic compounds (VOCs), and defense regulatory proteins play a crucial role (Yu et al., 2022; Zhu et al., 2022). Plants may respond forcefully to attacks from pathogens by priming their leaf tissues with the aid of ISR, rather than activating genes associated with defense directly (Li et al., 2020; Kim & Felton, 2013). As plant diseases remain a major concern in agriculture, the elucidation of microbial dynamics in soil in response to an attack, the role of environmental factors affecting these dynamics, with the molecular pathways involved in this plant-microbe-environment interaction, must be studied to gain a better understanding of the plant-pathogen interactions. This approach is expected to facilitate a better management of crop diseases in the field. This chapter looks into the interplay in the soil microbiome, its function and its engineering for disease suppression.

4.2 Plant-Microbe Interactions in Microbial Assemblage

Microbial diversity within soil is staggering. Numerous microbial species live on plant surfaces and in their tissues, where they support a variety of vital functions, including resiliency to abiotic and biotic stressors, growth promotion, and nutrient uptake (Noman et al., 2021; Koza et al., 2022). Microbial colonization regulates the networks of processes that control different biological and chemical interactions both below- and above-ground. The plant genotype has also been identified as a significant factor in the colonization of microbial populations, in addition to the interactions (Chaudhry et al., 2021; Nadarajah & Abdul Rahman, 2021). Additionally, according to Wei et al. (2017), the type of soil (i.e., loam and clay soils) also has a decisive role in structuring the predominance of different rhizospheric microbial populations. As a result, interactions between microbes and plants, as well as between microbes and their environment, affect how the microbiome is assembled.

The modifications of the rhizospheric microbiome are thought to offer a potential solution for both climate-resilience and disease-suppressive soil systems that contribute to the host fitness (Abdul Rahman et al., 2021). Different microorganisms are engaged in the battle for plant-derived nutrients in the rhizosphere, which affects how they interact with the roots (Sharma et al., 2023). These rivalries are what determine whether relationships within the microbiome succeed or fail. Two methods by which the soil microbiome suppresses disease are through inhibitory microbes in the soil or priming of plants through ISR. To understand these interactions let's breakdown the processes to root exudation, the microbe-microbe interactions and plant-microbe interactions. The root and microbial exudates play in fact an important role in the microbe and plant interplays (Hartman & Tringe, 2019; DiLegge et al., 2022).

4.2.1 Root Secretions and Chemotaxis

In the microbial colonization process, many taxa use chemotaxis to recognize and react to signals originating from plants, such as sugars released by roots. Following the signal perception, microorganisms mobilize towards plants and stick to the roots surface by originating a biofilm (Scharf et al., 2016; Abdul Hamid & Nadarajah, 2022). In contrast to bulk soil, the microbial communities of the phyllosphere and rhizosphere possess an abundance of genes that are involved in biofilm formation, chemotaxis, flagella assembly, secretion, motility, and the two-component regulatory system (Noman et al., 2021). The large numbers of substrate transporters found in members of the phyla Proteobacteria and Firmicutes enables these bacteria to adapt to the nutrient-rich environment of the plants (Mendes et al., 2013). Chemotaxis is encouraged by microorganisms found on the root surface or through chemical substances present in root exudates (Feng et al., 2021; López-Farfán et al., 2019).

Plant developmental phases, genotypes, and environmental circumstances influence the changes in the composition of root exudates, resulting in a difficult characterization of these substances. Plant-plant interactions are influenced by a series of below-ground signals that include the availability of nutrients and water, the soil microbiota, and chemicals emitted by roots (exudates) (Ma et al., 2022; Sharma et al., 2023). Root exudates contain a wide variety of compounds, many of which are species-specific and change based on the above-ground and below-ground environments (Salem et al., 2022). At the intraspecific level, root exudates differ greatly as well. For instance, mutations affecting relevant biosynthetic pathway genes in certain accessions of *A. thaliana* can result in the loss or lack of crucial substances that facilitate recruitment of microorganisms to the root (Malka & Cheng, 2017). This very unique molecular signature is in line with the theory that soluble root exudates may help plants distinguish between individuals of the same genotype (Frerigmann & Gigolashvili, 2014). For instance, it was discovered that rice roots develop toward the roots of plants of related genotypes while distancing themselves from other different genotypes (Fang et al., 2013). However, a number of chemicals (some common and some rare) have been discovered in specific plant species. The root tips mostly release polysaccharides, which are also abundantly found in root caps and mucilage. Plant-friendly bacteria showed a capacity to recognize various molecules through particular receptors, specific for those substances (Tosif et al., 2021; Noman et al., 2021). These diverse metabolites may be secreted from root cells and transported to the soil where concentration gradients serve as a driving force.

Selected transmembrane proteins assist in the efflux and influx of these compounds (Sharma et al., 2023; Muratore et al., 2021). Primary metabolites such as organic acids, amino acids, and sugars are more easily secreted into the rhizosphere thanks to specialized carriers and channel proteins. For instance, plants use ALMT/malate and MATE/citrate transporters for organic acids, CAT and GDU transporters for amino acids, and members of the SWEET transporter family for sugar molecules (Sharma et al., 2023; Canarini et al., 2019). Plants also modify their root system architecture (RSA) in response to the concentration of primary metabolites in the root region, which in return controls the exudation process (Canarini et al., 2019; Sharma et al., 2023). Amino acids have a cascading effect on the RSA, which is critically shaped by changes in the root metabolite efflux or influx. RSA affects the pattern of root exudation through the involvement of specialized carriers. Further research on the genes responsible for determining a plant RSA may aid in a thorough knowledge of the root exudation mechanism (Chaparro et al., 2013; Han et al., 2023) (Fig. 4.1).

4.2.2 *Microbe-Microbe Interaction in Microbiome Assemblage*

It is not surprising that genes active in the soil microbiota play crucial roles in either competing or collaborating with other microbial species. These microorganisms may produce a wide range of chemicals, that may have a significant impact on the complex web of interactions among microbes (Berg et al., 2020). Researchers have identified several unique and different gene clusters that are in charge of the production of numerous chemicals, particularly in the context of bacteria that are connected with plants (Netzker et al., 2015; Abdalla et al., 2017). Genomes of plant-associated bacteria contain genetic components necessary for the synthesis of compounds that can affect bacteria both within and beyond species boundaries. The existence, diversity, and geographic distribution of distinct microbial communities within host plants are regulated by these substances, which include antibiotics and toxins (Zhang et al., 2023; Liu et al., 2023). The mechanisms directing the synthesis of these substances, and their regulatory functions in the dynamics of microbe-microbe interactions, are still open to questions. Further research is also necessary

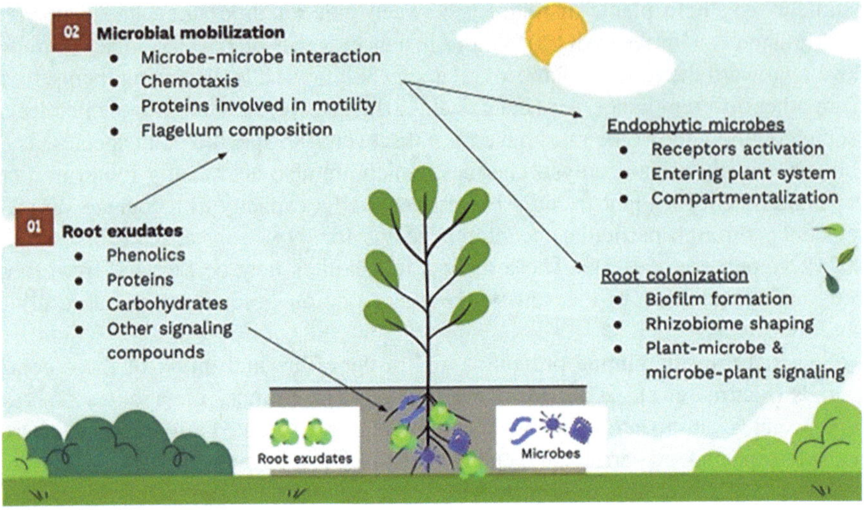

Fig. 4.1 The root exudates and the recruitment of microbes in soil and roots, and their effect on roots and other plant organs

to understand the complex molecular pathways behind these actions (Brown et al., 2020). Notably, specific functional characteristics that are connected to the suppression of infections, such as the presence of genes involved in the production of antifungal substances and the secretion of proteins, are extremely prevalent in rhizobiomes counteracting plant diseases (Mendes et al., 2011; Zhou et al., 2019a).

An enormous number of uncharacterized biosynthetic gene clusters (BGCs) are activated in response to pathogenic risks. These BGCs contain important gene sequences for key enzymes such as chitinases, polyketide synthases (PKSs), and non-ribosomal peptide synthases (NRPSs). They are crucial in causing the disease-suppressive activity observed in the endophytic root microbiome (Mendes et al., 2011). There are approximately a thousand BGCs among the several biosynthetic classes, found in the phyllospheric bacteria of *Arabidopsis*. These cover a variety of biosynthetic processes, such as terpene systems, post-translationally modified and ribosomally produced peptides, trans-AT PKSs, and NRPSs (Noman et al., 2021). The intricate connections that develop within a specific microbial habitat are largely driven by the complex interplay among these BGCs (Berendsen et al., 2018). There is an extensive distribution of antibiotic resistance genes within the microbiome, suggesting that the formation of the microbial community is governed by fierce competition among various microbial groups (Sengupta et al., 2013).

By detecting and producing signaling molecules such as homoserine lactone (HSL), quorum sensing (QS) enables bacteria to participate in intercellular communication (Xiao et al., 2022). Surprisingly, distinct bacterial taxa are able to produce the same signaling molecules, allowing for interactions with nearby microorganisms of other taxa that can either involve interference or collaboration (Prazdnova et al., 2022). For instance, the taxis of finger millet endophyte M6 towards *Fusarium graminearum* serves as an example of how QS has an impact on communication between organisms. This coordinated response causes a biofilm to form on the root hair, thereby trapping and preventing the entry of the endophytic pathogen (Mousa et al., 2016; Pan et al., 2015). The stimulation of the root colonization is achieved by controlling the root exudation of compounds, fueled by long-distance communication among various rhizospheric system components (Lyu & Smith, 2022; Upadhyay et al., 2022). Numerous reports have been published on the capacity of *Bacillus* and *Pseudomonas* species to produce compounds that are unique to these bacteria, thereby causing the systemic colonization of particular bacterial groups (Riera et al., 2017).

The QS mechanism, which enables certain signal molecules, such as AHLs, to allow intercellular bacterial gene expression, is a crucial aspect of microbial communication (Majdura et al., 2023). A wide variety of phenotypes are produced under the control of the QS system, many of which directly affect pathogenesis in economically significant bacterial pathogens. The fascinating tactic of interfering with or interrupting QS offers considerable potential as a method of reducing pathogenic infections in agriculture (Rutherford & Bassler, 2012). Quorum quenching, often known as QQ, has become widely recognized as a successful tactic (Rehman & Leiknes, 2018). Enzymatic destruction of AHL signal molecules is a common method in this field. For instance, several *Pseudomonas* strains showed a QQ activity capable to reduce the symptoms of soft rot (Zhou et al., 2022). By generating QS inhibitors, which may either degrade QS signals or compete for signal receptors, plants have additionally evolved a capacity to interfere with the QS system. The QS inhibitors also include antibiotics, and bioactive chemicals derived from marine organisms and fungi (Chen et al., 2019). The capacity of QS inhibitors to function below a minimum inhibitory concentration is a strong characteristic. Pathogens are able to continue developing without causing disease because to this special property (Bulgarelli et al., 2015; Carlström et al., 2019).

In addition to host plant signals, genetic resources for inter-microbiome competition are also essential for successful microbial colonization (Noman et al., 2021). In microbial communities within a variety of plant hosts, such as wheat, cucumber (Ofek-Lalzar et al., 2014) barley (Bulgarelli et al., 2015), and citrus (Xu et al., 2018), genes connected to the type VI secretion system (T6SS) have been discovered. This system releases harmful proteins to eliminate competitors (Carezzano et al., 2023).

The effector family known as Hyde1 in pathogenic *Acidovorax* spp. T6SS controls phyllospheric bacterial growth, suggesting a possible role in interbacterial competition for plant colonization (Noman et al., 2021). Microbial groups colonize by responding to environmental factors, plant signals, competition fitness, survival rates, and environmental conditions. However, integrated modeling and detailed research on the interactions between plant holobionts are needed to fully comprehend the complex biochemical underpinnings of microbial relationships. Recent research has shed light on the eco-evolutionary mechanisms that shape the composition of microbial communities. These include shifts in microbial niches that result in the selection of the best fitting populations for any given environment, ecological drift, and genetic diversification (Cordovez et al., 2019). In rhizospheric and phyllospheric communities, initial colonization is sparked by random events (Noman et al., 2021). Although core microbial communities are strong, plants can incorporate new species without affecting the structure of the existing microbiome. It is essential to comprehend the mechanisms underlying the assembly, dynamics, occurrence, and reactions to biotic and abiotic stimuli in order to emphasize the importance of microbiomes in the plant responses to challenging environments (Fadiji et al., 2023).

4.2.3 Plant-Pathogen Interactions: Role in Microbiome Assemblage

Pathogen and plant targeted molecules are responsible for the intricate processes that underlie plant–pathogen interactions. The chemicals that pathogens secrete have a substantial impact on their capacity to colonize, develop, and cause disease in the host (He et al., 2021). One noteworthy example is the infamous cause of potato late blight, *Phytophthora infestans*, which controls host defenses via a proteinaceous effector known as *AVR3a* (Bos et al., 2010). Similar to this, chemicals from plants are essential for identifying pathogens and triggering a variety of host defense mechanisms. These interactions frequently start in the apoplast, where plant receptors pick up cues from microbes. Pattern recognition receptors (PRRs) on the plant cell membranes often detect these cues, also known as pathogen-associated molecular patterns (PAMPs) (Lü et al., 2022; Abdul Malik et al., 2020). PAMPs include a wide variety of chemicals, such as bacterial flagellin, oomycete cellulose, and fungal chitin, which are all recognized by specific PRRs, such as chitin elicitor receptor kinase 1 (CERK1), flagellin-sensing 2 (FLS2), and cellulose binding elicitor lectin (CBEL) (Zhou et al., 2019b; Beck et al., 2014; Larroque et al., 2011). These interactions, in combination with the identification of PAMPs, are a critical component of the plant defense mechanism against potential intruders.

PAMP-triggered immunity, or PTI, is a first line of defense that plants employ when they detect certain chemicals from microbes, known as PAMPs, via the PRR specialized receptors. Pathogens evolved to secrete a plethora of effectors into host cells in order to avoid PTI that may be inhibited or neutralized by these effectors (Abdul Malik et al., 2020). The XA21 binding protein (XBP1) from *Xanthomonas oryzae* and the *AVR-Pii* from *Magnaporthe oryzae* are two common examples of these microbial effectors that impede the plant defense responses. In response, plants evolved a number of disease resistance (R) proteins that are able to identify the Avr proteins linked to pathogens (Abdul Malik et al., 2020). Effector-triggered immunity (ETI), also known as the second line of defense, is initiated by this interaction between the plant R proteins and the pathogen Avr proteins. ETI frequently causes localized cell death of both pathogen and plant cells in the location of infection. It is also quicker and stronger than PTI. Together, PTI and ETI form the innate immune system of plants, which enables them to recognize, fend off, and eliminate invasive pathogens (Ding et al., 2022; Abdul Malik et al., 2020).

The conditions under which plants and pathogens are grown, their genetic makeup, changes in the host physiology due to biotic or abiotic stresses, and the composition of microbiome associated with the host are all environmental factors that affect how plant and pathogens interact (Velásquez et al., 2018). For instance, the fungal pathogen *Fusarium oxysporum* was shown to infect cucumber plants in greenhouse conditions, whereas the bacterial pathogen *Xanthomonas campestris* causes a tomato disease (Noman et al., 2021). The plant microbiome, particularly the rhizobiome, acts as a protective barrier that disrupts the interactions between plants and pathogens, leading to positive effects on the plant health (Noman et al., 2021). Through resource competition, sequestering of necessary nutrients, i.e., zinc, and the production of pathogen-toxic substances like antifungal metabolites, microbial communities are able to restrict plant-pathogen-related activities (Cabot et al., 2019). Additionally, by encouraging systemic resistance in plants, beneficial plant-associated bacteria indirectly interrupt the communication between pathogens and their hosts (Zhu et al., 2022). Due to their useful benefits in preserving plant health, particularly under stressful environmental situations, native plant-associated microbial communities have recently received a great deal of interest from the scientific community.

4.3 Soil Microbiome and Plant Health

Each component of the microcosm often includes a number of beneficial traits that contribute to controlling physiological processes in host plants under various environmental circumstances. The unique characteristics displayed by various microbe groups within the microbiome have a substantial impact on plant health. Numerous variables, such as the variety of microorganisms, the habitat, and the host plant species, have an impact on these characteristics (Abdul Rahman et al., 2021). In addition to assisting in nutrient uptake, reducing environmental stressors, and offering protection against diseases, microorganisms have direct positive effects on the health of their host plants (Nadarajah & Abdul Rahman, 2021). The significance of plant microbiomes in protecting hosts from phytopathogens has long been the focus, where studies on development of "biocontrol" have intensified for use in green agriculture. Beneficial microorganisms are typically responsible for protecting plants against invading pathogens directly or through indirect means of host innate immune response elicitation (Manjunatha et al., 2022).

4.3.1 Direct Suppression of Plant Pathogens

There are many players in the dynamic world of the plant microbiota, including pathogenic, neutral, and beneficial microorganisms. A plant develops beneficial symbiotic interactions with microbial populations throughout its existence, but it also has to deal with diseases brought on by a wide variety of destructive microbes. Numerous plant species, particularly those that are economically significant, are harmed by soil-borne diseases, which result in reduced crop quality, yield and significant financial losses (Agrahari et al., 2020). *Sclerotinia sclerotiorum* (Hossain et al., 2023), *Phytophthora infestans* (Ivanov et al., 2021), *Pythium* spp. (Arora et al., 2021), and *Botrytis cinerea* (Pandit et al., 2022) are noteworthy examples of these soil-borne fungal diseases. These pathogens may survive for long periods of time in the soil by acquiring different dormant forms (such as oospores, chlamydospores, resting mycelia, and sclerotia) until they acquire activation cues from their respective hosts. Flavonoids and organic acids found in the root exudates of legumes significantly increase *B. cinerea* sporulation and spore germination (Pascale et al., 2020). When soil-borne infections invade plants, seedlings frequently experience stunted growth, root or collar rot, wilting, and damping-off. Some of these pathogens can infect a variety of hosts, making traditional management methods useless. It has been determined that specific bacterial communities that live in soil have a detrimental effect on plant health. Some examples of well researched soil-borne bacteria that threaten plants are *Pseudomonas syringae* (Sundin et al., 2016), which causes bacterial speck disease, and *Xanthomonas campestris* (Vicente & Holub, 2013), the causal agent of bacterial blight.

The complex interactions that take place in the rhizosphere diverse microbial community determine the extent of plant pathogen interaction. Beneficial microbes engage with pathogens in soil through resource and space competition, and the secretion of antimicrobial compounds (Li et al., 2021). Existing pathogens contribute to the development of additional detrimental microbial colonies within the soil by releasing effector proteins that suppress the activity of beneficial rhizosphere bacteria (Niu et al., 2020). Plants have benefited from nutrient uptake, resistance to environmental challenges, and the capacity to resist diseases due to coevolution with their microbial counterparts, over many million years. Disease prevention and nutrient mobilization are greatly aided by communal assemblies of beneficial microorganisms connected to hosts (Malgioglio et al., 2022). For instance, *Streptomyces* spp. can inhibit plant pathogens through antagonistic interactions and antibiosis. The genetics hosts and pathogens, the dynamics of the pathogen populations, the variety and composition of plant microbiota, as well as other abiotic and biotic stressors, all influence how well diseases are suppressed in soil (Bonaterra i Carreras et al., 2022; Dow et al., 2023). Further research is required to fully understand the complex interactions linking potential antagonists and disease-causing phytopathogens, even though disease suppression often arises from the combined efforts of several microorganisms rather than individual ones (Chen et al., 2023). The effectiveness of disease-suppressive soils has been linked to a few simple mechanisms, including the synthesis of antimicrobial metabolites and volatile chemicals by hostile bacteria (Noman et al., 2021).

Antibiosis is one such strategy, which involves the creation of antimicrobial metabolites to inhibit the growth and proliferation of other species. Xanthobaccins and kanosamines are two prominent examples of antibiotics that inhibit plant diseases. These substances, released in soil by different *Bacillus* species, inhibit the development of diseases such as bacterial wilt and root rot (Saeed et al., 2021). Additionally, some *Burkholderia* species produce pyrrolnitrin and phenazines that successfully inhibit the development of pathogenic *Fusarium* species (Xu et al., 2020). Another fascinating illustration is provided by the bacterium *Pseudomonas* spp., which secretes volatile organic chemicals to inhibit plant diseases and promote plant development. The possible impact of substances like phenazine and pyoluteorin on reducing the spread of phytopathogens has been highlighted by earlier studies (Bonaterra i Carreras et al., 2022; Wu et al., 2011). The dynamic competition for resources, survival, and colonization opportunities between plant-beneficial and detrimental microorganisms influences the growth of pathogens in soil. If, however, a disease manages to bypass the rhizobiome, root endophytes provide an additional layer of defense against the invading pathogens. They mobilize microbial communities in response to invasion, activating their genetic machinery to produce defense-related enzymes and metabolites (Uroz et al., 2009).

Fusarium wilt was reduced when several crops, including beans, spinach, and potato, were grown in disease-resistant soils enriched with bacterial phyla such as Actinobacteria and Bacteroidetes (Noman et al., 2021; Chen et al., 2023). Changing the microbial communities structure through various management approaches, such as intercropping and the application of organic amendments, can accomplish broad or targeted disease suppression (Ayangbenro et al., 2022). Our comprehension of the presence of particular gene clusters that coordinate pathogen-inhibitory effects has improved thanks to genomic research, which may someday result in the identification of brand-new antibiotics (Kalia et al., 2019). Plant-associated microbes provide an important source of potential antagonists with the ability to combat phytopathogens. These microbial communities interact directly with pathogens, preventing their spread (Fenta et al., 2023). Additionally, it is acknowledged that these communities are crucial accelerators fostering efficient biocontrol in shared ecological niches, through competition for resources.

4.3.2 Microbiome Role in Plant Immunity

Recent findings underscore the involvement of plant-associated microbiomes in shaping plant health. They exhibit beneficial attributes that enhance plant immune responses against both biotic and abiotic environmental stressors (Li et al., 2022; Haichar et al., 2021). These microbiomes essentially help their host plants to build resistance to diseases by altering their defense mechanisms. Microbe-induced immune reactions improve plants response to pathogens, leading to a notable enhancement in disease resistance and tolerance (Nishad et al., 2020). Bacterial antagonists from genera such as *Bacillus*, *Streptomyces*, and *Pseudomonas* showed a remarkable biocontrol potential against *Magnaporthe oryzae*, eliciting the expression of key defense genes such as *PR1*, *PR5*, and *NPR1* in rice seedlings, and effectively guarding against rice blast disease (Law et al., 2017). Similarly, root-associated *Rhizobium* and *Serratia* spp. decreased disease incidences in *M. oryzae*-infected rice plants by activating JA- and ETH-dependent systemic resistance (ISR) responses (Spence et al., 2014).

Pathogen attacks trigger shifts in the root exudation patterns of the host plants, facilitating the colonization by specific, resistance-inducing microbiota. Through the modulation of the SA-pathway, various microbial populations in the endosphere and episphere were able to activate defensive mechanisms against this pathogen (Di et al., 2017). Furthermore, in another controlled field trial, the root-associated microbiome was able to confer resistance in strawberry plants against two soil-borne fungal diseases, *Macrophomina phaseolina* and *Verticillium dahliae* (Lazcano et al., 2021). Similarly, field data showed a link between microbiome-induced resistance in maize plants and suppression of the disease caused by *Rhizoctonia solani* (Balint-Kurti et al., 2010; Noman et al., 2021). While certain microbes could induce plant immune responses in a controlled greenhouse environment, many fail in field conditions. This mismatch is frequently related to the microorganisms failure to flourish and establish themselves within the rhizosphere in unfavorable environments, resulting in reduced protection in field situations (Fiodor et al., 2021). Overall, the interplay between environmental conditions and soil health emerges as pivotal in shaping beneficial plant-microbe interactions.

The underlying molecular mechanism driving rhizobacteria-mediated ISR is shared by a variety of plant species and represents a unique aspect of inducible immunity (Van Wees et al., 1999). Certain cell surface components of biocontrol bacteria, such as lipopolysaccharides or peptidoglycan, can serve as triggers for initiating ISR in plants (Abdul Malik et al., 2020). Moreover, plants can evoke an ISR response when directly exposed to compounds secreted by beneficial bacteria, including the volatile compound dimethyl disulfide, as well as compounds like phenazines, cyclic lipopeptides, and exopolysaccharides (Sánchez-López et al., 2016). Moreover, transcription factors such as MYB72 are thought to play a role in controlling ISR in plants. (Van der Ent et al., 2008).

The modulation of ISR is complex and requires the interplay between various componentsthat evolved to harness associated microbes as signals to regulate and refine the plant immune system. Notably, plants rely on microbes during their initial growth stages to establish contact with soil, which is vital for their survival (Yu et al., 2022). ISR-inducing strains, on the other hand, control the release of antimicrobial substances from roots. Changes in the composition of root exudates are crucial to direct root microbiomes and promote the recruitment of advantageous microbial groups into the rhizosphere (Eichmann et al., 2021; Li et al., 2021).

However, in order to establish a symbiotic relationship with their host, bacteria eliciting ISR must get past the host plant immune system. Beneficial microorganisms use pathogen-like tactics to suppress this immune reaction (Yu et al., 2022).

For example, according to Yuejun et al. (2017), the AM fungus *Funneliformis mosseae*, which induces ISR, evades the host immune reactions and encourages root colonization through the synthesis of the SP7 effector.

The MiSSP7 effector is also used by the ectomycorrhizal fungus to inhibit SA-mediated immune responses in plants (Noman et al., 2021). Additionally, *Arabidopsis thaliana* showed reduced immunological responses after colonization by *Trichoderma* sp. or *Bacillus megaterium* (Ortíz-Castro et al., 2008; Noman et al., 2021). Elicitors and signals of plant-beneficial microorganisms trigger the immune response systemically, but downregulate locally this responses to aid in colonization. However, further research is needed to fully understand the complex molecular mechanisms underlying mutualistic plant-microbe interactions (Pršić & Ongena, 2020).

4.4 Functional Benefits of Microbiome Engineering on Plant Health

The plant microbiota play a crucial role in determining their host capacity to thrive and mature across diverse environmental conditions. Innovative techniques enhancing the plant stress resilience, the defense against diseases, and the absorption of nutrients have been shown to bolster these attributes. Advancing technologies in microbiome engineering encompass practices such as soil enhancement, construction of synthetic microbial communities, and customization of host-dependent microbiomes.

4.4.1 Amendments in Microbiome Engineering

Maintaining a consistent diversity of beneficial bacteria is critical to enhance soil health to improve crop resistance to environmental challenges and achieve higher yields. Organic soil formulations that can be utilized to support the establishment of functioning microbial populations include compost, organic residues, charcoal, organic wastes, and peat (Pathma & Sakthivel, 2012). Pathogenic organisms can be effectively handled utilizing methods such as biofumigation, an organic soil conditioning approach. Organic soil conditioners have the advantage of enriching the soil with a variety of beneficial and interconnected microbial species that may contribute to various activities more efficiently than chemical fertilizers (Tian-Yu et al., 2022). Characterizing the activities of beneficial microbial groups in response to certain organic amendments is critical for gaining a better understanding of soil health biochemistry and developing long-term strategies. Optimizing the application of organic soil amendments for various crops and soil types also promotes sustainable soil health management.

Salicylic acid and a variety of other metabolites found in root exudates are examples of plant-oriented signaling molecules that have a substantial impact on the dynamics and structure of the microbiome (Noman et al., 2021). This finding emphasizes the possibility of purposeful modification of plant microbiomes via targeted application of microbe-stimulating substances. An example is the release of phenolic compounds like coumarins from roots as an illustration. In addition to altering the microbial environment around the roots, coumarins also encourage the growth of advantageous rhizobacteria (Yang et al., 2023). Consequently, this has been shown in numerous studies to be a powerful defense mechanism that protects plants from soil-borne diseases (Chen et al., 2023; Mendes et al., 2013). The use of a distinct class of root exudates known as benzoxazinoids is another noteworthy aspect of this phenomenon. These substances are essential for drawing beneficial bacterial and fungal microbial populations into the rhizosphere and enabling their colonization. This complex interplay serves as a natural defense system, shielding maize plants from the attacks by herbivorous insects (Neal et al., 2012). As we learn more about how plants and microbes interact, it becomes clearer that these signaling molecules direct a complex web of ecological interactions that give plants the tools they need to survive in a variety of challenging environments. Expanded study into the many functions of root exudates will be beneficial for the investigation of innovative directions, such as host-specific biofertilizers. Salicylic acid, flavonoids, terpenoids, and phytoalexins are a few examples that highlight the potential of these substances to influence microbiome engineering and facilitate chemical interactions through symbiotic partnerships between particular signaling molecules and microbial taxa. This approach may set new paths for ground-breaking tactics to improve the plants resilience and health.

4.4.2 Microbial Consortia and Microbiome Engineering

Much like the principles of synthetic biology, microbiome engineering involves the use of artificial microbial consortia to finely manipulate the structure and role of plant microbiomes. This technique allows for the construction of consortia holding attributes crucial for optimal plant growth and development, even in challenging circumstances. Embracing such an approach presents a promising solution to various issues associated with conventional biofertilizers. These problems include the inability to effectively compete with native microorganisms during field trials, low performance within specific local environments, and compatibility issues with the host plants (Orukotan et al., 2023). Creating an optimal microbial consortium is a systematic process that encompasses various stages. These include choosing the microbial source, isolating and growing the core microbiota, identifying functionally active microbial groups, optimizing microbe-to-microbe interactions, and assessing the effectiveness of the resulting consortium (Cao et al., 2022).

In the rhizosphere, countless microorganisms engage in intricate interplays with other microbes, acting as crucial components in functional consortia. An example is the synergy between plant growth-promoting rhizobacteria (PGPR) and AM fungi in terms of ecosystem dynamics and nutrient availability (Hakim et al., 2021). Building upon this concept, specific microbial strains can be intentionally introduced into soils to reshape the composition of microbial communities. A well-constructed artificial microbial consortium is widely acknowledged for its ability to enhance plant stress resilience, harnessing microbial interactions to fortify plants *vs* environmental challenges (Maciag et al., 2023). An example is the synergistic introduction of the AM fungus *Glomus versiforme* and the plant-beneficial bacterium *B. rhizosphaerae* into the maize rhizosphere. In addition to promoting maize development, this partnership also encouraged the growth of a beneficial microbial community, which was especially important when dealing with salt or heavy metal stress (Ujvári et al., 2021).

In a similar view, cotton plants were fortified against growth restrictions and the soil-borne pathogen *F. oxysporum* using a potent microbial combination based on *Streptomyces* and *Bacillus* spp. (Murty & Shim, 2021). This consortium boosted disease resistance while also promoting plant development. Additionally, a noticeable change in the composition of the bacterial population was brought about by the introduction of *Rhizobium radiobacter* into the rhizosphere of potato plants. Other beneficial PGPR, such as *Bradyrhizobium* and *Streptomyces* spp. also became more prevalent as a result of these alterations in the rhizobiome. Potato yield, antioxidant levels, phytochemicals, phosphorus content, and root nodulation all significantly increased as a result of microbial community transformation (Naqqash et al., 2016).

Additionally, a synergistic strategy involving the simultaneous introduction of *B. amyloliquefaciens* and *Burkholderia phytofirmans* into the rhizosphere of *Capsicum* plants accelerated growth and stimulated the production of secondary metabolites like volatile organic compounds, renowned for their antagonistic effects against the pathogen *Alternaria alternata* (Soliman et al., 2023). These consortia make use of bacterial strains that have 1-aminocyclopropane-1-carboxylic acid deaminase genes. In addition to having powerful antibacterial effects, these tailored microbial blends also served as plant growth promoters (Glick & Nascimento, 2021). In a similar view, co-inoculation with strains of *Trichoderma* spp. dramatically improved resistance of pepper to the destructive disease *Phytophthora capsici*. By adjusting the activity of antioxidant enzymes and encouraging the build-up of disease-preventive phenolic compounds in the face of a pathogenic onslaught, this consortium coordinated a calibrated response against the pathogen (Sang et al., 2013; Santos Hernández et al., 2023). These examples highlight the great potential of utilizing microbial partnerships to improve plant health and resilience under various environmental stressors. In order to delve deeper into the molecular specifics of how they prevent diseases, additional research is required to comprehend the delicate interplay between these interactions and diseases. A practical strategy is required to ensure the lifespan of synthetic microbial communities across succeeding plant generations. To preserve their effectiveness and influence, artificial microbial consortia must be continuously applied at regular intervals.

4.4.3 Host Role in Microbiome Engineering

Beneficial microorganisms are typically found in the rhizosphere, where roots serve as gatekeepers, allowing only beneficial microbes to enter and act as endophytes (Liu et al., 2017). Plants simultaneously discharge several bacterial species into the rhizosphere, but it is still unclear how beneficial or harmful microorganisms enter and leave the plant holobiont (Lombardi et al., 2021). Further evidence show that roots can directly use related microorganisms as a source of nitrogen (Dellagi et al., 2020). To determine whether plants show preferences for particular microorganisms for beneficial interactions, additional in-depth investigation is necessary. Importantly, the dynamics and capabilities of these microenvironments are strongly influenced by plant genomes (Xiong et al., 2021). In certain accessions of *Arabidopsis*, *Pseudomonas fragi* WCS417r increased the plant biomass output (Wintermans et al., 2016). Changes in phyllospheric microbial diversity were caused by changes in the PTI signaling system and MIN7 vesicle-trafficking route in *Arabidopsis*, demonstrating a genetic link between cell wall integrity regulation, plant defense, and phyllospheric bacterial diversity (Machado Wood et al., 2021).

At the microbiome level, unique plant genotypes support a variety of disease-suppressive and advantageous microorganisms, affecting microbial diversity through changes in the metabolite released from the roots (Ha et al., 2021). Notably, the rhizobiomes of several bacterial taxa, i.e., *Enterobacter* and *Kosakonia*, are more prevalent and sustained resistance to Fusarium wilt on specific banana cultivars (Noman et al., 2021). However, only 0.7% of the operational taxonomic units (OTUs) found in the rhizosphere were in common among bean plants, and bean genotypes also had a substantial impact on this composition (Ayangbenro et al., 2017). Studies showed that there is a genetic connection between the variety of the epiphytic microbial population, the plant and its level of resistance (Noman et al., 2021). T.

The potential to increase agricultural outputs through the selection and development of "microbe-friendly" crops is enormous. Genetic alterations in plants to entice and stabilize functioning microorganisms within their microcosm are now considered, due to the understanding of these advantageous interactions. In order to make this happen, 'designer plants' may be created that secrete hormones and exudates that encourage the colonization by and recruitment of advantageous microbiomes.

Genes involved in the construction of advantageous microbiomes may be discovered by examining wild species or their relatives. For instance, modern domesticated bean accessions showed comparatively large amounts of Proteobacteria and Actinobacteria, but wild bean accessions showed abundant *Bacteroidetes*. Phyllospheric microbial diversity was strongly correlated with host genotype in a number of tomato accessions. The endophytic microbial diversity was strongly correlated with the host genetic diversity also in wheat plants (Noman et al., 2021). It is possible that changes in microbial communities are linked to changing plant genotypes and modified root architecture. Comprehensive research is necessary to clarify how plants control and promote specific microbiome colonization as well as how host-genotype-microbiome interactions recruit beneficial microbial communities, for desired features (Dastogeer et al., 2020). It appears easier to maintain healthy, useful microbiomes by keeping track of any real time change in the diversity of microorganisms, in the field.

4.4.4 Plant Manipulation and Microbiome Engineering

Comprehensive and long-term strategies must be implemented to protect the world food resources to address global and environmental issues such as shifting climate patterns, desertification, soil salinization, new and emerging diseases, and crop resistance to various biological and environmental stress factors (Çakmakçı et al., 2023). Innovative plant manipulation approaches utilizing modern molecular biology techniques, biochemical tactics, induced genetic mutations, selective breeding, and crop variety development have considerably improved plant resilience to a variety of environmental alterations. High-throughput gene expression profiling, the identification of intricate protein networks on a global scale, molecular modeling, and their interaction with environmental variations can all lead to improved agricultural traits and higher crop yields (Ahmad, 2023; Naqvi et al., 2022).

Researchers have unlocked the possibility for site-specific genome changes across the whole plant kingdom, through precision gene-editing technologies such as TALENs and CRISPR/Cas systems. The use of CRISPR/Cas systems for targeted crop trait modification has received a lot of attention. For example, CRISPR/Cas9 was used to precisely disrupt the carotenoid cleavage dioxygenase (*ccd7*) gene in rice, which is required for strigolactone production (Butt et al., 2018). This genetic change resulted in more tillering in *ccd7* mutant rice plants, revealing the accuracy of CRISPR/Cas systems in changing the plant architecture. In related research, CRISPR/Cas9 allowed production of novel eIF4G alleles conferring resistance to RTSV in the IR64 rice lines, via targeted mutagenesis.

Macovei et al. (2018) introduced mutations to the *eIF4G* gene via CRISPR/Cas9, paving the way for the production of RTSV-resistant rice cultivars. Furthermore, the CRISPR/Cas9 system has been demonstrated to be effective in inducing beneficial alterations in endogenous genes (Yeap et al., 2021). Researchers were able to construct favorable mutant alleles with increased oleic acid content in *Brassica napus* by altering the agronomically crucial fatty acid desaturase 2 gene (*FAD2*). This result has substantial implications for crop improvement and sustainability. These cutting-edge plant genome engineering approaches hold enormous potential for creating and sustaining a new "Green Revolution".

As a result, a thorough understanding of different approaches from soil science, microbial ecology, and plant biology promises opportunities for improving the phytomicrobiome. Nonetheless, the effective execution of such manipulation tactics faces significant obstacles due to present technical limits, which impede accelerated development of microbiome solutions. Some of the factors that contribute to these

Table 4.1 Soil environment and plant factors influencing the microbiome shaping

Factors	Summary	References
Host influence	The composition of the microbial communities in the rhizosphere, roots, leaves, and seeds, as well as plant phenotypes, are greatly influenced by the host genotype. Additionally, the microbial communities in the rhizosphere and roots are further influenced by root exudates.	Chaudhry et al. (2021), Fang et al. (2013), Frerigmann and Gigolashvili (2014), Ha et al. (2021), Nadarajah and Abdul Rahman (2021), Noman et al. (2021) and Rolfe et al. (2019)
	Nutrient status of P and Fe can alter microbial communities.	Han et al. (2023), He et al. (2017) and Lazcano et al. (2021)
	Phytohormones play a role in defense and can shape the soil microbiome in a host-dependent manner.	Eichmann et al. (2021), Yu et al. (2022) and Zhu et al. (2022)
Soil and abiotic factors	Soil physico-chemical properties and environment affect the microbiome	Fadiji et al. (2023), Salem et al. (2022)
Soil and biotic factors	Plant diseases are associated with distinct microbial communities; suppressive soils may include a higher concentration of microorganisms that are disease suppressive. Plant health may be impacted by early variations in the composition of the soil microbiome.	Bonaterra i Carreras et al. (2022), Dow et al. (2023) and Noman et al. (2021)

limitations are: host genotype variations, root exudates composition, root architecture, induced systemic reactions, competition from native soil microflora at the root-soil interface, limited microbial colonization, unfavorable environments, transient effects, lack of plant-root interface recognition, and the absence of standardized protocols and tools for *in situ* manipulation. Furthermore, the use of synthetic biology approaches holds promise for the identification of plasmids, transposable elements and novel phages that can be used for *in situ* microbiome engineering.

These findings have a potential to promote the formation and spread of healthy microbiomes, opening the door for revolutionary changes in agriculture and environmental management (Bakker et al., 2020). Table 4.1 presents some main factors that influence the microbiome shaping.

4.5 Conclusions and Future Perspectives

The interactions between plants and their associated microbiomes have been studied quantitatively, exposing the underlying mechanisms, thanks to the integration of experimental biology, multi-omics methods, and computational biology. Important bacterial and fungal groups that commonly engage in symbiotic relationships with plants have been identified thanks to the application of these approaches to a wide range of crops and model plant species. However, additional investigation into the diversity of microorganisms is needed to identify the functional communities of bacteria uniquely adapted to important crops. For this goal, comprehensive approaches must be developed to identify the core microbiota and explicate their activities within host plants, particularly for microbiomes of critical economic and ecological value. Studies on the microbiome using metagenome and other sequence-related methods have discovered key players with functional genes in charge of colonizing plants, boosting plant fitness features, and influencing the composition and diversity of plant-microbiota communities. The effect of microbiomes on plant fitness has been clarified thanks to these methodologies, although a sizable fraction of taxa are still unclassified and their variability is still unknown. Large-scale longitudinal studies on the dynamics of plant-microbiome interactions considering the host age and temporal fluctuations are crucial to fill this gap. A deeper comprehension of the dynamic interplay between plants, microbiomes, and challenging environmental conditions is essential to pave the way for engineering resilient microbial consortia with predictable and consistent outcomes. Furthermore, the fusion of experimental methodologies with modeling approaches will expedite scientific progress, addressing methodological and technical challenges. We will be able to engineer and improve the activities of complex microbial consortia with accuracy and consistency by using integrative methodologies that benefit of knowledge proceeding from a variety of scientific areas. We shall be better prepared to use engineered microbial consortia for sustainable and improved plant production in the face of continuously changing environmental conditions as our understanding of the subtleties of plant-microbiome-environment interactions grows.

References

Abdalla, M. A., Sulieman, S., & McGaw, L. J. (2017). Microbial communication: A significant approach for new leads. *South African Journal of Botany, 113*, 461–470.

Abdul Hamid, N. W., & Nadarajah, K. (2022). Microbe related chemical signalling and its application in agriculture. *International Journal of Molecular Sciences, 23*, 8998.

Abdul Malik, N. A., Kumar, I. S., & Nadarajah, K. (2020). Elicitor and receptor molecules: Orchestrators of plant defense and immunity. *International Journal of Molecular Sciences, 21*(3), 963.

Abdul Rahman, N. S. N., Abdul Hamid, N. W., & Nadarajah, K. (2021). Effects of abiotic stress on soil microbiome. *International Journal of Molecular Sciences, 22*, 9036.

Agrahari, R. K., Singh, P., Koyama, H., & Panda, S. K. (2020). Plant-microbe interactions for sustainable agriculture in the postgenomic era. *Current Genomics, 21*, 168–178.

Ahmad, M. (2023). Plant breeding advancements with "CRISPR-Cas" genome editing technologies will assist future food security. *Frontiers in Plant Science, 14*, 1133036.

Arora, H., Sharma, A., Sharma, S., Haron, F. F., Gafur, A., Sayyed, R. Z., & Datta, R. (2021). *Pythium* damping-off and root rot of *Capsicum annuum* L.: Impacts, diagnosis, and management. *Microorganisms, 9*, 823.

Ayangbenro, A. S., & Babalola, O. O. (2017). A new strategy for heavy metal polluted environments: a review of microbial biosorbents. *International journal of environmental research and public health, 14*(1), 94.

Ayangbenro, A. S., Chukwuneme, C. F., Ayilara, M. S., Kutu, F. R., Khantsi, M., Adeleke, B. S., et al. (2022). Harnessing the rhizosphere soil microbiome of organically amended soil for plant productivity. *Agronomy, 12*, 3179.

Bakker, P. A., Berendsen, R. L., Van Pelt, J. A., Vismans, G., Yu, K., Li, E., et al. (2020). The soil-borne identity and microbiome-assisted agriculture: Looking back to the future. *Molecular Plant, 13*, 1394–1401.

Balint-Kurti, P., Simmons, S. J., Blum, J. E., Ballaré, C. L., & Stapleton, A. E. (2010). Maize leaf epiphytic bacteria diversity patterns are genetically correlated with resistance to fungal pathogen infection. *Molecular Plant-Microbe Interactions, 23*, 473–484.

Beck, M., Wyrsch, I., Strutt, J., Wimalasekera, R., Webb, A., Boller, T., & Robatzek, S. (2014). Expression patterns of flagellin sensing 2 map to bacterial entry sites in plant shoots and roots. *Journal of Experimental Botany, 65*, 6487–6498.

Berendsen, R. L., Vismans, G., Yu, K., Song, Y., de Jonge, R., Burgman, W. P., et al. (2018). Disease-induced assemblage of a plant-beneficial bacterial consortium. *The ISME Journal, 12*, 1496–1507.

Berg, G., Grube, M., Schloter, M., & Smalla, K. (2014). Unraveling the plant microbiome: Looking back and future perspectives. *Frontiers in Microbiology, 5*, 148.

Berg, G., Rybakova, D., Fischer, D., Cernava, T., Vergès, M. C. C., Charles, T., et al. (2020). Microbiome definition re-visited: Old concepts and new challenges. *Microbiome, 8*, 1–22.

Bonaterra i Carreras, A., Badosa Romañó, E., Daranas Boadella, N., Francés Ortega, J., Roselló Prados, G., & Montesinos Seguí, E. (2022). Bacteria as biological control agents of plant diseases. *Microorganisms, 10*, 1759.

Bos, J. I., Armstrong, M. R., Gilroy, E. M., Boevink, P. C., Hein, I., Taylor, R. M., et al. (2010). *Phytophthora infestans* effector AVR3a is essential for virulence and manipulates plant immunity by stabilizing host E3 ligase CMPG1. *Proceedings of the National Academy of Sciences of the United States of America, 107*, 9909–9914.

Brown, S. P., Grillo, M. A., Podowski, J. C., & Heath, K. D. (2020). Soil origin and plant genotype structure distinct microbiome compartments in the model legume *Medicago truncatula*. *Microbiome, 8*, 1–17.

Bulgarelli, D., Garrido-Oter, R., Münch, P. C., Weiman, A., Dröge, J., Pan, Y., et al. (2015). Structure and function of the bacterial root microbiota in wild and domesticated barley. *Cell Host & Microbe, 17*, 392–403.

Butt, H., Jamil, M., Wang, J. Y., Al-Babili, S., & Mahfouz, M. (2018). Engineering plant architecture via CRISPR/Cas9-mediated alteration of strigolactone biosynthesis. *BMC Plant Biology, 18*, 1–9.

Cabot, C., Martos, S., Llugany, M., Gallego, B., Tolrà, R., & Poschenrieder, C. (2019). A role for zinc in plant defense against pathogens and herbivores. *Frontiers in Plant Science, 10*, 1171.

Çakmakçı, R., Salık, M. A., & Çakmakçı, S. (2023). Assessment and principles of environmentally sustainable food and agriculture systems. *Agriculture, 13*, 1073.

Canarini, A., Kaiser, C., Merchant, A., Richter, A., & Wanek, W. (2019). Root exudation of primary metabolites: Mechanisms and their roles in plant responses to environmental stimuli. *Frontiers in Plant Science, 10*, 157.

Cao, Z., Yan, W., Ding, M., & Yuan, Y. (2022). Construction of microbial consortia for microbial degradation of complex compounds. *Frontiers in Bioengineering and Biotechnology, 10*, 1051233.

Carezzano, M. E., Paletti Rovey, M. F., Cappellari, L. D. R., Gallarato, L. A., Bogino, P., et al. (2023). Biofilm-forming ability of phytopathogenic bacteria: A review of its involvement in plant stress. *Plants, 12*, 2207.

Carlström, C. I., Field, C. M., Bortfeld-Miller, M., Müller, B., Sunagawa, S., & Vorholt, J. A. (2019). Synthetic microbiota reveal priority effects and keystone strains in the *Arabidopsis* phyllosphere. *Nature Ecology & Evolution, 3*, 1445–1454.

Chaparro, J. M., Badri, D. V., Bakker, M. G., Sugiyama, A., Manter, D. K., & Vivanco, J. M. (2013). Root exudation of phytochemicals in *Arabidopsis* follows specific patterns that are developmentally programmed and correlate with soil microbial functions. *PLoS One, 8*, e55731.

Chaudhry, V., Runge, P., Sengupta, P., Doehlemann, G., Parker, J. E., & Kemen, E. (2021). Shaping the leaf microbiota: Plant–microbe–microbe interactions. *Journal of Experimental Botany, 72*, 36–56.

Chen, J., Wang, B., Lu, Y., Guo, Y., Sun, J., Wei, B., et al. (2019). Quorum sensing inhibitors from marine microorganisms and their synthetic derivatives. *Marine Drugs, 17*, 80.

Chen, W., Modi, D., & Picot, A. (2023). Soil and phytomicrobiome for plant disease suppression and management under climate change: A review. *Plants, 12*, 2736.

Cheng, Y. T., Zhang, L., & He, S. Y. (2019). Plant-microbe interactions facing environmental challenge. *Cell Host & Microbe, 26*, 183–192.

Cordovez, V., Dini-Andreote, F., Carrión, V. J., & Raaijmakers, J. M. (2019). Ecology and evolution of plant microbiomes. *Annual Review of Microbiology, 73*, 69–88.

Dastogeer, K. M., Tumpa, F. H., Sultana, A., Akter, M. A., & Chakraborty, A. (2020). Plant microbiome–an account of the factors that shape community composition and diversity. *Current Plant Biology, 23*, 100161.

Dellagi, A., Quillere, I., & Hirel, B. (2020). Beneficial soil-borne bacteria and fungi: A promising way to improve plant nitrogen acquisition. *Journal of Experimental Botany, 71*, 4469–4479.

Di, X., Gomila, J. O., & Takken, F. L. (2017). Involvement of salicylic acid, ethylene and jasmonic acid signalling pathways in the susceptibility of tomato to *Fusarium oxysporum*. *Molecular Plant Pathology, 18*, 1024–1035.

DiLegge, M. J., Manter, D. K., & Vivanco, J. M. (2022). Soil microbiome disruption reveals specific and general plant-bacterial relationships in three agroecosystem soils. *PLoS One, 17*, e0277529.

Ding, L. N., Li, Y. T., Wu, Y. Z., Li, T., Geng, R., Cao, J., et al. (2022). Plant disease resistance-related signaling pathways: Recent progress and future prospects. *International Journal of Molecular Sciences, 23*, 16200.

Dow, L., Gallart, M., Ramarajan, M., Law, S. R., & Thatcher, L. F. (2023). *Streptomyces* and their specialised metabolites for phytopathogen control–comparative *in vitro* and *in planta* metabolic approaches. *Frontiers in Plant Science, 14*, 1151912.

Eichmann, R., Richards, L., & Schäfer, P. (2021). Hormones as go-betweens in plant microbiome assembly. *The Plant Journal, 105*, 518–541.

Fadiji, A. E., Yadav, A. N., Santoyo, G., & Babalola, O. O. (2023). Understanding the plant-microbe interactions in environments exposed to abiotic stresses: An overview. *Microbiological Research, 271*, 127368.

Fang, S., Clark, R. T., Zheng, Y., Iyer-Pascuzzi, A. S., Weitz, J. S., Kochian, L. V., et al. (2013). Genotypic recognition and spatial responses by rice roots. *Proceedings of the National Academy of Sciences of the United States of America, 110*, 2670–2675.

Feng, H., Fu, R., Hou, X., Lv, Y., Zhang, N., Liu, Y., et al. (2021). Chemotaxis of beneficial rhizobacteria to root exudates: The first step towards root–microbe rhizosphere interactions. *International Journal of Molecular Sciences, 22*, 6655.

Fenta, L., Mekonnen, H., & Kabtimer, N. (2023). The exploitation of microbial antagonists against postharvest plant pathogens. *Microorganisms, 11*, 1044.

Fiodor, A., Singh, S., & Pranaw, K. (2021). The contrivance of plant growth promoting microbes to mitigate climate change impact in agriculture. *Microorganisms, 9*, 1841.

Frerigmann, H., & Gigolashvili, T. (2014). MYB34, MYB51, and MYB122 distinctly regulate indolic glucosinolate biosynthesis in *Arabidopsis thaliana*. *Molecular Plant, 7*, 814–828.

Glick, B. R., & Nascimento, F. X. (2021). Pseudomonas 1-aminocyclopropane-1-carboxylate (ACC) deaminase and its role in beneficial plant-microbe interactions. *Microorganisms, 9*(12), 2467.

Ha, J., Gao, Y., Zhang, R., Li, K., Zhang, Y., Niu, X., et al. (2021). Diversity of the bacterial microbiome associated with the endosphere and rhizosphere of different cassava (*Manihot esculenta* Crantz) genotypes. *Frontiers in Microbiology, 12*, 729022.

Haichar, F. E. Z., Cernava, T., Liu, J., & Timm, C. M. (2021). Novel insights into the response of the plant microbiome to abiotic factors. *Frontiers in Plant Science, 12*, 607874.

Hakim, S., Naqqash, T., Nawaz, M. S., Laraib, I., Siddique, M. J., Zia, R., et al. (2021). Rhizosphere engineering with plant growth-promoting microorganisms for agriculture and ecological sustainability. *Frontiers in Sustainable Food Systems, 5*, 617157.

Han, L., Zhang, H., Bai, X., & Jiang, B. (2023). The peanut root exudate increases the transport and metabolism of nutrients and enhances the plant growth-promoting effects of *Burkholderia pyrrocinia* strain P10. *BMC Microbiology, 23*, 1–16.

Hartman, K., & Tringe, S. G. (2019). Interactions between plants and soil shaping the root microbiome under abiotic stress. *The Biochemical Journal, 476*, 2705–2724.

He, Y., Jiang, C., Yang, H., Wang, Y., & Zhong, Z. (2017). Arbuscular mycorrhizal fungal composition affects the growth and nutrient acquisition of two plants from a karst area. *Sains Malaysiana, 46*, 1701–1708.

He, D. C., He, M. H., Amalin, D. M., Liu, W., Alvindia, D. G., & Zhan, J. (2021). Biological control of plant diseases: An evolutionary and eco-economic consideration. *Pathogens, 10*, 1311.

Hossain, M. M., Sultana, F., Li, W., Tran, L. S. P., & Mostofa, M. G. (2023). Sclerotinia sclerotiorum (Lib.) de Bary: Insights into the pathogenomic features of a global pathogen. *Cells, 12*(7), 1063.

Ivanov, A. A., Ukladov, E. O., & Golubeva, T. S. (2021). Phytophthora infestans: An overview of methods and attempts to combat late blight. *Journal of Fungi, 7*(12), 1071.

Jamil, F., Mukhtar, H., Fouillaud, M., & Dufossé, L. (2022). Rhizosphere signaling: Insights into plant–rhizomicrobiome interactions for sustainable agronomy. *Microorganisms, 10*, 899.

Kalia, V. C., Patel, S. K., Kang, Y. C., & Lee, J. K. (2019). Quorum sensing inhibitors as antipathogens: Biotechnological applications. *Biotechnology Advances, 37*, 68–90.

Kim, J., & Felton, G. W. (2013). Priming of antiherbivore defensive responses in plants. *Insect Science, 20*, 273–285.

Koza, N. A., Adedayo, A. A., Babalola, O. O., & Kappo, A. P. (2022). Microorganisms in plant growth and development: Roles in abiotic stress tolerance and secondary metabolites secretion. *Microorganisms, 10*, 1528.

Larroque, M., Ramirez, D., Lafitte, C., Borderies, G., Dumas, B., & Gaulin, E. (2011). Expression and purification of a biologically active *Phytophthora parasitica* cellulose binding elicitor lectin in *Pichia pastoris*. *Protein Expression and Purification, 80*, 217–223.

Law, J. W. F., Ser, H. L., Khan, T. M., Chuah, L. H., Pusparajah, P., Chan, K. G., et al. (2017). The potential of *Streptomyces* as biocontrol agents against the rice blast fungus, *Magnaporthe oryzae* (*Pyricularia oryzae*). *Frontiers in Microbiology, 8*, 3.

Lazcano, C., Boyd, E., Holmes, G., Hewavitharana, S., Pasulka, A., et al. (2021). The rhizosphere microbiome plays a role in the resistance to soil-borne pathogens and nutrient uptake of strawberry cultivars under field conditions. *Scientific Reports, 11*, 3188.

Lebeis, S. L. (2014). The potential for give and take in plant–microbiome relationships. *Frontiers in Plant Science, 5*, 287.

Li, P., Lu, Y. J., Chen, H., & Day, B. (2020). The lifecycle of the plant immune system. *Critical Reviews in Plant Sciences, 39*, 72–100.

Li, J., Wang, C., Liang, W., & Liu, S. (2021). Rhizosphere microbiome: The emerging barrier in plant-pathogen interactions. *Frontiers in Microbiology, 12*, 772420.

Li, Y. H., Yang, Y. Y., Wang, Z. G., & Chen, Z. (2022). Emerging function of ecotype-specific splicing in the recruitment of commensal microbiome. *International Journal of Molecular Sciences, 23*, 4860.

Liu, H., Carvalhais, L. C., Crawford, M., Singh, E., Dennis, P. G., Pieterse, C. M., & Schenk, P. M. (2017). Inner plant values: Diversity, colonization and benefits from endophytic bacteria. *Frontiers in Microbiology, 8*, 2552.

Liu, Q., Cheng, L., Nian, H., Jin, J., & Lian, T. (2023). Linking plant functional genes to rhizosphere microbes: A review. *Plant Biotechnology Journal, 21*, 902–917.

Lombardi, N., Woo, S. L., Vinale, F., Turrà, D., & Marra, R. (2021). The plant holobiont, volume II: Impacts of the rhizosphere on plant health. *Frontiers in Plant Science, 12*, 809291.

López-Farfán, D., Reyes-Darias, J. A., Matilla, M. A., & Krell, T. (2019). Concentration dependent effect of plant root exudates on the chemosensory systems of *Pseudomonas putida* KT2440. *Frontiers in Microbiology, 10*, 78.

Lü, P., Liu, Y., Yu, X., Shi, C. L., & Liu, X. (2022). The right microbe-associated molecular patterns for effective recognition by plants. *Frontiers in Microbiology, 13*, 1019069.

Lyu, D., & Smith, D. L. (2022). The root signals in rhizospheric inter-organismal communications. *Frontiers in Plant Science, 13*, 5328.

Ma, W., Tang, S., Dengzeng, Z., Zhang, D., Zhang, T., & Ma, X. (2022). Root exudates contribute to belowground ecosystem hotspots: A review. *Frontiers in Microbiology, 13*, 937940.

Machado Wood, A. K., Panwar, V., Grimwade-Mann, M., Ashfield, T., Hammond-Kosack, K. E., & Kanyuka, K. (2021). The vesicular trafficking system component MIN7 is required for minimizing Fusarium graminearum infection. *Journal of Experimental Botany, 72*, 5010–5023.

Maciag, T., Kozieł, E., Rusin, P., Otulak-Kozieł, K., Jafra, S., & Czajkowski, R. (2023). Microbial consortia for plant protection against diseases: More than the sum of its parts. *International Journal of Molecular Sciences, 24*, 12227.

Macovei, A., Sevilla, N. R., Cantos, C., Jonson, G. B., Slamet-Loedin, I., Čermák, T., et al. (2018). Novel alleles of rice eIF4G generated by CRISPR/Cas9-targeted mutagenesis confer resistance to Rice tungro spherical virus. *Plant Biotechnology Journal, 16*, 1918–1927.

Majdura, J., Jankiewicz, U., Gałązka, A., & Orzechowski, S. (2023). The role of quorum sensing molecules in bacterial–plant interactions. *Metabolites, 13*, 114.

Malgioglio, G., Rizzo, G. F., Nigro, S., Lefebvre du Prey, V., Herforth-Rahmé, J., Catara, V., & Branca, F. (2022). Plant-microbe interaction in sustainable agriculture: The factors that may influence the efficacy of PGPM application. *Sustainability, 14*, 2253.

Malka, S. K., & Cheng, Y. (2017). Possible interactions between the biosynthetic pathways of indole glucosinolate and auxin. *Frontiers in Plant Science, 8*, 2131.

Manjunatha, L., Rajashekara, H., Uppala, L. S., Ambika, D. S., Patil, B., Shankarappa, K. S., et al. (2022). Mechanisms of microbial plant protection and control of plant viruses. *Plants, 11*, 3449.

Mendes, R., Kruijt, M., De Bruijn, I., Dekkers, E., Van Der Voort, M., Schneider, J. H., et al. (2011). Deciphering the rhizosphere microbiome for disease-suppressive bacteria. *Science, 332*, 1097–1100.

Mendes, R., Garbeva, P., & Raaijmakers, J. M. (2013). The rhizosphere microbiome: Significance of plant beneficial, plant pathogenic, and human pathogenic microorganisms. *FEMS Microbiology Reviews, 37*, 634–663.

Mousa, W. K., Shearer, C., Limay-Rios, V., Ettinger, C. L., Eisen, J. A., & Raizada, M. N. (2016). Root-hair endophyte stacking in finger millet creates a physicochemical barrier to trap the fungal pathogen *Fusarium graminearum. Nature Microbiology, 1*, 1–12.

Muratore, C., Espen, L., & Prinsi, B. (2021). Nitrogen uptake in plants: The plasma membrane root transport systems from a physiological and proteomic perspective. *Plants, 10*, 681.

Murty, L. D., & Shim, W. B. (2021). *Streptomyces* and *Bacillus* species utilize volatile organic compounds to impact *Fusarium oxysporum* f. sp. *vasinfectum* race 4 (Fov4) virulence and suppress Fusarium wilt in Pima cotton. *BioRxiv*, 2021–2010.

Nadarajah, K., & Abdul Rahman, N. S. N. (2021). Plant–microbe interaction: Aboveground to belowground, from the good to the bad. *International Journal of Molecular Sciences, 22*, 10388.

Naqqash, T., Hameed, S., Imran, A., Hanif, M. K., Majeed, A., & van Elsas, J. D. (2016). Differential response of potato toward inoculation with taxonomically diverse plant growth promoting rhizobacteria. *Frontiers in Plant Science, 7*, 144.

Naqvi, R. Z., Siddiqui, H. A., Mahmood, M. A., Najeebullah, S., Ehsan, A., Azhar, M., et al. (2022). Smart breeding approaches in post-genomics era for developing climate-resilient food crops. *Frontiers in Plant Science, 13*, 972164.

Neal, A. L., Ahmad, S., Gordon-Weeks, R., & Ton, J. (2012). Benzoxazinoids in root exudates of maize attract *Pseudomonas putida* to the rhizosphere. *PLoS One, 7*, e35498.

Netzker, T., Fischer, J., Weber, J., Mattern, D. J., König, C. C., Valiante, V., et al. (2015). Microbial communication leading to the activation of silent fungal secondary metabolite gene clusters. *Frontiers in Microbiology, 6*, 299.

Nishad, R., Ahmed, T., Rahman, V. J., & Kareem, A. (2020). Modulation of plant defense system in response to microbial interactions. *Frontiers in Microbiology, 11*, 1298.

Niu, B., Wang, W., Yuan, Z., Sederoff, R. R., Sederoff, H., Chiang, V. L., & Borriss, R. (2020). Microbial interactions within multiple-strain biological control agents impact soil-borne plant disease. *Frontiers in Microbiology, 11*, 585404.

Noman, M., Ahmed, T., Ijaz, U., Shahid, M., Azizullah, L. D., & Song, F. (2021). Plant–microbiome crosstalk: Dawning from composition and assembly of microbial community to improvement of disease resilience in plants. *International Journal of Molecular Sciences, 22*, 6852.

Ofek-Lalzar, M., Sela, N., Goldman-Voronov, M., Green, S. J., Hadar, Y., & Minz, D. (2014). Niche and host-associated functional signatures of the root surface microbiome. *Nature Communications, 5*, 4950.

Olanrewaju, O. S., Ayangbenro, A. S., Glick, B. R., & Babalola, O. O. (2019). Plant health: Feedback effect of root exudates-rhizobiome interactions. *Applied Microbiology and Biotechnology, 103*, 1155–1166.

Ortíz-Castro, R., Valencia-Cantero, E., & López-Bucio, J. (2008). Plant growth promotion by *Bacillus megaterium* involves cytokinin signaling. *Plant Signaling & Behavior, 3*, 263–265.

Orukotan, K. E., Elughi, G. N., Abimbola, B. S., Akinyosoye, A. D., Ahuekwe, E. F., & Oziegbe, O. (2023). Plant microbiome engineering: Principles methods and current trends. In *Biotechnological approaches to sustainable development goals* (pp. 251–267). Springer Nature.

Pan, D., Mionetto, A., Tiscornia, S., & Bettucci, L. (2015). Endophytic bacteria from wheat grain as biocontrol agents of *Fusarium graminearum* and deoxynivalenol production in wheat. *Mycotoxin Research, 31*, 137–143.

Pandit, M. A., Kumar, J., Gulati, S., Bhandari, N., Mehta, P., Katyal, R., et al. (2022). Major biological control strategies for plant pathogens. *Pathogens, 11*, 273.

Pascale, A., Proietti, S., Pantelides, I. S., & Stringlis, I. A. (2020). Modulation of the root microbiome by plant molecules: The basis for targeted disease suppression and plant growth promotion. *Frontiers in Plant Science, 10*, 1741.

Pathma, J., & Sakthivel, N. (2012). Microbial diversity of vermicompost bacteria that exhibit useful agricultural traits and waste management potential. *Springer Plus, 1*, 1–19.

Prazdnova, E. V., Gorovtsov, A. V., Vasilchenko, N. G., Kulikov, M. P., Statsenko, V. N., Bogdanova, A. A., et al. (2022). Quorum-sensing inhibition by Gram-positive bacteria. *Microorganisms, 10*, 350.

Pršić, J., & Ongena, M. (2020). Elicitors of plant immunity triggered by beneficial bacteria. *Frontiers in Plant Science, 11*, 594530.

Rehman, Z. U., & Leiknes, T. (2018). Quorum-quenching bacteria isolated from Red Sea sediments reduce biofilm formation by *Pseudomonas aeruginosa. Frontiers in Microbiology, 9*, 1354.

Riera, N., Handique, U., Zhang, Y., Dewdney, M. M., & Wang, N. (2017). Characterization of antimicrobial-producing beneficial bacteria isolated from Huanglongbing escape citrus trees. *Frontiers in Microbiology, 8*, 2415.

Rolfe, S. A., Griffiths, J., & Ton, J. (2019). Crying out for help with root exudates: Adaptive mechanisms by which stressed plants assemble health-promoting soil microbiomes. *Current Opinion in Microbiology, 49*, 73–82.

Rutherford, S. T., & Bassler, B. L. (2012). Bacterial quorum sensing: Its role in virulence and possibilities for its control. *Cold Spring Harbor Perspectives in Medicine, 2*, a012427.

Saeed, Q., Xiukang, W., Haider, F. U., Kučerik, J., Mumtaz, M. Z., Holatko, J., et al. (2021). Rhizosphere bacteria in plant growth promotion biocontrol and bioremediation of contaminated sites: A comprehensive review of effects and mechanisms. *International Journal of Molecular Sciences, 22*, 10529.

Salem, M. A, Wang, J. Y., & Al-Babili, S. (2022). Metabolomics of plant root exudates: From sample preparation to data analysis. *Frontiers in Plant Science, 13*, 1062982.

Sánchez-López, Á. M., Baslam, M., De Diego, N., Muñoz, F. J., Bahaji, A., Almagro, G., et al. (2016). Volatile compounds emitted by diverse phytopathogenic microorganisms promote plant growth and flowering through cytokinin action. *Plant, Cell and Environment, 39*, 2592–2608.

Sang, M. K., Shrestha, A., Kim, D. Y., Park, K., Pak, C. H., & Kim, K. D. (2013). Biocontrol of Phytophthora blight and anthracnose in pepper by sequentially selected antagonistic rhizobacteria against *Phytophthora capsici*. *The Plant Pathology Journal, 29*, 154.

Santos Hernández, M., Diánez Martínez, F. J., Sánchez Montesinos, B., Huertas, V., Esteban García, A. B., Moreno-Gavira, A., et al. (2023). Biocontrol of diseases caused by *Phytophthora capsici* and *Pparasitica* in pepper plants. *Journal of Fungi, 9*, 360.

Santoyo, G., Urtis-Flores, C. A., Loeza-Lara, P. D., Orozco-Mosqueda, M. D. C., & Glick, B. R. (2021). Rhizosphere colonization determinants by plant growth-promoting rhizobacteria (PGPR). *Biology, 10*, 475.

Scharf, B. E., Hynes, M. F., & Alexandre, G. M. (2016). Chemotaxis signaling systems in model beneficial plant–bacteria associations. *Plant Molecular Biology, 90*, 549–559.

Sengupta, S., Chattopadhyay, M. K., & Grossart, H. P. (2013). The multifaceted roles of antibiotics and antibiotic resistance in nature. *Frontiers in Microbiology, 4*, 47.

Sharma, I., Kashyap, S., & Agarwala, N. (2023). Biotic stress-induced changes in root exudation confer plant stress tolerance by altering rhizospheric microbial community. *Frontiers in Plant Science, 14*, 1132824.

Soliman, S. A., Abdelhameed, R. E., & Metwally, R. A. (2023). *In vivo* and *in vitro* evaluation of the antifungal activity of the PGPR *Bacillus amyloliquefaciens* RaSh1 (MZ945930) against *Alternaria alternata* with growth promotion influences on *Capsicum annuum* Lplants. *Microbial Cell Factories, 22*, 70.

Spence, C., Alff, E., Johnson, C., Ramos, C., Donofrio, N., Sundaresan, V., & Bais, H. (2014). Natural rice rhizospheric microbes suppress rice blast infections. *BMC Plant Biology, 14*, 1–17.

Sundin, G. W., Castiblanco, L. F., Yuan, X., Zeng, Q., & Yang, C. H. (2016). Bacterial disease management: Challenges experience innovation and future prospects: Challenges in bacterial molecular plant pathology. *Molecular Plant Pathology, 17*, 1506–1518.

Tosif, M. M., Najda, A., Bains, A., Kaushik, R., Dhull, S. B., Chawla, P., & Walasek-Janusz, M. (2021). A comprehensive review on plant-derived mucilage: characterization, functional properties, applications, and its utilization for nanocarrier fabrication. *Polymers, 13*(7), 1066.

Ujvári, G., Turrini, A., Avio, L., & Agnolucci, M. (2021). Possible role of arbuscular mycorrhizal fungi and associated bacteria in the recruitment of endophytic bacterial communities by plant roots. *Mycorrhiza, 31*, 527–544.

Upadhyay, S. K., Srivastava, A. K., Rajput, V. D., Chauhan, P. K., Bhojiya, A. A., Jain, D., et al. (2022). Root exudates: Mechanistic insight of plant growth promoting rhizobacteria for sustainable crop production. *Frontiers in Microbiology, 13*, 916488.

Uroz, S., Dessaux, Y., & Oger, P. (2009). Quorum sensing and quorum quenching: The yin and yang of bacterial communication. *ChemBioChem, 10*, 205–216.

Van der Ent, S., Verhagen, B. W., Van Doorn, R., Bakker, D., Verlaan, M. G., Pel, M. J., et al. (2008). MYB72 is required in early signaling steps of rhizobacteria-induced systemic resistance in *Arabidopsis*. *Plant Physiology, 146*, 1293–1304.

Van Wees, S. C., Luijendijk, M., Smoorenburg, I., Van Loon, L. C., & Pieterse, C. M. (1999). Rhizobacteria-mediated induced systemic resistance (ISR) in *Arabidopsis* is not associated with a direct effect on expression of known defense-related genes but stimulates the expression of the jasmonate-inducible gene Atvsp upon challenge. *Plant Molecular Biology, 41*, 537–549.

Velásquez, A. C., Castroverde, C. D. M., & He, S. Y. (2018). Plant–pathogen warfare under changing climate conditions. *Current Biology, 28*, R619–R634.

Vicente, J. G., & Holub, E. B. (2013). *Xanthomonas campestris* pv*campestris* (cause of black rot of crucifers) in the genomic era is still a worldwide threat to brassica crops. *Molecular Plant Pathology, 14*, 2–18.

Wei, Z., Hu, X., Li, X., Zhang, Y., Jiang, L., Li, J., et al. (2017). The rhizospheric microbial community structure and diversity of deciduous and evergreen forests in Taihu Lake area China. *PLoS One, 12*, e0174411.

Wintermans, P. C., Bakker, P. A., & Pieterse, C. M. (2016). Natural genetic variation in *Arabidopsis* for responsiveness to plant growth-promoting rhizobacteria. *Plant Molecular Biology, 90*, 623–634.

Wu, D. Q., Ye, J., Ou, H. Y., Wei, X., Huang, X., He, Y. W., & Xu, Y. (2011). Genomic analysis and temperature-dependent transcriptome profiles of the rhizosphere originating strain *Pseudomonas aeruginosa* M18. *BMC Genomics, 12*, 1–17.

Xiao, Y., Zou, H., Li, J., Song, T., Lv, W., Wang, W., et al. (2022). Impact of quorum sensing signaling molecules in gram-negative bacteria on host cells: Current understanding and future perspectives. *Gut Microbes, 14*, 2039048.

Xiong, Q., Hu, J., Wei, H., Zhang, H., & Zhu, J. (2021). Relationship between plant roots rhizosphere microorganisms and nitrogen and its special focus on rice. *Agriculture, 11*, 234.

Xu, J., Zhang, Y., Zhang, P., Trivedi, P., Riera, N., Wang, Y., et al. (2018). The structure and function of the global citrus rhizosphere microbiome. *Nature Communications, 9*, 4894.

Xu, Z., Wang, M., Du, J., Huang, T., Liu, J., Dong, T., & Chen, Y. (2020). Isolation of *Burkholderia* sp HQB-1 a promising biocontrol bacteria to protect banana against Fusarium wilt through phenazine-1-carboxylic acid secretion. *Frontiers in Microbiology, 11*, 605152.

Yang, Y., Xu, J., Li, Y., He, Y., Yang, Y., Liu, D., & Wu, C. (2023). Effects of coumarin on rhizosphere microbiome and metabolome of *Lolium multiflorum*. *Plants, 12*, 1096.

Yeap, W. C., Khan, N. C. M., Jamalludin, N., Muad, M. R., Appleton, D. R., & Kulaveerasingam, H. (2021). An efficient clustered regularly interspaced short palindromic repeat (CRISPR)/ CRISPR-associated protein 9 mutagenesis system for oil palm (Elaeis guineensis). *Frontiers in Plant Science, 12*, 773656.

Yu, Y., Gui, Y., Li, Z., Jiang, C., Guo, J., & Niu, D. (2022). Induced systemic resistance for improving plant immunity by beneficial microbes. *Plants, 11*, 386.

Yuejun H., Jiang, C., Yang, H., Wang, Y., & Zhong, Z. (2017). Arbuscular mycorrhizal fungal composition affects the growth and nutrient acquisition of two plants from a karst area. Penerbit Universiti Kebangsaan Malaysia.

Zhang, Q., Linke, V., Overmyer, K. A., Traeger, L. L., Kasahara, K., Miller, I. J., et al. (2023). Genetic mapping of microbial and host traits reveals production of immunomodulatory lipids by *Akkermansia muciniphila* in the murine gut. *Nature Microbiology, 8*, 424–440.

Zhou, D., Jing, T., Chen, Y., Wang, F., Qi, D., Feng, R., et al. (2019a). Deciphering microbial diversity associated with Fusarium wilt-diseased and disease-free banana rhizosphere soil. *BMC Microbiology, 19*, 1–13.

Zhou, Z., Zhao, Y., Bi, G., Liang, X., & Zhou, J. M. (2019b). Early signalling mechanisms underlying receptor kinase-mediated immunity in plants. *Philosophical Transactions of the Royal Society B, 374*(1767), 20180310.

Zhou, Z., Wu, X., Li, J., Zhang, Y., Huang, Y., Zhang, W., et al. (2022). A novel quorum quencher, *Rhodococcus pyridinivorans* XN-36, is a powerful agent for the biocontrol of soft rot disease in various host plants. *Biological Control, 169*, 104889.

Zhu, L., Huang, J., Lu, X., & Zhou, C. (2022). Development of plant systemic resistance by beneficial rhizobacteria: Recognition initiation elicitation and regulation. *Frontiers in Plant Science, 13*, 952397.

Chapter 5
Rhizosphere Microbiome and Plant Tolerance Against Biotic Agents

Arun Kumar, Vikanksha, and Jatinder Singh

Abstract The rhizosphere is a focal point for microbial relationships due to the exudates released by roots. They are the mainstay of microbial nutrition and play a significant role in influencing the population density and functions of many microorganisms. The rhizosphere is a complex environment that supports a diverse range of species. While some of them have a neutral impact on the plant, others are attracted to the rhizosphere and may either have harmful or beneficial impacts. The organisms that affect plant development and their general well-being include pathogenic fungi, oomycetes, bacteria, and nematodes. The majority of soilborne pathogens have developed adaptations that allow them to thrive and persist in bulk soil. However, it is inside the rhizosphere that they actively engage with the plant, establishing a parasitic connection. The rhizosphere is a dynamic environment where the pathogens effectively infect the plant, including a diverse array of microflora and microfauna components engaged in intricate interactions.

Keywords Microbiome · Pathogenic microorganisms · Rhizosphere

5.1 Introduction

The rhizosphere is a distinct ecological niche that is actively shaped by plants. It serves as a habitat for both plant pathogenic and beneficial microorganisms. These microbes play a significant role in influencing the plant development and overall health. The rhizosphere organisms include several bacteria, fungi, nematodes, protozoa, algae, and microarthropods (Chialva et al., 2022). Numerous members within these groups have a neutral impact on the plant. However, they play a crucial role in the intricate food web that capitalizes on the substantial carbon fixation by the

A. Kumar · Vikanksha · J. Singh (✉)
Lovely Professional University, Phagwara, Punjab, India
e-mail: jatinder.19305@lpu.co.in

P. Mathur, S. Roy (eds.), *Plant Microbiome and Biological Control*, Sustainability in Plant and Crop Protection 20,
https://doi.org/10.1007/978-3-031-75845-4_5

photosynthesis and the subsequent transfer of metabolites into the rhizosphere, often referred to as rhizodeposits (Bang et al., 2018). The rhizosphere is inhabited by both detrimental and advantageous species that influence the plant metabolism. The microorganisms that have a negative impact on plant growth and wellness involve pathogenic fungi, oomycetes, bacteria, and nematodes. Conversely, there are species that provide key benefits, such as nitrogen-fixing bacteria, endo- and ectomycorrhizal fungi, as well as plant growth-promoting rhizobacteria (PGPR) and other endophytic fungi. The abundance and variety of harmful and beneficial bacteria are influenced by the amount and composition of the organic compounds released by roots, as well as by the resulting microbial relationships that take place in the rhizosphere (Hassan et al., 2019).

The study of the mechanisms that govern the rhizosphere structuring, dynamics, and functionality has garnered the attention of researchers from several fields. This research area has a potential for the advancement of novel approaches aimed at enhancing plants growth and their overall well-being. This review primarily includes the epidemiology, namely the geographic and chronologic elements of soilborne pathogens, as well as the financial impact of soil borne illnesses (Zhang et al., 2021). The focus is on the processes that govern the relationships among soilborne pathogens and beneficial microbes, with particular emphasis on the aggressive and safeguarding behaviors. We will also examine the direct favorable impacts of rhizosphere microorganisms on plant growth and development (Pegg et al., 2019). In conclusion, we review the impact of farming procedures on communities of pathogens and antagonists, as well as the strategies used to modify these procedures to foster soil suppressiveness and enhance plant growth and overall well-being (Spiegel et al., 2022).

The rhizospheric region is a specialized habitat for microorganisms, influenced by the root ability to withstand phytopathogens. It offers a diverse range of advantages to the well-being of plants, either via direct and indirect means (Sessitsch et al., 2023). The role of soil microorganisms in acting as a major source of nutrients for plants is well acknowledged by the scientific community. This is due to their involvement in many nutrient metabolism activities, including nitrogen cycling, organic material breakdown, and mineral erosion. When combined, these factors have a significant impact on the plant development (Dudáš et al., 2022). Therefore, the relationships between roots and microorganisms play a crucial role in facilitating the success of plants in their environment. In order to comprehend the reactions of ecosystems to shifting surroundings, an in-depth knowledge about the location and roles of such a variety of microbial life is essential. In addition, root exudates consist of a wide range of chemical compounds, such as amino acids, sugars, siderophores, and catalysts. They have a potential to enhance the functions of the microbial populations and to induce notable alterations in the composition of the rhizospheric populations (Dubey et al., 2019). These changes depend on specific plant species and genotypes. The microorganisms linked to plants play a crucial role in enhancing plant growth and increasing their ability to withstand various abiotic and biotic stress factors.

Many functions of the soil microbial communities i.e., involving carbon sequestration and biogeochemical turnover, are also significant to address worldwide problems related to global warming (Choudhary et al., 2016). Improving the capacity to anticipate the functioning of root microorganisms based on dynamical deviations is of utmost importance to develop targeted microbiome treatments that enhance crop yield and durability (Guo et al., 2021). More extensive genomic and physiological categorizations of plant-associated microorganisms, their genetic components, and plants-microbes relationships are required to accurately determine specific beneficial functions. Certain microbes may establish helpful or harmful associations with their plants of residence. The microbiomes associated with plants often have advantageous effects at the social level, helping farmers to safeguard plants from diseases and promoting their development, overall well-being, and resilience to expected changes (Xiang et al., 2022).

5.2 Rhizosphere Microbiome

In the rhizosphere, plant, soil, and microbe interactions are linked (Fig. 5.1), resulting in a unique and active ecosystem (Pii et al., 2015). Understanding the intricate nature of these regulations and operations requires a thorough understanding of the interactions between these three components. Rhizosphere systems manage the environment because of the conditions they develop. Since sustainable agriculture relies on precise management of microorganisms, a robust and thriving rhizosphere is essential for crop production because of the bioactive compounds it contains, released by its various microbial communities (Rekadwad, 2020).

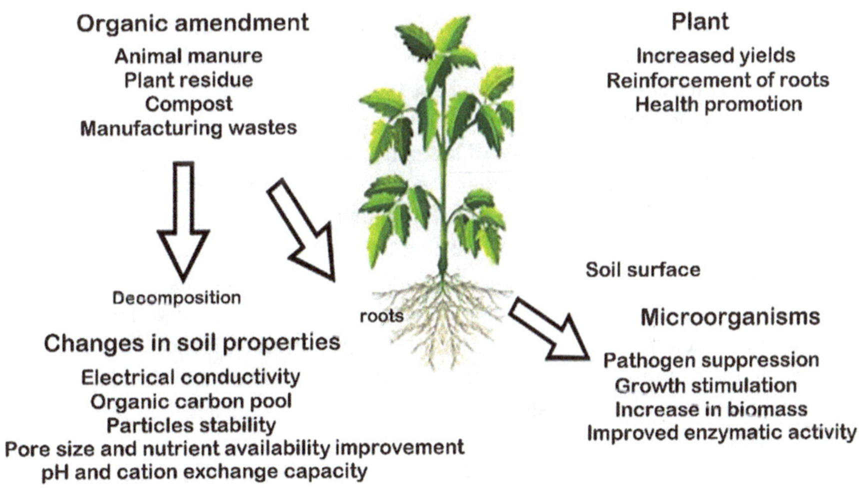

Fig. 5.1 The impact of organic amendments on soil, plant and microorganisms

Rhizodeposition is a second significant event where plants emit a variety of molecules into the soil, including vitamins, carbohydrates, secondary metabolites, amino acids, etc. They are known as root exudates, affecting the rhizosphere and, subsequently, the soil chemistry (Prashar et al., 2014). The third factor is the ability of rhizospheric microorganisms to migrate from roots towards either the endosphere (internal tissues) or the phyllosphere (aerial regions). The presence of various soil textures creates an eclectic atmosphere that promotes cooperation of a broad spectrum of species, such as bacteria, archaea, fungi, oomycetes, viruses, and various protists. These taxa engage in intricate trophic webs and substitution systems, interacting with one another.

The adverse effects of rhizosphere bacteria, namely soil-borne diseases, on plant development, yield, and agriculture produce have been extensively investigated over the course of many decades. Nevertheless, the presence of beneficial microorganisms, such as collaborative bacteria, may facilitate the development of plants via diverse mechanisms. These include the augmentation of nutrients accessibility, the synthesis of plant hormones, and the reinforcement of resistance towards both biotic and abiotic stress factors (Jayaraman et al., 2021).

The rhizosphere may be divided into two distinct sections, namely the ectorhizosphere and the endorhizosphere. Nevertheless, several ecosystems are inhabited by a diverse range of microorganisms, and their interactions with roots have been extensively studied. Previous studies demonstrated that, among the vast array of plant-associated taxa, only a limited number of species are harmful, whereas the majority of them are engaged in advantageous relationships that sustain plant survival and fitness. Endophytes have historically been undervalued; nevertheless, they are now receiving significant interest due to their capacity for N_2 fixation (Bag et al., 2022).

Multiple investigations provided evidence of the existence of endophytes within the root nodules of various agricultural plants, of which *Rhizobium* spp. are the most investigated and known. Additionally, non-nodulating endophytic populations, including i.e., *Microbacterium trichothecenolyticumn*, *Brevibacillus choshinensis*, *Endobacter medicaginis*, and *Micromonospora* spp., have been also found to inhabit nodules (Kumar & Dubey, 2020). The taxonomic differences among endophytic bacteria that inhabit various sections of the plant, such as leaves, stems, and leguminous nodules, have been elucidated by means of metagenomic methodologies (Papik et al., 2020). However, there is still a paucity of information on endophytic microbes that are often discovered in association with other agricultural crops. The physiological aspects of root-associated microbial communities are central in enhancing the plant health and resilience to diseases and stressors. This is often achieved via selection and delivery of diverse micronutrients. Hence, it is essential to prioritize the examination and understanding of the makeup and development of the microbiome in various divisions, in order to harness their abilities in full (Sani & Yong, 2021).

Rhizosphere bacteria may protect plants from many diseases via various mechanisms, including antagonism, niche occupation, rivalry for resources, and other effects related to the microbial diversity. Moreover, the mycorrhizal association between fungi and plants is a classical example that has been shown to provide many advantages to both organisms. The benefits include improved nutrient availability and increased resilience towards stress and pathogenic species. Mycorrhizal fungi facilitate the relationship of plants with various soil microorganisms, such as pathogens and mycorrhizosphere mutualists. These mutualists not only produce vitamins but also provide protection against antagonistic organisms. As an example, ectomycorrhizal (EcM) fungi offer significant defense against soil-borne diseases by the process of enveloping feeder roots and locally acidifying the soil. The plant species also plays a fundamental role in these interactions. For example, wheat has shown a particular attraction towards advantageous rhizosphere bacterial taxa. These include members of genera *Chitinophaga*, *Pseudomonas*, *Chryseobacterium*, and *Flavobacterium*, as well as a collection of microbes that promote plant growth and/or fix nitrogen, such as species of *Pedobacter*, *Variovorax*, and *Rhizobium*. Together, these bacteria exhibited a detrimental behavior against the soil-borne pathogen *Rhizoctonia solani* AG8. Moreover, *Janthinobacterium* showed significant antagonistic effects *vs* soil-borne pathogens such as *Pythium ultimum*, *R. solani* AG8, and *R. oryzae* under laboratory conditions (Li et al., 2021). The bacterium also demonstrated infection inhibitory properties towards *R. solani* AG8 when applied to soil. As a further example, cucumber plants sensitive to Fusarium wilt gather helpful microorganisms, such as members of Comamonadaceae and Xanthomonadaceae, to regulate the disease through an increased secretion of organic acids.

The *in vitro* growth of *F. oxysporum* was reduced in a growth suppression experiment, specifically including species of Comamonadaceae, *Pseudomonas*, and *Stenotrophomonas*. Although flavobacterial isolates did not exhibit direct antibacterial properties against *Ralstonia solanacearum* on solid substrates, it is plausible that the monosaccharide carriers present in flavobacteria facilitate the absorption of monosaccharides (Singh et al., 2023). This absorption process may subsequently decrease the quantity of carbohydrates that the lectin of *R. solanacearum* connects to, ultimately leading to a reduction in tomato disease beneath experimental circumstances.

The aforementioned data indicate that a rhizosphere microbiome has a potential to include advantageous microorganisms which may directly impede the insurgence of diseases, thus improving crop yields. In summary, rhizosphere microorganisms may exert defense mechanisms with significant influences on the plant well-being, when faced with one or more pathogens (Jha, 2023). These benefits are achieved by direct antagonism and/or the augmentation of host-induced systemic resistance (ISR), both of which promote disease suppression. Hence, comprehending the impact of plants on the advantageous composition of the rhizosphere species, the interactions among these microorganisms, and their final effect on plant defense has a significant importance in crop production (Kumawat et al., 2022).

5.3 Rhizosphere Engineering

The relationship of chemical communications and the exchange of chemicals between plants and soil microorganisms take place primarily in the soil microecosystem, comprising plants and soil microorganisms (Zhou et al., 2009). Several kinds of pollutants are concentrated in the soil due to the heavy use of chemical fertilizers and urban dwellers, degrading and altering the rhizospheric communities, and thus compromising crop production and quality. According to this perspective, restoring the rhizosphere microbiome can be an excellent approach to enhance the performance of the plant rhizosphere (Mendes et al., 2013; Rao, 2007). The rhizosphere microbiota is made up of living bacteria that support improved soil health and plant development while stimulating tolerance to biotic (phytopathogens) and abiotic stress factors such drought, salt, and soil pollution(Del Carmen et al., 2018; Kumar & Dubey, 2020). Even so, as a result of their varying interactions with other species, a complete microbiome may not include all of the aforementioned functions. Therefore, strengthening or adding a few taxa can promote the performance of the rhizosphere as a whole.

Different plants that interact with microorganisms in the rhizosphere have varying capacities for withstanding biotic and abiotic stress, optimal growth ranges at various altitudes or latitudes, and occasionally nutrient-poor circumstances that enable them to thrive in hostile environments. Therefore, a good substitute for improving the rhizosphere is to transfer those potential microorganisms, their genetic material, or their genes (for example, genes coding for *Bacillus thuringiensis* Cry toxins) to the desired rhizosphere or to microorganisms that lack those functions (Kolstø et al., 2009; Santoyo et al., 2017). However, the actual comprehension of the manipulation and management of the rhizosphere microbiota remains significantly constrained. Bioinoculation is widely regarded as an optimal and successful approach for manipulating the microbiota. Numerous solutions have been introduced to the global sales, which developed consortia of useful microorganisms such as PGPR and AM fungi.

Most bacteria are typically obtained by conventional laboratory culture methods, which may not accurately replicate the chemical circumstances found in soil. Bioinoculants often provide very favorable outcomes when tested in controlled laboratory and greenhouse environments. However, there is not enough proof to back up the notion that these microorganisms possess the ability to rival one another and operate effectively, since they do not consistently reproduce the results in controlled conditons, when tested in the farm natural soil. A significant proportion of the aforementioned inoculants exhibited failure when subjected to agricultural field settings, due to their susceptibility to predation and competing with indigenous bacteria for scarce nutrients (Berninger et al., 2018). To be considered successful, it is crucial that bioinoculants possess the capability to establish symbiotic or synergic relationships with different microbiota in close proximity. This ability enables these bacteria to replicate the robust and interconnected network seen in their indigenous rhizosphere soils. The underlying concept of this methodology is to introduce a diverse range of microorganisms to enhance plant functionality and confer comprehensive tolerance to both abiotic and biotic stresses (Sánchez-Cañizares et al., 2017).

The current progress in artificial microbial instruments and gene editing techniques provide a unique opportunity to manipulate the microbiome to achieve certain functionalities. Hence, the manipulation of plant characteristics and crop breeding are two approaches that may be used to design the rhizosphere. These strategies will be briefly examined in this chapter. By comprehending the fundamental structure of roots as well as the particular exudates produced by hosts and other plant features while selecting beneficial microorganisms, it appears possible to modify crops to possess these qualities via gene editing technologies such as CRISPR (Trivedi et al., 2021). Therefore, this particular technique has more potential since it replicates the connections that facilitate the advantageous microorganisms, contributing to the holobiont improvement. In the foreseeable future, it is anticipated that advancements in engineering methodologies will enable deliberate manipulation of the rhizosphere (Voolstra et al., 2021). The achievement of such effective engineering of the rhizosphere microbiome necessitates the adoption of a systematic strategy. By comprehending the fundamental mechanisms governing the formation of the rhizosphere, it is possible to seek a general ecological viability and efficacy to sustain agricultural outputs. This may be achieved through the emulation and management of advantageous symbiotic interactions observed among soils, microorganisms, and plants (Pradhan et al., 2022).

5.4 Soil Pathogens and Their Economic Impact

Soilborne plant diseases are a significant constraint in the development of commercial harvests in the majority of farming settings. In contrast to diseases that target the foliage and stems, they exhibit a higher level of resistance to oversight and control measures (Bruehl, 1987). Soilborne pathogens have evolved to be well-suited for survival and proliferation in soil environments. Plants often grow and thrive in bulk soil, but it is in the rhizosphere that the pathogen initiates contact with the plant and starts its parasitic association. The rhizosphere is a site with intricate interactions involving the diverse variety of microflora and microfauna, which may ultimately impact the progression and outcome of the pathogen infection (Mitter et al., 2013).

The classification of plant pathogens consists of four primary categories. However, in the context of soil, only two of them have significant importance, namely fungi (including true fungi and oomycetes) and nematodes. Limited numbers of bacterial groups are extended for periods of time. This is likely due to the fact that non-spore producing bacteria are not well adapted to the soil environment. Bacteria need a wound or natural orifice to infiltrate the plant and induce infection. A number of viruses have the capability to infect plant roots too. Similar to bacteria, these pathogens need a wound to initiate infection and are mostly spread by vectors (Finlay & Thorn, 2019).

Nematodes are invertebrates that represent one of the most abundant animal lineages on Earth. Most soil nematodes exhibit a free-living lifestyle, where they feed on bacteria, fungi, or other nematodes or even insects. However, there are some lineages that evolved as plant parasites. Certain taxa get nutrients by feeding externally on the root surface, known as migratory ectoparasites. Other species enter the root and wander inside its recesses while eating, referred to as migratory endoparasites. Additionally, some nematodes establish a feeding site within the root where they remain for development, known as sedentary endoparasites.

The primary emphasis of this study is on fungi and oomycetes, which are recognized as the most significant soilborne pathogens (Barnes et al., 2009). Fungi belong to the eukaryotic realm. They are characterized by a filamentous structure with a multicellular nature. They are heterotrophic, as they get nutrients from organic sources. Additionally, fungi are capable of producing a complex collection of interconnected structures (Gupta et al., 2017). The fungal hyphae, referred to as the mycelium, have the ability to assimilate nutrients from the adjacent substrate. The Oomycetes have a morphological resemblance to fungi, however they are chronologically associated with brown algae. These organisms generate motile spores known as zoospores and possess cellulose in their cellular walls, in contrast to real fungi which have chitin. However, it is important to note that the processes of parasitism and the resulting illnesses exhibit similarities to those seen in true fungi (Dick, 2013). As a result, it is appropriate to address them jointly in this context.

The majority of soilborne fungi exhibit a necrotrophic lifestyle, whereby they use enzymes and poisons to induce host tissue death, prior to hyphal invasion. These fungi do not rely on live cells for nutrition acquisition. The majority of biotrophic pathogens, including Rusts and powdery mildews, affect the aerial parts of plants and rely on the presence of live cells (Spanu et al., 2010). Several root pathogens, including i.e., *Phytophthora sojae*, have a semibiotrophic lifestyle. It is noteworthy that a limited number of root pathogens exhibit a biotrophic lifestyle. Several examples involve basal zoosporic fungi and Oomycetes, including *Plasmodiophora brassicae* and *Plasmopara halstedii*.

Most necrotrophic pathogens exhibit a broad spectrum of host specificity, making them generalists in nature. In contrast, biotrophic pathogens developed in close association with certain host plant species, resulting in a narrower range of host compatibility. Hence, it is often seen that necrotrophic infections lack a discernible population framework, whereas the plants do not possess distinct single-gene barriers. The soil environmental conditions often do not provide favourable circumstances for fungal development (Doehlemann et al., 2017). Exceptionally freezing temperatures and icy terrain, for example, as well as highly dry circumstances, may have a significant impact on the pathosystem. Pathogens, however, evolved survival capabilities in soil due to their resistance. Crop residues are also a significant factor to consider. When the environmental circumstances are conducive and a seed or root comes into proximity with the inactive propagules, they trigger the stimulation of the fungus. The process of germination occurs via the influence of root or seed exudates, which chemotactically stimulate growth towards the plant (Palti, 2012). A germination tube, a structure formed by some fungi during the process of germination, initiates infection.

In Oomycetes, the zoospores adhere to the external surface of the root, infiltrate the cells of the epidermal layer located at the root tip, secondary roots, and root hairs, or assault the root system. The growing branches and radicles of seedlings are observed. Some rapidly proliferating pathogens, such as species of *Pythium*, have the ability to rapidly increase in numbers, targeting the seeds and embryos even before their emergence.

Fungi represent a distinct group of eukaryotes (Perrine-Walker, 2020). The process by which penetration occurs, despite intact cell membranes, involves the use of cell wall-degrading enzymes and physiological turgor pressure to establish colonies inside the root cortex. The majority of soilborne fungi primarily target immature, young tissues rather than mature woody roots. Once the the roots are colonized and these fungi extensively spread throughout the cortex, the hyphae proceed to generate spores. The mycelium has the ability to propagate by extending its growth inside or outside the root, as well as by spreading to neighboring tissues. Plants that are infected by soil fungal pathogens may exhibit a variety of illnesses and characteristics (Gandhi & Mehta, 2011).

The disease symptoms seen below- and aboveground may lack distinguishing characteristics or resemble those induced by abiotic causes. Examples of environmental factors that may have detrimental effects on organisms include drought, physiological stress, and nutritional deficiencies that lead to reduced crop yield and quality. These pathogens are often present in the soil and may infect seeds, seedlings, and mature plants. They can be spread by contaminated soil, water, or plant debris. Soilborne pathogens can have a significant impact on agricultural productivity and can be challenging to manage. Effective control measures include crop rotation, the use of resistant varieties, and/or the use of fungicides, if available. The movement of pathogens into the stem basal region leads to the development of crown rot and wilting symptoms. Fungi have the ability to colonize the collar of perennial trees, resulting in their girdling. Additionally, the fruit of these trees may host a fungal inoculum, leading to rotting. Nevertheless, the predominant ailment is root rot. The elimination of root development on a particular axis is achieved by causing the death of the tips (Jaouad et al., 2020). The destruction of delicate feeder roots and root hairs results in a reduction of the plant capacity to absorb water and essential nutrients. Consequently, these factors diminish growth, hinder subsequent development, increase the plant susceptibility to drought, and provide inadequate nutrition levels (Nadarajan & Sukumaran, 2021).

5.5 Factors Controlling Rhizosphere Microbiome

The composition of the rhizosphere bacteria usually differs from that of the non-rhizospheric, bulk soil. This distinction suggests that the development of the microbial population in the rhizosphere is not an arbitrary event, being instead influenced by the host plants. Throughout the course of plant development, a significant proportion (5–21%) of the carbon fixed by photosynthesis is released into the rhizosphere micro-domain via the roots (Gutschick, 2012). This secretion plays a crucial role as a nutrition supply for the microbes that live in soil, hence influencing the composition of the rhizosphere. Presently, there is a prevailing consensus that roots exhibit a strong and intimate association with microbial populations (Yadav et al., 2017).

The quality and amounts of root exudates, which are complex fluids released by roots, are primarily influenced by the plant genotype/species, its growth stage, as well as by further abiotic and biotic stress factors. Root exudates include sugars, organic acids that attract and select microorganisms in the rhizosphere, leading to changes in the makeup and organization of these microbial communities. Several studies highlighted the importance of the exudates in mediating interactions between plants and microorganisms. For example, it has been shown that long-chain fatty acids and amino acids have significant implications in the attraction of possible advantageous microorganisms, such as *Pseudomonas* spp. (Nwachukwu et al., 2021). A recent study has reported four short-chain organic acids (citric, pyruvic, succinic, and fumaric acids), that enhanced members of Comamonadaceae, a potentially helpful bacteria group. Additionally, it has been observed that malic acid produced by roots can attract advantageous *Bacillus* spp. to the rhizosphere. The proportion and breakdown of sugars and phenolics in root exudates have a significant impact on the microbial makeup of soil. Furthermore, the introduction of coumaric acid, a phenolic compound, in soil has been discovered to impact microbial populations in the rhizosphere of cucumber plants. In addition, it has been shown that phenolics have a potential to enhance the population of bacteria or plant growth-promoting rhizobacteria (PGPR) in the rhizosphere of *Arabidopsis*. Furthermore, *Arabidopsis* is capable of synthesizing several specialized triterpenes that play a crucial role in orchestrating the establishment of distinct root microbiota (Kshetri et al., 2019). This ability allows *Arabidopsis* to actively shape and customize the ecosystem of microbes surrounding its roots. A further example is given by maize plants that emit benzoxazinoids via their roots, which play a crucial role in enhancing the growth and defense of plants, by influencing the composition of the rhizosphere microbiota.

Flavonoids have been recognized as essential signaling molecules in the root-rhizosphere system, mediating the relationship between roots and different microbes (Hassan & Mathesius, 2012). These include rhizobia, mycorrhizal fungi, root bacterial and fungal pathogens, pests such as insects, and nematodes. For example, in maize the introduction of flavones produced from roots has been shown to enhance the presence of Oxalobacteraceae species in the rhizosphere, leading to improved growth and nitrogen uptake. This suggests that relationships among roots and microbes mediated by flavonoids may potentially play a role in regulating embryonic procedures in host plants. In addition, formation of reactive oxygen species (ROS) via the receptor kinase FERONIA was found to regulate the abundance of beneficial *Pseudomonas* sp. in the rhizosphere microbiome (Ali et al., 2022). The defense-related phytohormones salicylic acid (SA) and jasmonic acid (JA) are involved in the regulation of the rhizosphere microbial communities in *Arabidopsis*. Studies have shown that the removal of genes responsible for SA or JA production led to changes in the rhizosphere microbial community. Moreover, the accumulation of SA or JA by biotrophic or necrotrophic pathogens, respectively, suggested that virus invasion may influence the development of the rhizosphere microbiome via the action of phytohormones (Chen et al., 2020).

In conclusion, root exudates act as chemical agents in modulating the composition of rhizosphere microbial communities. This modulation is contingent upon several factors such as the particular host plant and prevailing surroundings, particularly the presence of pathogenic attacks (Horwath & Paul, 2015). The impact of exudates on the communities of bacteria in the rhizosphere exhibits significant variability, complexity, and dynamism which still remain to be completely comprehended (Cesco et al., 2012).

5.6 Relationship of Plants Immunity and Pathogens Differentiation with the Rhizosphere Microbiome

The use of certain soil microorganisms to enhance the overall well-being and productivity of crops is being effectively employed to enhance pest control and crop production, by manipulating the soil microbiome population. The advent of genome sequencing technology has led to a resurgence of studies on the structure and function of bacterial communities, in relation to the regulation of plant production and soil health. Consequently, there has been a notable increase in the study of various bacterial, fungal, and archaeal constituents of the soil microbiome (Moran, 2022). Consistent with previous research, data showed that plant types in various stages of succession exhibit distinct ecophysiological characteristics that can have a substantial influence on soil natural attributes. For instance, in the early stages of succession, rapid-growing plant species generate a readily decomposable carbon base that promotes dominance of some microbial components (Mensah, 2015). In contrast, species with sluggish growth rates that dominate later stages of regeneration characterized by high lignin content, promote the proliferation of lignin-degrading fungi. The different connections found among an ensemble of microorganisms or inoculated strain and the host system have been expanding. They include abiotic soil conditions, the host plant inoculated and soil type, the micro-and macrobiota that surround them, as well as broader environmental circumstances. This scientific tackle provides insights for the development of improved and efficient microbe usages in the context of microbiome functions (Silveira et al., 2015).

The presence of microbiomes connected to plants may have advantageous effects on the hosts. Nevertheless, the composition of the communities of bacteria in the rhizosphere, as well as the efficacy of colonization, can be regulated by several variables, including ecological conditions, the genetic code of the host, and its physical state. There is a strong correlation between several plant characteristics such as its output, infection repression, development reaction, and blooming phenotype, as well as the presence of specific plant microbiota (Mendes et al., 2013). Consequently, the microbiota alteration or the impact of agricultural procedures have an effect on the plant traits as well as its production efficiency. Understanding both the biological and mechanical aspects of the relationship among microorganisms, vegetation, surrounding environments, farming procedures and management is crucial to achieve farming advantages from plant microbes (Bisen et al., 2015).

5.7 Beneficial Microorganisms *vs* Soilborne Diseases

The rhizosphere serves as both a habitat and a site for disease causal agents, wherein soilborne species may establish a parasitic association with roots. Nevertheless, the rhizosphere is a dynamic environment whereby a diverse variety of microflora and microfauna engage in interactions, ultimately impacting the progression and result of pathogenic infections (Usman et al., 2018). Numerous beneficial microbes in the rhizosphere have the ability to prevent the development or functioning of soilborne pathogenic fungi, oomycetes, bacteria, and/or nematodes. The impact and consequences of advantageous rhizosphere microbes on plant development and well-being have been extensively demonstrated in the case of bacteria species. These bacteria are often found in several environments, including soil, water, and the human body including fungi from the Deuteromycetes, such as *Trichoderma, Gliocladium*, and non-pathogenic *F. oxysporum* (Nihorimbere et al., 2011).

The relationships between soilborne pathogens and their environment primarily include three kinds of "conversations", including spatial, metabolic, and others. The colonization of the rhizosphere and subsequent successful establishment within the root zone of a pathogen is a necessary condition for achieving effectiveness. The phenomena seen in biocontrol bacteria may be partially attributed to the regulation of various antagonistic characteristics and chemicals, which are influenced by cell-density dependent control, or quorum detection. Furthermore, interspecific competition may act as a biocontrol system, particularly for the organic molecules required by seeds or for multiplication and root colonization by the pathogen (Agaras et al., 2015). Competition may also occur for micronutrients, particularly iron, crucial for development and physical performance. The competition for soluble ferric iron revolves on the development and/or use of chelators known as siderophores, which possess a high affinity for iron (Potrykus et al., 2013).

5.8 Rhizosphere Microorganisms: Benefits for Plants

In addition to their biocontrol activities, rhizosphere microorganisms have the potential to directly enhance plant growth and promote their overall health. Frequently, this is one of the many means by which these microorganisms

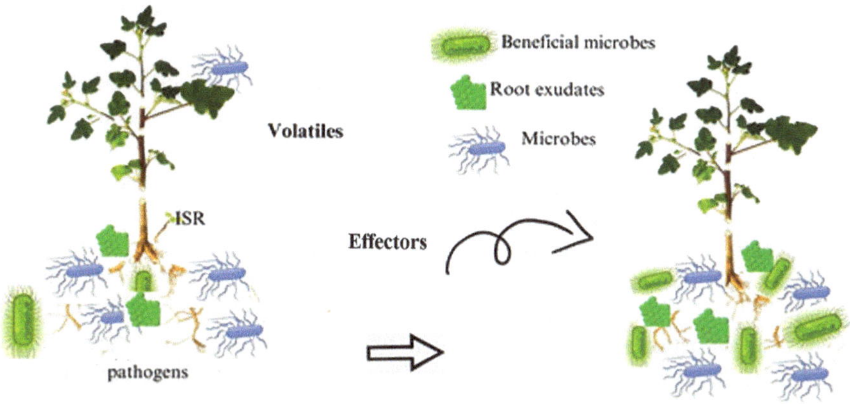

Fig. 5.2 A conceptual model of pathogen-mediated rhizosphere microbial recruitment for plant protection. First, previous crop plants release root exudates in soil, manipulating soil microbial dynamics or specifically recruiting beneficial species, after recognizing pathogens invasion. Microbial recruitment could effectively modulate immune responses and then directly inhibit pathogens as well as ISR. The surrounding plants possibly recruit rhizosphere microbiota after receiving aerial signals (volatiles) from diseased plants. Successful pathogens use effectors, or other strategies, to break down the rhizosphere microbes barrier

contribute to the promotion of plant health. Initially, it should be noted that phytostimulatory and biofertilizing microorganisms have a potential to enhance plant health via the process of fortification. Furthermore, it has been shown that several rhizosphere bacteria have the ability to elicit a systemic response in plants, leading to the engagement of defense systems (Venturi & Keel, 2016).

This defense mechanism does not require a close relationship between the microorganisms and the pathogen in question. Furthermore, it has been shown that ISR has the potential to be efficacious in real-world settings, including both field circumstances and industrial greenhouses (Fig. 5.2). ISR has many parallels with SAR, which represents the plant defensive response activated upon encountering pathogens (Sergeant & Renaut, 2010). However, ISR also displays distinct variations from SAR. ISR has been shown to include the signaling molecules jasmonate and ethylene, as demonstrated in experimental assays conducted on particular *Arabidopsis* mutants. In contrast to SAR, ISR is often not reliant on salicylates. However, it is worth noting that some beneficial bacteria have the ability to trigger a salicylate-dependent pathway inside plants (Rayapuram, 2007).

The number of cellular components found on the outermost layer of biocontrol bacteria, including lipopolysaccharides and flagella, has been identified as potential ISR triggers. ISR also happens when plants are exposed to certain beneficial substances generated by bacteria. Examples of such substances include the volatile 2,3-butanediol, the siderophore pyoverdine, and DAPG.

5.9 Impact of Rhizosphere Bacteria on Plant Nutrition

The rhizosphere microbiome has a substantial impact on the nutritional status of plants. Prominent examples are the nitrogen-fixing rhizobia and the mycorrhizal fungi that enhance phosphorus absorption. The significance of symbionts, including mycorrhizal fungi, in facilitating the transfer of nutrients and minerals from soil to the plant has been extensively acknowledged and supported by various studies (Meena et al., 2018). Additionally, these symbionts play a crucial role in soil structuring, and are also instrumental in the decrease of pathogens. The recognition and documentation of these functions have been well-established in the literature. In addition to *Rhizobium* and *Bradyrhizobium*, many additional bacteria that are capable of nitrogen fixation and reside in the rhizosphere have been documented (Khan et al., 2022a, b). As an example, the examination of the cowpea rhizosphere demonstrated a substantial genetic variation in symbiotic rhizobial species throughout the western Amazon region. According to the findings of glasshouse tests and analysis of the 16S rRNA gene sequencing, it was shown that *Bradyrhizobium*, *Rhizobium*, *Burkholderia*, and *Achromobacter* spp. have the ability to form nodules on cowpea plants and are effective in the biological nitrogen fixation (Moura et al., 2022).

Considering the many studies conducted on the process of nitrogen fixation by rhizobia, the effective introduction of this legume-specific relationship on different crop with farming significance has not yet been accomplished. A more comprehensive knowledge is needed on the cell-mediated reactions elicited by *Rhizobium* and mycorrhizal fungi, despite their apparent similarities, to realize this long-standing aspiration (Pankievicz et al., 2019).

Rhizosphere bacteria have also the ability to enhance the absorption of certain minerals. Iron is found in large quantities in soil. However, when the soil pH is moderate to alkaline, iron mostly resides in the form of insoluble ferric oxide, that is not accessible for microbial development. Bacteria utilize multiple processes to control levels of iron inside cells, in response to its limited availability and the potential toxicity of high levels of free iron. Those processes involve the production of siderophores. Plants, on the host side, use two strategies in response to iron restriction. The first one is to improve the dissolution of inorganic iron in the rhizosphere. The second mechanism involves the release of phytosiderophores, which are then carried into the root by a specialized uptake mechanism, as shown in rice (Fazeli-Nasab et al., 2022).

A number of investigations also suggested an alternative approach to iron acquisition by plants, which involves the consumption of iron bound to siderophores produced by bacteria. This phenomenon was shown in studies on pseudomonads, whereby they facilitated iron uptake via the production of siderophores, benefiting both graminaceous and dicotyledonous plant species. Rhizoferrin, a siderophore generated by the fungus *Rhizopus arrhizus*, also showed a high level of effectiveness in transporting iron to plants. The effectiveness of this process has been determined to be similar with that of artificial chelates (Gupta & Prakash, 2019). The activation of the plant endogenous iron absorption mechanism by rhizobacteria, such as *Bacillus subtilis* GB03, has also been shown. This isolate was shown to increase the gene expression of the *Arabidopsis* Fe-deficiency-induced transcription factor 1 (FIT1), leading to the induction of the ferric reductase FRO2 and the iron transporter IRT1. This observation provided more comprehensive insights into the process through which rhizosphere microorganisms impact the absorption of iron by plants (Krishna et al., 2023).

The majority of rhizobacterial species are classified as organotrophs, meaning that they get energy by the digestion of organic molecules. The restricted accessibility and quantity of biodegradable organic substances in the majority of soils, along with the scarcity of carbon, is a primary constraint for soil bacteria development (Xiang et al., 2022). Bacterial populations are of utmost importance in the process of liberating nutritive ions from soil minerals, not only crucial for their own nutritional needs but also for plants nutrition. Mineral metabolizing bacteria identified in diverse habitats, including the rhizosphere and the ecto-mycorrhizosphere, show a potential to enhance plant growth in soils that are deficient in nutrients.

5.10 Rhizosphere Microbiome *vs* Nematodes and Other Pathogens

Plant-parasitic nematodes (PPN) are a significant contributor to inadequate nutrition in plants and are responsible for substantial agricultural losses. They have been observed to concentrate on attacking the root systems of plants, employing multiple methods to spread (Nicol et al., 2011). These processes consist of penetrating the roots and consuming the local nutrients present, impeding the plant ability to absorb water and nutrients. This process ultimately makes the roots also vulnerable to potential pathogens eventually present in soil (Fig. 5.3). As a consequence, nematode parasitism can reduce agricultural and horticultural crops productivity, resulting in low yields (Koenning et al., 2004).

The primary emphasis of study in the field of crop production has concentrated on two prominent categories of PPN, namely the cyst nematodes (*Heterodera* and *Globodera* spp.) and root-knot nematodes (*Meloidogyne* spp.), due to their significant global, economic impact (Pires et al., 2022). Nevertheless, an alternative perspective exists that regards the occurrence of other nematodes, namely the free-living species present in soil throughout any crop cultivation, an indicator of soil characteristics associated with its nutrient availability. Since the '70 s, such nematodes that live freely have been used as indicators to assess key ecologic and soil conditions. Within the rhizosphere, populations of rhizobacteria may deploy multiple tactics that successfully contrast both PPN (Mhatre et al., 2019). These effects can be broadly categorized into two primary activities: *i*) direct antagonism, involving the generation of metabolites with both hazardous and advantageous properties, and *ii*) indirect processes. The latter encompass regulating nematode conduct, altering the chemical make-up of root excretions, releasing of nematode repelling agents by the

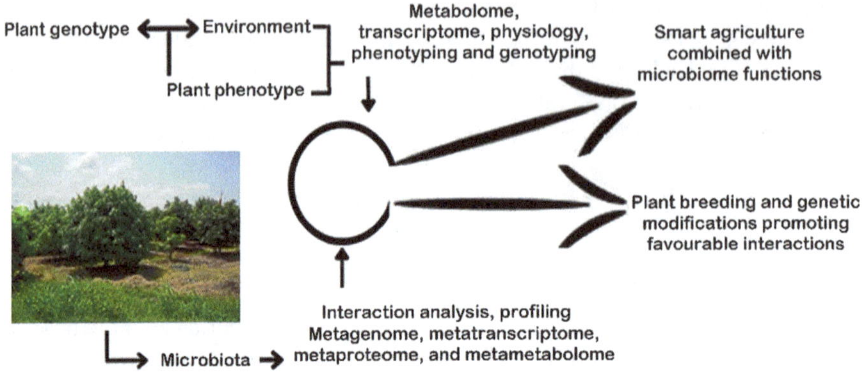

Fig. 5.3 In-depth knowledge on the interactions of holobionts and their functions, leading to knowledge-based approaches for microbiome applications, harnessing the potential of the plant-soil-microbe interaction in: (*i*) the selections of plant type or varieties favoring beneficial plant-microbe interactions, (*ii*) microbial inoculant complementation and (*iii*) agricultural management practices that enrich microbiota with desirable bacteria (Compant et al., 2010)

plant, competing for vital nutrients at root locations, and ultimately leading to general resistance (Yatoo et al., 2021).

The rhizosphere microbiome has the ability to control the PPN proliferation via many mechanisms, including the synthesis of phytohormones, nitrogen fixation, dissolution of phosphorus and potassium, the secretion of siderophores which have iron, and the generation of nitrogen gas in the form of ammonia (Kour et al., 2019). This occurrence includes the ability of rhizobacteria to counteract PPN by hindering their developing, growth, surviving, and development. Rhizobacteria achieve this goal through the production of toxins, metabolic by-products, and proteins that are of bacterial origin (Prasad et al., 2020). Ammonifiers have been shown to impede the proliferation of nematodes throughout the breakdown of nitrogenous organic materials. In addition, it has been shown that secondary metabolites such as 2–4 diacetylphloroglucinol, which are produced by *Pseudomonas fluorescens*, might potentially inhibit the PPN proliferation (Aioub et al., 2022). This inhibiting impact was particularly enhanced when these compounds were combined with neem cake.

In addition, it has been shown that some rhizobacteria have the ability to produce hydrogen cyanide (HCN), which could effectively eliminate contaminants in the rhizosphere. This process has shown potential in creating a favorable root microenvironment that promotes improved plant development (Collange et al., 2011). Chickpea, a prominent legume in several regions globally, has root galling caused by *M. javanica*. However, studies have shown that the presence of *Azospirillium*, *Azotobacter*, *Rhizobium*, and the mycorrhizal species *Glomus* sp. may effectively reduce the occurrence of root galling (Alloush et al., 2000). A different kind of direct antagonism may be seen in the synthesis of lytic enzymes, including chitinases, peroxidase, phenylalanine ammonia-lyase, dehydrogenase, lipases, β glucanase, proteases, and phosphatases (Khan et al., 2022a, b). Chitinase synthesis has been seen in *Corynebacterium paurometabolus*, which showed inhibitory effects on nematode eggs development, as well as reproductive processes, ultimately leading to a decrease in population size. The presence of three bacterial strains, namely *Stenotrophomonas maltophilia*, *Bacillus mycoides*, and *Pseudomonas* sp., resulted in a 56–74% decrease in potato cyst nematode (PPN) populations in potato (Mahmud et al., 2021). Additionally, plants with nematocidal properties exhibited a PPN reduction, likely due to the presence of hydrolytic enzymes, which have antifungal properties, as well as release of hydrogen cyanide (HCN).

Direct antagonisms consist of ISR, which may be considered as a common strategy used by plants to enhance their defense mechanisms and bolster their ability to fight towards a wide range of pests and illnesses (Oliveira et al., 2016). This heightened protective response is triggered by external inducing substances,and confers protection from a range of pathogens (see previous paragraphs). The ISR response involves the activation of various defense amino acids, which subsequently trigger the induction of resistance, with the synthesis of phytoalexins and additional phenolic compounds (Boubakri, 2020).

In outdoor environments, the presence of native and/or introduced rhizobacteria has the potential to increase nematode resistance/tolerance, achieved through the induction of plant systemic defense mechanisms, which involve processes such as cell wall thickening, deposition of callus, and building up of phenolic compounds. Additionally, some rhizobacteria may also contribute to the biosynthesis and up-regulation of phytoalexins, lipopolysaccharides, siderophores, SA, and JA (Pinski et al., 2019).

5.11 Conclusions

In order to reduce negative environmental consequences without compromising food security and standards, modern agriculture should focus on sustainable technologies and procedures. Rhizosphere biotechnology and microbiome engineering are both growing into topics that may have significant impacts in the agriculture industry. Future studies should, however, concentrate on modifying rhizospheric bacteria, creating advantageous consortia that persist in the field for several generations. In addition to studies on their impacts on diverse crops, more investigation is needed on top-down and bottom-up tactics, methodologies, and omics approaches. To reduce the use of chemical fertilizers/pesticides and increase crop output and quality, a combination of soil alterations, plant breeding and genetic manipulation methods, and targeted engineering of rhizospheric bacteria with microbiome engineering/biotechnology, can be applied. The feasibility of effectively integrating microbiome engineering with biotechnology across a range of additional applications, such as bioremediation, human health, etc., should be examined, as well as the practical implementation of such approaches.

References

Agaras, B. C., Scandiani, M., Luque Fernández, L., Farina, F., Carmona, M., & Valverde, C. (2015). Quantification of the potential biocontrol and direct plant growth promotion abilities based on multiple biological traits distinguish different groups of *Pseudomonas* spp. isolates. *Biological Control, 90*, 173–186.

Aioub, A. A., Elesawy, A. E., & Ammar, E. E. (2022). Plant growth promoting rhizobacteria (PGPR) and their role in plant-parasitic nematodes control: A fresh look at an old issue. *Journal of Plant Diseases and Protection, 129*, 1305–1321.

Ali, S., Tyagi, A., Park, S., Mir, R. A., Mushtaq, M., et al. (2022). Deciphering the plant microbiome to improve drought tolerance: Mechanisms and perspectives. *Environmental and Experimental Botany, 201*, 104933.

Alloush, G. A., Zeto, S. K., & Clark, R. B. (2000). Phosphorus source, organic matter, and arbuscular mycorrhiza effects on growth and mineral acquisition of chickpea grown in acidic soil. *Journal of Plant Nutrition, 23*, 1351–1369.

Bag, S., Mondal, A., Majumder, A., Mondal, S. K., & Banik, A. (2022). Flavonoid mediated selective cross-talk between plants and beneficial soil microbiome. *Phytochemistry Reviews, 21*, 1739–1760.

Bang, C., Dagan, T., Deines, P., Dubilier, N., Duschl, W. J., et al. (2018). Metaorganisms in extreme environments: Do microbes play a role in organismal adaptation ? *Zoology, 127*, 1–19.

Barnes, R. S. K., Calow, P. P., Olive, P. J., Golding, D. W., & Spicer, J. I. (2009). *The invertebrates: A synthesis*. Wiley. 512 pp.

Berninger, T., González López, Ó., Bejarano, A., Preininger, C., & Sessitsch, A. (2018). Maintenance and assessment of cell viability in formulation of non-sporulating bacterial inoculants. *Microbial Biotechnology, 11*, 277–301.

Bisen, K., Keswani, C., Mishra, S., Saxena, A., Rakshit, A., & Singh, H. B. (2015). Unrealized potential of seed biopriming for versatile agriculture. *Nutrient use efficiency: From basics to advances* (pp. 193–206).

Boubakri, H. (2020). *Induced resistance to biotic stress in plants by natural compounds: Possible mechanisms in priming-mediated stress and cross-stress tolerance in crop plants* (pp. 79–99). Academic.

Bruehl, G. W. (1987). *Soilborne plant pathogens* (pp. xv+–368pp).

Cesco, S., Mimmo, T., Tonon, G., Tomasi, N., Pinton, R., et al. (2012). Plant-borne flavonoids released into the rhizosphere: Impact on soil bio-activities related to plant nutrition a review. *Biology and Fertility of Soils, 48*, 123–149.

Chen, X., Marszałkowska, M., & Reinhold-Hurek, B. (2020). Jasmonic acid, not salicyclic acid restricts endophytic root colonization of rice. *Frontiers in Plant Science, 10*, 1758.

Chialva, M., Lanfranco, L., & Bonfante, P. (2022). The plant microbiota: Composition, functions, and engineering. *Current Opinion in Biotechnology, 73*, 135–142.

Choudhary, D. K., Kasotia, A., Jain, S., Vaishnav, A., Kumari, S., et al. (2016). Bacterial-mediated tolerance and resistance to plants under abiotic and biotic stresses. *Journal of Plant Growth Regulation, 35*, 276–300.

Collange, B., Navarrete, M., Peyre, G., Mateille, T., & Tchamitchian, M. (2011). Root-knot nematode (*Meloidogyne*) management in vegetable crop production: The challenge of an agronomic system analysis. *Crop Protection, 30*, 1251–1262.

Compant, S., Clément, C., & Sessitsch, A. (2010). Plant growth-promoting bacteria in the rhizo- and endosphere of plants: Their role, colonization, mechanisms involved and prospects for utilization. *Soil Biology and Biochemistry, 42*, 669–678.

Del Carmen, O.-M. M., Del Carmen, R.-G. M., Glick, B. R., et al. (2018). Microbiome engineering to improve biocontrol and plant growth-promoting mechanisms. *Microbiological Research, 208*, 25–31.

Dick, M. W. (2013). *Straminipilous fungi: Systematics of the Peronosporomycetes including accounts of the marine straminipilous protists, the plasmodiophorids and similar organisms* (p. 670). Springer Science & Business Media.

Doehlemann, G., Ökmen, B., Zhu, W., & Sharon, A. (2017). Plant pathogenic fungi. *Microbiology Spectrum, 5*(10), 1128.

Dubey, A., Malla, M. A., Khan, F., Chowdhary, K., Yadav, S., et al. (2019). Soil microbiome: A key player for conservation of soil health under changing climate. *Biodiversity and Conservation, 28*, 2405–2429.

Dudáš, M., Pjevac, P., Kotianová, M., Gančarčíková, K., Rozmoš, M., et al. (2022). Arbuscular mycorrhiza and nitrification: Disentangling processes and players by using synthetic nitrification inhibitors. *Applied and Environmental Microbiology, 88*, e01369–e01322.

Fazeli-Nasab, B., Piri, R., & Rahmani, A. F. (2022). Assessment of the role of rhizosphere in soil and its relationship with microorganisms and element absorption. In R. Soni, D. C. Suyal, & R. Goel (Eds.), *Plant protection: From chemicals to biologicals* (pp. 225–252). De Gruyter.

Finlay, R. D., & Thorn, R. G. (2019). *The fungi in soil. Modern soil microbiology* (e ed., pp. 65–90). CRC Press.

Gandhi, S. K., & Mehta, N. (2011). Management of fungal and bacterial diseases in vegetable crops. In *Breeding and protection of vegetables [Waktu dan tempat pertemuan tidak diketahui]* (pp. 376–440). New India Publishing Agency.

Guo, S., Xiong, W., Hang, X., Gao, Z., Jiao, Z., et al. (2021). Protists as main indicators and determinants of plant performance. *Microbiome, 9*, 1–11.

Gupta, C., & Prakash, D. (2019). Enhancement of active constituents of medicinal plants through the use of microbes. In A. Varma, S. Tripathi, & R. Prasad (Eds.), *Plant biotic interactions: State of the art* (pp. 227–241).

Gupta, A., Gupta, R., & Singh, R. L. (2017). Microbes and environment. In R. L. Singh (Ed.), *Principles and applications of environmental biotechnology for a sustainable future* (pp. 43–84). Springer.

Gutschick, V. P. (2012). *A functional biology of crop plants* (p. 232). Springer Science & Business Media.

Hassan, S., & Mathesius, U. (2012). The role of flavonoids in root–rhizosphere signalling: Opportunities and challenges for improving plant–microbe interactions. *Journal of Experimental Botany, 63*, 3429–3444.

Hassan, M. K., McInroy, J. A., & Kloepper, J. W. (2019). The interactions of rhizodeposits with plant growth-promoting rhizobacteria in the rhizosphere: A review. *Agriculture, 9*, 142.

Horwath, W., & Paul, E. A. (2015). Carbon cycling: The dynamics and formation of organic matter. *Soil Microbiology, Ecology and Biochemistry, 4*, 339–382.

Jaouad, M., Moinina, A., Ezrari, S., & Lahlali, R. (2020). Key pests and diseases of citrus trees with emphasis on root rot diseases: An overview. *Moroccan Journal of Agricultural Sciences, 1*.

Jayaraman, S., Naorem, A. K., Lal, R., Dalal, R. C., Sinha, N. K., Patra, A. K., & Chaudhari, S. K. (2021). Disease-suppressive soils — Beyond food production: A critical review. *Journal of Soil Science and Plant Nutrition, 21*, 1437–1465.

Jha, D. K. (2023). Problems and prospects of utilization of bacterial endophytes for the management of plant diseases. *Indian Phytopathology, 76*, 3–20.

Khan, A. R., Mustafa, A., Hyder, S., Valipour, M., Rizvi, Z. F., et al. (2022a). *Bacillus* spp. as bioagents: Uses and application for sustainable agriculture. *Biology, 11*, 1763.

Khan, N. F., Rasool, A., Mansoor, S., Saleem, S., Baba, T. R., et al. (2022b). Potential applications of Rhizobacteria as eco-friendly biological control, plant growth promotion and soil metal bioremediation. In V. S. Meena, M. Choudhary, R. P. Yadav, & S. Kumari Meena (Eds.), (pp. 104–170). Sustainable crop production – Recent advances, IntechOpen.

Koenning, S. R., Wrather, J. A., Kirkpatrick, T. L., Walker, N. R., Starr, J. L., & Mueller, J. D. (2004). Plant-parasitic nematodes attacking cotton in the United States: Old and emerging production challenges. *Plant Disease, 88*, 100–113.

Kolstø, A. B., Tourasse, N. J., & Økstad, O. A. (2009). What sets *Bacillus anthracis* apart from other *Bacillus* species ? *Annual Review of Microbiology, 63*, 451–476.

Kour, D., Rana, K. L., Yadav, N., Yadav, A. N., Kumar, et al. (2019). Rhizospheric microbiomes: biodiversity, mechanisms of plant growth promotion, and biotechnological applications for sustainable agriculture. In A. Kumar & V. S. Meena (Eds.), *Plant growth promoting rhizobacteria for agricultural sustainability: from theory to practices* (pp. 19–65). Springer Singapore.

Krishna, T. P. A., Ceasar, S. A., & Maharajan, T. (2023). Biofortification of crops to fight anemia: Role of vacuolar iron transporters. *Journal of Agricultural and Food Chemistry, 71*, 3583–3598.

Kshetri, L., Naseem, F., & Pandey, P. (2019). Role of *Serratia* sp. as biocontrol agent and plant growth stimulator, with prospects of biotic stress management in plant. In R. Z. Sayyed (Ed.), *Plant growth promoting rhizobacteria for sustainable stress management: Vol. 2: Rhizobacteria in biotic stress management* (pp. 169–200). Springer Nature Singapore.

Kumar, A., & Dubey, A. (2020). Rhizosphere microbiome: Engineering bacterial competitiveness for enhancing crop production. *Journal of Advanced Research, 24*, 337–352.

Kumawat, K. C., Nagpal, S., & Sharma, P. (2022). Potential of plant growth-promoting rhizobacteria-plant interactions in mitigating salt stress for sustainable agriculture: A review. *Pedosphere, 32*(2), 223–245.

Li, J., Wang, C., Liang, W., & Liu, S. (2021). Rhizosphere microbiome: The emerging barrier in plant-pathogen interactions. *Frontiers in Microbiology, 12*, 772420.

Mahmud, K., Missaoui, A., Lee, K., et al. (2021). Rhizosphere microbiome manipulation for sustainable crop production. *Current Plant Biology, 27*, 100210.

Meena, R. S., Vijayakumar, V., Yadav, G. S., & Mitran, T. (2018). Response and interaction of *Bradyrhizobium japonicum* and arbuscular mycorrhizal fungi in the soybean rhizosphere. *Plant Growth Regulation, 84*, 207–223.

Mendes, R., Garbeva, P., & Raaijmakers, J. M. (2013). The rhizosphere microbiome: Significance of plant beneficial, plant pathogenic, and human pathogenic microorganisms. *FEMS Microbiology Reviews, 37*, 634–663.

Mensah, A. K. (2015). Role of revegetation in restoring fertility of degraded mined soils in Ghana: A review. *International Journal of Biodiversity and Conservation, 7*, 57–80.

Mhatre, P. H., Karthik, C., Kadirvelu, K., Divya, K. L., Venkatasalam, E. P., et al. (2019). Plant growth promoting rhizobacteria (PGPR): A potential alternative tool for nematodes bio-control. *Biocatalysis and Agricultural Biotechnology, 17*, 119–128.

Mitter, B., Brader, G., Afzal, M., Compant, S., Naveed, M., et al. (2013). Advances in elucidating beneficial interactions between plants, soil, and bacteria. *Advances in Agronomy, 121*, 381–445.

Moran, E. F. (2022). *Human adaptability: An introduction to ecological anthropology* (4th ed., p. 460). Routledge.

Moura, F. T., Ribeiro, R. A., Helene, L. C. F., Nogueira, M. A., & Hungria, M. (2022). So many rhizobial partners, so little nitrogen fixed: The intriguing symbiotic promiscuity of common bean (*Phaseolus vulgaris* L.). *Symbiosis, 86*, 169–185.

Nadarajan, S., & Sukumaran, S. (2021). Chemistry and toxicology behind chemical fertilizers. In F. B. Lewu, T. Volova, S. Thomas, & K. R. Rakhimol (Eds.), *Controlled release fertilizers for sustainable agriculture* (pp. 195–229). Academic.

Nicol, J. M., Turner, S. J., Coyne, D. L., Nijs, L. D., Hockland, S., & Maafi, Z. T. (2011). Current nematode threats to world agriculture. In J. Jones, G. Gheysen, & C. Fenoll (Eds.), *Genomics and molecular genetics of plant-nematode interactions* (pp. 21–43). Springer.

Nihorimbere, V., Ongena, M., Smargiassi, M., & Thonart, P. (2011). Beneficial effect of the rhizosphere microbial community for plant growth and health. *Biotechnology, Agronomy, Society and Environment, 15*(2).

Nwachukwu, C. M., Wang, C., & Wetterlund, E. (2021). Exploring the role of forest biomass in abating fossil CO_2 emissions in the iron and steel industry–The case of Sweden. *Applied Energy, 288*, 116558.

Oliveira, M. D. M., Varanda, C. M. R., & Félix, M. R. F. (2016). Induced resistance during the interaction pathogen x plant and the use of resistance inducers. *Phytochemistry Letters, 15*, 152–158.
Palti, J. (2012). *Cultural practices and infectious crop diseases* (Vol. 9). Springer Science & Business Media.
Pankievicz, V., Irving, T. B., Maia, L. G., & Ané, J. M. (2019). Are we there yet? The long walk towards the development of efficient symbiotic associations between nitrogen-fixing bacteria and non-leguminous crops. *BMC Biology, 17*, 1–17.
Papik, J., Folkmanova, M., Polivkova-Majorova, M., Suman, J., & Uhlik, O. (2020). The invisible life inside plants: Deciphering the riddles of endophytic bacterial diversity. *Biotechnology Advances, 44*, 107614.
Pegg, K. G., Coates, L. M., O'Neill, W. T., & Turner, D. W. (2019). The epidemiology of Fusarium wilt of banana. *Frontiers in Plant Science, 10*, 1395.
Perrine-Walker, F. (2020). *Phytophthora palmivora*—Cocoa interaction. *Journal of Fungi, 6*, 167.
Pii, Y., Mimmo, T., Tomasi, N., et al. (2015). Microbial interactions in the rhizosphere: Beneficial influences of plant growth-promoting rhizobacteria on the nutrient acquisition process. A review. *Biology and Fertility of Soils, 51*, 403–415.
Pinski, A., Betekhtin, A., Hupert-Kocurek, K., Mur, L. A., & Hasterok, R. (2019). Defining the genetic basis of plant–endophytic bacteria interactions. *International Journal of Molecular Sciences, 20*, 1947.
Pires, D., Vicente, C. S., Menéndez, E., Faria, J. M., Rusinque, L., et al. (2022). The fight against plant-parasitic nematodes: Current status of bacterial and fungal biocontrol agents. *Pathogens, 11*, 1178.
Potrykus, J., Stead, D., MacCallum, D. M., Urgast, D. S., Raab, A., et al. (2013). Fungal iron availability during deep seated candidiasis is defined by a complex interplay involving systemic and local events. *PLoS Pathogens, 9*, e1003676.
Pradhan, D., Jaiswal, A. K., & Jaiswal, S. (2022). Emerging technologies for the production of nanocellulose from lignocellulosic biomass. *Carbohydrate Polymers, 285*, 119258.
Prasad, B. N., Subramanyam, B., Lakshmipathi, R. N., Ansari, R. A., Rizvi, R., Sumbul, A., & Rachmi, C. M. (2020). Utilization of beneficial microorganisms in sustainable control of phytonematodes. In R. A. Ansari, R. Rizvi, & I. Mahmood (Eds.), *Management of phytonematodes: Recent advances and future challenges* (pp. 317–337). Spinger Nature.
Prashar, P., Kapoor, N., & Sachdeva, S. (2014). Rhizosphere: Its structure, bacterial diversity and significance. *Reviews in Environmental Science and Bio/Technology, 13*, 63–77.
Rao, D. L. N. (2007). Microbial diversity, soil health and sustainability. *Journal. Indian Society of Soil Science, 55*, 392.
Rayapuram, C. (2007). *Molecular and ecological analysis of LOX3-and NPR1-dependent defense responses in plant-herbivore interactions*. Doctoral dissertation,. Jena University.
Rekadwad, B. (Ed.). (2020). *Microbial systematics: Taxonomy, microbial ecology, diversity*. CRC Press.
Sánchez-Cañizares, C., Jorrín, B., Poole, P. S., & Tkacz, A. (2017). Understanding the holobiont: The interdependence of plants and their microbiome. *Current Opinion in Microbiology, 38*, 188–196.
Sani, M. N. H., & Yong, J. W. (2021). Harnessing synergistic biostimulatory processes: A plausible approach for enhanced crop growth and resilience in organic farming. *Biology, 11*, 41.
Santoyo, G., Pacheco, C. H., Salmerón, J. H., et al. (2017). The role of abiotic factors modulating the plant-microbe-soil interactions: Toward sustainable agriculture. A review. *Spanish Journal of Agricultural Research, 15*, 13.
Sergeant, K., & Renaut, J. (2010). Plant biotic stress and proteomics. *Current Proteomics, 7*, 275–297.
Sessitsch, A., Wakelin, S., Schloter, M., Maguin, E., Cernava, T., et al. (2023). Microbiome interconnectedness throughout environments with major consequences for healthy people and a healthy planet. *Microbiology and Molecular Biology Reviews, 87*, e00212–e00222.

Silveira, M. H. L., Morais, A. R. C., da Costa Lopes, A. M., Olekszyszen, D. N., Bogel-Łukasik, R., et al. (2015). Current pretreatment technologies for the development of cellulosic ethanol and biorefineries. *ChemSusChem, 8*, 3366–3390.

Singh, S., Aghdam, S. A., Lahowetz, R. M., & Brown, A. M. (2023). Metapangenomics of wild and cultivated banana microbiome reveals a plethora of host-associated protective functions. *Environmental Microbiome, 18*, 36.

Spanu, P. D., Abbott, J. C., Amselem, J., Burgis, T. A., Soanes, D. M., et al. (2010). Genome expansion and gene loss in powdery mildew fungi reveal tradeoffs in extreme parasitism. *Science, 330*, 1543–1546.

Spiegel, Y., Oren, O., & Ehlers, R. U. (2022). Critical terms during development and commercialization of microbial agents for the control of plant parasitic nematodes. In R. A. Sikora, J. Desaeger, & L. Molendijk (Eds.), *Integrated nematode management: State-of-the-art and visions for the future* (pp. 446–453). CABI.

Trivedi, P., Mattupalli, C., Eversole, K., & Leach, J. E. (2021). Enabling sustainable agriculture through understanding and enhancement of microbiomes. *The New Phytologist, 230*, 2129–2147.

Usman, M., Nangere, M. G., & Oyetola, S. O. (2018). Functional significance of plant-associated microbes in soil fertility management. *International Journal of Research and Review, 5*, 29–35.

Venturi, V., & Keel, C. (2016). Signaling in the rhizosphere. *Trends in Plant Science, 21*, 187–198.

Voolstra, C. R., Suggett, D. J., Peixoto, R. S., Parkinson, J. E., Quigley, K. M., et al. (2021). Extending the natural adaptive capacity of coral holobionts. *Nature Reviews Earth & Environment, 2*, 747–762.

Xiang, L., Harindintwali, J. D., Wang, F., Redmile-Gordon, M., Chang, S. X., et al. (2022). Integrating biochar, bacteria, and plants for sustainable remediation of soils contaminated with organic pollutants. *Environmental Science & Technology, 56*, 16546–16566.

Yadav, G., Vishwakarma, K., Sharma, S., Kumar, V., Upadhyay, N., et al. (2017). Emerging significance of rhizospheric probiotics and its impact on plant health: Current perspective towards sustainable agriculture. In V. Kumar, M. Kumar, S. Sharma, & R. Prasad (Eds.), *Probiotics and plant health* (pp. 233–251). Springer Nature.

Yatoo, A. M., Ali, M. N., Baba, Z. A., & Hassan, B. (2021). Sustainable management of diseases and pests in crops by vermicompost and vermicompost tea. A review. *Agronomy for Sustainable Development, 41*, 1–26.

Zhang, H., Serwah Boateng, N. A., Ngolong Ngea, G. L., Shi, Y., Lin, H., et al. (2021). Unravelling the fruit microbiome: The key for developing effective biological control strategies for postharvest diseases. *Comprehensive Reviews in Food Science and Food Safety, 20*, 4906–4930.

Zhou, Q., Li, K., Jun, X., Bo, L., et al. (2009). Role and functions of beneficial microorganisms in sustainable aquaculture. *Bioresource Technology, 100*, 3780–3786.

Chapter 6
Rhizospheric Microbiome: Biodiversity, Significance, and Prospects for Biotechnological Advancements

Elif Özlem Arslan Aydoğdu, Nahdhoit Ahamada Rachid, and Nihal Doğruöz Güngör

Abstract The rhizosphere, the thin space around the root surface, is an important microenvironment housing a highly diversified microbial community. It includes of three zones named endorhizosphere, ectorhizosphere, and rhizosphere. Each zone harbors microorganisms with defined functions like nutrient cycling, disease protection, and pathogenicity. The rhizosphere microbiome is a critical community since it is implicated in the interactions linking roots (plant), soil, and microorganisms. This community comprises bacteria, fungi, oomycetes, viruses, and other harmful and beneficial microorganisms. While the harmful organisms are the causal agents of diseases, soil drought, and other negative impacts on plants and the environment, the beneficial species balance these stresses by producing secondary metabolites. These include antibiotics and other bioactive compounds such as hydrolytic enzymes and volatile organic compounds active in plant growth promotion, plant immuno-system improvement, and local environment bioremediation. Studies for sustainable agriculture significantly considered the rhizosphere microorganisms, and showed the importance of using this community to enhance crop development and yield, as a response to food crisis. This chapter focuses on the diversity, significance, mechanisms and modes of action of rhizosphere microorganisms, particularly in plant protection, environmental bioremediation, and sustainable agriculture.

Keywords Bioactive compounds · NRPs · PGPR · PKs · Rhizosphere microbiome

E. Ö. Arslan Aydoğdu · N. Doğruöz Güngör (✉)
Department of Biology, Faculty of Science, Istanbul University, Istanbul, Turkey
e-mail: eoarslan@istanbul.edu.tr; ndogruoz@istanbul.edu.tr

N. Ahamada Rachid
Institute of Graduate Studies in Sciences, Istanbul University, Istanbul, Turkey
e-mail: nahamadarachid@gmail.com

P. Mathur, S. Roy (eds.), *Plant Microbiome and Biological Control*, Sustainability in Plant and Crop Protection 20,
https://doi.org/10.1007/978-3-031-75845-4_6

Abbreviations

ACCD	1-aminocyclopropane-1-carboxylate deaminase
AMF	Arbuscular mycorrhizal fungi
AMGs	Auxiliary metabolic genes
BGCs	Biosynthesis gene clusters
CO_2	Carbon dioxide
EMF	Ectomycorrhizal fungi
GA_3	Gibberellic acid
IAA	Indole-3-acetic acid
IC50	Half-maximal inhibitory concentration
IRB	Iron-reducing bacteria
ISR	Inducing stemic resistance
MeHg	Methylated mercury
NRPs	Non-ribosomal polypeptides
PGPB	Plant growth-promoting bacteria
PGPR	Plant growth promoting rhizobacterial
PKs	Polyketides
ROS	Reactive oxygen species
SRB	Sulfate-reducing bacteria

6.1 Introduction

The rhizosphere microbiome is a community of microorganisms that live in the thin region surrounding roots, known as the rhizosphere. These microorganisms interact with the roots and influence nutrients uptake, disease resistance, and plant growth. The rhizosphere microbiome is a complex niche shaped by root exudates, siderophores, and antimicrobial compounds released by plants, including significant microbial components of soil such as bacteria, fungi, nematodes, algae, oomycetes, and viruses (Gu et al., 2020; Meraklı & Memon, 2020; Yang et al., 2023b, d). The rhizosphere is essential in providing various ecosystem services such as nutrients cycling and carbon absorption. In this layer, there is a selective enrichment of some soil microorganisms which provide a maximum benefit to plants. For this reason, plant secretions that function as nutrients for microorganisms as well as antimicrobial secondary metabolites such as flavonoids have a functional role (Kumar & Dubey, 2020). While carbohydrates were initially mentioned as nutritive secretions, later research proved that amino acids are the most critical chemical determinants in this environment (Venturi & Keel, 2016).

6.2 The Impact of Biotic and Abiotic Factors on the Rhizosphere Microbiome

Several factors, such as the soil type, agricultural techniques, environmental conditions, and vegetation diversity, influence the composition of the rhizosphere microbiome. Plants influence the rhizosphere microbiome composition by releasing a distinctive array of compounds and exudates into the soil, thereby nurturing distinct communities of microorganisms. With its unique properties and characteristics, the soil serves as a sculptor of microbial life by providing a mix of physical and chemical conditions that actively shape microbial diversity and behaviors (Sanon et al., 2009; Tkacz et al., 2015; Chen et al., 2022a, b, c).

The agricultural management practices further cause significant changes in the rhizosphere microbiome. The practices of crop rotation, tillage, fertilization, and the application of pesticides yield both direct and indirect impacts on this intricate microbial ecosystem (Breitkreuz et al., 2021; Yang et al., 2023d). For instance, applying pesticides directly influences the abundance and variety of specific microbial groups, while tillage practices indirectly mold the microbiome structure by altering soil textures and nutrient availability (Huang et al., 2021; Jacquiod et al., 2021). Moreover, combining environmental factors such as temperature, moisture, pH, and nutrient availability, harmonizes the rhizosphere microbiome composition and functions (Munoz-Ucros et al., 2021; Afridi et al., 2022). These environmental cues act as conductors, guiding the performance of the microbial communities in the rhizosphere. The type and frequency of soil cultivation, the combination of salinity, major elements such as nitrogen, and the proportion of pollutants such as heavy metals are also factors that directly affect the microbial diversity. All these abiotic factors are also stress sources that affect the plant growth. Recent studies have revealed that the rhizosphere microbiomes assist plants to develop resistance against these stress factors (Breitkreuz et al., 2021; Munoz-Ucros et al., 2021; Afridi et al., 2022).

Macroscopic and microscopic organisms also play an active role in soil by shaping the microbiota. For example, a study on tomato plants revealed that the diversity and numbers of bacteria in the rhizosphere decreased during *Ralstonia solanacearum* infestation, with a lack of apparent differentiation between pathogenic and non-pathogenic bacterial species. The prevailing theory suggests that this shift in the flora was a direct result of the plant undergoing a state of stress, leading to the down-regulation of certain genes. This, in turn, culminated in alterations or reductions in the nutrient content available to the rhizosphere bacteria (Wei et al., 2018). In such a situation, the protective function of the rhizosphere microbiome for the plant should not be ignored. Similarly, Kong et al. (2016), showed that the distribution of bacteria in the rhizobial flora of chili plants changed even in a short period of 1 week under insect infestation as biocontrol species, especially *Pseudomonas* sp., became dominant. In addition to *Pseudomonas* sp., three taxa, including Caulobacteraceae, Enterobacteriaceae, and Flavobacteriaceae, and three genera, including *Achromobacter*, *Janthinobacterium*, and *Stenotrophomonas* became dominant in the rhizosphere. A further study supporting this view proved that some bacterial species added to soybean soil help the plant to protect itself and be less affected by physical and biological stresses, such as drought and insect infestation, respectively (Komatsu et al., 2023).

6.3 Complexities of Microbial Interactions in Plant-Microbe Systems

Interactions between plants and microorganisms in soil occur across three distinct layers: the endorhizosphere, ectorhizosphere, and rhizoplane. The endorhizosphere is the innermost layer where microorganisms closely associate with the surface of roots and even enter root tissues (Nazir et al., 2016; Mahmud et al., 2021). These microorganisms are often considered mutualists as they can provide direct benefits to the plant, such as improved nutrients uptake and protection against pathogens. The second layer, the rhizoplane, is the root surface, and hosts a specific microbial community. These microorganisms are near the root but do not invade its tissues (Reinhold-Hurek et al., 2015). They can affect the nutrient absorption and cycling. The last layer, the ectorhizosphere, is the outermost part of the rhizosphere, extending into the surrounding soil. Here, microorganisms interact with the root exudates and the soil matrix. This layer is critical for the rhizosphere overall functioning, as it influences the exchange of nutrients and of signaling compounds between the plant and the surrounding soil environment. These complex and dynamic layers interact with each other and the plant host, influencing the soil environment and the plant-microbe interactions (Bashir et al., 2016).

A diverse and robust rhizosphere microbiome promotes plant growth and stress resilience. A flourishing and varied microbiome within the rhizosphere is crucial to enhance plant growth and enable them to withstand various stress factors (Qu et al., 2020; Bano et al., 2021; Kumawat et al., 2022).

Furthermore, it is important to note that not all interactions between these microorganisms and plants are beneficial. The plant-microbe relationship can range from commensal to mutualistic or parasitic, depending on various factors (Raaijmakers et al., 2009; Pii et al., 2015). Understanding these relationships is of paramount importance, as it holds the potential to drive advancements in productive agricultural practices and contribute, on a wider scale, to address food crises. Consequently, research in this field has steadily increased in recent years.

6.4 The Role of Rhizospheric Bacteria: Functions and Significance in Plant Growth and Development

Nowadays, the importance of high-yielding agriculture cannot be overstated, particularly in light of the rapid rise in global population growth and an expected sharp increase of the demand for food. Many factors, such as wrong agricultural practices, climate change, drought, and land degradation, affect the productivity of agricultural lands. It is crucial to recognize that the increasingly urgent issue of global climate change presents a significant and immediate danger to both the farming industry and the broader field of nutrition (Afridi et al., 2022; Kumawat et al., 2022; Vidal et al., 2022).

Part of the solution to the problem is to increase sustainable agricultural practices by improving soil fertility through organic means. For this purpose, rhizosphere management practices are becoming increasingly widespread. It is known that even small changes in the microbial communities in the rhizosphere may affect yield (Sun et al., 2021). A practical approach to bolster the rhizosphere is to employ microbial fertilizers that transform essential nutrients such as N, K, and P into plant-usable forms. Conventional farming practices rely on NPK fertilizers in low cost, inorganic forms, which however ultimately deteriorate farmland productivity. When these fertilizers are used, the soil pH, irrigation type, rainwater content, and microbial content can convert elements into compounds unavailable for the plant. For this reason, the soil needs microorganisms to transform these elements into usable forms (Glaser & Lehr, 2019).

Plant growth-promoting rhizobacteria (PGPR) are the most investigated among the rhizosphere microbiomes components. They promote plant growth and protection from some plant pathogens. Intracellular symbiotic PGPRs promote nodulation in roots and colonize these nodules. Many non-rhizobial endophytic taxa, such as members of *Arthrobacter*, *Pseudomonas*, *Micromonospora*, *Mycobacterium*, *Agrobacter*, *Pantoea*, *Bacillus*, and *Enterobacter*, can colonize the nodules, in addition to rhizobial bacteria like *Rhizobium, Azorhizobium*, and *Sinorhizobium* (Leite et al., 2017; Benitez et al., 2017; Hakim et al., 2021). Some species of *Bacillus*, *Pseudomonas*, *Serratia*, and *Azobacter* belong to the so-called extracellular PGPRs and are found in the soil close to the roots. In addition to PGPRs, other important bacterial components of the rhizosphere include actinomycetes (Muleta & Assefa, 2018; Qu et al., 2020).

The rhizosphere microbiome, as a holobiont, is particularly important in understanding the role of microorganisms that evolved and specialized together with plants. The microorganisms, which have a tremendous taxonomic and functional diversity thanks to their reproductive cycles and rapid evolution with related environmental adaptations, form unique holobiont communities. These are shaped by biotic and abiotic environmental factors reflecting the needs of each plant species or even individual microbes, when interacting with plants. Among these communities, culturable bacterial species are undoubtedly the most studied. Thanks to the swift advancements in molecular techniques, a more comprehensive range of bacteria can now be detected and analyzed. Nevertheless, the emphasis and methodology of these investigations differ. The main reason is that if cultured, it is possible to determine the functions of these bacteria or their relationship with the plant. Still, it is also possible to support the development of other plants by inoculating microbiomes in which these bacteria are not included.

The rhizosphere microbiome is pivotal in bolstering plant productivity and ensuring sustainability within agricultural systems. As mentioned earlier, studies have proven that rhizosphere microorganisms play a crucial role in enhancing nutrient absorption, promoting plant growth, strengthening disease resistance, and augmenting soil fertility (Dries et al., 2021). Plants establish a symbiotic relationship with microorganisms that bestow advantageous effects, ultimately leading to a discerning microbial consortium within the soil compartments. Consequently, the rhizosphere biome gradually distinguishes itself from the broader soil microbiome in which the plant is situated. These advantageous microorganisms play a crucial role in enhancing nutrient uptake by solubilizing minerals and generating enzymes that break down organic matter, thus rendering nutrients more readily available to plants. The rhizosphere layer in which the bacteria are located also determines their relationship with the plant. Such culturable flora represents the forefront resource to harness the potential of crop microbiomes to improve food security.

Utilizing state-of-the-art technology, scientists aim to improve plant growth and combat plant diseases by investigating the biological functions of plant microbiomes in depth. For example, in the root systems of legume plants, mycorrhizal fungi and rhizobacteria cooperate with host plants to overcome nutrient deficiencies in the soil, including NPK and other elements (Cordovez et al., 2019; Oldroyd & Leyser, 2020). Especially in the case of bacteria responsible for N_2 fixation, studies on this subject increased rapidly in recent years as the use of chemical fertilizers decreased. Some plant-associated taxa such as *Rhizobium*, *Bacillus*, *Azospirillum*, *Pantoea*, *Streptomyces*, *Flavobacterium*, and *Pseudomonas* play an essential role in nitrogen fixation by enriching soil through symbiotic relationships with roots. Similarly, also in other crops, it was possible to observe the potential of endophytic bacteria in i.e., banana plants to develop biofertilizers, thus promoting sustainable productions (Ngamau et al., 2014). Similar studies can be found also in soybean (Vargas-Día et al., 2019), maize (Jha, 2022), canola (Jamalzadeh et al., 2021), or ornamental plants such as *Amaranthus* (Yashaswini et al., 2021) and halophilic succulents (Dragojević et al., 2023). Biofertilizers are thus recognized for their safety and environmental friendliness, offering a viable alternative to agrochemicals, such as synthetic fertilizers and pesticides.

Moreover, the rhizosphere bacteria can produce phytohormones that promote plant growth and development. For example, some bacteria can stimulate root growth and increase nutrient uptake by producing auxin. In addition, some bacteria can produce cytokinins and gibberellins that promote shoot growth and flowering. On the other hand, other rhizosphere bacteria can degrade phytohormones, thus regulating plant growth and development. Furthermore, other rhizosphere bacteria may have biocontrol functions that help protect plants against soil-borne pathogens (Banerjee et al., 2018; Zhao et al., 2018; Patel et al., 2019; Anzalone et al., 2022).

The rhizosphere bacteria support plant growth and soil health through their potent enzymatic properties. They exhibit various enzymatic activities vital for nutrient cycling, organic matter breakdown, and overall ecosystem stability. Some of the key enzymatic properties include cellulase production, nitrogen fixation, phosphatase activity, nitrification and denitrification capabilities, siderophore synthesis, organic matter decomposition, phytohormone generation, as well as the production of antimicrobial and other extracellular enzymes. Understanding the enzymatic properties of rhizosphere bacteria is essential as it provides insights into their contributions to nutrient cycling, organic matter breakdown, and overall soil. For example, bacterial cellulases in the rhizosphere are functional in breaking down cellulose, a major component of plant cell walls, into simpler sugars that plants can take up.

Within the rhizosphere, the bacterial enzymes that hold significant importance are responsible for nitrogen fixation, phosphate solubilization, and K solubilization. Nitrogen fixation plays a crucial role in the rhizosphere as certain bacteria can convert atmospheric nitrogen into a usable form for plant growth and development (Purahong et al., 2019). In addition to their enzymatic properties, rhizosphere bacteria also play a significant role in the availability and cycling of other essential nutrients, such as solubilized P and K, making them more accessible to plants (Gupta et al., 2020; Inbaraj, 2021).

The rhizosphere microbiome allows the release of a wide range of compounds, including phytohormones, siderophores, and antimicrobial metabolites, that are essential for the plant health and vitality. Among them the siderophores stand out as specialized molecules that facilitate the absorption of iron, a vital nutrient that is essential for various physiological processes and growth.

Produced by rhizosphere bacteria, these molecules serve as an essential nutrient for both the plant and the bacterial community, highlighting the network of intricate and mutually beneficial relationships between plants and associated microorganisms (Hassan & Ajaj, 2022). On the other hand, antimicrobial metabolites or enzymes protect plants because of their inhibitory effects on pathogens (Verma et al., 2018; Gao et al., 2022). Plant growth and development are heavily influenced by phytohormones, a group of organic compounds that includes auxins, cytokinins, and gibberellins. Plant hormones, also known as phytohormones, regulate several physiological processes such as cell differentiation, elongation, and nutrient distribution. Auxins promote cell elongation and apical dominance, whereas cytokinins enhance cell division and prevent senescence. Gibberellins, on the other hand, promote stem elongation and seed germination. The complex interactions among these hormones are crucial for the overall growth and development of plants (Sodiq et al., 2021; Fernandes et al., 2022).

6.5 The Role of Rhizospheric Fungi: Functions and Significance in Plant Growth and Development

The microbiome provides a healthy and sustainable environment for the plant and contributes to its overall longevity and productivity, thanks to its complex neighborhood. Although bacteria are the most studied rhizosphere elements, other microorganisms are also known to be vital. Fungi are another group of microorganisms whose role in the rhizosphere is increasingly recognized, due to their high adaptability in soil ecosystems and their different metabolic products.

Fungi are widespread in soil and can grow at different pH and temperature. They are renowned for their dominance in the rhizosphere microbiome and superior performance in biogeochemical cycles, especially in arid environments and soils with acidic or alkaline properties. For example, fungi and Archaea (the second prokaryote Kingdom) are more functional than bacteria in acidic soils where semi-arid crops are grown, and solubilize phosphate better than bacteria under acidic soil conditions (Pretscher et al., 2018; Farhadi et al., 2023). Recent research has shown that fungi play a vital role in the plant rhizosphere beyond their physical presence.

Unlike bacteria, using fungi in plant growth is also possible by direct extraction. Research has shown that using yeast extracts derived from *Saccharomyces cerevisiae* can effectively enhance plant growth through an environmentally friendly approach. This highlights the importance of harnessing these natural resources potential to sustain agricultural productivity (. Yeasts are a natural source of cytokines that stimulate cell division and growth, as well as protein, nucleic acid, and chlorophyll II synthesis. Furthermore, their extracts improve plant growth by activating photosynthesis, increasing CO_2 usage, increasing protein and vitamin B contents, and producing natural plant growth regulators (e.g., auxins, gibberellin, and cytokinins) (Haider et al., 2021; Hamed et al., 2022; Farhadi et al., 2023).

Root mutualistic fungi play an essential role in maintaining soil fertility through many mechanisms. In addition to mycorrhizal fungi, fungal components of the rhizosphere microbiome play a crucial role in promoting plant growth and development, through increased nutrient acquisition. Furthermore, fungal enzymes in the rhizosphere can convert complex organic compounds such as cellulose and lignin into simpler forms that plants can easily absorb. Mycorrhizae are among the oldest and most well-documented symbiotic associations between fungi and plants, common in many plant species. Arbuscular mycorrhizal fungi (AMF) are essential elements of the rhizosphere microflora in natural environments and significantly influence nutrient cycling. They help to absorb plant nutrients, especially P, which can be present in an unavailable form in soil. AMF provide an important contribution to sustainable agricultural practices by promoting the growth and productivity of host plants (Rashid et al., 2016; Wen-Ya et al., 2021).

By altering the morphology or anatomical structure of roots and improving the physical and chemical properties of the rhizosphere environment, AMF also stimulate the plant to produce different metabolitesthat sustain photosynthesis, plant growth, and resistance. In addition, AMF suppress colonization by pathogenic bacteria and fungi since the flora competition is also involved. Studies carried out on many different plant species such as *Citrus reticulata*, *Fragaria ananassa*, *Cucumis sativus*, *Zea mays*, *Impatiens glandulifera*, and *Solanum tuberosum* showed that AMF reduce the impact of not only on competitive mycoflora but also of many diseases and pests, including nematodes (Manresa-Grao et al., 2022; Cockerton et al., 2020; Ravnskov et al., 2020; Zhang et al., 2023; Razak & Gange, 2023; Valdebenito et al., 2022). In summary, AMF colonization reduces the need of fertilizers, insecticides, and fungicides supporting plant growth and inter-plant communication, as well as protect the plant from parasites, strengthening its immune response.

Apart from the AMF, another critical group of fungi are ectomycorrhizal fungi (EMF). These play an active role in the element cycles in the soil, especially in the N and P balance, and also help to protect the plant from environmental stress (Kuyper & Suz, 2023). In a study conducted on *Pinus tabulaeformis* under drought stress, EMF promoted more prolonged survival of seedlings and increased their biomass. The study also found that the plant increased the production of soluble sugars in the presence of EMF under drought stress, and increased their ratio to starch in all tissues (Wang et al., 2021). EMF can increase plant resistance not only during drought but also in other circumstances. Sun et al. (2022) found that EMF increased Cd tolerance by strengthening the antioxidant defense response in oak (*Quercus acutissima* Carruth.) seedlings. The study showed that, in seedlings with EMF-inoculated soil, the level of Cd (which stimulates the formation of reactive oxygen species (ROS) that cause oxidative stress in plants) increases, the production of enzymes responsible for antioxidant activity increases. In addition to Cd, EMF play an essential role in strengthening tolerance to Al and other heavy metals (Dagher et al., 2020; Dhalaria et al., 2020; Liu et al., 2020, 2022; Riaz et al., 2021). Zhu et al. (2022), suggested that EMF may accelerate the restoration of vegetation in areas contaminated by mining activities. As a result of the study conducted in the restoration area, it was reported that EMF diversity varied depending on the plant species and that there was a positive correlation between soil parameters, important for plant growth, and many EMF phyla or genera.

In summary, rhizosphere fungal biota, which is shaped by many biotic and abiotic factors, play a vital role in plants health, both through the breakdown of complex soil compounds and in the inactivation of heavy metals that adversely affect plants growth. These vital components of the rhizosphere are active in communication, promoting the secretion of plant metabolites and defense against pathogens. In addition, their tolerance to environmental stresses suggests that fungi will have a more active role than bacteria in many environments.

6.6 Other Rhizosphere Microbiome Components

The rhizosphere microbiome is composed of various microorganisms, including bacteria, fungi, viruses, archaea, protozoa, and algae (Coats & Rumpho, 2014). Recent research has revealed that Archaea, a distinct group of microorganisms, play a critical role in plant development and ecosystems functioning. Specifically, the Archaea present in the rhizosphere microbiome are essential in facilitating the interaction between plants and microbes. Rhizosphere Archaea have been found to play a role in the nitrogen cycle as they can participate in processes such as nitrification and denitrification. Furthermore, the rhizosphere Archaea may contribute to nutrient cycling by participating in sulfur oxidation and methane production (Liu et al., 2016).

Research on the rhizosphere microbiome has primarily focused on bacteria and fungi, while studies on Archaea have been limited. Understanding the role of these microorganisms in the rhizosphere microbiome is crucial for understanding nutrient cycling and overall soil health. In addition, these organisms — of which extremophilic examples are frequently encountered with their adaptation to stress conditions — are active in almost every cycle. In this way, it is expected that they are one of the crucial elements of the rhizosphere microbiome in projects such as the reclamation of heavy metal contaminated areas. Ma et al. (2019) examined archaeal and bacterial diversity in Hg-contaminated rice fields. Methylated Hg (MeHg), which is highly toxic to living organisms, is generally produced in nature from elemental or inorganic mercury through biotic and abiotic processes. Sulfate-reducing bacteria (SRB), iron-reducing bacteria (IRB), and methanogens are particularly effective in biotic pathways. However, organisms with high MeHg resistance usually have enzymes to cleave the C-Hg bond. The diversity of bacteria and Archaea in soil showed the occurrence of the dsrB (SRB), merA (Hg-resistant microorganisms), hgcA (Hg-methylating microorganisms), and mcrA (methanogens) genes. Both Archaea and bacterial species present in the sampled soil environments were found to be active in Hg methylation, being provided with a MeHg resistance potential.

Archaea, which utilize alternative energy sources, can exist in aerobic and anaerobic environments. They can regulate the elemental cycle in terrestrial and aquatic rhizospheres and, in turn, the distribution of other microorganisms in the community. Therefore, they also indirectly regulate plant growth. The phyla Euryarchaeota and Thaumarchaeota are the most frequently studied archaea and, therefore, the most well-known. Their members are known to regulate element cycles through various metabolisms, ranging from methane producers to sulfur and ammonia oxidizers or P solubilizers. In addition, they are becoming increasingly the centre of attention in sustainable agriculture with the production of phytohormones, such as indole-3-acetic acid (IAA) and zeatin, siderophores, and bioremediation functions (Anderson et al., 2011; Haroon et al., 2013; Jiao et al., 2019; Zhang et al., 2019). Another important archaeal metabolite recently identified is 1-aminocyclopropane-1-carboxylate deaminase (ACCD). This enzyme is responsible for the degradation of ACC, the precursor compound of ethylene, a phytohormone produced under biotic and abiotic stress conditions, converted to α-ketobutyric acid and ammonia. The genetic locus of the enzyme, previously thought to have been exclusive to bacterial organisms, has been found in Archaea, including *Archaeoglobus fulgidus*, *Pyrococcus abyssi* and *P. furiosus* (Fujino et al., 2004; Naitam & Kaushik, 2021).

Another essential component of the rhizosphere is represented by viruses.. Their significance in soil ecosystems became increasingly evident when considering they represent one-fifth of the biomass in the oceans with an estimated 70-fold population in terrestrial environments (Muscatt et al., 2022). Viruses play a crucial role in all cycles, and as such are also defined as the most significant genetic reservoir (Kimura et al., 2008; Buee et al., 2009). Besides their functions in elementary cycles, viruses such as the phages (pathogens of bacteria) are important in transferring genetic information with their lysogenic role. The rhizosphere of agricultural soils boasts a significantly richer phage diversity, which can be attributed to the large bacterial populations thriving in these areas (Ashelford et al., 2003; Pratama et al., 2020). It should also be noted that the phages can support plant growth, especially by suppressing the populations of pathogenic bacteria (Yang et al., 2023c). Metagenomic data indicate that viruses can influence C, N, S, and phosphate biogeochemical cycles, as well as the photosynthetic capacity by reprogramming their hosts metabolism through the expression of auxiliary metabolic genes (AMGs) (Roux et al., 2019; Middelboe & Brussaard, 2017; Bi et al., 2021).

Apart from phages, another important virus group is the archaeal viruses. Studies have revealed that they have more complex morphologic arrangements than phages and other viruses (Prangishvili et al., 2006). However, studies on the soil virosphere need to be more extensive, as this area needs to be studied more in detail to better understand their rhizosphere functions. The potential of viruses as biological agents such as herbicides, insecticides, mainly thanks to their host-specificity, has led to a better understanding of their distribution and functionsin soil ecosystems.

Another essential component of the rhizosphere is given by the protozoa. These organisms are unicellular eukaryotes identified as primary consumers of bacteria and fungi in soil. They also include plant and animal pathogenic species. They are thought to provide homeostasis by suppressing the bacterial and fungal populations that can get out of control due to high amounts of plant-derived nutrient inputs in soil (Gao et al., 2019). While protozoa, which feed on pathogenic microorganisms, contribute to plant health in this way, they also contribute to plant development by increasing soil mineralization.

Encouraging the production of secondary metabolites in bacteria and fungi that are helpful to plants is one way to improve plant growth. Additionally, this process induces a stress on predators, leading to increased secretion of beneficial compounds. Research has shown that protozoa play a role in stimulating the production of phytohormones and microbial-derived antibacterial agents and regulators such as siderophores (Weidner et al., 2017).

Microalgae and cyanobacteria are a further rhizosphere group worth mentioning. These eukaryotic and prokaryotic groups are responsible for photosynthesis and contribute to the C and N cycles in soil, supporting development of plants. It is also known that microalgae support beneficial bacterial populations in soil, regulating important parameters for homeostasis such as pH, salinity, and electrolyte balance. They also increase seed germination rates (Renuka et al., 2018). The algae can also produce various phytohormones such as auxins, cytokinins, gibberellins, jasmonic acid, abscisic acid, and ethylene (Tan et al., 2021). In addition to microalgae, which can be found in the rhizosphere microbiome, marine microalgae can also be used to sustain plants development, thanks to their extremely rich nutrient content as biofertilizers (Wally et al., 2013; Lv et al., 2020; Neveux et al., 2020).

Cyanobacteria, like microalgae, also contribute to plant development and shaping of the rhizosphere microbiome. They were initially thought to be effective in the wetland rhizosphere, but have been proven to exist in all types of soils, where they are functional. These organisms are involved in many activities such as increasing C content of soil, phosphate transformation, fixing atmospheric N_2, nutrient mobilization, synthesis of some phytohormone derivatives, and even remediation applications. They were found to support the development of various crops, including rice, wheat, corn and tomato (Jäger et al., 2010; Gayathri et al., 2015; Hussain et al., 2015; Mohan & Kumar, 2019). Cyanobacteria, which release plant growth-promoting substances such as auxin, gibberellins, cytokinin, ABA, several vitamins and amino acids, as well as compounds such as polypeptides of antibacterial nature, are also functional in suppressing plant pathogens in soil. The use of cyanobacteria in agricultural lands is not limited to their role in supporting plant growth, as they can also reduce the emission of greenhouse gases. For this reason, cyanobacteria inoculation into soil should become increasingly common among sustainable agricultural practices (Srivastava et al., 2021).

6.7 Significance of Rhizospheric Microbiome

The discovery of new microbial metabolites is essential for the control of resistant pathogens, diseases, and toxic compounds. For this reason, scientists search and isolate microorganisms in many different environments (i.e., caves, sea organisms, deep sea etc.) and evaluate their activities (Rachid & Güngör, 2023; Aydoğdu et al., 2023, Yang et al., 2023a). Recently, the rhizospheric microbiome has been added to these environments (Kour et al., 2020; Niazi et al., 2023). The rhizosphere is one of the most complex ecosystems with a broad range of specialized interactions among soil, microorganisms and plants. These special areas are natural microenvironments where microorganisms are very dense. In addition, the rhizosphere interface shows a highly interactive microenvironment, with high levels of metabolic activities and functions. The metabolic and catabolic functions of rhizospheric microorganisms indicate that they are a valuable reservoir of bioactive compounds with a potential for huge industrial and biotechnological falllouts (Fig. 6.1) (Mokrani & El-Hafid, 2020). From agriculture, environment, to various industries, the significance of this microbiome is outstanding.

6.7.1 Therapeutic Uses of Rhizosphere Microbes

Research on this topic is mostly on the effects of rhizospheric microbial diversity on plant health (Kour et al., 2020). Studies on bioactive substances, including secondary metabolites produced by rhizosphere microorganisms, for therapeutic and industrial applications have recently started to increase, and are expected to provide many new metabolites. Apsari et al. (2019) investigated the antimicrobial effects of actinomycetes isolated from rhizosphere soil on human pathogenic bacteria. These authors tested the supernatants of actinomycetes cultured in liquid medium and crude extracts on urinary tract infection bacteria (*Acinetobacter calcoaceticus*, *Klebsiella pneumoniae*, *Citrobacter braakii* and *Escherichia coli*) and found that both the supernatant and extract were effective against these bacteria.

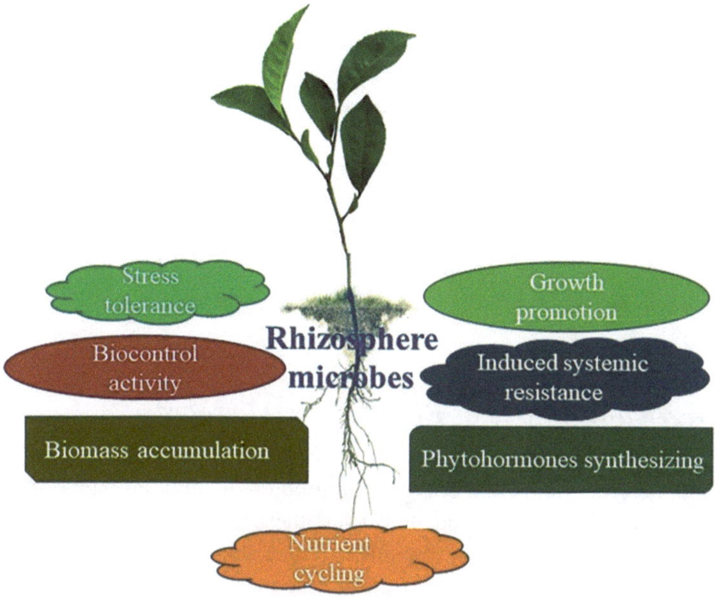

Fig. 6.1 Biotechnological application of rhizosphere microorganisms

Aspergillus niger AK6 isolate showed moderate antimicrobial activity against pathogens such as *Staphylococcus aureus*, *Shigella flexneri*, *Bacillus subtilis*, *Klebsiella pneumonia*, *Escherichia coli*, and *Candida albicans*, It showed strong antifungal activity (with 47.2, 59.4 and 64.1% inhibition) against *Sclerotium rolfsii*, *Cercospora canescens* and *Fusarium sambucinum* phytopathogens, respectively. In addition, *Aspergillus niger* AK6 extract also exhibited anticancer activity against MCF-7 human tumor cell line, with an IC50 value of 102.01 μg/mL. This study shows that rhizosphere microorganisms have a potential not only for agricultural applications but also for medical uses as new antifungal, antimicrobial and anticancer drugs (Niazi et al., 2023).

6.7.2 *Significances of Rhizosphere Microbes in the Environment*

Contamination of water and soil with toxic metals affects soil fertility, water quality, agricultural production and thus the ecosystem. Heavy metal pollution in the ecosystem seriously threatens human health, with a global pollution burden that is increasing. Heavy metal pollution is the result of intensive use of fungicides and herbicides, including industrialization, mining, industrial production and rapid urbanization (Kumar et al., 2023). Heavy metals spread in the environment originate from both natural (volcanic eruptions, forest fires, weathering of rocks, etc.)

and anthropogenic (industrial waste, agricultural fertilizers, mining, burning of fuels, etc.) sources (Klonowska et al., 2020; Xia et al., 2018; Liang et al., 2017; Cheng et al., 2018; Wu et al., 2020).

One of the methods used to reduce heavy metal pollution or convert it into harmless forms is phytoremediation. Studies have shown that rhizosphere microorganisms are largely beneficial in phytoremediation, by increasing metal solubility, producing plant growth promoters, organic acids, and polysaccharides. It has also been shown that the combination of bioremediation and phytoremediation can be an effective approach for replanting contaminated soils (Anerao et al., 2022; Pouresmaieli et al., 2022; Liu et al., 2022; Racić et al., 2023; Jiao et al., 2023).

6.7.3 Agricultural Use of Rhizosphere Microbes

To enhance the crop yields and contrast the food crisis, high amounts of chemical fertilizers and pesticides are used. However, these chemicals are harmful to the environment, humans, and other organisms since they enter the food chain and interfere with the environmental microflora. The need of biological productions is hence necessary and urgent. According to their interaction with the host plants, the rhizosphere microorganisms can be classified as harmful and beneficial (Yu et al., 2019). Figure 6.2 illustrates the many advantages provided by rhizospheric microbiomes. Beneficial microorganisms promote the growth of plants, increase crop yield or/and protect the host from diseases and pathogens (Qu et al., 2020; Li et al., 2021a). The rhizosphere microbial community, depending on its composition, host species, root genotypes and phenotypes, as well as environmental factors, can inhibit pathogens through different mechanisms. These include antagonism or reinforcement of the plant immune system by inducing systemic resistance (ISR) in the host (Qu et al., 2020; Li et al., 2021a, b). It is reported that theorganic acids, sugars, secondary metabolites, and amino acids present in the root exudates control the

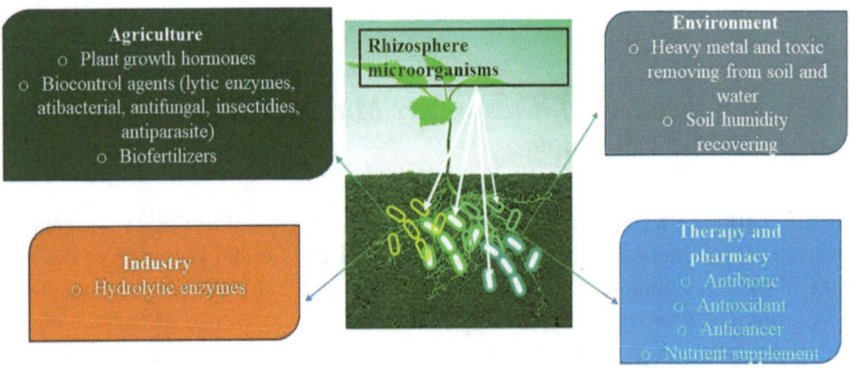

Fig. 6.2 Potential significance of rhizosphere microorganisms

nutrition and recruitment of soil microbiome to the plant (Preece & Peñuelas, 2020; Sun et al., 2021; de Faria et al., 2021). Even though actual knowledge on the recruitment of specific taxa is still limited, the diversity of rhizosphere microbiome appears defined by the root exudates (Zhalnina et al., 2018; Sun et al., 2021).

Plants benefit from the different biological compounds secreted by rhizosphere microbes. Extra-cellular lytic enzymes, secondary metabolites such as antibiotics, plant growth hormones like IAA and GA3, volatile organic compounds, and siderophores are among the bioactive compounds that act significantly in plant growth promotion and biocontrol (Subrahmanyam et al., 2020). Microbial extracellular enzymes are known for their capacity in the lysis of polymers present in the pathogens cell walls, membranes, or other cellular parts. These include chitin, one of the main natural polymers, constituent of insect shells and fungal cell walls (Karlsson & Stenlid, 2008). For example, bacterial strains and a *Trichoderma hazarnium* isolate proceeding from the rhizosphere of sugarcane showed an antifungal activity *vs* the sugarcane fungal pathogen *Colletotrichum falcatum,* which is attributed to their chitinase secretion ability (Viswanathan et al., 2003). Furthermore, authors observed an increase secretion of N-acetylglucosaminidase and β-1,3- glucanase when *Trichoderma hazianum* strain T5 was inoculated on a medium containing chitin or hyphaeof the fungal pathogen. These observations evidenced the mycolytic effect of these extracellular enzymes, characterized by high lytic effects (Viswanathan et al., 2003).

Atif et al. (2023), demonstrated the insecticidal activity of the culture filtrate of high chitinase producing *Streptomyces* spp. strains, isolated from eggplants rhizosphere soil. It was shown that these actinobacteria reduced the number of root-not nematode *Meloidogyne incognita* and inhibited egg hatching. The highest chitinase producing strain showed the highest egg hatching suppression, enhancing plant growth, and reducer nematode numbers. *Stenotrophomonas maltophilia* W81, isolated from the rhizosphere of sugar beet, inhibited the growth of the fungal pathogen *Pythium ultimum.* This protective effect was associated to the bacterium protease production (Dunne et al., 1997). In another study, Feng et al. (2022), showed that isolates of *Streptomyces, Pseudomonas, Saccharothrix*, and *Nocardiopsis*, proceeding from the rhizosphere of resistant potato plants, inhibited the agent of potato late blight, *Phytophthora infestans*. The cellulase and catalase activities observed for these isolates were thought to sustain the inhibitory activity. In addition, the density changes were observed for some microbial species after treatment with these strains (Feng et al., 2022). *Trichoderma asperellum* was also shown to act as a potential biocontrol agent *vs* soil-borne *Fusarium* spp. pathogens (Win et al., 2021). This antagonistic activity was attributed to the production of cell wall lytic enzymes, including chitinase and β-1,3-glucanase in cultures containing the phytopathogen. Nine weeks after application of the beneficial fungi on banana plantlets previously infected by *Fusarium oxysporum*, the disease severity was reduced by 70% (Win et al., 2021).

Antagonism of rhizospheric microorganisms towards phytopathogens is also attributed to secondary metabolites with antimicrobial and pesticide-like compounds. Several fungi and bacteria inhibited growth of phytopathogens by producing different antimicrobial compounds. Antimicrobial properties of *Streptomyces* spp. isolated from black pepper rhizosphere was attributed to their biocontrol activity. The culture filtrate of some *Streptomyces* isolates showed antifungal properties *vs* two severe fungal pathogens such as *Phytophthora capsici* and *Sclerotium rolfsii*. Compounds like brefeldine A, salfredine B11, fusaric acid, and dermadin were some of the antibiotics identified from some bacterial extracts. Other antifungal compounds identified consisted of isonitrinic acid E, harianopyridone 1 and 2, enniatin B, and brevianamide F (Anusree et al., 2019).

Trichoderma isolates proceeding from the rhizosphere of maize, banana, and cassava were investigated for their effects *vs* some strains of *Fusarium* isolated from diseased rice plants and oak leaves. The antifungal activity was observed after dual culture experiments, and *Trichoderma* culture filtrate assays. In the dual culture trial *T. koningiopsis* showed high inhibitory effects *vs F. tricinctum* F2B by 31.17%, whereas *T. erinaceum* inhibited *F. proliferatum* 296 by 47.37%. Like other *Trichoderma* species, the tested fungal strains are considered to act as bio-agents *vs* the pathogenic *Fuasarium* strains (Olowe et al., 2022).

Metagenomic analysis of rhizospheric microbes may help to identify biosynthesis gene clusters (BGCs) encoding secondary metabolites of diverse functions, including non-ribosomal polypeptides (NRPs), polyketides (PKs), terpenes, and alkaloids (Pattnaik & Busi, 2019; Zhou et al., 2021).. In a study by Zhou et al. (2021), rhizospheric *Bacillus* strains were screened for antimicrobial activities against phytopathogenic bacteria, fungi, and oomycetes. Antagonism by some of these strains was shown *vs Verticillium dahliae*, *Pseudomonas syringae*, *Phytophthora infestans*, *Botrytis cinerea, Rhizoctonia solani,* and *Erwinia carotovora*. The identification of 120 BGCs encoding secondary metabolites in 10 *Bacillus* strains, including bacteriocins, terpenes, PKs-NRPs, PKs, and NRPs led to the identification of genes encoding known compounds such as fencing, petrobactin, bacillary, and subtilin. Since these *Bacillus* spp.were isolated from tomato rhizosphere soil, they were considered as suitable candidates for manufacturing biocontrol agents and products for tomato disease control. Moreover, they represent a reservoir of novel antimicrobial and other secondary metabolites (Zhou et al., 2021). Khan et al. (2021) also identified, by high-performance liquid chromatography, lipopeptide surfactin from the culture extract of *Bacillus thuringiensis*. This secondary metabolite showed significant antagonism against the phytopathogenic fungus *Fusarium graminearum*. The *Bacillus* sp. was isolated from the rhizosphere of maize (*Zea mays*). Use of surfactin was considered to increase yields of maize and other cereal crops, which are often affected by this pathogen (Khan et al., 2021).

Eight *Trichoderma* isolates obtained from rhizosphere soil of solanaceous crops viz. tomato, brinjal and chili, were investigated for their antimicrobial activities against fungal and bacterial pathogens. Results showed that all isolates had antagonistic effects *vs* the tested phytopathogenic fungi *Fusarium oxysporum* f. sp. *lycopersici*, *Colletotrichum gloeosporioides*, *Rhizoctonia solani*, and the bacterium *Ralstonia solanacearum*. The highest antimicrobial effect was observed for the *T. harzianum* strains NBG and MC2. Among the antibacterial, antifungal and antibiotics identified from the culture extracts of the NBG strain, authors found trichodin A, viridiol, aternariol monomethyl ether, 5-hydroxyvertinolide, pyridoxatin, and Isoechinulin A. Those antimicrobials identified from the culture extracts of MC2 strain included 3-hydroxytrichodenone C, viridin, cyclopenol, gliotoxin, cremenolid, and pyridoxatin (Rahman et al., 2023).

Elshafie et al. (2023) isolated *Streptomyces* sp. (Act1), *Streptomyces atratus* (Act2), and *Arthrobacter humicola* (Act3) from the rhizosphere of basil, *Ocimum basilicum* L. They showed a high reduction in disease symptoms of "black leg" necrotic stem lesions caused by *Fusarium tabacinum* in this plant. In addition, the antimicrobial activity of cell-free culture filtrates obtained from these isolates was determined. The presence of 20 different components in the chemically characterised filtrate revealed that some important secondary metabolites such as Rhizocticin A, Eponemycin and N-Acetyl-l-histidinol were synthesised.

The potential of *Trichoderma asperelloides, T. asperellum* and *T. longibrachiatum* isolated from the rhizosphere of coffee (*Coffea arabica*) plants in Ethiopia was investigated for the biocontrol of coffee wilt disease caused by *Fusarium xylarioides* on coffee plants. Data showed that *T. asperellum* and *T. longibrachiatum* reduced the mycelial growth of *F. xylarioides* by over 80%. Thus, these isolates play a major role in reducing the disease severity with an applied potential in the biocontrol of coffee wilt disease (Mulatu et al., 2022).

Plant growth promotion attributed to the rhizosphere microbes is realized through different mechanisms as previously introduced. To enhance the crops production and resistance against abiotic stresses, microorganisms present in the rhizosphere areas can respond to the plant needs by solubilizing minerals such as K and phosphate, fixing nitrogen in roots, producing siderophores, or secreting plant hormones (de Faria et al., 2021; Subrahmanyam et al., 2020; Pattnaik & Busi, 2019). Plant growth hormones have different functions, increasing root surface and length, to facilitate the nutrient uptake from soil, seed germination, and the development of stem, leaf, flower, and fruit (Subrahmanyam et al., 2020).

One of the most important hormones in plants is IAA. It consists of an auxin, naturally secreted by plants or by microorganisms for the plants. IAA accelerates plant growth by expanding the root surface and length (Tsavkelova et al., 2005; Larekeng et al., 2019; Ahemad & Khan, 2012). Different rhizosphere microorganisms have been reported to produce phytohormones. Different species of *Rhizopus, Fusarium, Trichoderma, Gliocladium,* and *Aspergillus*, isolated from the rhizosphere of *Swietenia mahagoni* stands, were observed with IAA production ability. The highest production was observed in the genus *Trichoderma* (Larekeng et al., 2019). To improve the significant production in hydroponic farming systems, plant growth promoting fungi are reported to be used under normal and stress conditions (Tarroum et al., 2022). Tarroum et al. (2022) observed the production of IAA in the *Penicillium olsonii* isolated from the rhizosphere of the halophytic plant *Aeluropus lithoralis*. Its plant growth-promoting effect was shown by treating tobacco plants in a hydroponic farming system with both fungal mycelium and a cell-free culture filtrate. The resultsshowed a significantly increased leaf area, with higher dry weight, height, and total pigment (chlorophyll) content. Additionally, the K content in the leaves of treated plants increased, while the Na accumulation in roots decreased. The fungus also induced the expression of salt stress related genes in the treated plant, and allowed the reduction of the synthetic fertilizer inputs by 50% (Tarroum et al., 2022). Also some bacteria isolated from the rhizosphere of pigeon pea, wheat, and maize plants including *Pantoea dispersa* JRBHU3, *Enterobacter hormaechei* JRBHU9, and *Bacillus subtilis* JRBHU11 showed high IAA production (Prasad et al., 2022).

After IAA, gibberellin (GA3) is the second phytohormone studied in the rhizosphere microbiome. It is thought to participate in plant growth mechanisms by enhancing the nodule formation, seed germination, as well as flower, fruit, stem, and leaf differentiation (Subrahmanyam et al., 2020; Kour et al., 2020).

Other phytohormones include cytokinins, that are derived from adenine. Apart higher plants, these hormones are mostly secreted by algae, fungi, and bacteria (Li et al., 2021a, b). Their mechanisms of action consist of plant growth promotion, improvement of plant physiological activities, and regulation of growth under abiotic stresses. In this context, they participate in the development of the shoot, root, chloroplast, and vascular system. They also play a role in cell multiplication, nutrient distribution, and control of the plant at high or low temperature, salts, drought, and other environmental stresses (Li et al., 2021a, b; Subrahmanyam et al., 2020; Kour et al., 2020). Rhizospheric fungi and bacteria which synthesize these phytohormones include *Penicillium citrium*, *Aspergillus fumigatus*, *Azospirillum lipoferum*, and *Proteus vulgaris* (Gu et al., 2023; Khan et al., 2011; Creus et al., 2004; Bhattacharyya et al., 2015).

Nutrients uptake and distribution are fundamental for growth, development, and yields. Nitrogen is an essential element needed by plants. However, few plants can fix the atmospheric nitrogen in their bodies (Sharma et al., 2022). This element is often provided as chemical fertilizers but the consequences that they cause. i.e. environmental pollution, promoted the search of alternative biological pathways. The biological way to offer nitrogen to plants is to use microorganisms that can provide nitrate or ammonium, the two forms of nitrogen directly assumed by plants. Some microorganisms have nitrogenase enzymes system, including the presence of the *nif*H gene, among others (Singh et al., 2020). Several bacterial isolates obtained from the rhizosphere of sugarcane showed the ability to fix N_2. Furthermore, the *nif*H gene was identified in the isolates, with the most N-fixing being *Bacillus megaterium* and *B. mycoides* (Singh et al., 2020). Karthika et al. (2020) showed that *Bacillus* sp. KTMA4 obtained from tomato rhizosphere had different plant growth promoting traits, including nitrogen fixation, as shown by high N-accumulation in the *Bacillus* treated plant (Karthika et al., 2020). In addition, *Erwinia rhapontici* EU-B1SP1 and *Pseudomonas gessardi* EU-LWNA-25 isolated, from cereal crops rhizosphere, were observed to fix nitrogen and solubilize phosphate (Devi et al., 2022). Yeast strains isolated from cherry tomato and identified as *Candida* spp. also showed a N-fixation ability by producing ammonium under nitrogen free environments (Chen et al., 2022a, b, c). All these rhizosphere microbes can hence be used as potential bio-fertilizers.

The role of siderophores in plnt growth is remarkable. These compounds provide iron, an essential element. Their structures are suitable to chelate iron from organic and inorganic iron compounds present in soil (Sarwar et al., 2022; El-Maraghy et al., 2020). The ferric-siderophore complexes are uptaken by roots and then provide iron to both plants and other herbivores, including humans (Sarwar et al., 2022). These molecules are also play fundamental in the bioremediation of toxic heavy metals. It is reported that iron and other heavy metals are in competition with siderophores (Tóth et al., 2016; Khaksar et al., 2016; El-Maraghy et al., 2020). Microbial siderophores are mostly reported, and rhizosphere microbial siderophores are not excluded. Solubilization of mineral nutrients is a physiological activity that provides essential minerals to plants. Even if these minerals are required in low concentrations, their absence can affect the development of crops and reduce yields (Kaul et al., 2019). Phosphate is reported to be the second essential nutrient element after nitrogen. It plays a vital role in plant physiology since it is a component of biological compounds such as co-enzymes, nucleic acids, and phospholipids (Kaul et al., 2019; Kumar & Dubey, 2020). Bacteria with phosphate-solubilizing ability are reported to rely on different mechanisms of P-solubilisation. Rhizobacteria can produce organic acids that chelate mineral ions bound to precipitated phosphorous, or induce the dissolution of phosphate by locally decreasing the environmental pH (Rashid et al., 2004).

Other elements are required for sustainable agriculture such as Zn and K. Rhizosphere fungi and bacteria may solubilize these minerals from organic and inorganic sources in soil, since plants are enabled to uptake them directly. Species identified with siderophores producing capacity and suggested as plant growth enhancers, for reducing toxic heavy metal, or for synthesis of bio-fertilizers include: *Bacillus subtilis, B. halotolerans*, and *B. safensis* isolated from *Arachis hypogaea* (Sarwar et al., 2022); *B. subtilis* LSBS2 isolated from the rhizosphere of sesame plants (Nithyapriya et al., 2021); *Trichosporon ovoides* YEAST-6 and *Yarrowia lipolytica* YEAST-16, isolated from the rhizosphere of *Triticum aestivum* (El-Maraghy et al., 2020); *Brevibacillus brevis* GZDF3, isolated from rhizosphere soil of *Pinellia ternate* (Sheng et al., 2020).

Species recognized as phosphate solubilizing include: *Trichosporon ovoides* IFM63839 and *Saccharomyces cerevisiae* FI25-1F, obtained from *Triticum aestivum* rhizosphere (El-Maraghy et al., 2020); *Erwinia perscinia* EUA3K3 (Devi et al., 2022), *Rhizobium* sp. Mp1-Ha7, and *Phyllobacterium ifriqiyense* Mp1-Ha8 (Qingwei et al., 2023). Zn-phosphate solubilizing rhizospheric fungi include *Aspergillus chevalieri, Fusarium moniliforme*, and *Trichoderma harzianum* (Abdel-Ghany et al., 2019). *Halomonas aquamaria* EUB2ANL2 was observed to hold a K solubilization ability. Plants treated with a consortium including P-solubilizer, K-solubilizer, and N-fixing rhizospheric bacteria showed higher plant growth traits than those of plants receiving only one species, or the untreated control (Devi et al., 2022) (Table 6.1).

Table 6.1 Functions and application of rhizosphere microorganisms

Microorganisms	Group	Plant	Function	Bioactive Compounds[a]	Application area	References
Bacillus tequilensis PKDN31 Bacillus licheniformis PKDL10	Bacteria	Tomato	Enzyme production	4, 5, 6, 7	Agriculture, Industry	Karthika et al. (2022)
Trichoderma viride VTV7	Fungi	Tea	Enzyme production	1, 3, 4, 5, 6	Agriculture, Industry	Kolandasamy et al. (2023)
Bacillus subtilis VBS3 Pseudomonas fluorescens VPF5 Streptomyces griseus VSG4	Bacteria					
Trichoderma gamsii	Fungi	Black gram	Enzyme production	1, 3, 5	Agriculture, Industry	Valan Arasu et al. (2023)
Trichoderma spp.	Fungi	Olive	Enzyme production	2, 6, 7	Agriculture, Industry	Abdenaceur et al. (2022)
Streptomyces sp. CMAA 1653 Streptomyces sp. CMAA 1527	Bacteria	*Deschampsia antarctica*	Antitumour	Actinomycin V	Pharmaceutical	Silva et al. (2020)
				Cinerubin B		
Aspergillus carneus ABRF4	Fungi	ND	Antibacterial	ND	Pharmaceutical	Sahu et al. (2023)
			Anticancer	Trans-1,3-dimethyl—cyclohexane		
Actinobacteria Rizo Pp Ac-11	Bacteria	Polygala sp.	Antimicrobial	ND	Pharmaceutical	Cruz (2018)
Bacillus vietnamensis Pseudomonas monteilii MN759447 P. fluorescens	Bacteria	*Zingiber officinale*	Antifungal	2,4-bis (1,1-dimethylethyl) phenol	Pharmaceutical	John et al. (2021)
	Bacteria	*D. sissoo*	Biological control	Pseudomonine, salicylic acid, kynurenic acid (siderophores)	Agriculture	Srivastava et al. (2022)
	Bacteria	*Pisum sativum*	Reduced Pb concentration	ND	Bioremediation	Shabaan et al. (2021)

Bacillus altitudinis AS19	Bacteria	*Paris polyphylla* var. *yunnanensis*	Plant growth	Hydroxamate type-siderophore; phosphate-solubilization	Agriculture	Wang et al. (2022)
			Antimicrobial activity	ND	Pharmaceutical	
Aspergillus niger *Microdochium bolleyi* *Westerdikeya centenaria*	Fungi	*Baccharis scandens, Solanum chilense*	Plant growth	IAA, siderophore	Agriculture	Fuentes-Quiroz et al. (2023)
				IAA		
				iaa, siderophore, phosphate solubilization		
Proteus sp., Pseudomonas sp., E. meliloti,	Bacteria	*Medicago sativa*	Decreased Cu, Zn, Pb uptake	ND	Bioremediation	Raklami et al. (2021)
Glomus sp., Sclerocystis sp., Acaulospora sp.	Fungi					
Acinetobacter beijerinckii *Raoultella planticola*	Bacteria	*Chlorophytum comosum*	Plant growth, heavy metal removing	IAA, salicylic acid (siderophore), p-solubilization	Agriculture, Bioremediation	Hussain et al. (2022)
Aspergillus niger, Penicillium chrysosporium	Fungi	*Vicia faba* L.	Cd, Pb removing	ND	Bioremediation	El-Mahdy et al. (2021)
Talaromyces pinophilus	Fungi	*Triticum aestivum*	reduced Cd, Ni, Cu, and Zn	ND	Bioremediation	El-Shahir et al. (2021)
Pseudomonas koreensis MU2	Bacteria	ND	Plant growth	GA1 and GA3	Agriculture	Kang et al. (2019)
			Mineral or metal solubilization, organic matter degradtion	Citric, malicand tartaric acid	Agriculture, Bioremediation	
Rhizobium sp.	Bacteria	Rice	Plant growth; Silicate, P, K, and Zn solubilization	IAA, ACC deaminase	Agriculture	Chandrakala et al. (2019)

[a] ND: Not defined; 1 = chitinase; 2 = pectinase; 3 = cellulase; 4 = glucanase; 5 = proteinase; 6 = lipase; 7 = amylase

6.8 Conclusion and Future Prospects

Changing world conditions, such as population growth and climate change, are putting strong pressure on the development of agricultural productions. Main social goal should be to maintain crop yields during droughts and to develop new strategies to increase yields to meet growing food demand. Many studies have been carried out on the interaction of growth-promoting microorganisms with plants and their expected benefits. However, more research is needed on these microbes. The microorganisms found in the rhizosphere are specialized, due to specific groups of metabolites secreted by the root, hence their composition may also be different. These microbial communities may protect plants from phytopathogens, improving soil health by removing or breaking down heavy metals and toxic substances that inhibit plant growth. By this way beneficial species may enhance growth more efficiently, enabling the plant to resist stress conditions i.e., drought, salinity etc.

Even if studies on these microbial communities show promising results, their field application is still limited, especially in developing countries. Aim of these studies is principally how to fight food crisis. From this point of view, the application of such strategies should be globalized by multiplying studies in the different world regions. In addition, the specificity of these microorganisms depends on soil structure and plant phenotype and genotype. The studies on the diversity of such microbial consortia should be multiplied, depending on these two factors, relying on new and low-cost methodologies. However, exploring the diversity of these useful microorganisms cannot be limited only to their discovery. Sustainable methods should be also revised and proposed for their conservation and protection.

Applications for sustainable agriculture require also the study of rhizosphere microorganisms other than bacteria and fungi, as it is necessary to better understand their interaction with plants. Furthermore, the rhizospheric microbes include both harmful and beneficial taxa. In this view, to improve crop development and yield, the impact of rhizosphere harmful organisms need to be explored in more detail, as understanding their mechanisms on plants can help us to find new sustainable strategies for their management.

Understanding the mode of action of each bioactive compound produced by rhizospheric microorganisms can facilitate the synthesis of biofertilizers without applying the microorganisms directly to the field, to decrease the likelihood of inducing undesired changes in microbial diversity, especially for plant growth promoters. The analysis of microbial culture filtrates through chemical analyses revealed that these microorganisms produce diverse unknown bioactive compounds. For this reason, studying these metabolites should be also the focus of future studies. Therefore, the rhizosphere microbial communities are recognized as the most important resource in improving sustainable agricultural productions. Moreover, due to their high content of bioactive metabolites, biotechnological fallouts such as pharmaceutical and industrial uses, may also be possible. Research in this area should be intensified as indicated by the promising results of the few studies conducted in this context.

References

Abdel-Ghany, T. M., Mohamed, Z. H., Al Abboud, M. A., Helmy, E. A., Al-Rajhi, A. M., & Shater, A. R. M. (2019). Solubilization of inorganic phosphate by rhizospheric fungi isolated from soil cultivated with *Sorghum bicolor* L. *BioResources, 14*, 5521–5532.

Abdenaceur, R., Farida, B. T., Mourad, D., Rima, H., Zahia, O., & Fatma, S. H. (2022). Effective biofertilizer *Trichoderma* spp. isolates with enzymatic activity and metabolites enhancing plant growth. *International Microbiology, 25*, 817–829.

Afridi, M. S., Fakhar, A., Kumar, A., Ali, S., Medeiros, F. H., et al. (2022). Harnessing microbial multitrophic interactions for rhizosphere microbiome engineering. *Microbiological Research, 127199*, 127199.

Ahemad, M., & Khan, M. S. (2012). Effect of fungicides on plant growth promoting activities of phosphate solubilizing *Pseudomonas putida* isolated from mustard (*Brassica compestris*) rhizosphere. *Chemosphere, 86*, 945–950.

Anderson, I., Scheuner, C., Göker, M., Mavromatis, K., Hooper, S. D., et al. (2011). Novel insights into the diversity of catabolic metabolism from ten haloarchaeal genomes. *PLoS One, 6*, e20237.

Anerao, P., Kaware, R., Kumar Khedikar, A., Kumar, M., & Singh, L. (2022). Phytoremediation of persistent organic pollutants: Concept challenges and perspectives. In V. Kumar, M. P. Shah, & S. K. Shahi (Eds.), *Phytoremediation technology for the removal of heavy metals and other contaminants from soil and water* (pp. 375–404). Elsevier.

Anusree, T., Suseela Bhai, R., Ahammed Shabeer, T. P., & Oulkar, D. (2019). *Streptomyces* spp. from black pepper rhizosphere: A boundless reservoir of antimicrobial and growth promoting metabolites. *Journal of Biologically Active Products from Nature, 9*(1), 1–23.

Anzalone, A., Mosca, A., Dimaria, G., Nicotra, D., Tessitori, M., et al. (2022). Soil and soilless tomato cultivation promote different microbial communities that provide new models for future crop interventions. *International Journal of Molecular Sciences, 23*(15), 8820.

Apsari, P. P., Budiarti, S. R. I., & Wahyudi, A. T. (2019). Actinomycetes of rhizosphere soil producing antibacterial compounds against urinary tract infection bacteria. *Biodiversity, 20.*

Ashelford, K. E., Day, M. J., & Fry, J. C. (2003). Elevated abundance of bacteriophage infecting bacteria in soil. *Applied and Environmental Microbiology, 69*, 285–289.

Atif, A. M., Elzamik, F. I., Mohamed, G. M., Al-Quwaie, D. A., Ashkan, M. F., et al. (2023). Biological control of the root-knot nematode (*Meloidogyne incognita*) on eggplants with various chitinase-producing Streptomyces strains. *European Journal of Plant Pathology, 167*, 1–24.

Aydoğdu, E. Ö. A., Kesiktaş, M., Sanli, N. O., Güngör, N. D., Sancar, S., Yildiz, T., & Yemişken, E. (2023). Preliminary study on antimicrobial activities of skin mucus from by-catch of species. *Oceanological and Hydrobiological Studies, 52*, 137–146.

Banerjee, G., Gorthi, S., & Chattopadhyay, P. (2018). Beneficial effects of bio-controlling agent *Bacillus cereus* IB311 on the agricultural crop production and its biomass optimization through response surface methodology. *Anais da Academia Brasileira de Ciências, 90*(suppl 1), 2149–2159.

Bano, S., Wu, X., & Zhang, X. (2021). Towards sustainable agriculture: Rhizosphere microbiome engineering. *Applied Microbiology and Biotechnology, 105*, 7141–7160.

Bashir, O., Khan, K. A., Hakeem, K. R., Mir, N. A., Rather, G. H., & Mohi-ud-din, R. (2016). Soil microbe diversity and root exudates as important aspects of rhizosphere ecosystem. In K. R. Hakeem & M. S. Akhtar (Eds.), *Plant, soil and microbes: Volume 2: Mechanisms and molecular interactions* (pp. 337–357). Springer.

Benitez, M. S., Osborne, S. L., & Lehman, R. M. (2017). Previous crop and rotation history effects on maize seedling health and associated rhizosphere microbiome. *Scientific Reports, 7*, 15709.

Bhattacharyya, D., Garladinne, M., & Lee, Y. H. (2015). Volatile indole produced by rhizobacterium Proteus vulgaris JBLS202 stimulates growth of Arabidopsis thaliana through auxin, cytokinin, and brassinosteroid pathways. *Journal of Plant Growth Regulation, 34*, 158–168.

Bi, L., Yu, D. T., Du, S., Zhang, L. M., Zhang, L. Y., Wu, C. F., Xiong, C., Han, L. L., & He, J. Z. (2021). Diversity and potential biogeochemical impacts of viruses in bulk and rhizosphere soils. *Environmental Microbiology, 23*(2), 588–599.

Breitkreuz, C., Herzig, L., Buscot, F., Reitz, T., & Tarkka, M. (2021). Interactions between soil properties, agricultural management and cultivar type drive structural and functional adaptations of the wheat rhizosphere microbiome to drought. *Environmental Microbiology, 23*(10), 5866–5882.

Buee, M., De Boer, W., Martin, F., Van Overbeek, L., & Jurkevitch, E. (2009). The rhizosphere zoo: An overview of plant-associated communities of microorganisms, including phages, bacteria, archaea, and fungi, and of some of their structuring factors. *Plant & Soil, 321*, 189–212.

Chandrakala, C., Voleti, S. R., Bandeppa, S., Sunil Kumar, N., & Latha, P. C. (2019). Silicate solubilization and plant growth promoting potential of *rhizobium* sp. isolated from rice rhizosphere. *SILICON, 11*(6), 2895–2906.

Chen, J. M., Feng, W. M., Yan, H., Liu, P., Zhou, G. S., et al. (2022a). Explore the interaction between root metabolism and rhizosphere microbiota during the growth of Angelica sinensis. *Frontiers in Plant Science, 13*, 1005711.

Chen, R. Y., Jiang, W., Fu, S. F., & Chou, J. Y. (2022b). Screening, evaluation, and selection of yeasts with nitrogen fixation and ammonia production ability from the cherry tomato (*Lycopersicon esculentum* var. *cerasiforme*) rhizosphere as a potential bio-fertilizer. *Rhizosphere, 23*, 100580.

Chen, S., Sun, Y., Wei, Y., Li, H., & Yang, S. (2022c). Different rhizosphere soil microbes are recruited by tomatoes with different fruit color phenotypes. *BMC Microbiology, 22*(1), 210.

Cheng, Z., Chen, L. J., Li, H. H., Lin, J. Q., Yang, Z. B., et al. (2018). Characteristics and health risk assessment of heavy metals exposure via household dust from urban area in Chengdu, China. *Science of the Total Environment, 619–620*, 621–629.

Coats, V. C., & Rumpho, M. E. (2014). The rhizosphere microbiota of plant invaders: An overview of recent advances in the microbiomics of invasive plants. *Frontiers in Microbiology, 5*, 368.

Cockerton, H. M., Li, B., Stavridou, E., Johnson, A., Karlström, A., Armitage, A. D., Martinez-Crucis, A., Galiano-Arjona, L., Harrison, N., Barber-Pérez, N., Cobo-Medina, M., & Harrison, R. J. (2020). Genetic and phenotypic associations between root architecture, arbuscular mycorrhizal fungi colonisation and low phosphate tolerance in strawberry (Fragaria × ananassa). *BMC Plant Biology, 20*(1), 154.

Cordovez, V., Dini-Andreote, F., Carrión, V. J., & Raaijmakers, J. M. (2019). Ecology and evolution of plant microbiomes. *Annual Review of Microbiology, 73*, 69–88.

Creus, C. M., Sueldo, R. J., & Barassi, C. A. (2004). Water relations and yield in Azospirillum-inoculated wheat exposed to drought in the field. *Canadian Journal of Botany, 82*(2), 273–281.

Cruz, F.D.P.N. (2018) Isolation of the endophytic and rhizospheric microbiome associated to *Polygala spp.*: Evaluation of the biotechnological potential and antimicrobial activity.

Dagher, D. J., Pitre, F. E., & Hijri, M. (2020). Ectomycorrhizal fungal inoculation of *Sphaerosporella brunnea* significantly increased stem biomass of *Salix miyabeana* and decreased lead, tin, and zinc, soil concentrations during the phytoremediation of an industrial landfill. *Journal of Fungi, 6*(2), 87.

de Faria, M. R., Costa, L. S. A. S., Chiaramonte, J. B., Bettiol, W., & Mendes, R. (2021). The rhizosphere microbiome: Functions, dynamics, and role in plant protection. *Tropical Plant Pathology, 46*, 13–25.

Devi, R., Kaur, T., Kour, D., Yadav, A. N., & Suman, A. (2022). Potential applications of mineral solubilizing rhizospheric and nitrogen fixing endophytic bacteria as microbial consortium for the growth promotion of chilli (*Capsicum annum* L.). *Biologia, 77*(10), 2933–2943.

Dhalaria, R., Kumar, D., Kumar, H., Nepovimova, E., Kuča, K., et al. (2020). Arbuscular Mycorrhizal fungi as potential agents in ameliorating heavy metal stress in plants. *Agronomy, 10*(6), 815.

Dragojević, M., Stankovic, N., Djokic, L., Raičević, V., & Jovičić-Petrović, J. (2023). Endorhizosphere of indigenous succulent halophytes: A valuable resource of plant growth promoting bacteria. *Environmental Microbiology, 18*(1), 1–16.

Dries, L., Bussotti, S., Pozzi, C., Kunz, R., Schnell, S., et al. (2021). Rootstocks shape their microbiome—Bacterial communities in the rhizosphere of different grapevine rootstocks. *Microorganisms, 9*(4), 822.

Dunne, C., Crowley, J. J., Moënne-Loccoz, Y., Dowling, D. N., Bruijn, S., & O'Gara, F. (1997). Biological control of *Pythium ultimum* by *Stenotrophomonas maltophilia* W81 is mediated by an extracellular proteolytic activity. *Microbiology, 143*(12), 3921–3931.

El-Mahdy, O. M., Mohamed, H. I., & Mogazy, A. M. (2021). Biosorption effect of *Aspergillus Niger* and *Penicillium chrysosporium* for Cd- and Pb-contaminated soil and their physiological effects on *Vicia faba* L. *Environmental Science and Pollution Research, 28*(47), 67608–67631.

El-Maraghy, S. S., Tohamy, T. A., & Hussein, K. A. (2020). Expression of SidD gene and physiological characterization of the rhizosphere plant growth-promoting yeasts. *Heliyon, 6*(7), e04384.

Elshafie, H. S., De Martino, L., Formisano, C., Caputo, L., De Feo, V., & Camele, I. (2023). Chemical identification of secondary metabolites from rhizospheric Actinomycetes using LC-MS analysis: In silico antifungal evaluation and growth-promoting effects. *Plants, 12*(9), 1869.

El-Shahir, A. A., El-Tayeh, N. A., Ali, O. M., Abdel Latef, A. A. H., & Loutfy, N. (2021). The effect of endophytic Talaromyces pinophilus on growth, absorption and accumulation of heavy metals of Triticum aestivum grown on sandy soil amended by sewage sludge. *Plants, 10*(12), 2659.

Farhadi, D., Asghari, H. R., Baradaran Firouzabadi, M., & Abbaspour, A. (2023). Improved sweet corn growth, yield and rhizosphere enzymes by application of *Funneliformis mosseae, Piriformospora indica* and yeast extract. *Gesunde Pflanzen, 75*, 1–13.

Feng, S., Jin, L., Tang, S., Jian, Y., & Li, Z. (2022). Combination of rhizosphere bacteria isolated from resistant potato plants for biocontrol of potato late blight. *Pest Management Science, 78*(1), 166–176.

Fernandes, R. H., Silva, D. J. H. D., Delazari, F. T., Vital, C. E., & Lopes, E. A. (2022). Is hormonal analysis a predictive tool for grafting success in tomato? *Bragantia, 81*.

Fuentes-Quiroz, A., Herrera, H., Ortiz, J., Arriagada, C., & Jorquera-Fontena, E. (2023). Rhizosphere-inhabiting fungi isolated from native plants of the Atacama desert affect leaf traits of 'Chardonnay'grapevines (*Vitis vinifera* L.). *Rhizosphere, 27*, 100715.

Fujino, A., Ose, T., Yao, M., Tokiwano, T., Honma, M., Watanabe, N., & Tanaka, I. (2004). Structural and enzymatic properties of 1-aminocyclopropane-1-carboxylate deaminase homologue from *Pyrococcus horikoshii. Journal of Molecular Biology, 341*(4), 999–1013.

Gao, Z., Karlsson, I., Geisen, S., Kowalchuk, G., & Jousset, A. (2019). Protists: Puppet masters of the rhizosphere microbiome. *Trends in Plant Science, 24*(2), 165–176.

Gao, J. L., Khan, M. S., Sun, Y. C., Xue, J., Du, Y., et al. (2022). Characterization of an endophytic antagonistic bacterial strain *Bacillus halotolerans* LBG-1-13 with multiple plant growth-promoting traits, stress tolerance, and its effects on lily growth. *BioMed Research International, 2022*, 5960004–5960012.

Gayathri, M., Kumar, P. S., Prabha, A. M. L., & Muralitharan, G. (2015). In vitro regeneration of Arachis hypogaea L. and Moringa oleifera Lam. Using extracellular phytohormones from Aphanothece sp. MBDU 515. *Algal Research, 7*, 100–105.

Glaser, B., & Lehr, V. I. (2019). Biochar effects on phosphorus availability in agricultural soils: A meta-analysis. *Scientific Reports, 9*(1), 9338.

Gu, S., Wei, Z., Shao, Z., Friman, V. P., Cao, K., et al. (2020). Competition for iron drives phytopathogen control by natural rhizosphere microbiomes. *Nature Microbiology, 5*(8), 1002–1010.

Gu, K., Chen, C. Y., Selvaraj, P., Pavagadhi, S., Yeap, Y. T., et al. (2023). *Penicillium citrinum* provides transkingdom growth benefits in Choy Sum (*Brassica rapa* var. parachinensis). *Journal of Fungi, 9*(4), 420.

Gupta, A., Rico-Medina, A., & Caño-Delgado, A. I. (2020). The physiology of plant responses to drought. *Science, 368*(6488), 266–269.

Haider, I., Raza, M. A. S., Iqbal, R., Ahmad, S., Aslam, M. U., et al. (2021). Alleviating the drought stress in wheat (*Triticum aestivum* L.) by foliar application of amino acid and yeast. *Pakistan Journal of Agricultural Research, 34*(1).

Hakim, S., Naqqash, T., Nawaz, M. S., Laraib, I., Siddique, M. J., et al. (2021). Rhizosphere engineering with plant growth-promoting microorganisms for agriculture and ecological sustainability. *Frontiers in Sustainable Food Systems, 5*, 617157.

Hamed, S. M., El-Gaml, N. M., & Eissa, S. T. (2022). Integrated biofertilization using yeast with cyanobacteria on growth and productivity of wheat. *Beni-Suef University Journal of Basic and Applied Sciences, 11*(1), 112.

Haroon, M. F., Hu, S., Shi, Y., Imelfort, M., Keller, J., et al. (2013). Anaerobic oxidation of methane coupled to nitrate reduction in a novel archaeal lineage. *Nature, 500*(7464), 567–570.

Hassan, A. A., & Ajaj, D. F. (2022). Isolation and molecular identification of *Lactobacillus* bacteria and evaluation of their efficacy in inhibiting the pathogenic fungus *Pythium aphanidermatum*. *Tikrit Journal for Agricultural Sciences, 21*(4), 40–53.

Huang, W., Sun, D., Lu, Y., Dai, S., Chen, L., & An, Y. (2021). Effects of pesticide–fertilizer combinations on the rhizosphere microbiome of sugarcane: A preliminary study. *Sugar Tech, 23*, 571–579.

Hussain, A., Shah, S. T., Rahman, H., Irshad, M., & Iqbal, A. (2015). Effect of IAA on in vitro growth and colonization of Nostoc in plant roots. *Frontiers in Plant Science, 6*, 46.

Hussain, A., Shah, M., Hamayun, M. H., Iqbal, A., Qadir, M. F., et al. (2022). Phytohormones producing rhizobacteria alleviate heavy metals stress in soybean through multilayered response. *Microbiological Research, 266*, 127237.

Inbaraj, M. P. (2021). Plant-microbe interactions in alleviating abiotic stress—A mini review. *Frontiers in Agronomy, 3*, 667903.

Jacquiod, S., Raynaud, T., Pimet, E., Ducourtieux, C., Casieri, L., et al. (2021). Changes in wheat rhizosphere microbiota in response to chemical inputs, plant genotype and phenotypic plasticity. *BioRxiv*, 2021–2005.

Jäger, K., Bartók, T., Ördög, V., & Barnabás, B. (2010). Improvement of maize (*Zea mays* L.) anther culture responses by algae-derived natural substances. *South African Journal of Botany, 76*(3), 511–516.

Jamalzadeh, A., Darvishnia, M., Khodakaramian, G., Bazgir, E., & Zafari, D. (2021). Genetic diversity and plant growth-promoting activity of the dominant bacteria from canola plants in Western Iran. *Egyptian Journal Of Biological Pest Control, 31*, 1–9.

Jha, Y. (2022). Enhanced cell viability with induction of pathogenesis related proteins against aspergillus Niger in maize by Endo-Rhizospheric bacteria. *Jordan Journal of Biological Sciences, 15*(1), 139–147.

Jiao, S., Xu, Y., Zhang, J., & Lu, Y. (2019). Environmental filtering drives distinct continental atlases of soil archaea between dryland and wetland agricultural ecosystems. *Microbiome, 7*(1), 1–13.

Jiao, G., Huang, Y., Dai, H., Gou, H., Li, Z., et al. (2023). Responses of rhizosphere microbial community structure and metabolic function to heavy metal coinhibition. *Environmental Geochemistry and Health, 1*, 22–6198.

John, J., Mallikarjunaswamy, G. E., & Noushad, N. (2021). Probiotic rhizospheric *Bacillus* sp. from *Zingiber officinale* Rosc. Displays antifungal activity against soft rot pathogen *Pythium* sp. *Current Plant Biology, 27*, 100217.

Kang, S. M., Adhikari, A., Lee, K. E., & Park, Y. G. (2019). Gibberellin producing rhizobacteria *Pseudomonas koreensis* MU2 enhance growth of lettuce (*Lactuca sativa*) and Chinese cabbage (*Brassica rapa, chinensis*). *Journal of Microbiology, Biotechnology and Food Sciences, 2*.

Karlsson, M., & Stenlid, J. (2008). Comparative evolutionary histories of the fungal chitinase gene family reveal non-random size expansions and contractions due to adaptive natural selection. *Evolutionary Bioinformatics Online, 4*, 47–60.

Karthika, S., Midhun, S. J., & Jisha, M. S. (2020). A potential antifungal and growth-promoting bacterium Bacillus sp. KTMA4 from tomato rhizosphere. *Microbial Pathogenesis, 142*, 104049.

Karthika, S., Remya, M., Varghese, S., Dhanraj, N. D., Sali, S., Rebello, S., Jose, S. M., & Jisha, M. S. (2022). *Bacillus tequilensis* PKDN31 and *Bacillus licheniformis* PKDL10 – As double headed swords to combat *Fusarium oxysporum* f. sp. *lycopersici* induced tomato wilt. *Microbial Pathogenesis, 172*, 105784.

Kaul, S., Sharma, S., Apra, & Dhar, M. K. (2019). Phosphate-solubilising fungi and their potential role in sustainable agriculture. *Biofertilizers for Sustainable Agriculture and Environment*, 371–393.

Khaksar, G., Treesubsuntorn, C., & Thiravetyan, P. (2016). Effect of endophytic *Bacillus cereus* ERBP inoculation into non-native host: Potentials and challenges for airborne formaldehyde removal. *Plant Physiology and Biochemistry, 107*, 326–336.

Khan, A. L., Hamayun, M., Kim, Y. H., Kang, S. M., Lee, J. H., & Lee, I. J. (2011). Gibberellins producing endophytic Aspergillus fumigatus sp. LH02 influenced endogenous phytohormonal levels, isoflavonoids production and plant growth in salinity stress. *Process Biochemistry, 46*(2), 440–447.

Khan, M., Salman, M., Shah, S. H., & Israr, M. (2021). *In vitro* antifungal potential of surfactin isolated from rhizospheric *Bacillus thuringiensis* Berliner 1915 against maize (*Zea mays* L.) fungal phytopathogen *Fusarium graminearum* Schwabe. *Acta Agriculturae Slovenica, 117*(4), 1–7.

Kimura, M., Jia, Z. J., Nakayama, N., & Asakawa, S. (2008). Ecology of viruses in soils: Past, present and future perspectives. *Soil Science & Plant Nutrition, 54*(1), 1–32.

Klonowska, A., Moulin, L., Ardley, J. K., Braun, F., Gollagher, M. M., et al. (2020). Novel heavy metal resistance gene clusters are present in the genome of *Cupriavidus neocaledonicus* STM 6070, a new species of *Mimosa pudica* microsymbiont isolated from heavy-metal-rich mining site soil. *BMC Genomics, 21*(1), 214.

Kolandasamy, M., Mandal, A. K. A., Balasubramanian, M. G., & Ponnusamy, P. (2023). Multifaceted plant growth-promoting traits of indigenous rhizospheric microbes against *Phomopsis theae*, a causal agent of stem canker in tea plants. *World Journal of Microbiology and Biotechnology, 39*(9), 237.

Komatsu, K. J., Esch, N. L., Bloodworth, K. J., Burghardt, K. T., McGurrin, K., et al. (2023). Rhizobial diversity impacts soybean resistance, but not tolerance, to herbivory during drought. *Basic and Applied Ecology, 66*, 31–39.

Kong, H. G., Kim, B. K., Song, G. C., Lee, S., & Ryu, C. M. (2016). Aboveground whitefly infestation-mediated reshaping of the root microbiota. *Frontiers in Microbiology, 7*, 1314.

Kour, D., Kaur, T., Devi, R., Rana, K. L., Yadav, N., et al. (2020). Biotechnological applications of beneficial microbiomes for evergreen agriculture and human health. In *New and future developments in microbial biotechnology and bioengineering* (pp. 255–279). Elsevier.

Kumar, A., & Dubey, A. (2020). Rhizosphere microbiome: Engineering bacterial competitiveness for enhancing crop production. *Journal of Advanced Research, 24*, 337–352.

Kumar, N., Verma, D. K., Marwein, B. M., Gupta, N., Barman, B., et al. (2023). Sustainable remediation of the toxic impact of heavy metals in soil and water ecosystem using diverse approaches and microalgal cultivation. *Octa Journal of Biosciences, 10*(2), 62–76.

Kumawat, K. C., Razdan, N., & Saharan, K. (2022). Rhizospheric microbiome: Bio-based emerging strategies for sustainable agriculture development and future perspectives. *Microbiological Research, 254*, 126901.

Kuyper, T. W., & Suz, L. M. (2023). Do ectomycorrhizal trees select ectomycorrhizal fungi that enhance phosphorus uptake under nitrogen enrichment? *Forests, 14*(3), 467.

Larekeng, S. H., Restu, M., Tunggal, A., & Susilowati, A. (2019). Isolation and identification of rhizospheric fungus under mahoni (*Swietenia mahagoni*) stands and its ability to produce IAA (Indole Acetid Acid) hormones. In *IOP conference series: Earth and environmental science*. (vol. 343, No. 1, pp. 012051. IOP Publishing.

Leite, J., Fischer, D., Rouws, L. F., Fernandes-Júnior, P. I., Hofmann, A., Kublik, S., ... & Radl, V. (2017). Cowpea nodules harbor non-rhizobial bacterial communities that are shaped by soil type rather than plant genotype. *Frontiers in Plant Science, 7*, 2064.

Li, J., Wang, C., Liang, W., & Liu, S. (2021a). Rhizosphere microbiome: The emerging barrier in plant-pathogen interactions. *Frontiers in Microbiology, 12*, 772420.

Li, S. M., Zheng, H. X., Zhang, X. S., & Sui, N. (2021b). Cytokinins as central regulators during plant growth and stress response. *Plant Cell Reports, 40*, 271–282.

Liang, J., Feng, C., Zeng, G., Gao, X., Zhong, M., et al. (2017). Spatial distribution and source identification of heavy metals in surface soils in a typical coal mine city, Lianyuan, China. *Environmental Pollution, 225*, 681–690.

Liu, D., Tago, K., Hayatsu, M., Tokida, T., Sakai, H., et al. (2016). Effect of elevated CO_2 concentration, elevated temperature and no nitrogen fertilization on methanogenic archaeal and methane-oxidizing bacterial community structures in paddy soil. *Microbes and Environments, 31*(3), 349–356.

Liu, H., Chen, H., Ding, G., Li, K., & Ren, Q. (2020). Identification of candidate genes conferring tolerance to aluminum stress in *Pinus massoniana* inoculated with ectomycorrhizal fungus. *BMC Plant Biology, 20*, 1–13.

Liu, B., Dong, P., Zhang, X., Feng, Z., Wen, Z., et al. (2022). Identification and characterization of eight metallothionein genes involved in heavy metal tolerance from the ectomycorrhizal fungus *Laccaria bicolor*. *Environmental Science and Pollution Research, 29*(10), 14430–14442.

Lv, J., Liu, S., Feng, J., Liu, Q., Guo, J., et al. (2020). Effects of microalgal biomass as biofertilizer on the growth of cucumber and microbial communities in the cucumber rhizosphere *Turk. Journal of Botany, 44*(2), 167–177.

Ma, M., Du, H., Sun, T., An, S., Yang, G., & Wang, D. (2019). Characteristics of archaea and bacteria in rice rhizosphere along a mercury gradient. *Science of The Total Environment, 650*(Pt 1), 1640–1651.

Mahmud, K., Missaoui, A., Lee, K., Ghimire, B., Presley, H. W., & Makaju, S. (2021). Rhizosphere microbiome manipulation for sustainable crop production. *Current Plant Biology, 27*, 100210.

Manresa-Grao, M., Pastor-Fernández, J., Sanchez-Bel, P., Jaques, J. A., Pastor, V., & Flors, V. (2022). Mycorrhizal symbiosis triggers local resistance in citrus plants against spider mites. *Frontiers in Plant Science, 13*, 867778.

Meraklı, N., & Memon, A. R. (2020). Role of plant growth promoting bacteria (pgpr) in plant growth and development: Soil-plant relationship. *Turkish Journal of Agriculture – Food Science and Technology, 8*, 2590–2602.

Middelboe, M., & Brussaard, C. P. (2017). Marine viruses: Key players in marine ecosystems. *Viruses, 9*(10), 302.

Mohan, A., & Kumar, B. (2019). Plant growth promoting activities of cyanobacteria growing in rhizosphere of agriculturally fertile soil. *IOSR Journal of Biotechnology and Biochemistry, 5*(4), 28–36.

Mokrani, S., & El-Hafid, N. (2020). Rhizospheric microbiome: Biodiversity, current advancement and potential biotechnological applications. In *Advances in plant microbiome and sustainable agriculture: Diversity and biotechnological applications* (pp. 39–60). Springer.

Muleta, A., & Assefa, F. (2018). Isolation and screening of antibiotic producing actinomycetes from rhizosphere and agricultural soils. *African Journal of Biotechnology, 17*(22), 700–715.

Mulatu, A., Megersa, N., Abena, T., Kanagarajan, S., Liu, Q., et al. (2022). Biodiversity of the genus *Trichoderma* in the rhizosphere of coffee (*Coffea arabica*) plants in Ethiopia and their potential use in biocontrol of coffee wilt disease. *Crops, 2*(2), 120–141.

Munoz-Ucros, J., Zwetsloot, M. J., CuellarGempeler, C., & Bauerle, T. L. (2021). Spatiotemporal patterns of rhizosphere microbiome assembly: From ecological theory to agricultural application. *Journal of Applied Ecology, 58*(5), 894–904.

Muscatt, G., Hilton, S., Raguideau, S., Teakle, G., Lidbury, I. D., et al. (2022). Crop management shapes the diversity and activity of DNA and RNA viruses in the rhizosphere. *Microbiome, 10*(1), 1–16.

Naitam, M. G., & Kaushik, R. (2021). Archaea: An agro-ecological perspective. *Current Microbiology, 78*(7), 2510–2521.

Nazir, N., Kamili, A. N., Zargar, M. Y., Khan, I., Shah, D., et al. (2016). Effect of root exudates on rhizosphere soil microbial communities. *Journal of Research and Development, 16*(9).
Neveux, N., Nugroho, A. A., Roberts, D. A., Vucko, M. J., & de Nys, R. (2020). Selecting extraction conditions for the production of liquid biostimulants from the freshwater macroalga *Oedogonium intermedium*. *Journal of Applied Phycology, 32*, 539–551.
Ngamau, C. N., Matiru, V. N., Tani, A., & Muthuri, C. W. (2014). Isolation and identification of endophytic bacteria of bananas (Musa spp.) in Kenya and their potential as biofertilizers for sustainable banana production. *African Journal of Microbiology Research, 6*(34), 6414–6422.
Niazi, S. K., Basavarajappa, D. S., Kumaraswamy, S. H., Bepari, A., Hiremath, H., et al. (2023). GC-MS based characterization, antibacterial, antifungal and anti-oncogenic activity of ethyl acetate extract of *Aspergillus Niger* strain AK-6 isolated from rhizospheric soil. *Current Issues in Molecular Biology, 45*(5), 3733–3756.
Nithyapriya, S., Lalitha, S., Sayyed, R. Z., Reddy, M. S., Dailin, D. J., et al. (2021). Production, purification, and characterization of bacillibactin siderophore of *Bacillus subtilis* and its application for improvement in plant growth and oil content in sesame. *Sustainability, 13*(10), 5394.
Oldroyd, G. E. D., & Leyser, O. (2020). A plant's diet, surviving in a variable nutrient environment. *Science, 368*(6486), eaba0196.
Olowe, O. M., Nicola, L., Asemoloye, M. D., Akanmu, A. O., Sobowale, A. A., & Babalola, O. O. (2022). Characterization and antagonistic potentials of selected rhizosphere *Trichoderma* species against some *Fusarium* species. *Frontiers in Microbiology, 13*, 985874.
Patel, P., Shah, R., Joshi, B., Ramar, K., & Natarajan, A. (2019). Molecular identification and biocontrol activity of sugarcane rhizosphere bacteria against red rot pathogen *Colletotrichum falcatum*. *Biotechnology Reports, 21*, e00317.
Pattnaik, S. S., & Busi, S. (2019). Rhizospheric fungi: Diversity and potential biotechnological applications. In *Recent advancement in white biotechnology through fungi: volume 1: Diversity and enzymes perspectives* (pp. 63–84). Springer.
Pii, Y., Mimmo, T., Tomasi, N., Terzano, R., Cesco, S., & Crecchio, C. (2015). Microbial interactions in the rhizosphere: Beneficial influences of plant growth-promoting rhizobacteria on nutrient acquisition process. A review. *Biology and Fertility of Soils, 51*, 403–415.
Pouresmaieli, M., Ataei, M., Forouzandeh, P., Azizollahi, P., & Mahmoudifard, M. (2022). Recent progress on sustainable phytoremediation of heavy metals from soil. *Journal of Environmental Chemical Engineering, 10*, 108482.
Prangishvili, D., Forterre, P., & Garrett, R. A. (2006). Viruses of the archaea: A unifying view. *Nature Reviews. Microbiology, 4*(11), 837–848.
Prasad, J. K., Dey, R., & Raghuwanshi, R. (2022). Exopolysaccharide-producing rhizospheric bacteria enhance yield via promoting wheat (*Triticum aestivum* L.) growth at early stages. *Microbiology, 91*(6), 757–769.
Pratama, A. A., Terpstra, J., de Oliveria, A. L. M., & Salles, J. F. (2020). The role of rhizosphere bacteriophages in plant health. *Trends in Microbiology, 28*(9), 709–718.
Preece, C., & Peñuelas, J. (2020). A return to the wild: Root exudates and food security. *Trends in Plant Science, 25*(1), 14–21.
Pretscher, J., Fischkal, T., Branscheidt, S., Jäger, L., Kahl, S., et al. (2018). Yeasts from different habitats and their potential as biocontrol agents. *Fermentation, 4*(2), 31.
Purahong, W., Sadubsarn, D., Tanunchai, B., Wahdan, S. F. M., Sansupa, C., et al. (2019). First insights into the microbiome of a mangrove tree reveal significant differences in taxonomic and functional composition among plant and soil compartments. *Microorganisms, 7*(12), 585.
Qingwei, Z., Lushi, T., Yu, Z., Yu, S., Wanting, W., et al. (2023). Isolation and characterization of phosphate-solubilizing bacteria from rhizosphere of poplar on road verge and their antagonistic potential against various phytopathogens. *BMC Microbiology, 23*, 221.
Qu, Q., Zhang, Z., Peijnenburg, W. J. G. M., Liu, W., Lu, T., et al. (2020). Rhizosphere microbiome assembly and its impact on plant growth. *Journal of Agricultural and Food Chemistry, 68*(18), 5024–5038.

Raaijmakers, J. M., Paulitz, T. C., Steinberg, C., Alabouvette, C., & Moënne-Loccoz, Y. (2009). The rhizosphere: A playground and battlefield for soilborne pathogens and beneficial microorganisms. *Plant & Soil, 321*, 341–361.

Rachid, N. A., & Güngör, N. D. (2023). Screening of bioactive compounds for biomedical and industrial uses from actinobacteria isolated from the Parsık Cave, Turkey: Identifying novel compounds from extreme environments. *Johnson Matthey Technology Review, 67*(2), 159–170.

Racić, G., Vukelić, I., Kordić, B., Radić, D., Lazović, M., Nešić, L., & Panković, D. (2023). Screening of native *Trichoderma* species for nickel and copper bioremediation potential determined by FTIR and XRF. *Microorganisms, 11*(3), 815.

Rahman, M., Borah, S. M., Borah, P. K., Bora, P., Sarmah, B. K., et al. (2023). Deciphering the antimicrobial activity of multifaceted rhizospheric biocontrol agents of solanaceous crops viz., *Trichoderma harzianum* MC2, and *Trichoderma harzianum* NBG. *Frontiers in Plant Science, 14*, 1141506.

Raklami, A., Tahiri, A. I., Bechtaoui, N., Abdelhay, E. G., Pajuelo, E., et al. (2021). Restoring the plant productivity of heavy metal-contaminated soil using phosphate sludge, marble waste, and beneficial microorganisms. *Journal of Environmental Sciences, 99*, 210–221.

Rashid, M. T., Khalil, S., Ayub, N., Alam, S., & Latif, F. (2004). Organic acids production and phosphate solubilization by phosphate solubilizing microorganisms (PSM) under *in vitro* conditions. *Pakistan Journal of Biological Sciences, 7*, 187–196.

Rashid, M. I., Mujawar, L. H., Shahzad, T., Almeelbi, T., Ismail, I. M., & Oves, M. (2016). Bacteria and fungi can contribute to nutrients bioavailability and aggregate formation in degraded soils. *Microbiological Research, 183*, 26–41.

Ravnskov, S., Cabral, C., & Larsen, J. (2020). Mycorrhiza induced tolerance in *Cucumis sativus* against root rot caused by *Pythium ultimum* depends on fungal species in the arbuscular mycorrhizal symbiosis. *Biological Control, 141*, 104133.

Razak, N. A., & Gange, A. C. (2023). Multitrophic interactions between arbuscular mycorrhizal fungi, foliar endophytic fungi and aphids. *Microbial Ecology, 85*(1), 146–156.

Reinhold-Hurek, B., Bünger, W., Burbano, C. S., Sabale, M., & Hurek, T. (2015). Roots shaping their microbiome: Global hotspots for microbial activity. *Annual Review of Phytopathology, 53*, 403–424.

Renuka, N., Guldhe, A., Prasanna, R., Singh, P., & Bux, F. (2018). Microalgae as multi-functional options in modern agriculture: Current trends, prospects and challenges. *Biotechnology Advances, 36*(4), 1255–1273.

Riaz, M., Kamran, M., Fang, Y., Wang, Q., Cao, H., et al. (2021). Arbuscular mycorrhizal fungi-induced mitigation of heavy metal phytotoxicity in metal contaminated soils: A critical review. *Journal of Hazardous Materials, 402*, 123919.

Roux, S., Adriaenssens, E. M., Dutilh, B. E., Koonin, E. V., Kropinski, A. M., et al. (2019). Minimum information about an uncultivated virus genome (MIUViG). *Nature Biotechnology, 37*(1), 29–37.

Sahu, M. K., Yeeravallı, R., Das, A., & Jha, H. (2023). Secondary metabolites of rhizospheric fungal isolate *Aspergillus carneus* ABRF4 regulate the antibacterial and anti-proliferative activity against cancer cells. *NUS Biological Sciences, 15*(2).

Sanon, A., Andrianjaka, Z. N., Prin, Y., Bally, R., Thioulouse, J., et al. (2009). Rhizosphere microbiota interfers with plant-plant interactions. *Plant & Soil, 321*, 259–278.

Sarwar, S., Khaliq, A., Yousra, M., & Sultan, T. (2022). Iron biofortification potential of siderophore producing rhizobacterial strains for improving growth, yield and iron contents of groundnut. *Journal of Plant Nutrition, 45*(15), 2332–2347.

Shabaan, M., Asghar, H. N., Akhtar, M. J., Ali, Q., & Ejaz, M. (2021). Role of plant growth promoting rhizobacteria in the alleviation of lead toxicity to *Pisum sativum* L. *International Journal of Phytoremediation, 23*(8), 837–845.

Sharma, T., Dhar, M. K., & Kaul, S. (2022). Rhizomicrobes: The underground life for sustainable agriculture. In *Re-visiting the rhizosphere eco-system for agricultural sustainability* (pp. 167–180). Springer Nature.

Sheng, M. M., Jia, H. K., Zhang, G. Y., Zeng, L. N., Zhang, T. T., et al. (2020). Siderophore production by rhizosphere biological control bacteria *Brevibacillus brevis* GZDF3 of *Pinellia ternata* and its antifungal effects on *Candida albicans*. *Journal of Microbiology and Biotechnology, 30*(5), 689–699.

Silva, L. J., Crevelin, E. J., Souza, D. T., Lacerda-Júnior, G. V., de Oliveira, V. M., et al. (2020). Actinobacteria from Antarctica as a source for anticancer discovery. *Scientific Reports, 10*(1), 13870.

Singh, R. K., Singh, P., Li, H. B., Song, Q. Q., Guo, D. J., et al. (2020). Diversity of nitrogen-fixing rhizobacteria associated with sugarcane: A comprehensive study of plant-microbe interactions for growth enhancement in *Saccharum* spp. *BMC Plant Biology, 20*(1), 220.

Sodiq, A., Setiawati, M., Santosa, D., & Widayat, D. (2021). Molecular identification of isolates from local microorganisms as potential biofertilizer. *Sains Tanah – Journal of Soil Science and Agroclimatology, 18*(2), 188–193.

Srivastava, R., Kanda, T., Yadav, S., Mishra, R., & Atri, N. (2021). Cyanobacteria in rhizosphere: Dynamics, diversity, and symbiosis. In S. K. Dubey & S. K. Verma (Eds.), *Plant, soil and microbes in tropical ecosystems, rhizosphere biology* (pp. 51–69). Springer.

Srivastava, P., Sahgal, M., Sharma, K., Enshasy, H. A. E., Gafur, A., et al. (2022). Optimization and identification of siderophores produced by *Pseudomonas monteilii* strain MN759447 and its antagonism toward fungi associated with mortality in *Dalbergia sissoo* plantation forests. *Frontiers in Plant Science, 13*, 984522.

Subrahmanyam, G., Kumar, A., Sandilya, S. P., Chutia, M., & Yadav, A. N. (2020). Diversity, plant growth promoting attributes, and agricultural applications of rhizospheric microbes. In *Plant microbiomes for sustainable agriculture* (pp. 1–52). Springer.

Sun, H., Jiang, S., Jiang, C., Wu, C., Gao, M., & Wang, Q. (2021). A review of root exudates and rhizosphere microbiome for crop production. *Environmental Science and Pollution Research International, 28*(39), 54497–54510.

Sun, W., Yang, B., Zhu, Y., Wang, H., Qin, G., & Yang, H. (2022). Ectomycorrhizal fungi enhance the tolerance of phytotoxicity and cadmium accumulation in oak (*Quercus acutissima* Carruth.) seedlings: Modulation of growth properties and the antioxidant defense responses. *Environmental Science and Pollution Research International, 29*(5), 6526–6537.

Tan, C. Y., Dodd, I. C., Chen, J. E., Phang, S. M., Chin, C. F., et al. (2021). Regulation of algal and cyanobacterial auxin production, physiology, and application in agriculture: An overview. *Journal of Applied Phycology, 33*(5), 2995–3023.

Tarroum, M., Romdhane, W. B., Al-Qurainy, F., Ali, A. A. M., Al-Doss, A., et al. (2022). A novel PGPF *Penicillium olsonii* isolated from the rhizosphere of *Aeluropus littoralis* promotes plant growth, enhances salt stress tolerance, and reduces chemical fertilizers inputs in hydroponic system. *Frontiers in Microbiology, 13*, 996054.

Tkacz, A., Cheema, J., Chandra, G., Grant, A., & Poole, P. S. (2015). Stability and succession of the rhizosphere microbiota depends upon plant type and soil composition. *The ISME Journal, 9*, 2349–2359.

Tóth, G., Hermann, T., Da Silva, M. R., & Montanarella, L. (2016). Heavy metals in agricultural soils of the European Union with implications for food safety. *Environment International, 88*, 299–309.

Tsavkelova, E. A., Cherdyntseva, T. A., & Netrusov, A. (2005). Auxin production by bacteria associated with orchid roots. *Microbiology, 74*, 46–53.

Valan Arasu, M., Vijayaraghavan, P., Al-Dhabi, N. A., Choi, K. C., & Moovendhan, M. (2023). Biocontrol of *Trichoderma gamsii* induces soil suppressive and growth-promoting impacts and rot disease-protecting activities. *Journal of Basic Microbiology, 63*, 801–813.

Valdebenito, A., Nahuelcura, J., Santander, C., Cornejo, P., Contreras, B., et al. (2022). Physiological and metabolic effects of the inoculation of arbuscular mycorrhizal fungi in *Solanum tuberosum* crops under water stress. *Plants, 11*, 2539.

Vargas-Día, A. A., Ferrera-Cerrato, R., Silva-Rojas, H. V., & Alarcón, A. (2019). Isolation and evaluation of endophytic bacteria from root nodules of *Glycine max* L. (Merr.) and their potential use as biofertilizers. *The Spanish Journal of Agricultural Research, 17*, e1103–e1103.

Venturi, V., & Keel, C. (2016). Signaling in the rhizosphere. *Trends in Plant Science, 21*, 187–198.

Verma, S. K., Kingsley, K. L., Bergen, M. S., Kowalski, K. P., & White, J. F. (2018). Fungal disease prevention in seedlings of rice (*Oryza sativa*) and other grasses by growth-promoting seed-associated endophytic bacteria from invasive *Phragmites australis*. *Microorganisms, 6*(1), 21.

Vidal, C., González, F., Santander, C., Pérez, R., Gallardo, V., et al. (2022). Management of rhizosphere microbiota and plant production under drought stress: A comprehensive review. *Plants, 11*, 2437.

Viswanathan, R., Sundar, A. R., & Premkumari, S. M. (2003). Mycolytic effect of extracellular enzymes of antagonistic microbes to *Colletotrichum falcatum*, red rot pathogen of sugarcane. *World Journal of Microbiology and Biotechnology, 19*, 953–959.

Wally, O. S., Critchley, A. T., Hiltz, D., Craigie, J. S., Han, X., et al. (2013). Regulation of phytohormone biosynthesis and accumulation in *Arabidopsis* following treatment with commercial extract from the marine macroalga *Ascophyllum nodosum*. *Journal of Plant Growth Regulation, 32*, 324–339.

Wang, J., Zhang, H., Gao, J., Zhang, Y., Liu, Y., & Tang, M. (2021). Effects of ectomycorrhizal fungi (*Suillus variegatus*) on the growth, hydraulic function, and non-structural carbohydrates of Pinus tabulaeformis under drought stress. *BMC Plant Biology, 21*(1), 171.

Wang, Y., Zhang, G., Huang, Y., Guo, M., Song, J., et al. (2022). A potential biofertilizer—Siderophilic bacteria isolated from the rhizosphere of *Paris polyphylla* var. *yunnanensis*. *Frontiers in Microbiology, 13*, 870413.

Wei, Z., Hu, J., Yin, S., Xu, Y., Jousset, A., et al. (2018). *Ralstonia solanacearum* pathogen disrupts bacterial rhizosphere microbiome during an invasion. *Soil Biology and Biochemistry, 118*, 8–17.

Weidner, S., Latz, E., Agaras, B., Valverde, C., & Jousset, A. (2017). Protozoa stimulate the plant beneficial activity of rhizospheric pseudomonads. *Plant & Soil., 410*, 509–515.

Wen-Ya, M. A., Qiang-Sheng, W. U., Yong-Jie, X. U., & Kamil, K. U. Č. A. (2021). Exploring mycorrhizal fungi in walnut with a focus on physiological roles. *Notulae Botanicae Horti Agrobotanici Cluj-Napoca, 49*(2), 12363–12363.

Win, T. T., Bo, B., Malec, P., Khan, S., & Fu, P. (2021). Newly isolated strain of *Trichoderma asperellum* from disease suppressive soil is a potential bio-control agent to suppress Fusarium soil borne fungal phytopathogens. *The Plant Pathology Journal, 103*, 549–561.

Wu, M. L., Cheng, H., Zhao, H., Sun, F. L., Wang, Y. T., et al. (2020). Distribution patterns and source identification for heavy metals in Mirs Bay of Hong Kong in China. *Ecotoxicology, 29*(6), 762–770.

Xia, F., Qu, L., Wang, T., Luo, L., Chen, H., et al. (2018). Distribution and source analysis of heavy metal pollutants in sediments of a rapid developing urban river system. *Chemosphere, 207*, 218–228.

Yang, J., Song, Y., Zhou, Z., Huang, Y., Wang, S., et al. (2023a). Sulfoxanthicillin from the deep-sea derived *Penicillium* sp. SCSIO sof101: An antimicrobial compound against gram-positive and-negative pathogens. *The Journal of Antibiotics, 76*(3), 113–120.

Yang, K., Fu, R., Feng, H., Jiang, G., Finkel, O., et al. (2023b). RIN enhances plant disease resistance via root exudate-mediated assembly of disease-suppressive rhizosphere microbiota. *Molecular Plant, 16*(9), 1379–1395.

Yang, K., Wang, X., Hou, R., Lu, C., Fan, Z., et al. (2023c). Rhizosphere phage communities drive soil suppressiveness to bacterial wilt disease. *Microbiome, 11*(1), 1–18.

Yang, S., Liu, H., Xie, P., Wen, T., Shen, Q., & Yuan, J. (2023d). Emerging pathways for engineering the rhizosphere microbiome for optimal plant health. *Journal of Agricultural and Food Chemistry, 71*(11), 4441–4449.

Yashaswini, M. S., Nysanth, N. S., & Anith, K. N. (2021). Endospore-forming bacterial endophytes from Amaranthus spp. improve plant growth and suppress leaf blight (*Rhizoctonia solani* Kühn) disease of Amaranthus tricolor L. *Rhizosphere, 19*, 100387.

Yu, K., Pieterse, C. M., Bakker, P. A., & Berendsen, R. L. (2019). Beneficial microbes going underground of root immunity. *Plant, Cell & Environment, 42*, 2860–2870.

Zhalnina, K., Louie, K. B., Hao, Z., Mansoori, N., da Rocha, U. N., et al. (2018). Dynamic root exudate chemistry and microbial substrate preferences drive patterns in rhizosphere microbial community assembly. *Nature Microbiology, 3*(4), 470–480.

Zhang, R., Neu, T. R., Li, Q., Blanchard, V., Zhang, Y., et al. (2019). Insight into interactions of thermoacidophilic archaea with elemental sulfur: Biofilm dynamics and EPS analysis. *Frontiers in Microbiology, 10*, 896.

Zhang, M., Shi, Z., Lu, S., & Wang, F. (2023). AMF inoculation alleviates molybdenum toxicity to maize by protecting leaf performance. *Journal of Fungi, 9*(4), 479.

Zhao, J., Liu, J., Liang, H., Huang, J., Chen, Z., et al. (2018). Manipulation of the rhizosphere microbial community through application of a new bio-organic fertilizer improves watermelon quality and health. *PLoS One, 13*, e0192967.

Zhou, L., Song, C., Li, Z., & Kuipers, O. P. (2021). Antimicrobial activity screening of rhizosphere soil bacteria from tomato and genome-based analysis of their antimicrobial biosynthetic potential. *BMC Genomics, 22*, 1–14.

Zhu, W., Ding, C., Zhu, K., Zhang, W., Liang, D., et al. (2022). Characterization of ectomycorrhizal fungal communities associated with tree species on an iron tailings deposit undergoing restoration. *Environmental Science and Pollution Research International, 29*, 84396–84409.

Chapter 7
Rhizobiome Microbials in the Protection of Plants from Biotic Stress

Supriya Sharma, N. Boopathi, R. K. Mohana Pradeep, Shanmugam Velmurugan, V. Jayanth, and V. Bhuvaneswari

Abstract The Rhizosphere is a dynamic and complex niche with a rich microbial diversity nourished by organic compounds released by roots. The rhizobiome is the microbial community present in the soil surrounding the root zone under the influence of root exudates. Plant health is a crucial factor for global crop production which is threatened by various abiotic and biotic factors. The use of pesticides for biotic stress management leaves environmental footprints and affects the ecosystem functioning. The rhizobiome has a unique functional role in determining plant health in terms of growth and disease protection, which ultimately reduces crop yield losses and enhances production. Rhizospheric microorganisms from healthy plants are exploited to manage diverse diseases, and further induce plant resistance mechanisms. This chapter elucidates the concept of rhizobiome, rhizobiome recruitment by plants, and its role in biotic stress management.

Keywords Biotic stress · Induced systemic resistance · Plant health · Rhizobiome

S. Sharma (✉) · V. Bhuvaneswari
Department of Plant Pathology, Assam Agricultural University, Jorhat, Assam, India
e-mail: supriya.sharma@aau.ac.in

N. Boopathi · R. K. Mohana Pradeep
Department of Plant Pathology, Centre for Plant Protection Studies, Tamil Nadu Agricultural University, Coimbatore, Tamil Nadu, India

S. Velmurugan
Division of Plant Pathology, ICAR- Indian Agricultural Research Institute, New Delhi, India

V. Jayanth
Division of Entomology, ICAR- Indian Agricultural Research Institute, New Delhi, India

P. Mathur, S. Roy (eds.), *Plant Microbiome and Biological Control*, Sustainability in Plant and Crop Protection 20,
https://doi.org/10.1007/978-3-031-75845-4_7

Abbreviations

ETS	Effector-Triggered Susceptibility
MTI	MAMP-triggered immunity
mVAMPs	Microbial volatile-associated microbial patterns (mVAMPs)
MVOCs	Microbial Volatile Organic Compounds
PRRs	Plasma membrane Pattern Recognition Receptors
RLKs	Receptor-Like Kinases
RLPs	Receptor-Like Proteins

7.1 Background: Rhizosphere Microbial Communities

Plants evolved microbe interactions that are inherent to their environmental adaptation. Unravelling these interactions is now possible thanks to the rapid development of computational biology, sequencing technology and -omics analyses. This progress has led to some exciting revelations about the "*Rhizosphere microbiome*", known as the second genome of plants, collectively containing bacteria, fungi, and oomycetes, closely related to plant growth and health (Cai et al., 2017; Wu et al., 2018). The beneficial microorganisms function as an indirect layer of the plant immune system by acting as a barrier towards pathogen invasion or inducing systemic resistance. Thus, several microbial groups show potential functions related to probiotics and plant protection, attracting the attention of the research community. However, there is still a dearth of proven research on how the rhizosphere microbial communities influence growth and resistance in plants (Hacquard et al., 2017).

Rhizosphere microorganisms exhibit dichotomous behaviour, *viz.,* they can be deleterious to plant growth and development and thereby threaten agricultural production (Yin et al., 2021). In alternative, they may promote plant growth by mobilizing nutrients present in soil, by producing hormones and thus increasing the tolerance towards biotic or abiotic stress (Jacoby et al., 2017). With emphasis on beneficial rhizosphere microbes, the mechanism or strategies for protecting plants or crops may be directly attributed to antagonism, niches and resource competition, or microbial diversity (Kwak et al., 2018; Yin et al., 2021). For symbiotic fungi, mycorrhizae could benefit plants by providing enhanced nutrient access and tolerance to stress or pathogens (Smith & Read, 2010). Mycorrhizal fungi also mediate plant interactions with other soil microbes, including pathogens and mycorrhizosphere mutualists that produce vitamins (Tedersoo et al., 2020).

The rhizosphere is an organized unit composed of the root and its associated microorganisms The interaction dynamics in this limited soil region are directly influenced by the living root and its associated microbiome, that is the ensemble of the microbial genomes therein present. Thus, numerous processes occur in parallel in the rhizosphere, originating a unique and active niche. The soil microflora, which is as diverse as numerous, has the ability to modulate the plant immune responses against biotic and abiotic stresses. On the other hand, plants are also able to shape their rhizosphere microbiome by recruiting beneficial microbial communities and excreting root-type specific metabolites. These include amino acids and proteins, organic acids, carbohydrates, sugars, vitamins and mucilages, which constitute a large part of the plant fixed carbon. Meanwhile, plants and beneficial microbes exhibit their abilities to avoid excessive immune responses in their reciprocal symbiosis. Substantial lines of evidence also highlighted potential pathogens that might be secreting proteins/effectors to overcome, for their advantage, the emerging root peripheral barriers. Notably, plant-beneficial microbes, including nitrogen-fixing rhizobia, plant growth promoting rhizobacteria (PGPR), and mycorrhiza can also modify plant volatiles to induce a defense response (Kong et al., 2021). Recent studies also revealed that plants can recruit specialized associated microbiomes as an adaptation strategy for growth promotion and protection, antagonizing pathogens or modulating their immune system (Wen et al., 2021; Yin et al., 2021).

Overall, beneficial microbial communities are the active component in this multitrophic complex interaction in the rhizosphere, and their power and potential to effectively suppress diseases and thereby increase productivity are being realized through advanced molecular tools. This chapter will elucidate the evolution, host microbiome recruitment and related benefits in managing plant health.

7.2 Evolution of Metaorganisms: The Microbe—Plant Interactions

Plants are exposed to both aerial and soil environments with diverse microorganisms some of which are engaged in close associations with the root, thus together forming a metaorganism. This concept implies the complementary interdependence of the multicellular host with its associated fungi, bacteria, archaea and other eukaryotic organisms (Hassani et al., 2019). Such intimate harmony with the host requires specific adaptations that permit the microbes to deal with plant defenses and colonize the plant tissue successfully during its growth and development (Friesen et al., 2011). Plants are highly selective in recruiting their microbiota, a process that never occurs randomly (Porras-Alfaro & Bayman, 2011; Muller et al., 2016). The recruitment agent, mucilages/exudates from the roots, release flavonoids, organic acids, amino acids, cutin monomers, hormones, sugars, nutrients and polyphenols, which are entangled in changing the plant-microbe cross talks and microbial gene expression (Patel et al., 2015; Egamberdieva et al., 2017; Compant et al., 2021). The compounds secreted by roots were initially thoroughly enlisted by Vives-Peris et al. (2020), modified from previous lists given by Dakora and Phillips (2002) and Dennis et al. (2010).

The number of rhizospheric microbes may be much less in young plantlets as compared to mature plants. The radical apex that emerges from the seed interacts with bulk soil microbes to originate its own rhizosphere microbiome. The hosts have well-adapted mechanisms to release key metabolites in the vicinity of the root, including many molecules involved in active and passive transports (Baetz & Martinoia, 2014). Diffusion is responsible for the spread of low molecular-weight compounds, including amino acids, phenolics, carboxylic acids and sugars (Badri & Vivanco, 2009). Ionic channels are responsible for releasing specific carbohydrates and carboxylates such as oxalate and malate, which are carried over across the membranes, mediated by proteins. Two different anionic channels have been identified, namely SLow Anion Channels (SLACs known as S-type) and QUick Anion Channels (QUACs named as R-type). SLACs are activated in several seconds and QUACs within a few milliseconds (Dreyer et al., 2012).

Under phosphate and iron stress conditions, some plants prefer to increase the secretion of malate, citrate, and oxalate to enhance the root zone with organic carbon content, which could attract further beneficial microorganisms (Vives-Peris et al., 2020). If there is presence of toxic Al^{3+} ions in soil, the aluminium activated malate transporters (ALMT) release organic acids, mainly malate, which chelate and inactivate toxic aluminium ions (Sharma et al., 2016). Few exudates, viz., flavonoids, citrate, oxalate coumarins, malate, camalexin, ethylene, malic acids and benzoxazinoids, can be specific taxon recruiters (Arif et al., 2020). For vesicle transport, roots release high molecular weight metabolites stored in vesicles, a mechanism named exocytosis. The exuded molecules proceed from the Golgi apparatus or endoplasmic reticulum, and may protect roots from phytopathogens (Weston et al., 2012).

The mutualistic symbiotic *Rhizobium*-legumes associations are most widely documented among the plant kingdom, and several studies provided deep insights into the intrinsic mechanisms of induced disease resistance and growth promotion conferred by microbial communities (Xin et al., 2016; Cao et al., 2017; Poole et al., 2018).

7.3 How Do the Plants Recruit Their Rhizobiome?

Being immobile, plants access to the microbiomes from different environments is restricted (Zhu, 2016). However, they recruit their own rhizobiome "guards" by diverse mechanisms. The non-pathogenic microorganisms associated with plants protect against invading harmful pathogens and enhance their growth and fitness through the secretion of metabolites (Haney et al., 2015). It is clear that not all the interactions are beneficial since certain microorganisms can attack the host system for their survival (Badet & Croll, 2020). Rhizobiome recruitment by plants is indeed fascinating, since it depends on various ecological interactions with co-occurring microorganisms (Haney et al., 2015).

7.3.1 Plant Invite to Recruit the Rhizobiome

Root activities influence the microbial dynamics in the surrounding soil or rhizosphere (Li et al., 2014). The rhizosphere serves as a major habitat for microorganisms, including fungi, bacteria, archaea, oomycetes, viruses, and other macro-organisms, including nematodes and insects (Hartmann et al., 2008). Endophytic microbiome reservoirs present in seeds or other planting materials initiate the successional microbiome assembly, which ultimately builds the rhizobiome community of plants (Girsowicz et al., 2019). These planting materials contain a variety of endophytic microbes that are transferred inherently from mother plants (Kong et al., 2019). Resident microbiota and bulk soil surrounding the root region play a vital role in forming specific microbiomes in the rhizosphere, together with the seed microbiome. Bakker et al. (2015) reported that the formation of a microbiome in the rhizosphere is a dynamic and complex process that mainly depends on the soil type and ron esources such as root exudates and released carbon compounds. Roots organize the rhizosphere community assembly through rhizodeposition and carbon source enrichment in their rhizospheric region (Bais et al., 2006). The higher availability of energy and carbon-rich compounds hence promote the rhizosphere colonization by beneficial microbes (Nuccio et al., 2020) (Fig. 7.1).

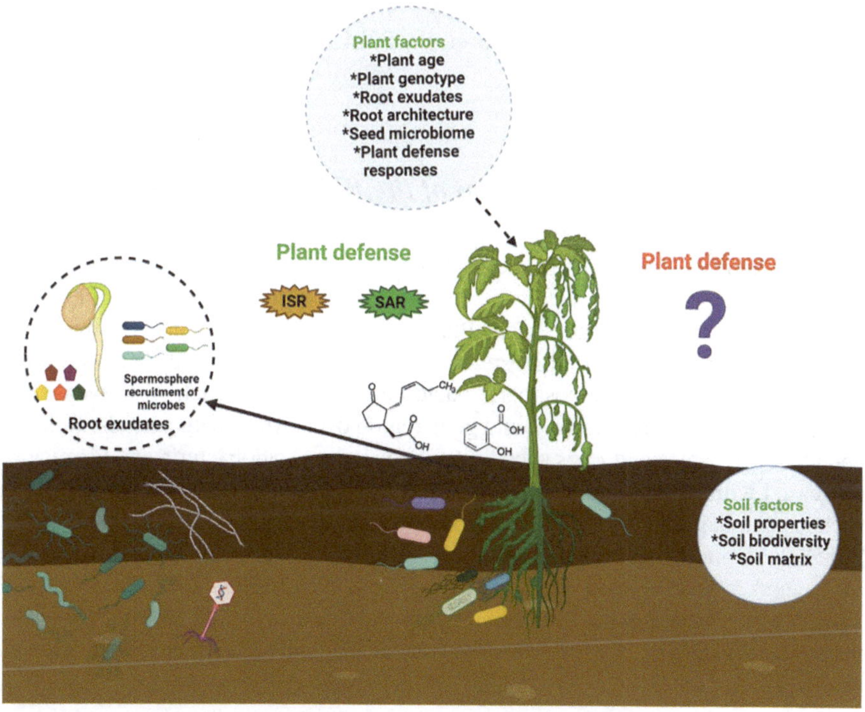

Fig. 7.1 Factors influencing plant rhizobiome recruitment and their benefits for plant health

The composition of the root exudate varies with different developmental stages and plant genetic makeups. During germination, the spermosphere assembly is highly influenced by exudates released by seed components, mainly containing metabolites and low molecular weight compounds (Schiltz et al., 2015). Aulakh et al. (2001) reported that the rice microbiome composition varies with root exudates released at different growth stagesDuring earlier stages, the composition of root exudates showed abundance ofsugars and alcohols, and it further decreased in the advancing plant stages, while the exudates in later stages were abundant with amino acids and phenolics. Rhizo-microbiome and root exudation patterns in relation to the physiological status of the host plant are strongly inter-related, influencing the recruitment in the rhizosphere (Chaparro et al., 2013).

Rhizodeposition is one of the driving forces for the recruitment of microbial rhizosphere communities. The rhizosphere composition is largely influenced by the genetic makeup of the host plant (Berendsen et al., 2012). Its microbial composition varies with plant species in different regions and environments, which elucidates that recruitment and assembly of the rhizobiome depend on the host plant genotype (Li et al., 2018; Matthews et al., 2019). Some plants can interact specifically with selective groups of microorganisms. Legumes secret phenolic compounds which activate the expression of nodulation genes in symbiotic nitrogen fixing bacteria (Rhizobia) resulting in formation of root nodulesExamples of this symbiotic association are *Phaseolus* spp. with *Rhizobium* spp., *Medicago* species with *Sinorhizobium* spp., *Lotus* spp. with *Mesorhizobium* spp., *Glycine* spp. with *Bradyrhizobium* spp. etc. (Abd-Alla et al., 2014; Li et al., 2020; Lopez-Gomez et al., 2012).

7.3.2 Progressive Recruitment of the Rhizosphere Microbiome

Being a complex and dynamic belowground region, the rhizosphere shapes the microbiome community structure due to the complex interactions for competition and survival (Gamalero et al., 2020). Plant species may also re-shape the microbial community inhabited in the rhizosphere. The onset of root growth in *Avena fatua* showed a change in the rhizospheric bacterial community, which was revealed by sequencing the 16S rRNA ribosomal gene region over two growth periods. Relative abundance of bacterial groups such as Actinobacteria, Acidobacteria and Firmicutes decreased with a reduction in diversity as roots specifically recruited bacterial taxa such as Alphaproteobacteria, Betaproteobacteria and Bacteriodetes (Shi et al., 2015). In tomato, microbiome assembly differed between the modern genotype Moneymaker and its wild relative *Solanum pimpinellifolium* LA1578. The wild tomato was enriched in taxa such as *Rhizobium*, *Acidovorax* and *Massilia*, whereas the modern genotype was enriched by species of genus *Pseudomonas* (Cordovez et al., 2021).

7.4 Intruders Hold off: Rhizobiome in Action

The rhizobiome is a complex ecosystem of microorganisms that colonize the root zone, where they form a front line of defence against harmful pathogens and promote plant growth. The rhizobiome community consists of beneficial actinomycetes, bacteria, and fungi that are involved in a dynamic interplay between plants and pathogens by producing lytic enzymes, antibiotics, siderophores, and other secondary metabolites (SAR and ISR related) that inhibit the growth of pathogenic invaders. However, these microorganisms can stimulate the plant innate immune system, enabling it to recognize pathogenic threats and fight them more effectively through MTI. This is an example of the remarkable ability of these microenvironments to evolve and select mechanisms that promote plant health and vigor, in the face of pathogenic challenges.

7.4.1 Lytic Enzymes

Lytic or cell wall-degrading enzymes are formidable weapons produced by PGPR that play a key role in the constant battle between plants and their pathogens, by promoting their defense mechanisms. When a plant encounters a pathogenic threat, it can release a plethora of lytic enzymes, namely lipases, cellulases, phosphatases, proteases, glucanases, chitosanases and chitinases to destroy the integrity of the pathogen cell wall, leading to its demise. For example, Homma (1984) exploited the potential of *Polyangium* sp. to lyse hyphae of *Rhizoctonia solani* and conidia of the severe rice pathogen, *Cochliobolus miyabeanus* by perforating their cell walls.

Li et al. (2019) deciphered a novel outer membrane vesicle (OMV)-derived β-1,6-glucanase GluM from the myxobacterium *Corallococcus* sp. strain EGB. This bacterium was capable of disassembling the integrity of the fungal cell wall of *Magnaporthe oryzae* leading to terminate its infection in rice plants. Adaikpoh (2021) observed that increased transcription of a putative lytic CHAP amidase triggered production of lytic enzymes to support the micro-predatory response of *Archangium* sp. strain CB G35 (myxobacteria) against plant pathogens in response to the phytohormone MeJA. Gao et al. (2008) detected a 41 kDa chitosanase from *B. cereus* strain D-11 with a deacetylation level of 70–100% in the cleavage of chitosan in *R. solani* AG-1. Muslim et al. (2016) identified a novel strain of *B. licheniformis* with the highest antibiofilm activity of chitosanases active against *Klebsiella pneumoniae* and *Pseudomonas aeruginosa*.

Saber et al. (2015) demonstrated that *B. subtilis* ATCC 11774 has higher chitinase activity acting as an efficient biocontrol agent of *R. solani*, the causal agent of potato black scurf disease. Win et al. (2021) demonstrated a distinct strain of *Trichoderma asperellum* from a Panama wilt contaminated field of banana, that secretes higher mycolytic enzymes viz., chitinase and β-1,3 glucanase, active against pathogenic *Fusarium oxysporum*, *F. fujikuroi*, *F. cantenulatum* and *F.*

tricinctum. By deploying these lytic enzymes in the rhizobiome, plants can strategically halt their pathogens progression and sustain their own survival.

7.4.2 Induced Systemic Resistance

The root microbiome, a thriving community of beneficial microorganisms (bacteria, fungi and actinomycetes) residing in the host-root zone, has emerged as a key player in bolstering the plant innate defenses against deleterious pathogens. Dating back to the late'90 s, ISR represents a heightened state of plant defense mechanism. It was first reported in carnation for non-pathogenic, root colonizing, plant growth promoting *Pseudomonas fluorescences* strain WCS417 against *F. oxysporum* f. sp. *dianthi* (Van Peer et al., 1991) and for *Colletotrichum orbiculare* in cucumber (Wei et al., 1991). Subsequently, ISR responses were found against various pathogens in *Arabidopsis* (Bakker et al., 2003), bean (Meziane et al., 2005), eucalyptus (Ran et al., 2005), radish (Leeman et al., 1995), rice (De Vleesschauwer et al., 2006), tobacco (De Meyer et al., 1999), and tomato (Audenaert et al., 2002).

In this context, ISR has also been described for many PGPR belonging to genera *Bacillus* (Kloepper et al., 2004), *Serratia* (Schuhegger et al., 2006), *Streptomyces* (Wang et al., 2020), *Lysobacter* (Kilic-Ekici & Yuen, 2003) and for plant growth-promoting fungi (PGPF) of genus *Trichoderma* (Viterbo et al., 2005; Segarra et al., 2009), non-pathogenic strains of *Fusarium oxysporum* (Alabouvette et al., 2007), *Pythium oligandrum* (Takenaka et al., 2003), *Penicillium oxalicum* (De Cal et al., 1997), and *Penicillium chrysogenum* (Thuerig et al., 2006). Other reports include *Phoma* sp., *Colletotrichum magnum* (Redman et al., 1999), *Piriformospora indica* (Molitor & Kogel, 2009), *Rhizoctonia* sp. (Olson & Benson, 2007) and AMF (Khan et al., 2010). However, ISR is a fascinating phenomenon characterized by the activation of the plant innate defense specifically in response to a pathogenic attack. It saves the plant energy resources by avoiding passive defense expenditures. This mechanism, known as "priming," represents an energy-efficient evolutionary strategy by activating cellular defenses resulting in more efficient and robust resistance to deleterious plant pathogens.

Microbial Volatile Organic Compounds (MVOCs) released by the root-microbiome serve as chemical messengers that orchestrate a systemic response to fend off invasive plant pathogens. The low molecular weight 6-pentyl-α-pyrone (6PP) volatile compound released by *T. asperellum* lsmT5 activates a multi-layered defense mechanism through a cascade of signalling pathways in *Arabidopsis vs Botrytis cinerea* and *Alternaria brassicicola* (Kottb et al., 2015). In Addition, these MVOCs may also function as Microbe Associated Molecular Patterns (MAMPs) and so-called microbial volatile-associated microbial patterns (mVAMPs). When the plants recognize and respond to mVAMPs through receptors, they activate Mitogen-Activated Protein Kinase (MAPK) cascades, which in turn trigger distinct responses. This intricate process results in diverse and tailored plant responses to volatile cues. However, there is still a paucity of concrete evidences to substantiate the existence of mVAMPs in activating MAPK cascades for defense against pathogens.

7.4.3 Harmonizing Plant-Rhizobiome Immunity

Root microbial populations comprise a plethora of microorganisms, which include both beneficial and detrimental species. Consequently, plants have developed an in-built complex detection mechanism to discern these microbes, modulating their signalling cascades, and initiate various downstream defense responses. The initial defense mechanism involves the recognition of a myriad of MAMPs, chemical cues emanating from diverse beneficial root microbiomes. These chemical cues are termed "elicitors" and are detected by surface or plasma membrane Pattern Recognition Receptors (PRRs) anchored with Receptor-Like Kinases (RLKs) and Receptor-Like Proteins (RLPs). Upon detection, these receptors activate MTI (MAMP-triggered immunity). This process encompasses a spectrum of first-line defense mechanisms, which include ion fluxes involving the influx of Ca^{2+} and H^+, as well as the efflux of K^+ and Cl^-.

These responses also entail oxidative bursts, protein phosphorylation mediated by MAPK cascades, reinforcement of the cell wall, and synthesis of phytoalexins through various pathways such as the shikimic acid, acetate, and mevalonate pathways. Also, late-stage defenses involve callose deposition, activation of defense genes, and biosynthesis of ethylene and camalexin. Subsequently, pathogens also co-evolved certain countermeasures such as the delivery of bacterial effector proteins directly into the host cell cytoplasm via different types of secretion systems, for example the Type III secretion system (TTSS) for Gram-negative bacteria, to circumvent MTI, leading to Effector-Triggered Susceptibility (ETS). This dynamic interplay between plant immunity and susceptibility toward invasive pathogens has been conceptualized as the Zig-Zag model (Jones & Dangl, 2006). Extensive research has been conducted to explore the initiation of MAMP-triggered immunity (MTI) in plants, which has been achieved through the use of bacterial-derived small peptides such as flg22 (flagellin) and elf18 (elongation factor), which are perceived by FLS2 and EFR receptors, respectively. Moreover, bacterial glycoconjugates, such as peptidoglycan (PGN) and lipopolysaccharides (LPS) derived from the outer membranes of Gram-positive and Gram-negative bacteria, have also been identified as MAMP molecules capable of stimulating plant innate immunity. Furthermore, fungal chitin and its oligomers, perceived by CERK1 and LYK5 receptors, trigger systemic defense responses in both monocot and dicot plants.

LPS are the primary constituents of the outer membrane of Gram-negative bacteria and play various roles in the interactions between plants and microbes. However, the mechanism of perception of the LPS receptor still remains obscure. LPS and its components such as lipid A, the inner core, and the side chain of the O antigen act as MAMP cues to induce innate immune responses against invasive pathogens in plants. Van Peer et al. (1991) and Leeman et al. (1995) deciphered how

LPS of growth-promoting rhizobacteria *P. fluorescens* strains WCS417 and WCS374 (O antigen) elicit systemic resistance (independent of PR gene expression) to Fusarium wilt, in carnation and radish. Subsequently, (Mishina & Zeier 2006, 2007) observed systemic defense responses by inducing the expression and accumulation of PR genes and *flavin-dependent monooxygenase 1* in *Arabidopsis*, in response to *P. aeruginosa* and other non-host bacteria. The adapted rhizobiome exerts selective pressure on their conserved MAMP cues to evade recognition by host receptors, for effective colonization. Silipo et al. (2011) discovered a novel bicyclic monosaccharide (bradyrhizose) in LPS (O-antigen) of *Bradyrhizobium* sp. BTAi1, masking their recognition by the host receptor, in different plant families.

7.5 Elucidating the Role of Rhizosphere Microorganisms Against Biotic Stress

Plants harbour their microbial reservoir in the rhizosphere by recruiting a broad range of diverse microorganisms. The pathogenic microbes in turn establish infection by crossing the physical and natural defence system of plants. In these processes the plants are protected, with different outcomes, by microbes of antagonistic nature recruited in the rhizosphere and by the disease resistance they induce.

7.5.1 Rhizobiome vs *Phytopathogens*

In recent decades, rhizospheric microbes with antimicrobial and plant growth promotion activity have been explored and commercially utilized for application in different crops. The application of rhizospheric microorganisms as microbial inoculants or biocontrol agents helps in managing soil borne plant diseases in an eco-friendly manner, in an alternative to chemical management. For example, different *Trichoderma* spp. have been reported for their effectiveness in managing root rot diseases of tomato (Olowe et al., 2022). Mishra et al. (2022) observed that the use of *Pseudomonas flourescens* JM-1 suppressed the maize ear rot disease caused by *F. moniliforme* by producing 2,4-diacetylphloroglucinol (DAPG). These authors improved the growth parameters in maize by enhancing the production of indolacetic acid (IAA), and the Zn and phosphate solubilization.

Secondary metabolites produced by bacteria that belong to a diverse class of low molecular weight compounds are termed antimicrobials. However, most bacteria are produced in less quantities, and are ineffective compared with other species used as biocontrol agents (Köhl et al., 2019). A biocontrol agent is considered effective when it produces enough quantities of antibiotics active against plant pathogens (Junaid et al., 2013). Broad-spectrum antimicrobial activity has been reported in bacterial genera such as *Bacillus*, *Agrobacterium*, *Pantoea*, *Serratia*, *Pseudomonas*, *Streptomyces* and *Stenotrophomonas* (Table 7.1).

Table 7.1 Bio-agents for managing different phytopathogens

Causal organism	Disease	Host	PGPR agents	References
Fungal pathogens				
Bipolaris maydis	Southern corn leaf blight (SCLB)	Maize	*Bacillus subtilis, Funneliformis mosseae*	Begum et al. (2019) and Zhu et al. (2021)
Rhizoctonia solani	Banded leaf and sheath blight (BLSB)	Maize, Soybean	*Rhizoctonia* spp. and hypovirulent *R. solani, Pseudomonas* spp., *Pseudomonas aeruginosa, P. fluorescens, Gliocladium, Trichoderma, Bacillus subtilis, Paenibacillus polymyxa, Chryseobacterium gleum, Streptomyces spp,*	Herr (1995), Ajayi-Oyetunde and Bradley (2018), Singh et al. (2020), Rai and Singh (2018), Fiers et al. (2012), and Begum et al. (2019)
Colletotrichum graminicola	Anthracnose stalk rot, Anthracnose leaf blight	Maize, sorghum, wheat, barley, Ryegrass, bluegrass	*Sporobolomyces roseus, Streptomyces ganmycicus*	Zhu et al. (2021) and Back et al. (1999)
Puccinia polysora	Southern rust	Maize	*Trichoderma* spp.	Zhu et al. (2021) and Veenstra et al. (2019)
Fusarium verticillioides, Fusarium Graminearum	Maize ear and stalk rot, Fusarium head blight	Maize, wheat	*Trichoderma asperellum, Bacillus subtilis, Bacillus velezensis, Trichoderma viride, Trichoderma harzianum, Pseudomonas fluorescens*	Ye et al. (2012), Ramesha et al. (2018), Arici and Tuncel (2020), Zhu et al. (2021) and Tom and Patel (2021)
Pythium inflatum and *P. aphanidermatum*	Pythium stalk rot	Cucumber, soybean, cotton, cucurbits, pepper	*Streptomyces* Spp., *Bacillus subtilis*	Abd-Elsalam (2020), Hassanisaadi et al. (2021), Olawuyi et al. (2014) and Zhu et al. (2021)
Bacterial pathogens				
Dickeya zeae	Rice, maize, banana	Bacterial soft rot, stalk rot, foot rot	*Pseudomonas aureofaciens, P. putida, P. chlororaphis, P. protegens, P. fluorescens, P. kilonensis*	Begum et al. (2019), Hu et al. (2018), Zhu et al. (2021), and dos Santos et al. (2020)

(continued)

Table 7.1 (continued)

Causal organism	Disease	Host	PGPR agents	References
Clavibacter michiganensis subsp. *nebraskensis*	Maize, sorghum	Bacterial wilt and blight	*Pseudomonas* and *Bacillus* spp.	Ikley et al. (2015), Zhu et al. (2021) and Goyal and Anvitha (2021)
Xanthomonas vasicola pv.*vasculorum, X. campestris*	Maize, banana, sugarcane, sorghum	Bacterial leaf streak, bacterial leaf spot	*Rhizobium, Azotobacter, Azospirillum*	Coutinho et al. (2015), Lalitha (2017), Damicone et al. (2018) and Ocimati et al. (2018)

7.5.2 Rhizobiome Against Insect Pests and Plant Parasitic Nematodes

The term "rhizobiome" typically refers to the community of microorganisms, such as bacteria and fungi, that reside in the rhizosphere, which is the thin soil volume directly surrounding roots. The rhizobiome plays a crucial role in plant health and can indirectly affect insect pests and nematode management by influencing plant defenses and attracting or repelling insect pests. However, when discussing pest management, the focus is often on how the rhizobiome influences plants and their interactions with pests.

Microbe-plant interactions can begin at the earliest stage, even during seed formation, with many endophytes being transmitted to the seeds through their parental line. After the seed germinates in soil, the colonization of roots by various microbial species commences. The majority of these plant-associated microbes can be found in the rhizosphere, a narrow soil layer influenced by roots and heavily impacted by the release of exudates. This microbial community associated with roots, known as the rhizosphere microbiome, plays a crucial role in promoting plant health and overall survival (Bakker et al., 2013; Mendes et al., 2013).

Beneficial root-associated microbes offer significant potential in environmentally safe pest control strategies, both above- and belowground. The use of certain species or strains of non-pathogenic bacteria and fungi in agricultural soils has been a long-standing practice. These microbes can be applied to stimulate plant growth or act as biocontrol agents against plant pathogens or insects. Here we briefly discuss some ways in which the rhizobiome can impact insect pest management.

7.5.2.1 Plant Health and Resilience

A diverse and balanced rhizobiome not only promotes plant health through improved nutrient uptake, better soil structure, and pest protection but may also enhance plants resistance to nematode damages. These microorganisms support plants by utilizing various biochemical mechanisms, including nitrogen fixation, phosphorus solubilization, potassium mobilization, zinc solubilization, and the mobilization of phosphates and micronutrients (Kumar, 2019). As a result, healthy plants become more robust against insect pest and nematode attacks, recover more effectively from damage, and exhibit increased resistance to insect and nematode-induced damage.

7.5.2.2 Indirect Pest Control

Certain microorganisms within the rhizobiome have the ability to generate compounds or enzymes that can be harmful to nematodes or impede their activity. In particular, some fungi and bacteria are capable of producing nematicidal compounds that contribute to nematode population management. Moreover, these same microorganisms in the rhizobiome can produce secondary metabolites or compounds that serve as deterrents or inhibitors against insect pests. These bioactive compounds may act as natural insecticides, reducing the necessity for chemical pesticides.

7.5.2.3 Enhanced Plant Defenses

Rhizobiome microorganisms can trigger the plant defense mechanisms, such as the production of secondary metabolites or the induction of systemic resistance. These defenses can make the plant less attractive to pests and better able to withstand their attacks. Induced systemic resistance represents an indirect approach to biocontrol. Certain plants activate systemic resistance by means of phytohormones i.e., ethylene and jasmonic acid, while others trigger it through the salicylic acid (SA) pathway (Mendes et al., 2013).

7.5.2.4 Attracting Natural Enemies

Through their rhizobiome, some plants can release volatile organic compounds that attract natural enemies of insect pests. Plant characteristics like noxious secondary compounds can harm herbivores, whereas plant volatiles induced by herbivores influence the herbivores and, in turn, indirectly impact the plants by attracting their natural enemies (Stam et al., 2014; Turlings & Erb, 2018).

7.5.2.5 Competition and Exclusion

Beneficial microorganisms within the rhizobiome can compete with plant-parasitic nematodes, restraining their growth and establishment while limiting their access to essential nutrients and resources.

7.5.2.6 Root Exudates and Nematode-Attracting Microorganisms

Root exudates support distinct microbial populations in the rhizosphere, thereby attracting different nematode species (Wang et al., 2023). These nematode-attracting microorganisms can be used strategically to trap nematodes, preventing them from infesting susceptible crops and serving as bait.

Researchers and farmers are exploring strategies to utilize rhizobiomes for insect pest and plant-parasitic nematode control. These strategies include introducing beneficial microbes and adopting practices that enhance the rhizobiome diversity and health. These methods are integrated into pest management programs aiming at reducing chemical pesticide use by combining biological, cultural, and physical approaches. Table 7.2 lists some of the plant-associated microbes, for which the microbial metabolites active in biocontrol of multiple pests are characterized.

It is important to acknowledge that the effectiveness of rhizobiome-based pest control can vary depending on factors such as plant species, insect pests, and environmental conditions. Ongoing research is essential to better understand the intricate rhizobiome interactions and to optimize practical applications.

For insect pests and plant-parasitic nematodes, practices like crop rotation, cover cropping, and organic soil amendments are recommended to foster a diverse and beneficial rhizobiome. Using specific insect and nematode-suppressive microorganisms as biocontrol agents can also effectively manage, promoting sustainable and eco-friendly agricultural practices, while reducing the need for chemical pesticides.

7.6 Effect of Rhizobiome in Improving Plant Health

The plant-microbiome-soil interactions occur at the rhizosphere region, which serves as the entry portal for the plants to absorb minerals, nutrients, and for phytohormone production, iron acquisition acting as the prime line of defense against various biotic and abiotic stress factors (Zhang et al., 2017).

7.6.1 Nutrient Acquisition

Plant-microbe interactions trigger the essential functions of the rhizosphere microbiome in improving plant nutrition. PGPR can improve plant nutrient absorption either by enhancing the root system architecture or by increasing the availability of soil nutrients. For example, the nitrogen use efficiency (NUE) was found to be better in *Oryza indica* than in *O. japonica* varieties. This is due to the higher proportion of N-cycle related bacteria recruitment by the host NRT1.1B gene of *O. indica* varieties (Zhang et al., 2019). Strawberry associated rhizobiome, including *Arthrobacter, Nocardioides, Gaiella,* Gp6 and Gp16 Acidobacteria, were able to fix atmospheric nitrogen similar to *Burkholderia* and *Rhizobium* (Lazcano et al., 2021).

Table 7.2 Rhizobiome guarding plants from multiple pests

Microorganism	Pest	Active metabolites	References
Bacillus amyloliquefaciens AG1	*Tuta absoluta*	Biosurfactant	Khedher et al. (2015)
Bacillus atrophaeus L193	*Rhopalosiphum padi*	Biosurfactant	Rodríguez et al. (2018)
Bacillus subtilis PTB185	*Aulacorthum solani* *Aphis gossypii*	Chitinase	Kahia et al. (2021)
Bacillus subtilis SPB1	*Ectomyelois ceratoniae* *Spodoptera littoralis*	Biosurfactant	Ghribi et al. (2012) and Mnif et al. (2013)
Bacillus subtilis V26	*Tuta absoluta*	Biosurfactant	Ben Khedher et al. (2020)
Bacillus thuringiensis strains	*Plutella xylostella*	BT toxin	Wang et al. (2020)
Bacillus thuringiensis CMB26	*Pieris rapae crucivora*	Lipopeptide	Kim et al. (2004)
Brevibacillus laterosporus Lak1210	*Plutella xylostella*	Chitinase	Prasanna et al. (2013)
Paenibacillus elgii HOA73	*Plutella xylostella*, *Meloidogyne incognita*	Crude enzyme, Gelatinase, Chitinase	Neung et al. (2014) and Nguyen et al. (2013)
Paenibacillus polymyxa BMP-11	*Tribolium castaneum*	1-Octen-3-ol Benzothiazole	Zhao et al. (2011)
Streptomyces tanaschiensis	*Musca domestica*, *Locusta migratoria*	Flavensomycin	Craveri and Giolitti (1957a, b)
Photorhabdus temperate M1021	*Galleria mellonella*	Benzaldehyde	Ullah et al. (2015)
Pseudomonas chlororaphis PCL1391	*Spodoptera littoralis*, *Heliothis virescens*, *Plutella xylostella*, *galleria mellonella*	Potent insect toxin, HCN, lipopeptide, fit toxin	Ruffner et al. (2013) and Flury et al. (2017)
Pseudomonas chlororaphis O6	*Meloidogyne hapla*, *Myzus persicae*	Hydrogen cyanide (HCN), cyclic Lipopeptides	Lee et al. (2011), Kang et al. (2018) and Kang et al. (2019)
Pseudomonas chlororaphis PA23	*Caenorhabditis elegans*	Pyrrolnitrin, HCN	Nandi et al. (2015)
Azospirillum brasilense	*Diabrotica speciosa*	NA	Santos et al. (2014)
Glomus intraradices (also known as *Rhizophagus intraradices*)	*Diabrotica virgifera virgifera*	NA	Dematheis et al. (2013)
Glomus spp.	*Otiorhynchus sulcatus*	NA	Bennett et al. (2013)

Rhizobium spp. are nitrogen-fixing bacteria that establish symbiotic associations with legumes. However, recent studies showed that they can also naturally colonize the non-legume crops rhizosphere, namely in hosts such as corn, barley, sunflower and rice. They can be also successfully inoculated in tomato, lettuce, pepper and radish rhizosphere, improving the plant growth significantly (Alami et al., 2000; García-Fraile et al., 2012). The *Burkholderia* genus includes numerous diazotrophic species found to play vital role in nitrogen fixation in corn, sugarcane and tomato (Caballero-Mellado et al., 2007; Van Deynze et al., 2018).

Yu et al. (2021) reported that root-derived falvone exudates predominantly improved the enrichment of Oxalobacteraceae in the root zone, which then promoted LRT-1 mediated lateral root development and enhanced nitrogen acquisition and growth of maize varieties. In addition to N uptake, the plant rhizobiome also enhanced iron (Fe) and phosphorus (P) under iron limited and phosphorus deprivation situations, respectively. During phosphorus starvation conditions, the rhizosphere microbiome of *Arabidopsis thaliana* activated the master transcriptional regulator PSR (PHR1 mediated phosphate starvation response) genes, which in turn promoted the acquisition of inorganic phosphate (Castrillo et al., 2017).

7.6.2 Phytohormone Production

Phytohormones alter plant physiological activities at very tiny concentrations acting as chemical stimulants. They play a vital role in plant behavior by influencing physiological and developmental processes. Consistently, the plant hormones have been classified into five classes namely, auxin, gibberellins (GA_3), cytokinine, abscissic acid (ABA) and ethylene (ET) (Oleńska et al., 2020). Furthermore, jasmonate (JA), salicylic acid (SA) and brassinosteroids are known to play a remarkable role in plant growth promotion under stress conditions (Wong et al., 2015; Kang et al., 2016).

A wide range of rhizobacteria is potentially involved in phytohormone production and facilitates communication among plant cells (Maheshwari et al., 2015). Indol-3-acetic acid (IAA) produced by bacteria improves the surface area and root length for sufficient uptake of nutrients and water. More than 80% of rhizosphere microbes are proficient of synthesizing and releasing IAA (Oleńska et al., 2020). In the cucumber—*B. amyloliquefaciens* strain SQR9 system, inoculation of auxin-producing PGPR secretes high amount of tryptophan through cucumber roots, thereby increasing the auxin synthesis (Liu et al., 2016). Several case studies were conducted to improve root elongation, growth, and yield by IAA-producing rhizobacteria (Ali et al., 2014; Majeed et al., 2015). The application of IAA-producing *Bthuringiensis*, *B. simplex*, *Enterobacer* sp. and *Pseudomonas* strains showed a beneficial effect on the growth and development of wheat crops under drought stress conditions (Raheem et al., 2018). In addition, bacteria-derived auxins are mostly involved in salinity, drought and heavy metal stress conditions (Kudoyarova et al., 2019). Maheshwari et al. (2015) reported cytokinin production in various bacterial genera *viz.*, *Agrobacterium*, *Escherichia*, *Bacillus*, *Pseudomonas*, *Methylobacterium*,

Klebsiella and *Proteus*. Cytokinin biosynthesis pathway involves the transfer of adenosine monophosphate (AMP) and dimethylallyl pyrophosphate (DMAPP) into N6-isopentenyl adenosine monophosphate. But in the case of bacteria, cytokinin biosynthesis occurs by the formation of AMP from hydroxyl dimethyl butenyl diphosphate (HMBDP) (Wong et al., 2015).

7.6.3 Siderophore Supply to Roots

Iron is present abundantly in soil but not in a form that is accessible to plants or microorganisms. Rhizosphere microbes are known to produce siderophores (Mehnaz, 2013). These are low molecular weight, high-affinity, iron-chelating compounds that have the potential to supply iron to host plants (Hider & Kong, 2010). The rhizobiome synthesized siderophores show high affinity towards rhizospheric Fe^{3+} ions, which increases the iron uptake and simultaneously may retard growth of pathogenic taxa. The iron-chelating rhizobial microbes colonize entirely the roots, eliminating all other deleterious microflora from their rhizosphere habitat. *Pseudomonas* spinoculated plants showed improved iron sequestration, and inhibited disease development (Sahu & Sindhu, 2011).

Pyoverdine and pyochelin are the major siderophores produced by *Pseudomonas* strains. They help to enhance uptake of iron and retard the growth and sporulation of pathogens due to competition for iron (Shen et al., 2013). Several studies were showed siderophore producing rhizobiome including *Bacillus, Bradyrhizobium, Rhizobium, Klebsiella, Streptomyces* and *Serratia* (Mustafa et al., 2019). The coumarin scopoletin secretion by the root-specific MYB72 regulated biosynthesis is an excellent iron chelator with selective antimicrobial activity. It recruits specific bacterial taxa for rhizobacteria-mediated iron accumulation, under iron stress conditions (Stringlis et al., 2018; Harbort et al., 2020).

7.7 Rhizobiome Engineering to Boost Plant Immune System

The rhizomicrobiome plays a crucial role in plant health directly and indirectly. The concept of rhizobiome engineering was coined by O'Connel and co-workers in 1996Rhizobiome engineering alters rhizobiome composition, which could affect plant health (O'Connell et al., 1996). In other words, rhizobiome engineering manipulates the microbiome assembly in the rhizosphere to obtain a desired effect on plants, such as stress resilience, ameliorated response *vs* toxicity, and so on. It is important to understand the factors that influence the rhizobiome assembly for its effective manipulation. They include thc plant species that may selectively recruit specific microbes, the development of the type and quantity of root exudates released by the plants, the competitive and cooperative nature of soil microbes. Also soil pH, nutrient status, organic matter content, temperature and moisture are involved in shaping the rhizosphere microbiome assembly (Musilova et al., 2016).

Plant-based, soil-based, and microbe-based approaches can alter the microbiome assembly of the rhizosphere. In plant-based approaches, a regulatory role of the host plant in recruiting the rhizosphere microbe is utilized. The plant physiology is altered using resistance-triggering elicitors (chemical or biological) to bring changes in the root exudates, which impact the composition of rhizosphere microbes. For instance, the foliar application of salicylic acid and acibenzolar-s-methyl on pine trees increased the chitinase content in root exudation. As a result, the microbes utilizing chitin as a food source were drastically reduced in numbers (Mannaa et al., 2020). Applying *Pseudomonas koreensis* and *Lysobacter enzymogenes* on pine foliage altered the rhizosphere microbiome, yielding a defense against pine wilt pathogen (Han et al., 2021).

In rhizobiome engineering by plant breeding programs the cultivars with specific root exudates are integrated with a breeding program to enrich the rhizosphere with beneficial microbes, while suppressing phytopathogens (Bakker et al., 2012). A compositional and functional shift in the rhizosphere microbial community was observed between a resistant and susceptible cultivar. In this system the resistant cultivar rhizosphere was enriched with beneficial microbes belonging to families Bacillaceae, Pseudomonadaceae, Cytophagaceae and Solibacteraceae (Mendes et al., 2018). Genetic modification of plants aims at producing specific root exudates that could be utilized by a special class of microbes, resulting in their enrichment in the rhizosphere. In nature we can see that *Agrobacterium* transforms the plant to produce opines, which is only catabolized by *Agrobacterium* (Tzfira & Citovsky, 2006). The same could be replicated too, but the prerequisite is to identify the gene that synthesizes a specific metabolites. This could be done using genome-wide association studies (GWAS) and QTL analysis (Deng et al., 2021). Genome editing techniques like zinc finger nucleases, TALEN, and CRISPR/Cas system can engineer plants to exudate metabolites to enrich beneficial microbes in the rhizosphere (Kumar & Dubey, 2020).

The target in soil-based approach for rhizobiome engineering is to modify the soil and edaphic factors to impact the composition of the rhizosphere microbiome (Lori et al., 2017). The use of soil amendments such as biochar (O'Neill et al., 2009), organic manure, lime (Chen et al., 2022), green manure, animal manure (Khatri et al., 2023), silicon nanoparticles (Rajput et al., 2021), nanosilver (Sillen et al., 2015) and nano-chitosan (Agri et al., 2022) increased the rhizosphere microbial population, diversity and functionality, suppressing phytopathogens more than the conventional soil without amendments.

The microbe-based approach for engineering rhizobiome includes methods such as single strain inoculation, consortia, syn-com inoculation, and microbiome transplantation. The classic single-strain inoculation method is well-used for biocontrol, biofertilizers, and biopesticides. After testing its efficacy for the above-mentioned role, the strain is isolated from a large community of microbes and selected as a useful genotype. Applying this strain is expected to improve plant health and to enhance the rhizosphere microbiome compositional and functional diversity.

Application of *Bacillus velezensis* and *Pseudomonas fluorescens* to control the bacterial wilt pathogen *Ralstonia solanacearum*, resulted in a rhizosphere microbiome shift by enriching three core members present in suppressive soils, namely Actinobacteria, *Arthrobacter* and *Gaiella* (Elsayed et al., 2020).

In contrast to the use of a single strain as a biocontrol agent, multi-species or synthetic communities of microbes are used for plant protection, growth promotion and stress resilience. Four bacterial strains used for dressing cotton seeds increased their growth by indirectly influencing the rhizosphere microbial composition (Kaur et al., 2022). Biocontrol methods of inoculation based on consortia do not involve the majority of unculturable microbes that are naturally present in soil. Nevertheless, rhizosphere microbiome transplantation includes a whole set of microbiomes, including unculturable taxa (Park et al., 2023). Recently, a suitable systematic protocol for transplanting rhizosphere microbiome was proposed, with an indication of enhanced plant defense against soil-borne pathogens after transplantation (Jiang et al., 2022).

7.8 Conclusion

Attaining sustainable food production and decreasing hunger is essential to achieve the important Sustainable Development Goal (SDG) of 2030, which can only be reached via effective plant health management. Using chemicals to manage plant health imposes greater risk for human beings, animals and the environmentHence, the beneficial microbes residing in the plant rhizosphere must be better exploited for ecofriendly management of biotic stress caused by plant pathogens. Rhizobiome shapes the plant phenotype to withstand against various biotic and abiotic stress factors. Broad spectrum formulations of efficient beneficial microbes with antagonistic nature, proceeding from the rhizobiome, may provide an alternate for chemical use in plant disease management. The application of chemical fertilizers imbalances the soil microbial community. Instead, applying the organic amendments helps to maintain the native microbiome, which further increases the richness and diversity of species in the plant rhizosphere.

References

Abd-Alla, M. H., El-Enany, A. W. E., Nafady, N. A., Khalaf, D. M., & Morsy, F. M. (2014). Synergistic interaction of *Rhizobium leguminosarum* bvviciae and arbuscular mycorrhizal fungi as a plant growth promoting biofertilizers for faba bean (*Vicia faba* L.) in alkaline soil. *Microbiological Research, 169*, 49–58.

Abd-Elsalam, K. A. (2020). Host plants and specificity of the genus *Pythium*. In *Pythium* (pp. 162–175). CRC Press.

Adaikpoh, B. I. (2021). *Responses of Myxobacteria to plant-related chemosignals suggests their roles in the rhizobiome*. Doctoral dissertation:. University of Mississippi.

Agri, U., Chaudhary, P., Sharma, A., & Kukreti, B. (2022). Physiological response of maize plants and its rhizospheric microbiome under the influence of potential bioinoculants and nanochitosan. *Plant and Soil, 474*(1), 451–468.

Ajayi-Oyetunde, O. O., & Bradley, C. A. (2018). *Rhizoctonia solani*: Taxonomy, population biology and management of rhizoctonia seedling disease of soybean. *Plant Pathology, 67*, 3–17.

Alabouvette, C., Olivain, C. L., Haridon, F., Aimé, S., & Steinberg, C. (2007). Using strains of *Fusarium oxysporum* to control Fusarium wilts: Dream or reality? In *Novel biotechnologies for biocontrol agent enhancement and management* (pp. 157–177). Springer.

Alami, Y., Achouak, W., Marol, C., & Heulin, T. (2000). Rhizosphere soil aggregation and plant growth promotion of sunflowers by an exopolysaccharide-producing *Rhizobium* spstrain isolated from sunflower roots. *Applied and Environmental Microbiology, 66*, 3393–3398.

Ali, S., Hameed, S., Imran, A., Iqbal, M., & Lazarovits, G. (2014). Genetic, physiological and biochemical characterization of *Bacillus* sp. strain RMB7 exhibiting plant growth promoting and broad spectrum antifungal activities. *Microbial Cell Factories, 13*, 1–15.

Arici, S. E., & Tuncel, Z. N. (2020). Antifungal activity of useful microorganisms against a phytopathogenic fungus on maize. *Emerging Materials Research, 9*, 743–749.

Arif, I., Batool, M., & Schenk, P. M. (2020). Plant microbiome engineering: Expected benefits for improved crop growth and resilience. *Trends in Biotechnology, 38*, 1385–1396.

Audenaert, K., Pattery, T., Cornelis, P., & Höfte, M. (2002). Induction of systemic resistance to *Botrytis cinerea* in tomato by *Pseudomonas aeruginosa* 7NSK2: Role of salicylic acid, pyochelin, and pyocyanin. *Molecular Plant-Microbe Interactions, 15*, 1147–1156.

Aulakh, M. S., Wassmann, R., Bueno, C., Kreuzwieser, J., & Rennenberg, H. (2001). Characterization of root exudates at different growth stages of ten rice (*Oryza sativa* L.) cultivars. *Plant Biology, 3*, 139–148.

Back, P. A., Landschoot, P. J., & Huff, D. R. (1999). Variation in pathogenicity, morphology, and RAPD marker profiles in *Colletotrichum graminicola* from turfgrasses. *Crop Science, 39*, 1129–1135.

Badet, T., & Croll, D. (2020) The rise and fall of genes: origins and functions of plant pathogen pangenomes. *Current Opinion in Plant Biology, 56*, 65–73.

Badri, D. V., & Vivanco, J. M. (2009). Regulation and function of root exudates. *Plant, Cell & Environment, 32*, 666–681.

Baetz, U., & Martinoia, E. (2014). Root exudates: The hidden part of plant defense. *Trends in Plant Science, 19*, 90–98.

Bais, H. P., Weir, T. L., Perry, L. G., Gilroy, S., & Vivanco, J. M. (2006). The role of root exudates in rhizosphere interactions with plants and other organisms. *Annual Review of Plant Biology, 57*, 233–266.

Bakker, P. A. H. M., Ran, L. X., Pieterse, C. M. J., & Van Loon, L. C. (2003). Understanding the involvement of rhizobacteria-mediated induction of systemic resistance in biocontrol of plant diseases. *Canadian Journal of Plant Pathology, 25*, 5–9.

Bakker, M. G., Manter, D. K., Sheflin, A. M., Weir, T. L., & Vivanco, J. M. (2012). Harnessing the rhizosphere microbiome through plant breeding and agricultural management. *Plant and Soil, 360*, 1–13.

Bakker, P. A., Doornbos, R. F., Zamioudis, C., Berendsen, R. L., & Pieterse, C. M. (2013). Induced systemic resistance and the rhizosphere microbiome. *The Plant Pathology Journal, 29*, 136.

Bakker, M. G., Chaparro, J. M., Manter, D. K., & Vivanco, J. M. (2015). Impacts of bulk soil microbial community structure on rhizosphere microbiomes of *Zea mays*. *Plant and Soil, 392*, 115–126.

Begum, N., Qin, C., Ahanger, M. A., Raza, S., Khan, M. I., et al. (2019). Role of arbuscular mycorrhizal fungi in plant growth regulation: Implications in abiotic stress tolerance. *Frontiers in Plant Science, 10*, 1068.

Ben Khedher, S., Boukedi, H., Laarif, A., & Tounsi, S. (2020). Biosurfactant produced by *Bacillus subtilis* V26: a potential biological control approach for sustainable agriculture development. *Organic Agriculture, 10*, 117–124.

Bennett, A. E., Macrae, A. M., Moore, B. D., Caul, S., & Johnson, S. N. (2013). Early root herbivory impairs arbuscular mycorrhizal fungal colonization and shifts defence allocation in establishing *Plantago lanceolata. PLoS One, 8*, e66053.

Berendsen, R. L., Pieterse, C. M., & Bakker, P. A. (2012). The rhizosphere microbiome and plant health. *Trends in Plant Science, 17*(8), 478-486.

Caballero-Mellado, J., Onofre-Lemu, J., Estrada-De Los Santos, P., & Martínez-Aguilar, L. (2007). The tomato rhizosphere, an environment rich in nitrogen-fixing *Burkholderia* species with capabilities of interest for agriculture and bioremediation. *Applied and Environmental Microbiology, 73*, 5308–5319.

Cai, F., Pang, G., Miao, Y., Li, R., Li, R., et al. (2017). The nutrient preference of plants influences their rhizosphere microbiome. *Applied Soil Ecology, 110*, 146–150.

Cao, Y., Halane, M. K., Gassmann, W., & Stacey, G. (2017). The role of plant innate immunity in the legume-rhizobium symbiosis. *Annual Review of Plant Biology, 68*, 535–561.

Castrillo, G., Teixeira, P. J. P. L., Paredes, S. H., Law, T. F., De Lorenzo, L., et al. (2017). Root microbiota drive direct integration of phosphate stress and immunity. *Nature, 543*(7646), 513–518.

Chaparro, J. M., Badri, D. V., Bakker, M. G., Sugiyama, A., Manter, D. K., & Vivanco, J. M. (2013). Root exudation of phytochemicals in Arabidopsis follows specific patterns that are developmentally programmed and correlate with soil microbial functions. *PloS One, 8*(2), e55731.

Chen, D., Wang, X., Carrion, V. J., Yin, S., Yue, Z., Liao, Y., Dong, Y., & Li, X. (2022). Acidic amelioration of soil amendments improves soil health by impacting rhizosphere microbial assemblies. *Soil Biology and Biochemistry, 167*, 108599.

Compant, S., Cambon, M. C., Vacher, C., Mitter, B., Samad, A., & Sessitsch, A. (2021). The plant endosphere world–bacterial life within plants. *Environmental Microbiology, 23*(4), 1812–1829.

Cordovez, V., Rotoni, C., Dini-Andreote, F., Oyserman, B., Carrión, V. J., & Raaijmakers, J. M. (2021). Successive plant growth amplifies genotype-specific assembly of the tomato rhizosphere microbiome. *Science of the Total Environment, 772*, 144825.

Coutinho, T. A., Van der Westhuizen, L., Roux, J., McFarlane, S. A., & Venter, S. N. (2015). Significant host jump of *Xanthomonas vasicola* from sugarcane to a *Eucalyptus grandis* clone in South Africa. *Plant Pathology, 64*, 576–581.

Craveri, R., & Giolitti, G. (1957a). An antibiotic with fungicidal and insecticidal activity produced by Streptomyces. *Nature, 179*(4573), 1307–1307.

Craveri, R., & Giolitti, G. (1957b). An antibiotic with fungicidal and insecticidal activity produced by *Streptomyces. Nature, 179*, 1307.

Dakora, F. D., & Phillips, D. A. (2002). Root exudates as mediators of mineral acquisition in low-nutrient environments. In: Food security in nutrient-stressed environments: Exploiting plants' genetic capabilities. *Plant and Soil, 245*, 201–213.

Damicone, J., Cevallos, F., & Olson, J. (2018). First report of bacterial leaf streak of corn caused by *Xanthomonas vasicola* pv*vasculorum* in Oklahoma. *Plant Pathology, 102*, 437–437.

De Cal, A., Pascual, S., & Melgarejo, P. (1997). Involvement of resistance induction by *Penicillum oxalicum* in the biocontrol of tomato wilt. *Plant Pathology, 46*, 72–79.

De Meyer, G., Audenaert, K., & Höfte, M. (1999). *Pseudomonas aeruginosa* 7NSK2-induced systemic resistance in tobacco depends on in planta salicylic acid accumulation but is not associated with PR1a expression. *European Journal of Plant Pathology, 105*, 513–517.

De Vleesschauwer, D., Cornelis, P., & Höfte, M. (2006). Redox-active pyocyanin secreted by Pseudomonas aeruginosa 7NSK2 triggers systemic resistance to *Magnaporthe grisea* but enhances *Rhizoctonia solani* susceptibility in rice. *Molecular Plant-Microbe Interactions, 19*, 1406–1419.

Dematheis, F., Kurtz, B., Vidal, S., & Smalla, K. (2013). Multitrophic interactions among Western corn Rootworm, *Glomus intraradices* and microbial communities in the rhizosphere and endorhiza of maize. *Frontiers in Microbiology, 4*, 357.

Deng, S., Caddell, D. F., Xu, G., Dahlen, L., Washington, L., et al. (2021). Genome wide association study reveals plant loci controlling heritability of the rhizosphere microbiome. *The ISME Journal, 15*, 3181–3194.

Dennis, P. G., Miller, A. J., & Hirsch, P. R. (2010). Are root exudates more important than other sources of rhizodeposits in structuring rhizosphere bacterial communities? *FEMS Microbiology Ecology, 72*, 313–327.

Dos Santos, R. M., Diaz, P. A. E., Lobo, L. L. B., & Rigobelo, E. C. (2020). Use of plant growth-promoting rhizobacteria in maize and sugarcane: Characteristics and applications. *Frontiers in Sustainable Food Systems, 4*, 136.

Dreyer, I., Gomez-Porras, J. L., Riaño-Pachón, D. M., Hedrich, R., & Geiger, D. (2012). Molecular evolution of slow and quick anion channels (SLACs and QUACs/ALMTs). *Frontiers in Plant Science, 3*, 263.

Egamberdieva, D., Wirth, S. J., Alqarawi, A. A., Abd Allah, E. F., & Hashem, A. (2017). Phytohormones and beneficial microbes: Essential components for plants to balance stress and fitness. *Frontiers in Microbiology, 8*, 2104.

Elsayed, T. R., Jacquiod, S., Nour, E. H., Sørensen, S. J., & Smalla, K. (2020). Biocontrol of bacterial wilt disease through complex interaction between tomato plant, antagonists, the indigenous rhizosphere microbiota, and *Ralstonia solanacearum*. *Frontiers in Microbiology, 10*, 2835.

Fiers, M., Edel-Hermann, V., Chatot, C., Le Hingrat, Y., Alabouvette, C., & Steinberg, C. (2012). Potato soil-borne diseases. A review. *Agronomy for Sustainable Development, 32*, 93–132.

Flury, P., Vesga, P., Péchy-Tarr, M., Aellen, N., Dennert, F., et al. (2017). Antimicrobial and insecticidal: Cyclic lipopeptides and hydrogen cyanide produced by plant-beneficial pseudomonas strains CHA0, CMR12a, and PCL1391 contribute to insect killing. *Frontiers in Microbiology, 8*, 100.

Friesen, M. L., Porter, S. S., Stark, S. C., von Wettberg, E. J., Sachs, J. L., & Martinez-Romero, E. (2011). Microbially mediated plant functional traits. *Annual Review of Ecology, Evolution, and Systematics, 42*, 23–46.

Gamalero, E., Bona, E., Todeschini, V., & Lingua, G. (2020). Saline and arid soils: Impact on bacteria, plants, and their interaction. *Biology, 9*, 116.

Gao, X. A., Ju, W. T., Jung, W. J., & Park, R. D. (2008). Purification and characterization of chitosanase from *Bacillus cereus* D-11. *Carbohydrate Polymers, 72*, 513–520.

García-Fraile, P., Carro, L., Robledo, M., Ramirez-Bahena, M. H., Flores-Felix, J. D., et al. (2012). *Rhizobium* promotes non-legumes growth and quality in several production steps: Towards a biofertilization of edible raw vegetables healthy for humans. *PLoS One, 7*, e38122.

Ghribi, D., Abdelkefi-Mesrati, L., Boukedi, H., Elleuch, M., Ellouze-Chaabouni, S., & Tounsi, S. (2012). The impact of the *Bacillus subtilis* SPB1 biosurfactant on the midgut histology of *Spodoptera littoralis* (Lepidoptera: Noctuidae) and determination of its putative receptor. *Journal of Invertebrate Pathology, 109*, 183–186.

Girsowicz, R., Moroenyane, I., & Steinberger, Y. (2019). Bacterial seed endophyte community of annual plants modulated by plant photosynthetic pathways. *Microbiological Research, 223*, 58–62.

Goyal, K., & Anvitha, D. (2021). Wilting: The root cause of amateur death. *International Journal of All Research Education & Scientific Methods, 9*, 634–638.

Hacquard, S., Spaepen, S., Garrido-Oter, R., & Schulze-Lefert, P. (2017). Interplay between innate immunity and the plant microbiota. *Annual Review of Phytopathology, 55*, 565–589.

Han, G., Mannaa, M., Kim, N., Jeon, H. W., Jung, H., et al. (2021). Response of pine rhizosphere microbiota to foliar treatment with resistance-inducing bacteria against pine wilt disease. *Microorganisms, 9*, 688.

Haney, C. H., Samuel, B. S., Bush, J., & Ausubel, F. M. (2015). Associations with rhizosphere bacteria can confer an adaptive advantage to plants. *Nature Plants, 1*, 1–9.

Harbort, C. J., Hashimoto, M., Inoue, H., Niu, Y., Guan, R., et al. (2020). Root-secreted coumarins and the microbiota interact to improve iron nutrition in Arabidopsis. *Cell Host & Microbe, 28*, 825–837.

Hartmann, A., Rothballer, M., & Schmid, M. (2008). Lorenz Hiltner, a pioneer in rhizosphere microbial ecology and soil bacteriology research. *Plant and Soil, 312*, 7–14.
Hassani, M. A., Ozkurt, E., Seybold, H., Dagan, T., & Stukenbrock, E. H. (2019). Interactions and coadaptation in plant metaorganisms. *Annual Review of Phytopathology, 57*, 483–503.
Hassanisaadi, M., Shahidi Bonjar, G. H., Hosseinipour, A., Abdolshahi, R., Ait Barka, E., & Saadoun, I. (2021). Biological control of *Pythium aphanidermatum*, the causal agent of tomato root rot by two *Streptomyces* root symbionts. *Agronomy, 11*, 846.
Herr, L. J. (1995). Biological control of *Rhizoctonia solani* by binucleate *Rhizoctonia* sppand hypovirulent *Rsolani* agents. *Crop Protection, 14*, 179–186.
Hider, R. C., & Kong, X. (2010). Chemistry and biology of siderophores. *Natural Product Reports, 27*, 637–657.
Homma, Y. (1984). Perforation and lysis of hyphae of *Rhizoctonia solani* and conidia of *Cochliobolus miyabeanus* by soil myxobacteria. *Phytopathology, 74*, 1234–1239.
Hu, M., Li, J., Chen, R., Li, W., Feng, L., et al. (2018). *Dickeya zeae* strains isolated from rice, banana and clivia rot plants show great virulence differentials. *BMC Microbiology, 18*, 1–15.
Ikley, J. T., Wise, K. A., & Johnson, W. G. (2015). Annual ryegrass (*Lolium multiflorum*), Johnsongrass (*Sorghum halepense*), and large crabgrass (*Digitaria sanguinalis*) are alternative hosts for *Clavibacter michiganensis* subsp. *nebraskensis*, causal agent of goss's wilt of corn. *Weed Science, 63*, 901–909.
Jacoby, R., Peukert, M., Succurro, A., Koprivova, A., & Kopriva, S. (2017). The role of soil microorganisms in plant mineral nutrition — Current knowledge and future directions. *Frontiers in Plant Science, 8*, 1617.
Jiang, G., Zhang, Y., Gan, G., Li, W., Wan, W., Jiang, Y., ... & Dini-Andreote, F. (2022). Exploring rhizo-microbiome transplants as a tool for protective plant-microbiome manipulation. *ISME Communications*, *2*(1), 10.
Jones, J. D., & Dangl, J. L. (2006). The plant immune system. *Nature, 444*(7117), 323–329.
Junaid, J. M., Dar, N. A., Bhat, T. A., Bhat, A. H., & Bhat, M. A. (2013). Commercial biocontrol agents and their mechanism of action in the management of plant pathogens. *International Journal of Modern Plant and Animal Sciences, 1*, 39–57.
Kahia, M., Nguyen, T. T. A., McCune, F., Naasz, R., Antoun, H., & Fournier, V. (2021). Insecticidal effect of *Bacillus pumilus* PTB180 and *Bacillus subtilis* PTB185 used alone and in combination against the foxglove aphid and the melon aphid (Hemiptera: Aphididae). *Canadian Entomologist, 153*, 726–740.
Kang, M. S., Hamayun, M., Waqas, M., Kim, J., Shinwari, Z. K., & Lee, I. (2016). *Burkholderia* sp KCTC 11096BP modulates pepper growth and resistance against *Phytophthora capsici*. *Pakistan Journal of Botany, 48*, 1965–1970.
Kang, B. R., Anderson, A. J., & Kim, Y. C. (2018). Hydrogen cyanide produced by *Pseudomonas chlororaphis* O6 exhibits nematicidal activity against *Meloidogyne hapla*. *The Plant Pathology Journal, 34*, 35.
Kang, B. R., Anderson, A. J., & Kim, Y. C. (2019). Hydrogen cyanide produced by *Pseudomonas chlororaphis* O6 is a key aphicidal metabolite. *Canadian Journal of Microbiology, 65*, 185–190.
Kaur, S., Egidi, E., Qiu, Z., Macdonald, C. A., Verma, J. P., et al. (2022). Synthetic community improves crop performance and alters rhizosphere microbial communities. *Journal of Sustainable Agriculture and Environment, 1*, 118–131.
Khan, M. H., Meghvansi, M. K., Panwar, V., Gogoi, H. K., & Singh, L. (2010). Arbuscular mycorrhizal fungi-induced signalling in plant defence against phytopathogens. *Journal of Phytology, 2*, 53–69.
Khatri, S., Dubey, S., Shivay, Y. S., Jelsbak, L., & Sharma, S. (2023). Organic farming induces changes in bacterial community and disease suppressiveness against fungal phytopathogens. *Applied Soil Ecology, 181*, 104658.
Khedher, S. B., Boukedi, H., Kilani-Feki, O., Chaib, I., Laarif, A., et al. (2015). *Bacillus amyloliquefaciens* AG1 biosurfactant: Putative receptor diversity and histopathological effects on *Tuta absoluta* midgut. *Journal of Invertebrate Pathology, 132*, 42–47.

Kilic-Ekici, O., & Yuen, G. Y. (2003). Induced resistance as a mechanism of biological control by *Lysobacter enzymogenes* strain C3. *Phytopathology, 93*, 1103–1110.
Kim, P. I., Bai, H., Bai, D., Chae, H., Chung, S., et al. (2004). Purification and characterization of a lipopeptide produced by *Bacillus thuringiensis* CMB26. *Journal of Applied Microbiology, 97*, 942–949.
Kloepper, J. W., Ryu, C. M., & Zhang, S. (2004). Induced systemic resistance and promotion of plant growth by *Bacillus* spp. *Phytopathology, 94*, 1259–1266.
Köhl, J., Kolnaar, R., & Ravensberg, W. J. (2019). Mode of action of microbial biological control agents against plant diseases: Relevance beyond efficacy. *Frontiers in Plant Science, 10*, 845.
Kong, H. G., Song, G. C., & Ryu, C. M. (2019). Inheritance of seed and rhizosphere microbial communities through plant–soil feedback and soil memory. *Environmental Microbiology Reports, 11*, 479–486.
Kong, H. G., Song, G. C., Sim, H. J., & Ryu, C. M. (2021). Achieving similar root microbiota composition in neighbouring plants through airborne signalling. *The ISME Journal, 15*, 397–408.
Kottb, M., Gigolashvili, T., Großkinsky, D. K., & Piechulla, B. (2015). *Trichoderma* volatiles effecting *Arabidopsis*: From inhibition to protection against phytopathogenic fungi. *Frontiers in Microbiology, 6*, 995.
Kudoyarova, G., Arkhipova, T., Korshunova, T., Bakaeva, M., Loginov, O., & Dodd, I. C. (2019). Phytohormone mediation of interactions between plants and non-symbiotic growth promoting bacteria under edaphic stresses. *Frontiers in Plant Science, 10*, 1368.
Kumar, V. V. (2019). Influence of the rhizospheric microbiome in plant health management. In V. Kumar, R. Prasad, M. Kumar, & D. Choudhary (Eds.), *Microbiome in plant health and disease: Challenges and opportunities* (pp. 215–230). Springer.
Kumar, A., & Dubey, A. (2020). Rhizosphere microbiome: Engineering bacterial competitiveness for enhancing crop production. *Journal of Advanced Research, 24*, 337–352.
Kwak, M. J., Kong, H. G., Choi, K., Kwon, S. K., Song, J. Y., et al. (2018). Rhizosphere microbiome structure alters to enable wilt resistance in tomato. *Nature Biotechnology, 36*, 1100–1109.
Lalitha, S. (2017). Plant growth–promoting microbes: A boon for sustainable agriculture. In A. Dhanarajan (Ed.), *Sustainable agriculture towards food security* (pp. 125–158). Springer.
Lazcano, C., Boyd, E., Holmes, G., Hewavitharana, S., Pasulka, A., & Ivors, K. (2021). The rhizosphere microbiome plays a role in the resistance to soil-borne pathogens and nutrient uptake of strawberry cultivars under field conditions. *Scientific Reports, 11*, 3188.
Lee, J. H., Ma, K. C., Ko, S. J., Kang, B. R., Kim, I. S., & Kim, Y. C. (2011). Nematicidal activity of a nonpathogenic biocontrol bacterium, *pseudomonas chlororaphis* O6. *Current Microbiology, 62*, 746–751.
Leeman, M., Van Pelt, J. A., Den Ouden, F. M., Heinsbroek, M., Bakker, P. A. H. M., & Schippers, B. (1995). Induction of systemic resistance against *Fusarium* wilt of radish by lipopolysaccharides of *Pseudomonas fluorescens*. *Phytopathology, 85*, 1021–1027.
Li, X., Rui, J., Mao, Y., Yannarell, A., & Mackie, R. (2014). Dynamics of the bacterial community structure in the rhizosphere of a maize cultivar. *Soil Biology and Biochemistry, 68*, 392–401.
Li, Y., Kong, Y., Teng, D., Zhang, X., He, X., et al. (2018). Rhizobacterial communities of five co-occurring desert halophytes. *PeerJ, 6*, e5508.
Li, Z., Ye, X., Liu, M., Xia, C., Zhang, L., et al. (2019). A novel outer membrane β-1, 6-glucanase is deployed in the predation of fungi by myxobacteria. *The ISME Journal, 13*, 2223–2235.
Li, R., Feng, Y., Chen, H., Zhang, C., Huang, Y., et al. (2020). Whole-genome sequencing of *Bradyrhizobium diazoefficiens* 113-2 and comparative genomic analysis provide molecular insights into species specificity and host specificity. *Frontiers in Microbiology, 11*, 576800.
Liu, Y., Chen, L., Zhang, N., Li, Z., Zhang, G., et al. (2016). Plant-microbe communication enhances auxin biosynthesis by a root-associated bacterium, *bacillus amyloliquefaciens* SQR9. *Molecular Plant-Microbe Interactions, 29*, 324–330.
Lopez-Gomez, M., Sandal, N., Stougaard, J., & Boller, T. (2012). Interplay of flg22-induced defence responses and nodulation in *Lotus japonicus*. *Journal of Experimental Botany, 63*, 393–401.

Lori, M., Symnaczik, S., Mäder, P., De Deyn, G., & Gattinger, A. (2017). Organic farming enhances soil microbial abundance and activity—A meta-analysis and meta-regression. *PloS One, 12*, e0180442.

Maheshwari, D. K., Dheeman, S., & Agarwal, M. (2015). Phytohormone-producing PGPR for sustainable agriculture. In D. Maheshwari (Ed.), *Bacterial metabolites in sustainable agroecosystem* (Sustainable development and biodiversity, vol) (Vol. 12, pp. 159–182). Springer.

Majeed, A., Abbasi, M. K., Hameed, S., Imran, A., & Rahim, N. (2015). Isolation and characterization of plant growth-promoting rhizobacteria from wheat rhizosphere and their effect on plant growth promotion. *Frontiers in Microbiology, 6*, 198.

Mannaa, M., Han, G., Jeon, H. W., Kim, J., Kim, N., et al. (2020). Influence of resistance-inducing chemical elicitors against pine wilt disease on the rhizosphere microbiome. *Microorganisms, 8*, 884.

Matthews, A., Pierce, S., Hipperson, H., & Raymond, B. (2019). Rhizobacterial community assembly patterns vary between crop species. *Frontiers in Microbiology, 10*, 581.

Mehnaz, S. (2013). Secondary metabolites of *Pseudomonas aurantiaca* and their role in plant growth promotion. In *Plant microbe symbiosis: Fundamentals and advances* (pp. 373–393). Springer India.

Mendes, R., Garbeva, P., & Raaijmakers, J. M. (2013). The rhizosphere microbiome: Significance of plant beneficial, plant pathogenic, and human pathogenic microorganisms. *FEMS Microbiology Reviews, 37*, 634–663.

Mendes, L. W., Mendes, R., Raaijmakers, J. M., & Tsai, S. M. (2018). Breeding for soil-borne pathogen resistance impacts active rhizosphere microbiome of common bean. *The ISME Journal, 12*, 3038–3042.

Meziane, H., Van Der Sluis, I., Van Loon, L. C., Höfte, M., & Bakker, P. A. (2005). Determinants of *Pseudomonas putida* WCS358 involved in inducing systemic resistance in plants. *Molecular Plant Pathology, 6*, 177–185.

Mishina, T. E., & Zeier, J. (2006). The *Arabidopsis* flavin-dependent monooxygenase FMO1 is an essential component of biologically induced systemic acquired resistance. *Plant Physiology, 141*, 1666–1675.

Mishina, T. E., & Zeier, J. (2007). Pathogen-associated molecular pattern recognition rather than development of tissue necrosis contributes to bacterial induction of systemic acquired resistance in *Arabidopsis*. *Plant Journal, 50*, 500–513.

Mishra, J., Mishra, I., & Arora, N. K. (2022). 2, 4-Diacetylphloroglucinol producing *Pseudomonas fluorescens* JM-1 for management of ear rot disease caused by *Fusarium moniliforme* in *Zea mays* L. *3. Biotechnology, 12*, 138.

Mnif, I., Elleuch, M., Chaabouni, S. E., & Ghribi, D. (2013). *Bacillus subtilis* SPB1 biosurfactant: Production optimization and insecticidal activity against the carob moth *Ectomyelois ceratoniae*. *Crop Protection, 50*, 66–72.

Molitor, A., & Kogel, K. H. (2009). Induced resistance triggered by *Piriformospora indica*. *Plant Signaling and Behaviour, 4*, 215–216.

Muller, D. B., Vogel, C., Bai, Y., & Vorholt, J. A. (2016). The plant microbiota: Systems-level insights and perspectives. *Annual Review of Genetics, 50*, 211–234.

Musilova, L., Ridl, J., Polivkova, M., Macek, T., & Uhlik, O. (2016). Effects of secondary plant metabolites on microbial populations: Changes in community structure and metabolic activity in contaminated environments. *International Journal of Molecular Sciences, 17*, 1205.

Muslim, S. N., Al-Kadmy, I. M., Hussein, N. H., Ali, A. N. M., Taha, B. M., et al. (2016). Chitosanase purified from bacterial isolate *Bacillus licheniformis* of ruined vegetables displays broad spectrum biofilm inhibition. *Microbial Pathogenesis, 100*, 257–262.

Mustafa, S., Kabir, S., Shabbir, U., & Batool, R. (2019). Plant growth promoting rhizobacteria in sustainable agriculture: From theoretical to pragmatic approach. *Symbiosis, 78*, 115–123.

Nandi, M., Selin, C., Brassinga, A. K. C., Belmonte, M. F., Fernando, W. D., et al. (2015). Pyrrolnitrin and hydrogen cyanide production by *Pseudomonas chlororaphis* strain PA23

exhibits nematicidal and repellent activity against *Caenorhabditis elegans*. *PLoS One, 10*, e0123184.
Neung, S., Nguyen, X. H., Naing, K. W., Lee, Y. S., & Kim, K. Y. (2014). Insecticidal potential of *Paenibacillus elgii* HOA73 and its combination with organic sulfur pesticide on diamond-back moth, *Plutella xylostella*. *Journal of Korean Society for Applied Biological Chemistry, 57*, 181–186.
Nguyen, X. H., Naing, K. W., Lee, Y. S., Jung, W. J., Anees, M., & Kim, K. Y. (2013). Antagonistic potential of *Paenibacillus elgii* HOA73 against the root-knot nematode, *Meloidogyne incognita*. *Nematology, 15*, 991–1000.
Nuccio, E., Starr, E., Karaoz, U., Brodie, E. L., Zhou, J., et al. (2020). Niche differentiation is spatially and temporally regulated in the rhizosphere. *The ISME Journal, 14*, 999–1014.
O'Connell, K. P., Goodman, R. M., & Handelsman, J. (1996). Engineering the rhizosphere: Expressing a bias. *Trends in Biotechnology, 14*, 83–88.
O'Neill, B., Grossman, J., Tsai, M., Gomes, J. E., Lehmann, J., et al. (2009). Bacterial community composition in Brazilian Anthrosols and adjacent soils characterized using culturing and molecular identification. *Microbial Ecology, 58*, 23–35.
Ocimati, W., Were, E., Groot, J. C., Tittonell, P., Nakato, G. V., & Blomme, G. (2018). Risks posed by intercrops and weeds as alternative hosts to *Xanthomonas campestris* pv*musacearum* in banana fields. *Frontiers in Plant Science, 9*, 1471.
Olawuyi, O. J., Odebode, A. C., Oyewole, I. O., Akanmu, A. O., & Afolabi, O. (2014). Effect of arbuscular mycorrhizal fungi on *Pythium aphanidermatum* causing foot rot disease on pawpaw (*Carica papaya* L.) seedlings. *Archives of Phytopathology and Plant Protection, 47*, 185–193.
Oleńska, E., Małek, W., Wójcik, M., Swiecicka, I., Thijs, S., & Vangronsveld, J. (2020). Beneficial features of plant growth-promoting rhizobacteria for improving plant growth and health in challenging conditions: A methodical review. *Science of the Total Environment, 743*, 140682.
Olowe, O. M., Nicola, L., Asemoloye, M. D., Akanmu, A. O., & Babalola, O. O. (2022). Trichoderma: Potential bio-resource for the management of tomato root rot diseases in Africa. *Microbiological Research, 257*, 126978.
Olson, H. A., & Benson, D. M. (2007). Induced systemic resistance and the role of binucleate *Rhizoctonia* and *Trichoderma hamatum* 382 in biocontrol of *Botrytis* blight in geranium. *Biological Control, 42*, 233–241.
Park, I., Seo, Y. S., & Mannaa, M. (2023). Recruitment of the rhizo-microbiome army: Assembly determinants and engineering of the rhizosphere microbiome as a key to unlocking plant potential. *Frontiers in Microbiology, 14*, 1163832.
Patel, J. S., Singh, A., Singh, H. B., & Sarma, B. K. (2015). Plant genotype, microbial recruitment and nutritional security. *Frontiers in Plant Science, 6*, 608.
Poole, P., Ramachandran, V., & Terpolilli, J. (2018). Rhizobia: From saprophytes to endosymbionts. *Nature Reviews. Microbiology, 16*, 291–303.
Porras-Alfaro, A., & Bayman, P. (2011). Hidden fungi, emergent properties: endophytes and microbiomes. *Annual Review of Phytopathology, 49*, 291–315.
Prasanna, L., Eijsink, V. G., Meadow, R., & Gåseidnes, S. (2013). A novel strain of *Brevibacillus laterosporus* produces chitinases that contribute to its biocontrol potential. *Applied Microbiology and Biotechnology, 97*, 1601–1611.
Raheem, A., Shaposhnikov, A., Belimov, A. A., Dodd, I. C., & Ali, B. (2018). Auxin production by rhizobacteria was associated with improved yield of wheat (*Triticum aestivum* L.) under drought stress. *Archives of Agronomy and Soil Science, 64*, 574–587.
Rai, D., & Singh, S. K. (2018). Is banded leaf and sheath blight a potential threat to maize cultivation in Bihar. *International Journal of Current Microbiology and Applied Sciences, 7*, 671–683.
Rajput, V. D., Minkina, T., Feizi, M., Kumari, A., Khan, M., et al. (2021). Effects of silicon and silicon-based nanoparticles on rhizosphere microbiome, plant stress and growth. *Biology, 10*, 791.
Ramesha, V., Naik, L. K., & Somasekhara, M. (2018). Effect of different biocontrol agents against stalk rot of maize (*Fusarium solani*). *Trends in Biosciences, 11*, 3288–3292.

Ran, L. X., Li, Z. N., Wu, G. J., Van Loon, L. C., & Bakker, P. A. H. M. (2005). Induction of systemic resistance against bacterial wilt in *Eucalyptus urophylla* by fluorescent *Pseudomonas* spp. *European Journal of Plant Pathology, 113*, 59–70.

Redman, R. S., Freeman, S., Clifton, D. R., Morrel, J., Brown, G., & Rodriguez, R. J. (1999). Biochemical analysis of plant protection afforded by a nonpathogenic endophytic mutant of *Colletotrichum magna*. *Plant Physiology, 119*, 795–804.

Rodríguez, M., Marín, A., Torres, M., Béjar, V., Campos, M., & Sampedro, I. (2018). Aphicidal activity of surfactants produced by *Bacillus atrophaeus* L193. *Frontiers in Microbiology, 9*, 3114.

Ruffner, B., Péchy-Tarr, M., Ryffel, F., Hoegger, P., Obris, C., et al. (2013). Oral insecticidal activity of plant-associated *pseudomonads*. *Environmental Microbiology, 15*, 751–763.

Saber, W. I., Ghoneem, K. M., Al-Askar, A. A., Rashad, Y. M., Ali, A. A., & Rashad, E. M. (2015). Chitinase production by *Bacillus subtilis* ATCC 11774 and its effect on biocontrol of *Rhizoctonia* diseases of potato. *Acta Biologica Hungarica, 66*, 436–448.

Sahu, G. K., & Sindhu, S. S. (2011). Disease control and plant growth promotion of green gram by siderophore producing *Pseudomonas* sp. *Research Journal of Microbiology, 6*, 735.

Santos, F., Penaflor, M. F. G., Paré, P. W., Sanches, P. A., Kamiya, A. C., et al. (2014). A novel interaction between plant-beneficial rhizobacteria and roots: Colonization induces corn resistance against the root herbivore *Diabrotica speciosa*. *PLoS One, 9*, e113280.

Schiltz, S., Gaillard, I., Pawlicki-Jullian, N., Thiombiano, B., Mesnard, F., & Gontier, E. (2015). A review: What is the spermosphere and how can it be studied? *Journal of Applied Microbiology, 119*, 1467–1481.

Schuhegger, R., Ihring, A., Gantner, S., Bahnweg, G., Knappe, C., et al. (2006). Induction of systemic resistance in tomato by N-acyl-L-homoserine lactone-producing rhizosphere bacteria. *Plant, Cell & Environment, 29*, 909–918.

Segarra, G., Van der Ent, S., Trillas, I., & Pieterse, C. M. J. (2009). MYB72, a node of convergence in induced systemic resistance triggered by a fungal and a bacterial beneficial microbe. *Plant Biology, 11*, 90–96.

Sharma, T., Dreyer, I., Kochian, L., & Piñeros, M. A. (2016). The ALMT family of organic acid transporters in plants and their involvement in detoxification and nutrient security. *Frontiers in Plant Science, 7*, 1488.

Shen, X., Hu, H., Peng, H., Wang, W., & Zhang, X. (2013). Comparative genomic analysis of four representative plant growth-promoting rhizobacteria in *Pseudomonas*. *BMC Genomics, 14*, 1–20.

Shi, S., Nuccio, E., Herman, D. J., Rijkers, R., Estera, K., et al. (2015). Successional trajectories of rhizosphere bacterial communities over consecutive seasons. *MBio, 6*, e00746–e00715.

Silipo, A., Leon, M. R., Erbs, G., Lanzetta, R., Parrilli, M., et al. (2011). A unique bicyclic monosaccharide from the *Bradyrhizobium* lipopolysaccharide and its role in the molecular interaction with plants. *Angewandte Chemie International Edition, 50*, 12610–12612.

Sillen, W. M., Thijs, S., Abbamondi, G. R., Janssen, J., Weyens, N., et al. (2015). Effects of silver nanoparticles on soil microorganisms and maize biomass are linked in the rhizosphere. *Soil Biology and Biochemistry, 91*, 14–22.

Singh, S., Singh, U. B., Malviya, D., Paul, S., Sahu, P. K., et al. (2020). Seed biopriming with microbial inoculant triggers local and systemic defense responses against *Rhizoctonia solani* causing banded leaf and sheath blight in maize (*Zea mays* L.). *International Journal of Environmental Research and Public Health, 17*, 1396.

Smith, S. E., & Read, D. J. (2010). *Mycorrhizal symbiosis*. Academic press.

Stam, J. M., Kroes, A., Li, Y., Gols, R., van Loon, J. J., et al. (2014). Plant interactions with multiple insect herbivores: From community to genes. *Annual Review of Plant Biology, 65*, 689–713.

Stringlis, I. A., Yu, K., Feussner, K., de Jonge, R., Van Bentum, S., et al. (2018). MYB72-dependent coumarin exudation shapes root microbiome assembly to promote plant health. *PNAS, 115*, E5213–E5222.

Takenaka, S., Nishio, Z., & Nakamura, Y. (2003). Induction of defense reactions in sugar beet and wheat by treatment with cell wall protein fractions from the mycoparasite *Pythium oligandrum*. *Phytopathology, 93*, 1228–1232.

Tedersoo, L., Bahram, M., & Zobel, M. (2020). How mycorrhizal associations drive plant population and community biology. *Science, 367*(6480), eaba1223.

Thuerig, B., Binder, A., Boller, T., Guyer, U., Jiménez, S., Rentsch, C., & Tamm, L. (2006). An aqueous extract of the dry mycelium of Penicillium chrysogenum induces resistance in several crops under controlled and field conditions. *European Journal of Plant Pathology, 114*, 185–197.

Tom, A., & Patel, S. (2021). Dynamic association of *Fusarium verticillioides* with maize and its biological control. *The Pharma Innovation Journal, 10*, 1010–1019.

Turlings, T. C., & Erb, M. (2018). Tritrophic interactions mediated by herbivore-induced plant volatiles: Mechanisms, ecological relevance, and application potential. *Annual Review of Entomology, 63*, 433–452.

Tzfira, T., & Citovsky, V. (2006). *Agrobacterium*-mediated genetic transformation of plants: Biology and biotechnology. *Current Opinion in Biotechnology, 17*, 147–154.

Ullah, I., Khan, A. L., Ali, L., Khan, A. R., Waqas, M., et al. (2015). Benzaldehyde as an insecticidal, antimicrobial, and antioxidant compound produced by *Photorhabdus temperata* M1021. *Journal of Microbiology, 53*, 127–133.

Van Deynze, A., Zamora, P., Delaux, P. M., Heitmann, C., Jayaraman, D., et al. (2018). Nitrogen fixation in a landrace of maize is supported by a mucilage-associated diazotrophic microbiota. *PLoS Biology, 16*, e2006352.

Van Peer, R., Niemann, G. J., & Schippers, B. (1991). Induced resistance and phytoalexin accumulation in biological control of *Fusarium* wilt of carnation by *pseudomonas* sp.strain WCS 417. *Phytopathology, 81*, 728–734.

Veenstra, A., Rafudeen, M. S., & Murray, S. L. (2019). *Trichoderma asperellum* isolated from African maize seed directly inhibits *Fusarium verticillioides* growth in vitro. *European Journal of Plant Pathology, 153*, 279–283.

Viterbo, A., Harel, M., Horwitz, B. A., Chet, I., & Mukherjee, P. K. (2005). *Trichoderma* mitogen-activated protein kinase signalling is involved in induction of plant systemic resistance. *Applied and Environmental Microbiology, 71*, 6241–6246.

Vives-Peris, V., De Ollas, C., Gómez-Cadenas, A., & Pérez-Clemente, R. M. (2020). Root exudates: From plant to rhizosphere and beyond. *Plant Cell Reports, 39*, 3–17.

Wang, M., Geng, L., Sun, X., Shu, C., Song, F., & Zhang, J. (2020). Screening of *Bacillus thuringiensis* strains to identify new potential biocontrol agents against *Sclerotinia sclerotiorum* and *Plutella xylostella* in *Brassica campestris* L. *Biological Control, 145*, 104262.

Wang, X., Wang, C., Chen, R., Wang, W., Wang, D., & Tian, X. (2023). Plant genotype shapes the soil nematode community in the rhizosphere of tomatoes with different resistance *to Meloidognye incognita*. *Plants, 12*, 1528.

Wei, G., Kloepper, J. W., & Tuzun, S. (1991). Induction of systemic resistance of cucumber to *Colletotrichum orbiculare* by select strains of plant growth-promoting rhizobacteria. *Phytopathology, 81*, 1508–1512.

Wen, T., Zhao, M., Yuan, J., Kowalchuk, G. A., & Shen, Q. (2021). Root exudates mediate plant defense against foliar pathogens by recruiting beneficial microbes. *Soil Ecology Letters, 3*, 42–51.

Weston, L. A., Ryan, P. R., & Watt, M. (2012). Mechanisms for cellular transport and release of allelochemicals from plant roots into the rhizosphere. *Journal of Experimental Botany, 63*, 3445–3454.

Win, T. T., Bo, B., Malec, P., Khan, S., & Fu, P. (2021). Newly isolated strain of *Trichoderma asperellum* from disease suppressive soil is a potential bio-control agent to suppress *Fusarium* soil borne fungal phytopathogens. *Journal of Plant Pathology, 103*, 549–561.

Wong, W. S., Tan, S. N., Ge, L., Chen, X., & Yong, J. W. H. (2015). The importance of phytohormones and microbes in biofertilizers. In D. Maheshwari (Ed.), *Bacterial metabolites in sustain-*

able agroecosystem (Sustainable development and biodiversity, Vol) (Vol. 12, pp. 105–158). Springer.
Wu, L., Chen, J., Xiao, Z., Zhu, X., Wang, J., et al. (2018). Barcoded pyrosequencing reveals a shift in the bacterial community in the rhizosphere and rhizoplane of *Rehmannia glutinosa* under consecutive monoculture. *International Journal of Molecular Sciences, 19*, 850.
Xin, X. F., Nomura, K., Aung, K., Velásquez, A. C., Yao, J., et al. (2016). Bacteria establish an aqueous living space in plants crucial for virulence. *Nature, 539*(7630), 524–529.
Ye, Y. F., Li, Q. Q., Gang, F. U., Yuan, G. Q., Miao, J. H., & Wei, L. I. N. (2012). Identification of antifungal substance (Iturin A2) produced by *Bacillus subtilis* B47 and its effect on southern corn leaf blight. *Journal of Integrative Agriculture, 11*, 90–99.
Yin, C., Casa Vargas, J. M., Schlatter, D. C., Hagerty, C. H., Hulbert, S. H., & Paulitz, T. C. (2021). Rhizosphere community selection reveals bacteria associated with reduced root disease. *Microbiome, 9*, 1–8.
Yu, P., He, X., Baer, M., Beirinckx, S., Tian, T., Moya, Y. A., & Hochholdinger, F. (2021). Plant flavones enrich rhizosphere Oxalobacteraceae to improve maize performance under nitrogen deprivation. *Nature Plants, 7*, 481–499.
Zhang, R., Vivanco, J. M., & Shen, Q. (2017). The unseen rhizosphere root–soil–microbe interactions for crop production. *Current Opinion in Microbiology, 37*, 8–14.
Zhang, J., Liu, Y. X., Zhang, N., Hu, B., Jin, T., et al. (2019). NRT1.1B is associated with root microbiota composition and nitrogen use in field-grown rice. *Nature Biotechnology, 37*, 676–684.
Zhao, L. J., Yang, X. N., Li, X. Y., Wei, M. U., & Feng, L. I. U. (2011). Antifungal, insecticidal and herbicidal properties of volatile components from *Paenibacillus polymyxa* strain BMP-11. *Agricultural Sciences in China, 10*, 728–736.
Zhu, J. K. (2016). Abiotic stress signaling and responses in plants. *Cell, 167*, 313–324.
Zhu, M., Tong, L., Xu, M., & Zhong, T. (2021). Genetic dissection of maize disease resistance and its applications in molecular breeding. *Molecular Breeding, 41*, 32.

Chapter 8
Harnessing Rhizosphere Mycobiome for Sustainable Agriculture: Perspectives and Prospects

Sukanya Das, Laishram Shantikumar Singh, Amrit Kumar, and Jintu Rabha

Abstract Global agriculture faces the dual challenge of feeding a growing population while preserving the environment. Regrettably, traditional farming methods have led to significant environmental damage. Consequently, there is an increased emphasis on investigating the beneficial effects of plant-associated microorganisms in crops as a green and sustainable alternative. One such promising area of study is the rhizosphere mycobiome, a dynamic and intricate community of fungi inhabiting the root zone, which holds immense potential for sustainable agriculture. These fungi play a multifaceted role in enhancing crop productivity, managing pests and pathogens, and promoting soil health. Mycorrhizal symbiosis and rhizosphere fungi, in particular, are crucial in facilitating nutrient uptake and improving soil structure. Moreover, rhizospheric fungi exhibit biocontrol potential against plant pathogens through multiple mechanisms. They also contribute to stress tolerance in crops, fortifying their resilience to environmental challenges such as drought and disease outbreaks. The evolving landscape of mycobiome-based agriculture and emerging trends encourage further research and innovation in this field for sustainable farming practices. The exploration of the rhizosphere mycobiome offers a promising pathway for inspiring innovative research and propelling the advancement of sustainable agricultural practices.

Keywords Agricultural sustainability · Biological control · Nutrient cycling · Plant growth promotion · Rhizosphere fungi

S. Das · J. Rabha (✉)
Department of Botany, Gauhati University, Guwahati, Assam, India
e-mail: jinturabha@gauhati.ac.in

L. S. Singh
Department of Microbiology, Assam Down Town University, Guwahati, Assam, India

A. Kumar
Department of Botany, Nowgong College (Autonomous), Nagaon, Assam, India

P. Mathur, S. Roy (eds.), *Plant Microbiome and Biological Control*, Sustainability in Plant and Crop Protection 20,
https://doi.org/10.1007/978-3-031-75845-4_8

8.1 Introduction

Agriculture is an ever-evolving landscape, facing the dual challenges of a growing global population and the pressing need for environmental sustainability. Given these circumstances, the quest for innovative and holistic solutions is more imperative than ever. Traditional agricultural practices, characterized by intensive chemical inputs, monocultural cropping systems, and relentless utilization of finite natural resources, have undeniably driven remarkable increases in food production over the past few decades (Singh & Singh, 2017; Patel et al., 2020). However, they also exacted a substantial toll on the environment, leaving many challenges, including soil degradation, biodiversity loss, nutrient imbalances, and escalating environmental concerns (Vlek et al., 2017).

While scientists have long been intrigued by the complex interactions linking plants and microorganisms, recent years have witnessed a surge in the prioritization of research focusing on plant-associated microbiomes. The plant microbiome, consisting of a diverse population of bacteria, fungi, and other microorganisms that live within and around plant tissues, is crucial in influencing plant health, growth, and overall well-being (Vandenkoornhuyse et al., 2015). In this context, the rhizosphere mycobiome emerges as an intriguing and promising solution. It offers a natural, resilient, and ecologically sustainable frontier for pursuing more environment-friendly and sustainable agricultural practices (Kandasamy et al., 2021; Siegieda et al., 2023). This chapter delves into the current understanding of the rhizosphere mycobiome and its potential applications in agriculture. It explores the mechanisms through which fungi impact plant growth and health, including their roles in services such as nutrient cycling, disease suppression, and stress tolerance. We hope this exploration will inspire further research in this promising field and contribute to our collective pursuit of sustainable agricultural practices.

8.2 The Rhizosphere and Its Importance in Plant-Microbe Interactions

The rhizosphere is a narrow volume of soil that surrounds and is influenced by the roots (McNear Jr., 2013). It is a dynamic and highly active environment where complex interactions occur among plants, soil, and a multitude of microorganisms, including bacteria, fungi, and archaea. The term "rhizosphere" was first coined by the German scientist Lorenz Hiltner in 1904 to describe the volume of soil influenced by root exudates, where microbial populations differ from those in bulk soil. This region is characterized by a unique and dynamic microenvironment, playing a pivotal role in plant health, growth, and nutrition (Berendsen et al., 2012).

The significance of the rhizosphere in plant-microbe interactions cannot be overstated. It serves as a hotspot for microbial activity due to the release of organic compounds by roots, which include sugars, amino acids, organic acids, and secondary metabolites (Singh & Mukerji, 2006). These root exudates act as a rich source of energy and nutrients for soil microorganisms, leading to increased microbial biomass and diversity in the rhizosphere, compared to bulk soil. In return, these microorganisms can benefit plants, in processes involving nutrient acquisition, disease suppression, and stress tolerance (Brahmaprakash et al., 2017).

8.3 The Rhizosphere Mycobiome: Fungi in the Rhizosphere

Among the different groups of rhizosphere-associated microorganisms, fungi constitute a significant portion of the microbial biomass and play crucial roles in soil health and plant productivity. This vast and diverse group of organisms comprises the rhizosphere mycobiome. The rhizosphere mycobiome serves as a treasure trove of biological activity. It is a hotbed for symbiotic associations between fungi and plants, which co-evolved over million years (Arnold et al., 2010). These symbiotic relationships are beneficial and often result essential for plant growth and development. Mycorrhizal fungi, for instance, enhance plant nutrient uptake by extending their hyphal networks into the soil, effectively increasing the root surface area. In return, they receive photosynthetically derived carbon compounds from the plant (Sharma & Jha, 2014). However, the role of the rhizosphere mycobiome extends beyond nutrient acquisition. Fungi in the rhizosphere also contribute to disease suppression through competition for resources, production of antagonistic metabolites, or inducing systemic resistance in plants against a broad spectrum of pathogens. Furthermore, certain groups of fungi can enhance plant tolerance to abiotic stresses such as drought and salinity. Considering these multifaceted benefits, harnessing the rhizosphere mycobiome for sustainable agriculture appears to be a logical step forward. However, this is easier said than done. Numerous factors, including plant genotype, soil type, climate, and agricultural practices, influence the rhizosphere mycobiome (Moussa et al., 2023). As elaborated in the subsequent sections, exploring and comprehending these interactions is crucial for effectively manipulating the rhizosphere mycobiome to enhance crop productivity.

8.3.1 Rhizosphere Mycobial Ecology

The mycobial ecology of the rhizosphere encompasses a diverse array of fungal interactions that profoundly influence plant-microbe dynamics and soil ecosystems. At the forefront of these interactions are mycorrhizal fungi, notably arbuscular mycorrhizal fungi (AMF) and ectomycorrhizal (ECM) fungi, which form mutualistic associations with roots. Concurrently, endophytic fungi colonize roots internally, providing benefits such as improved nutrient acquisition, stress tolerance, and protection against pathogens (Deka et al., 2017; Das & Narzary, 2017). Saprotrophic fungi play a vital role in the rhizosphere by decomposing organic matter, contributing to nutrient cycling and soil organic matter decomposition (Nichols & Wright, 2004). This process involves breaking down complex organic compounds, thus releasing nutrients that plants can subsequently absorb. However, the rhizosphere has its challenges, as pathogenic fungi can also take residence, posing threats to plant health and crop productivity. Understanding and managing these pathogenic fungi become imperative in sustaining healthy crops and optimizing yields. Additionally, certain fungi in the rhizosphere act as biocontrol agents, suppressing the growth of plant pathogens and safeguarding plants from diseases (Thambugala et al., 2020). Studying the diverse fungal communities in the rhizosphere is crucial for gaining insights into their roles and contributions to the complex network of interactions. Examining rhizosphere fungal diversity involves culture-based and culture-independent approaches. Culture-based studies rely on cultivating fungi in laboratory conditions, while culture-independent methods, such as DNA sequencing, directly analyze genetic material from environmental samples, providing a comprehensive view of the fungal community taxonomically classified, without the need for cultivation. These investigations enhance our knowledge of the intricate dynamics at play and pave the way for informed agricultural practices to harness the symbiotic relationships between plants and fungi.

8.3.2 Fungal Diversity in the Rhizosphere: Case Studies

The root-associated domain hosts a remarkable diversity of fungi, encompassing various species with unique roles and interactions. The intricate relationship between plants and the diverse fungi in the rhizosphere highlights the significance of fungal diversity in the functioning and resilience of terrestrial ecosystems. Numerous studies have investigated the fungal diversity in the rhizosphere of various crop plants. Zhang et al. (2017) studied field fungal populations in wheat rhizospheres and bulk soils in the North China Plain. The rhizosphere exhibited lower fungal diversity and a distinct community structure compared to bulk soil. Their study demonstrated that the composition of fungal communities in wheat fields was influenced by both present and past environmental factors, with their effects varying depending on the proximity to roots. In a metagenomic-based similar study, Noor et al. (2021) compared the fungal rhizosphere communities of the following plants: *Rhazya stricta*, *Enneapogon desvauxii*, *Citrullus colonynthis*, *Senna italica*, and *Zygophyllum simplex*. Based on results, they concluded that the prevalent fungal phyla in all samples were Ascomycota and Basidiomycota. The rhizosphere of the examined plants contained only saprotrophic and pathogenic fungi, such as *Aspergillus*, *Fusarium*, *Penicillium*, *Ceratobasidium*, *Myrothecium*, *Cochliobolus*, *Phoma*, and *Pestalotiopsis*. Notably, Glomeromycota were not detected in their study.

Kazerooni et al. (2017) studied fungal diversity and composition in the rhizosphere of tomato plants grown in conventional and desert farming systems in Oman, by using 454-pyrosequencing and culture-based techniques. The results showed that

25% of fungal classes were shared between the two techniques, with pyrosequencing recovering more taxa. They found that although the fungal diversity in conventional and desert farming systems was comparable, the composition of fungal classes and taxa differed. Pyrosequencing revealed Microsporidetes and Dothideomycetes as common fungal classes, while culture-based technique revealed Eurotiomycetes as the most abundant class. They also observed that most fungal species in the rhizosphere of tomato were saprotrophs. The application of fertilizers has also been reported to affect the rhizosphere fungal diversity. In a similar study, Zhang et al. (2018) investigated the geographic distribution and community assembly of fungi in soybean soil and rhizosphere from 43 sites in China. Their results demonstrated that the rhizosphere samples had lower alpha diversity with unique fungal community structures. The composition of the fungal communities was reported to be influenced by historical processes as well as dispersal limitations. The study also demonstrates that dispersal limitation significantly influences rare sub-communities more significantly than abundant ones. These findings helped unravel the biogeography of root-associated fungi communities in agricultural soils.

Semenov et al. (2019) compared bacterial and fungal diversity in soil under potato, maize, and white mustard cultivation with mineral and organic fertilizers. Organic fertilizers increased the diversity of beneficial fungi such as *Trichoderma harzianum*, which is known as an antagonist of several plant pathogens. Fertilization significantly impacted fungal communities, with organic fertilizers proving more beneficial for fungal diversity, in both the soil and the root environment, than mineral fertilizers. Besides fertilization, various farming management practices are also known to effect fungal diversity in soil. Wang et al. (2017) examined how long-term conservation tillage and conventional tillage affected rhizosphere fungal communities during wheat growth in northern China. Their findings revealed that zero tillage enhances fungal alpha diversity in both the rhizosphere and the bulk soil, whereas plough tillage promoted more stable communities. Moreover, the researchers also discovered that variations in nutritional resources between the rhizosphere and bulk soils can select for distinct species of fungi, boosting community diversity during plant growth. According to this study, long-term changes in tillage regimes may result in unique soil fungal ecology, potentially influencing other elements of soil functioning.

Plants significantly influence the microbial communities that inhabit the soil immediately surrounding roots. In ecological studies, representative plants intentionally introduced into a given environment serve as indicators of ecological conditions, allowing researchers to assess the impact of specific factors on plant-microbe interactions. Schöps et al. (2020) investigated the rhizosphere mycobiomes of phytometer and resident plants of common grassland species, *Dactylis glomerata* L. *sensu stricto* and *Plantago lanceolata* L., along a land-use intensity gradient. Despite some fungal taxa exhibiting plant specificity, the phytometer plants hosted a comparable rhizosphere mycobiome as resident plants. They concluded that the rhizosphere of plant species exhibited distinct fungal richness, with *P. lanceolata* having higher richness than *D. glomerata*, and that land-use intensity significantly impacted the mycobiome. Gomes et al. (2003) carried out a culture-independent analysis of fungal populations in tropical soil and rhizosphere of two maize cultivars. They observed a consistent rhizosphere effect on fungal community composition throughout the plant development stages. The rhizospheres of young maize plants exhibited the presence of Ascomycetes (order Pleosporales), while the rhizosphere of senescent maize plants showed different Ascomycetes members and basidiomycetous yeasts.

Several reports highlighted the positive correlation between crop health and fungal diversity. Kandasamy et al. (2021) conducted a study in southwestern Ontario to investigate the impact of fungal diversity on maize crop production. Aerial infrared imaging identified low and high-productive zones on 10 farms, with higher yields linked to increased fungal phylogenetic and alpha diversity. While specific fungal species correlated with productivity, no universal markers could be found across farms. Soil texture and chemistry, including clay, sand, phosphorus saturation, cation exchange capacity, and aluminium, significantly influenced both yield and mycobiome diversity. High-yielding sites harboured species such as *Penicillium* spp., *Trichoderma*, *Chalara fungorum*, and *Gibellulopsis*, while low-yielding locations shared species such as *Fusarium* spp., *Pythium*, *Setophoma terrestris*, and *Neonectria*.

In a recent study, Siegieda et al. (2023) investigated the role of biological indicators in monitoring soil microbial diversity. They revealed characteristic differences between the mycobiomes of various soil compartments, such as the bulk and rhizosphere, and plant niches such as roots and shoots. Their research was based on observations from 13 organic strawberry farms, both healthy and diseased. Data indicated that fungal trophic mode composition is more essential in shaping plant health than mycobiome alpha diversity. Moreover, healthy soil niches and roots exhibited fewer core taxa but a greater diversity, compared to unhealthy ones. The research also emphasized the importance of uncommon fungal species and highlighted the stability and consistency of fungal networks in healthy strawberry farms, particularly in the roots-shoot axis of sick strawberry plants.

8.4 Role of Fungal Community in the Rhizosphere

Fungi in the rhizosphere enhance plant growth and support sustainable agriculture by promoting plant health, facilitating nutrient acquisition, and protecting against pathogens. They also aid in nutrient cycling and enhance soil fertility (Fig. 8.1).

8.4.1 Plant Growth Promoting Fungi in the Rhizosphere and Agricultural Sustainability

The term Plant growth-promoting fungi (PGPF) was coined to describe a diverse group of fungi that significantly improve plant growth and support sustainable agriculture (Devi et al., 2020; El-Maraghy et al., 2020). These fungi, predominantly

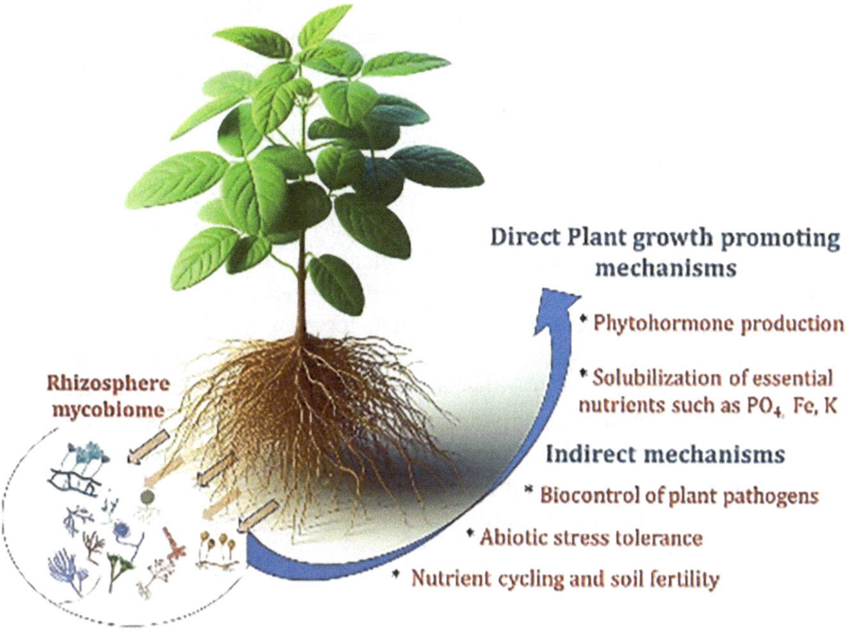

Fig. 8.1 Diagrammatic representation of the beneficial roles played by fungi in the rhizosphere

residing in the rhizosphere, are pivotal in enhancing plant health, nutrient acquisition, and protection against pathogens. Various plant beneficial mechanisms employed by PGPF, including mycorrhizal fungi, are particularly noteworthy. They rely on mutualistic relationships, facilitating nutrient acquisition, enhancing the absorption of essential nutrients, such as phosphorus and nitrogen, from soil. This is achieved through the extension of fungal hyphae into the soil, increasing the root nutrient absorption capacity and making nutrients more available to the plant (Aleklett et al., 2021). Rhizosphere fungi play a crucial role as prominent phosphate solubilizers in soil. Phosphate deficiency in soil is a common difficulty for plant growth, as this element is necessary in many physiological processes.

Despite being present in many soils, phosphorus often exists in insoluble forms that are unavailable to plants. This constraint, however, can be alleviated by certain fungi known as phosphate solubilizers, which play a crucial role in increasing plant phosphorus availability (Kaur & Sudhakara Reddy, 2017). Particularly, the phosphate solubilizing fungi found in the rhizosphere have the ability to transform the insoluble forms of phosphorus into soluble ones, thereby enhancing their accessibility to plants. The secretion of organic acids, enzymes, and other substances, that break down the bonds that keep phosphorus in its insoluble state, is involved in this process (Sindhu et al., 2014). These fungi solubilize phosphorus, converting it into a form that roots may easily absorb, resulting in enhanced nutrient uptake (Khan et al., 2010). Root-dwelling fungi can also improve plant growth through the release of phytohormones in the rhizosphere.

Phytohormones are chemical substances produced naturally by plants that regulate growth and development. Some rhizospheric fungi have been found to produce these hormones, thereby influencing plant growth (Rabha et al., 2014; Sarabia et al., 2018; Das et al., 2023). AMF have also been shown to produce cytokinin, auxin, and gibberellin (Pons et al., 2020). Rhizospheric fungi produce siderophores, iron-binding molecules that are crucial in making iron available to plants. These siderophores dissolve and translocate iron in the soil to the roots. Furthermore, a few root-associated fungi have also been reported to play a significant role in potassium solubilization, a process that is crucial for plant growth and yield (Teotia et al., 2016). Fungi like *Aspergillus* spp., and *Aspergillus terreus*, can solubilize this insoluble potassium into forms that are accessible for plant uptake. They do this by excreting organic acids, which solubilize the unavailable potassium. The use of these potassium-solubilizing microbes as biofertilizers may be a sustainable solution to increasing crop productivity and reducing the reliance on chemical fertilizers. These microbes may also protect against phytopathogens, thereby playing a significant role in plant nutrition and overall health. Table 8.1 shows a list of recently reported fungal species with an attributed role in plant growth promotion through various mechanisms.

8.4.2 Mycorrhizal Symbiosis in Sustainable Agriculture

Mycorrhizal symbiosis is a critical and intricate mutualistic association between roots and fungi. The fungus facilitates the absorption of essential minerals from soil, whereas the plant provides the fungus with carbohydrates produced through photosynthesis. Mycorrhizal networks increase the reach of the root systems, improving nutrients uptake and water absorption thus promoting plant growth and overall resilience to environmental stresses. This symbiosis is critical to ecosystem functioning because it influences plant community dynamics, nutrient cycling, and soil structure (Carteron et al., 2022). Mycorrhizal associations are important components of terrestrial ecosystems and contribute to the ecological succession of different plant species (Harrison, 1997). A more robust and eco-friendly sustainable agricultural system can be developed by utilizing the complex interaction between mycorrhizal fungi and plants. It can provide a comprehensive strategy that places an emphasis on biodiversity preservation, long-term soil health, and the development of agricultural systems that are both economical and environmentally sound.

The hyphae of mycorrhizal fungi extended into the soil form a vast network that greatly increases the root system. This network improves nutrient absorption and is frequently more effective than roots alone (Shi et al., 2023). Mycorrhizal fungi have been shown to increase plant uptake of macronutrients such as phosphorus and nitrogen by up to 80% (van Der Heijden et al., 2015). For phosphate uptake, mycorrhizal fungi use their extensive extra-radical hyphae to reach phosphate that would otherwise be inaccessible to plants, found in soil nutrient patches outside the rhizosphere. These extra-radical hyphae also produce low-molecular-weight organic

Table 8.1 Plant growth promotion exhibited by rhizospheric fungi and mechanisms of action

Fungi	Plant species	Mechanisms	Reference
Trichoderma harzianum	*Persea bombycina*	IAA, Phosphate solubilization	Rabha et al., (2014)
Trichoderma spp.	*Arabidopsis thaliana*	IAA, Volatile Organic Compounds (VOCs)	Nieto-Jacobo et al., (2017)
Aspergillus sp.	Wheat (*Triticum aestivum*) and chickpea (*Cicer arietinum*)	Gibberelic acid, IAA, siderophore, phosphate	Pandya et al., (2018)
Candida railenensis, *Crypococcus flavus*, *Solicoccozyma aeria* and *Clavispora lusitaniae*	Maize (*Zea mays* L.)	Phosphate	Sarabia et al., (2018)
Brettanomyces naardensis and AMF	Sunflower (*Helianthus annuus* L.)	IAA, phosphate and ammonia	Nafady et al., (2019)
T. harzianum and *T. virens*	*Pinus sylvestris* var. *mongolica*	-	Halifu et al., (2019)
Talaromyces funiculosus	Chilli (*Capsicum annuum* L.)	IAA, siderophore, phosphate	Naziya et al., (2020)
Mortierella spp.	Multiple plant species	IAA, phosphate, siderophore, ACC deaminase	Ozimek & Hanaka, (2020)
Aspergillus niger, Penicillium sp. and Rhizopus oryzae	Wheat (*Triticum aestivum*) var. PBW 343	IAA, siderophore, phosphate, ammonia	Imran et al., (2021)
Penicillium aurantiogriseum	Sesame (*Sesamum indicum* L.)	IAA, phosphate, siderophore	Zhao et al. (2021b)
T. afroharzianum	Tomato (*Solanum lycopersicum* L.)	IAA, phosphate, siderophore, ACC deaminase	Zhao et al. (2021a)
Purpureocillium lilacinum, Metarhizium marquandii	Maize (*Zea mays*), bean (*Phaseolus vulgaris*) and soybean (*Glycine max*)	Phosphate, IAA	Baron et al., (2020)
Penicillium olsonii	*Aeluropus littoralis*	IAA	Tarroum et al., (2022)
Piriformospora indica	Tomato	Induced higher levels of putrescine, auxin (indole-3-acetic acid) and gibberellin (GA4 and GA7) levels in host.	Kundu et al., 2022
Papiliotrema laurentii, Saitozyma podzolica	Brinjal (*S. melongena*)	IAA production, phosphate, ammonia, etc.	Das et al., 2023
Pochonia chlamydosporia	Tomato and Barley (*Hordeum vulgare*)	Defense induction	Maciá-Vicente et al., 2009 and Pentimone et al., 2019

acids, which help to dissolve phosphorus from non-available sources like iron and calcium phosphate precipitates (Shi et al., 2023).

Recently, the comprehensive mechanism of interaction, and the pathways involved in mycorrhizal-assisted plant phosphate uptake, have been elucidated (Shi et al., 2023). Therefore, we refrain from discussing it further in this chapter. The mycorrhizal fungi were also reported to collaborate with phosphate-solubilizing bacteria (PSB), where PSB convert non-available phosphorous compounds into orthophosphate, which mycorrhizal fungi take up and impart to their host plant (Nacoon et al., 2020). Furthermore, the co-inoculation of the PSB *Bacillus megaterium* and the AMF *Funneliformis mosseae* were reported to improve nutritional quality and promote shoot growth in alfalfa (Liu et al., 2020). Similarly, co-inoculation of *Klebsiella variicola* and AMF *Rhizophagus intraradices*, in combination with rock phosphate, promoted growth and enhanced inulin production in *Helianthus tuberosus* (Nacoon et al., 2020).

The hyphal network formed by mycorrhizal fungi acts as a natural extension of the root system, increasing its specific surface area also for water absorption (Püschel et al., 2020). An enhanced water uptake is crucial in water-stressed environments, contributing to the plant resilience during periods of drought. The mycorrhizal association thus becomes a key strategy for mitigating the impact of climate variability on crop productions. Besides, mycorrhizal fungi play a pivotal role in soil structure improvement. The hyphae produce glomalin, a glycoprotein that helps bind soil particles into stable aggregates (Singh et al., 2022). This aggregation enhances soil structure, promoting water infiltration, aeration, and root exploration. Improved soil structure is fundamental to mitigate soil erosion, enhancing fertility, and fostering a conducive environment for plant growth (Singh et al., 2022).

8.4.3 *Fungal Biocontrol Agents in the Rhizosphere*

The pursuit of sustainable agriculture encouraged a growing interest in harnessing natural mechanisms to manage plant diseases while minimizing environmental impact (Kumar et al., 2021). Rhizosphere fungi play a crucial role in the biocontrol of pathogens that threaten plant health and productivity. These fungi exhibit a range of mechanisms that act as a powerful arsenal against invading pathogens, as discussed below.

8.4.3.1 Antagonism

Antagonism is a key mechanism employed by rhizospheric fungi in biocontrol. These fungi use antagonistic activities to combat and suppress the growth of plant pathogens in the soil, contributing to effective disease management in the rhizosphere (Vallance et al., 2009). The capacity to counteract pathogens stems primarily from producing robust defensive chemicals. These compounds, ranging from antibiotics to secondary metabolites, serve as molecular weapons that disrupt the growth and development of pathogenic organisms (Das et al., 2023). Through an intricate interplay of biochemical pathways, these fungi synthesize and release compounds that exhibit selective toxicity, targeting specific pathogens while sparing beneficial microbes and the host plant (Raaijmakers et al., 2009).

Several rhizospheric fungi have been identified for their remarkable antagonistic activities. Notable examples include species from different genera i.e., *Trichoderma* and *Penicillium*, renowned for suppressing a broad spectrum of soil-borne pathogens. In particular, many *Trichoderma* spp., a group of fungi commonly found in soil, are known for their biocontrol capabilities and ability to suppress growth of other phytopathogenic fungi. The antifungal activities are attributed to the production of various secondary metabolites (SMs), including terpenes, pyrones, gliotoxin, gliovirin, and peptaibols. Each group of secondary metabolites plays a specific role in the antagonistic interaction between *Trichoderma* spp. and phytopathogens. Also *Penicillium* is a source of novel bioactive molecules, which can be used as antifungal agents. The antifungal potential has been reported *vs* a diverse group of phytopathogenic species. Filtrates of *Penicillium* isolates showed high antifungal effects on mycelial growth of *Fusarium oxysporum, F. solani, Macrophomina phaseolina, Aspergillus japonicus* var. *aculeatus* and *Cladosporium cladosporioides* (Khokhar et al., 2011). Recent studies have also highlighted the potential of yeasts as biocontrol agents (Das et al., 2023; de Oliveira et al., 2023; Bosqueiro et al., 2023). The biocontrol effect of these fungi has been harnessed in agricultural practices worldwide, offering sustainable alternatives to conventional, chemical-based, disease management strategies.

8.4.3.2 Competition for Resources

Competition for limited resources is a fundamental driving force in the rhizosphere microcosm, shaping the dynamics between plant-associated microorganisms. Rhizospheric fungi, equipped with an intricate set of adaptive strategies, engage in resource competition to establish dominance and suppress the proliferation of potential plant pathogens. This activity relies on mechanisms such as nutrient scavenging strategies, which ensure their nutritional requirements and strategically deprive potential pathogens of the resources needed for growth. The competition for nutrients, such as nitrogen and phosphorus, creates an environment where rhizosphere fungi gain a competitive edge, limiting the establishment and virulence of pathogenic organisms. Siderophores are low molecular weight organic compounds produced by certain rhizospheric fungi such as *Trichoderma*, *Beauveria* and *Metarhizium* spp. (Ghosh et al., 2017), *Aspergillus niger* (Osman et al., 2018), *Aureobasidium pullulans* (Di Francesco & Baraldi, 2021). The siderophore-producing ability enhances the competitive advantage of rhizospheric fungi by scavenging the available limited iron in the rhizosphere, and contributes to the overall suppression of pathogenic species. Besides, siderophores are also known to directly influence the growth of various pathogens (Leong, 1986; Ghosh et al., 2020; Di Francesco & Baraldi, 2021).

8.4.3.3 Induced Systemic Resistance

Induced Systemic Resistance (ISR) mediated by rhizospheric fungi plays a pivotal role in the biocontrol of plant pathogens, offering a sustainable and environment-friendly approach to disease management. This sophisticated defense mechanism involves the activation of systemic resistance in plants through intricate signaling pathways and molecular crosstalk between the plant and beneficial rhizospheric fungi (Ghosh & Panja, 2021; see also Chaps. 1 and 3 in this Volume). By recognizing specific microbial molecules, known as elicitors, these fungi initiate a cascade of events that prime the plant immune system for a heightened and more efficient response against a broad spectrum of pathogens (Pieterse et al., 2014). The signaling pathways involved in ISR are mediated by key signaling molecules, including salicylic acid, jasmonic acid, and ethylene. Recognition of these elicitors leads to the activation of defense-related genes, producing antimicrobial compounds, reinforcing cell walls, and activating enzymes involved in defense activities within the plant (Pieterse et al., 2014). This priming effect fortifies the plant, that is then ready to repel invading pathogens. *Trichoderma* spp. are widely known for their ability to induce systemic resistance in plants against a wide range of pathogens (Nawrocka & Małolepsza, 2013).

8.4.4 Nutrient Cycling and Soil Fertility

The sustainable productivity of all ecosystems, including agricultural and natural environments, depends on soil quality (Azcón-Aguilar & Barea, 2015). The rhizosphere is a hotspot for the microbial activity and plays an important role in nutrient cycle activities (Reinhold-Hurek et al., 2015). Understanding the composition and function of the rhizosphere mycobiome is a rapidly developing field of research with implications for sustainable agriculture and ecosystem functioning. In this regard, various diversity-based studies have been conducted on the rhizosphere microflora, mostly focusing on the diversity of the bacterial components that received much attention (Prashar et al., 2014; Beckers et al., 2016; Lei et al., 2019; Ling et al., 2022). However, studies have shown that the rhizosphere mycobiome is also highly diverse and plays a crucial role in plant health and nutrient cycling (Xu et al., 2012; Kandasamy et al., 2021; Semenov et al., 2022). Fungi in the rhizosphere can form long-term, mutualistic associations with plants, providing enhanced nutrient uptakes and protection from pathogens (Moussa et al., 2023).

The decomposition of organic matter is a fundamental ecological process that fuels ecosystem nutrients cycling. As key decomposers, fungi can break down complex organic compounds, releasing essential nutrients into the ecosystem (Condron et al., 2010). Many fungi utilize a plethora of enzymes to depolymerize recalcitrant organic substrates. Extracellular enzymes, such as cellulases, ligninases, and proteases, act synergistically to hydrolyze complex polymers into simpler, assimilable compounds (Oyewole et al., 2022). These enzymes play a crucial role in the breakdown of plant litter, facilitating the release of carbon, nitrogen, and other essential elements (Kjøller & Struwe, 2002). According to Liang et al. (2023), fungal guilds such as plant pathogens, mycorrhizal fungi, and saprotrophs are highly prevalent in soils, particularly within the rhizosphere. Higher diversity of saprotrophic fungi has been shown to increase decomposition rates when assessing leaf mass losses (Fang et al., 2020). The mycelial network of fungi extends into the surrounding environment, actively foraging for organic matter and nutrients. This hyphal exploration increases the decomposition efficiency and enhances nutrient mobilization from the substrate (Aleklett et al., 2021). The extensive mycelial network acts as a conduit for nutrient transport, connecting decomposing material with the broader ecosystem. In addition to their decomposing activities, manysymbiotic relationships with plants enable efficient nutrient transfer between fungi and plants, enhancing nutrient uptake and promoting plant growth (Jha et al., 1993). The mycorrhizal network serves as a nutrient exchange conduit, further influencing ecosystem nutrient dynamics. Various environmental factors influence fungal decomposition rates, including temperature, moisture, and substrate quality (Bradford et al., 2014). Understanding these factors is critical for predicting the impact of global environmental changes on fungal decomposition dynamics and nutrient cycling (van der Wal et al., 2013).

8.5 Mycobiome Signaling and Plant Health

The coevolution of plants with the surrounding microflora opened up new possibilities for studying the dynamics related to plant growth and disease control. Since the emergence of plants from aquatic to terrestrial ecosystems during the course of evolution, fossil records indicated the early establishment of arbuscular mycorrhizal fungi, at a time when the soil was in a poor nutritional status for promoting plant growth and development (Redecker et al., 2000). In the long run, these groups of fungi, both mycorrhizal or free-living, carried out repeated cycling of nutrients, degradation of soil organic matter, and carbon distribution. They thus changed the organic matter dynamics and the microbial communities in successive trophic levels, within the early food webs (Anal et al., 2020). These plant-associated fungal communities are now cosmopolitan in distribution. They occur in a variety of habitats, viz. soil, dead organic matter, living plant host (endophytes), the surface of leaves (phyllosphere) or roots (mycorrhizae), where they play an important role by sustaining or promoting tolerance to biotic and abiotic stresses, by enhancing plant growth and, in general, in any ecological management (Rai et al., 2019; Hardoim et al., 2015; Anal et al., 2020).

The mycobiome components associated with plants may be grouped into five categories: saprotrophic, pathogenic, epiphytic, endophytic and mycorrhizal fungi. These components maintain their relative fitness and diversity either through cross-protection, parasitism, ISR induction in the host plant, nutrients acquisition, competition through the action of catabolic enzymes and activation of hexose transporters genes (Hardoim et al., 2015; Cai et al., 2015; Bajpai et al., 2019; Anal et al., 2020). Out of all the fungi associated with plants, the mycorrhizal association is the most predominant one. The majority of the ascomycetes and basidiomycetes involved in symbiotic associations are mycorrhizae, primarily associated with dominant trees in forest ecosystem, thus shaping the forest structure (Bajpai et al., 2019).

Recent studies related to the symbiotic associations between plants and fungi indicated the presence of effector molecules such as cysteine-rich secreted proteins i.e., MISSP7, that mediate the exchange of nutrients between the two partners (Martin et al., 2008; Bonfante & Genre, 2010). However, in absence of an effector molecule, symbiosis may be mediated by the expression of genes which encode proteins related to the metabolism of indole-3-acetic acid and root morphogenesis (Bajpai et al., 2019). Beneficial fungi primarily utilize various fungal effector molecules to limit plant defense responses (Pozo et al., 2021). Various hormonal effectors, including Ca^{2+}, reactive oxygen species, and nitric oxide signatures, regulate mutualistic symbiosis (Pozo et al., 2021). Molecular mechanisms related to symbiosis include the differential expression of protein coding genes, as in *Laccaria bicolor* and *Tuber melanosporum*, due to the presence of transposable elements (Martin et al., 2008; Riccioni et al., 2008; Bajpai et al., 2019). These elements occupy a significant portion of the plant genome and evolved to encode enzymes targeted to cell wall, to avoid the plant defense responses during symbiosis (Bonfante & Genre, 2010; Bajpai et al., 2019). Another example includes sucrose synthase genes, whose inhibition reduces plant height, weight, and seed yield (Baier et al., 2010; Bajpai et al., 2019). Transportation of major macro-elements such as nitrogen is mediated either directly by releasing ammonium ions or by the assimilation of ammonium into amino acids in the hyphae, due to the expression of fungal genes involved in glutamine synthase and arginine breakdown (Bonfante & Genre, 2010; Bajpai et al., 2019). In the mutual association between the roots and fungi, bioactive molecules viz. strigolactones or flavonoids are firstly released by roots. These molecules are specifically detected by the fungus that responds by releasing low molecular weight, partially lipophilic molecules viz. the Myc factor, that is detected by the Myc factor receptor in roots (Bajpai et al., 2019; Pozo et al., 2021). On receptor activation, the expression of a membrane-bound steroid binding protein occurs, which increases the synthesis of secondary messenger Ca^{2+} required for subsequent colonization (Bonfante & Genre, 2010; Kuhn et al., 2010; Bajpai et al., 2019).

In mycorrhizal symbiosis, downstream elements participating in rapid signal transduction are localized in the plant cell nucleus through a common signal transduction pathway, called symbiosis (SYM) pathway (Bajpai et al., 2019). Several plant genes — viz. LjSYMRK in *Lotus japonicus* and MtDM12 in *Medicago truncata* — encode a receptor kinase to perceive signals from fungi (Markmann et al., 2008; Bajpai et al., 2019). Around 500 genes are expressed differentially and involved in cell wall synthesis during mycorrhizal colonization (Bajpai et al., 2019).

Chitin, a major element of fungal cell wall, elicits plant defense responses. On the other hand, pattern recognition receptors mediate in plants the recognition of potential aggressors through multiple defense barriers (Pozo et al., 2021). On successful completion of a pre-symbiotic phase (receptor and signal interaction), physical contact between the two partners occurs, which results in proliferation and branching of both root tissues and hyphae. The hyphae next develop a hyphopodium, mediated by differential expression of genes involved in cell wall remodelling. The hyphopodium is a mycelial structure which adheres to the root epidermis by undergoing repeated branching on the cell wall of the root epidermal cells (Bajpai et al., 2019; Anal et al., 2020). Formation of hyphopodium is followed by a resting period of 4–6 h during which the plant prepares for subsequent colonization by the fungus, by producing a pre-penetration apparatus (PPA) and developing an interface compartment for the fungus (Bajpai et al., 2019; Anal et al., 2020). At the point of contact, aggregation of the cytoplasm occurs, membrane-bound organelles become concentrated in the pre-penetration apparatus and hyphal penetration occurs by the development of thick columns which invade the plant tissue and repeatedly branch to form arbuscules within the inner root cortex region (Genre et al., 2005; Paszowski, 2006; Bajpai et al., 2019). These arbuscules function as haustoria and help in nutrient exchange between plants and fungi. The Vapyrin gene has been reported to mediate the process, and its inhibition resulted in a significant decrease in epidermis penetration, ABC transporters and arbuscules formation (Pumplin et al., 2010; Zhang et al., 2010; Bajpai et al., 2019).

8.6 Enhancing Crop Resilience to Abiotic Stress

The rhizospheric mycobiome plays a crucial role in enhancing the resistance of crops to various abiotic stressors such as drought, salinity, and extreme temperatures (Sharma et al., 2021; Chandra et al., 2022). Mycorrhizal associations, by forming extensive hyphal networks, aid nutrient absorption and water uptake, essential for plant survival in arid and nutrient-poor conditions (Bouskout et al., 2022). Moreover, certain fungal species within the rhizosphere are known to produce bioactive compounds. These molecules can act as elicitors, triggering the plant stress response mechanisms and bolstering its resilience against environmental challenges (Pattnaik & Busi, 2019; Imran et al., 2021). Various fungi have been reported to produce ACC (1-aminocyclopropane-1-carboxylate) deaminase, an enzyme that significantly induces abiotic stress resistance in plants (Nadeem et al., 2010). These fungi may lower the inhibitory concentrations of ethylene, a plant hormone that increases under stress, allowing plants to better cope with stressful conditions. The enzyme ACC deaminase present in these fungi cleaves ACC, the precursor of ethylene, thereby lowering the ethylene level and facilitating plant growth (Adedayo & Babalola, 2023). Furthermore, ACC deaminase-positive plant growth-promoting fungal strains can intensively promote the growth and development of crops cultivated under various abiotic stresses conditions (Nadeem et al., 2010; Ripa et al., 2019; Zhang et al., 2019; Rauf et al., 2021).

The rhizospheric mycobiome also contributes to the modulation of plant hormonal balance, specifically auxins and cytokinins (Sharma et al., 2017; Sharma et al., 2021, These hormones are pivotal for plant growth and stress adaptation. Fungal symbionts can positively influence these hormonal pathways, promoting root development and enhancing the plant's ability to withstand stress (Ho-Plágaro & García-Garrido, 2022). In addition, the secretion of metabolites by rhizospheric fungi can act as signaling molecules (Dlamini et al., 2022). These molecules prime the plant for an anticipatory response to impending stressors (Sharma & Kumawat, 2022). Harnessing the potential of the rhizospheric mycobiome in inducing abiotic stress resistance offers promising prospects for optimizing yield and quality of crops (Pozo et al., 2021). Moreover, it contributes to the pursuit of sustainable agricultural practices in the face of a changing climate (Aich et al., 2022). Understanding these complex interactions and their implications for plant health is a crucial area of ongoing research. The International LIFE MycoRestore consortium prompted the utilization of fungal biodiversity to increase resilience towards climatic changes and biological pathogens, leading to increased plant health (Diez-Hermano et al., 2022).

8.7 Current Challenges and Future Prospects

Mycobiome rhizosphere research is crucial to advance agricultural sustainability, providing insights into the complex relationship between fungi and roots. The rhizosphere houses diverse fungal communities. However, understanding this mycobiome is challenging due to the vast diversity of fungal species and the limitations of traditional culture-based methods. High-throughput sequencing offers a solution but requires extensive bioinformatics expertise and databases for interpretation.

Despite these challenges, mycobiome research presents promising opportunities in sustainable agriculture. Technological advances allow for deeper exploration of the mycobiome, uncovering novel fungal species and their functional roles. Integrating multi-omics approaches provides a holistic understanding of the mycobiome, identifying key fungal contributors to plant health and soil ecosystems. This knowledge has practical implications for sustainable agricultural practices, offering a base to develop eco-friendly alternatives to conventional agrochemicals.

Translating laboratory findings into practical applications requires cost-effective methods for mass production, formulation, and delivery of beneficial fungal strains. However, the lack of well-defined regulatory frameworks and potential safety concerns may hinder or delay the adoption of these novel strains by farmers.

Despite these hurdles, the field of rhizosphere mycobiome research offers many opportunities for discovery and innovation. Deciphering the "black box" of fungal interactions is a critical frontier. Understanding how fungi communicate with each other, or with other soil microbes and plants is key to maximizing the benefits of these networks. Exploring the mycobiome role in supporting broader ecosystem services will be highly informative to define innovative strategies aiming at building resilient and sustainable agricultural systems.

References

Adedayo, A. A., & Babalola, O. O. (2023). Fungi that promote plant growth in the rhizosphere boost crop growth. *Journal of Fungi, 9*, 239. https://doi.org/10.3390/jof9020239

Aich, A., Dey, D., & Roy, A. (2022). Climate change resilient agricultural practices: A learning experience from indigenous communities over India. *PLOS Sustainability and Transformation, 1*(7), e0000022. https://doi.org/10.1371/journal.pstr.0000022

Aleklett, K., Ohlsson, P., Bengtsson, M., & Hammer, E. C. (2021). Fungal foraging behaviour and hyphal space exploration in micro-structured soil chips. *The ISME Journal, 15*(6), 1782–1793.

Anal, A. K. D., Rai, S., Singh, M., & Solanki, M. K. (2020). Plant mycobiome: Current research and applications. In *Phytobiomes: Current insights and future vistas* (pp. 81–104). Springer Nature.

Arnold, A. E., Lamit, L. J., Gehring, C. A., Bidartondo, M. I., & Callahan, H. (2010). Interwoven branches of the plant and fungal trees of life. *The New Phytologist, 185*, 874–878.

Azcón-Aguilar, C., & Barea, J. M. (2015). Nutrient cycling in the mycorrhizosphere. *Journal of Soil Science and Plant Nutrition, 15*(2), 372–396.

Baier, M. C., Keck, M., Gödde, V., Niehaus, K., Küster, H., & Hohnjec, N. (2010). Knockdown of the symbiotic sucrose synthase MtSucS1 affects arbuscule maturation and maintenance in mycorrhizal roots of *Medicago truncatula. Plant Physiology, 152*, 1000–1014.

Bajpai, A., Rawat, S., & Johri, B. N. (2019). Fungal diversity: Global perspective and ecosystem dynamics. In T. Satyanarayana, B. N. Johri, & S. K. Das (Eds.), *Microbial diversity in ecosystem sustainability and biotechnological applications* (pp. 83–113). Springer.

Baron, N. C., de Souza Pollo, A., & Rigobelo, E. C. (2020). *Purpureocillium lilacinum* and *Metarhizium marquandii* as plant growth-promoting fungi. *Peer J, 8*, e9005.

Beckers, B., Op De Beeck, M., Thijs, S., Truyens, S., Weyens, N., Boerjan, W., & Vangronsveld, J. (2016). Performance of 16s rDNA primer pairs in the study of rhizosphere and endosphere bacterial microbiomes in metabarcoding studies. *Frontiers in Microbiology, 7*, 650.

Berendsen, R. L., Pieterse, C. M., & Bakker, P. A. (2012). The rhizosphere microbiome and plant health. *Trends in Plant Science, 17*(8), 478–486.

Bonfante, P., & Genre, A. (2010). Mechanisms underlying beneficial plant-fungus interactions in mycorrhizal symbiosis. *Nature Communications, 1*, 48.

Bosqueiro, A. S., Bizarria, R., Jr., & Rosa-Magri, M. M. (2023). Biocontrol of post-harvest tomato rot caused by *Alternaria arborescens* using *Torulaspora indica. Biocontrol Science and Technology, 33*(2), 115–132.

Bouskout, M., Bourhia, M., Al Feddy, M. N., Dounas, H., Salamatullah, A. M., Soufan, W., et al. (2022). Mycorrhizal fungi inoculation improves *Capparis spinosa*'s yield, nutrient uptake and photosynthetic efficiency under water deficit. *Agronomy, 12*, 149.

Bradford, M. A., Wood, S. A., Bardgett, R. D., Black, H. I., Bonkowski, M., et al. (2014). Discontinuity in the responses of ecosystem processes and multifunctionality to altered soil community composition. *Proceedings of the National Academy of Sciences of the United States of America, 111*(40), 14478–14483.

Brahmaprakash, G. P., Sahu, P. K., Lavanya, G., Nair, S. S., Gangaraddi, V. K., & Gupta, A. (2017). Microbial functions of the rhizosphere. In D. P. Singh, H. B. Singh, & R. Prabha (Eds.), *Plant-microbe interactions in agro-ecological perspectives: Volume 1: Fundamental mechanisms, methods and functions* (pp. 177–210). Springer.

Cai, F., Chen, W., Wei, Z., Pang, G., Li, R., Ran, W., & Shen, Q. (2015). Colonization of *Trichoderma harzianum* strain SQR-T037 on tomato roots and its relationship to growth, nutrient availability and soil microflora. *Plant and Soil, 388*, 337–350.

Carteron, A., Vellend, M., & Laliberté, E. (2022). Mycorrhizal dominance reduces local tree species diversity across US forests. *Nature Ecology & Evolution, 6*, 370–374. https://doi.org/10.1038/s41559-021-01634-6

Chandra, P., Rai, A. K., Sundha, P., Basak, N., & Kaur, H. (2022). Rhizospheric soil–plant-microbial interactions for abiotic stress mitigation and enhancing crop performance. In *Soil health and environmental sustainability: Application of geospatial technology* (pp. 593–614). Springer. https://doi.org/10.1007/978-3-031-09270-1_26

Condron, L., Stark, C., O'Callaghan, M., Clinton, P., & Huang, Z. (2010). The role of microbial communities in the formation and decomposition of soil organic matter. In *Soil microbiology and sustainable crop production* (pp. 81–118). Springer.

Das, S., & Narzary, D. (2017). Diversity study on endophytic fungi associated with *Oroxylum indicum* and their interactions with some phytopathogens. *Journal of Advnaced Plant Science, 9*(2), 27–41.

Das, S., Rabha, J., & Narzary, D. (2023). Assessment of soil yeasts *Papiliotrema laurentii* S-08 and *Saitozyma podzolica* S-77 for plant growth promotion and biocontrol of Fusarium wilt of brinjal. *Journal of Applied Microbiology, 134*(11), lxad252.

de Oliveira, A. J., Ono, M. A., Suguiura, I. M. D. S., Zucareli, C., Garcia, E. B., Olchanheski, L. R., & Ono, E. Y. S. (2023). Potential of yeasts as biocontrol agents against *Fusarium graminearum* in vitro and on corn. *Journal of Applied Microbiology, 134*(12), lxad296.

Deka, D., Tayung, K., & Jha, D. K. (2017). Harnessing fungal endophytes for plant and human health. In D. K. Maheshwari (Ed.), *Endophytes: Biology and Biotechnology* (Vol. 1, pp. 59–98). Springer.

Devi, R., Kaur, T., Kour, D., Rana, K. L., Yadav, A., & Yadav, A. N. (2020). Beneficial fungal communities from different habitats and their roles in plant growth promotion and soil health. *Microbial Biosystems, 5*(1), 21–47.

Di Francesco, A., & Baraldi, E. (2021). How siderophore production can influence the biocontrol activity of *Aureobasidium pullulans* against *Monilinia laxa* on peaches. *Biological Control, 152*, 104456.

Diez-Hermano, S., Ahmad, F., Niño-Sanchez, J., Benito, A., Hidalgo, E., et al. (2022). Health condition and mycobiome diversity in Mediterranean tree species. *Frontiers in Forests and Global Change, 5*, 1056980. https://doi.org/10.3389/ffgc.2022.1056980

Dlamini, S. P., Akanmu, A. O., & Babalola, O. O. (2022). Rhizospheric microorganisms: The gateway to a sustainable plant health. *Frontiers in Sustainable Food Systems, 6*, 925802. https://doi.org/10.3389/fsufs.2022.925802

El-Maraghy, S. S., Tohamy, T. A., & Hussein, K. A. (2020). Role of plant-growth promoting fungi (PGPF) in defensive genes expression of *Triticum aestivum* against wilt disease. *Rhizosphere, 15*, 100223.

Fang, M., Liang, M., Liu, X., Li, W., Huang, E., & Yu, S. (2020). Abundance of saprotrophic fungi determines decomposition rates of leaf litter from arbuscular mycorrhizal and ectomycorrhizal trees in a subtropical forest. *Soil Biology and Biochemistry, 149*, 107966.

Genre, A., Chabaud, M., Timmers, T., Bonfante, P., & Barker, D. G. (2005). Arbuscular mycorrhizal fungi elicit a novel intracellular apparatus in *Medicago truncatula* root epidermal cells before infection. *Plant Cell, 17*, 3489–3499.

Ghosh, S. K., & Panja, A. (2021). Different mechanisms of signaling pathways for plant protection from diseases by fungi. In *Biocontrol agents and secondary metabolites* (pp. 591–630). Woodhead Publishing.

Ghosh, S. K., Banerjee, S., & Sengupta, C. (2017). Bioassay, characterization and estimation of siderophores from some important antagonistic fungi. *Journal of Biopesticides, 10*(2), 105–112.

Ghosh, S. K., Bera, T., & Chakrabarty, A. M. (2020). Microbial siderophore–A boon to agricultural sciences. *Biological Control, 144*, 104214.

Gomes, N. C. M., Fagbola, O., Costa, R., Rumjanek, N. G., Buchner, A., Mendona-Hagler, L., & Smalla, K. (2003). Dynamics of fungal communities in bulk and maize rhizosphere soil in the tropics. *Applied and Environmental Microbiology, 69*(7), 3758–3766.

Halifu, S., Deng, X., Song, X., & Song, R. (2019). Effects of two Trichoderma strains on plant growth, rhizosphere soil nutrients, and fungal community of *Pinus sylvestris* var. *mongolica* annual seedlings. *Forests, 10*, 758. https://doi.org/10.3390/f10090758

Hardoim, P. R., van Overbeek, L. S., Berg, G., Pirttila, A. M., Company, S., et al. (2015). The hidden world within plants: Ecological and evolutionary considerations for defining functioning of microbial endophytes. *Microbiology & Molecular Biology Reviews, 79*, 293–320.

Harrison, M. J. (1997). The arbuscular mycorrhizal symbiosis. In G. Stacey & N. T. Keen (Eds.), *Plant-microbe interactions*. Springer. https://doi.org/10.1007/978-1-4615-6019-7_1

Ho-Plágaro, T., & García-Garrido, J. M. (2022). Molecular regulation of arbuscular mycorrhizal Symbiosis. *International Journal of Molecular Sciences, 23*, 5960. https://doi.org/10.3390/ijms23115960

Imran, M., Abulreesh, H. H., Monjed, M. K., Elbanna, K., & Samreen, & Ahmad, I. (2021). Multifarious functional traits of free-living rhizospheric fungi, with special reference to *Aspergillus* spp. isolated from north Indian soil, and their inoculation effect on plant growth. *Annals of Microbiology, 71*, 1–17. https://doi.org/10.1186/s13213-021-01643-7

Jha, D. K., Sharma, G. D., & Mishra, R. R. (1993). Mineral nutrition in the tripartite interaction between Frankia, Glomus and Alnus at different soil phosphorus regimes. *New Phytologist, 123*(2), 307–311.

Kandasamy, S., Weerasuriya, N., Subramanian, G., Thorn, R. G., Patterson, G., Ali, S., & Lazarovits, G. (2021). Disentangling the association of corn root mycobiome with plant productivity and the importance of soil physicochemical balance in shaping their relationship. *Frontiers in Sustainable Food Systems, 5*, 617332.

Kaur, G., & Sudhakara Reddy, M. (2017). Role of phosphate-solubilizing fungi in sustainable agriculture. In T. Satyanarayana, S. Deshmukh, & B. Johri (Eds.), *Developments in fungal biology and applied mycology* (pp. 391–412). Springer.

Kazerooni, E. A., Maharachchikumbura, S. S. N., Rethinasamy, V., Al-Mahrouqi, H., & Al-Sadi, A. M. (2017). Fungal diversity in tomato rhizosphere soil under conventional and desert farming systems. *Frontiers in Microbiology, 8*, 1462. https://doi.org/10.3389/fmicb.2017.01462

Khan, M. S., Zaidi, A., Ahemad, M., Oves, M., & Wani, P. A. (2010). Plant growth promotion by phosphate solubilizing fungi–current perspective. *Archives of Agronomy and Soil Science, 56*(1), 73–98.

Khokhar, I., Mukhtar, I., & Mushtaq, S. (2011). Antifungal effect of Penicillium metabolites against some fungi. *Archives of Phytopathology and Plant Protection, 44*(14), 1347–1351.

Kjøller, A. H., & Struwe, S. (2002). Fungal communities, succession, enzymes, and decomposition. In *Enzymes in the environment: Activity, ecology and applications* (Vol. 1, pp. 267–284). Marcel Dekker.

Kuhn, H., Kuster, H., & Requena, N. (2010). Membrane steroid-binding protein 1 induced by a diffusible fungal signal is critical for mycorrhization in *Medicago truncatula*. *New Phytologist, 185*, 716–733.

Kumar, A., Rabha, J., & Jha, D. K. (2021). Antagonistic activity of lipopeptide-biosurfactant producing *Bacillus subtilis* AKP, against *Colletotrichum capsici*, the causal organism of anthracnose disease of chilli. *Biocatalysis and Agricultural Biotechnology, 36*, 102133.

Kundu, A., Mishra, S., Kundu, P., Jogawat, A., & Vadassery, J. (2022). *Piriformospora indica* recruits host-derived putrescine for growth promotion in plants. *Plant Physiology, 188*(4), 2289–2307.

Lei, S., Xu, X., Cheng, Z., Xiong, J., Ma, R., Zhang, L., et al. (2019). Analysis of the community composition and bacterial diversity of the rhizosphere microbiome across different plant taxa. *Microbiology, 8*(6), e00762.

Leong, J. (1986). Siderophores: Their biochemistry and possible role in the biocontrol of plant pathogens. *Annual Review of Phytopathology, 24*(1), 187–209.

Liang, S., Arraiano-Castilho, R., Neuenkamp, L., et al. (2023). Rhizosphere fungal community assembly varied across functional guilds in a temperate forest. *Ecological Processes, 12*, 6. https://doi.org/10.1186/s13717-023-00417-0

Ling, N., Wang, T., & Kuzyakov, Y. (2022). Rhizosphere bacteriome structure and functions. *Nature Communications, 13*(1), 836.

Liu, J., Liu, X., Zhang, Q., Li, S., Sun, Y., Lu, W., & Ma, C. (2020). Response of alfalfa growth to arbuscular mycorrhizal fungi and phosphate-solubilizing bacteria under different phosphorus application levels. *AMB Express, 10*, 1–13.

Maciá-Vicente, J. G., Rosso, L. C., Ciancio, A., Jansson, H.-B., & Lopez-Llorca, L. V. (2009). Colonisation of barley roots by endophytic *fusarium equiseti* and *Pochonia chlamydosporia*: Effects on plant growth and disease. *Annals of Applied Biology, 155*, 391–401.

Markmann, K., Giczey, G., & Parniske, M. (2008). Functional adaptation of a plant receptor-kinase paved the way for the evolution of intracellular root symbioses with bacteria. *PLoS Biology, 6*, 497–506.

Martin, F., Aerts, A., Ahrén, D., et al. (2008). The genome of Laccaria bicolor provides insights into mycorrhizal symbiosis. *Nature, 452*(7183), 88–92.

McNear, D. H., Jr. (2013). The rhizosphere — Roots, soil and everything in between. *Nature Education Knowledge, 4*, 1.

Moussa, T. A., Mohamed, A. H., & Zaky, M. S. (2023). Rhizosphere mycobiome: Roles, diversity, and dynamics. In *Plant mycobiome: Diversity, interactions and uses* (pp. 47–61). Springer International Publishing.

Nacoon, S., Jogloy, S., Riddech, N., Mongkolthanaruk, W., Kuyper, T. W., & Boonlue, S. (2020). Interaction between phosphate solubilizing bacteria and arbuscular mycorrhizal fungi on growth promotion and tuber inulin content of *Helianthus tuberosus* L. *Scientific Reports, 10*(1), 4916.

Nadeem, S. M., Zahir, Z. A., Naveed, M., & Arshad, M. (2010). Microbial ACC-deaminase: Prospects and applications for inducing salt tolerance in plants. *Critical Reviews in Plant Sciences, 29*(6), 360–393.

Nafady, N. A., Hashem, M., Hassan, E. A., Ahmed, H. A., & Alamri, S. A. (2019). The combined effect of arbuscular mycorrhizae and plant-growth-promoting yeast improves sunflower defense against *Macrophomina phaseolina* diseases. *Biological Control, 138*, 104049.

Nawrocka, J., & Małolepsza, U. (2013). Diversity in plant systemic resistance induced by Trichoderma. *Biological Control, 67*(2), 149–156.

Naziya, B., Murali, M., & Amruthesh, K. N. (2020). Plant Growth-Promoting Fungi (PGPF) instigate plant growth and induce disease resistance in *Capsicum annuum* L. upon infection with *Colletotrichum capsici* (Syd.) Butler & Bisby. *Biomolecules, 10*, 41. https://doi.org/10.3390/biom10010041

Nichols, K. A., & Wright, S. F. (2004). Contributions of fungi to soil organic matter in agroecosystems. In C. R. C. Press (Ed.), *Soil organic matter in sustainable agriculture* (pp. 179–198).

Nieto-Jacobo, M. F., Steyaert, J. M., Salazar-Badillo, F. B., Nguyen, D. V., Rostás, M., Braithwaite, M., et al. (2017). Environmental growth conditions of *Trichoderma* spp. affects indole acetic acid derivatives, volatile organic compounds, and plant growth promotion. *Frontiers in Plant Science, 8*, 102.

Noor, S. O., Al-Zahrani, D. A., Hussein, R. M., Baeshen, M. N., Moussa, T. A., Abo-Aba, S. M., et al. (2021). Assessment of fungal diversity in soil rhizosphere associated with Rhazya stricta and some desert plants using metagenomics. *Archives of Microbiology, 203*, 1211–1219.

Osman, Y., Mowafy, A. M., Gebreil, A. S., & Hamed, S. M. (2018). Siderophore production by rhizosphere inhabiting bacteria and fungi. *Journal of Plant Production, 9*(9), 717–721.

Oyewole, O. A., Saidu, M. M., Idris, A. D., Yakubu, J. G., & Bello, A. B. (2022). Communities of microbial enzymes and biodegradation of persistent environmental pollutants. In *Ecological interplays in microbial enzymology* (pp. 247–277). Springer Nature.

Ozimek, E., & Hanaka, A. (2020). Mortierella species as the plant growth-promoting fungi present in the agricultural soils. *Agriculture, 11*(1), 7.

Pandya, N. D., Desai, P. V., Jadhav, H. P., & Sayyed, R. Z. (2018). Plant growth promoting potential of *Aspergillus* sp. NPF7, isolated from wheat rhizosphere in South Gujarat, India. *Environmental Sustainability, 1*(3), 245–252.

Paszowski, U. (2006). A journey through signaling in arbuscular mycorrhizal symbioses. *Tansley Reviews in New Phytologist, 172*, 35–46.

Patel, S. K., Sharma, A., & Singh, G. S. (2020). Traditional agricultural practices in India: An approach for environmental sustainability and food security. *Energy, Ecology and Environment, 5*, 253–271.

Pattnaik, S. S., & Busi, S. (2019). Rhizospheric fungi: Diversity and potential biotechnological applications. In A. N. Yadav, S. Mishra, S. Singh, & A. Gupta (Eds.), *Recent advancement in white biotechnology through fungi: Volume 1: Diversity and enzymes perspectives* (pp. 63–84). Springer.

Pentimone, I., Colagiero, M., Ferrara, M., Nigro, F., & Ciancio, A. (2019). Time-dependent effects of *Pochonia chlamydosporia* endophytism on gene expression profiles of colonized tomato roots. *Applied Microbiology and Biotechnology, 103*, 8511–8527.

Pieterse, C. M., Zamioudis, C., Berendsen, R. L., Weller, D. M., Van Wees, S. C., & Bakker, P. A. (2014). Induced systemic resistance by beneficial microbes. *Annual Review of Phytopathology, 52*, 347–375.

Pons, S., Fournier, S., Chervin, C., Bécard, G., Rochange, S., Frei Dit Frey, N., et al. (2020). Phytohormone production by the arbuscular mycorrhizal fungus *Rhizophagus irregularis*. *PLoS One, 15*(10), e0240886. https://doi.org/10.1371/journal.pone.0240886

Pozo, M. J., Zabalgogeazcoa, I., de Vazquez Aldana, B. R., & Martinez-Medina, A. (2021). Untapping the potential of plant mycobiomes for applications in agriculture. *Current Opinion in Plant Biology, 60*, 102034. https://doi.org/10.1016/j.pbi.2021.102034

Prashar, P., Kapoor, N., & Sachdeva, S. (2014). Rhizosphere: Its structure, bacterial diversity and significance. *Reviews in Environmental Science and Bio/Technology, 13*, 63–77.

Pumplin, N., Mondo, S. J., Topp, S., Starker, C. G., Gantt, J. S., & Harrison, M. J. (2010). *Medicago truncatula* Vapyrin is a novel protein required for arbuscular mycorrhizal symbiosis. *The Plant Journal, 61*(3), 482–494.

Püschel, D., Bitterlich, M., Rydlová, J., & Jansa, J. (2020). Facilitation of plant water uptake by an arbuscular mycorrhizal fungus: A Gordian knot of roots and hyphae. *Mycorrhiza, 30*, 299–313.

Raaijmakers, J. M., Paulitz, T. C., Steinberg, C., Alabouvette, C., & Moënne-Loccoz, Y. (2009). The rhizosphere: A playground and battlefield for soilborne pathogens and beneficial microorganisms. *Plant and Soil, 321*, 341–361. https://doi.org/10.1007/s11104-008-9568-6

Rabha, J., Jha, D. K., Sarma, H. K., Acharya, A., Chakraborty, U., & Chakraborty, B. N. (2014). Growth improvement and biochemical responses of *Persea bombycina* to native rhizosphere *Trichoderma harzanium* isolate PBR1. *Journal of Botanical Society of Bengal, 68*, 149–157.

Rai, S., Slanki, M. K., Solanki, A. C., & Surapathrudu, K. (2019). *Trichoderma* as biocontrol agent: Molecular prospectus and application. In R. A. Ansari & I. Mahmood (Eds.), *Plant health under biotic stress*. Springer. https://doi.org/10.1007/978-981-13-6040-4_7

Rauf, M., Awais, M., Ud-Din, A., Ali, K., Gul, H., Rahman, M. M., Hamayun, M., & Arif, M. (2021). Molecular mechanisms of the 1-Aminocyclopropane-1-Carboxylic Acid (ACC) deaminase producing *Trichoderma asperellum* MAP 1 in enhancing wheat tolerance to waterlogging stress. *Frontiers in Plant Science, 11*, 614971. https://doi.org/10.3389/fpls.2020.614971

Redecker, D., Kodner, R., & Graham, L. E. (2000). Glomalean fungi from the Ordovician. *Science, 289*, 1920–1921. https://doi.org/10.1126/science.289.5486.1920

Reinhold-Hurek, B., Bünger, W., Burbano, C. S., Sabale, M., & Hurek, T. (2015). Roots shaping their microbiome: Global hotspots for microbial activity. *Annual Review of Phytopathology, 53*, 403–424.

Riccioni, C., Belfiori, B., Rubini, A., Passeri, V., Arcioni, S., & Paolocci, F. (2008). *Tuber melanosporum* outcrosses: Analysis of the genetic diversity within and among its natural populations under this new scenario. *New Phytologist, 180*(2), 466–478.

Ripa, F. A., Cao, W. D., Tong, S., & Sun, J. G. (2019). Assessment of plant growth promoting and abiotic stress tolerance properties of wheat endophytic fungi. *BioMed Research International, 6105865.*

Sarabia, M., Cazares, S., González-Rodríguez, A., Mora, F., Carreón-Abud, Y., & Larsen, J. (2018). Plant growth promotion traits of rhizosphere yeasts and their response to soil characteristics and crop cycle in maize agroecosystems. *Rhizosphere, 6*, 67–73.

Schöps, R., Goldmann, K., Korell, L., Bruelheide, H., Wubet, T., & Buscot, F. (2020). Resident and phytometer plants host comparable rhizosphere fungal communities in managed grassland ecosystems. *Scientific Reports, 10*, 919.

Semenov, M. V., Nikitin, D. A., Stepanov, A. L., & Semenov, V. M. (2019). The structure of bacterial and fungal communities in the rhizosphere and root-free loci of gray forest soil. *Eurasian Soil Science, 52*, 319–332.

Semenov, M. V., Krasnov, G. S., Semenov, V. M., & van Bruggen, A. (2022). Mineral and organic fertilizers distinctly affect fungal communities in the crop rhizosphere. *Journal of Fungi, 8*, 251.

Sharma, B., & Jha, D. K. (2014). Arbuscular mycorrhizal fungal association in some commonly occurring economically important plants of Kaziranga National Park, Assam, India. *Indian Phytopathology, 67*(3), 243–251.

Sharma, B., & Kumawat, K. C. (2022). Functional diversity of microbes in rhizosphere: A key player for soil health conservation under changing climatic conditions. In A. Vaishnav, S. Arya, & D. K. Choudhary (Eds.), *Plant stress Mitigators*. Springer. https://doi.org/10.1007/978-981-16-7759-5_22

Sharma, R., Minakshi, S & Chauhan, A. (2017). Rhizosphere microbiome and its role in plant growth promotion. In: Kalia, V., Shouche, Y., Purohit, H., Rahi, P. (eds) Mining of microbial wealth and metagenomics, 29–56 Springer.

Sharma, S., Chandra, D., & Sharma, A. K. (2021). Rhizosphere plant–microbe interactions under abiotic stress. In V. V. S. R. Gupta & A. K. Sharma (Eds.), *Rhizosphere biology: Interactions between microbes and plants* (pp. 195–216). Springer. https://doi.org/10.1007/978-981-15-6125-2_10

Shi, J., Wang, X., & Wang, E. (2023). Mycorrhizal symbiosis in plant growth and stress adaptation: From genes to ecosystems. *Annual Review of Plant Biology, 74*, 569–607.

Siegieda, D., Panek, J., & Frąc, M. (2023). Plant and soil health in organic strawberry farms–Greater importance of fungal trophic modes and networks than α-diversity of the mycobiome. *Applied Soil Ecology, 188*, 104925.

Sindhu, S. S., Phour, M., Choudhary, S. R., & Chaudhary, D. (2014). Phosphorus cycling: Prospects of using rhizosphere microorganisms for improving phosphorus nutrition of plants. In N. Parmar & A. Singh (Eds.), *Geomicrobiology and biogeochemistry* (pp. 199–237). Springer.

Singh, G., & Mukerji, K. G. (2006). Root exudates as determinant of rhizospheric microbial biodiversity. In K. G. Mukerji, C. Manoharachary, & J. Singh (Eds.), *Microbial activity in the rhizoshere* (pp. 39–53). Springer Berlin Heidelberg.

Singh, R., & Singh, G. S. (2017). Traditional agriculture: A climate-smart approach for sustainable food production. *Energy, Ecology and Environment, 2*, 296–316.

Singh, P., Mugunthan, J., Patel, R., & Mukherjee, S. (2022). Mycorrhizal product glomalin: A proficient agent of nutrient sequestration and soil fertility restoration under jeopardized agroecosystem. In *Microbes and microbial biotechnology for green remediation* (pp. 781–794). Elsevier.

Tarroum, M., Romdhane, W. B., Al-Qurainy, F., Ali, A. A. M., Al-Doss, A., Fki, L., & Hassairi, A. (2022). A novel PGPF *Penicillium olsonii* isolated from the rhizosphere of *Aeluropus littoralis* promotes plant growth, enhances salt stress tolerance, and reduces chemical fertilizers inputs in hydroponic system. *Frontiers in Microbiology, 13*, 996054.

Teotia, P., Kumar, V., Kumar, M., Shrivastava, N., & Varma, A. (2016). Rhizosphere microbes: Potassium solubilization and crop productivity–Present and future aspects. In V. Meena, B. Maurya, J. Verma, & R. Meena (Eds.), *Potassium solubilizing microorganisms for sustainable agriculture* (pp. 315–325). Springer.

Thambugala, K. M., Daranagama, D. A., Phillips, A. J., Kannangara, S. D., & Promputtha, I. (2020). Fungi vs. fungi in biocontrol: An overview of fungal antagonists applied against fungal plant pathogens. *Frontiers in Cellular and Infection Microbiology, 10*, 604923.

Vallance, J., Le Floch, G., Déniel, F., Barbier, G., Lévesque, C. A., & Rey, P. (2009). Influence of *Pythium oligandrum* biocontrol on fungal and oomycete population dynamics in the rhizosphere. *Applied and Environmental Microbiology, 75*, 4790–4800.

van Der Heijden, M. G., Martin, F. M., Selosse, M. A., & Sanders, I. R. (2015). Mycorrhizal ecology and evolution: The past, the present, and the future. *New Phytologist, 205*, 1406–1423.

van der Wal, A., Geydan, T. D., Kuyper, T. W., & De Boer, W. (2013). A thready affair: Linking fungal diversity and community dynamics to terrestrial decomposition processes. *FEMS Microbiology Reviews, 37*, 477–494.

Vandenkoornhuyse, P., Quaiser, A., Duhamel, M., Le Van, A., & Dufresne, A. (2015). The importance of the microbiome of the plant holobiont. *New Phytologist, 206*, 1196–1206.

Vlek, P. L., Khamzina, A., & Tamene, L. D. (2017). *Land degradation and the sustainable development goals: Threats and potential remedies* (CIAT Publication no. 440). The International Center for Tropical Agriculture.

Wang, Z., Li, T., Wen, X., Liu, Y., Han, J., Liao, Y., & DeBruyn, J. M. (2017). Fungal communities in rhizosphere soil under conservation tillage shift in response to plant growth. *Frontiers in Microbiology, 8*, 1301. https://doi.org/10.3389/fmicb.2017.01301

Xu, L., Ravnskov, S., Larsen, J., & Nicolaisen, M. (2012). Linking fungal communities in roots, rhizosphere, and soil to the health status of *Pisum sativum*. *FEMS Microbiology Ecology, 82*(3), 736–745.

Zhang, Q., Blaylock, L. A., & Harrison, M. J. (2010). Two *Medicago truncatula* half-ABC transporters are essential for arbuscule development in arbuscular mycorrhizal symbiosis. *Plant Cell, 22*, 1483–1497.

Zhang, K., Adams, J. M., Shi, Y., Yang, T., Sun, R., et al. (2017). Environment and geographic distance differ in relative importance for determining fungal community of rhizosphere and bulk soil. *Environmental Microbiology, 19*, 3649–3659.

Zhang, J., Zhang, B., Liu, Y., Guo, Y., Shi, P., & Wei, G. (2018). Distinct large-scale biogeographic patterns of fungal communities in bulk soil and soybean rhizosphere in China. *Science of the Total Environment, 644*, 791–800.

Zhang, S., Gan, Y., & Xu, B. (2019). Mechanisms of the IAA and ACC-deaminase producing strain of *Trichoderma longibrachiatum* T6 in enhancing wheat seedling tolerance to NaCl stress. *BMC Plant Biology, 19*, 22. https://doi.org/10.1186/s12870-018-1618-5

Zhao, J., Liu, T., Liu, W. C., Zhang, D. P., Dong, D., Wu, H. L., et al. (2021a). Transcriptomic insights into growth promotion effect of *Trichoderma afroharzianum* TM2-4 microbial agent on tomato plants. *Journal of Integrative Agriculture, 20*(5), 1266–1276.

Zhao, X., Ni, Y., Zhao, H., Liu, X., He, B., et al. (2021b). Plant growth-promoting ability and control efficacy of *Penicillium aurantiogriseum* 44M-3 against sesame Fusarium wilt disease. *Biocontrol Science and Technology, 31*, 1314–1329.

Chapter 9
Boosting Plant Immunity: The Functional Role and Mechanism of Arbuscular Mycorrhizal Fungi in Resistance

Maunata Ghorui, Shouvik Chowdhury, Prakash Balu, Keshab Das, and Kiran Sunar

Abstract In plant-microorganism rhizosphere interactions, the Arbuscular Mycorrhizal Fungi (AMF) play a pivotal role in fortifying plant defense mechanisms *vs* deleterious pathogens. This intricate interplay is reinforced by the plant innate capacity to discriminate between symbiotic AMF and pathogenic entities, underpinned by complex molecular signalling cascades and receptor-mediated responses. AMF colonization initiates a series of events encompassing shifts in gene expression, modulation of protein profiles, and alterations in metabolite composition, mediated by bioactive compounds such as strigolactones and Myc-factors. Successful recognition culminates in substantial cellular reorganization, notably the formation of arbuscules within plant cells, signifying a harmonious symbiotic union. Concomitantly, plants have evolved discerning receptor complexes tailored to differentiate between AMF and pathogenic incursions, orchestrating a meticulous equilibrium between immune responses and symbiotic associations. AMF-mediated biocontrol represents a pivotal facet of this mutualistic partnership. These fungi function as biocontrol agents, shielding plants against an array of soil-borne adversaries, including bacteria, nematodes, and phytopathogenic fungi. The intricate mechanisms underpinning AMF-mediated biocontrol encompass induced systemic resistance, alterations in root morphological traits, biochemical modifications, and perturbations within the rhizospheric microbial community. This review elucidates the multifaceted molecular, physiological, and ecological dimensions of the symbiotic interplay between plants and AMF, unveiling the remarkable mechanisms augmenting plant defense strategies and restraining soil-borne pathogens.

M. Ghorui · S. Chowdhury · P. Balu
Department of Biotechnology, School of Life Sciences, Vels Institute of Science, Technology and Advanced Studies (VISTAS), Chennai, Tamil Nadu, India

K. Das · K. Sunar (✉)
Department of Botany, Balurghat Mahila Mahavidyalaya Balurghat, Dakshin Dinajpur, West Bengal, India

P. Mathur, S. Roy (eds.), *Plant Microbiome and Biological Control*, Sustainability in Plant and Crop Protection 20,
https://doi.org/10.1007/978-3-031-75845-4_9

Keywords Arbuscular Mycorrhizal fungi · Biocontrol agent · Induced systemic resistance · Systemic acquired resistance · Mycorrhiza induced resistance

Abbreviations

AMF	Arbuscular Mycorrhizal Fungi
BCA	Biocontrol Agent
ISR	Induced Systemic Resistance
SAR	Systemic Acquired Resistance
MIR	Mycorrhiza-Induced Resistance
PGPR	Plant Growth Promoting Rhizobacteria
BF	Branching factor
CSSP	Common Symbiosis Signalling Pathway
ABC	ATP-binding cassette
CO	Chito oligosaccharides
LCO	Lipo-chito-oligosaccharides
HRGP	Hydroxyproline-Rich Glycoprotein
GRSP	*Glomus*-related Soil Protein
IPM	Integrated Pest Management

9.1 Introduction

In the intricate tapestry of the Earth ecosystems, a silent partnership unfolds beneath the soil surface, hidden from the naked eye but profoundly influential in the survival of numerous plants. This alliance involves an ancient group of microorganisms known as Arbuscular Mycorrhizal Fungi (AMF). Across the globe, from the rainforests to the arid deserts, these remarkable fungi forge mutualistic connections with the roots of around 80% of terrestrial plants (Weng et al., 2022), forming a bond that has endured for hundreds of million years.

While the intricacies of plant-microbe interactions have fascinated scientists for generations, the role, and mechanisms by which AMF bolster a plant ability to fend off diseases remain a subject of intense research and fascination. Our voyage shall embark on a journey through the molecular conversations between AMF and host plants, elucidating the transformations instigated within plant tissues. A foray into the augmentation of nutrient acquisition, modulation of hormonal networks, and reinforcement of physical barriers will reveal the multifaceted strategies orchestrated by AMF to augment plant resistance against pathogenic intruders.

Extensive use of chemical pesticides has led to pesticide-resistant pathogens along with the decline of natural predators, exacerbating pest and disease issues (Huang et al., 2011). Chemical pesticides also pollute air, water, soil, and food, raising food safety concerns (Razak & Gange, 2021; Hage-Ahmed et al., 2019). Environmental scientists and plant pathologists are increasingly focused on green, eco-friendly disease and pest control methods, with biological control gaining prominence due to its efficiency, low resource requirements, and environmental safety (Hildebrandt et al., 2007; Gianinazzi et al., 2010; van Driesche et al., 2010; Lee et al., 2013; Begum et al., 2019). This includes using parasitic and predatory insects, as well as microorganisms, for disease and pest control (Obrycki et al., 2009).

AMF are a prevalent and significant fungal group in mycorrhizal associations, having a crucial role in inhibiting soil-borne pathogens (Berg, 2009; Veresoglou & Rillig, 2012; Allsup et al., 2021). More than 30 AMF species have demonstrated efficacy in managing soil-borne diseases (Miransari, 2011; Jung et al., 2012). The numerous benefits offered by AMF to plants have encouraged extensive research into their potential as BCA (Vishwakarma et al., 2022). Research indicates that AMF can influence host plants by altering root structure, enhancing the rhizosphere environment, competing with pathogens, and activating plant defense mechanisms. They have been shown to mitigate damage from various pathogens in several plant species, reducing the need for pesticides (Gianinazzi et al., 2010; Lee et al., 2013; Li et al., 2018; Begum et al., 2019; Razak & Gange, 2021; Allsup et al., 2021).

9.2 Plant Mechanisms for Distinguishing Beneficial Microorganisms from Pathogens

From the plant perspective, mycorrhizal fungi and other beneficial organisms like *Rhizobium* are considered alien "non-self" organisms, which can potentially trigger an immune response that the fungus has to overcome for successful interaction. The colonization process by AMF is like the infection by biotrophic pathogens, leading to similar initial defence responses (Pieterse et al., 2012).

9.2.1 Arbuscular Mycorrhizal Fungi

AMF spore germination initiates non-symbiotic growth. As the pre-symbiotic stage progresses, the fungi start to rely on signals released by roots for a continued growth (Besserer et al., 2006). The interaction between AMF and plants begins with chemical signalling in the pre-symbiotic stage, leading to changes in gene expression, protein profiles, and metabolite composition (Kaur & Suseela, 2020). Strigolactones, which are the branching factor (BF) produced in the root cortex, serve as primary signals in the common symbiosis signalling pathway (CSSP) (Genre & Bonfante, 2020). These strigolactones (2′ epi-5-deoxystrigol or epi-5DS) generated during symbiosis, are released into the rhizosphere via PDR-type ATP-binding cassette (ABC) transporters. Further hyphal growth requires flavonoids, while AMF also release chemical signals collectively known as "Myc-factors," of which chito oligosaccharides (COs) and lipo-chito-oligosaccharides (LCOs) are the main types (Kaur & Suseela, 2020). Successful recognition leads to cellular reorganization, including changes in AMF shape, movement, and increased mitochondrial density, driven by strigolactones (Besserer et al., 2006). Internalization of hyphae occurs, forming hyphopodia, and further, arbuscules develop within cortical cells as hyphae spread longitudinally (Kobae et al., 2018).

9.2.2 *Pathogen*

Interestingly, pathogenic fungi also respond to these chitin-related compounds, but plants have evolved specific molecular patterns on their receptor complexes for microbe identification. The LysM-RLK complex initially recognizes all chitin-related molecules. In rice, *OsCERK1* activated during systemic acquired resistance (SAR) is involved in pathogen recognition. Then a receptor-like protein OsCEBiP directly binds to pathogen-associated COs. When encountering Myc-COs, OsMYR1, another LysM-RLK, is produced by the plant to facilitate interaction with Myc-LCOs and Nod-factors, indirectly verified by OsSYMRK (Genre & Bonfante, 2020).

9.3 AMF-Mediated Biocontrol Against Different Types of Microorganisms

The symbiosis between plants and AMF plays a crucial role in enhancing plant defences against soil-borne pests and diseases (Sanchez-Bel et al., 2016) (Fig. 9.1). As AMF establish themselves within the host plants, they serve as an initial defence against soil-borne diseases (Cruz & Ishii, 2012; Li et al., 2021). AMF can serve as a biocontrol agent (BCA) to safeguard host plants from diseases caused by soil pathogens (Veresoglou & Rillig, 2012). In a study, the disease index and incidence of *Ralstonia solanacearum*, inoculated with *Glomus rhizogenes* and *Funneliformis. mosseae*, were reduced by 9.7% and 49.8%, respectively, compared to the control without AMF inoculation (Aguk et al., 2018).. The interactions among AMF, plants, pathogenic bacteria, and other microorganisms are highly intricate and influenced by different experimental conditions and materials researchers use. Consequently, the disease resistance conferred by AMF in plants can vary (Vos et al., 2013) (Fig. 9.2).

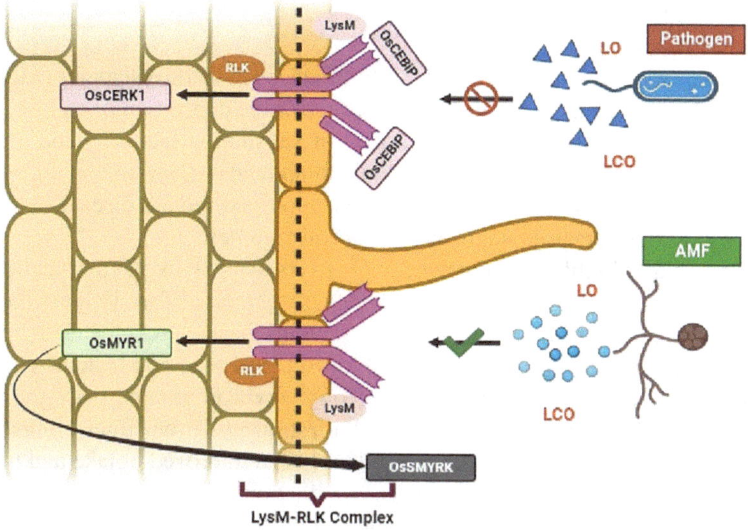

Fig. 9.1 Role of different types of signalling molecules to distinguish between pathogens and symbiotic microorganisms by host plants. When encountering a pathogen, the host plant activates *OsCERK1* through *OsCEBiP*, leading to SAR. In the presence of Myc-factors (*LO*/*LCO*) from AMF, plant activates *OsMYR1*, and *OsSYMRK* verifies *LO*/*LCO* interactions, triggering a common symbiosis pathway

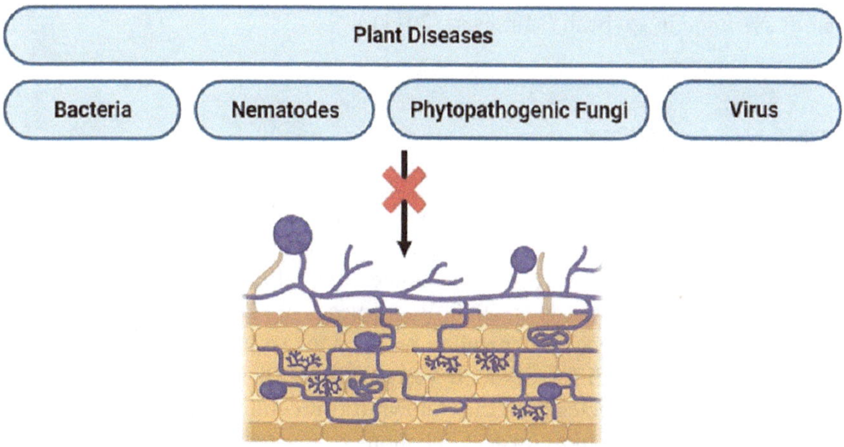

Fig. 9.2 Rhizosphere interaction and AMF-mediated biocontrol of microorganisms. AMF-colonized plants exhibit heightened resistance to bacteria, nematodes, viruses and phytopathogenic fungi

9.3.1 *Bacteria*

AMF colonization of host plants leads to mycorrhizal-induced resistance (MIR) and the activation of various immune responses, such as the deposition of callose, thickening of the cell walls, and the production of substances like ethylene, reactive oxygen species (ROS), and antimicrobial compounds (Dowarah et al., 2021). Research has demonstrated that AMF colonization can reduce disease symptoms caused by various bacterial pathogens in tropical legumes.

Several studies highlighted the beneficial effects of AMF in plant protection. For instance, the pathogen *Pseudomonas syringae* pv. *glycinea* (Psg) induces bacterial blight in soybean plants. It produces effector proteins that suppress the plant immune system, resulting in symptoms like leaf chlorosis, necrosis, pod lesions, and stem discoloration (Xin et al., 2018). Inoculating soybean plants with the AMF *Entrophospora infrequens* reduced Psg colonization by seven times compared to non-AMF inoculated plants (Malik et al., 2016). The improved uptake and transfer of nitrogen, along with increased biomass production facilitated by *E. infrequens*, were suggested to play a crucial role in enhancing the immunity of soybean plants.

Mulberry inoculated with *G. fasciculatum* or *F. mosseae* significantly reduced the incidence of *Pseudomonas syringae* (Kamble & Agre, 2014). Grapevines inoculated with AMF decreased *P. fluorescens* density on the root surface, diminishing the likelihood of reinfection of grape plants (Jung et al., 2012). The colonization of AMF on apple seedling roots lowered their susceptibility to actinomycete infections, consequently reducing disease incidence (Jung et al., 2012). *Funneliformis mosseae* effectively prevented plant infection by reducing the population of pathogenic *P. syringae* in soybean (Miransari, 2011).

9.3.2 *Nematodes*

Numerous studies have reported the ability of AMF to alleviate plant-parasitic nematode attacks. AMF can parasitize nematode cysts preventing nematode damages in crops such as oats, soybeans, cotton, cucumbers, tomato, kidney bean, alfalfa, citrus, and peach (De Sá & Campos, 2020; Rodrigues et al., 2021). The impact of root exudates on nematodes varies depending on the nematode species, so the level of protection provided will differ (Sikder & Vestergård, 2020). For instance, *Heterodera cajani* is a nematode species primarily known for infecting pigeon peas, but it can also affect other legumes like cowpeas and mung beans, resulting in symptoms such as root galling, stunted growth, and leaf chlorosis (Maurya et al., 2018). Similarly, *Meloidogyne incognita* induces root galling and leads to reduced growth and leaf chlorosis in its host plants (Xiang et al., 2018). AMF showed an inhibitory effect on root-knot nematodes in tomato, resulting in fewer galls on the roots of mycorrhizal plants compared to the control, and a significant decrease in the nematode parasitism rate (Vos et al., 2013). Specific inhibitory effects were observed, with half of the larvae perishing after 1 day of culture (Liu et al., 2012).

Vos et al. (2012) conducted pot and field experiments to investigate the inhibitory impact of AMF on *M. incognita* and found that AMF promoted the growth of infested plants and reduced the root-knot index. Pre-inoculating tobacco with *F. mosseae* significantly decreased the number of *Thielaviopsis basicola* propagules and *Meloidogyne incognita* juveniles, and enhanced plant resistance to nematodes (Liu et al., 2012). AMF symbionts enhance plant nematode tolerance but may not eliminate nematode damage, especially when they occur at high densities. The inoculation of chickpea plants with various AMF species, including *Glomus fasciculatum* (syn. *Rhizophagus fasciculatus*), *G. constrictum* (syn. *S. constrictum*), *G. intraradices* (syn. *R. intraradices*), *G. margarita*, or *Acaulospora* spp., when infected with *M. incognita*, resulted in increased plant height, greater fresh and dry weight, improved yield, higher chlorophyll content, and a reduction in nematode population densities. Additionally, these AMF-inoculated plants exhibited lower values of the root galling index, compared to plants without AMF inoculation (Siddiqui & Akhtar, 2006; Akhtar & Siddiqui, 2007).

9.3.3 Phytopathogenic Fungi

AMF use has received extensive attention as a biological control method against various phytopathogenic fungi (Lin et al., 2021). *Phytophthora*, a widely used model pathogenic fungus, has been employed extensively to study AMF-mediated plant disease control (Krzyzaniak et al., 2021). AMF species like *G. fasciculatum*, *E. etunicatum*, *G. macrocarpum*, *G. margarita*, *G. heterogama*, and *G. calospora* have demonstrated the ability to mitigate diseases caused by pathogenic fungi in various crops such as barley, peanut, soybean, banana, cotton, kidney bean, onion, tobacco, citrus, peach, poplar, strawberries, red clover, and ginseng (Steinkellner et al., 2012; Bubici et al., 2019; Eke et al., 2020; Aljawasim et al., 2020; Guzman et al., 2021). The AMF species *Rhizophagus irregularis* (syn. *Glomus intraradices*) inhibited the growth of the pathogenic fungus *Fusarium oxysporum*, suggesting that the chemical equilibrium established by mycorrhizae curtails the growth and reproduction of pathogenic fungi. Tomato seedling roots infected with *F. oxysporum* for 20 days showed, after inoculation with *G. polyphylla* and *G. macrocarpum*, a reduction in disease severity by 78% and 75%, respectively (Steinkellner et al., 2012).

Pea plants infected by *Aphanomyces euteiches* revealed that the establishment of a full AMF symbiosis is crucial for plant defence against pathogens (Slezack et al., 2000). A study revealed that AMF-colonized cotton plants displayed greater resistance to *Thielaviopsis basicola* infection compared to those with non-mycorrhizal roots (Schönbeck & Dehne, 1977). Subsequent investigations revealed an inverse relationship between the degree of mycorrhizal infection and the chlamydospore yield of *T. basicola* (Baltruschat & Schönbeck, 1975). *Phytophthora sojae* is a pathogen that specifically infects soybean, causing damping off in seedlings, with root rot and stem lesions in mature plants (Dorrance, 2018). Inoculation with *G. intraradices* reduced oxidative damage in soybean plants infected with *P. sojae* by decreasing H_2O_2 and increasing jasmonic acid (JA) content, glutathione reductase activity, and nitrogen and carbon metabolism compared to non-inoculated controls (Li et al., 2013).

Charcoal root rot, which is caused by the wide-ranging fungus *Macrophomina phaseolina*, is a soilborne disease that infects the roots and lower stems of over 500 plant species, including legumes such as peanut, soybean, and chickpea (Marquez et al., 2021). Several different AMF species have demonstrated the ability to alleviate the symptoms caused by *M. phaseolina* in legumes. In the case of chickpea plants, the inoculation with individual AMF species such as *G. fasciculatum* (syn. *Rhizophagus fasciculatus*), *G. constrictum* (syn. *S. constrictum*), *R. irregularis* (syn. *G. intraradices*), *Gigaspora margarita*, or *Acaulospora* spp. resulted in a reduction in the severity of root rot and an increase in plant growth, chlorophyll content, and the number of pods compared to non-AMF plants (Siddiqui & Akhtar, 2006; Akhtar & Siddiqui, 2007). For soybean plants infected with *M. phaseolina*, the AMF species *R. irregularis* was found to upregulate pathogenesis and disease-resistant genes while increasing lignin production. This response led to a reduction in disease incidence and severity, ultimately improving plant growth and yield (Spagnoletti et al., 2020).

Various wilt diseases affecting legumes are caused by fungi, including *Fusarium* wilt in pigeon peas due to infection with *F. nudum*. Symptoms of this infection include chlorosis, drooping leaves and stems, and wilting. Inoculation of pigeon pea with AMF species *R. fasciculatus* (syn *G. fasciculatum*) significantly reduced wilting severity, increasing plant height, shoot dry weight, and phosphorous content, compared to non-AMF plants (Siddiqui & Mahmood, 1996).

Another wilt infection that results in severe root rot and seedling damping off, caused by *Aphanomyces euteiches*, affects a wide range of legume plants. Inoculating pea with AMF species *R. fasciculatus* (syn. *G. fasciculatum*) and *R. irregularis* (syn. *G. intraradices*) significantly reduced spore production and root-rot severity compared to non-AMF plants (Garzo et al., 2018; Wu et al., 2018).

The combined application of AMF and chitosan nanoparticles significantly enhanced the resistance of faba bean plants to root rot caused by *Rhizoctonia solani*. This approach not only prevented but also effectively cured the disease (El-Gazzar et al., 2023).

9.3.4 Viruses

The impact of AMF symbiosis on virus infections in potato plants appears closely tied to the specific viral strain which are single-strand (ss) RNA(+), constituting the largest group of RNA viruses (Elsharkawy et al., 2012). Mycorrhizal colonization had a beneficial effect in attenuating the disease caused by the Tomato Yellow Leaf Curl Sardinia virus (TYLCSV), a ssDNA virus belonging to the *Geminiviridae* family, responsible for one of the most severe tomato viral diseases (Maffei et al., 2013). These different outcomes in virus interactions can be attributed to the unique characteristics of the viruses, such as their location within cells and phloem restriction.

9.4 Mechanism of AMF-Mediated Biocontrol

9.4.1 Induced Systemic Resistance (ISR)

ISR is a plant defence mechanism triggered by factors like virulence, avirulence, non-pathogenic microorganisms, changes in environmental conditions, or chemicals (Avis et al., 2008; Liu & Wang, 2016). AMF triggers ISR in plants as a biocontrol mechanism (Xavier & Boyetchko, 2003). In a split-root experiment, a reduction in disease symptoms in the non-mycorrhizal portion of the root system was observed, compared to AMF colonized roots (Pozo et al., 2010; Schouteden et al., 2015). This phenomenon is referred to as "mycorrhizal autoregulation", indicating that once a plant undergoes mycorrhization, it restricts additional colonization of its roots by AMF or other pathogenic fungi to minimize its energy expenditure (Vierheilig et al., 2008).

Upon forming symbiotic relationships with plants, AMF can stimulate the production of various signalling molecules, including nitric oxide, JA, salicylic acid, ethylene, hydrogen peroxide, abscisic acid, calcium ions, and sugar signals (Schmitz & Harrison, 2014). These signalling molecules play crucial roles in plant-AMF recognition, mycorrhizal symbiosis establishment, and activation of plant defense mechanisms. Mycorrhizal-induced JA can activate plant defense systems, enhancing plant resistance to pathogens (Metwally & Al-Amri, 2020). AMF colonization is closely linked to the accumulation of nitric oxide in plants. Inoculation of *Gi. margarita* in *Alfalfa truncula ta* resulted in a significant increase in nitric oxide levels. Specifically, the leaves exhibited a 3.3-fold increase, while the roots showed a 1.9-fold increase compared to the control treatment. This highlights the ability of AMF to stimulate the accumulation of nitric oxide, contributing to systemic resistance (He et al., 2010).

Studies have revealed that JA and ethylene generally confer resistance against saprotrophic pathogens, while salicylic acid inhibits biotrophic pathogens. JA and ethylene are associated with ISR, while SA is linked to SAR when plants face pathogenic infections (Hause et al., 2007). Nitric oxide has garnered significant attention in plant cell signalling research, serving as a molecule involved in processes related to plant defence systems, signalling, and gene expression (Calcagno et al., 2012). AMF induce significant changes in defence-related gene transcripts in *Medicago truncatula* and resulted in both local and systemic defence responses in plants against damage caused by *Xanthomonas campestris*.

The mechanisms by which AMF enhance host plant disease resistance are diverse, with effects stemming from individual mechanisms or the synergistic interplay of multiple mechanisms. The types of disease resistance induced by AMF can be either systemic or local (Pozo et al., 2002; Tabin et al., 2009). *Glomus versiforme* triggered systemic resistance to *R. solanacearum* in tomato plants root-splitting experiments, leading to a significant increase in phenolics content in the roots (Zhu & Yao, 2004). Salicylic acid has been identified as the primary signalling molecule through which AMF induce systemic disease resistance in plants (Dugassa et al., 1996).

9.5 AMF Interaction with Rhizosphere Microbiome: Integrated Disease Control

Once colonized by AMF, plants activate a robust defence strategy called MIR (Cameron et al., 2013) (Fig. 9.3). AMF contributes to disease and pest suppression through ISR (Jung et al., 2012; Song et al., 2015). MIR shares characteristics with pathogen-induced SAR and non-pathogenic *Rhizobacterium*-ISR (Nishad et al., 2020). Utilizing biocontrol microorganisms i.e., plant growth-promoting bacteria such as *Pseudomonas* and *Bacillus* spp. in conjunction with AMF can enhance the plant partial resistance to pathogens through various mechanisms, including induced ISR and MIR (Cameron et al., 2013) (Fig. 9.4). Several instances of such interactions are given below.

9.5.1 Cumulative Effects of AMF and Biocontrol Agents

The combination of AMF and Plant Growth-Promoting Rhizobacteria (PGPR) has emerged as a promising strategy to enhance plant growth and disease resistance. Beneficial microorganisms, including PGPR, Plant Growth-Promoting Fungi (PGPF), AMF, and biocontrol agents (BCA), when applied individually or as consortia, enhance various rhizosphere processes. They promote plant growth, mobilize essential nutrients, improve root nodulation, enhance photosynthesis and gas

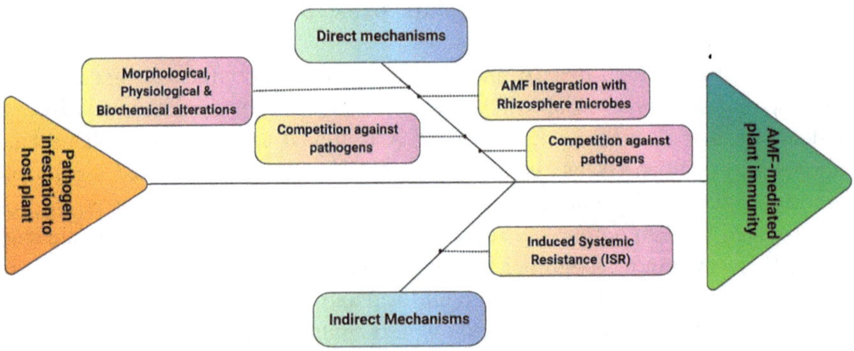

Fig. 9.3 Mechanisms of AMF-mediated biocontrol against microorganisms. AMF enhance plant immunity through a variety of both direct and indirect mechanisms involved in developing host immunity. Direct mechanisms include morphological, physiological and biochemical alterations in the host and rhizosphere. AMF and pathogens compete each other to overcome this response in the rhizosphere. AMF also establishe a positive interaction with other beneficial rhizosphere species. Indirect mechanisms involve induction and activation by AMF of defense related pathways leading to ISR. AMF entry in the host tissue will induce ISR at the early stages of symbiosis

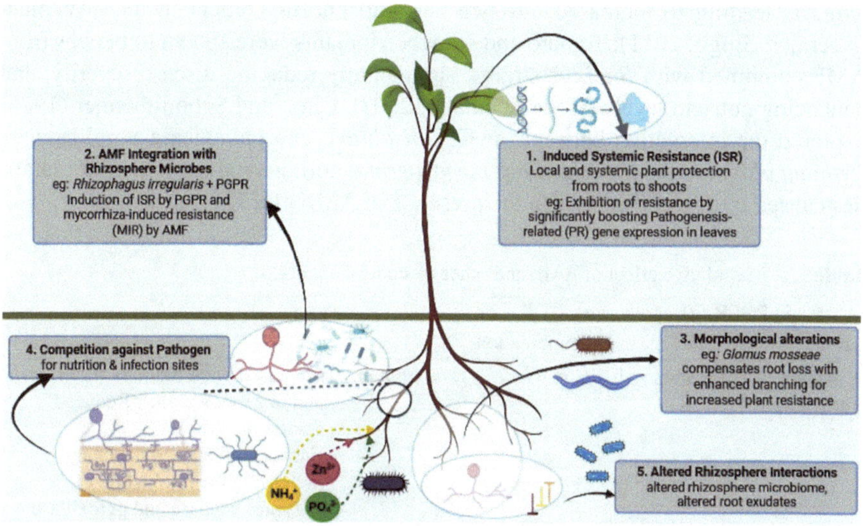

Fig. 9.4 Schematic illustration of AMF-mediated biocontrol mechanisms targeting microorganisms, including ISR, synergy with diverse rhizosphere microbes, morphological modifications, competitive exclusion, and modification of rhizosphere interactions. Underground interactions include: competition, altered rhizosphere interactions, morphological alterations as well as AMF integration with rhizosphere microorganisms. Aboveground interaction includes ISR induction

exchange, mitigate salinity and drought stress, and induce resistance against phytopathogens (Sunar et al., 2023).

Ziane et al. (2023) investigated resistance of date palm against Fusarium wilt, where a consortium of 31 AMF species combined with PGPR consortia demonstrated a strong antagonistic effect against *Fusarium* growth and effectively stimulated the plant natural defenses. The combination of *F. mosseae*, *G. fasciculatum*, and *Bacillus* spp., were effective in inhibiting Fusarium root wilt of tomato (Devi et al., 2022).

The combination of bio-agents *F. mosseae* and *Trichoderma harzianum* protected cucumber plants (*Cucumis sativus*) from the fungal pathogen *Alternaria alternata*, the causal agent of cucumber wilt. Their combination had a synergistic, positive impact on the cucumber plant biomass (both fresh and dry root weight), and on disease severity as well, by enhancing plant resistance with increased production of enzymes such as catalase and peroxidase (Matrood et al., 2020). Similarly, various combinations involving AMF (*A. laevis*) and PGPR (*T. viride*) significantly reduced disease incidence, with some combinations completely preventing the Tomato wilt disease caused by *F. oxysporum* (Tanwar et al., 2013). In *Cucumis melo*, the combination of AMF, *T. harzianum*, and their joint application reduced *Fusarium* wilt disease incidence by more than 60% (Martínez-Medina et al., 2011). Individual inoculation with *G. sinosum*, *Gi. albida*, or *P. fluorescens*, as well as their combined inoculation, reduced disease incidence by up to 69.2% in *Phaseolus vulgaris* leading to increased nitrogen and phosphorus content in the host plant (Neeraj & Singh, 2011). Tomato and strawberry plants were shown to benefit from AMF combined with *Bacillus* strains, significantly reducing disease severity, and enhancing nutrient levels (Hoda & Ghalia, 2007). Chou and Schmitthenner (1974) explored the interaction between AMF, *Rhizobium*, and pathogenic fungi such as *Pythium ultimum* and *Phytophthora megasperma*, noting a reduction in plant fatalities caused by *P. megasperma* in the presence of AMF (Table 9.1).

Table 9.1 Cumulative effect of AMF and other biocontrol agents

AMF and PGPR (*Trichoderma)*			
Host	Disease	Combination	Impact
Cucumis sativus	Cucumber wilt	*F. mosseae* + *T. harzianum*	Increased plant biomass (both fresh and dry root weight), Reduced disease severity Enhanced plant resistance by increasing the production of enzymes catalase and peroxidase (Matrood et al., 2020)
		AMF AMF + T. *harzianum*	Aboveground biomass increased by 11.5% Aboveground biomass increased by 69.7% (Tanwar et al., 2013)
Tomato	*F. oxysporum*	*F. mosseae* + *A. laevis* AMF + *T. viridi* *F. mosseae* + *A. laevis* + *T. viride*	Reduction in disease incidence to 20% Reduction to 10% Completely prevented the disease (Tanwar et al., 2013)
Cucumis melo	Fusarium wilt	AMF *T. harzianum* AMF + *T. harzianum*	Reduced disease incidence by 25–50% Reduced disease incidence by 60% Reduced disease incidence by >60% (Martínez-Medina et al., 2011)
AMF and PGPR (*Pseudomonas*)			
Phaseolus vulgaris		Individual inoculation with *G. sinosum*, *Gi. albida*, or *P. fluorescens* Combined inoculation	Reduced the incidence rate by 50.5–52.8% Reduced the incidence rate by 68.9–69.2%. Exhibited higher nitrogen and phosphorus content (Neeraj & Singh, 2011)
Tomato	Fusarium wilt	*G. sinuosum* + *P. fluorescens*	More effective than single inoculation (Akköprü & Demir, 2005; Srivastava et al., 2010)
Papaya	Root rot by *F. oxysporum*	Combined inoculation	Proved more effective (Martínez-Medina et al., 2011).
	Elaeis guineensis basal rot	*G. intraradices* or *G. clarum* *G. intraradices* or *G. clarum* + *P. aeruginosa*	Reduced the severity by 15–17%, Reduced severity by 57–80% showing excellent control against plant root diseases (Parvin et al., 2020).

(continued)

Table 9.1 (continued)

AMF and PGPR (*Trichoderma)*			
Host	Disease	Combination	Impact
AMF and PGPR (*Bacillus)*			
Tomato	Fusarium root wilt (*F. oxysporum* f. sp. *lycopersici*)	*F. mosseae* + *G. fasciculatum* + *B. velezensis* + *Bacillus* spp.	Safe and effective biocontrol agents against *Fusarium* wilt of tomato with a maximum inhibition of 58.43 and 55.68% (Devi et al., 2022)
	F. oxysporum	AMF + *Bacillus* AMF	73.6–82.1% 34.1% (Cai et al., 2021)
Tomato	Fusarium root rot	*G. mosseae* + *B. subtilis*	Reduced the disease severity by 85.0–93.4% but also enhanced plant nutrient levels (NPK, Ca, Mg, Fe, Zn), leaf pigment content, total soluble sugar, total soluble protein, and total free amino acids (Hoda & Ghalia, 2007)
Strawberry	Fusarium wilt	*Glomus* + *B. subtilis*	Increased fresh weight of strawberries by 61.7–90.9% compared to single inoculation (Tahmatsidou et al., 2006)
Strawberry	Powdery mildew	*F. mosseae* + *G. caledonium* + *G. sinuosum* + *B. subtilis*	Reduced disease severity by 12% (Lowe et al., 2012)
Cotton	Verticillium wilt	*G. versiforme* *G. versiforme* + *B. vallismortis*	Reduced disease severity by 37.7% Reduced disease severity by 63.3% (Zhang et al., 2012)
AMF and other BCAs			
Tomato	Fusarium crown and root rot	*F. mosseae* + *R. intraradices* + *T. virens l40012* + *T. harzianum l40015* + *B. subtilis PS1-3* + *P. fluorescens PS2-6* + *F. oxysporum* f. sp. *radicis-lycopersici*	Reduced the disease severity by 85.0–93.4% but also enhanced plant nutrient levels (NPK, Ca, Mg, Fe, Zn), leaf pigment content, total soluble sugar, total soluble protein, and total free amino acids (Hoda & Ghalia, 2007)
Date palm	Fusarium wilt (*Fusarium. oxysporum* f. sp. *Albedinis*)	AMF consortium (31 species) + PGPR consortia (4 strains)	Bacterial strains *Bacillus cereus* 263AG5' of the PGPR consortium inhibited *Fusarium* growth (60.59%). High synergistic effect between PGPR and AMF could be attributed to the association of their antagonistic effect with their strong capacity for stimulation of plant natural defenses (Ziane et al., 2023)

9.5.2 Histological Changes

9.5.2.1 Morphological Alterations

During pathogen invasion of the plant cell wall, AMF induce a range of advantageous changes in the host root system, with increasing root growth, thickening, enhanced branching, accelerated lignification of the cell wall, thickening of the root tip epidermis, and development of more cell layers which collectively halt the infection (Boutaj et al., 2019; Basyal & Emery, 2021). Mycorrhization of *F. mosseae* and *E. etunicatum* in cotton exhibited increased root xylem structure when infested with pathogenic *Verticillium dahliae*. This resulted in distorted palisade tissue and vessels, the production of gelatinous substances within the vessels, cell deformation and contraction, color deepening, and significant thickening of cell walls. These alterations, together with lignification, a decrease in vacuole count, the disappearance of mitochondrial inner folds, and structural changes in the root system, collectively enhancde the host resistance to *V. dahliae* (Boutaj et al., 2019). AMF establishes a network of mycelium, arbuscules, vesicles and spores within the root epidermis and endodermis. This network impedes the penetration of pathogens into root cell tissues and, consequently, the progress of infection. These changes were observed in the anatomical structure of tomato roots, that impacted the infection dynamics of *Phytophthora* (Pozo et al., 2002).

Additionally, alterations in the root systems of mycorrhizal plants leads to the production of hydroxyproline-rich glycoproteins (HRGPs) (Balestrini & Bonfante, 2014). HRGP, a linear sugar-protein compound on the plant cell wall, reinforces the strength of the cell wall, increasing its resistance to degradation by proteases, cellulases, and hemicellulases secreted by the pathogens during infection. HRGP also acts as a lectin during pathogenic invasion, immobilizing pathogens on the cell wall and preventing their penetration into the host cells. Moreover, some substances present on the AMF cell wall can inhibit pathogenic bacteria. For instance, certain *Rhizopahgus* AMF strains contain β-1,3-glucan on their extra-root hyphae, germ tube inner walls, and spore cell walls, suggesting that AMF can serve as a barrier against pathogenic microorganisms. By modifying root anatomy and producing structural substances, AMF enhance host plant resistance to pests and diseases, effectively achieving an indirect biological control (Artursson et al., 2006; Scervino et al., 2007; Silvestri et al., 2020; Ma et al., 2021).

9.5.2.2 Biochemical Alterations

AMF also enhance plant disease resistance by regulating host secondary metabolites (Elliott et al., 2021). They produce beneficial secondary metabolites such as phytochemicals, calloses, alkaloids, and phenols on their hyphae along with phenolic compounds i.e., flavonoids and phenolic carboxylic acids. They are crucial for plant signalling and defence systems, helping plants resist disease-causing conditions (Tahat et al., 2012). Phytoprotectins, produced during plant-pathogen interactions, are linked to disease resistance and act as barriers around infected cells (Jaiti et al., 2008). *Funneliformis mosseae* induces a phytotoxin stress response in roots, enhancing disease resistance. Plant inoculation with *R. intraradices* promoted callose deposition in cucumber roots, protecting them from *Colletotrichum orbiculare* poisoning (Bais et al., 2006).

AMF colonization increases vinblastine content in rose leaves, boosting resistance to biotic stress (Martínez-Medina et al., 2011). AMF also increase flavonoids near invasion sites during infection. Mycorrhizal plants, like cotton, accumulate phenolic compounds, enhancing resistance to pathogens like *V. dahliae* (Lioussanne et al., 2008). *Funneliformis mosseae* significantly reduced disease incidence and severity in strawberries, increasing total polyphenol and ascorbic acid contents (Chandanie et al., 2009). *Glomus versiforme* induces systemic resistance to *Ralstonia solanacearum* in tomato, increasing phenolics content in roots (Lioussanne et al., 2008). Although some studies showed no significant changes in phenolic compounds in AMF-vaccinated *Phoenix dactylifera* during *F. oxysporum* infection, hydroxycinnamic acid derivatives was found to help resist chlorosis in date palms (Jaiti et al., 2008).

9.5.2.3 Molecular Alterations

Cameron et al. (2013) introduced a four-phase spatial and temporal model to elucidate the comprehensive process underlying the mycorrhizal association and MIR. MIR stems from direct plant responses to arbuscular mycorrhizal colonization and indirect responses to mycorrhizosphere rhizobacteria, suggesting ISR induced by bacteria in MIR. Systemic priming of ethylene- and JA-inducible defenses are commonly observed in both ISR and MIR (Van der Ent et al., 2009). JA-dependent defence priming during MIR is attributed to ISR induced by rhizobacteria (Cameron et al., 2013).

(i) *Root exudate release and AMF attraction*: in the pre-symbiotic phase, plants release root exudates containing sugars, amino acids, phenolic compounds, and other secondary metabolites including strigolactones which signal AMF towards roots (Akiyama et al., 2005).
(ii) *Plant immune response to AMF*: upon AMF colonization or pathogen infection, roots increase salicylic acid (SA) production, accumulate defensive compounds, hydrolytic enzymes, ROS, and activate the phenylpropanoid pathway, although at a lesser extent than during pathogen interactions (Fester & Hause, 2005; Scheler et al., 2013). This response is temporally limited in AMF symbiosis but prolonged in pathogen interactions, suggesting an AMF control over immunity (García-Garrido & Ocampo, 2002). During mycorrhization initiation, plants recognize AMF microbe-associated molecular patterns (MAMPs) via pattern recognition receptors (PRRs), boosting SA levels and immune activation (Zhang & Zhou, 2010).

(iii) *Immune suppression by AMF and mycorrhizosphere development*: in the subsequent phase, AMF suppress plant immunity, akin to pathogenic fungi, using mechanisms like calcium/calmodulin kinase DMI3 that repress early defence genes (Siciliano et al., 2007; de Jonge et al., 2011). AMF also induce abscisic acid (ABA) production, hindering SA-dependent defences and promoting colonization (Ton et al., 2009). Despite SA suppression, the initial SA response may create long-SAR signals, potentially forming the basis for SA-based defence (Heil & Ton, 2008; Conrath, 2006). Inhibiting some SA-regulated responses is necessary for successful AMF association, as observed in other mutualistic relationships (Dumas-Gaudot et al., 2000). AMF secreted effectors disrupt the host immune system by suppressing SA-dependent defences (Kloppholz et al., 2011). Additionally, upregulation of phytohormones like JA is crucial for establishing and maintaining the AMF mutualistic association (Hause et al., 2007).

(iv) *Mycorrhizosphere development and ISR*: in the final phase, a successful mycorrhizal association enhances photosynthate transport to roots and affects sugar-dependent signalling pathways, potentially leading to the recruitment of beneficial rhizobacteria and mycorrhizosphere development (Smith & Smith, 2011). AMF, once established, recruit root-associated bacteria, including PGPR, to boost immune responses against pathogens.

9.5.2.4 Altered Rhizosphere Interactions

After forming a symbiotic relationship with plants, AMF changes the permeability of the root cell membranes and alter root exudates, affecting the rhizosphere physico-chemical properties and its microbial population. AMF hyphae penetrate soil pores and secrete substances i.e. the *Glomus* related soil protein (GRSP), organic acids, and polyamines. They promote formation of soil aggregates, improve local pH, water stability, aeration, and permeability, and enhance redox potential. These changes support healthy plant growth and resistance to diseases (Weaver & Wehunt, 1975; Tatsumi et al., 2020).

Root exudates from AMF-symbiotic plants directly impact the growth, development, and reproduction of other soil fungi, nematodes, and bacteria, leading to alterations in the soil microbial community in terms of structure, nature, quantity, and spatial distribution (Cordier et al., 1996; Whipps, 2004; Tabin et al., 2009). Tomato root exudates in association with AMF were found to immobilize nematodes, reducing their invasion into the root system, and repel *Phytophthora nicotianae* zoospores, preventing them from accessing roots (Lioussanne et al., 2008; Liu et al., 2012). AMF also foster beneficial interactions with soil microorganisms, particularly those antagonistic to soil-borne pathogens. This stimulation boosts the number of beneficial microorganisms in the rhizosphere, such as *Gliocladium*, *Streptomyces*, *Trichoderma*, fungi, nitrogen-fixing and phosphorus-dissolving bacteria, including *Pseudomonas fluorescence*, *Bacillus* spp., and other PGPR and Actinomycetes (Chandanie et al., 2009). Beneficial microorganisms help reduce pathogen numbers and the likelihood of plant infection, indirectly enhancing plant resistance to pathogens. PGPR also strengthen the symbiotic relationship between AMF and plants (Cordier et al., 1996).

9.5.2.5 Competition

AMF and soil-borne pathogens share the same ecological niche and invasion sites in the soil rhizosphere. In general, they compete spatially, with the primary biocontrol effect being the reduction of initial infection and re-infection of root epidermal pathogens. *Funneliformis mosseae* reduced infection by *Phytophthora nicotianae* in both colonized and adjacent uninfected root systems by restricting the pathogen invasion in arbuscular cells (Cordier et al., 1996). AMF-colonized roots and adjacent uninfected roots had significantly fewer nematodes, *Heterodera glycines* (Whipps, 2004). A study reported competition for infection sites between AMF and pathogenic bacteria in *Aquilaria agallocha* (Zhou et al., 2020). Inoculation of *G. fasciculatum* also inhibited the development of *Pythium aphanidermatum* in *A. agallocha* roots and reduced plant morbidity and damping-off symptoms (Zhou et al., 2020). AMF structures such as vesicles, hyphae and arbuscules often parasitize nematode galls. Inoculatin of olive planting stocks with *Glomus* spp. significantly reduced root galling by root-knot nematodes and increased growth by 88.9% (Castillo et al., 2006).

9.6 Conclusions

AMF have emerged as a promising BCA in agriculture, offering a range of strengths and opportunities. One of their key strengths lies in their eco-friendliness, as they reduce the reliance on chemical pesticides that can harm beneficial organisms and ecosystems. Furthermore, AMF not only protect plants against soil-borne pathogens but also enhance nutrient uptake, leading to their overall improved health and productivity. AMF versatility is another advantage, as they can form symbiotic relationships with a wide variety of host species, making them adaptable to different agricultural systems. Additionally, their use aligns well with sustainable agricultural practices, contributing to soil health and long-term ecosystem stability.

AMF-based biocontrol also has its weaknesses and faces certain threats. The effectiveness can vary depending on environmental conditions, host plants, and pathogens. The slow establishment of AMF populations in soil can delay their effectiveness in controlling diseases. To fully leverage their potential, there is an opportunity to develop commercial biological products containing specific AMF strains for targeted disease control. Integrating AMF into Integrated Pest Management strategies can enhance disease management while minimizing chemical pesticide applications. Ongoing research and innovation may uncover new AMF strains with enhanced biocontrol properties and better adaptation to various conditions.

However, challenges remain, including the need to raise awareness among farmers and agricultural practitioners about the benefits of AMF applications. Competition with chemical pesticides and other biological control agents, which may already have an established market and presence, is another hurdle. In conclusion, while AMF show promise as biocontrol agents in agriculture, their successful integration into farming practices will depend on continued research, innovation, and increased awareness within the agricultural community. This can lead to more sustainable and effective disease management, while minimizing negative environmental impacts.

Acknowledgement The authors are grateful to Symbiotic Sciences Pvt. Ltd., India for the support extended during this study.

References

Aguk, J. A., Karanja, N., Schulte-Geldermann, E., Bruns, C., Kinyua, Z., & Parker, M. (2018). Control of bacterial wilt (*Ralstonia solanacearum*) in potato (*Solanum tuberosum*) using rhizobacteria and arbuscular mycorrhiza fungi. *African Journal of Food Agriculture Nutrition and Development, 18*, 13372–13388.

Akhtar, M. S., & Siddiqui, Z. A. (2007). Glomus intraradices, Pseudomonas alcaligenes, and Bacillus pumilus: Effective agents for the control of root-rot disease complex of chickpea (Cicer arietinum L.). *Journal of General Plant Pathology, 74*, 53–60.

Akiyama, K., Matsuzaki, K., & Hayashi, H. (2005). Plant sesquiterpenes induce hyphal branching in arbuscular mycorrhizal fungi. *Nature, 435*, 824–827.

Akköprü, A., & Demir, S. (2005). Biological control of Fusarium wilt in tomato caused by *Fusarium oxysporum* f. sp. *lycopersici* by AMF *Glomus intraradices* and some rhizobacteria. *Journal of Phytopathology, 153*, 544–550.

Aljawasim, D. B., Khaeim, M. H., & Manshood, A. M. (2020). Assessment of arbuscular mycorrhizal fungi (*Glomus* spp.) as potential biocontrol agents against damping-off disease *Rhizoctonia solani* on cucumber. *Journal of Crop Protection, 9*, 141–147.

Allsup, C. M., Lankau, R. A., & Paige, K. N. (2021). Herbivory and soil water availability induce changes in arbuscular mycorrhizal fungal abundance and composition. *Microbial Ecology, 84*, 141–152.

Artursson, V., Finlay, R. D., & Jansson, J. K. (2006). Interactions between arbuscular mycorrhizal fungi and bacteria and their potential for stimulating plant growth. *Environmental Microbiology, 8*, 1–10.

Avis, T. J., Gravel, V., Antoun, H., & Tweddell, R. J. (2008). Multifaceted beneficial effects of rhizosphere microorganisms on plant health and productivity. *Soil Biology and Biochemistry, 40*, 1733–1740.

Bais, H. P., Weir, T. L., Perry, L. G., Gilroy, S., & Vivanco, J. M. (2006). The role of root exudates in rhizosphere interactions with plants and other organisms. *Annual Review of Plant Biology, 57*, 233–266.

Balestrini, R., & Bonfante, P. (2014). Cell wall remodeling in mycorrhizal symbiosis: A way towards biotrophism. *Frontiers in Plant Science, 5*, 237.

Baltruschat, H., & Schönbeck, F. (1975). The influence of endotrophic mycorrhiza on the infestation of tobacco by *Thielaviopsis basicola*. *Journal of Phytopathology, 84*, 172–188.

Basyal, B., & Emery, S. M. (2021). An arbuscular mycorrhizal fungus alters switchgrass growth root architecture and cell wall chemistry across a soil moisture gradient. *Mycorrhiza, 31*, 251–258.

Begum, N., Qin, C., Ahanger, M. A., Raza, S., Khan, M. I., et al. (2019). Role of arbuscular mycorrhizal fungi in plant growth regulation: Implications in abiotic stress tolerance. *Frontiers in Plant Science, 10*, 1068.

Berg, G. (2009). Plant-microbe interactions promoting plant growth and health: Perspectives for controlled use of microorganisms in agriculture. *Applied Microbiology and Biotechnology, 84*, 11–18.

Besserer, A., Puech-Pagès, V., Kiefer, P., Gomez-Roldan, V., Jauneau, A., et al. (2006). Strigolactones stimulate arbuscular mycorrhizal fungi by activating mitochondria. *PLoS Biology, 4*, e226.

Boutaj, H., Meddich, A., Wahbi, S., Moukhli, A., El Alaoui-Talibi, Z., et al. (2019). Effect of Arbuscular Mycorrhizal fungi on Verticillium wilt development of olive trees caused by *Verticillium dahliae. Research Journal of Biotechnology, 14*, 8.

Bubici, G., Kaushal, M., Prigigallo, M. I., Gómez-Lama Cabanás, C., & Mercado-Blanco, J. (2019). Biological control agents against Fusarium wilt of banana. *Frontiers in Microbiology, 10*, 616.

Cai, X., Zhao, H., Liang, C., Li, M., & Liu, R. (2021). Effects and mechanisms of symbiotic microbial combination agents to control tomato Fusarium crown and root rot disease. *Frontiers in Microbiology, 12*, 629793.

Calcagno, C., Novero, M., Genre, A., Bonfante, P., & Lanfranco, L. (2012). The exudate from an arbuscular mycorrhizal fungus induces nitric oxide accumulation in *Medicago truncatula* roots. *Mycorrhiza, 22*, 259–269.

Cameron, D. D., Neal, A. L., van Wees, S. C., & Ton, J. (2013). Mycorrhiza induced resistance: More than the sum of its parts? *Trends in Plant Science, 18*, 539–545.

Castillo, P., Nico, A. I., Azcón-Aguilar, C., Del Río, R. C., Calvet, C., et al. (2006). Protection of olive planting stocks against parasitism of root-knot nematodes by arbuscular mycorrhizal fungi. *Plant Pathology, 55*, 705–713.

Chandanie, W. A., Kubota, M., & Hyakumachi, M. (2009). Interactions between the arbuscular mycorrhizal fungus *Glomus mosseae* and plant growth-promoting fungi and their significance for enhancing plant growth and suppressing damping-off of cucumber (*Cucumis sativus* L). *Applied Soil Ecology, 41*, 336–341.

Chou, L. G., & Schmitthenner, A. F. (1974). Effect of rhizobium japonicum and Endogone mosseae on soybean root rot caused by Pythium ultimum and Phytophthora megasperma var. sojae. *Plant Disease Reports, 58*, 221–225.

Conrath, U. (2006). Systemic acquired resistance. *Plant Signaling & Behavior, 1*, 179–184.

Cordier, C., Gianinazzi, S., & Gianinazzi-Pearson, V. (1996). Colonisation patterns of root tissues by *Phytophthora nicotianae* var *parasitica* related to reduced disease in mycorrhizal tomato. *Plant and Soil, 185*, 223–232.

Cruz, A. F., & Ishii, T. (2012). Arbuscular mycorrhizal fungal spores host bacteria that affect nutrient biodynamics and biocontrol of soil-borne plant pathogens. *Biology Open, 1*, 52–57.

de Jonge, R., Bolton, M. D., & Thomma, B. P. H. J. (2011). How filamentous pathogens co-opt plants: The ins and outs of fungal effectors. *Current Opinion in Plant Biology, 14*, 400–406.

De Sá, C. S. B., & Campos, M. A. S. (2020). Arbuscular mycorrhizal fungi decrease *Meloidogyne enterolobii* infection of guava seedlings. *Journal of Helminthology, 94*, e183.

Devi, N. O., Devi, R., Debbarma, M., Hajong, M., & Thokchom, S. (2022). Effect of endophytic *Bacillus* and arbuscular mycorrhiza fungi (AMF) against Fusarium wilt of tomato caused by *Fusarium oxysporum* f sp *lycopersici. Egyptian Journal of Biological Pest Control, 32*, 1.

Diagne, N., Ndour, M., Djighaly, P. I., Ngom, D., Ngom, M. C. N., Ndong, G., Svistoonoff, S., & Cherif-Silini, H. (2020). Effect of plant growth promoting rhizobacteria (PGPR) and arbuscular mycorrhizal fungi (AMF) on salt stress tolerance of *Casuarina obesa* (Miq). *Frontiers in Sustainable Food Systems, 4*, 601004.

Dorrance, A. E. (2018). Management of *Phytophthora sojae* of soybean: A review and future perspectives. *Canadian Journal of Plant Pathology, 40*, 210–219.

Dowarah, B., Gill, S. S., & Agarwala, N. (2021). Arbuscular mycorrhizal fungi in conferring tolerance to biotic stresses in plants. *Journal of Plant Growth Regulation, 41*, 1429–1444.

Dugassa, G. D., Vonalten, H., & Schonbeck, F. (1996). Effects of arbuscular mycorrhizal (AM) on health of *linumusita tissimuml* infected by fungal pathogens. *Plant and Soil, 185*, 173–182.
Dumas-Gaudot, E., Gollotte, A., Cordier, C., Gianinazzi, S., & Gianinazzi Pearson, V. (2000). Modulation of host defence systems. In Y. Kapulnick & D. D. Douds Jr. (Eds.), *Arbuscular mycorrhizas: Physiology and function* (pp. 173–200). Kluwer Academic Publishers.
Eke, P., Adamou, S., Fokom, R., Dinango Nya, V., Tsouh Fokou, P. V., et al. (2020). Arbuscular mycorrhizal fungi alter antifungal potential of lemongrass essential oil against *Fusarium solani* causing root rot in common bean (*Phaseolus vulgaris* L). *Heliyon, 6*, e05737.
El-Gazzar, N., KMA, E.-H., Teama Safaa, A. M., & Rabie, G. H. (2023). Enhancing *Vicia faba's* immunity against *Rhizoctonia solani* root rot diseases by arbuscular mycorrhizal fungi and nano chitosan. *BMC Plant Biology, 23*(1), 403.
Elliott, A. J., Daniell, T. J., Cameron, D. D., & Field, K. J. (2021). A commercial arbuscular mycorrhizal inoculum increases root colonization across wheat cultivars but does not increase assimilation of mycorrhiza-acquired nutrients. *Plants, People, Planet, 3*, 588–599.
Elsharkawy, M. M., Shimizu, M., Takahashi, H., & Hyakumachi, M. (2012). The plant growth-promoting fungus *Fusarium equiseti* and the arbuscular mycorrhizal fungus *Glomus mosseae* induce systemic resistance against *Cucumber* mosaic virus in cucumber plants. *Plant and Soil, 361*, 397–409.
Fester, T., & Hause, G. (2005). Accumulation of reactive oxygen species in arbuscular mycorrhizal roots. *Mycorrhiza, 15*, 373–379.
García-Garrido, J. M., & Ocampo, J. A. (2002). Regulation of the plant defense response in arbuscular mycorrhizal symbiosis. *Journal of Experimental Botany, 53*, 1377–1386.
Garzo, E., Rizzo, E., Fereres, A., & Gomez, S. K. (2018). High levels of arbuscular mycorrhizal fungus colonization on *Medicago truncatula* reduces plant suitability as a host for pea aphids (*Acyrthosiphon pisum*). *Insect Science, 27*, 99–112.
Genre, A., & Bonfante, P. A. (2020). A rice receptor for mycorrhizal fungal signals opens new opportunities for the development of sustainable agricultural practices. *Molecular Plant, 13*, 181–183.
Gianinazzi, S., Gollotte, A., Binet, M. N., van Tuinen, D., Redecker, D., & Wipf, D. (2010). Agroecology: The key role of arbuscular mycorrhizas in ecosystem services. *Mycorrhiza, 20*, 519–530.
Guzman, A., Montes, M., Hutchins, L., DeLaCerda, G., Yang, P., et al. (2021). Crop diversity enriches arbuscular mycorrhizal fungal communities in an intensive agricultural landscape. *The New Phytologist, 231*, 447–459.
Hage-Ahmed, K., Rosner, K., & Steinkellner, S. (2019). Arbuscular mycorrhizal fungi and their response to pesticides. *Pest Management Science, 75*(3), 583–590.
Hause, B., Mrosk, C., Isayenkov, S., & Strack, D. (2007). Jasmonates in arbuscular mycorrhizal interactions. *Phytochemistry, 68*, 101–110.
He, Z. Q., Li, H. X., & Tang, H. R. (2010). Effect of arbuscular mycorrhizal fungi on endogenous in cucumber after *Rhizoctonia solani* inoculation. *Chinese Science Bulletin, 26*, 187–190.
Heil, M., & Ton, J. (2008). Long-distance signalling in plant defence. *Trends in Plant Science, 13*, 264–272.
Hildebrandt, U., Regvar, M., & Bothe, H. (2007). Arbuscular mycorrhiza and heavy metal tolerance. *Phytochemistry, 68*, 139–146.
Hoda, H., & Ghalia, A. (2007). Effect of arbuscular mycorrhizal fungus *Glomus mosseae* and *Glomus subtilis* on the infection of tomato roots by *Fusarium oxysporum*. *Egyptian Journal of Microbiology, 42*, 29–49.
Huang, N. X., Enkegaard, A., Osborne, L. S., Ramakers, P. M. J., Messelink, G. J., et al. (2011). The banker plant method in biological control. *Critical Reviews in Plant Sciences, 30*, 259–278.
Jaiti, F., Kassami, M., Meddich, A., & El Hadrami, I. E. (2008). Effect of arbuscular mycorrhization on the accumulation of hydroxycinnamic acid derivatives in data palm seedlings challenged with *Fusarium oxysporum* f sp *albedinis*. *Journal of Phytopathology, 156*(11–12), 641–646.

Jung, S. C., Martinez-Medina, A., Lopez-Raez, J. A., & Pozo, M. J. (2012). Mycorrhiza-induced resistance and priming of plant defenses. *Journal of Chemical Ecology, 38*, 651–664.

Kamble, V. R., & Agre, D. G. (2014). New report on AMF colonization in root parasite *Striga gesnerioides* and its host *Lepidagathis hamiltoniana* from high altitude region of Maharashtra. *International Multidisciplinary Research Journal, 3*, 27–31.

Kaur, S., & Suseela, V. (2020). Unraveling Arbuscular mycorrhiza-induced changes in plant primary and secondary metabolome. *Metabolites, 10*, 335.

Kloppholz, S., Kuhn, H., & Requena, N. (2011). A secreted fungal effector of *Glomus intraradices* promotes symbiotic biotrophy. *Current Biology, 21*, 1204–1209.

Kobae, Y., Kameoka, H., Sugimura, Y., Saito, K., Ohtomo, R., Fujiwara, T., & Kyozuka, J. (2018). Strigolactone biosynthesis genes of Rice are required for the punctual entry of Arbuscular Mycorrhizal fungi into the roots. *Plant & Cell Physiology, 59*(3), 544–553.

Krzyzaniak, Y., Magnin-Robert, M., Randoux, B., Fontaine, J., & Lounès-Hadj Sahraoui, A. L. (2021). *Combined use of beneficial bacteria and arbuscular mycorrhizal fungi for the biocontrol of plant cryptogamic diseases: Evidence methodology and limits in symbiotic soil microorganisms* (pp. 429–468). Springer.

Lee, E. H., Eo, J. K., Ka, K. H., & Eom, A. H. (2013). Diversity of arbuscular mycorrhizal fungi and their roles in ecosystems. *Mycobiology, 41*(3), 121–125.

Li, H. Y., Liu, R. J., Shu, H., & Li, Y. (2005). *Chib1* and *PAL5* directly involved in the defense responses induced by the arbuscular mycorrhizal fungus *Glomus fasciculatus* against nematode. *Mycosystema, 24*(3), 385–393.

Li, Y., Liu, Z., Hou, H., Hong, L., Zhu, X., Li, X., He, X., & Tian, C. (2013). Arbuscular mycorrhizal fungi-enhanced resistance against *Phytophthora sojae* infection on soybean leaves is mediated by a network involving hydrogen peroxide jasmonic acid and the metabolism of carbon and nitrogen. *Acta Physiologiae Plantarum, 35*(12), 3465–3475.

Li, F., Guo, Y., Christensen, M. J., Gao, P., Li, Y., & Duan, T. (2018). An arbuscular mycorrhizal fungus and *Epichloë festucae* var *lolii* reduce *Bipolaris sorokiniana* disease incidence and improve perennial ryegrass growth. *Mycorrhiza, 28*(2), 159–169.

Li, Y., Duan, T., Nan, Z., & Li, Y. (2021). Arbuscular mycorrhizal fungus alleviates alfalfa leaf spots caused by *Phoma medicaginis* revealed by RNA-seq analysis. *Journal of Applied Microbiology, 130*(2), 547–560.

Lin, P., Zhang, M., Wang, M., Li, Y., Liu, J., & Chen, Y. (2021). Inoculation with arbuscular mycorrhizal fungus modulates defense-related genes expression in banana seedlings susceptible to wilt disease. *Plant Signaling & Behavior, 16*(5), 1884782.

Lioussanne, L., Jolicoeur, M., & St-Arnaud, M. (2008). Mycorrhizal colonization with *Glomus intraradices* and development stage of transformed tomato roots significantly modify the chemotactic response of zoospores of the pathogen *Phytophthora nicotianae*. *Soil Biology and Biochemistry, 40*(9), 2217–2224.

Liu, W., & Wang, G.-L. (2016). Plant innate immunity in rice: A defense against pathogen infection. *National Science, 3*(3), 295–308.

Liu, R. J., Dai, M., Wu, X., Li, M., & Liu, X. Z. (2012). Suppression of the root-knot nematode *Meloidogyne incognita* (Kofoid and White) Chitwood. *Mycorrhiza, 22*(4), 289–296.

Lowe, A., Rafferty-McArdle, S. M., & Cassells, A. C. (2012). Effects of AMF- and PGPR-root inoculation and a foliar chitosan spray in single and combined treatments on powdery mildew disease in strawberry. *Agricultural and Food Science, 21*(1), 28–38.

Ma, X., Li, X., & Ludewig, U. (2021). Arbuscular mycorrhizal colonization outcompetes root hairs in maize under low phosphorus availability. *Annals of Botany, 127*(1), 155–166.

Maffei, G., Miozzi, L., Fiorilli, V., Novero, M., Lanfranco, L., & Accotto, G. P. (2013). The arbuscular mycorrhizal symbiosis attenuates symptom severity and reduces virus concentration in tomato infected by tomato yellow leaf curl Sardinia virus (TYLCSV). *Mycorrhiza, 24*(3), 179–186.

Malik, R. J., Dixon, M. H., & Bever, J. D. (2016). Mycorrhizal composition can predict foliar pathogen colonization in soybean. *Biological Control, 103*, 46–53.

Marquez, N., Giachero, M. L., Gallou, A., Debat, H. J., Cranenbrouck, S., et al. (2018). Transcriptional changes in mycorrhizal and nonmycorrhizal soybean plants upon infection with the fungal pathogen *Macrophomina phaseolina*. *MPMI, 31*, 842–855.

Marquez, N., Giachero, M. L., Declerck, S., & Ducasse, D. A. (2021). *Macrophomina phaseolina*: General characteristics of pathogenicity and methods of control. *Frontiers in Plant Science, 12*, 634397.

Martínez-Medina, A., Roldán, A., & Pascual, J. A. (2011). Interaction between arbuscular mycorrhizal fungi and *Trichoderma harzianum* under conventional and low input fertilization field condition in melon crops: Growth response and Fusarium wilt biocontrol. *Applied Soil Ecology, 47*(2), 98–105.

Matrood, A. A. A., Khrieba, M. I., & Okon, O. G. (2020). Synergistic interaction of *Glomus mosseae* T. and *Trichoderma harzianum* R. In the induction of systemic resistance of *Cucumis sativus* L. to *Alternaria alternata* (Fr.) K. *Plant Science Today, 7*, 101–108.

Maurya, A. K., Simon, S., John, V. T., & Lal, A. A. (2018). Survey of pigeon pea wilt caused by cyst nematode (*Heterodera cajani*) in Trans Yamuna and Ganga Taluks of Allahabad District. *International Journal of Current Microbiology and Applied Sciences, 7*(06), 799–802.

Metwally, R. A., & Al-Amri, S. M. (2020). Individual and interactive role of *Trichoderma viride* and arbuscular mycorrhizal fungi on growth and pigment content of onion plants. *Letters in Applied Microbiology, 70*, 79–86.

Miransari, M. (2011). Interactions between arbuscular mycorrhizal fungi and soil bacteria. *Applied Microbiology and Biotechnology, 89*(4), 917–930.

Neeraj, K., & Singh, K. (2011). Organic amendments to soil inoculated arbuscular mycorrhizal fungi and *Pseudomonas fluorescens* treatments reduce the development of root-rot disease and enhance the yield of *Phaseolus vulgaris* L. *European Journal of Soil Biology, 47*, 288–295.

Nishad, R., Ahmed, T., Rahman, V. J., & Kareem, A. (2020). Modulation of plant defense system in response to microbial interactions. *Frontiers in Microbiology, 11*, 1298.

Obrycki, J. J., Harwood, J. D., Kring, T. J., & O'Neil, R. J. (2009). Aphidophagy by Coccinellidae: Application of biological control in agroecosystems. *Biological Control, 51*(2), 244–254.

Parvin, W., Govender, N., Othman, R., Jaafar, H., Rahman, M., & Wong, M. Y. (2020). Phenazine from *Pseudomonas aeruginosa* UPMP3 induced the host resistance in oil palm (*Elaeis guineensis* Jacq)-Ganoderma boninense pathosystem. *Scientific Reports, 10*(1), 15621.

Pieterse, C. M. J., Van der Does, D., Zamioudis, C., Leon-Reyes, A., & Van Wees, S. C. M. (2012). Hormonal modulation of plant immunity. *Annual Review of Cell and Developmental Biology, 28*, 489–521.

Pozo, M. J., Cordier, C., Dumas-Gaudot, E., Gianinazzi, S., Barea, J. M., & Azcón-Aguilar, C. (2002). Localized versus systemic effect of arbuscular mycorrhizal fungi on defence responses to *Phytophthora infection* in tomato plants. *Journal of Experimental Botany, 53*, 525–534.

Pozo, M. J., Jung, S. C., & López-Ráez, J. A. (2010). Impact of arbuscular mycorrhizal symbiosis on plant response to biotic stress: The role of plant defence mechanisms. In H. Koltai & Y. Kapulnik (Eds.), *Arbuscular mycorrhizas: Physiology and function* (pp. 193–207). Springer.

Razak, N. A., & Gange, A. C. (2021). Multitrophic interactions between arbuscular mycorrhizal fungi foliar endophytic fungi and aphids. *Microbial Ecology*, 1–11.

Rodrigues, E., Silva, M. T., Calandrelli, A., Miamoto, A., Rinaldi, L. K., et al. (2021). Pre-inoculation with arbuscular mycorrhizal fungi affects essential oil quality and the reproduction of root lesion nematode in *Cymbopogon citrates*. *Mycorrhiza, 31*(5), 613–623.

Sanchez-Bel, P., Troncho, P., Gamir, J., Pozo, M. J., Camañes, G., et al. (2016). The nitrogen availability interferes with mycorrhiza-induced resistance against *Botrytis cinerea* in tomato. *Frontiers in Microbiology, 7*, 1598.

Scervino, J. M., Ponce, M. A., Erra-Bassells, R., Bompadre, J., Vierheilig, H., Ocampo, J. A., & Godeas, A. (2007). The effect of flavones and flavonols on colonization of tomato plants by *arbuscular mycorrhizal* fungi of the genera *Gigaspora* and *Glomus*. *Canadian Journal of Microbiology, 53*, 702–709.

Scheler, C., Durner, J., & Astier, J. (2013). Nitric oxid and reactive oxygen species in plant biotic interactions. *Current Opinion in Plant Biology, 16*, 534–539.

Schmitz, A. M., & Harrison, M. J. (2014). Signaling events during initiation of arbuscular mycorrhizal symbiosis. *Journal of Integrative Plant Biology, 56*, 250–261.

Schönbeck, F., & Dehne, H. W. (1977). Damage to mycorrhizal and nonmycorrhizal cotton seedlings by *Thielaviopsis basicola. Plant Disease Report, 61*, 266–276.

Schouteden, N., De Waele, D., Panis, B., & Vos, C. M. (2015). Arbuscular Mycorrhizal fungi for the biocontrol of plant-parasitic nematodes: A review of the mechanisms involved. *Frontiers in Microbiology, 6*, 1280.

Siciliano, V., Genre, A., Balestrini, R., Cappellazzo, G., deWit, P. J. G. M., & Bonfante, P. (2007). Transcriptome analysis of arbuscular mycorrhizal roots during development of the prepenetration apparatus. *Plant Physiology, 144*, 1455–1466.

Siddiqui, Z. A., & Akhtar, M. S. (2006). Biological control of root-rot disease complex of chickpea by AM fungi. *Archives of Phytopathology, 39*, 389–395.

Siddiqui, Z. A., & Mahmood, I. (1996). Biological control of Heterodera cajani and Fusarium udum on pigeon pea by Glomus mosseae, Trichoderma harzianum and Verticillium chlamydosporium. *Israel Journal of Plant Sciences, 44*, 49–56.

Sikder, M. M., & Vestergård, M. (2020). Impacts of root metabolites on soil nematodes. *Frontiers in Plant Science, 10*, 1792.

Silvestri, A., Pérez-Tienda, J., & López-Ráez, J. A. (2020). Arbuscular mycorrhizal fungal gene expression analysis by real-time PCR. *Methods in Molecular Biology, 2146*, 157–170.

Slezack, S., Dumas-Gaudot, E., Paynot, M., & Gianinazzi, S. (2000). Is a fully established arbuscular mycorrhizal symbiosis required for bioprotection of *Pisum sativum* roots against *Aphanomyces euteiches*? *Molecular Plant-Microbe Interactions, 13*, 238–241.

Smith, S. E., & Smith, F. A. (2011). Roles of arbuscular mycorrhizas in plant nutrition and growth: New paradigms from cellular to ecosystem scales. *Annual Review of Phytopathology, 62*, 227–250.

Song, Y., Chen, D., Lu, K., Sun, Z., & Zeng, R. (2015). Enhanced tomato disease resistance primed by arbuscular mycorrhizal fungus. *Frontiers in Plant Science, 6*, 786.

Spagnoletti, F. N., Cornero, M., Chiocchio, V. M., Lavado, R. S., & Roberts, I. N. (2020). Arbuscular mycorrhiza protects soybean plants against *Macrophomina phaseolina* even under nitrogen fertilization. *European Journal of Plant Pathology, 156*, 839–849.

Srivastava, R., Khalid, A., Singh, U. S., & Sharma, A. K. (2010). Evaluation of *arbuscular mycorrhizal* fungus *fluorescent Pseudomonas* and *Trichoderma harzianum* formulation against *Fusarium oxysporum* f sp lycopersici for the management of tomato wilt. *Biological Control, 53*, 24–31.

Steinkellner, S., Hage-Ahmed, K., García-Garrido, J. M., Illana, A., Ocampo, J. A., & Vierheilig, H. (2012). A comparison of wild-type old and modern tomato cultivars in the interaction with the arbuscular mycorrhizal fungus *Glomus mosseae* and the tomato pathogen *Fusarium oxysporum* f. sp. *lycopersici. Mycorrhiza, 22*, 189–194.

Sunar, K., Das, K., Rai, A. K., & Gurung, S. A. (2023). Beneficial microbial consortia and their role in sustainable agriculture under climate change conditions. In P. Mathur, R. Kapoor, & S. Roy (Eds.), *Microbial symbionts and plant health: Trends and applications for changing climate* (pp. 41–73). Springer.

Tabin, T., Arunachalam, A., Shrivastava, K., & Arunachalam, K. (2009). Effect of arbuscular mycorrhizal fungi on damping-off disease in *Aquilaria agallocha* Roxb. Seedlings. *Tropical Ecology, 50*, 243–248.

Tahat, M. M., Sijam, K., & Othman, R. (2012). Ultrastructural changes of tomatoes (Lycopersicon esculentum) root colonized by *Glomus mosseae* and *Ralstonia solanacearum. African Journal of Biotechnology, 11*, 6681–6686.

Tahmatsidou, V., O'Sullivan, J., Cassells, A. C., Voyiatzis, D., & Paroussi, G. (2006). Comparison of AMF and PGPR inoculants for the suppression of *Verticillium wilt* of strawberry (*Fragaria × ananassa* cv Selva). *Applied Soil Ecology, 32*(3), 316–324.

Tanwar, A., Aggarwal, A., & Panwar, V. (2013). Arbuscular mycorrhizal fungi and *Trichoderma viride* mediated Fusarium wilt control in tomato. *Biocontrol Science and Technology, 23*(5), 485–498.

Tatsumi, C., Hyodo, F., Taniguchi, T., Shi, W., Koba, K., et al. (2020). Arbuscular mycorrhizal community in roots and nitrogen uptake patterns of understory trees beneath ectomycorrhizal and non-ectomycorrhizal overstory trees. *Frontiers in Plant Science, 11*, 583585.

Ton, J., Flors, V., & Mauch-Mani, B. (2009). The multifaceted role of ABA in disease resistance. *Trends in Plant Science, 14*, 310–317.

Van der Ent, S., Van Wees, S. C. M., & Pieterse, C. M. J. (2009). Jasmonate signaling in plant interactions with resistance-inducing beneficial microbes. *Phytochemistry, 70*, 1581–1588.

van Driesche, R. G., Carruthers, R. I., Center, T., Hoddle, M. S., Hough-Goldstein, J., et al. (2010). Classical biological control for the protection of natural ecosystems. *Biological Control, 54*(S1), 2–33.

Veresoglou, S. D., & Rillig, M. C. (2012). Suppression of fungal and nematode plant pathogens through arbuscular mycorrhizal fungi. *Biology Letters, 8*, 214–217.

Vierheilig, H., Steinkellner, S., Khaosaad, T., & Garcia-Garrido, J. M. (2008). The biocontrol effect of mycorrhization on soilborne fungal pathogens and the autoregulation of the AM symbiosis: One mechanism two effects? In A. Varma (Ed.), *Mycorrhiza* (pp. 307–320). Springer.

Vishwakarma, S. K., Ilyas, T., Malviya, D., Shafi, Z., Shahid, M., et al. (2022). Arbuscular mycorrhizal fungi (AMF) as potential biocontrol agents. In U. B. Singh, P. K. Sahu, H. V. Singh, P. K. Sharma, S. K. Sharma, et al. (Eds.), *Rhizosphere microbes. Microorganisms for sustainability 40* (pp. 197–222). Springer.

Vos, C., Geerinckx, K., Mkandawire, R., Panis, B., De Waele, D., & Elsen, A. (2012). Arbuscular mycorrhizal fungi affect both penetration and further life stage development of root-knot nematodes in tomato. *Mycorrhiza, 22*, 157–163.

Vos, C., Schouteden, N., van Tuinen, D., Chatagnier, O., Elsen, A., et al. (2013). Mycorrhiza-induced resistance against the root-knot nematode *Meloidogyne incognita* involves priming of defense gene responses in tomato. *Soil Biology and Biochemistry, 60*, 45–54.

Weaver, D. J., & Wehunt, E. J. (1975). Effect of soil pH on susceptibility of peach to *Pseudomonas syringae*. *Phytopathology, 65*, 984–989.

Weng, W., Yan, J., Zhou, M., Yao, X., Gao, A., Ma, C., Cheng, J., & Ruan, J. (2022). Roles of arbuscular mycorrhizal fungi as a biocontrol agent in the control of plant diseases. *Microorganisms, 10*, 1266.

Whipps, J. M. (2004). Prospects and limitations for mycorrhizas in biocontrol of root pathogens. *Canadian Journal of Botany, 82*, 1198–1227.

Wu, L., Chang, K. F., Conner, R. L., Strelkov, S. E., Fredua-Agyeman, R., et al. (2018). *Aphanomyces euteiches*: A threat to canadian field pea production. *Engineering, 4*, 542–551.

Xavier, L. J. C., & Boyetchko, S. M. (2003). Arbuscular mycorrhizal fungi in plant disease control. In D. K. Arora (Ed.), *Fungal biotechnology in agricultural food and environmental applications* (pp. 183–194). CRC Press.

Xiang, N., Lawrence, K., & Donald, P. A. (2018). Biological control potential of plant growth-promoting rhizobacteria suppression of *Meloidogyne incognita* on cotton and *Heterodera glycines* on soybean. *Annual Review of Phytopathology, 166*(7–8), 449–458.

Xin, X., Kvitko, B. H., & He, S. Y. (2018). *Pseudomonas syringae*: What it takes to be a pathogen. *Nature Reviews. Microbiology, 16*, 316–328.

Zhang, J., & Zhou, J. M. (2010). Plant immunity triggered by microbial molecular signatures. *Molecular Plant, 3*, 783–793.

Zhang, G., Raza, W., Wang, X., Ran, W., & Shen, Q. (2012). Systemic modification of cotton root exudates induced by arbuscular mycorrhizal fungi and *Bacillus vallismortis* HJ-5 and their effects on *Verticillium wilt* disease. *Applied Soil Ecology, 61*, 85–91.

Zhou, J., Chai, X., Zhang, L., George, T. S., Wang, F., & Feng, G. (2020). Different arbuscular mycorrhizal fungi cocolonizing on a single plant root system recruit distinct microbiomes. *mSystems, 5*, e00929-20.

Zhu, H. H., & Yao, Q. (2004). Localized and systemic increase of phenols in tomato roots induced by *Glomus versiforme* inhibits *Ralstonia solanacearum*. *Journal of Phytopathology, 152*, 537–542.

Ziane, S. O., Talibi, Z. E. A., Koraichi, S. I., Douira, A., Amir, S., Meddich, A., & Modafar, C. E. (2023). Synergistic effects of arbuscular mycorrhizal fungi associated to plant growth-promoting rhizobacteria in suppression of soil-borne Fusarium wilt of date palm. *Biocatalysis and Agricultural Biotechnology, 51*, 102753.

Chapter 10
Phytostimulation and Induction of Systemic Resistance by Plant Growth Promoting Fungi

Balwant Singh, Shivangi Tripathi, Santvana Tyagi, Abhishek Tripathi, Priya Dubey, and Shailendra Kumar

Abstract In the present era, the world needs high production of crops to fulfill the demand for food. For this reason, farmers use chemicals and fertilizers to increase crop production, but this increase in yield causes damage to the environment. To prevent the harmful effects of chemicals and fertilizers, eco-friendly approaches are needed to ensure environmental and human health. For the innovative view of sustainable agriculture, biofertilizers are important replacements of chemicals for growing demands and crop production. Use of beneficial microorganisms is a new strategy to improve plant growth and crop productivity. The role of microbial populations, particularly fungi, is pivotal in solubilizing and mineralizing soil nutrients essential for plant growth. Among these microbial populations, certain fungi such as *Trichoderma*, *Penicillium*, *Aspergillus*, *Fusarium*, *Phoma*, or *Piriformospora* may act as plant growth-promoting fungi (PGPF). They showed effectiveness in significantly boosting plant growth, improving germination, seedling strength, biomass

B. Singh (✉)
Department of Botany, K. S. Saket PG College Ayodhya, Dr. Ram Manohar Lohia Avadh University, Ayodhya, Uttar Pradesh, India

S. Tripathi
Department of Bioscience, Integral University, Lucknow, Uttar Pradesh, India

S. Tyagi
Department of Molecular Biology and Genetic Engineering, G. B. Pant University of Agriculture and Technology, Pantnagar, Uttarakhand, India

A. Tripathi
Department of Biotechnology, Institute of Engineering and Technology, Abdul Kalam Technical University, Lucknow, Uttar Pradesh, India

P. Dubey
Department of Bioscience, Integral University, Lucknow, Uttar Pradesh, India

S. Kumar
Department of Microbiology, Dr. Ram Manohar Lohia Avadh University, Ayodhya, Uttar Pradesh, India

P. Mathur, S. Roy (eds.), *Plant Microbiome and Biological Control*, Sustainability in Plant and Crop Protection 20,
https://doi.org/10.1007/978-3-031-75845-4_10

production, root hair development, photosynthetic efficiency, flowering, and overall yields. Furthermore, PGPFs exert control over foliar and root pathogens by inducing systemic resistance (ISR) in the host plants. The exceptional capabilities of some PGPFs originate from their ability to enhance nutrient absorption, stimulate the production of phytohormones and alter plant gene expression by activating signaling pathways. This chapter offers a succinct overview of the interactions between plants and their associated PGPFs, shedding light on their role in fostering enhanced plant growth and inducing systemic resistance against invading pathogens.

Keywords Fungi · ISR · PGPF · Plant growth · Sustainable agriculture

Abbreviations

AAO	ABA Aldehyde Oxidase
AM	Arbuscular Mycorrhizae
CHI	Chitinases
ECM	Ecto-Mycorrhizae
ET	Ethylene
GLU	Glucanases
HCN	Hydrogen Cyanide
IAA	Indole Acetic Acid
ISR	Induced Systemic Resistance
JA	Jasmonate
NCED	9-Cis-Epoxycarotenoid Dioxygenase
PAL	Phenylalanine Ammonia-Lyase
PGPF	Plant Growth Promoting Fungi
PGPM	Plant-Growth-Promoting Microorganisms
PGPR	Plant Growth-Promoting Rhizobacteria
POX	Peroxidases
ROS	Reactive Oxygen Species
SA	Salicylic Acid
SAR	Systemic Acquired Resistance
TMV	Tobacco Mosaic Virus
TNV	Tobacco Necrosis Virus
ZEP	Zeathanxin Epoxidase

10.1 Introduction

A large class of non-pathogenic fungi is found in plants called "Plant growth-promoting Fungi" (PGPF). They induce systemic resistance and phytostimulation which are two important characteristics of the complex interactions between plants and helpful microbes. PGPF mediate beneficial changes or enhance plant development, health, and protect against different stress types. The majority of PGPF originates from the soil or roots of other hosts. PGPF can inhibit plant diseases in several ways, such as by producing antibiotics and competing with pathogens for resources i.e., food and space. A few of the fungi classified as PGPF belong to the Basidiomycota (*Rhizoctonia*, *Limonomyces*, *Rhodotorula* and Sterile fungi) and Zygomycota (*Rhizopus* and *Mucor*), while the majority are classified in the phylum Ascomycota (*Aureobasidium*, *Aspergillus*, *Cladosporium*, *Chaetomium*, *Colletotrichum*, *Trichoderma*, *Exophiala*, *Penicillium*, *Fusarium*, *Gliocladium*, *Phoma*, *Purpureocillium*, *Phomopsis*, and *Talaromyces*) (Debbarma et al., 2021; Hossain et al., 2017). Plant fitness can be significantly impacted by fungi or bacteria found around roots. The plant growth-promoting rhizobacteria (PGPR) or PGPF are useful microorganisms that inhabit the rhizosphere (Vílchez et al., 2020). They encourage the adaptive development of vigorous root systems and improve the ability of their hosts to absorb nutrients and water (Verbon & Liberman, 2016). PGPF are non-pathogenic, naturally occurring saprotrophs that contribute to soil fertility, boost plant development and trigger a frontline defense against pathogenic infections (Hossain et al., 2017; Zhang et al., 2016). The ability of PGPF to colonize roots is the primary mechanism in preventing pathogens infection. In addition to promoting disease resistance, the PGPF capacity to solubilize phosphates and generate indole acetic acid (IAA), siderophores, cellulase, chitinase, and other compounds can directly or indirectly improve growth and development of the host plant (Jogaiah et al., 2013; Muslim et al., 2019; Naziya et al., 2020).

Despite the accumulation of lignin, callose, phenols, and other compounds, PGPF alter the plant cell walls to induce systemic resistance (ISR). This alteration prevents invasive pathogens from entering and growing (Zhou et al., 2018). Furthermore, PGPF cause plants to produce more defense-related enzymes, such as β-1,3 glucanase, chitinase, peroxidases (POX), and phenylalanine ammonia-lyase (PAL), that directly contribute to defense systems. The reported effects demonstrate the diverse roles that PGPF play in strengthening plants against disease threats, including both structural and enzymatic alterations (Naziya et al., 2020).

The prevalent use of chemical applications, including pesticides and fertilizers, has adverse effects on the environment and soil microflora. Dependence on inorganic chemical fertilizers poses threats to human health and the environment, leading to public concerns about agrochemical use. Moreover, long-term chemical applications may result in plant pathogenic fungi developing resistance. A shift towards biological-based organic fertilizers is gaining attetion, to address these concerns. The rhizosphere microbiome plays a crucial role in natural agriculture ecosystems, influencing crop quality through services like organic matter decomposition, nutrient acquisition, water absorption, weed control, and pest or disease biocontrol. Application of beneficial microorganisms emerged in the lase decade as a new strategy to enhance plant health and productivity. This review emphasizes the potential of PGPF inoculants in improving crops, showcasing the diverse activities of rhizosphere free-living fungi, and advocating for the use of microbes as biocontrol agents and biofertilizers (El-Maraghy et al., 2021).

Innovative strategies aiming at sustainable plant defense systems with improved physiological immunity and soil mineral bioavailability focus on the exploitation of growth-stimulating microbes. These species function as bio-nutrients, enhancing the nutrition content of crops and improving the soil in an environmentally beneficial way. These biological agents have the ability to secrete enzymes that aid in the intake of nutrients without creating pollution to the environment, hence increasing productivity and producing high-value crops. A protective physiological condition called "stimulated resistance," brought on by environmentally beneficial stimuli, is essential for fending off a wide range of plant diseases, including pathogenic fungi.

Through mechanisms such as systemic resistance stimulation, plant nutrition enhancement, and toxicity to pathogens, PGPF play a role in this process. PGPF may release substances like hydrogen cyanide (HCN), acting against a range of infections. Furthermore, PGPF generates IAA, which stimulates plant development and solubilizes organic phosphates, improving the health of plants. These fungi may be used to counteract severe pathogens, i.e., Fusarium which have a detrimental impact on yields, strengthening the plants immune systems and increasing their resilience to biological challenges (Abd Alhakim et al., 2022).

Global agricultural crops are seriously threatened by fungal diseases, which can result in serious harm and large financial losses. They not only reduce agricultural outputs but also jeopardize the quality of produced goods by releasing toxic chemicals such as mycotoxins. These metabolites can cause poisoning and other major health problems in both humans and animals who consume contaminated food. Fungal pathogens are extremely destructive and common, accounting for around 80% of crop diseases. It is difficult to control soil-borne fungi because of the intricate nature of the soil ecosystem and the quick emergence of new infection paths (Abd Alhakim et al., 2022; Attia et al., 2022). Scientific data emphasize the plant innate capacity to protect themselves from infections through resistance mechanisms, particularly acting against the backdrop of climatic changes that increase their vulnerability. Understanding and utilizing plant defenses is crucial in the face of changing environmental challenges, as this resistance might act synthetically through phenotypic or anatomic traits, or chemically through the presence of chemicals limiting the pathogen growth. The interactions between plants and PGPF encompass phytostimulation and systemic resistance induction. Unraveling the intricacies of these interactions holds great promise for developing sustainable agricultural strategies that foster plant resilience and productivity.

10.2 Plant Growth-Promoting Fungi

Fungi are the second most diverse (after insects) group of organisms with heterotrophic modes of life such as saprotrophic, parasitic, and symbiotic. (Singh & Singh, 2023). Specialized symbiotic fungi which form a biological relationship with higher plants' roots fostering its growth is termed PGPF. They form useful associations with plants based on a mutualistic performance and induce plant growth as well as production through different mechanisms.

10.2.1 Nature of PGPF

Numerous studies have been conducted to investigate the attributes of PGPF, elucidating their mechanisms of action, and exploring their potential application in agriculture and ecosystem management. Phosphorus growth promoting fungi (PGPFs) encompass a wide range of non-pathogenic species that establish symbiotic relationships with plants, hence promoting growth and enhancing overall health (Hossain et al., 2017). The misclassification of fungal species as PGPF can be attributed to their inherent biological degree of dissimilarity. Extensive scholarly investigations yielded results indicating significant variations in taxonomy, habitat distribution, physiology, and plant-interaction properties among the fungi categorized as PGPF. Bent (2006) presented evidence that challenged the efficacy of the PGPF designation. These findings indicate that the PGPF do not induce plant growth consistently, hence contradicting their purported general benefits. After a thorough analysis of the effects of various environmental variables on the growth and development of certain plant species, it became apparent that the role of a fungus may not exhibit regular patterns of influence. An analogous contention might concern the categorization of specific fungi that elicit plant development as plant growth-promoting species (Jahagirdar et al., 2013).

10.2.2 Host Specificity

The concept of host specificity in PGPF refers to the tendency of these fungi to establish mutualistic associations with specific plant species or groups. Certain species of PGPF exhibit a broad spectrum of host compatibility and are capable of forming associations with various host species (García et al., 2005). Conversely, there are other PGPF that exhibit a more restricted or specific host preference. The host specificity of PGPF is influenced by various factors.

Different types of mycorrhizal associations, such as arbuscular mycorrhizae (AM) and ectomycorrhiza (ECM), exhibit varying degrees of host specificity. For example, specific fungi forms symbiotic associations primarily with plants in the family Fabaceae, commonly referred to as legumes. In contrast, other fungi display affiliations with a broader range of plant taxa (García-Garza et al., 1997; Harman et al., 2004).

The PGPF host specificity can be modulated by a range of environmental factors, such as soil pH, nutrient concentrations, and climatic conditions. Several fungi possess the ability to thrive under specific environmental conditions that are conducive to the growth and reproduction of particular plant species. Empirical evidence has substantiated the existence of mutualistic connections between plants and PGPF, with a given degree of versatility (Samuels & Hebbar, 2015). These associations confer significant benefits to the plants involved. Based on a comprehensive body of documented evidence, it is commonly recognized that the fungal genera *Aspergillus*, *Fusarium*, *Penicillium*, *Piriformospora*, *Phoma*, and *Trichoderma* are prominent examples of PGPF (Navi & Bandyopadhyay, 2002).

Furthermore, plant growth-promoting rhizobacteria (PGPR) play a significant role in alleviating abiotic stresses. The prerequisite for the manifestation of positive effects of the majority of PGPF on plants depends on host colonization process.

Several species of *Trichoderma*, *Aspergillus*, *Penicillium*, and certain endophytes have been employed as agents to augment growth and development of plants in the agricultural sector (Ahmad & Baker, 1987). These fungi are widely recognized as plant growth promoters.

10.2.3 Composition

PGPF demonstrate an ability to inhibit the spread of plant diseases in the rhizosphere. This conclusion is supported by the promotion of mineral solubilization, as well as by the synthesis of hydrolytic enzymes and plant hormones. The assemblage comprises the mineral elements nitrogen (N), phosphorus (P), and iron (Fe). Furthermore, the establishment of an equilibrium, the competition for saprotrophic colonization, and mycoparasitism are additional mechanisms that contribute to the advancement of systemic resistance (Lewis & Papavizas, 1991). The convergence of several signalling pathways ultimately leads to the development of systemic resistance. The phenomenon, discussed by Hyakumachi (1994), pertains to the recurrent formation of colonies by PGPF within the root system. The results of this colonizing endeavor are enhanced development, fostered growth, and strengthened defense capabilities.

The scientific literature has reported a diverse array of features in multiple species of *Trichoderma*, as evidenced by the studies conducted by García-Garza et al. (1997), Harman et al. (2004), and García et al. (2005). The afore mentioned characteristics encompassed their ability to improve plant development, optimize nutrient uptake, alter the rhizosphere, and decrease the density of dangerous fungi by strengthening defense mechanisms. Verma et al. (2007) and Samuels and Hebbar (2015) conducted extensive studies about the use of bio-fungicides and agricultural products derived from several species of *Trichoderma*. In their study, Navi and Bandyopadhyay (2002) conducted a comprehensive collection of commercially available biocontrol agents specifically developed for the purpose of managing fungal plant diseases. The inventory comprised a varied assortment of species, including members of genera *Ampelomyces*, *Candida*, *Coniothyrium*, *Gliocladium*, and *Talaromyces*.

10.2.4 Diversity

The classification of PGPF is extremely heterogeneous. A significant proportion of fungi that are categorized as PGPF are predominantly classified within the taxonomic phylum Ascomycota (Kredics et al., 2014). The compilation comprises the following genera: *Talaromyces*, *Aureobasidium*, *Chaetomium*, *Cladosporium*, *Colletotrichum*, *Exophiala*, *Penicillium*, *Trichoderma*, *Fusarium*, *Gliocladium*, *Phoma*, *Phomopsis*, and *Purpureocillium*. With the exception of *Limonomyces*, *Rhodotorula*, *Rhizoctonia*, and non-reproductive fungi, Basidiomycota include only a limited number of PGPF species (Verma et al., 2007).

Furthermore, distinct PGPF species can be identified in the taxonomic phylum Zygomycota, more precisely in genera *Mucor* and *Rhizopus*. Nevertheless, despite their genetic similarity, these fungi are devoid of the essential virulence characteristics required to infect a substantial proportion of the plant hosts from which they can be obtained. Sterile fungi include also some PGPF. They correspond to a unique class of mycelial fungi that are characterized by the lack of an ability to produce spores. The taxonomic group Oomycota is predominantly comprised of organisms that have gained notoriety for causing significant harm to vegetation. However, it is crucial to note that a minority of the species comprising this assemblage have been classified as non-pathogenic (Thines & Kamoun, 2010). *Pythium oligandrum* and *Phytophthora cryptogea*, two non-pathogenic oomycetes, colonize as PGPF the root ecosystems (Attitalla et al., 2001; Benhamou et al., 2012).

10.3 Phytostimulation

Plants establish intricate relationships in the rhizosphere with a variety of microorganisms, particularly bacteria and fungi. These interactions may strengthen and change their metabolism, encouraging growth and development by producing low molecular weight volatile components, solubilizing soil phosphorus, and synthesizing plant hormones like IAA (Jogaiah et al., 2013; Hossain et al., 2017). PGPF may promote root and shoot growth, seed germination, seedling vigor, increase photosynthetic efficacy, and thus agricultural output. When compared to mycorrhizae, PGPF include more varied taxonomic groups. They present a wider variety of host plants involved in a non-obligate mutualism. For this reason, even though symbiotic mycorrhizal fungi are known to promote plant growth, they are distinct from PGPF (Bent, 2006). In comparison to mycorrhiza, PGPF are more diverse in terms of taxonomy and have more complex relationships with plants. To improve seed germination, seedling vigour, growth, flowering, and productivity, they rely on a variety of methods (Hossain et al., 2017). PGPF inhibit phytopathogens in addition to having beneficial effects on plant development through their root associations. PGPFs such as *Trichoderma* spp. have beneficial effects, but may also function as biocontrol agents (Kaewchai, 2009). Moreover, PGPF are also essential in the formulation of biofertilizers and myco-fungicides (Vujanovic & Goh, 2012).

Plant growth is influenced by rhizosphere microorganisms through both direct and indirect mechanisms (Fig. 10.1). The genera *Penicillium* and *Trichoderma* provide examples of (PGPF that promote plant growth and create systemic resistance to diseases in a variety of host species (El-Maraghy et al., 2021; Romera et al., 2019). When used as a biological control agent, *T. harzianum* not only inhibits infections but also encourages plant development, which lowers production costs, minimizes environmental effects, and requires less chemical fungicide usage (El-Maraghy et al., 2021).

Many PGPF have different ways of treating infections and stimulating plant development (El-Maraghy et al., 2021). Micronutrients and phosphate are solubilized by *T. harzianum* and non-pathogenic *Fusarium* species, increasing plant accessibility (Hussein & Joo, 2015). *Fusarium* wilt is efficiently managed in a variety of crops by other *F. oxysporum* strains. Cucumbers develop a systemic resistance to *Colletotrichum orbiculare* due to *Phoma* sp. and *P. simplicissimum* (Muslim et al., 2019). Further examples include the endophyte *Piriformospora indica* tath stimulates the growth of host plants. The effectiveness of *Phoma* sp., *P. simplicissimum,* and *Glomus mosseae* rely on boosting cucumber systemic resistance against anthracnose.

Mannitol and other sugar alcohols have an impact on the formation of rhizosphere fungal colonies (Frąc et al., 2018). All things considered, PGPF provide a variety of advantages by managing pathogen diversity, stress conditions, and

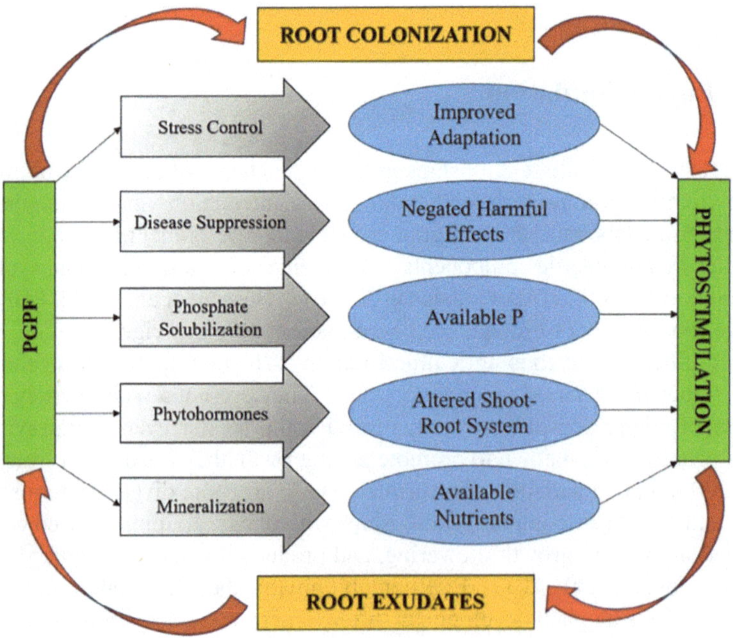

Fig. 10.1 Beneficial associations between plant and PGPF (phytostimulation cycle through PGPF). (Adapted from Hossain & Sultana, 2020)

nutrient availability within many geographic agricultural landscapes. These results advance our knowledge on the critical function that PGPF play in efficient and sustainable plant management (El-Maraghy et al., 2021).

Trichoderma sp. has shown the capacity to promote plant development in addition to its biocontrol properties. The release of metabolites that regulate plant growth, such as phytohormones, and the increased saprotrophic activity of *Trichoderma* in soil underpin this effect (Baker, 1988). Additionally, the control of minor pathogens may also play a role in increased root growth and nutrient absorption (Ousley et al., 1993). Moreover, other *Trichoderma* isolates demonstrated biocontrol against *Sclerotinia* on cabbage. Furthermore, a field trial was conducted on lettuce to test *Trichoderma longipile* 6Sr4 (Rabeendran et al., 2000). Improvements in the photosynthetic machinery have occasionally been connected to increases in the growth of specific plants through interactions with PGPF. In greenhouse trials, the application of *T. longipile* and *T. tomentosum* increased the area of cabbage leaves by 58–71% (Hossain et al., 2017).

Abiotic stress factors such as heat, salinity, and drought can damage mitochondria and chloroplasts, reduce water-use efficiency, and decrease rates of photosynthesis and nutrient uptake in crop plants, all of which can dramatically reduce growth and yield (Cramer et al., 2011). An early response of plants to these challenges is the activation of abscisic acid (ABA) to regulate stomatal closure. Enzyme-encoding genes like zeathanxin epoxidase (*ZEP* or *ABA1*), ABA aldehyde oxidase (*AAO3*), 9-cis-epoxycarotenoid dioxygenase (*NCED*), and *ABA3* are responsible for synthesizing ABA. Abiotic stress in plants regulates the expression of these genes (Lee & Luan, 2012). Anion channels located in the plasma membrane are activated by elevated ABA levels, which also raise the cytosolic Ca^{2+} concentration. Increased ABA also increases the synthesis of reactive oxygen species (ROS), such as H_2O_2, which function as a signalling mediator to cause stomatal closure (Khan et al., 2015).

PGPF may positively affect many of these metabolic pathways, leading to increased leaf efficiency. *Penicillium* species increased the amount of chlorophyll in cucumber and chili leaves. Cucumber and Indian spinach showed a considerable increase in leaf area and leaf chlorophyll content following soil amendment with *Aspergillus* spp. (Babu et al., 2015). *Fusarium* sp. PPA1 and PPF1, increased the rate of photosynthesis and the net accumulation of carbohydrates in plants, respectively, improving the amount, area and chlorophyll content of leaves (Hossain et al., 2017; Islam et al., 2014).

10.4 Induced Systemic Resistance

Exploration into the biological regulation of plant diseases in soil unveiled a significant breakthrough: the identification of a specific form of induced resistance. This phenomenon occurs when certain PGPR colonize roots, triggering a process known as induced systemic resistance (ISR). ISR is acknowledged for its ability to decrease the occurrence and severity of a variety of fungal, bacterial, viral, nematode, and oomycete diseases across a broad spectrum of plant species. In contrast to innate defense mechanisms, ISR is seen as a cost-effective approach. It accomplishes this by efficiently managing the allocation of resources in plant defense, establishing a more optimal balance between resource investment and potential threats. Consequently, this leads to a reduction in the physiological costs incurred by plants. Therefore, ISR serves as an efficient defense mechanism against invasive pathogens. Induced resistance falls into two main categories: induced systemic resistance (ISR) and systemic acquired resistance (SAR) (Murali et al., 2021). This categorization is determined by the inherent nature of the inducer and the regulatory pathways for defense signaling. ISR offers a constructive approach to contrast infection levels in plants caused by detrimental pathogens.

The activation of ISR upon pathogen invasion is systematically evident in plant regions spatially removed from the initial inducing site (Tarroum et al., 2022). The treatment of roots with PGPF shows a systemic impact by eliciting resistance in leaves and shoots. The implementation of this approach through PGPF treatment is recognized as a secure and cost-effective strategy for enhancing the growth and productivity of crops (Debbarma et al., 2021). ISR functions as a mechanism to hinder the invasion of plant diseases orchestrated by specific phytopathogens. The initiation of ISR after a phytopathogen attack is consistently observed in plant organs spatially distant from the initial inducing site. Conversely, the capacity of PGPF acting on roots results in the induction of resistance in the shoots covering the upper portion of the plant (Debbarma et al., 2021).

The systematic instigation of resistance through the application of PGPF stands out as a secure and highly efficient strategy to elevate plants growth. The extensive deployment of PGPF, encompassing fungi like *Trichoderma* sp. or *Penicillium* sp. elicits ISR impeding the proliferation of soil-borne phytopathogens (Murali et al., 2021). Elicitors, exemplified by glycoproteins derived from *Phytophthora cinnamomi, Cladosporium fulvum, Colletotrichum lindemuthianum* and *Rhizoctonia solani* showcase their prowess in inducing resistance in crop plants (Debbarma et al., 2021). Notably, in tobacco plants, an eliciting protein extracted from *Phytophthora* sp. was employed to activate defense mechanisms, exemplifying the versatility of these elicitors (Attia et al., 2022). Furthermore, a proteinaceous non-enzymatic elicitor sourced from *T. virens* not only effectively triggered defense responses in plants but also stimulated ISR against foliar phytopathogens. Potent elicitors, such as oligosaccharides, isolated from PGPF, amplify the activation of plant defense mechanisms against phytopathogens. Fungi responsible for soft rot diseases in tomato and other crops, such as *F. oxysporum*, generate cerebrosides, primarily glycosphingolipids. These cerebrosides have proven their mettle by demonstrating resistance against wilt and soft rot diseases in tomato plants (Attia et al., 2022). Notably, in greenhouse experiments, the application of cerebrosides also effectively impeded the proliferation of anthracnose disease in chili plants, underscoring their potential in crop protection and disease management (Fig. 10.2).

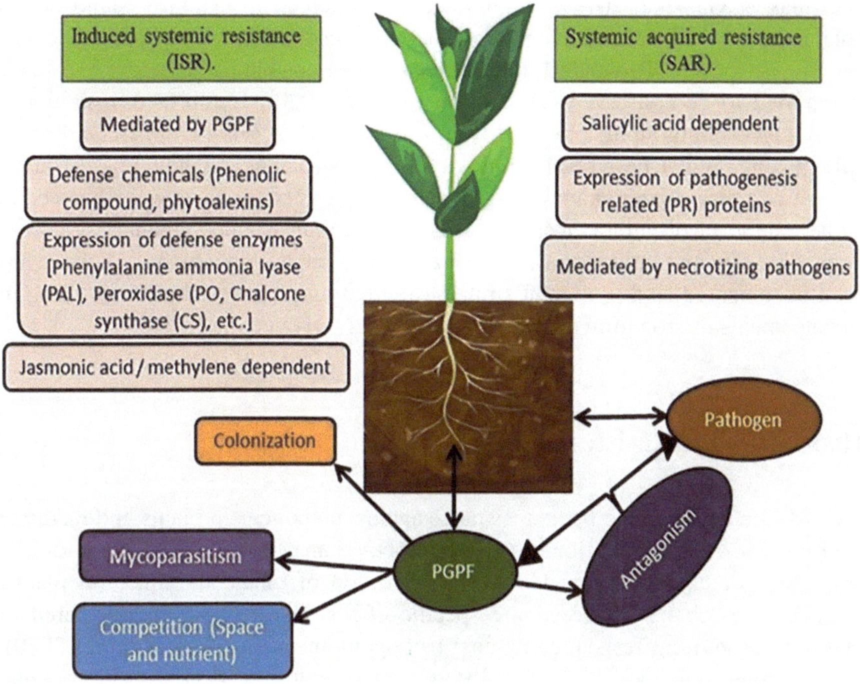

Fig. 10.2 The resistance mechanism provided by PGPF against invading pathogens

10.5 Systemic Acquired Resistance (SAR)

(SAR delineates a distinct signal transduction pathway marked by the synthesis of pathogenesis-related proteins, playing a pivotal role in fortifying the plant defense mechanisms *vs* diverse pathogens. In tobacco plants, the induction of SAR significantly mitigated the expression of disease symptoms triggered by an array of pathogenic microorganisms. These encompass *Phytophthora parasitica*, *Cercospora nicotianae*, *Peronospora tabacina*, as well as *Tobacco Mosaic Virus* (TMV), *Tobacco Necrosis Virus (*TNV), *Pseudomonas syringae* pv. *tabaci*, and *Erwinia carotovora*. The introduction of *Trichoderma harzianum* resulted in the synthesis of PR proteins in pepper plants when subjected to infection by *Phytophthora capsici*. Metabolites originating from diverse fungi show a significant capability to instill resistance within plants.

A well-established connection exists between the expression of elicitor-inducible pathogenesis-related (PR) proteins and disease resistance. An array of glycoprotein elicitors derived from *P. cinnamomi, Cladosporium fulvum, Colletotrichum lindemuthianum* and *R. solani*, showcased remarkable capabilities in inducing resistance in plants (Debbarma et al., 2021). For example, a 10 kDa protein elicitor (elicitin) isolated from *Phytopthora* sp. triggered defense responses in tobacco plants

(Woźniak & Gałązka, 2019). Oligosaccharides sourced from PGPF stand out as potent elicitors capable of stimulating defense responses in crop plants against pathogenic species. The underlying mechanism responsible for inducing resistance is reported to be conserved (Hossain & Sultana, 2020). Cerebrosides, predominantly glycosphingolipids obtained from *Fusarium oxysporum* f. sp. *lycopersici*, a wilt-causing fungus, have been identified for their proficiency in inducing resistance against wilt disease in tomato plants. Treatment with cerebrosides was observed to reduce the level of anthracnose infection in chili plants under greenhouse conditions. The effective utilization of bio-preparations containing fungal elicitors has been documented as a proficient strategy in eliciting plant resistance, resulting in enhancements in crop growth, yield and quality (Naziya et al., 2020).

10.6 Disease Resistance

(PGPMs are employed to boost resistance against pathogens in plants and facilitate their growth. The introduction of these microbes is an eco-friendly strategy for disease management, enduring due to the activation of innate immunity in plants. Limited research has explored the specific effectiveness of PGPF compared to PGPR in promoting resistance against pathogens invasion (Naziya et al., 2020). PGPF, being nonpathogenic, naturally exist as saprotrophs, decomposing organic matter. The primary mechanism preventing pathogen invasion involves the ability of PGPF to establish root colonization, concurrently aiding in nutrient assimilation. The capabilities of PGPF encompass various activities such as phosphate solubilization, IAA production, siderophore production, and the synthesis of enzymes like cellulase and chitinase, all of which directly or indirectly foster and improve resistance to diseases (Mukherjee et al., 2023).

10.7 Plant Defense Signalling

Plants produce various defense enzymes, including signal molecules and PR proteins, to execute antimicrobial processes during both local and systemic defense responses. The investigation of PGPF-mediated resistance revealed specific defense enzymes (Woźniak & Gałązka, 2019). Despite the distinct functions of phytohormones such as ethylene (ET), jasmonate (JA), and salicylic acid (SA), defense responses activated by both pathogenic and beneficial microbes control the processes (Fig. 10.2). In the context of local and systemic defense responses, plants generate a broad array of enzymes, PR proteins, and signal molecules to induce antimicrobial activity. The transcript accumulation pattern of defense enzymes during PGPF-mediated resistance has received limited attention. Defense responses, whether triggered by beneficial or parasitic microorganisms, regulate defense signaling networks, involving plant hormones like JA and ET, along with the accumulation of specific proteins (thaumatin) (Murali et al., 2021). This stimulation leads to the production of defense-related enzymes, mainly GLU (glucanases) and CHI (chitinases), which directly or indirectly hinder the growth and proliferation of pathogens. Before regulating defense signalling pathways, PGPF induce or increase the synthesis of secondary compounds (lignin, callose) and the production of antioxidant enzymes, protecting plants from various biotic and abiotic stresses.

Certain species of *Trichoderma* have been reported to decompose cellulose in plants, stimulating ISR activities by inducing ET or JA defense signalling. Cross-communication between phytohormone tracts modulate plant defense responses based on phytopathogen invasion. Several studies have demonstrated the potential of the SA-dependent response following treatment with *Trichoderma* sp. When tomato plants were treated with *Trichoderma*, there was an observed increase in the expression of JA-responsive genes, leading to the activation of systemic resistance against the phytopathogen *Botrytis cinerea* (Murali et al., 2021). *Trichoderma*, by eliciting ISR, actively participated in both ET and JA signalling. Additionally, certain strains of *Trichoderma* have been reported to stimulate both SA and JA signaling (Abbas et al., 2022). Some *Trichoderma* species were also found to break down plant cellulose through enzymatic activities (Sarangi et al., 2021) thereby inducing ISR in various plants such as beans, corn, cucumbers, limes, and tobacco. This induction is achieved by triggering ET or JA defense signaling pathways (Fig. 10.3).

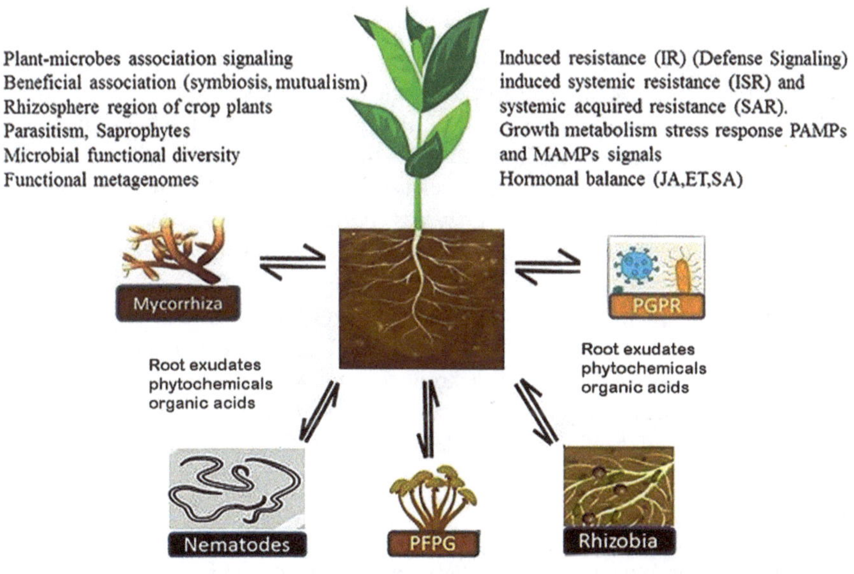

Fig. 10.3 Microbes and plant growth-promoting fungi elicit resistance and trigger plant defense signaling

10.8 Conclusions

Within the field of sustainable agriculture, there has been a lot of interest in investigating biological approaches to improve plant growth and reduce the impact of pathogens. PGPF represent a promising means of phytostimulation and generation of systemic resistance in plants. PGPF stimulates plant development through phytostimulation, which is closely related to the dynamic interactions taking place in the rhizosphere. Such symbiotic connections with plant roots give rise to an environment that is advantageous to both parties. The complex network of relationships includes the generation of hormones, the exchange of nutrients, and the regulation of physiological functions in plants. The uptake of nutrients by roots is facilitated by PGPF, as a crucial component of phytostimulation.

Plant development and growth are directly facilitated by the consequent boost in the availability of nutrients such as phosphates. Furthermore, the PGPF impact on plant hormones, such as auxins, cytokinins, and gibberellins, plays a pivotal role in phytostimulation. These fungi are known to produce IAA, that influences various aspects of plant growth, including cell elongation and root development. The production of plant hormones by PGPF serves as a regulatory mechanism, modulating plant physiology to promote optimal growth conditions. Additionally, PGPF contribute to the synthesis of secondary metabolites, which can enhance plant resilience to environmental stressors. An important part of the PGPF agricultural value lies in their capacity develop systemic resistance beyond phytostimulation. Systemic resistance entails initiating the plant innate immune system, which offers an enhanced barrier against harmful intruders. To accomplish this, PGPF induce ISR, that entails the build up of substances such as lignin, callose, and phenols, that modify the plant cell walls. These alterations function as barriers, preventing the pathogens entry and hindering their development and spread within the root tissues.

In addition to its effect on the cell wall, PGPF also play a part in the activation of enzymes linked to defense. They facilitate the greater accumulation of several enzymes, including chitinase, β-1,3 glucanase, PAL, and POX. The plant defense mechanisms rely heavily on these enzymes. For example, PAL has a role in the production of phenolic chemicals, which support the cell walls and have antibacterial properties. Fungal cell walls are the target of chitinase and β-1,3 glucanase, which strengthen the plant defenses against harmful fungi. The combined effect of these effectors emphasizes how all-encompassing PGPF mediate induction of resistance. Plants relying on PGPF have a stronger immune system that is ready to fight against pathogens by boosting physical obstacles and triggering enzymatic defenses. This supports the idea of a sustainable and ecologically friendly agriculture by increasing the plant resistance to diseases and lowering the need for chemical pesticides.

The consequences of resistance system induction and phytostimulation mediated by PGPF go beyond specific plants to larger environmental and agricultural settings. Using the potential of PGPF gives a strategic advantage in agroecosystems, where maintaining a delicate balance between crop outputs and environmental sustainability is crucial. Reducing reliance on artificial fertilizers and pesticides is consistent with agroecological concepts, encouraging environmentally beneficial techniques that are also commercially feasible. Furthermore, PGPF improve soil fertility and health, supporting agricultural sustainability. As non-pathogenic saprotrophs, these fungi aid in the breakdown of organic material, in the cycling of nutrients, and in the enhancement of soil structure. Plant health and ecosystem resilience are impacted in a cascade manner by PGPF management of soil fertility. Therefore, PGPF integration into agricultural methods has potential to promote sustainable and all-encompassing land management strategies. However, the status of research in this area requires further investigation and breakthroughs that can be achieved. Many studies emphasized PGPF to stimulate plant growth and develop resistance to specific phytopathogens. However, actual knowledge of the specificity of PGPF-plant interactions and their efficacy in a variety of species and environmental circumstances is noticeably lacking.

In addition, there are formulation, distribution, and persistence issues in using PGPF practically in field settings. Creating efficient formulations that preserve the PGPF activity and viability is essential for successful applications in agricultural systems. Additionally, to maximize the long-term benefits of PGPF, it is crucial to comprehend the aspects that affect their adaptability and persistence in various soil settings. An area of growing attention is the function of PGPF in lowering the effects of abiotic stress factors on plants. There is an opportunity to investigate how these fungi support plant resistance against abiotic stresses like drought, salinity, and severe temperatures, even though the interactions between PGPF and plants are well-documented in the context of biotic stress. Deciphering these intricacies will offer a more sophisticated comprehension of the multiplicity of functions that PGPF can fulfill to augment crops performance in distinct environmental settings.

In summary, research on phytostimulation and induction of systemic resistance by PGPF is an important step toward sustainable and resilient agriculture. The complex interactions herein described highlight a broad range of possibilities. In addition to addressing current agricultural issues, the PGPF capacity to boost plant growth and strengthen systemic resistance is in line with more general objectives of protecting the environment and conserving soil. Unlocking the full potential of PGPF holds promise for making agriculture a more harmonious and environmentally conscientious activity, as research advances in this subject. Our understanding of cultivation could be completely transformed by incorporating these discoveries into workable agricultural plans that promote long-term resilience, environmental preservation, and productivity.

References

Abbas, A., Mubeen, M., Zheng, H., Sohail, M. A., Shakeel, Q., Solanki, M. K., et al. (2022). *Trichoderma* sp. genes involved in the biocontrol activity against *Rhizoctonia solani*. *Frontiers in Microbiology, 13*, 1–22.

Abd Alhakim, A., Hashem, A., Abdelaziz, A. M., & Attia, M. S. (2022). Impact of plant growth promoting fungi on biochemical defense performance of tomato under fusarial infection. *Egyptian Journal of Chemistry, 65*, 291–301.

Ahmad, J. S., & Baker, R. (1987). Rhizosphere competence of *Trichoderma harzianum*. *Phyopathology, 77*, 182–189.

Attia, M. S., Abdelaziz, A. M., Al-Askar, A. A., Arishi, A. A., Abdelhakim, A. M., & Hashem, A. H. (2022). Plant growth-promoting fungi as biocontrol tool against Fusarium wilt disease of tomato plant. *Journal of Fungai, 8*, 775.

Attitalla, I. H., Johnson, P., Brishammar, S., & Quintanilla, P. (2001). Systemic resistance to Fusarium wilt in tomato induced by *Phytophthora cryptogea. Journal of Phytopathology, 149*, 373–380.

Babu, A. G., Kim, S. W., Yadav, D. J., Hyum, U., Adhikari, M., & Lee, Y. S. (2015). *Penicillium menonorum*: A novel fungus to promote growth and nutrient management in cucumber plants. *Mycobiology, 43*, 49–56.

Baker, R. (1988). *Trichoderma* spp. as plant-growth stimulants. *Critical Reviews in Biotechnology, 7*, 97–106.

Benhamou, N., le Floch, G., Vallance, J., Gerbore, J., Grizard, D., et al. (2012). *Pythium oligandrum*: An example of opportunistic success. *Microbiology, 158*, 2679–2694.

Bent, E. (2006). Induced systemic resistance mediated by plant growth-promoting rhizobacteria (PGPR) and fungi (PGPF). In S. Tuzun & E. Bent (Eds.), *Multigenic and induced systemic resistance in plants* (pp. 225–258). Springer.

Cramer, G. R., Urano, K., Delrot, S., Pezzotti, M., & Shinozaki, K. (2011). Effects of abiotic stress on plants: A systems biology perspective. *BMC Plant Biology, 11*, 163.

Debbarma, M., Rajesh, T., Devi, R. K. T., Thakuria, D., Thakur, N. S. A., et al. (2021). Morphological characterization of plant growth promoting fungi (PGPF) isolated from maize rhizosphere in Meghalaya. *The Pharma Innovation Journal, 10*, 570–575.

El-Maraghy, S. S., Tohamy, A. T., & Hussein, K. A. (2021). Plant protection properties of the plant growth-promoting fungi (PGPF): Mechanisms and potentiality. *Current Research in Environmental & Applied Mycology, 11*(1), 391–415.

Frąc, M., Hannula, S. E., Bełka, M., & Jędryczka, M. (2018). Fungal biodiversity and their role in soil health. *Frontiers in Microbiology, 9*, 707.

García, B. F. J., Santamarina, M. P., & Roselló, J. (2005). *Trichoderma*: mecanismos de control. *Phytoma, 172*, 106–107.

García-Garza, J. A., Reeleder, R. D., & Paulitz, T. C. (1997). Degradation of sclerotia of *Sclerotinia sclerotiorum* by fungus gnats (*Bradysia copropila*) and the biocontrol fungi *Trichoderma* sp. *Soil Biology and Biochemistry, 29*, 123–129.

Harman, G., Hotel, C. R., Viterbo, A., Chet, I., & Lorito, M. (2004). *Trichoderma* species opportunistic, avirulent plant symbionts. *Nature Reviews, 2*, 43–56.

Hossain, M. M., & Sultana, F. (2020). Application and mechanisms of plant growth promoting fungi (PGPF) for phytostimulation. *Organic Agriculture*, IntechOpen. https://doi.org/10.5772/intechopen.92338

Hossain, M. M., Sultana, F., & Islam, S. (2017). Plant growth-promoting fungi (PGPF): Phytostimulation and induced systemic resistance. In D. Singh, H. Singh, & R. Prabha (Eds.), *Plant-microbe interactions in agro-ecological perspectives* (pp. 135–191). Springer.

Hussein, K. A., & Joo, J. H. (2015). Isolation and characterization of rhizomicrobial isolates for phosphate solubilization and indole acetic acid production. *Journal of the Korean Society for Applied Biological Chemistry, 58*, 847–855.

Hyakumachi, M. (1994). Plant-growth-promoting fungi from turf grass rhizosphere with potential for disease suppression. *Soil Microorganism, 44*, 53–68.

Islam, S., Akanda, A. M., Prova, A., Sultana, F., & Hossain, M. M. (2014). Growth promotion effect of *Fusarium* sp. PPF1 from Bermuda grass (*Cynodon dactylon*) rhizosphere on Indian spinach (*Basella alba*) seedlings are linked to root colonization. *Archives of Phytopathology and Plant Protection, 47*, 2319–2331.

Jahagirdar, S., Sachin, K., Ramesh, B., Basavaraja, G. T., & Yogesh, B. (2013). Biointensive management of Asian soybean rust caused by Phakopsora pachyrhizi. *Syd Resources and Crops, 14*, 182–188.

Jogaiah, S., Abdelrahman, M., Tran, L. S., & Shin-Ichi, I. (2013). Characterization of rhizosphere fungi that mediate resistance in tomato against bacterial wilt disease. *Journal of Experimental Botany, 64*, 3829–3842.

Kaewchai, S. (2009). Mycofungicides and fungal biofertilizers. *Fungal Diversity, 38*, 25–50.

Khan, A. L., Waqas, M., & Lee, I. J. (2015). Resilience of *Penicillium resedanum* LK6 and exogenous gibberellin in improving *Capsicum annuum* growth under abiotic stresses. *Journal of Plant Research, 128*, 259–268.

Kredics, L., Hatvani, L., Naeimi, S., Körmöczi, P., Manczinger, L., et al. (2014). Biodiversity of the genus *Hypocrea/Trichoderma* in different habitats. In V. K. Gupta, M. Schmoll, A. Herrera-Estrella, R. S. Upadhyay, I. Druzhinina, & M. G. Tuohy (Eds.), *Biotechnology and biology of Trichoderma* (pp. 3–24). Elsevier.

Lee, S. C., & Luan, S. (2012). ABA signal transduction at the crossroad of biotic and abiotic stress responses. *Plant, Cell & Environment, 35*, 53–60.

Lewis, J. A., & Papavizas, G. C. (1991). Biocontrol of plant diseases: The approach of tomorrow. *Crop Protection, 10*, 95–105.

Mukherjee, B., Roy, S., Parvin, N., Tarafdar, S., & Dutta, S. (2023). Characterization of a potent plant growth promoting fungal strain *Aspergillus fumigatus* MCC 1721 with special reference to idole-3-acetic acid production. *Plant Science Today, 10*, 210–223.

Murali, M., Naziya, B., Ansari, M. A., Alomary, M. N., Al-Yahya, S., Almatroudi, A., et al. (2021). Bioprospecting of rhizosphere-resident fungi: Their role and importance in sustainable agriculture. *Journal of Fungi, 7*, 314.

Muslim, A., Hyakumachi, M., Kageyama, K., & Suwandi, S. (2019). Induction of systemic resistance in cucumber by hypovirulent binucleate *Rhizoctonia* against anthracnose caused by *Colletotrichum orbiculare*. *Tropical Life Sciences Research, 30*, 109–122.

Navi, S. S., & Bandyopadhyay, R. (2002). Biological control of fungal plant pathogens. In J. M. Waller, J. M. Lenné, & S. J. Waller (Eds.), *Plant pathologists pocketbook* (3rd ed., pp. 354–365). CAB International.

Naziya, B., Murali, M., & Amruthesh, K. N. (2020). Plant Growth-Promoting Fungi (PGPF) instigate plant growth and induce disease resistance in *Capsicum annuum* L. upon infection with *Colletotrichum capsici* (Syd.) Butler & Bisby. *Biomolecules, 10*, 41.

Ousley, M. A., Lynch, J. M., & Whipps, J. M. (1993). Effect of *Trichoderma* on plant growth: A balance between inhibition and growth promotion. *Microbial Ecology, 26*, 277–285.

Rabeendran, N., Moot, D. J., Jones, E. E., & Stewart, A. (2000). Inconsistent growth promotion of cabbage and lettuce from *Trichoderma* isolates. *New Zealand Plant Protection, 53*, 143–146.

Romera, F. J., García, M. J., Lucena, C., Martínez-Medina, A., Aparicio, M. A., Ramos, J., et al. (2019). Induced systemic resistance (ISR) and Fe deficiency responses in dicot plants. *Frontiers in Plant Science, 10*, 287.

Samuels, G. J., & Hebbar, P. K. (2015). *Trichoderma: Identification and agricultural applications* (pp. 7–34). The American Phytopathological Society.

Sarangi, S., Swain, H., Adak, T., Bhattacharyya, P., Mukherjee, A. K., Kumar, G., & Mehetre, S. T. (2021). *Trichoderma*-mediated rice straw compost promotes plant growth and imparts stress tolerance. *Environmental Science and Pollution Research, 28*, 44014–44027.

Singh, B., & Singh, V. K. (2023). Diversity of wood-inhabiting macrofungi from district Ayodhya, Uttar Pradesh, India. *Kavak, 59*, 1–11.

Tarroum, M., Romdhane, W. B., Al-Qurainy, F., Ali, A. A. M., Al-Doss, A., Fki, L., & Hassairi, A. (2022). A novel PGPF *Penicillium olsonii* isolated from the rhizosphere of *Aeluropus littoralis* promotes plant growth, enhances salt stress tolerance, and reduces chemical fertilizers inputs in hydroponic system. *Frontiers in Microbiology, 13*, 996054.

Thines, M., & Kamoun, S. (2010). Oomycete-plant coevolution: Recent advances and future prospects. *Current Opinion in Plant Biology, 13*, 427–433.

Verbon, E. H., & Liberman, L. M. (2016). Beneficial microbes affect endogenous mechanisms controlling root development. *Trends in Plant Science, 21*, 218–229.

Verma, M., Brar, S. K., Tyagi, R., Surampalli, R., & Valero, J. (2007). Antagonistic fungi, *Trichoderma* sp.: Panoply of biological control. *Biochemical Engineering Journal, 37*, 1–20.
Vílchez, J. I., Yang, Y., He, D., Zi, H., Peng, L., Lv, S., et al. (2020). DNA demethylases are required for myo-inositolmediated mutualism between plants and beneficial rhizobacteria. *Nature Plants, 6*, 983–995.
Vujanovic, V., & Goh, Y. K. (2012). qPCR quantification of *Sphaerodes mycoparasitica* biotrophic mycoparasite interaction with *Fusarium graminearum*: *In vitro* and *in planta* assays. *Archives of Microbiology, 194*, 707–717.
Woźniak, M., & Gałązka, A. (2019). The rhizosphere microbiome and its beneficial effects on plants - current knowledge and perspectives. *Advances in Microbiology, 58*, 59–69.
Zhang, S. W., Gan, Y. T., & Xu, B. L. (2016). Application of plant-growth-promoting fungi *Trichoderma longibrachiatum* T6 enhances tolerance of wheat to salt stress through improvement of antioxidative defense system and gene expression. *Frontiers in Plant Science, 7*, 1405.
Zhou, L. S., Tang, K., & Guo, S. X. (2018). The plant growth-promoting fungus (PGPF) *Alternaria* sp. A13 markedly enhances *Salvia miltiorrhiza* root growth and active ingredient accumulation under greenhouse and field conditions. *International Journal of Molecular Sciences, 19*(1), 270.

Chapter 11
Functional Role and Mechanism of PGPRs, AMFs and Endophytes in Improving Host Resistance Against Plant Pathogens

Prasann Kumar and Debjani Choudhury

Abstract Plant pathogens pose a significant threat to agricultural productivity and food security worldwide. To combat these diseases, researchers have explored the potential of beneficial plant-associated microorganisms, including Plant Growth-Promoting Rhizobacteria (PGPRs), Arbuscular Mycorrhizal Fungi (AMFs), and Endophytes. This Chapter provides an overview of the functional mechanisms by which these microorganisms enhance host resistance *vs* plant pathogens. PGPRs, through multiple mechanisms such as antimicrobial compound production, nutrient and space competition, and systemic resistance induction, promote plant growth and confer protection against pathogens. AMFs establish mutualistic associations with plant roots, enhancing nutrient uptake, inducing systemic defence responses, and directly suppressing pathogen growth. Endophytes within plant tissues produce antimicrobial compounds, induce systemic resistance, and compete with pathogens, thus boosting host resistance. Understanding the functional roles and mechanisms of PGPRs, AMFs, and endophytes in enhancing host resistance provides valuable insights for sustainable disease management strategies, reducing reliance on chemical pesticides, and ensuring global food security. Further research is needed to optimize the application of these beneficial microorganisms in agricultural systems.

Keywords Defence · Environment · Microorganism · Mycorrhiza · Zero hunger · No poverty

P. Kumar (✉)
Department of Agronomy, School of Agriculture, Lovely Professional University, Phagwara, Punjab, India

D. Choudhury
Department of Plant Pathology, School of Agriculture, Lovely Professional University, Phagwara, Punjab, India

P. Mathur, S. Roy (eds.), *Plant Microbiome and Biological Control*, Sustainability in Plant and Crop Protection 20,
https://doi.org/10.1007/978-3-031-75845-4_11

Abbreviations

PGPRs	Plant Growth-Promoting Rhizobacteria
AMFs	Arbuscular Mycorrhizal Fungi
DNA	Deoxyribose Nucleic Acid
PTI	Pattern-Triggered Immunity
ETI	Effective-Triggered Immunity
PAMPS	Pathogen-Associated Molecular Patterns
HGT	Horizontal Gene Transfer

11.1 Introduction

Plant diseases caused by pathogenic microorganisms significantly challenge agricultural productivity and food security worldwide. Conventional approaches to disease management, such as chemical pesticides, have limitations in terms of environmental impact and the development of resistance in susceptible plants Sayanti Mandal et al., 2023). As a result, growing interest has been in exploring the potential of beneficial plant-associated microorganisms as sustainable alternatives to enhance host resistance against plant pathogens. Among these microorganisms, Plant Growth-Promoting Rhizobacteria (PGPRs), Arbuscular Mycorrhizal Fungi (AMFs), and Endophytes have emerged as promising candidates due to their functional roles and mechanisms in improving host resistance. PGPRs are a diverse group of bacteria that naturally inhabit the rhizosphere and the internal root tissues (Bhanse et al., 2022). They establish beneficial associations with plants and have been extensively studied for their ability to promote plant growth and development. In addition to their growth-promoting effects, PGPRs have been shown to play a crucial role in enhancing host resistance against a wide range of plant pathogens (Turan et al., 2022; Veerapagu et al., 2023). These beneficial bacteria employ various mechanisms to suppress growth and colonization of pathogens. They produce antimicrobial compounds, such as antibiotics and siderophores, that directly inhibit the growth of bacteria or fungi. PGPRs also compete with pathogens for essential nutrients and space, limiting their proliferation (Abdelsattar et al., 2023; Kaur & Garg, 2022). Furthermore, these microorganisms can induce systemic resistance in plants, triggering a cascade of defence responses that enhance the plant ability to withstand a pathogen attack. This systemic resistance is often mediated by activating specific signalling pathways and modulating defence-related gene expression.

AMFs are a group of mutualistic fungi that form symbiotic associations with the roots of most terrestrial plant species. They establish intimate relationships with the plant, facilitating the exchange of nutrients between the two partners. AMFs are vital in improving nutrient uptake, particularly phosphorus, which is often limited in soil (Joshi et al., 2021; Mahmud & Chong, 2021; Sharma et al., 2021).

Beyond nutrient acquisition, AMFs have been found to contribute significantly to enhance host resistance against pathogens. They employ multiple mechanisms to confer protection to their host plants. AMFs also compete with pathogens for nutrients and space, limiting the availability of resources necessary for pathogen growth. Additionally, these fungi can directly suppress a pathogen proliferation through mycoparasitism, where they directly attack and consume other fungi (Ezrari et al., 2022; Reddy & Saravanan, 2013; Verma et al., 2023). Furthermore, AMFs induce systemic defence responses in plants, priming them for enhanced resistance against future pathogen attacks. This priming effect involves the modulation of plant hormone signalling pathways and the production of defence-related enzymes and compounds.

Endophytes are bacteria, fungi, and other microbial species that colonize the internal tissues of plants without causing apparent harm. They have gained attention for their multifunctional roles in plant health and disease management (Bukhat et al., 2020; Jariyal & Soni, 2021). These microorganisms can produce secondary metabolites with antimicrobial properties, which directly inhibit the growth of plant pathogens within the host tissues. Additionally, endophytes induce systemic resistance in plants by stimulating the production of defense-related signaling molecules and activating defense mechanisms. Furthermore, they can compete with pathogens for resources within the plant tissues, further limiting the pathogen colonization and spread. Understanding the functional roles and mechanisms of PGPRs, AMFs, and endophytes in improving host resistance against plant pathogens provides valuable insights for the development of sustainable and eco-friendly strategies for disease management in agriculture (Bukhat et al., 2020; Jariyal & Soni, 2021; Reddy & Saravanan, 2013). Harnessing the potential of these beneficial microorganisms can reduce reliance on chemical pesticides, enhance plant health, and ensure global food security. However, further research is needed to explore the complex interactions between these microorganisms, plants, and pathogens, as well as to optimize their application in diverse agricultural systems.

11.2 Research on Microbes Functional Role and Mechanisms

While there has been significant research on the functional roles and mechanisms of PGPRs, AMFs, and endophytes in enhancing host resistance against plant pathogens, there are still several research gaps that need to be addressed (da Silva et al., 2022; Ghorbanpour et al., 2018; Kumawat et al., 2022; Panda et al., 2019; Sahab et al., 2021).

1. Elucidating specific mechanisms of action: while the general mechanisms of PGPRs, AMFs, and endophytes in improving host resistance are understood to some extent, there is a need for more in-depth investigations into the specific molecular and biochemical pathways involved. Identifying and characterizing

the genes, proteins, and metabolites responsible for pathogen suppression, induction of systemic resistance, and other beneficial effects can provide valuable insights for practical application.
2. Exploring the diversity and specificity of microorganism-plant interactions: the effectiveness of PGPRs, AMFs, and endophytes can vary depending on the microorganism and host plant combination. More research is needed to understand the factors influencing the diversity and specificity of these interactions. This includes exploring the impact of different microbial strains or species on host resistance and determining the plant traits that influence their colonization and beneficial effects.
3. Unravelling the signalling mechanisms and cross-talk between microorganisms and plants: the communication and signalling between microorganisms and plants play a crucial role in inducing defence responses and enhancing host resistance. Investigating the molecular dialogues, signalling pathways, and cross-talk between the microorganisms and the host plant can provide insights into the underlying mechanisms of beneficial interactions. Understanding these complex interactions will contribute to developing targeted approaches to optimize the use of PGPRs, AMFs, and endophytes in disease management.
4. Scaling up from laboratory studies to field application: most studies investigating functional roles and mechanisms of these microrganisms have been conducted under controlled laboratory conditions. However, there is a gap in understanding how beneficial microorganisms perform in diverse field environments, where factors such as soil types, agronomic practices, and plant communities can influence their efficacy. Research focusing on field-scale trials and long-term monitoring is crucial for evaluating practical applicability and sustainability in real-world agricultural systems.
5. Integration with other disease management strategies: while PGPRs, AMFs, and endophytes show promise in enhancing host resistance, there is a need to explore their potential integration with other disease management strategies. Investigating their compatibility with biocontrol agents, plant genetic resistance, and cultural practices can help develop integrated approaches that maximize disease suppression while minimizing reliance on chemical pesticides.

Closing these research gaps will provide a deeper understanding of the functional roles and mechanisms of PGPRs, AMFs, and endophytes in improving host resistance against plant pathogens. This knowledge can guide the development of effective and sustainable disease management strategies that minimize the use of synthetic chemicals and contribute to the long-term sustainability of agricultural systems (Bhatt et al., 2022; Can et al., 2022; Ghorbanpour et al., 2018; Mahmud et al., 2021a; Mishra et al., 2022; Sharma, 2022; Singh et al., 2023).

11.3 Research Strategy and Solutions in Microbes Functional Role and Mechanism

To address the research gaps in the functional role and mechanism of PGPRs, AMFs, and endophytes in improving host resistance against plant pathogens, several strategies and solutions can be implemented:

1. Integration of -omics approaches: employing high-throughput techniques, such as genomics, transcriptomics, proteomics, and metabolomics, can provide a comprehensive understanding of the functional mechanisms underlying the interactions between microorganisms and plants. These approaches can identify specific genes, proteins, and metabolites involved in pathogen suppression, systemic resistance induction, and other beneficial effects. Integrating multiple -omics data sets can offer a system-level understanding of the complex interactions and signalling pathways involved in host-microbe interactions.
2. Comparative studies and meta-analyses: conducting comparative studies across different microorganism-plant combinations can help identify common and unique mechanisms of action. Meta-analyses can aggregate data from multiple studies to provide a broader perspective on the effectiveness and consistency of PGPRs, AMFs, and endophytes in improving host resistance. These approaches can help identify key factors influencing the functional roles and mechanisms and elucidate the broader applicability of these microorganisms in disease management.
3. Functional validation of candidate genes and metabolites: Once potential genes and metabolites involved in pathogen suppression and defence induction have been identified through omics approaches, functional validation experiments can be conducted. Gene knockout or overexpression studies and targeted metabolite profiling can help confirm the roles of specific genes and metabolites in the beneficial effects of PGPRs, AMFs, and endophytes. CRISPR-Cas9 technology can be utilized for targeted gene manipulation to validate their functions.
4. Field-scale trials and long-term monitoring: to bridge the gap between laboratory studies and field applications, conducting field-scale trials under diverse environmental conditions is essential. Long-term monitoring of plant-microbe interactions can provide insights into the persistence, stability, and efficacy of PGPRs, AMFs, and endophytes in real-world agricultural systems. Such trials can evaluate the effects of these microorganisms on disease suppression, plant growth, and crop yield over multiple growing seasons, considering variations in soil types, cropping systems, and agronomic practices.
5. Collaboration and interdisciplinary research: addressing the research gaps in the functional role and mechanism of PGPRs, AMFs, and endophytes requires collaboration among researchers from various disciplines, including microbiology, plant pathology, genetics, agronomy, and ecology. Integrating diverse expertise and perspectives can facilitate a holistic study of complex interactions and mechanisms. Collaborative efforts can also enhance the translation of research

findings into practical applications and foster the development of integrated disease management strategies that combine multiple approaches for optimal pathogen control.

6. Field surveys and diagnostic tools: conducting extensive field surveys to identify specific microorganisms and their abundance in different crop systems can provide insights into their potential roles in disease suppression. The development of advanced diagnostic tools, such as DNA-based techniques and high-throughput sequencing, can enable rapid and accurate identification and quantification of beneficial microorganisms in field samples. These tools can aid in assessing the distribution, diversity, and dynamics of PGPRs, AMFs, and endophytes and their associations with plant health and disease resistance.

By implementing these strategies and solutions, researchers can make significant progress in filling the research gaps and advancing our understanding of the functional roles and mechanisms of PGPRs, AMFs, and endophytes in improving host resistance against plant pathogens (Gupta et al., 2022; Sahu et al., 2019; Samanta et al., 2023; Sharma et al., 2021; Tokas et al., 2023). This knowledge can contribute to develop sustainable and effective disease management strategies, leading to enhanced crop productivity, reduced environmental impact, and global food security.

11.4 Host Resistance Against Pathogens and Challenges

Host resistance against plant pathogens refers to the ability of plants to withstand or limit the impact of pathogenic microorganisms, such as bacteria, fungi, viruses, or pests such as nematodes, that can cause severe yield losses. It is an essential component of plant defence mechanisms and is crucial in maintaining plant health and productivity. Host resistance can be achieved through various mechanisms, including physical barriers, chemical defences, and immune responses (Sabir, 2022; Sakthieaswari et al., 2022; Shahid & Mehnaz, 2020; Singh et al., 2019; Tyagi et al., 2023; Vimal et al., 2017).

Physical Barriers Plants have structures that act as barriers towards invasion by pathogens. The cell wall serves as the first line of defence, providing a physical barrier to hinder a pathogen penetration. Additionally, plants may possess specialized structures like trichomes, waxes, and thorns that deter pathogens attachment and movement. Furthermore, the stomatal pores on leaves can close in response to attacks, limiting the entry of pathogens into the plant tissues (Sati et al., 2020).

Chemical Defences Plants produce various chemical compounds that can inhibit growth and development of pathogens. These chemical defences include antimicrobial compounds, such as phytoalexins, produced in response to attacks. Other defence compounds, like terpenoids, phenolics, and alkaloids, can also have antimicrobial properties. Additionally, plants can produce volatile compounds that attract beneficial organisms or repel pathogens (Afridi et al., 2022; Chandra et al., 2021; De Corato, 2020; Goss et al., 2017; Meena et al., 2017).

Immune Responses Plants possess sophisticated immune systems recognising and responding to pathogen invasion. Plant immune responses activate two main defence pathways: pattern-triggered immunity (PTI) and effector-triggered immunity (ETI). PTI is a generalized response triggered by the recognition of pathogen-associated molecular patterns (PAMPs) by plant pattern recognition receptors (PRRs). ETI, conversely, is a specific response activated when plants recognize specific pathogen effectors through intracellular resistance (R) proteins. Both PTI and ETI result in the activation of defence responses, including the production of antimicrobial compounds, reinforcement of cell walls, and programmed cell death at the infection site (Basile et al., 2020; Kumar & Dubey, 2020; Li et al., 2023; Mahmud et al., 2021b; Mon et al., 2021; Yang et al., 2021).

Despite host resistance mechanisms, plant pathogens evolved strategies to overcome or evade plant defences. Challenges in achieving effective host resistance against plant pathogens include:

1. *Pathogen variability.* Pathogens exhibit high genetic diversity and the ability to evolve rapidly, leading to the emergence of new races or strains that can overcome host resistance mechanisms. This requires continuous monitoring and identification of new pathogen strains and the development of resistant plant varieties that can counter these evolving pathogens.
2. *Lack of durable resistance.* Some resistance mechanisms employed by plants are only effective against specific strains or races of pathogens. Over time, the pathogens populations can evolve and overcome these resistance mechanisms, leading to resistance breakdown. Developing plant varieties with durable resistance, which remains effective over an extended period, is a significant challenge.
3. *Complexity of plant-pathogen interactions.* The interactions between plants and pathogens involve multiple molecular and biochemical processes. Understanding these intricate interactions and the underlying resistance mechanisms requires interdisciplinary research efforts and the integration of various scientific disciplines.
4. *Compatibility with agronomic practices.* Developing resistant plant varieties and implementing host resistance strategies must consider compatibility with existing agronomic practices. Resistance traits, such as yield, quality, and agronomic performance, should not negatively impact desirable agronomic traits. Ensuring that resistance mechanisms can be effectively integrated into diverse cropping systems is essential.
5. *Trade-offs and costs of resistance.* In some cases, activating defence responses and deploying resistance mechanisms can produce a cost in terms of plant growth and development. The allocation of resources to defence mechanisms may result in reduced plant growth and yield. Balancing the trade-offs between defence and growth is challenging in achieving effective and sustainable resistance.

Addressing these challenges requires a multidisciplinary approach, including plant breeding, genetics, genomics, molecular biology, and crop management strategies (Aslam et al., 2022; Bhandari et al., 2022; Johnson et al., 2016; Khalid et al., 2021; Sharma et al., 2020). By enhancing our understanding of plant-pathogen interactions, developing durable resistance, and implementing integrated disease management practices, we can overcome these challenges and improve host resistance against plant pathogens, leading to more sustainable and resilient agricultural systems (Table 11.1).

Table 11.1 Host-resistance mechanisms, limitations and challenges

Host Resistance Mechanisms	Challenges and Limitations
Physical barriers	Variability in effectiveness across different pathogens
Cuticle thickening	Pathogens evolve mechanisms to penetrate thicker cuticles.
Cell wall fortification	Adaptation of pathogens to breach reinforced cell walls
Stomatal closure	Pathogen manipulation of stomatal opening and closure
Trichome production	Pathogen adaptation to trichome-mediated defence
Chemical defenses	Evolution of resistance in pathogens towards plant-produced toxins
Phytoalexin production	Pathogens evolving detoxification mechanisms
Antimicrobial peptides	Development of resistance mechanisms in pathogens
Secondary metabolites	Variation in effectiveness against different pathogen species
Immune responses	Pathogens evolve effectors to suppress plant immune responses.
Pattern-triggered immunity (PTI)	Pathogens evolving to avoid recognition by plant PRRs
Effector-triggered immunity (ETI)	Rapid evolution of pathogen effectors to overcome ETI
Hypersensitive response (HR)	Pathogens evolving mechanisms to avoid detection and subsequent HR
Systemic acquired resistance (SAR)	Trade-offs between defence activation and resource allocation
RNA interference (RNAi)	Pathogen evasion through RNAi suppression
Defence signaling pathways	Crosstalk and potential trade-offs with plant growth and development
Induced resistance	Limited effectiveness against highly virulent pathogens
Genetic resistance	Breakdown of resistance due to pathogen adaptation
Durable resistance	Identification and deployment of resistance genes with long-lasting effectiveness
Gene stacking	Ensuring compatibility and stability of multiple resistance genes
Compatible agronomic practices	Integration of resistance strategies with other agricultural practices
Resistance breeding	Time-consuming process and potential loss of other desirable traits
Deployment of resistant varieties	The rapid evolution of new pathogen races that overcome resistance
Disease management strategies	The integrated approach is required to mitigate pathogen evolution and reduce selection pressure.

11.5 Modelling Host Resistance Against Plant Pathogens

The following list of equations offer simplified mathematical representations of various aspects of host resistance (R). However, it is essential to note that host resistance is a complex phenomenon influenced by numerous factors, and that these equations provide simplified models to capture specific aspects of the process. The actual mathematical representation of host resistance may require more complex and context-specific models, integrating empirical data and rigorous analyses.

1. Resistance Index: $R = (G - P) / (G + P)$. This equation represents the resistance index, taking into account the difference between genetic factors (G) and pathogen virulence (P) normalized by their sum.
2. Phenotypic Plasticity: $R = G \cdot (1 + E)$. This equation considers the phenotypic plasticity of host resistance, where the genetic factors (G) are multiplied by the environmental influence (E).
3. Defence Protein Expression: $R = G \cdot \exp.(-k \cdot D)$. This equation represents the expression of defence proteins (D), where genetic factors (G) decrease exponentially with increasing defence protein expression.
4. Host-Pathogen Interaction: $R = G / (P + 1)$. This equation captures the host-pathogen interaction, where genetic factors (G) are divided by the pathogen virulence (P) plus 1.
5. Quantitative Trait Loci (QTL). Model: $R = G1 + G2 + ... + Gn$. This equation represents the QTL model, where host resistance (R) is the sum of the effects of multiple genetic factors (G1, G2, ..., Gn).
6. Adaptation Rate: $R = G / (1 + k \cdot P)$. This equation considers the adaptation rate, where genetic factors (G) are divided by 1 plus the product of a constant (k) and pathogen virulence (P).
7. Pathogen Population Size: $R = G / \log(N + 1)$. This equation relates R to the logarithm of the pathogen population size (N), with genetic factors (G) in the numerator.
8. Resource Allocation: $R = G / (P + R)$. This equation represents resource allocation, where host resistance (R) is divided by the sum of a pathogen virulence (P) and host resistance itself.
9. Cost of Resistance: $R = G / (P + C)$. This equation considers the cost of resistance (C), where R is divided by the sum of pathogen virulence (P) and the cost of resistance.
10. Evolutionary Game Theory: $R = (G \cdot P) / (G + P)$. This equation applies evolutionary game theory, where R is the product of genetic factors (G) and pathogen virulence (P) divided by their sum.
11. Induced Defense Signaling: $R = G \cdot S$. This equation represents the influence of induced defense signaling (S) on R, where genetic factors (G) are multiplied by the level of defense signaling.
12. Fitness Trade-off: $R = G / (1 + F)$. This equation captures the fitness trade-off between R and another trait (F), where genetic factors (G) are divided by 1 plus the fitness trait.

13. Spatial Dynamics: R = G / (P + D + S). This equation considers spatial dynamics, where R is divided by the sum of pathogen virulence (P), dispersal ability (D), and spatial structure (S).
14. Host-Pathogen Coevolution: R = (G · exp. (-k · P)) / (1 + exp. (-k · P)). This equation represents the coevolutionary dynamics between hosts and pathogens, where genetic factors (G) decrease exponentially with increasing pathogen virulence (P), following a logistic growth function.
15. Cost-Benefit Analysis: R = G – (C · P). This equation applies cost-benefit analysis, where host resistance (R) is the difference between genetic factors (G) and the product of a cost factor (C) and pathogen virulence (P).
16. Quantitative Resistance: R = (G · D) / (P + D). This equation represents quantitative resistance, where genetic factors (G) and defense mechanisms (D) are divided by the sum of pathogen virulence (P) and defense mechanisms.
17. Environmental Sensitivity: R = G · E^k. This equation captures the environmental sensitivity of host resistance, where genetic factors (G) are multiplied by the environmental influence (E) raised to a power (k).
18. Regulatory Network: R = G / (1 + k · P · D). This equation represents a regulatory network, where R is equal to genetic factors (G) divided by 1 plus the product of a constant (k), pathogen virulence (P), and defence mechanisms (D).
19. Coexistence Model: R = (G · P) / (G + P) · (1 – C). This equation represents a coexistence model, where host resistance (R) is the product of genetic factors (G) and pathogen virulence (P) divided by their sum, adjusted by a competition factor (C).
20. Dominance Model: R = G / (P + D). This equation represents a dominance model, where R is equal to genetic factors (G) divided by the sum of pathogen virulence (P) and dominance factors (D) related to genetic interactions.
21. Trade-off Selection: R = (G1 · P1) / (G2 + P2). This equation represents trade-off selection, where R is the product of genetic factors (G1) and pathogen virulence (P1) divided by the sum of other genetic factors (G2) and pathogen virulence (P2).
22. Epidemiological Model: R = G / (1 + k · P · I). This equation considers an epidemiological model, where R is equal to genetic factors (G) divided by 1 plus the product of a constant (k), pathogen virulence (P), and infection rate (I).
23. Plant Growth Model: R = G · (1 – P) · exp. (-k · G). This equation combines host resistance (R) with plant growth dynamics, where R is equal to genetic factors (G) and pathogen virulence (P) affect resistance, and resistance is modulated by a decay constant (k).
24. Costly Signaling Theory: R = G / (P + S). This equation applies costly signaling theory, where R is divided by the sum of pathogen virulence (P) and the cost of signaling (S).
25. Game Theory Model: R = (G · P) / (G + P + C). This equation represents a game theory model, where R is the product of genetic factors (G) and pathogen virulence (P) divided by their sum, adjusted by a cost factor (C).

26. Host Fitness: R = G / (P + F) This equation captures host fitness, where R is equal to genetic factors (G) divided by the sum of pathogen virulence (P) and a fitness factor (F).
27. Decision-Making Model: R = G / (P + E). This equation represents a decision-making model where R is equal to genetic factors (G) divided by the sum of pathogen virulence (P) and an environmental factor (E) influencing resistance.
28. Genetic Variation: R = G · V. This equation considers genetic variation (V) in host resistance, where genetic factors (G) are multiplied by the level of genetic variation.
29. Predator-Prey Model: R = (G · P) / (P + D). This equation represents a predator-prey model, where R is the product of genetic factors (G) and pathogen virulence (P) divided by the sum of pathogen virulence and predation (D).
30. Host-Pathogen Interaction Matrix: R = (G · P) / sqrt (G^2 + P^2). This equation considers a host-pathogen interaction matrix, where host resistance (R) is the product of genetic factors (G) and pathogen virulence (P) divided by the square root of their sum of squares.
31. Disease Progression: R = (G · D) / (P + D). This equation represents disease progression, where R is the product of genetic factors (G) and defence mechanisms (D) divided by the sum of pathogen virulence (P) and defence mechanisms.
32. Optimal Defense Strategy: R = G · (1 – P) / (1 + k · D). This equation represents an optimal defence strategy, where host resistance (R) is the product of genetic factors (G) and the difference between 1 and pathogen virulence (P), divided by 1 plus a constant (k) multiplied by defence mechanisms (D).
33. Genetic Recombination: R = (G1 + G2) / (P + D). This equation considers genetic recombination, where R is the sum of two genetic factors (G1 and G2) divided by the sum of pathogen virulence (P) and defence mechanisms (D).
34. Gene Expression: R = G · exp. (-k · E). This equation represents gene expression dynamics, where genetic factors (G) decrease exponentially with increasing environmental influence (E) based on a decay constant (k).
35. Gene Regulation Model: R = (G · exp.(-k · D)) / (1 + exp.(-k * D)). This equation represents a gene regulation model, where R is equal to genetic factors (G) that decrease exponentially with increasing defence mechanisms (D), following a logistic growth function.
36. Spatial Heterogeneity: R = G / (1 + k · (P + D) · S). This equation considers spatial heterogeneity, where host resistance (R) is divided by 1 plus the product of a constant (k), the sum of pathogen virulence (P) and defence mechanisms (D), and spatial structure (S).
37. Disease Threshold: R = 1 if P > T, else R = 0 This equation represents a disease threshold model, where host resistance (R) is 1 if the pathogen virulence (P) exceeds a certain threshold (T), and 0 otherwise.
38. Genetic Epistasis: R = (G1 + G2 + ... + Gn) / (1 + k · (P1 + P2 + ... + Pn)). This equation considers genetic epistasis, whcre R is the sum of multiple genetic factors (G1, G2, ..., Gn) divided by 1 plus a constant (k) multiplied by the sum of multiple pathogen virulence factors (P1, P2, ..., Pn).

39. Population Genetics Model: $R = (G \cdot \exp.(-k \cdot P \cdot N)) / (1 + \exp.(-k \cdot P \cdot N))$ This equation represents a population genetics model, where R is given by genetic factors (G) that decrease exponentially with increasing pathogen virulence (P) and population size (N), following a logistic growth function.
40. Host Defense Efficiency: $R = (G \cdot D) / (P + E)$. This equation represents the efficiency of host defence mechanisms (D), where host resistance (R) is the product of genetic factors (G) and defence mechanisms divided by the sum of pathogen virulence (P) and an environmental factor (E).
41. Coevolutionary Arms Race: $R = (G \cdot \exp.(-k \cdot P \cdot E)) / (1 + \exp.(-k \cdot P \cdot E))$. This equation captures the coevolutionary arms race between hosts and pathogens, where R is the product of genetic factors (G) and a decay function of pathogen virulence (P) and environmental influence (E), following a logistic growth function.
42. Optimal Resource Allocation: $R = G * (1 - P) / (1 + k * (D + E))$. This equation represents optimal resource allocation, where host resistance (R) is the product of genetic factors (G) and the difference between 1 and pathogen virulence (P), divided by 1 plus a constant (k) multiplied by the sum of defence mechanisms (D) and environmental factors (E).
43. Gene Flow: $R = (G1 + G2) / (1 + k \cdot (P1 + P2) \cdot F)$. This equation considers gene flow between populations, where R is the sum of two genetic factors (G1 and G2) divided by 1 plus a constant (k) multiplied by the sum of pathogen virulence (P1 and P2) and a gene flow factor (F).
44. Virulence Trade-off: $R = G / (P + V)$. This equation represents a virulence trade-off model, where host resistance (R) is divided by the sum of pathogen virulence (P) and a virulence trade-off factor (V).
45. Plant Defense Costs: $R = G / (P + C)$ This equation considers plant defense costs, where R is equal to a genetic factor divided by the sum of pathogen virulence (P) and the cost of defense (C).
46. Adaptive Immune Response: $R = (G \cdot A) / (P + A)$ This equation represents the adaptive immune response, where host resistance (R) is the product of genetic factors (G) and the adaptive immune response (A) divided by the sum of pathogen virulence (P) and the adaptive immune response.
47. Ecological Resilience: $R = G \cdot (1 - P) / (1 + k \cdot E)$, This equation captures ecological resilience, where R is the product of genetic factors (G) and the difference between 1 and pathogen virulence (P), divided by 1 plus a constant (k) multiplied by an ecological factor (E).
48. Temporal Dynamics: $R = G * (1—\exp. (-k \cdot t))$, This equation represents temporal dynamics of host resistance, where R is equal to genetic factors (G) multiplied by the difference between 1 and the exponential decay of time (t), with a decay constant (k).
49. Optimal Defense Allocation: $R = G / (P + D + E)$. This equation represents optimal defense allocation, where host resistance (R) is equal to genetic factors (G) divided by the sum of pathogen virulence (P), defense mechanisms (D), and an environmental factor (E).

50. Information Integration Model: $R = (G \cdot I) / (P + I)$. This equation represents an information integration model, where R is the product of genetic factors (G) and the information integration level (I) divided by the sum of pathogen virulence (P) and the information integration level.

11.6 Genomes and Plant-Endophyte Interactions

Genomic insights into plant-endophyte interactions have provided valuable information about the mechanisms and dynamics of this complex symbiotic relationship (Bind et al., 2023; Bosamia et al., 2020; Goss et al., 2023; Jing et al., 2022; Patwardhan et al., 2023). Endophytes are microorganisms that inhabit the internal tissues of plants without causing apparent disease symptoms. They can include bacteria, fungi, and other microbes. The genomic studies of these interactions have shed light on the genetic and molecular mechanisms underlying the mutualistic association between plants and endophytes.

1. *Diversity and Identification.* Genomic approaches have revealed the immense diversity of endophytic microorganisms associated with plants. High-throughput sequencing technologies, such as metagenomics and amplicon sequencing, have enabled the identification and classification of endophytes based on their DNA or RNA sequences. These studies have expanded our understanding of taxonomic composition and richness of endophytic communities, within different plant species and tissues.
2. *Mutualistic Benefits.* Genomic studies have unravelled functional traits of endophytes contributing to plant mutualistic benefits. Analysis of endophytic genomes has revealed the presence of genes involved in nutrient acquisition, production of plant growth-promoting compounds (e.g., phytohormones, enzymes), stress tolerance mechanisms, and defence against pathogens. These genomic insights provide a basis for understanding how endophytes enhance plant growth, nutrient uptake, stress tolerance, and protection against pathogens.
3. *Metabolic Interactions.* Genomic investigations have elucidated the metabolic interactions between plants and endophytes. Endophytic genomes encode diverse enzymes and metabolic pathways that contribute to nutrient cycling, synthesis of secondary metabolites, and modulation of plant hormone levels. Genomic analysis has uncovered the genetic basis of such metabolic interactions, highlighting the endophytes potential to enhance plant fitness and adaptability.
4. *Horizontal Gene Transfer.* Genomic studies have provided evidence for horizontal gene transfer (HGT) between endophytes and their host plants. HGT is the transfer of genetic material between distant organisms, not directly related by vertical gene transfer (i.e., from parent to offspring). Horizontal gene transfer between endophytes and plants can lead to the acquisition of new genetic traits, such as pathogen resistance genes or metabolic capabilities, which can significantly influence plant adaptation and evolution.

5. *Molecular Signaling and Communication.* Genomic analyses have unveiled the molecular mechanisms involved in the signaling and communication between plants and endophytes. These mechanisms often rely on the exchange of signaling molecules, such as small peptides or secondary metabolites. Genes associated with *quorum* sensing, biofilm formation, and host recognition have been identified in endophytic genomes, providing insights into the complex molecular dialogues during plant-endophyte interactions.
6. *Dynamic Interactions and Plasticity.* Genomic studies have highlighted the dynamic nature and plasticity of plant-endophyte interactions. Endophytes genomes often exhibit a high degree of genomic plasticity, including gene gain and loss, rearrangements, and genome size variation. This plasticity allows endophytes to adapt to diverse ecological niches and respond to changing environmental conditions, enabling them to establish and maintain symbiotic associations with plants.
7. *Application in Agriculture and Biotechnology.* Genomic insights into plant-endophyte interactions significantly affect agricultural and biotechnological applications. Understanding the genetic basis of beneficial endophyte-plant interactions can guide the selection and engineering of endophytes for improving crop productivity, nutrient uptake efficiency, and stress tolerance. Genomic information can be leveraged to develop targeted strategies for harnessing the potential of endophytes in sustainable agriculture and bioprospecting for novel bioactive compounds.

In conclusion, genomic studies of plant-endophyte interactions have provided deep insights into the genetic and molecular mechanisms underlying such mutualistic relationship (Aamir et al., 2020; Das et al., 2022a; Gayathri et al., 2023; Orozco-Mosqueda et al., 2022; Rathore & Srinivasulu, 2022). These studies have expanded our understanding of the symbiosis functional traits, metabolic interactions, and dynamic nature. Genomic insights hold immense promise for harnessing the potential of endophytes in agriculture, biotechnology, and environmental sustainability (Table 11.2).

Building a network model to represent the genomic insights of plant-endophyte interactions can be complex due to the many factors involved. However, it can provide a simplified example to illustrate the concept. Please note that a model may not capture the full complexity of a network of interactions (Table 11.2). A network graph may be produced using a library to represent the relationships between various genomic insights of plant-endophyte interactions, and factors associated with them. The "Genomic Insights" node represents the overarching concept of genomic discoveries in plant-endophyte interactions. Each subsequent node represents a specific genomic insight, as listed in Table 11.2. Edges are added to connect the "Genomic Insights" node with each specific insight node, illustrating their relationships and dependencies. A graphic network representation is, however, a simplified representation, and the actual network of genomic insights of plant-endophyte interactions is far more intricate and interconnected. This approach serves as a basic illustration of the network model. Further research and exploration are necessary to

Table 11.2 Description of genomic insights

Genomic study	Description
Diversity of endophytes	Genomic studies have revealed the immense taxonomic diversity of endophytic microorganisms associated with plants, including bacteria, fungi, and other microbes.
Functional traits	Genomic analysis uncovered genes involved in nutrient acquisition, production of plant growth-promoting compounds, stress tolerance mechanisms, and defense against pathogens.
Metabolic interactions	Endophytic genomes encode diverse enzymes and metabolic pathways involved in nutrient cycling, synthesis of secondary metabolites, and modulation of plant hormone levels.
Horizontal gene transfer (HGT)	Genomic evidence suggests that HGT occurs between endophytes and host plants, leading to the acquisition of new genetic traits, such as pathogen resistance genes or metabolic capabilities.
Molecular signaling	Genomic studies have identified genes associated with *quorum* sensing, biofilm formation, and host recognition in endophytic genomes, shedding light on the molecular dialogues between plants and endophytes.
Dynamic interactions	Endophytes genomes exhibit genomic plasticity, including gene gain and loss, rearrangements, and genome size variation, enabling endophytes to adapt to different ecological niches and respond to changing environmental conditions.
Functional genomics	Transcriptomic and proteomic analyses provide insights into gene expression patterns and protein profiles in endophytes during plant colonization, nutrient exchange, and modulation of plant physiology.
Community dynamics	Metagenomic studies allow the exploration of endophytic community dynamics, revealing shifts in the composition and abundance of endophytic species in response to various environmental factors and plant hosts.
Symbiotic gene expression	Genomic studies have revealed specific genes and gene clusters involved in establishing and maintaining symbiotic interactions between endophytes and plants, such as the expression of nodulation or mycorrhizal-related genes.
Biotechnological applications	Genomic information on endophytes can be leveraged to develop targeted strategies for agricultural biotechnology, including the selection and engineering of endophytes to improve crop productivity, nutrient uptake, and stress tolerance.

incorporate additional factors and interactions for a more comprehensive understanding of genomic insights in plant-endophyte interactions.

The following equations reflect different aspects of genomic insights in the study of plant-endophyte interactions. They represent various models and analyses used in the study of plant-endophyte interactions. Use of these models may help researchers to quantify and understand various aspects of endophytic genetics, genomics, and the dynamics of symbiotic relationships.

1. Gene Diversity: $GD = (H / N) \cdot \ln(N)$. This equation calculates the gene diversity (GD) based on the number of haplotypes (H) and the total number of individuals (N) in an endophytic population.

2. Shannon Diversity Index: SDI = − Σ(pi · ln(pi)). This equation calculates the Shannon Diversity Index (SDI) based on the proportional abundance (pi) of different endophytic species.
3. Simpson's Diversity Index: DI = 1 / Σ(pi^2). This equation calculates Simpson's Diversity Index (DI) based on the proportional abundance (pi) of different endophytic species.
4. Jaccard Index: JI = c / (a + b + c). This equation calculates the Jaccard Index (JI) to quantify the similarity (c) between two endophytic communities, considering the presence (a) and absence (b) of shared species.
5. Bray-Curtis Dissimilarity: BC = Σ(|ai − bi|) / (Σ(ai) + Σ(bi)). This equation calculates the Bray-Curtis Dissimilarity (BC) between two endophytic communities, considering the abundances (ai and bi) of shared species.
6. Endophyte Species Richness: SR = S / N. This equation calculates the endophyte species richness (SR) based on the number of species (S) divided by the total number of samples (N) in a study.
7. Relative Abundance: RA = (Ni / N) · 100 This equation calculates the relative abundance (RA) of an endophytic species (Ni) as a percentage of the total number of individuals (N).
8. Genome Size Estimation: GS = (k · C) / L This equation estimates the genome size (GS) of an endophytic organism based on the flow cytometry constant (k), the DNA content (C), and the length of a reference genome (L).
9. Gene Expression Analysis: FC = (E1 / E2). This equation calculates the fold change (FC) in gene expression levels (E1 and E2) between two conditions, such as endophyte-plant interaction and control.
10. Co-expression Network Analysis: C = cor(X, Y). This equation calculates the correlation coefficient (C) between the expression levels of two genes (X and Y) to identify co-expression patterns in endophyte-host interactions.
11. Differential Gene Expression: DEG = |log2(FC)| ≥ threshold. This equation identifies differentially expressed genes (DEGs) by comparing the fold change (FC) of gene expression levels against a predefined threshold.
12. Gene Ontology Enrichment: OR = (a · d) / (b · c) This equation calculates the odds ratio (OR) to evaluate the enrichment of specific Gene Ontology terms associated with endophyte-related genes.
13. Pathway Analysis: PA = Σ(corr(Xi, Yi)) / n. This equation calculates the average correlation (PA) between the expression profiles of genes (Xi and Yi) involved in specific pathways, providing insights into endophyte-host interactions.
14. Protein-Protein Interaction Network: PPI = Σ(d(Xi, Yi)) / n. This equation calculates the protein-protein interaction (PPI) score between proteins (Xi and Yi) to infer functional associations in endophyte-related processes.
15. Regulatory Network Inference: MI = ∑(P(xi, yi) · log(P(xi, yi) / (P(xi) · P(yi)))). This equation calculates the mutual information (MI) between gene pairs (xi and yi) to infer regulatory interactions in endophyte-host interactions.

16. Gene Co-expression Module Detection: $GS = \Sigma(corr(Xi, Mi)) / n$. This equation calculates the gene significance (GS) by summing the correlation between gene expression (Xi) and module eigengenes (Mi), aiding in identifying functionally related genes.
17. Genome-Wide Association Studies: $P = 1 / (1 + exp.(-\beta X))$. This equation models the probability (P) of endophyte-related traits based on the logistic regression coefficient (β) and genetic predictors (X) obtained from genome-wide association studies.
18. Quantitative Trait Locus Mapping: $LOD = (n / 4) \cdot log10(10^{(RSS0/n)} / 10^{(RSS1/n)})$. This equation calculates the logarithm of the odds (LOD) ratio for detecting quantitative trait loci (QTLs) associated with endophyte-related traits, where RSS0 and RSS1 are the residual sum of squares under null and alternative models, respectively.
19. Epigenetic Modification Analysis: $DM = |mean(X1) - mean(X2)|$. This equation calculates the DNA methylation difference (DM) between two conditions (X1 and X2) to identify epigenetic changes in endophyte-host interactions.
20. Genome-Wide DNA Methylation: $M = (Nme / N) \cdot 100$. This equation calculates the genome-wide DNA methylation percentage (M) based on the number of methylated sites (Nme) divided by the total number of sites (N).
21. Comparative Genomics: $ED = (1 - (2 \cdot c) / (a + b + 2 * c)) * 100$. This equation calculates the evolutionary distance (ED) between endophytic genomes based on the number of shared genes (c) and unique genes (a and b).
22. Phylogenetic Tree Construction: $d = -3/4 \cdot ln(1\text{–}4/3 \cdot (c / n))$. This equation estimates the genetic distance (d) between endophytic species based on the number of observed differences (c) among n loci.
23. Comparative Genomic Hybridization: $L = log2(T/C)$. This equation calculates the log2 ratio (L) of fluorescence intensities between test (T) and control (C) samples to identify genomic differences between endophytic strains.
24. Transposon Insertion Site Mapping: $PI = (Ni / N) \cdot 100$. This equation calculates the transposon insertion percentage (PI) at specific genomic loci (Ni) as a proportion of the total number of insertion sites (N).
25. Gene Duplication Analysis: $R = N / N0$. This equation calculates the duplication ratio (R) by dividing the number of duplicated genes (N) by the estimated number of ancestral genes (N0).
26. Comparative Genomic Analyses: $V = (c / a) / (d / b)$. This equation calculates the variation (V) in gene presence or absence among different endophytic species based on the number of shared genes (c) and unique genes (a and b).
27. Genome Rearrangement Analysis: $R = 1 - (2 \cdot b + c) / (a + b + c)$. This equation calculates the rearrangement distance (R) between endophytic genomes based on the number of breakpoints (b) and inversions (c) relative to the total number of syntenic blocks (a).
28. Comparative Genomic Synteny: $S = (2 \cdot a) / (2 \cdot a + b + c)$. This equation calculates the synteny score (S) between endophytic genomes based on the number of conserved gene orderings (a) compared to the total number of collinear genes (a, b, and c).

29. Gene Family Expansion/Contraction: FE = (N – N0) / N0. This equation calculates the gene family expansion (FE) or contraction as the difference in the number of genes (N) compared to the estimated number of ancestral genes (N0).
30. Genome-Wide Selection Pressure: ω = dN / dS. This equation calculates the ratio (ω) of nonsynonymous substitutions (dN) to synonymous substitutions (dS) to infer the selection pressure acting on endophytic genes.
31. Co-evolutionary Analysis: C = ∑(Pi · Qi). This equation calculates the co-evolutionary index (C) between genes (Pi and Qi) based on their evolutionary rates, enabling the identification of co-evolving endophytic and host genes.
32. Genomic Islands Detection: CI = (N – N0) / (n – n0). This equation calculates the composition index (CI) to identify genomic islands in endophytic genomes based on the difference in the number of genes (N – N0) and the difference in the number of insertion sites (n – n0).
33. Functional Annotation Enrichment: FE = (k / K) / (n / N). This equation calculates the functional enrichment (FE) of specific gene ontology terms based on the number of genes with functional annotations (k) relative to the total number of genes (K), and the number of differentially expressed genes (n) relative to the total number of genes (N).
34. Pathway Enrichment Analysis: PE = (k / K) / (n / N). This equation calculates the pathway enrichment (PE) based on the number of genes associated with a specific pathway (k) relative to the total number of genes (K), and the number of differentially expressed genes (n) relative to the total number of genes (N).
35. Network Analysis: CC = C / (n · (n – 1) / 2). This equation calculates the clustering coefficient (CC) of a gene co-expression network based on the number of connections (C) between genes relative to the maximum possible connections (n · (n – 1) / 2).
36. Regulatory Network Inference: GRN = W · X. This equation infers the gene regulatory network (GRN) by multiplying the gene expression matrix (X) with the weight matrix (W) obtained from network inference algorithms.
37. Transcription Factor Analysis: MR = (Ni / N) · 100. This equation calculates the motif representation (MR) of specific transcription factor binding sites (Ni) as a percentage of the total number of genes (N) in endophytic genomes.
38. Genome-Wide Association Network: GWN = A · R · X. This equation represents the genome-wide association network (GWN) by multiplying the adjacency matrix (A), the relatedness matrix (R), and the phenotype matrix (X) obtained from genome-wide association studies.
39. Functional Redundancy: FR = Σ(Wi) / n. This equation calculates the functional redundancy (FR) in endophytic genomes based on the sum of functional weights (Wi) assigned to each gene relative to the total number of genes (n).
40. Epistasis Analysis: EP = P(XY) – P(X) · P(Y). This equation calculates the epistatic interaction (EP) between two genetic loci (X and Y), based on their joint probability (P(XY)) and individual probabilities (P(X) and P(Y)).

41. Genetic Diversity: GD = $(1 - \Sigma(pi^2))$. This equation calculates the genetic diversity (GD) based on the proportion of different genotypes (pi) within an endophytic population.
42. Genetic Differentiation: FST = (HT – HS) / HT. This equation calculates the genetic differentiation (FST) among endophytic populations based on the total genetic variance (HT) and the within-population genetic variance (HS).
43. Phylogenetic Signal: PS = (M / V) – 1. This equation calculates the phylogenetic signal (PS) of endophytic traits based on the mean (M) and variance (V) of trait values across a phylogenetic tree.
44. Genomic Prediction: BP = $\sum(\beta i \cdot Xi)$. This equation represents the genomic prediction (BP) of a specific endophytic trait by summing the product of marker effects (βi) and marker values (Xi).
45. Genotype-Environment Interaction: GEI = $\Sigma((Yij - Y.)^2) - \Sigma((Yij - Yj.)^2)$. This equation calculates the genotype-environment interaction (GEI) for endophytic genotypes based on the sum of squared deviations from the overall mean (Y.) and the genotype-specific means (Yj.).
46. Selection Coefficient: S = (W – 1) / W. This equation calculates the selection coefficient (S) based on the fitness of endophytic genotypes (W) relative to the fittest genotype.
47. Genetic Linkage Mapping: LOD = $\log10(1 / (1 - e^{(-D)}))$.This equation estimates the logarithm of the odds (LOD) score for detecting genetic linkage between endophytic markers based on the recombination fraction (D).
48. Recombination Rate Estimation: $r = c / (2 \cdot N \cdot \mu)$. This equation estimates the recombination rate (r) between endophytic markers based on the number of recombination events (c), the adequate population size (N), and the mutation rate per generation (μ).
49. Polygenic Risk Score: PRS = $\sum(\beta i \cdot Xi)$. This equation calculates the polygenic risk score (PRS) for endophytic traits by summing the product of SNP effects (βi) and SNP genotypes (Xi).
50. Genetic Epistasis: G = (Y – Y1 – Y2 – Y3). This equation models the genetic epistasis (G) between three loci (Y, Y1, Y2, and Y3) in endophytic genomes.

11.7 Root Microbiome Inflection

The inflection of the root microbiome refers to the ability of plants to shape the composition and function of microbial communities residing in their root system. This intricate interaction between plants and the root microbiome appears crucial in plant growth promotion and disease management. The root microbiome consists of a diverse community of bacteria, fungi, archaea, and other microorganisms that colonize the root surface, hairs, and internal tissues (Bag et al., 2022; Cheng et al., 2019; Ghoghari et al., 2022; Kohler et al., 2009; Kotia et al., 2022; Kumar et al., 2021a; Malviya et al., 2021; Sethi, 2022). Plant growth promotion by the root microbiome is mediated through various mechanisms. Firstly, certain beneficial microorganisms in the root microbiome can enhance nutrient availability and uptake for the plant. For example, nitrogen-fixing bacteria can convert atmospheric nitrogen into plant-usable forms, such as ammonia, aiding plant nutrition.

Similarly, phosphate-solubilizing bacteria and mycorrhizal fungi can release phosphate from organic sources, making it accessible to the plant. These interactions improve plant nutrient status, leading to increased growth and development. In addition to nutrient acquisition, the root microbiome also influences plant growth through the production of phytohormones (Chakraborty et al., 2021; Das et al., 2022b; Kumar et al., 2021b; Kumari et al., 2022). Some microorganisms produce auxins, cytokinins, gibberellins, and other growth-promoting hormones that regulate plant growth processes, including cell division, elongation, and differentiation. These hormones stimulate root and shoot growth, enhance root branching, and promote plant vigour.

Moreover, the root microbiome can play a vital role in disease management. Certain microorganisms in the root microbiome can suppress the growth and activity of plant pathogens. They do so through various mechanisms, including competition for nutrients and space, production of antimicrobial compounds, and induction of plant systemic resistance. For instance, certain bacteria can produce antibiotics or siderophores that inhibit the growth of pathogens, while some fungi can parasitize or outcompete plant pathogens in the root environment. The root microbiome can also trigger the plant innate immune response, priming it for enhanced defence against future pathogen attacks. Plant genetic factors may shape the composition and function of the root microbiome. Plant species and genotypes have specific traits and root exudates that attract or select particular microbial communities (P. Kumar et al., 2020; Kumar et al., 2021a; Upadhyay et al., 2023).

Plant breeding and genetic engineering approaches can enhance the recruitment of beneficial microorganisms or modulate root exudation patterns to optimize plant-microbe interactions for growth promotion and disease resistance. The application of microbiome-based biostimulants and biocontrol agents in agriculture has gained considerable attention. These products contain specific microbial strains or consortia that can promote plant growth and protect against pathogens. By harnessing the beneficial interactions between plants and their root microbiome, it is possible to develop sustainable and environmentally friendly strategies for crop production, reducing the reliance on synthetic fertilizers and pesticides (Jing et al., 2022; Kumar et al., 2020; Kumari et al., 2022; Patwardhan et al., 2023; Upadhyay et al., 2023). Plant inflexion of the root microbiome has a profound impact on growth promotion and disease management, and on essential plant services, such as nutrient acquisition, hormone regulation, and disease suppression. Understanding and harnessing the complex interactions between plants and their root microbiome holds great potential for sustainable agriculture practices and improving crop productivity while minimizing chemical inputs.

11.8 Ecology of Plant Microbiomes

The ecology of plant microbiomes refers to the study of the complex interactions between plants and the diverse community of microorganisms that inhabit various plant tissues, including the rhizosphere (soil surrounding the roots), endosphere (internal tissues of the plant), phyllosphere (above-ground plant surfaces), and spermosphere (seed surface) (Aamir et al., 2020; Aslam et al., 2022; Chakraborty et al., 2021; Gayathri et al., 2023; Johnson et al., 2016; Kumar et al., 2021b; Orozco-Mosqueda et al., 2022; Sharma et al., 2020). The plant microbiome comprises bacteria, fungi, archaea, viruses, and other microorganisms, collectively called the plant-associated microbiota. Many factors, including plant species, genotype, developmental stage, environmental conditions, and soil characteristics, shape the microbiome ecology. These factors influence the microbiome composition, diversity, and function, ultimately impacting plant health, growth, and adaptation to environmental stressors.

1. *Composition and Diversity.* The plant microbiome is highly diverse and can vary significantly across host species, tissues, and environments. The microbial community composition is influenced by soil type, climate, and the rhizosphere and endosphere microbial reservoirs. Each plant species has a unique microbial signature, called the core microbiome, which consists of microorganisms consistently associated with a particular host species. Additionally, the plant microbiome exhibits temporal dynamics, with variations occurring during different developmental stages and in response to environmental changes.
2. *Assembly Processes.* The assembly of plant microbiomes is governed by ecological processes, including selection, drift, dispersal, and ecological filtering. Selection occurs when plants actively recruit or favour specific microorganisms based on their functional traits. Drift refers to random changes in microbial community composition over time. Dispersal involves the movement of microorganisms between plant hosts through various means, such as wind, water, and insect vectors. Ecological filtering occurs when environmental factors shape the composition of the plant microbiome by selecting microorganisms adapted to specific conditions.
3. *Rhizosphere Interactions.* The rhizosphere is the region of soil surrounding roots, where interactions between roots, soil, and microorganisms are particularly dynamic. Root exudates, consisting of various organic compounds, serve as a carbon source for microorganisms in the rhizosphere. Microorganisms can influence nutrient availability, cycling, and uptake, enhancing plant nutrient acquisition. Rhizosphere microorganisms can also promote plant growth by producing growth-promoting hormones, solubilization of minerals, and protection against pathogens.
4. *Endosphere Interactions.* The endosphere refers to the internal tissues of plants, including roots, stems, leaves, and reproductive organs. Endophytes are microorganisms that reside within the endosphere without causing apparent disease

symptoms. Endophytes can establish mutualistic associations with plants, providing benefits such as nutrient uptake, hormone regulation, stress tolerance, and disease resistance. They can also modulate plant physiology and metabolism by producing enzymes and secondary metabolites. The endosphere is a complex ecological niche, and the assembly of endophytic communities is influenced by host genotype, tissue type, and environmental conditions.

5. *Phyllosphere Interactions.* The phyllosphere encompasses the above-ground plant surfaces, including leaves, stems, and flowers. The phyllosphere microbiome is exposed to environmental conditions such as sunlight, temperature, and humidity. Microorganisms in the phyllosphere can influence plant health and interactions with pathogens and pests. They can contribute to leaf litter degradation, enhance nutrient cycling, and act as biocontrol agents against foliar pathogens. Leaf surface properties, such as trichomes, wax layers, and nutrient availability, influence the composition of the phyllosphere microbiome.
6. *Plant-Microbe Interactions.* Plant-microbe interactions in the microbiome can be classified as mutualistic, commensal, or pathogenic. Mutualistic interactions involve beneficial associations, such as nitrogen fixation by bacteria or mycorrhizal associations that enhance nutrient uptake. Commensal interactions involve microorganisms that benefit from the plant without harming or benefiting it significantly. Pathogenic interactions involve microbial pathogens that can cause diseases in plants. The balance and dynamics of these interactions shape the overall health and fitness of plants.
7. *Environmental Influences.* The ecology of plant microbiomes is influenced by environmental factors, including climate, soil type, land management practices, and pollution. Environmental conditions can shape microbial community structure, diversity, and function, influencing plant-microbe interactions and ecosystem processes. Climate change, in particular, can alter the composition and functioning of plant microbiomes, potentially affecting plant health and ecosystem stability.

Understanding the ecology of plant microbiomes has significant implications for agriculture, ecosystem functioning, and plant health management. It provides insights into sustainable agricultural practices, such as optimizing plant-microbe interactions for nutrient cycling, disease suppression, and stress tolerance (Bhatt et al., 2022; Can et al., 2022; Jing et al., 2022; Sahu et al., 2019; Samanta et al., 2023; Sharma, 2022; Singh et al., 2023; Tokas et al., 2023). Furthermore, harnessing the beneficial functions of the plant microbiome through microbiome-based approaches holds promise for improving crop productivity, reducing the use of synthetic inputs, and promoting environmental sustainability.

11.9 Mycorrhizosphere and Soil Environment

The mycorrhizosphere refers to the soil environment surrounding the roots of plants that form symbiotic associations with mycorrhizal fungi. Mycorrhizal fungi establish mutualistic relationships with roots, forming specialized structures called mycorrhizae (An Aslam et al., 2022; Bag et al., 2022; Basile et al., 2020; Kumar et al., 2020, 2021a, b; Kumari et al., 2022; Turan et al., 2022; Tyagi et al., 2023; Upadhyay et al., 2023). These associations play a vital role in plant growth, nutrient uptake, and stress management, making the mycorrhizosphere a valuable biostimulant in agricultural and ecological contexts.

1. *Nutrient Uptake Enhancement.* Mycorrhizal associations in the mycorrhizosphere significantly enhance plant nutrient uptake, especially for immobile phosphorus nutrients. Mycorrhizal fungi have extensive hyphal networks that extend beyond the root system, increasing the volume of soil explored for nutrient acquisition. The mycorrhizal hyphae effectively scavenge and absorb nutrients from a larger soil area, enabling plants to access otherwise inaccessible nutrients to their roots. This enhanced nutrient uptake capability of mycorrhizal associations contributes to improved plant growth and overall nutrient status.
2. *Phosphorus Solubilization and Acquisition.* Mycorrhizal fungi play a crucial role in phosphorus solubilization and acquisition. They produce enzymes, such as phosphatases, that break down organic phosphorus compounds into plant-usable forms. This enzymatic activity increases phosphorus availability in the soil, benefiting both mycorrhizal and non-mycorrhizal plants. The mycorrhizosphere acts as a hotspot for phosphorus cycling, as mycorrhizal fungi efficiently acquire and transfer phosphorus to their host plants, aiding in plant growth and development.
3. *Improved Water Uptake and Drought Tolerance.* Mycorrhizal associations in the mycorrhizosphere enhance plant water uptake, particularly under drought conditions. The extensive mycorrhizal hyphal network acts as a supplemental root system, increasing the soil exploration capacity and water-absorbing surface area. This enables plants to access water from deeper soil layers and areas beyond theroot reach. The mycorrhizosphere helps plants maintain water balance, enhances their ability to withstand drought stress, and improves overall water-use efficiency.
4. *Nutrient Cycling and Carbon Sequestration.* The mycorrhizosphere contributes to ecosystem nutrient cycling and carbon sequestration. Mycorrhizal fungi promote the decomposition of organic matter, releasing nutrients that plants subsequently take up. The mycorrhizal hyphal network acts as a conduit for nutrient transfer, facilitating the movement of nutrients between different plants in the ecosystem. Additionally, mycorrhizal associations enhance carbon sequestration in the soil, as mycorrhizal fungi allocate a significant portion of the photosynthetic products from their host plants to the soil, enriching its organic carbon pool.

5. *Enhanced Stress Management.* The mycorrhizosphere is critical in plant stress management, including resistance to biotic and abiotic stresses. Mycorrhizal associations confer increased resistance to diseases caused by various pathogens, including fungi, bacteria, and nematodes. The mycorrhizosphere also improves plant tolerance to abiotic stresses such as salinity, heavy metals, and extreme temperatures. Mycorrhizal fungi can enhance plant stress tolerance through various mechanisms, including producing stress-related hormones, synthesis of protective compounds, and modulation of plant defence responses.
6. *Biostimulant Effect.* The mycorrhizosphere acts as a biostimulant, promoting plant growth and development. The symbiotic association between mycorrhizal fungi and plant roots stimulates root development, increasing root biomass, root branching, and nutrient absorption capacity. This biostimulant effect results in enhanced plant growth, vigour, and productivity. Furthermore, mycorrhizal associations contribute to the establishment of sustainable and ecologically balanced agroecosystems by reducing the reliance on synthetic fertilizers and improving nutrient use efficiency.

In summary, the mycorrhizosphere has a biostimulant role with significant impacts on plant growth, nutrient uptake, and stress management. The mycorrhizal associations in the mycorrhizosphere enhance nutrient acquisition, especially for immobile nutrients like phosphorus, improve water uptake and drought tolerance, contribute to nutrient cycling and carbon sequestration, enhance stress management, and is a biostimulant for plant growth (Mahmud et al., 2021b; Mahmud & Chong, 2021; Malviya et al., 2021; Mandal et al., 2023; Meena et al., 2017; Mishra et al., 2022). Understanding and harnessing the potential of the mycorrhizosphere can significantly benefit agriculture, ecological restoration, and sustainable land management practices.

11.10 Compost with Disease Suppressive Features

Disease-suppressive compost enhances natural soil suppressiveness against soil-borne plant pathogens. This assumption refers to the phenomenon observed where applying disease-suppressive compost to soil, as it can enhance the soil inherent ability to suppress soil-borne plant diseases. This process involves transforming compost materials during composting, resulting in the development of beneficial microorganisms and the release of bioactive compounds that contribute to disease suppression.

The compost is an organic material produced by decomposing plant and animal residues under controlled conditions. Disease-suppressive compost refers to compost that has undergone specific composting techniques and management practices to enhance its disease-suppressive properties (Chandra et al., 2021; Cheng et al., 2019; da Silva et al., 2022; Das et al., 2022a, b; Dellagnezze & Sierra-Garcia, 2023; Goss et al., 2023; Jariyal & Soni, 2021; Jing et al., 2022). The composting process involves the microbial decomposition of organic matter, leading to heat generation, the breakdown of complex compounds, and the enrichment of nutrient content.

The disease-suppressive properties of compost are attributed to several mechanisms:

1. *Microbial Population Dynamics.* During the composting process, the high temperatures generated produce microenvironmental conditions that favour the proliferation of thermophilic microorganisms. These high-temperature conditions can eliminate or suppress plant pathogens in the compost materials. As the compost cools down, mesophilic microorganisms become dominant. These microbial populations contribute to disease suppression by outcompeting pathogens for resources and space, producing antagonistic compounds, and stimulating the plant defence mechanisms.
2. *Antagonistic Microorganisms.* Disease-suppressive compost contains a diverse range of beneficial species, including bacteria, fungi, and other microorganisms. Some of them possess antagonistic properties and can directly inhibit the growth and activity of plant pathogens. They do so through mechanisms such as the production of antibiotics, enzymes that degrade pathogen cell walls, competition for nutrients and space, and induction of systemic resistance in plants.
3. *Induced Systemic Resistance.* Disease-suppressive compost can induce systemic resistance in plants, priming them for enhanced defence against pathogens. Applying compost stimulates the plant innate immune system, triggering the production of defence-related compounds and the activation of defence pathways. This systemic resistance response can provide long-lasting protection against various pathogens, even beyond the initial contact with the compost.
4. *Bioactive Compounds.* Many composts contain various bioactive compounds, including organic acids, phenolic and volatile compounds, which can suppress growth of pathogens. These compounds can inhibit the pathogen spores germination, disrupt their cell membranes, interfere with pathogen signalling, and induce oxidative stress, leading to the papthogen inhibition or death.

The application of disease-suppressive compost to soil has several benefits for plant disease management:

1. *Disease Reduction.* Disease-suppressive compost can significantly reduce the incidence and severity of soil-borne plant diseases. The beneficial microorganisms and bioactive compounds in the compost collaborate to suppress the growth and activity of plant pathogens, limiting their ability to infect roots and cause disease.
2. *Increased Crop Health and Productivity.* Disease-suppressive compost enhances crop health and productivity by reducing disease pressure. Plants grown in compost-amended soil often exhibit improved vigour, increased root development, and enhanced nutrient uptake. This leads to healthier plants better equipped to withstand biotic and abiotic stresses, resulting in improved crop yields.
3. *Sustainable Disease Management.* Disease-suppressive compost offers a sustainable approach to disease management as it relies on utilising organic waste materials and harnesses the natural abilities of beneficial microorganisms. Recycling organic residues into compost and applying it to the soil reduces the need for chemical inputs, such as synthetic pesticides and fertilizers, thereby minimizing environmental impacts.

It is important to note that the effectiveness of disease-suppressive compost can vary depending on factors such as the compost quality, its application rate and timing, as well as crop-specific interactions. Additionally, the specific mechanisms of disease suppression may vary depending on the compost materials used and the compost and soil microbial community dynamics. The application of disease-suppressive compost to soil can also enhance its natural suppressiveness *vs* soil-borne plant pathogens (Malviya et al., 2021; Sayanti Mandal et al., 2023; Meena et al., 2017;). This is achieved through the presence of beneficial microorganisms, the production of bioactive compounds, and the induction of systemic resistance in plants. Disease-suppressive compost offers a sustainable approach to disease management, promoting crop health, productivity, and environmental stewardship.

11.11 Conclusion

In conclusion, PGPRs, AMFs, and endophytes play vital functional roles and employ diverse mechanisms to enhance host resistance against plant pathogens. These beneficial plant-microbe interactions have significant implications for sustainable agriculture and disease management. PGPRs can colonize the rhizosphere and stimulate plant growth and health through various mechanisms. They can directly inhibit the growth and activity of pathogens through competition for nutrients and space, production of antimicrobial compounds, and induction of systemic resistance in plants. Additionally, PGPRs can enhance nutrient availability and uptake, improve soil structure, and modulate plant hormone levels, all contributing to increased host resistance against pathogens. AMFs form mutualistic associations with roots, enhancing plant resistance against pathogens. AMFs improve nutrient acquisition, particularly phosphorus, for the host plant, thereby increasing its vigour and reducing disease susceptibility. They also induce systemic resistance in plants, activate defence-related genes, and produce antimicrobial compounds that can suppress the growth of pathogens. The extensive hyphal network of AMFs facilitates nutrient and information exchange between plants, enhancing their overall resilience.

Also endophytes offer unique mechanisms for improving host resistance against pathogens. They can produce antibiotics, enzymes, and secondary metabolites that directly inhibit pathogen growth or induce plant defence responses. They can also enhance plant stress tolerance, nutrient uptake, and fitness.

Endophytic colonization can result in long-term protection against pathogens by priming the plant immune system and triggering defence mechanisms. The functional roles and mechanisms of PGPRs, AMFs, and endophytes are not mutually exclusive, and their combined effects can synergistically impact plant health and disease resistance. These beneficial microorganisms can enhance nutrient availability, induce systemic resistance, improve plant growth, and suppress pathogen activity, reducing disease incidence and increasing crop productivity. Furthermore, their modes of action are often environmentally friendly, reducing the need for chemical inputs and promoting sustainable agricultural practices. However, much remains to learn about the intricate interactions between these microorganisms and their host plants. Further research is needed to elucidate specific mechanisms, optimize application methods, and explore the potential of harnessing these beneficial plant-microbe interactions for effective disease management strategies. Understanding the functional roles and mechanisms of PGPRs, AMFs, and endophytes offers promising avenues for developing sustainable and environmentally friendly approaches to enhance host resistance against plant pathogens, ultimately contributing to advancing agriculture and global food security.

11.12 Prospects for Future

The functional role and mechanism of PGPRs, AMFs, and endophytes in improving host resistance against plant pathogens offer promising prospects for sustainable agriculture and disease management. Here are some potential future directions and prospects for further exploration:

1. Enhanced understanding of microbial interactions: future research can focus on unravelling the intricate interactions between PGPRs, AMFs, and endophytes, as well as their interplay with host plants. Investigating the molecular and genetic mechanisms of these interactions will provide a deeper understanding of how these beneficial microorganisms enhance host resistance against pathogens.
2. Microbiome engineering: the ability to manipulate the composition and functions of plant-associated microbial communities holds a potential for disease management. Future research can explore strategies for targeted manipulation of the plant microbiome, such as developing microbiome-based biostimulants or engineering specific microbial consortia to maximize disease suppression and plant health.
3. Multi-trophic interactions: understanding the interactions between PGPRs, AMFs, endophytes, and other members of the plant microbiome, as well as the broader soil ecosystem, is crucial. Investigating the influence of microbial communities on disease suppression and plant health in the context of multi-trophic interactions will provide a more holistic understanding of microbial-mediated plant defence mechanisms.
4. Exploration of novel microbial strains: identifying and characterising novel microbial strains with enhanced disease-suppressive capabilities will be an essential prospect. Screening and isolating beneficial microorganisms from diverse environments, including extreme environments and native ecosystems, can lead to the discovery of novel strains with unique mechanisms for improving host resistance against pathogens.
5. Development of precision microbiome management: prospects include developing precision microbiome management approaches that allow for tailored and site-specific applications of beneficial microorganisms. This can involve precision agriculture techniques, such as site-specific application of microbial inoculants or targeted delivery systems, to optimize the establishment and function of PGPRs, AMFs, and endophytes in improving host resistance.

6. Integration of microbiome approaches with plant breeding: combining microbiome approaches with plant breeding programs can lead to crop varieties with enhanced microbial interactions and disease resistance. Prospects involve integrating knowledge of beneficial microbial traits and plant genetics to breed crops with optimized microbiome interactions for improved disease management and resilience.
7. Climate change adaptation: with changing environmental conditions due to climate change, exploring the potential of PGPRs, AMFs, and endophytes in enhancing plant resilience to abiotic stresses, such as drought, salinity, and extreme temperatures, is a critical prospect. Investigating the role of these microorganisms in mitigating climate change impacts on plant health and productivity will be crucial for sustainable agriculture.

In conclusion, the prospects for the functional role and mechanism of PGPRs, AMFs, and endophytes in improving host resistance against plant pathogens are promising. Continued research and innovation in these areas will contribute to developing sustainable and eco-friendly approaches for disease management, enhanced crop productivity, and advancing resilient agricultural systems.

Acknowledgements We would like to express our sincere gratitude to the Department of Agronomy for their support and assistance throughout the writing. The Department commitment to academic excellence and research has been instrumental in completing this endeavour.

Authors' Contribution All authors made significant contributions to the present study. Each author has participated sufficiently in the study, made intellectual contributions and is responsible for the work accuracy and integrity. The authors have collaborated closely, ensuring the completion of this work through collective effort, expertise, and dedication.

References

Aamir, M., Rai, K. K., Zehra, A., Dubey, M. K., Kumar, S., et al. (2020). Microbial bioformulation-based plant biostimulants: A plausible approach toward next generation of sustainable agriculture. In A. Kumar & E. K. Radhakrishnan (Eds.), *Microbial endophytes* (pp. 195–225). Woodhead Publishing.

Abdelsattar, A. M., Elsayed, A., El-Esawi, M. A., & Heikal, Y. M. (2023). Enhancing *Stevia rebaudiana* growth and yield through exploring beneficial plant-microbe interactions and their impact on the underlying mechanisms and crop sustainability. *Plant Physiology and Biochemistry, 198*, 107673.

Aslam, M. M., Okal, E. J., Idris, A. L., Qian, Z., Xu, W., et al. (2022). Rhizosphere microbiomes can regulate plant drought tolerance. *Pedosphere, 32*(1), 61–74.

Bag, S., Mondal, A., & Banik, A. (2022). Exploring tea (*Camellia sinensis*) microbiome: Insights into the functional characteristics and their impact on tea growth promotion. *Microbiological Research, 254*, 126890.

Basile, B., Rouphael, Y., Colla, G., Soppelsa, S., & Andreotti, C. (2020). Appraisal of emerging crop management opportunities in fruit trees, grapevines and berry crops facilitated by the application of biostimulants. *Scientia Horticulturae, 267*, 109330.

Bhandari, S., Kukreja, S., Kumar, V., Dey, A., & Goutam, U. (2022). Chapter 9 – Microbiome stimulants and their applications in crop plants. In J. Samuel, A. Kumar, & J. Singh (Eds.), *Relationship between microbes and the environment for sustainable ecosystem services, volume 1* (pp. 177–195). Elsevier.

Bhanse, P., Kumar, M., Singh, L., Awasthi, M. K., & Qureshi, A. (2022). Role of plant growth-promoting rhizobacteria in boosting the phytoremediation of stressed soils: Opportunities, challenges, and prospects. *Chemosphere, 303*, 134954.

Bhatt, K., Suyal, D. C., Kumar, S., Singh, K., & Goswami, P. (2022). New insights into engineered plant-microbe interactions for pesticide removal. *Chemosphere, 309*, 136635.

Bind, S., Bind, S., Kumar, A., & Chandra, D. (2023). Microbial cross talk: Below and above ground. In D. Chandra & P. Bhatt (Eds.), *Developments in applied microbiology and biotechnology* (pp. 213–226). Academic.

Bosamia, T. C., Barbadikar, K. M., & Modi, A. (2020). Genomic insights of plant endophyte interaction: Prospective and impact on plant fitness. In A. Kumar & E. K. Radhakrishnan (Eds.), *Microbial endophytes* (pp. 227–249). Woodhead Publishing.

Bukhat, S., Imran, A., Javaid, S., Shahid, M., Majeed, A., & Naqqash, T. (2020). Communication of plants with microbial world: Exploring the regulatory networks for PGPR mediated defense signaling. *Microbiological Research, 238*, 126486.

Can, H., Kal, U., Kayak, N., Dal, Y., & Turkmen, O. (2022). Chapter 10 – Use of microbial inoculants against biotic stress in vegetable crops: Physiological and molecular aspect. In M. Seymen, E. S. Kurtar, C. Erdinc, & A. Kumar (Eds.), *Developments in applied microbiology and biotechnology* (pp. 263–332). Academic.

Chakraborty, S., Kumar, P., Sanyal, R., Mane, A. B., Arvind Prasanth, D., et al. (2021). Unravelling the regulatory role of miRNAs in secondary metabolite production in medicinal crops. *Plant Gene, 27*, 100303.

Chandra, P., Wunnava, A., Verma, P., Chandra, A., & Sharma, R. K. (2021). Strategies to mitigate the adverse effect of drought stress on crop plants—Influences of soil bacteria: A review. *Pedosphere, 31*, 496–509.

Cheng, Y. T., Zhang, L., & He, S. Y. (2019). Plant-microbe interactions facing environmental challenge. *Cell Host & Microbe, 26*, 183–192.

da Silva, A. V., da Silva, M. K., Sampaio, E. B. T., Romanholo Ferreira, L. F. R., Zambrano Passarini, M. R., et al. (2022). Benefits of plant growth-promoting symbiotic microbes in climate change era. In A. Kumar, J. Singh, & L. F. Romanholo Ferreira (Eds.), *Microbiome under changing climate* (pp. 85–113). Woodhead Publishing.

Das, P. P., Singh, K. R. B., Nagpure, G., Mansoori, A., Singh, R. P., et al. (2022a). Plant-soil-microbes: A tripartite interaction for nutrient acquisition and better plant growth for sustainable agricultural practices. *Environmental Research, 214*, 113821.

Das, T., Saha, S. C., Sunita, K., Majumder, M., Ghorai, M., et al. (2022b). Promising botanical-derived monoamine oxidase (MAO) inhibitors: Pharmacological aspects and structure-activity studies. *South African Journal of Botany, 146*, 127–145.

De Corato, U. (2020). Disease-suppressive compost enhances natural soil suppressiveness against soil-borne plant pathogens: A critical review. *Rhizosphere, 13*, 100192.

Dellagnezze, B. M., & Sierra-Garcia, I. N. (2023). Chapter 8 – Immobilization of microbial inoculants for improving soil nutrient bioavailability. In V. K. Sharma, A. Kumar, M. R. Z. Passarini, S. Parmar, & V. K. Singh (Eds.), *Developments in applied microbiology and biotechnology* (pp. 161–181). Academic.

Ezrari, S., Radouane, N., Tahiri, A., El Housni, Z., Mokrini, F., et al. (2022). Dry root rot disease, an emerging threat to citrus industry worldwide under climate change: A review. *Physiological and Molecular Plant Pathology, 117*, 101753.

Afridi, M. S., Fakhar, A., Kumar, A., Ali, S., Medeiros, F. H. V., et al. (2022). Harnessing microbial multitrophic interactions for rhizosphere microbiome engineering. *Microbiological Research, 265*, 127199.

Gayathri, G., Gomathi, S., Ambikapathy, V., & Panneerselvam, A. (2023). Improvement of soil fertility through plant microbial interaction. In P. Swapnil, M. Meena, A. Harish, S. V. Marwal, & A. Zehra (Eds.), *Plant-microbe interaction – Recent advances in molecular and biochemical approaches* (pp. 281–301). Academic.

Ghoghari, N., Bharwad, K., Champaneria, A., & Rajkumar, S. (2022). Microbial consortia for augmentation of plant growth–revisiting the promising approach towards sustainable agriculture. In H. Singh & A. Vaishnav (Eds.), *New and future developments in microbial biotechnology and bioengineering* (pp. 231–256). Elsevier.

Ghorbanpour, M., Omidvari, M., Abbaszadeh-Dahaji, P., Omidvar, R., & Kariman, K. (2018). Mechanisms underlying the protective effects of beneficial fungi against plant diseases. *Biological Conservation, 117*, 147–157.

Goss, M. J., Carvalho, M., Brito, I., & Brígido, C. (2017). Diversity in Arbuscular Mycorrhizal fungi. In M. J. Goss, M. Carvalho, & I. Brito (Eds.), *Functional diversity of mycorrhiza and sustainable agriculture* (pp. 59–79). Academic.

Goss, M. J., Brito, I., & Carvalho, M. (2023). Management of soil biology for crop protection from biotic and abiotic stresses. In M. J. Goss & M. Oliver (Eds.), *Encyclopedia of soils in the environment* (2nd ed., pp. 447–468). Academic.

Gupta, A., Bano, A., Rai, S., Mishra, R., Singh, M., et al. (2022). Mechanistic insights of plant-microbe interaction towards drought and salinity stress in plants for enhancing the agriculture productivity. *Plant Stress, 4*, 100073.

Jariyal, M., & Soni, R. (2021). Role of microbial communities in restoration disturbed lands. In J. S. Singh, S. Tiwari, C. Singh, & A. K. Singh (Eds.), *Microbes in land use change management* (pp. 433–450). Elsevier.

Jing, J., Cong, W. F., & Bezemer, T. M. (2022). Legacies at work: Plant–soil–microbiome interactions underpinning agricultural sustainability. *Trends in Plant Science, 27*, 781–792.

Johnson, S. N., Benefer, C. M., Frew, A., Griffiths, B. S., Hartley, S. E., et al. (2016). New frontiers in belowground ecology for plant protection from root-feeding insects. *Applied Soil Ecology, 108*, 96–107.

Joshi, N., Parewa, H. P., Joshi, S., Sharma, J. K., Shukla, U. N., et al. (2021). Use of microbial biostimulants in organic farming. In V. S. Meena, S. K. Meena, A. Rakshit, J. Stanley, & C. Srinivasarao (Eds.), *Advances in organic farming* (pp. 59–73). Woodhead Publishing.

Kaur, H., & Garg, N. (2022). Chapter 24 – Microbial community in soil-plant systems: Role in heavy metal(loid) detoxification and sustainable agriculture. In R. C. Dubey & P. Kumar (Eds.), *Rhizosphere engineering* (pp. 471–498). Academic.

Khalid, M., Ur-Rahman, S., Hassani, D., Hayat, K., Zhou, P., & Hui, N. (2021). Advances in fungal-assisted phytoremediation of heavy metals: A review. *Pedosphere, 31*, 475–495.

Kohler, J., Hernández, J. A., Caravaca, F., & Roldán, A. (2009). Induction of antioxidant enzymes is involved in the greater effectiveness of a PGPR versus AM fungi with respect to increasing the tolerance of lettuce to severe salt stress. *Environmental and Experimental Botany, 65*, 245–252.

Kotia, A., Rutu, P., Singh, V., Kumar, A., Dhoke, S., et al. (2022). Rheological analysis of rice husk-starch suspended in water for sustainable agriculture application. *Materials Today: Proceedings, 50*, 1962–1966.

Kumar, A., & Dubey, A. (2020). Rhizosphere microbiome: Engineering bacterial competitiveness for enhancing crop production. *Journal of Advanced Research, 24*, 337–352.

Kumar, P., Kumar, T., Singh, S., Tuteja, N., Prasad, R., & Singh, J. (2020). Potassium: A key modulator for cell homeostasis. *Journal of Biotechnology, 324*, 198–210.

Kumar, P., Sharma, K., Saini, L., & Dey, S. R. (2021a). Chapter 8 – Role and behavior of microbial volatile organic compounds in mitigating stress. In A. Kumar, J. Singh, & J. Samuel (Eds.), *Volatiles and metabolites of microbes* (pp. 143–161). Academic.

Kumar, V., Dwivedi, P., Kumar, P., Singh, B. N., Pandey, D. K., et al. (2021b). Mitigation of heat stress responses in crops using nitrate primed seeds. *South African Journal of Botany, 140*, 25–36.

Kumari, P., Singh, J., & Kumar, P. (2022). Impact of bioenergy for the diminution of an ascending global variability and change in the climate. In A. Kumar, J. Singh, & L. F. Romanholo Ferreira (Eds.), *Microbiome under changing climate* (pp. 469–487). Woodhead Publishing.

Kumawat, K. C., Razdan, N., & Saharan, K. (2022). Rhizospheric microbiome: Bio-based emerging strategies for sustainable agriculture development and future perspectives. *Microbiological Research, 254*, 126901.

Li, X., Ren, Q., Liao, C., Wang, Q., Ren, M., et al. (2023). Chapter 24 – Plant-microbe interactions in wheat to deal with abiotic stress. In M. K. Khan, A. Pandey, M. Hamurcu, O. P. Gupta, & S. Gezgin (Eds.), *Abiotic stresses in wheat* (pp. 375–391). Academic.

Mahmud, M. S., & Chong, K. P. (2021). Formulation of biofertilizers from oil palm empty fruit bunches and plant growth-promoting microbes: A comprehensive and novel approach towards plant health. *Journal of King Saud University – Science, 33*, 101647.

Mahmud, A. A., Upadhyay, S. K., Srivastava, A. K., & Bhojiya, A. A. (2021a). Biofertilizers: A nexus between soil fertility and crop productivity under abiotic stress. *Current Research in Environmental Sustainability, 3*, 100063.

Mahmud, K., Missaoui, A., Lee, K., Ghimire, B., Presley, H. W., & Makaju, S. (2021b). Rhizosphere microbiome manipulation for sustainable crop production. *Current Plant Biology, 27*, 100210.

Malviya, M. K., Solanki, M. K., Li, C. N., Htun, R., Singh, R. K., et al. (2021). Sugarcane microbiome: Role in sustainable production. In M. K. Solanki, P. L. Kashyap, R. A. Ansari, & B. Kumari (Eds.), *Microbiomes and plant health* (pp. 225–242). Academic.

Mandal, S., Anand, U., López-Bucio, J., Radha, K. M., et al. (2023). Biostimulants and environmental stress mitigation in crops: A novel and emerging approach for agricultural sustainability under climate change. *Environmental Research, 233*, 116357.

Meena, V. S., Meena, S. K., Verma, J. P., Kumar, A., Aeron, A., et al. (2017). Plant beneficial rhizospheric microorganism (PBRM) strategies to improve nutrients use efficiency: A review. *Ecological Engineering, 107*, 8–32.

Mishra, P., Tiwari, S. K., Singh, A. K., & Malik, M. Z. (2022). Harnessing endophytic microbial diversity for stress alleviation: Current perspectives for crop improvement. In G. Santoyo, A. Kumar, M. Aamir, & S. Uthandi (Eds.), *Mitigation of plant abiotic stress by microorganisms* (pp. 149–174). Academic.

Mon, Y. Y., Bidabadi, S. S., Oo, K. S., & Zheng, S. J. (2021). The antagonistic mechanism of rhizosphere microbes and endophytes on the interaction between banana and *Fusarium oxysporum* f. sp. *cubense*. *Physiological and Molecular Plant Pathology, 116*, 101733.

Orozco-Mosqueda, M., Fadiji, A. E., Babalola, O. O., Glick, B. R., & Santoyo, G. (2022). Rhizobiome engineering: Unveiling complex rhizosphere interactions to enhance plant growth and health. *Microbiological Research, 263*, 127137.

Panda, S., Busatto, N., Hussain, K., & Kamble, A. (2019). Piriformospora indica-primed transcriptional reprogramming induces defense response against early blight in tomato. *Scientia Horticulturae, 255*, 209–219.

Patwardhan, S. B., Pandit, C., Pandit, S., Verma, D., Lahiri, D., Nag, M., et al. (2023). Illuminating the signalomics of microbial biofilm on plant surfaces. *Biocatalysis and Agricultural Biotechnology, 47*, 102537.

Rathore, M. S., & Srinivasulu, Y. (2022). Soil health sustenance through varied fertility management approaches for increasing microbial load under temperate climatic conditions: A practical approach. In J. Singh & D. Sharma (Eds.), *Microbial resource technologies for sustainable development* (pp. 319–331). Elsevier.

Reddy, C. A., & Saravanan, R. S. (2013). Chapter Three – Polymicrobial multi-functional approach for enhancement of crop productivity. In S. Sariaslani & G. M. Gadd (Eds.), *Advances in applied microbiology* (Vol. 82, pp. 53–113). Academic.

Sabir, A. (2022). Sustainable stress mitigation with microorganisms in viticulture. In M. Seymen, E. S. Kurtar, C. Erdinc, & A. Kumar (Eds.), *Developments in applied microbiology and biotechnology* (pp. 77–95). Academic.

Sahab, S., Suhani, I., & Singh, R. P. (2021). Valuing each patch of land: Utilizing plant-microbe interactions for the betterment of agriculture. In J. S. Singh, S. Tiwari, C. Singh, & A. K. Singh (Eds.), *Microbes in land use change management* (pp. 471–507). Elsevier.

Sahu, P. K., Singh, D. P., Prabha, R., Meena, K. K., & Abhilash, P. C. (2019). Connecting microbial capabilities with the soil and plant health: Options for agricultural sustainability. *Ecological Indicators, 105*, 601–612.

Sakthieaswari, P., Kannan, A., & Baby, S. (2022). Chapter 14 – Role of mycorrhizosphere as a biostimulant and its impact on plant growth, nutrient uptake and stress management. In H. B. Singh & A. Vaishnav (Eds.), *New and future developments in microbial biotechnology and bioengineering* (pp. 319–336). Elsevier.

Samanta, A., Banerjee, S., & Datta, S. (2023). The role of endophytes to boost the plant immunity. In M. Shah & D. Deka (Eds.), *Endophytic association: What, why and how* (pp. 199–223). Academic.

Sati, D., Pandey, S. C., Pande, V., Upreti, S., Gouri, V., et al. (2020). Toward an enhanced understanding of plant growth promoting microbes for sustainable agriculture. In S. De Mandal & P. Bhatt (Eds.), *Recent advancements in microbial diversity* (pp. 87–112). Academic.

Sethi, S. (2022). Utilization of beneficial fungal strain/bacterial strains in climate-resilient agriculture. In A. Kumar, J. Singh, & L. F. R. Ferreira (Eds.), *Microbiome under changing climate* (pp. 313–331). Woodhead Publishing.

Shahid, I., & Mehnaz, S. (2020). Chapter 4 – Microbial secondary metabolites: Effectual armors to improve stress survivability in crop plants. In J. S. Singh & S. R. Vimal (Eds.), *Microbial services in restoration ecology* (pp. 47–70). Elsevier.

Sharma, P. (2022). Role and significance of biofilm-forming microbes in phytoremediation -A review. *Environmental Technology and Innovation, 25*, 102182.

Sharma, V., Sharma, A., & Salwan, R. (2020). Overview and challenges in the implementation of plant beneficial microbes. In V. Sharma, R. Salwan, & L. K. T. Al-Ani (Eds.), *Molecular aspects of plant beneficial microbes in agriculture* (pp. 1–18). Academic.

Sharma, P., Kumar, T., Yadav, M., Gill, S. S., & Chauhan, N. S. (2021). Plant-microbe interactions for the sustainable agriculture and food security. *PLT Genes, 28*, 100325.

Singh, V. K., Shukla, A. K., & Singh, A. K. (2019). Chapter 7 – Impact of climate change on plant–microbe interactions under agroecosystems. In K. K. Choudhary, A. Kumar, & A. K. Singh (Eds.), *Climate change and agricultural ecosystems* (pp. 153–179). Woodhead Publishing.

Singh, S. K., Srikanth, G. S., Puranik, S., & Shukla, L. (2023). Chapter 9 – Chemical talk within plant holobiont: A fascinating conversation. In P. Swapnil, M. M. Harish, A. Marwal, S. Vijayalakshmi, & A. Zehra (Eds.), *Plant-microbe interaction – recent advances in molecular and biochemical approaches* (pp. 165–203). Academic.

Tokas, D., Singh, S., Yadav, R., & Singh, A. N. (2023). Plant-microbe interactions: Role in sustainable agriculture and food security in a changing climate. In P. Swapnil, M. Meena, A. M. Harish, S. Vijayalakshmi, & A. Zehra (Eds.), *Plant-microbe interaction – Recent advances in molecular and biochemical approaches* (pp. 363–391). Academic.

Turan, M., Argin, S., Bolouri, P., Arjumend, T., Ersoy, N., et al. (2022). Bioremediation of contaminated soil with plant growth rhizobium bacteria. In P. Pandey (Ed.), *Assisted phytoremediation* (pp. 265–284). Elsevier.

Tyagi, P., Singh, A., Saraswat, P., Chaturvedi, A., & Ranjan, R. (2023). Sustainable agricultural approach to study interaction of plants and microbes. In P. Swapnil, M. M. Harish, A. Marwal, S. Vijayalakshmi, & A. Zehra (Eds.), *Plant-microbe interaction – Recent advances in molecular and biochemical approaches* (pp. 331–361). Academic.

Upadhyay, S. K., Devi, P., Kumar, V., Pathak, H. K., Kumar, P., Rajput, V. D., & Dwivedi, P. (2023). Efficient removal of total arsenic ($As^{3+/5+}$) from contaminated water by novel strategies mediated iron and plant extract activated waste flowers of marigold. *Chemosphere, 313*, 137551.

Veerapagu, M., Jeya, K. R., & Sankaranarayanan, A. (2023). Role of plant growth-promoting microorganisms in phytoremediation efficiency. In P. Swapnil, M. M. Harish, A. Marwal, S. Vijayalakshmi, & A. Zehra (Eds.), *Plant-microbe interaction – Recent advances in molecular and biochemical approaches* (pp. 45–61). Academic.

Verma, A., Bind, S., & Bajeli, J. (2023). Inflection of the root microbiome by plants: Plant growth promotion and disease management. In D. Chandra & P. Bhatt (Eds.), *Developments in applied microbiology and biotechnology* (pp. 151–173). Academic.

Vimal, S. R., Singh, J. S., Arora, N. K., & Singh, S. (2017). Soil-plant-microbe interactions in stressed agriculture management: A review. *Pedosphere, 27*, 177–192.

Yang, T., Lupwayi, N., Marc, S. A., Siddique, K. H. M., & Bainard, L. D. (2021). Anthropogenic drivers of soil microbial communities and impacts on soil biological functions in agroecosystems. *Global Ecology and Conservation, 27*, e01521.

Chapter 12
Bacterial Secondary Metabolites (B-SMs): Empowering Agriculture Against Fungal Disease Challenges

Janki Ruparelia, Aniruddh Rabari, Chaitanya Kumar Jha, and R. Z. Sayyed

Abstract Bacterial secondary metabolites (B-SMs) are bioactive, low molecular weight compounds synthesized during the bacteria idio phase within their metabolic cycle. These specialized metabolites are dispensable for bacterial growth but are pivotal in eliciting secondary response mechanisms such as resistance, survival, and defense. Rhizobacteria, encompassing genera such as *Bacillus*, *Pseudomonas*, *Micrococcus*, *Streptomyces*, members of Actinomycetes such as *Nocardia* spp., are prolific producers of secondary metabolites characterized by their diverse physio-chemical and biotic properties. Among these metabolites are pigments, toxins, antibiotics, and antitumor agents. Notably, rhizobacterial B-SMs exhibit biocontrol activities, including potent antifungal, antiviral, and antibacterial effects. This Chapter reviews some secondary metabolites synthesized by various rhizobacterial species, particularly highlighting their antifungal properties, and offers insights into characterization methodologies and advanced analytical techniques.

Keywords Antifungal antibiotics · Bio-control · Disease management · Mode of action · Rhizobacteria · Sustainable agriculture

J. Ruparelia · A. Rabari
Department of Microbiology and Biotechnology, University School of Science, Gujarat University, Ahmedabad, India

Microbiology Department, Gujarat Arts and Science College, Ahmedabad, India

C. K. Jha (✉)
Microbiology Department, Gujarat Arts and Science College, Ahmedabad, India

R. Z. Sayyed
Department of Microbiology, PSGVP Mandal's S. I. Patil Arts, G. B. Patel Science and STKV Sangh Commerce College, Shahada, India

P. Mathur, S. Roy (eds.), *Plant Microbiome and Biological Control*, Sustainability in Plant and Crop Protection 20,
https://doi.org/10.1007/978-3-031-75845-4_12

Abbreviations

B-SMs	Bacterial secondary metabolites
FTIR	Fourier-transform infrared spectroscopy
GC	gas chromatography-mass spectrometry
GC-MS	liquid chromatography-mass spectrometry (LC-MS)
LPs	Lipopolypeptides
PKs	Polyketides
NRPs	Nonribosomal peptides

12.1 Introduction

Bacterial Secondary Metabolites (B-SMs) are bioactive molecules synthesized by bacteria during their secondary metabolism in the stationary (idio) phase. They are low molecular weight compounds and are also known as specialized metabolites. These compounds are not necessary for the growth of bacteria, but are needed to trigger secondary response mechanisms, i.e., resistance, survival, and defense. Many bacteria may directly or indirectly boost plant growth by synthesizing numerous kinds of B-SMs such as siderophores, phytohormones, etc. (Saranraj et al., 2022), that directly improve plant and soil health by providing numerous growth-promoting tactics. Moreover, B-SMs such as specialized enzymes, antibiotics, and toxins, indirectly sustain plant health through several biocontrol mechanisms acting *vs* many diseases. Many B-SMs are known for their biocontrol activities through antifungal, antiviral, antibacterial mechanisms and more, as described in this chapter (Pourakbar et al., 2021; Saranraj et al., 2022, 2023; Suriani et al., 2020).

Secondary metabolites (SMs) produced by pervasive microorganisms play a focal role in managing contagious diseases, as they hold antimicrobial potential (Kumar et al., 2021). Rhizobacteria produce considerable SM with countless diverse physiochemical and biotic properties. For example, *Bacillus*, *Pseudomonas*, *Micrococcus*, *Streptomyces*, *Actinomyces*, and *Nocardia* spp. produce numerous metabolites using specific metabolic pathways (Kadiri et al., 2023; Sevugapperumal et al., 2021). The B-SMs mode of action differs from one species to another: for example, extracted bacterial antibiotics can act upon specific fungi in a particular field or environment. Meanwhile, the change in environmental parameters may affect effectivity *vs* specificity. B-SMs act through various modes of action against phytopathogenic fungi, i.e., disruption of cell wall synthesis, inhibition of protein or nucleic acid synthesis, disruption of microtubules and/or of cell membranes (Fig. 12.1).

Detection of B-SMs could be very tedious, time-consuming, and expensive. Still, the analytical methods nowadays are helpful as they can quickly detect particular components (Jadhav et al., 2017). Several B-SMs could be detected by analytical methods such as TLC (Thin Layer Chromatography), HPLC (high-Performance Liquid Chromatography), GC-MS (Gas Chromatography-Mass Spectrometry), LC-MS (Liquid Chromatography-Mass Spectrometry), and more (Patel et al., 2016).

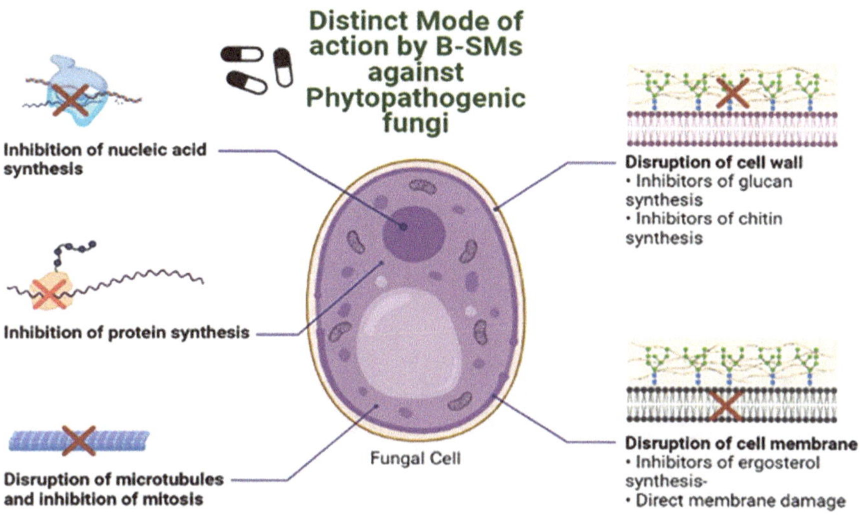

Fig. 12.1 Mode of action by B-SMs against phytopathogenic fungi. (Graphic source: BioRender)

12.2 Metabolite Groups

12.2.1 Antifungal B-SMs

Natural products (NPs) are a broad source of bioactive compounds with scientific relevance, as evidenced by the fact that half of existing drugs are NP or NP-derived (Goh et al., 2020). They can be further engineered for clinical, flavor, and agrochemical applications (Parkinson et al., 2015). Bacteria produce a variety of secondary metabolites, which have numerous applications in the field of agriculture and biocontrol, due to their proved potential as antibacterial, antifungal, anti-algal, and more (Fig. 12.2).

12.2.2 Macrolactones

The macrolactones have been extensively explored and used for their biological properties. They range in size from ten to more than twenty-membered rings.often comprising other chemical motifs. Macrocycles, particularly those with a macrolactone ring, are a diverse group of natural compounds, many of which exhibit fascinating bioactivity and have become a popular tool for finding new therapeutic targets for humans and plants. Macrolactones are long-chain hydroxyl acids with approx. Twelve atoms encompassing a ring-like structure. Amongst several beneficial characteristics of macrolactone, the ability to bind to challenging protein targets makes it appropriate as antifungal drugs. Cethromycin (macrolide),

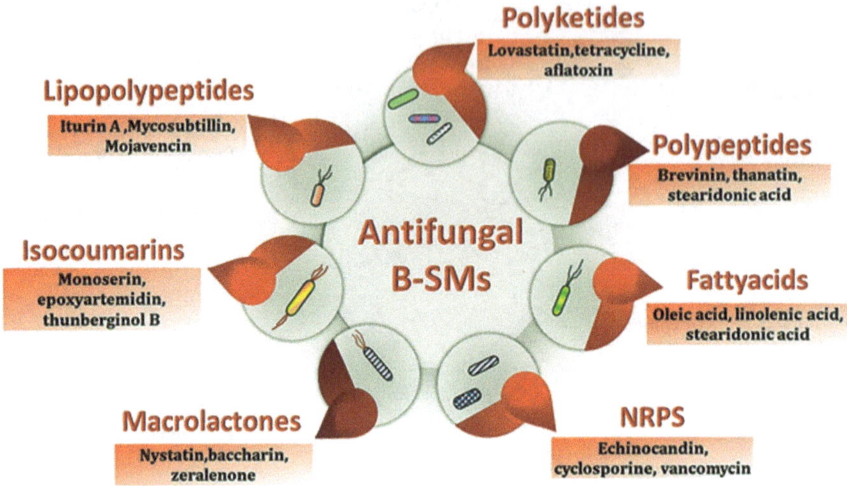

Fig. 12.2 Seven groups of antifungal secondary metabolites produced by bacteria

Warfarin (coumarin), Penicillin G (beta-lactam), Minocin (tetracycline), Vancomycin (glycopeptide), Phonol (anti-inflammatory), Amphotericin B (antifungal), Erythromycin (antibacterial), Abamectin (insecticidal), Temsirolimus (anticancer) and Epothilone B (anticancer) are some examples of bioactive macrolactones (Aldrich et al., 2005; Brzozowski & Wuest, 2017; Wu et al., 2019). Nevertheless, their dimensions and complicated assembly make them challenging to synthesize (Xue & Sherman, 2000; Zin et al., 2020).

12.2.3 Lipopolypeptides (LPs)

Lipopolypepetides comprise lipids and peptides assembled into a molecular structure (Hamley, 2015). Many bacterial species yield these molecules as a portion of their metabolism, specifically those of genera *Bacillus*, *Streptomyces*, and *Pseudomonas* (Coutte et al., 2017). A number of LPs are antibiotics, whereas additional LPs are receptor agonists (Hamley et al., 2014). Such examples of LPs include surfactin, iturin, and fengycin. Some LPs have solid antifungal and hemolytic properties (Maget-Dana & Peypoux, 1994). The activity of LPs is normally associated with the interactions of the plasma membrane (Nail Nasir et al., 2013). Ergosterol-containing monolayer constituents in the plasma membrane play an essential role in this interaction (Nasir & Besson, 2012). There is a general tendency, in the study of these molecules, considering that adding a lipid component of a particular length (i.e., C10–C12) to a lipopeptide will upsurge its antifungal activity (Meena & Kanwar, 2015). LPs with more C atoms (i.e., C-14 or C-16) in their lipid back are expected to characteristically show antifungal and antibacterial activity (Meena & Kanwar, 2015).

12.2.4 Fatty Acids

Fatty acid are generally joined in three clusters yielding a triglyceride molecule (Ollivier et al., 2003). Fatty acids deploy their antifungal activities by targeting diverse cellular functionalities comprising protein synthesis, fatty acid metabolism, and topoisomerase activity. However, fungal pathogens are less susceptible to develop resistance to antifungal fatty acids. The cell membrane is the most significant target of antifungal fatty acids (Pohl et al., 2011). Examples of antifungal fatty acids include auric, palmitic, linolenic, linoleic, oleic, stearic, and myristic acids (McGaw et al., 2002; Seidel & Taylor, 2004). Undecylenic acid is an unsaturated fatty acid that is the most commonly used antifungal agent also for human diseases, and is available in the market in the form of ointment, powder, or liquid (Naz et al., 2022; Saranraj et al., 2023).

12.2.5 Polyketides (PKs)

PKs are bio-active molecules isolated from many species of bacteria. Even though only a reduced amount of these antifungal compounds has been isolated by screening techniques, novel genomic approaches have delivered convincing data that the bio-synthesis of structurally varied PKs in nature is massive (Beletskaya et al., 2015; Oku et al., 2008). The bio-synthesis of PKs starts from acetyl-CoA, where the thioesters are catalyzed by polyketide synthase (PKS), an enzyme extremely homologous to the fatty acid synthase (FAS). Among the several types of PKs, type II PKS enzymes are mostly used to synthesize aromatic PKs, i.e., actinorhodin, tetracycline, and tetracenomycin (Korman et al., 2010). Many PKs are medically significant, with antifungal, antitumor, and immune-suppressive assets. Amphotericin B is a polyketide antifungal drug produced by *Streptomyces nodosus*, a gram-positive filamentous soil bacterium. It is thought to interact with fungal sterols to form transitory membrane-spanning channels that induce uncontrolled loss of ions and cellular components in fungi (Nikodinovic et al., 2003). Raman Microspectroscopy can detect the antibiotic amphotericin B produced by *Streptomyces nodosus* (Miyaoka et al., 2014). Moreover, they are involved in aiding antagonism for substrates and in communication among organisms (El-Bondkly, 2014).

12.2.6 Polypeptides

A polypeptide is a linear chain of more than 50 amino acids held together by peptide bonds. They are essential constituent of cells and tissues along with hormones, toxins, antibiotics, and enzymes (Hakim et al., 2015). Although some polypeptides possess primarily antifungal activity, i.e., lipopeptide (e.g., echinocandins) or histidine-rich (e.g., the linear histatins or branched HK), membrane-disrupting peptides (e.g., magainins, protegrins) possess a wider antimicrobial spectrum, including antifungal, antibacterial and antiviral activity (Fernández de Ullivarri et al., 2020). Their antifungal action includes inhibition of cell wall synthesis, depolarization of the cell membrane, inhibition of protein and nucleic acid synthesis (Reygaert, 2018). The antimicrobial mode of action of polypeptides depends upon the specific mechanism that relies on the type of microbial agent and polypeptide used.

12.2.7 Isocoumarins

Isocoumarins are the isomers of coumarins with an overturned lactone ring. Most natural isocoumarins possess a 3-alkyl (C1-C17) or a (un)substituted 3-phenyl ring on α-pyranone ring and 8-oxygenation on the benzene ring, due to their characteristic bio-synthetic origin (Patel et al., 2016; Saeed, 2016). They are abundant in microbes and in plants. The isocoumarins are considered for their amazing scaffold exploited in many pharmacological products (Saeed, 2016). Two new coumarin derivatives, 4,4′-dimethoxy-5,5′-dimethyl-7,7′-oxydicoumarin (1) and 7-(,-dimethylallyloxy)-5-methoxy-4-methylcoumarin were identified from roots of *Myoporum bontioides*. Their structures were determined using 1D and 2D NMR spectroscopy, and significantly suppressed *Fusarium oxysporum*, demonstrating antifungal properties (Wu et al., 2019).

12.2.8 Nonribosomal Peptides (NRPs)

These SMs are a class of peptide produced by a wide range of bacteria and fungi. Bacteria produce NRPs inside their cell using an enzyme known as nonribosomal peptide synthetase, found only in cyanobacteria, bacteria, and fungi (Oku et al., 2008; Sayari et al., 2019). Bio-chemical studies have shown the fundamental affinity of NRPs linked enzymes, with a much superior amount of catalytic tools for the combination of uncommon topographies, as compared to the ribosomal structure (Finking & Marahiel, 2004).

12.3 Characterisation of B-SMs

The nature and belongings of B-SMs are amplified by their physicochemical and biological characteristics. Reckoning impending applications and their deeds in a dissimilar environment may be performed by estimating the properties of secondary metabolites. The building block of metabolites may show physical similarities, however they can be distinguished through their association (linkage) patterns.

The characterization of B-SMs includes identifying the metabolite composition and its conformation, molecular weight, order and linkage pattern, chemical branching structure, and group elements (Lee, 2015). However, the structural complexity of B-SMs makes their characterization challenging to detect by a unique system. Thus, merging different methods is vital for their structural characterization (Pourakbar et al., 2021; Saranraj et al., 2023; Tyc et al., 2017) (Table 12.1).

Screening of antifungal bacteria is the first step to study and characterize particular B-SMs. Isolated bacteria are inoculated into respective medium for SMs production. However, fermentation conditions may directly affect SM synthesis with variances of up to 400-fold in the concentrations of SMs (Monaghan et al., 2011). Crude extraction can be performed by a rotary vapor device. Traditional and analytical methods could be used for characterization. Nowadays, to characterize and study SMs, omics approaches, i.e. metabolomics, are also very expedient (Chu et al., 2019).

Chromatographic techniques have been used to characterize metabolites for many years. Nevertheless, no single chromatographic technique is adequate to determine the particular metabolite present in the sample. Yet, the combination of chromatographic technique with mass analysis approaches has proven to be fruitful, i.e., LC-MS, GC-MS, HPLC-ES–MS, and more. TLC is performed to identify and purify the sub-fractions in the basic excerpt.

The SMs have proficient activities that can be further clarified and determined for further studies by other analytical techniques (Swathi et al., 2013). High-performance liquid chromatography (HPLC) analysis is used to estimate the number of SMs produced by microorganisms under various growth conditions, which is a less time-consuming although expensive procedure (approximately 1 min per assay). Where the measure of SMs production is centred on metabolite productivity, the index is computed from the electrospray mass spectra of samples injected right into a spectrometer (Higgs et al., 2001). The analysis HPLC- ELSD (evaporative light scattering detection) is preferred as it provides a quasi-universal estimation for the number of SMs confined in an extract. ELSD, therefore, is favored due to the non-selective nature of the detection of compounds. It provides compound facts regarding the molecular and ion fragments of SMs. This rapid technique to study SM productivity may be used to identify improved cultivation and extraction conditions by analyzing and classifying bulky collections of extracts. This method is also helpful in analyzing huge assemblies of extracts to classify subgroups of highly productive organisms for biological screening or additional study (Higgs et al., 2001).

Gas chromatography (GC) is the most appropriate system to determine the number of components and segments, in an intricate combination of volatile compounds (Padmanabhan, 2016). GC separates and detects tiny molecular weight molecules in the gas phase. The sample is vaporized in the injection port either as a gas or a liquid. Because it is difficult to vaporize bigger molecules, the chemicals studied are usually lower than 1000 Da. As it is exceedingly dependable and runs almost constantly, GC is ideal for environmental monitoring, industrial applications, and metabolites research (Jadhav, 2020). GC is commonly utilized in applications that require the detection of tiny, volatile compounds and non-aqueous solutions.

Table 12.1 Characterisation methods for antifungal B-SMs

Bacteria	B-BMs	Detection technique	Application	References
Bacillus endophyticus	Abscisic acid, salicylic acid	HPLC	Antifungal	Kousar et al. (2020)
Pseudomonas fluorescens N04, *P. koreensis* N19	Hydroxycinnamates, benzoates, flavonoids, glycoalkaloid	Ultra-high-performance liquid chromatography-mass spectrometry (UHPLC-MS)	Plants primed for enhanced defense responses	Mhlongo et al. (2020)
Pseudomonas chlororaphis O6	Phenazines and pyrrolnitrin	HPLC	Biocontrol of *Rhizoctonia solani* and *F. graminearum*	Park et al. (2011)
Pseudomonas fluorescens SS101	Cyclic lipopeptide massetolide A	Reverse-phase high-performance liquid chromatography (RP-HPLC)	Biocontrol of late blight caused by *Phytophthora infestans*	Tran et al. (2007)
Pseudomonas aeruginosa BRp3	1-hydroxy-phenazine	LCMS	Control of Bacterial Leaf Blight (BLB) in Basmati rice	Yasmin et al. (2017)
Paenibacillus alvei T19, *Lysinibacillus sphaericus* T22	Hydroxycinnamates, benzoates, flavonoids, glycoalkaloid	Ultra-high-performance liquid chromatography-mass spectrometry (UHPLC-MS)	Plants primed for enhanced defense responses	Mhlongo et al. (2020)
P. aeruginosa	Abscisic acid, salicylic acid	HPLC	Antifungal	Kousar et al. (2020)
Azotobacter chroococcum	Flavonoid content, phenolic content, and ABA	GC—flame ionization detector (GC-FID) and GC-MS	Drought resistance	Asghari et al. (2020)
B. amyloliquefaciens GA1	Cyclic lipopeptides surfactin, iturin A and fengycin	Mass spectrometry	Biocontrol agent, production of antibiotics	Arguelles-Arias et al. (2009)
Pseudomonas putida	Tropane alkaloids	GC-FID, GC-MS	Increased yield and water deficit stress induction	Ghorbanpour et al. (2013)
Leclercia adecarboxylata MO1	Indole-3-acetic-acid and ACC deaminase	GC-MS	Salinity stress tolerance	Kang et al. (2019)

(continued)

Table 12.1 (continued)

Bacteria	B-BMs	Detection technique	Application	References
Pseudomonas spp., *Azospirillum* spp.	Benzoxazinoids and diethyl phthalate	LC/MS	Enhanced total root surface, total root volume	Walker et al. (2012)
Azospirillum brasilense	Flavonoid content, phenolic content, and ABA	GC—flame ionization detector (GC-FID),GC-MS	Drought resistance	Asghari et al. (2020)
Bacillus amyloliquefaciens FZB42	Bacillomycin D, Difficidin, Macrolactin, and tripeptides	WGS	Biofertilizer and biocontrol potential	Chen Dekan and Schön (2009)
B. amyloliquefaciens NJN-6	Lipopeptides iturin A	HPLC	Suppressed *Fusarium* wilt	Yuan et al. (2013)
Pseudomonas fluorescens	Tropane alkaloids	GC-FID, GC-MS	Increased yield and water deficit stress induction	Ghorbanpour et al. (2013)
P. putida *P. fluorescens* *P. aeruginosa*	Phenazine and pyrrolnitrin	TLC	Antifungal activity against *Pythium ultimum*	Sharma et al. (2017)
P. fluorescens Pf 05	2,4-diacetyl-phloroglucinol (DAPG)	HPLC, NMR, MS	Antifungal metabolites against *Magnaporthe grisea* and *Rhizoctonia solani*	Prasanna Reddy et al. (2009)
Pseudomonas chlororaphis	WLIP and lahorenoic acid A production	HPLC	Biocontrol activity	Shahid et al. (2017)
Pseudomonas aurantiaca PB-St2	Lahorenoic acids A, B, C, phenazine-1-carboxylic acid, 2-hydroxy phenazine-1-carboxylic acid, and WLIP	1D and 2D NMR spectroscopic	Antifungal activity	Mehnaz et al. (2013)

Because the molecules do not need to evaporate, nonpolar molecules are separated using GC, whereas polar analytes may be separated using LC (Jadhav, 2020; Keyfi & Varasteh, 2016). When it is attached to mass spectrometry (GC-MS), additional information may be obtained about individually separated compound molecular mass and fundamental conformation. When high perseverance MS is used, efficient groups, and in definite belongings, molecular geometry and spatial isomerism may be identified.

FID is an excellent universal detector for organic compounds in GC, as it detects the quantity of C in a sample. Samples are then burnt in a hot hydrogen-air flame after they have passed through the column. Combustion releases C ions into the air. While the overall efficiency of the process is relatively low (only one out of every 10^5 carbon ions produces an ion in the flame), the total number of ions produced is proportionate to the amount of carbon in the sample. The current flowing through the ions is measured using electrodes. The entire sample is pyrolyzed, making FID a destructive detector. Non-flammable gases and water do not affect FID (Gnana Raja et al., 2014). This detector is highly sensitive to organic atoms (10–12 g/s = 1 pg/s, linear range, 10^6–10^7) but insensitive to tiny molecules such as N_2, NOx, H_2S, CO, CO_2, and H_2O. When the right amounts of hydrogen and air are mixed, the burning process produces no or few particles, resulting in a low background signal. Cations will be formed in the profluent stream if other carbon-containing substances are added to this stream. The more carbon atoms in a molecule, the more pieces are framed, and the more delicate the detector must be (Amagai et al., 2015; Napa et al., 2015).

12.4 Future Prospects

The predominance of drug-resistant fungal pathogens is a foremost universal health challenge. Therefore, innovative drugs with compelling antifungal activity and resistance are required. Considering this scenario for agriculture and soil health, organic farming is the green requirement to overcome the sturdiness of soil health and indirectly of human health. It allows farmers to overcome the side effects of chemical pesticides or chemical agents for disease management in agriculture. Antifungal B-SMs act as a biologibal resource to cure numerous kinds of fungal diseases by directly inhibiting or killing fungal pathogens. In this era, where organic farming is a central charm, only a few antifungal metabolites are known, and many secondary metabolites with biocontrol efficacy are yet to be explored. Combining one or more analytical methods with a next-generation -omics approach could be helpful to explore B-SM diversity and their molecular mechanisms.

12.5 Conclusion

B-SMs are vital bioactive natural substances that are directly synthesized from many bacteria. Most agriculturally essential SMs are derived from rhizobacteria, i.e., species of *Pseudomonas*, *Bacillus*, *Serratia*, *Micrococcus* and others. Actually, large amounts of crops yield are destroyed by fungal diseases and cured by chemical fungicide. These agents directly affect soil health and indirectly harm human health by the consumption of chemicals. B-SMs have a huge potential as biocontrol agents that can cure fungal diseases or inhibit the fungal pathogens on a particular crop. Moreover, B-SMs used as antifungal agents have proved also to be significant antibacterial, antiviral, and antinematodes. Some of the B-SMs, i.e., IAA, gibrellins, siderophore, etc., are also plant-promoting agents which directly or indirectly boost plant health.

Acknowledgments We gratefully acknowledge SODH (Scheme of developing high-quality research, Government of Gujarat, India), for fellowship, SSIP (student start-up & innovation policy, Government of Gujarat, India), and support from the Microbiology Department, Gujarat Arts and Science College, Ahmedabad, India.

Conflict of Interest The Authors declare no conflict of interest.

References

Aldrich, C. C., Venkatraman, L., Sherman, D. H., & Fecik, R. A. (2005). Chemoenzymatic synthesis of the polyketide macrolactone 10-deoxymethynolide. *Journal of the American Chemical Society, 127*, 8910–8911.

Amagai, T., Bai, H., Wang, Q., Miyake, Y., Noguchi, M., & Nakai, S. (2015). Determination of nicotine exposure using passive sampler and high performance liquid chromatography. *Pharmaceutica Analytica Acta, 6*, 7.

Arguelles-Arias, A., Ongena, M., Halimi, B., Lara, Y., Brans, A., Joris, B., & Fickers, P. (2009). Bacillus amyloliquefaciens GA1 as a source of potent antibiotics and other secondary metabolites for biocontrol of plant pathogens. *Microbial Cell Factories, 8*, 1–12.

Asghari, B., Khademian, R., & Sedaghati, B. (2020). Plant growth promoting rhizobacteria (PGPR) confer drought resistance and stimulate biosynthesis of secondary metabolites in pennyroyal (*Mentha pulegium* L.) under water shortage condition. *Scientia Horticulturae, 263*, 109132.

Beletskaya, T., Collins, C., & Murray, P. (2015). Polyketides as a source of chemical diversity. *Phycotoxins: Chemistry and Biochemistry, 2*, 381–405.

Brzozowski, R. S., & Wuest, W. M. (2017). Twelve-membered macrolactones: Privileged scaffolds for the development of new therapeutics. *Chemical Biology & Drug Design, 89*, 169–191.

Chen Dekan, X., & Schön, L. H. (2009). *Whole genome analysis of the plant growth-promoting Rhizobacteria Bacilllus amyloliquefaciens FZB42 with focus on its secondary metabolites*. Dissertation,. Humboldt-Universität. https://doi.org/10.18452/16095

Chu, S. H., Huang, M., Kelly, R. S., Benedetti, E., Siddiqui, J. K., Zeleznik, O. A., et al. (2019). Integration of metabolomic and other omics data in population-based study designs: An epidemiological perspective. *Metabolites, 9*, 117.

Coutte, F., Lecouturier, D., Dimitrov, K., Guez, J. S., Delvigne, F., Dhulster, P., & Jacques, P. (2017). Microbial lipopeptide production and purification bioprocesses, current progress and future challenges. *Biotechnology Journal, 12*(7), 1600566.

El-Bondkly, A. M. A. (2014). Sequence analysis of industrially important genes from Trichoderma. *Biotechnology and Biology of Trichoderma*, 377–392.

Fernández de Ullivarri, M., Arbulu, S., Garcia-Gutierrez, E., & Cotter, P. D. (2020). Antifungal peptides as therapeutic agents. *Frontiers in Cellular and Infection Microbiology, 10*, 518678.

Finking, R., & Marahiel, M. A. (2004). Biosynthesis of nonribosomal Peptides1. *Annual Review of Microbiology, 58*, 453–488.

Ghorbanpour, M., Hatami, M., & Khavazi, K. (2013). Role of plant growth promoting rhizobacteria on antioxidant enzyme activities and tropane alkaloid production of *Hyoscyamus Niger* under water deficit stress. *Turkish Journal of Biology, 37*, 350–360.

Gnana Raja, M., Geetha, G., & Sankaranarayanan, A. (2014). A concise study of organic volatile impurities in ten different marketed formulations by [GC/HS-FID/MS] gas chromatography technique. *Journal of Analytical & Bioanalytical Techniques, 5*, 4.

Goh, F., Zhang, M. M., Lim, T. R., Low, K. N., Nge, C. E., Heng, E., et al. (2020). Identification and engineering of 32 membered antifungal macrolactone notonesomycins. *Microbial Cell Factories, 19*, 1–13.

Hakim, M. A., Yang, S., & Lai, R. (2015). Centipede venoms and their components: Resources for potential therapeutic applications. *Toxins, 7*, 4832–4851.

Hamley, I. W. (2015). Lipopeptides: From self-assembly to bioactivity. *Chemical Communications, 51*, 8574–8583.

Hamley, I. W., Kirkham, S., Dehsorkhi, A., Castelletto, V., Reza, M., & Ruokolainen, J. (2014). Toll-like receptor agonist lipopeptides self-assemble into distinct nanostructures. *Chemical Communications, 50*, 15948–15951.

Higgs, R. E., Zahn, J. A., Gygi, J. D., & Hilton, M. D. (2001). Rapid method to estimate the presence of secondary metabolites in microbial extracts. *Applied and Environmental Microbiology, 67*, 371–376.

Jadhav, C. A. (2020). A review on gas chromatography. *World Journal of Pharmaceutical Research, 9*, 1246.

Jadhav, H. P., Shaikh, S. S., & Sayyed, R. Z. (2017). Role of hydrolytic enzymes of rhizoflora in biocontrol of fungal phytopathogens: An overview. In S. Mehnaz (Ed.), *Rhizotrophs: Plant growth promotion to bioremediation* (pp. 183–203). Springer.

Kadiri, M., Sevugapperumal, N., Nallusamy, S., Ragunathan, J., Ganesan, M. V., Alfarraj, S., Ansari, M. J., Sayyed, R. Z., Lim, H. R., & Show, P. L. (2023). Pan-genome analysis and molecular docking unveil the biocontrol potential of Bacillus velezensis VB7 against Phytophthora infestans. *Microbiological Research, 268*, 127277.

Kang, S. M., Shahzad, R., Bilal, S., Khan, A. L., Park, Y. G., Lee, K. E., Asaf, S., Khan, M. A., & Lee, I. J. (2019). Indole-3-acetic-acid and ACC deaminase producing *Leclercia adecarboxylata* MO1 improves *Solanum lycopersicum* L. growth and salinity stress tolerance by endogenous secondary metabolites regulation. *BMC Microbiology, 19*, 1–14.

Keyfi, F., & Varasteh, A. (2016). Development and validation of a GC-FID method for diagnosis of Methylmalonic Acidemia. *Reports of Biochemistry and Molecular Biology, 4*, 104–109.

Korman, T. P., Ames, B., & Tsai, S. C. (2010). 1.08 — Structural enzymology of polyketide synthase: The structure–sequence–function correlation. In L. Mander & H. W. Liu (Eds.), *Comprehensive natural products II: Chemistry and biology* (pp. 305–345). Elsevier.

Kousar, B., Bano, A., & Khan, N. (2020). PGPR modulation of secondary metabolites in tomato infested with *Spodoptera litura*. *Agronomy, 10*, 778.

Kumar, D., Kumar, S., & Kumar, A. (2021). Extraction and characterization of secondary metabolites produced by bacteria isolated from industrial wastewater. *Journal of Water Process Engineering, 40*, 101811.

Lee, M. K. (2015). *Single-molecule and super-resolution microscopy of bacterial cells*. PhD Dissertation, Stanford University. http://purl.stanford.edu/dt810gc1403

Maget-Dana, R., & Peypoux, F. (1994). Iturins, a special class of pore-forming lipopeptides: Biological and physicochemical properties. *Toxicology, 87*, 151–174.

McGaw, L. J., Jäger, A. K., & Van Staden, J. (2002). Isolation of antibacterial fatty acids from *Schotia brachypetala*. *Fitoterapia, 73*, 431–433.

Meena, K. R., & Kanwar, S. S. (2015). Lipopeptides as the antifungal and antibacterial agents: Applications in food safety and therapeutics. *BioMed Research International, 2015*, 473050.

Mehnaz, S., Saleem, R. S. Z., Yameen, B., Pianet, I., Schnakenburg, G., Pietraszkiewicz, H., et al. (2013). Lahorenoic acids A-C, ortho-dialkyl-substituted aromatic acids from the biocontrol strain pseudomonas aurantiaca PB-St2. *Journal of Natural Products, 76*, 135–141.

Mhlongo, M. I., Piater, L. A., Steenkamp, P. A., Labuschagne, N., & Dubery, I. A. (2020). Metabolic profiling of PGPR-treated tomato plants reveal priming-related adaptations of secondary metabolites and aromatic amino acids. *Metabolites, 10*, 210.

Miyaoka, R., Hosokawa, M., Ando, M., Mori, T., Hamaguchi, H. O., & Takeyama, H. (2014). In situ detection of antibiotic amphotericin B produced in *Streptomyces nodosus* using Raman microspectroscopy. *Marine Drugs, 12*, 2827–2839.

Monaghan, R. L., Polishook, J. D., Pecore, V. J., Bills, G. F., Nallin-Omstead, M., & Streicher, S. L. (2011). Discovery of novel secondary metabolites from fungi—Is it really a random walk through a random forest? *Canadian Journal of Botany, 73*(S1), 925–931.

Nail Nasir, M., Besson, F., & Deleu, M. (2013). Interactions des antibiotiques ituriniques avec la membrane plasmique. Apport des systèmes biomimétiques des membranes (synthèse bibliographique). *BASE (Biotechnology, Agronomy, Society and Environment), 17*, 505–516. https://popups.uliege.be/1780-4507/index.php?id=10280

Napa, D., Delhiraj, N., & Anbazhagan, S. (2015). A simple, isocratic and ultra-fast liquid chromatography/mass spectrometry method for the estimation of barnidipine in human plasma. *Pharmaceutica Analytica Acta, 6*, 7.

Nasir, M. N., & Besson, F. (2012). Interactions of the antifungal mycosubtilin with ergosterol-containing interfacial monolayers. *Biochimica et Biophysica Acta (BBA) – Biomembranes, 1818*, 1302–1308.

Naz, R., Khushhal, S., Asif, T., Mubeen, S., Saranraj, P., & Sayyed, R. Z. (2022). Inhibition of bacterial and fungal phytopathogens through volatile organic compounds produced by *Pseudomonas* sp. In R. Z. Sayyed & V. G. Uarrota (Eds.), *Secondary metabolites and volatiles of PGPR in plant-growth promotion* (pp. 95–118). Springer. https://doi.org/10.1007/978-3-031-07559-9_6

Nikodinovic, J., Barrow, K. D., & Chuck, J. A. (2003). High frequency transformation of the amphotericin-producing bacterium *Streptomyces nodosus. Journal of Microbiological Methods, 55*, 273–277.

Oku, N., Adachi, K., Matsuda, S., Kasai, H., Takatsuki, A., & Shizuri, Y. (2008). Ariakemicins A and B, novel polyketide-peptide antibiotics from a marine gliding bacterium of the genus *Rapidithrix. Organic Letters, 10*, 2481–2484.

Ollivier, D., Artaud, J., Pinatel, C., Durbec, J. P., & Guérère, M. (2003). Triacylglycerol and fatty acid compositions of French virgin olive oils. Characterization by chemometrics. *Journal of Agricultural and Food Chemistry, 51*, 5723–5731.

Padmanabhan, A. S. (2016). A comparative GC-MS analysis of bacterial secondary metabolites of *Pseudomonas* species. *The Pharma Innovation Journal, 5*, 84–89.

Park, J. Y., Oh, S. A., Anderson, A. J., Neiswender, J., Kim, J. C., & Kim, Y. C. (2011). Production of the antifungal compounds phenazine and pyrrolnitrin from *Pseudomonas chlororaphis* O6 is differentially regulated by glucose. *Letters in Applied Microbiology, 52*, 532–537.

Parkinson, E. I., Bair, J. S., Nakamura, B. A., Lee, H. Y., Kuttab, H. I., Southgate, E. H., et al. (2015). Deoxynybomycins inhibit mutant DNA gyrase and rescue mice infected with fluoroquinolone-resistant bacteria. *Nature Communications, 6*, 1–9.

Patel, S., Sayyed, R. Z., & Saraf, M. (2016). Bacterial determinants and plant defense induction: Their role as biocontrol agents in sustainable agriculture. In K. Hakeem & M. Akhtar (Eds.), *Plant, soil and microbes. Volume 2: Mechanisms and molecular interactions* (pp. 187–204). Springer. https://doi.org/10.1007/978-3-319-29573-2_9

Pohl, C. H., Thibane, V. S., Kock, J. L. F., & Thibane, V. S. (2011). Antifungal free fatty acids: A review. In A. Méndez-Vilas (Ed.), *Science against microbial pathogens: Communicating current research and technological advances* (*FORMATEX microbiology series 3*) (pp. 61–71).

Pourakbar, L., Moghaddama, S. S., El Enshasy, H. A., & Sayyed, R. Z. (2021). Antifungal activity of the extract of a macroalgae, *Gracilariopsis persica*, against four plant pathogenic fungi *in vitro*. *Plants, 10*, 1781.

Prasanna Reddy, B., Reddy, M. S., & Vijay Krishna Kumar, K. (2009). Characterization of antifungal metabolites of *Pseudomonas fluorescens* and their effect on mycelial growth of *Magnaporthe grisea* and *Rhizoctonia solani*. *International Journal of PharmTech Research–CODEN, 1*, 1490–1493.

Reygaert, W. C. (2018). An overview of the antimicrobial resistance mechanisms of bacteria. *AIMS Microbiology, 4*, 482–501.

Saeed, A. (2016). Isocoumarins, miraculous natural products blessed with diverse pharmacological activities. *European Journal of Medicinal Chemistry, 116*, 290–317.

Saranraj, P., Sayyed, R. Z., Kokila, M., Sudha, A., Sivasakthivelan, P., Devi, M. D., Naz, R., & Yasmin, H. (2022). Plant growth-promoting and biocontrol metabolites produced by endophytic *Pseudomonas fluorescence*. In R. Z. Sayyed & V. G. Uarrota (Eds.), *Secondary metabolites and volatiles of PGPR in plant-growth promotion* (pp. 349–381). Springer. https://doi.org/10.1007/978-3-031-07559-9_18

Saranraj, P., Sayyed, R. Z., Kokil, M., Salomi, V., Sivasakthivelan, P., Manigandan, M., & Mawar, R. (2023). Evolving concepts of biocontrol of phytopathogens by endophytic *Pseudomonas fluorescence*. In R. Mawar, R. Z. Sayyed, S. K. Sharma, & K. S. Sattiraju (Eds.), *Plant growth promoting microorganisms of arid region* (pp. 365–388). Springer. https://doi.org/10.1007/978-981-19-4124-5_17

Sayari, M., Van der Nest, M. A., Steenkamp, E. T., Soal, N. C., Wilken, P. M., & Wingfield, B. D. (2019). Distribution and evolution of nonribosomal peptide synthetase gene clusters in the Ceratocystidaceae. *Genes, 10*, 328.

Seidel, V., & Taylor, P. W. (2004). *In vitro* activity of extracts and constituents of *Pelagonium* against rapidly growing mycobacteria. *International Journal of Antimicrobial Agents, 23*, 613–619.

Sevugapperumal, N., Chinnaiah, S., Renukadevi, P., Nakkeeran, S., Saranya, N., Senthilraja, C., et al. (2021). Mining the genome of *Bacillus velezensis* VB7 (CP047587) for MAMP genes and non-ribosomal peptide synthetase gene clusters conferring antiviral and antifungal activity. *Microorganisms, 9*, 2511. https://doi.org/10.3390/microorganisms9122511

Shahid, I., Rizwan, M., Baig, D. N., Saleem, R. S., Malik, K. A., & Mehnaz, S. (2017). Secondary metabolites production and plant growth promotion by *Pseudomonas chlororaphis* and *P. aurantiaca* strains isolated from cactus, cotton, and Para grass. *Journal of Microbiology and Biotechnology, 27*, 480–491.

Sharma, P., Verma, P. P., & Kaur, M. (2017). Identification of secondary metabolites produced by fluorescent pseudomonads for controlling fungal pathogens of apple. *Indian Phytopathology, 70*, 452–456.

Suriani, N. L., Suprapta, D. N., Nazir, N., Parwanayoni, N. M. S., Darmadi, A. A. K., Dewi, D. A., et al. (2020). A mixture of piper leaves extracts and rhizobacteria for sustainable plant growth promotion and bio-control of blast pathogen of organic Bali rice. *Sustainability, 12*, 8490. https://doi.org/10.3390/SU12208490

Swathi, J., Narendra, K., Sowjanya, K. M., & Krishna, S. A. (2013). Biological characterisation of secondary metabolites from marine fungi microascus sps. *International Journal of Research in Pharmaceutical and Biomedical sciences, 4*, 754–759.

Tran, H., Ficke, A., Asiimwe, T., Höfte, M., & Raaijmakers, J. M. (2007). Role of the cyclic lipopeptide massetolide A in biological control of *Phytophthora infestans* and in colonization of tomato plants by *Pseudomonas fluorescens*. *The New Phytologist, 175*, 731–742.

Tyc, O., Song, C., Dickschat, J. S., Vos, M., & Garbeva, P. (2017). The ecological role of volatile and soluble secondary metabolites produced by soil bacteria. *Trends in Microbiology, 25*, 280–292.

Walker, V., Couillerot, O., von Felten, A., Bellvert, F., Jansa, J., Maurhofer, M., et al. (2012). Variation of secondary metabolite levels in maize seedling roots induced by inoculation with *Azospirillum*, *Pseudomonas* and *Glomus* consortium under field conditions. *Plant & Soil, 356*, 151–163.

Wu, Z., Chen, J., Zhang, X., Chen, Z., Li, T., She, Z., et al. (2019). Four new isocoumarins and a new natural tryptamine with antifungal activities from a mangrove endophytic fungus *Botryosphaeria ramosa* L29. *Marine Drugs, 17*, 88.

Xue, Y., & Sherman, D. H. (2000). Alternative modular polyketide synthase expression controls macrolactone structure. *Nature, 403*(6769), 571–575.

Yasmin, S., Hafeez, F. Y., Mirza, M. S., Rasul, M., Arshad, H. M. I., Zubair, M., & Iqbal, M. (2017). Biocontrol of bacterial leaf blight of rice and profiling of secondary metabolites produced by rhizospheric *Pseudomonas aeruginosa* BRp3. *Frontiers in Microbiology, 8*, 1895.

Yuan, J., Ruan, Y., Wang, B., Zhang, J., Waseem, R., Huang, Q., & Shen, Q. (2013). Plant growth-promoting rhizobacteria strain *Bacillus amyloliquefaciens* NJN-6-enriched bio-organic fertilizer suppressed Fusarium wilt and promoted the growth of banana plants. *Journal of Agricultural and Food Chemistry, 61*, 3774–3780.

Zin, P. P. K., Williams, G. J., & Ekins, S. (2020). Cheminformatics analysis and modeling with Macrolactone DB. *Scientific Reports, 10*, 1–12. https://doi.org/10.1038/s41598-020-63192-4

Chapter 13
Fungal Bio-stimulants: Cutting-Edge Bioinoculants for Sustainable Agriculture

M. Chaithra, K. Nishmitha, Pooja S. Patel, Debasis Mitra, R. Gangaraj, and R. Z. Sayyed

Abstract A more sustainable and agriculturally productive system is becoming increasingly necessary to maintain soil fertility and reduce the loss of biodiversity. Microbial bio-stimulants are cutting-edge technologies that can guarantee a high nutritional value in agricultural yield while overcoming the negative effects of environmental changes. This chapter presents an overview of fungal plant bio-stimulants and of their use in agriculture. Arbuscular Mycorrhizal Fungi (AMF) and *Trichoderma* are the two most important fungi that act as bio-stimulants in agriculture. Because of the ability of AMFs to function as bio-fertilizers, bio-protectors, or bio-degraders, these connections may affect plant productivity. Mycorrhizal fungi affect the diversity and quality of rhizosphere microflora, which changes the microbial activity in the rhizosphere as a whole. The benefits of synergistic interactions between AMF and other rhizosphere microorganisms, including rhizobia, non-symbiotic diazotrophs, phytohormone producers, and phosphate solubilizers, are

M. Chaithra (✉)
ICAR-Central Plantation Crop Research Institute, Research Center, Kahikuchi, Guwahati, Assam, India
e-mail: chaithra.m1@icar.gov.in

K. Nishmitha
Division of Plant Pathology, ICAR-Indian Agricultural Research Institute, New Delhi, India

P. S. Patel
Department of Plant Pathology, College of Agriculture, V. C. Farm, Mandya, Karnataka, India

D. Mitra
Department of Microbiology, Graphic Era (Deemed to be University), Dehradun, Uttarakhand, India

R. Gangaraj
ICAR-Indian Institute of Vegetable Research, Regional Research Station, Sargatia, Kushinagar, Uttar Pradesh, India
e-mail: gangaraj.r@icar.gov.in

R. Z. Sayyed
Department of Microbiology, PSGVP Mandal's S. I. Patil Arts, G. B. Patel Science and STKV Sangh Commerce College, Shahada, India

P. Mathur, S. Roy (eds.), *Plant Microbiome and Biological Control*, Sustainability in Plant and Crop Protection 20,
https://doi.org/10.1007/978-3-031-75845-4_13

widely acknowledged. The function of AMF in protecting plants from phytopathogenic bacteria and fungi is important. Despite these advantages, interactions between AMF and rhizosphere microflora in agriculture, horticulture, and forestry remain underutilized, partly because of the limitations imposed by their obligate biotrophic conditions and the absence of suitable technology for large-scale production. Nevertheless, further research is necessary, because the use of AMF-rhizosphere microflora combinations appears to be a potential method for promoting plant development against diseases and other environmental challenges.

Keywords Arbuscular mycorrhizal fungi · Beneficial microorganisms · Microbial bio-stimulants · Rhizosphere · *Trichoderma*

Abbreviations

AMF	Arbuscular mycorrhizal fungi
PSP	Phosphorous Solubilizing Bacterium
ABA	Abscisic acid
PGPR	Plant growth promoting rhizobacteria
IAA	Indole-3-acetic acid
JA	Jasmonic acid
SA	Salicylic acid

13.1 Introduction

Uncontrolled and excessive use of chemical inputs to boost agricultural output damages the soil and threatens the ecosystem, posing a major risk to human health (Nellemann et al., 2009). Therefore, it is crucial to transit to modern agriculture and search for new biotechnological developments that permit a reduction in the usage of chemical inputs without affecting crop yields and farmer revenue (Chbani et al., 2013). To maintain food biosecurity in a sustainable way, recent initiatives have focused on producing high-quality, nutrient-rich food using naturally available organisms/substances, called bio-stimulants, to control disease and enhance yields (Hamid et al., 2021; Basu et al., 2021; Gowtham et al., 2024; Raja, 2013; Benfatto et al., 2015).

Plant bio-stimulants are substances other than fertilizers that, when used appropriately may influence plant growth and ameliorate biotic and abiotic stress (Du Jardin, 2015; Al-Turki et al., 2023). Recently, many studies emphasized the use of bio-stimulants as an alternative method of plant disease control due to their non-toxic nature and cost-effectiveness (Lambert, 2006). According to Yakhin et al. (2017), bio-stimulants increase plant productivity due to emergent properties brought on by their constituent complexes, rather than as a sole result of the presence of essential nutrients, plant growth regulators, or plant protective compounds. Bio-stimulants also assist in lowering greenhouse gas emissions (through the better use of nitrogen). Through improved nutrient absorption, bio-stimulants aid in the preservation of non-renewable resources.

According to Calvo et al. (2014), bio-stimulants are not regarded as fertilizers because they do not provide sufficient nutrients. When used sparingly, bio-stimulants are substances other than fertilizers that stimulate plant growth (Kauffman et al., 2007). They can be categorized based on the effects they produce, their role, and the origin and nature of the materials employed in their production. Torre et al. (2016) identified five bio-stimulant categories: humic substances, algae extracts, protein and amino acid hydrolysates, inorganic salts, and microorganisms (beneficial bacteria and fungi). Yakhin et al. (2017) categorized humic compounds and PGPR as bio-fertilizers, phytostimulants, and bio-pesticides, respectively. According to Du Jardin (2015), bio-stimulants are a subset of bio-fertilizers. Other authors view microorganisms and algal extracts as bio-fertilizers (Torre et al., 2016). According to Alabouvette and Cordier (2018), single-strain fungi or bacteria, multiple fungi, multiple bacteria, and multiple fungi are examples of microbial bio-stimulants. The application of bio-stimulants encourages plant growth and crop yield by enhancing nutrient acquisition, phytohormone production, alleviating abiotic stress and inducing systemic resistance (Yakhin et al., 2017) (Fig. 13.1).

The use of bio-agents presents a viable and reliable substitute for traditional fungicides in the control of diseases. Consequently, the search for environmentally friendly alternatives to mitigate the widespread reliance on chemical pesticides has stimulated interest in the biological containment of plant diseases. Microbial biocontrol agents protect plants against diverse diseases through mechanisms such as antagonistic activity, competitive inhibition, and the synthesis of antimicrobial biomolecules (Kohl et al., 2019). For example, the application of *Aureobasidium pullulans* and *Bacillus amyloliquefaciens* reduced *Verticillum* wilt of olives by reducing mycelial growth rate by direct antagonism (Lopez-Moral et al., 2021). Tortora et al. (2012) showed that *Azospirillum brasilense* induced systemic resistance in strawberry plants against *Colletotrichum acutatum*.

Fig. 13.1 Fungal plant bio-stimulant and their effects on plants

Table 13.1 Effects of inoculation with plant bio-stimulants, on the agronomical, physiological, and biochemical performance of crops

Biostimulants	Crops	Properties	Functions	References
G. fasciculatum and *Rhizobium meliloti*	Alfalfa	Drought tolerance	Delay in leaf senescence and maintained or enhanced the number of stems under drought-stressed conditions	Giocoechea et al. (1995)
AMF	Onion and pepper	Salinity tolerance	Enhancement of plant growth in saline soil	Hirrel and Gerdemann (1980)
AMF and rhizobia	Legume plants	Soil disturbance tolerance	Enhanced soil fertility and quality	Requena et al. (2000)
AMF and rhizobacteria	Alfalfa	Toxicpollutants tolerance	Growth and heavy metal uptake by alfalfa	El-Kherbawy et al. (1989)
G. mosseae	Soybean	Bacterial disease control	Reduced *P. syringae* infection	Shalaby and Hanna (1998)
G. intraradices	Tomato	Fungal disease control	Reduced *Fusarium* crown and root rot of tomato	Caron et al. (1986)

In this chapter we discuss various microbial bio-stimulants, the interaction of mycorrhiza with other organism, and their influence on growth, emphasizing plant disease control (Table 13.1).

13.2 Microbial Plant Bio-stimulants

Several Plant Growth Promoting Rhizosphere (PGPR) species (*Azotobacter*, *Azospirillum*, and *Rhizobium* spp.) and fungal bio-stimulants, such as *Trichoderma* spp. and various endophytic fungi including mycorrhizal species, have been considered as microbial plant bio-stimulants for promising ways to ensure crop productivity under nutrient-challenged conditions (*i.e.*, N and/or P deficiency). These plant microbial bio-stimulants must be safe for the environment and effective at very low doses. These microorganisms, whether bacteria or fungi, comprise the rhizospheric, epiphytic, or endophytic microbiota, which can be employed as plant bio-stimulants and health promoters. Along with their effects on human and animal health, their activities are also supported by yet-to-be-understood genetic connections between plants and microbial species.

According to numerous studies, both plants and fungi coevolved with terrestrial plants, supporting the idea that the parasitism-mutualism continuum is advantageous for reducing ground and surface water contamination and boosting crop growth and productivity (Johnson & Graham, 2013). The principal mycorrhizal and *Trichoderma* spp. fungi used as fungal bio-stimulants have been shown to boost plant growth and nutrient uptake, improving plant defense in ecologically friendly and sustainable ways, thereby increasing crop productivity (Johnson & Gilbert, 2015) (Table 13.2).

Table 13.2 Bio-stimulants mediating plant disease control

Biostimulants	Disease/pathogen	Crop	References
Trichoderma harzianum T-22y	*Alternaria* spp., *Fusarium* spp., *Phytophthora* sp., *Neocosmospora solani*, *Rhizoctonia solani* and *Sclerotinia sclerotiorum*	Carrot	Patkowska et al. (2020)
Trichoderma harzianum G 227	*Fusarium* spp., *Phoma exigua* var. *exigua*, *R. solani* and *S. sclerotiorum*	Soybean	Pastucha and Patkowska (2005)
Trichoderma viride	*Rhizoctonia solani*	Soybean root rot	Haikal (2008)
T. harzianum	*Sclerotinia sclerotiorum*	Soybean stem rot	Rollan et al. (2007)
Glomus intraradices and *Glomus* spp.	*Fusarium oxysporum* f. sp. *cubense*	Banana	Jaizme-Vega et al. (1998)
G. versiforme	*Verticillium dahliae* and *Pseudomonas lacrymans*	Eggplant and cucumber	Li et al. (1997)
G. fasciculatum, *Gigaspora margarita*, *Acaulospora laevis*, and *Sclerocystis dussii*	*Sclerotium rolfsii*	Groundnut	Kulkarni et al. (1997)
G. fasciculatum and *Trichoderma harzianum*	Damping-off caused by *Pythium aphanidermatum* and black shank disease caused by *P. parasitica* var. *nicotianae*	Tobacco	Sreeramulu et al. (1998)
G. fasciculatum and *T. viride*	Root rot pathogen *Macrophomina phaseolina*	Safflower	Prashanthi et al. (1997)

13.2.1 Trichoderma

Several species from genus *Trichoderma* are becoming increasingly important in agriculture as plant microbial bio-stimulants (Simard et al., 2012). In addition, *Trichoderma* strains have demonstrated their bio-stimulatory function by promoting plant growth and developmental processes in harsh environments, leading to improved nutritional quality and yield (Saboor et al., 2021; Bastami et al., 2021; Najafi et al., 2021). Bio-priming maize seeds with *Trichoderma harzianum* alleviated cold stress in maize by increasing seedling emergence with higher chlorophyll a, b, and carotenoid contents. *Trichoderma* increased the production of cytokinin leading to stable chlorophyll synthesis in presence of a cold stress. Increased production of polyphenol oxidase and proline in bio-primed plants is an indicator that *Trichoderma* can mitigate cold stress in maize (Afrouz et al., 2023). Numerous studies revealed that *Trichoderma* spp. enhance root-to-shoot communication to perform their bio-stimulatory function. Additionally, they have been shown to increase plant productivity and growth by promoting the manufacture of several phytochemicals, including indole-3-acetaldehyde, indole-3-ethanol, and other volatile compounds, such as indole-3-carboxaldehyde. All these phytochemicals are known to have auxin-like properties that improve nutrients uptake, absorption, and solubilization. This boosts the root, shoot, and branch architecture, which improves the overall plant productivity (Colla et al., 2015).

Trichoderma stands out as the most widely employed biological control agent (BCA) against plant pathogens, earning a reputation as an effective antagonist of fungi causing plant diseases (Prasad et al., 2018; Abbas et al., 2017; Błaszczyk et al., 2014; Leelavathi et al., 2014). *Trichoderma* spp. have been reported to counteract the negative effects of various plant pathogens, including phytopathogenic fungi, and some bacterial diseases (Candido et al., 2013). Studies have shown that these fungi deploy secretions of antifungal metabolites, hydrolytic enzymes, boosting nutrient competition, mycoparasitism by coiling around the host, and developing appressoria to control pathogens (Zeilinger & Omann, 2007). These processes are stimulated by the biosynthesis of target metabolites, such as plant growth regulators, antibiotics, siderophores and lytic enzymes (especially chitinases, β-1,3-glucanases, β-1,6-glucanases), which completely degrade the cell walls of hyphae and spores (Sood et al., 2020; and Chet et al., 1981). The significant interest in the properties of *Trichoderma* isolates and the potential for their utilization in agriculture has led to the development of commercial products incorporating various *Trichoderma* species (Lopez-Bucio et al., 2015).

Trichoderma spp. are among the most studied and widely used sources of fungal bio-stimulants because of their bio-pesticide and bio-control properties, which help plants better withstand a variety of biotic and abiotic challenges. Various biotechnological enterprises have also used *Trichoderma* spp. commercially as a valuable source of enzymes (Nicolas et al., 2014). Therefore, using better crop management techniques, more plants should be treated with microorganisms to promote the interactions of beneficial fungi and utilize them to their maximum potential.

Trichoderma spp. also release bioactive fungal compounds that provide nutrients to the host plant (Behie & Bidochka, 2014). The bio-control and bio-pesticidal properties of these compounds have also been demonstrated (Nicolas et al., 2014). Numerous species of *Trichoderma* such as *T. harzianum, T. viridae, T. virens* and *T. asperellum* are reported to control *R. solani*, *Sclerotium rolfsii*, *Pythium aphanidermatium*, *Gaeumannomyces graminis* var. *tritici*, *Verticillium dahliae*, *Fusarium oxysporum*, and *Fusarium culmorum*, both under greenhouse and field conditions (Tyskiewicz et al., 2022). Similarly, Chaithra and Pankaja (2020) reported the antagonistic potential of different *Trichoderma* spp. against soil-borne, fungal plant pathogens. Different *Trichoderma* spp. isolated from soil and agricultural wastes, collected from different geographical locations, exhibited the highest potential against foliar fungal plant pathogens (Chaithra et al., 2022).

Application of *Trichoderma* isolates to vegetable crops enhanced organ growth, nutrient usage effectiveness, and resistance to abiotic stress (Colla et al., 2015). *Trichoderma* spp. have been reported to play a variety of roles in soil bioengineering, inducing systemic resistance in plants, improving rhizospheric competence *vs* soil-based competitors, producing siderophores, secreting many additional cellulosic enzymes, with the ability to neutralize toxic organic compounds of primary origin and secondary metabolites as well (Srivastava et al., 2022; Enshasy et al., 2020; Sood et al., 2020).

13.2.2 *Mycorrhizal Fungi*

Arbuscular mycorrhizal fungi (AMF, phylum Glomeromycota), includes 14 families and 29 genera with more than 200 species, classified in three classes (Glomeromycetes, Archaeosporomycetes, and Paraglomeromycetes), and five orders (Archaeosporales, Diversisporales, Gigasporales, Glomerales, and Paraglomerales) (Oehl et al., 2011). According to current estimates, more than 80% of land plants can create advantageous interactions with AMFs. Although many AMF species have been discovered, only a small number are used in farming. As a result, species from genera *Rhizophagus* and *Funneliformis*, which are generalist symbionts prevalent in virtually all soils and climate zones, are almost entirely present in the commercial inocula that are currently on the market. AMF hyphae penetrate the root cortical cells of plants to create a mutualistic interaction that results in highly branched structures known as arbuscules (Candido et al., 2013). The fungal companion is referred to as a mycorrhizal fungus, and the mix of fungus and root tissue is known as a mycorrhiza.

Mycorrhizal fungi fall into two categories: ectomycorrhizae (associated with beech, birch, fir, hemlock, larch, oak, pine, and spruce) and endomycorrhizae (associated with apple, ash, bayberry, cherry, dogwood, holly, juniper, turfgrass, and many herbaceous plants), which grow in and into the soil around the root cells. Arbuscular mycorrhizae (AM) and vesicular-arbuscular mycorrhizae (VAM) are two types of endomycorrhizae. Herbaceous and woody plants constitute a sizable portion of their combined host range.

The ability of plants to utilize soil resources is significantly improved by mycorrhizal root systems, which multiply the absorptive capacity of of roots by 10 to 1000 times. Strong chemicals (enzymes) released by mycorrhizal fungi into the soil make it difficult to extract nutrients such as phosphorus, iron, and other "tightly bonded" soil nutrients. This process of extraction is crucial to plant nutrition and explains why non-mycorrhizal plants need a lot of fertilizer to remain healthy. Most of this fertility is lost or wasted in non-mycorrhizal environments.

Mycorrhizal fungi are also involved in a wide range of other processes that aid the establishment and development of plants. Water uptake and storage rely on the same extensive network of fungal filaments, which are crucial for nutrient uptake (Mitra et al., 2020). Mycorrhizal plants experience much less drought stress in non-irrigated environments than non-mycorrhizal plants. The association of mycorrhizae with saffron alleviates saline stress by superoxide dismutase, catalase, and methyl jasmonate (MeJ) production (Hamidian et al., 2023). Incorporation of AMF along with bio-char enhanced growth, soil enzyme activities, and water stress tolerance in soybeans (Jabborova et al., 2022).

Suppression of diseases and pathogens is another advantage of mycorrhizal plants (Mitra et al., 2021a). A physical barrier that prevents the invasion of pathogens is present on mycorrhizal roots in the form of a mantle, which is a tight, interwoven, sock-like covering of thick filaments. Mycorrhizal fungi also fight pathogens and disease-causing organisms that infiltrate the root zone. Mycorrhizae induces physiological and biochemical changes in the host upon colonization, such as increased production of chitinases, β-1,3-glucanases, phytoalexins, phenylpropanoid pathway proteins, peroxidases, pathogenesis-related (PR) proteins, callose, hydroxyproline rich glycoproteins (HRGP) and phenolics (Gianinazzi-Pearson et al., 1994). A similar response was observed in wheat upon *Funneliformis mosseae* colonization against pathogenic *Gaeumannomyces graminis* (Falahian et al., 2007). Mycorrhizae competes with the pathogen for survival and photosynthates, decreasing its colonization due to higher carbon demand. Colonization of *Glomus intraradices* in maize improved absorption of essential nutrients such as phosphorus directly or indirectly correlated with *Fusarium verticillioides* resistance (Pozo & Azcon-Aguilar, 2007).

Mycorrhizal fungi also secrete certain antibiotics that immobilize and kill disease-causing organisms. For example, some mycorrhizal fungi shield pine trees from *Phytophora, Fusarium*, and *Rhizoctonia.* Additionally, mycorrhizal fungi enhance the soil structure. Humic substances and organic glues (extracellular polysaccharides) are produced by mycorrhizal filaments, which bind soil into aggregates and increase its porosity. This effect increases aeration, water circulation into the soil, root growth, distribution, porosity, and structure, with beneficial effects on plants growth. The capacity of mycorrhizal fungi to support soil structure in sandy or compacted soils may be more significant than their ability to find and mobilize nutrients.

The potential of soil to develop a mycorrhiza can be diminished by a variety of common practices: some of the practices that can reduce or eradicate these advantageous soil fungi include tillage, excessive use of fertilizer, topsoil removal, erosion, site preparation, road and home construction, fumigation, invasion of non-native plants, and leaving soils bare.

13.2.2.1 Mycorrhizal Diversity

Various mycorrhizal fungal species are present in natural habitats. Following any disturbance, the proportion and abundance of mycorrhizal species frequently changed. Mycorrhizal fungi in fact differ in their abilities and tolerances. Some are better at promoting drought resilience, whereas others may be better at warding off pathogens or tolerating extremes in soil temperature (Mitra et al., 2021a, b; Panneerselvam et al., 2019). It is unlikely that a single mycorrhizal fungus could provide all previously indicated benefits to all host species, and adapt to all environmental factors. This is because of the enormous range of soil, climatic and biotic profiles that characterize man-made habitats. For instance, mycorrhizal fungi linked to immature plants may exhibit different types and behaviors than those associated

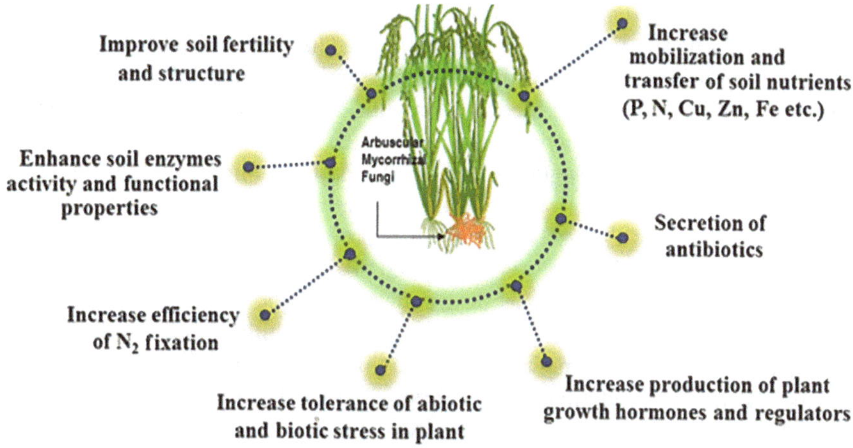

Fig. 13.2 Beneficial interactions of AMF with plants

with older plants. The mycorrhizal fungi required to support productivity across a long-lived plant may differ from those required to support seedling establishment in challenging environments. Diversity offers a level of buffering that is absent in settings with only one or a few species. A particular plant may be better to occupy several belowground niches and endure a variety of chemical and physical environments if it produces mycorrhizal fungi. Nevertheless, in addition to AMF, other fungi are known to colonize the roots of various plants, including *Trichoderma* spp. (Ascomycota) and *Sebacinales* (Basidiomycota). They do this without coming into direct contact with the plants, improving nutrient uptake and modifying their antioxidant defense mechanisms through a mechanism that is currently unknown (Mukherjee et al., 2012).

13.2.3 AMF-Microflora Interactions

Owing to the constant availability of nutrients near the roots, microorganisms in the rhizosphere coexist. Interactions occur as a result of the rhizosphere high bacterial density and AMF pervasiveness. Some of these interactions may be advantageous for plant development (Fig. 13.2).

13.2.3.1 Interactions Between AMF and Symbiotic Organisms

A diazotroph is a symbiotic bacterium capable to fix athmospheric nitrogen by the nitrogenase enzyme complex. The fixed nitrogen – is subsequently transmitted to the host. The direct involvement of the host in the nitrogen fixation process results from the fact that the site of nitrogen fixation differs among various diazotrophs.

The way the diazotroph and host interact also reveals the importance of the association for both parties participating in the symbiosis. This process is very important in the production of certain crops such as legumes, and ultimately in food production. Legumes can be given a second inoculation with AMF and the right *Rhizobium* sp. strains to increase their productivity. Regardless, the host plant has mostly benefited from the co-inoculating of legumes with AMF and the appropriate *Rhizobium* sp. strains. The choice of an appropriate *Rhizobium* sp. strain for inoculation purposes may not be as crucial as the choice of an appropriate AMF species to increase legume productivity. This is because nodulated plants profit from P uptake by AMF extraradical mycelium, which is necessary for the nitrogen fixation process.

The presence of many AMF species raises the possibility of variations in their capacity to promote development in the interactions with the host and the *Rhizobium* symbiont (Ianson & Linderman, 1993; Ibijbijen et al., 1996). Saxena et al. (1997) assessed *Cajanus cajani* for different AMF isolates: *Gigaspora etunicatum*, *G. aggregatum*, *G. deserticola*, *G. mosseae*, *G. margarita*, and *G. intraradices*. They discovered that the effects of various AMF species on nitrogen fixation and nodulation are significantly diverse. Additionally, these authors hypothesized that the variations were likely caused by a particular inter-endophyte interaction between the AMF species and *Rhizobium* sp. strain, which emerged under N-limiting conditions. The *Rhizobium* sp. strain uses the P that the AMF species mobilizes in an efficient and compatible tripartite symbiosis to help fix nitrogen. The specific AMF-*Rhizobium* sp. strain combination determines hence the effectiveness of such a tripartite symbiosis. This significant discovery indicates intense contact between the three symbiotic partners.

The tripartite interaction does not only enhance growth but also suppresses soilborne pathogens improving plant health. Gao et al. (2012) reported suppression of soybean crown rot disease due to enormous upregulation of pathogenesis-related (PR) protein (PR2, PR3, PR4, and PR10) upon co-inoculation with *Rhizobium* and AMF.

13.2.3.2 Interactions Between AMF and Non-Symbiotic Organisms

The co-inoculation of mycorrhizal hosts with different rhizobacteria has been studied by several researchers (Bagyaraj & Menge, 1978; Meyer & Linderman, 1986). For instance, the diazotroph *Acetobacter diazotrophicus* colonizes the aerial sections of sweet potato (*Ipomoea batatas* L.) and sugarcane (*Saccharum officinarum* L.) plants. Contrary to other diazotroph associations with AMF, *A. diazotrophicus* benefits sugarcane only when co-inoculated with AMF. The synergistic effects of co-inoculating sweet potato with *G. clarum* and *A. diazotrophicus* were evaluated by Paula et al. (1992). When co-inoculated with *G. clarum* spores, they discovered that *A. diazotrophicus* was found to infect aerial plant parts and increase tuber formation. The bacterium also promoted sporulation and intraradical AMF development.

To increase plant output in cereal crops, *Azospirillum* spp., a microaerophilic diazotroph, was coupled with AMF. Co-inoculation of *Azospirillum lipoferum* and

AMF reduced root-knot nematode, *Meloidogyne incognita* incidence by 80% in tomato. The co-inoculation reduced nematode parasitism by producing a wide variety of antibiotics, hydrolytic enzymes, organic compounds, HCN, and protease resulting in reduced egg hatching and killing juveniles (Ismaiel & Abdelaziz, 2019).

13.2.3.3 Interaction Between AMF and Phosphate Solubilizers

The utilization of AMF in transporting P following its release by a Phosphorous Solubilizing Bacterium (PSB) has received more attention than the ability of AMF to solubilize sparingly soluble phosphorus (Mitra et al. 2023a, b). The purpose of co-inoculating plants with AMF and PSB is to take advantage of the capacity of PSB to liberate phosphorus from bound or organic phosphorus sources in the soil. This phosphorus can then be taken up by AMF hyphae and successfully translocated to the host plant. For example, finger millet (*Eleusine coracana* L.) was co-inoculated with *G. fasciculatum* and the PSB *Bacillus circulans*. The plants were cultivated in a 1:1 sterilized soil:sand mixture supplemented with either radiolabeled superphosphate or tricalcium phosphate, and their response was evaluated. The plants inoculated with AMF removed more P from the soil, whereas plants co-inoculated with *G. fasciculatum* and *B. circulans* yielded more dry matter and P.

Kim et al. (1998) also reported synergistic interactions between PSB *Enterobacter agglomerans* and *G. etunicata*, resulting in enhanced P uptake in tomato plants. The only non-mycorrhizal fungal relationship that consistently improves host nutrition and growth is that between AMF and the phosphorus-solubilizing fungus, *Penicilium bilaii*. It appears that in some instances, host P nutrition significantly improved following an increase in P availability as a result of its solubilization by *P. bilaii* and movement to host roots by the AMF species. The synergistic interactions between AMF and *P. bilaiae* increased wheat's and field bean's yield (Kucey, 1987).

13.2.3.4 Interaction Between AMF and Phytohormone-Producing Rhizobacteria

AMF especially interact with *Pseudomonas* spp., a rhizobacterium that produces phytohormones. Meyer and Linderman (1986) investigated the effects of dual inoculation with an indigenous AMF mixed culture and *P. putida*, cultivated in non-sterile soil on subterranean clover (*Trifolium suhterraneum*). Co-inoculated plants considerably outperformed plants left un-inoculated or inoculated with the AMF combination or bacteria alone, in terms of shoot dry weight and mycorrhizal colonization. Additionally, co-inoculated plants had higher concentrations of Fe, Cu, Al, Zn, Co, and Ni. The nodulation of these plants by native soil rhizobia was improved. Amendment of soil with AMF and *P. fluorescens* reduced root rot caused by *Rhizoctonia solani* in French bean along with beneficial effect on yield (Song et al., 2015). The interaction of AMF with other soil microflora is worth investigating as it has enormous bio-control ability to suppress specific soil-borne pathogens.

13.2.4 Functions of AMF

13.2.4.1 Hormonal Regulation

(a) **Abscisic acid (ABA).** Increase in ABA in stressed plants after AMF or PGPR inoculation. According to Duc et al. (2018), *Septoglomus constrictum* colonized tomato roots showed down regulation of 9-cis-epoxy carotenoid dioxygenase (SlNCED) under drought stress, but not under normal or heat stress. Chitarra et al. (2016) showed that the type of mycorrhizal inoculum had an impact on ABA production, under the same stress conditions.

(b) **Indole-3-acetic acid (IAA).** Indole-3-acetic acid (IAA), the primary plant auxin, regulates cell functions like extension, division, and root growth. In water-scarce or salt-excess soils, IAA levels rise, with even greater increases in plants colonized by AMFs. Liu et al. (2018) found that drought-stressed plants inoculated with *Funneliformis mosseae* exhibited activated auxin-related genes (*PtYUC3*), up-regulation of inflow carriers (*PtABCB19*, *PtLAX2*), down-regulation of efflux carrier genes, and activation of other auxin-related genes (*PtYUC8*), and transport proteins (PtPIN1, PtPIN3). This suggests a complex interaction between IAA, AMFs, and the auxin pathway under environmental stress.

(c) **Jasmonic acid (JA) and salicylic acid (SA) biosynthesis.** AMF symbiosis and JA biosynthesis are inextricably linked. Under both normal and stressful conditions, AMF inoculation dramatically enhanced JA levels in durum wheat, maize, and cucumber plants. SA also stimulates the production of stress-related genes to preserve membrane stability and prevent oxidative damage. Although SA is known to play a crucial role in the control of disease signaling pathways (De Vleesschauwer et al., 2014), it also appears to play a role in the symbiotic connection between plant growth-promoting microorganisms (PGPMs) and plants. Tomato plants treated with AMF and JA induced resistance against *Fusarium* wilt, with increased total soluble sugars, free amino acids, and total soluble proteins in shoot and root (El-Khallal, 2007). Song et al. (2015) reported enhanced tomato resistance to early blight after mycorrhizal colonization due to priming of systemic defense response, and the JA signaling pathway is essential for mycorrhiza-primed disease resistance.

13.2.4.2 Antioxidant Mechanism

Both reactive oxygen species (ROS) and reactive nitrogen species (RNS) are produced under stress conditions to regulate several physiological processes and ensure plant life. However, owing to the high reactivity of both ROS and RNS, several biomolecules, including proteins, lipids, and nucleic acids, are vulnerable to their potent oxidative activity. Consequently, cells, DNA, RNA, and enzymes can sustain significant and irreparable damage, resulting in cell death. According to several investigations, plants inoculated with PGPMs exhibit greater ROS and RNS scavenging activities. In the tissues of inoculated plants, compared to non-inoculated plants, less hydrogen peroxide (H_2O_2) and malondialdehyde (MDA), markers of lipid peroxidation, were found.

13.2.4.3 Water Use Efficiency (WUE) and Photosynthetic Physiological Response Improvement

By increasing leaf water potential, leaf transpiration rate, stomata conductance, relative water content (RWC), or other factors that may have positive effects on water use efficiency (WUE), PGPM application has been shown in several studies to have an impact on plant water-related parameters. For instance, Barzana et al. (2012) showed that under both well-watered and drought-stressed conditions, the relative apoplastic water flow significantly increased in the roots of AMF-inoculated maize and tomato plants, compared to control plants.

13.2.4.4 Improvement of Nutrient Use Efficiency (NUE)

Nutrient utilization efficiency may be impacted by the inoculation of PGPMs due to potential contributions to mineral uptake, siderophore formation, atmospheric nitrogen fixation, and phosphate solubilization. Numerous studies have connected the improvement in root morphology observed after PGPM inoculation with an increase in mineral uptake. In particular, extra-radical hyphae generated by AMF inoculation may help plant roots use more soil, enhancing their ability to absorb nutrients. However, these events can also be explained by selective ion channel overexpression, which in some situations improves the K^+/Na^+ ratio and results in a crucial mechanism that promotes plant salinity tolerance (Santander et al., 2019).

13.2.4.5 Production of Ion-chelating Compounds (Siderophores)

Even though Fe is one of the most prevalent elements in soil, it is not easily accessible because it only exists in the insoluble form Fe^{3+}. As a result, certain bacteria secrete siderophores to scavenge iron from the mineral phases for assimilation. In this setting, bacterial siderophores can assist plants in absorbing the Fe required for several critical functions, including photosynthesis. In addition to increased nutrient availability, the iron-chelating siderophore also competes with the metabolism of pathogens and adversely affects their survival. For example, the siderophore-producing bacteria in the black rice rhizosphere inhibited incidence of the rice pathogen *Pyricularia oryzae* (Nabila & Kasiamdari, 2021).

These results imply that a better bio-stimulant impact is mostly determined by the inoculant inherent properties, which make the relationship with the host more favorable and hence more compatible. Utilizing this technique to implement unique sustainable agricultural practices that minimize the use of chemical fertilizers and enhance crop growth and productivity, metagenomics has emerged as a promising tool that can help identifying the benefits of such fungi.

13.2.5 Benefits of Plant Bio-stimulants in Agriculture

Farmers seeking enhanced harvestable yields may find plant bio-stimulants as a cost-effective and supplementary option, as these products indirectly influence growth by boosting nutrient uptake through soil-based microbes. Additionally, they enhance plant resistance to drought and salinity, contribute to soil erosion minimization, and support soil regeneration through an efficient mycelium network. Notably, bio-stimulants play a crucial role in carbon dioxide management by trapping the greenhouse CO_2 and transforming it into microbial biomass. They contribute to maintain the ecological health and productivity of ecosystems while being organic and leaving no undesirable residue or genetic alterations in crops. The potential reduction in the need for fertilizers and pesticides stems from the bio-stimulants ability to target indigenous soil microorganisms, facilitating natural processes that impact plant growth. Furthermore, plant bio-stimulants offer benefits in bio-fortification, emphasizing their multifaceted advantages in sustainable agriculture.

13.3 Conclusions

Extensive research has unequivocally demonstrated the potential and functionality of fungal plant bio-stimulants in serving as growth promoters, bio-protectors, and contributors to distinctive advantages such as salinity and drought resistance. It is imperative to recognize, however, that AMF represent just one subset among various microorganism groups inhabiting the rhizosphere. Consequently, there exist intricate interactions between AMF and other microbial groups, some of which hold significance in influencing crop yields. Leveraging these diverse connections becomes highly advantageous for enhancing productivity while minimizing reliance on excessive pesticides or fertilizers. Given the intrinsic qualities of AMF, capitalizing on these spontaneous correlations can be particularly beneficial when aiming at heightened yields, with reduced input levels.

References

Abbas, A., Jiang, D., & Fu, Y. (2017). *Trichoderma* spp. as antagonist of *Rhizoctonia solani*. *Journal of Plant Pathology & Microbiology, 8*, 402.

Afrouz, M., Sayyed, R. Z., Fazeli-Nasab, B., Piri, R., Almalki, W. H., & Fitriatin, B. N. (2023). Seed bio-priming with beneficial *Trichoderma harzianum* alleviates cold stress in maize. *PeerJ, 11*, e15644.

Alabouvette, C., & Cordier, C. (2018). Fertilite biologique des sols: des microorganismes utiles à la croissance des plantes. *Innovations Agronomiques. INRAE, 69*, ff10.15454/UIJG8Lff, hal02058232.
Al-Turki, A., Murali, M., Omar, A. F., Rehan, M., & Sayyed, R. Z. (2023). Exploring the recent advances in PGPR mediated resilience towards interactive effects of drought and salt stress in plants. *Frontiers in Microbiology, 14*, 1214845.
Bagyaraj, D. J., & Menge, J. A. (1978). Interaction between vesicular-arbuscular mycorrhizal fungi and *Azotobacter* and their effects on rhizosphere microflora and plant growth. *The New Phytologist, 80*, 567–573.
Barzana, G., Aroca, R., Paz, J. A., Chaumont, F., Martinez-Ballesta, M. C., Carvajal, M., & Ruiz-Lozano, J. M. (2012). Arbuscular mycorrhizal symbiosis increases relative apoplastic water flow in roots of the host plant under both well-watered and drought stress conditions. *Annals of Botany, 109*, 1009–1017.
Bastami, A., Amirnia, R., Sayyed, R. Z., & Enshasy, H. E. (2021). The effect of mycorrhizal fungi and organic fertilizers on quantitative and qualitative traits of two important *Satureja* species. *Agronomy, 11*, 1285. https://doi.org/10.3390/agronomy11071285
Basu, A., Prasad, P., Das, S. N., Kalam, S., Sayyed, R. Z., Reddy, M. S., & El Enshasy, H. (2021). Plant Growth Promoting Rhizobacteria (PGPR) as green bioinoculants: Recent developments, constraints, and prospects. *Sustainability, 13*, 1140. https://www.mdpi.com/2071-1050/13/3/1140
Behie, S. W., & Bidochka, M. J. (2014). Nutrient transfer in plant-fungal symbioses. *Trends in Plant Science, 19*, 734–740. https://doi.org/10.1016/j.tplants.2014.06.007
Benfatto, D., Matteo, R., Di Franco, F., San Lio, R. M., Ugolini, L., & Lazzeri, L. (2015). The use of bio-based liquid formulations in pest control of citrus groves. *Industrial Crops and Products, 75*, 42–47.
Błaszczyk, L., Siwulski, M., Sobieralski, K., Lisiecka, J. M., & Jedryczka, M. (2014). *Trichoderma* spp. – Application and prospects for use in organic farming and industry. *Journal of Plant Protection Research, 54*, 309–317.
Calvo, P., Nelson, L., & Kloepper, J. W. (2014). Agricultural uses of plant bio-stimulants. *Plant and Soil, 383*(1), 3–41. https://doi.org/10.1007/s11104-014-2131-8
Candido, V., Campanelli, G., D'Addabbo, T., Castronuovo, D., Renco, M., & Camele, I. (2013). Growth and yield promoting effect of articificial mycorrhization combined with different fertilizer rates on field-grown tomato. *Italian Journal of Agronomy, 8*, 168–174.
Caron, M., Fortin, J. A., & Richard, C. (1986). Effect of *Glomus intraradices* on the infection by *Fusarium oxysporum* f. sp. *radicis-lycopersici* in tomatoes over a 12-week period. *Canadian Journal of Botany, 64*, 552–556.
Chaithra, M., Pankaja, N. S., Venkatesh, & Benherlal, P. S. (2020). Antagonistic potentiality of *Trichoderma* isolates on soil borne fungal plant pathogens. *International Journal of Current Microbiology and Applied Sciences, 9*, 1042–1052.
Chaithra, M., Pankaja, N. S., Benherlal, P. S., & Mahesh, H. B. (2022). Micro-farming specific morpho-molecular characterization of *Trichoderma* isolates and its bio-control potential against foliar fungal plant pathogens. *Archives of Phytopathology and Plant Protection, 55*, 2250–2281.
Chbani, A., Mawlawi, H., & Zaouk, L. (2013). Evaluation of brown seaweed (*Padina pavonica*) as bio-stimulant of plant growth and development. *African Journal of Agricultural Research, 8*, 1155–1165.
Chet, I., Harman, G. E., & Baker, R. (1981). *Trichoderma hamatum*: Its hyphal interactions with *Rhizoctonia solani* and *Pythium* spp. *Microbial Ecology, 7*, 29–38.
Chitarra, W., Pagliarani, C., Maserti, B., Lumini, E., Siciliano, I., Cascone, P., et al. (2016). Insights on the impact of arbuscular mycorrhizal symbiosis on tomato tolerance to water stress. *Plant Physiology, 171*, 1009–1023.
Colla, G., Rouphael, Y., Di Mattia, E., El-Nakhel, C., & Cardarelli, M. (2015). Co-inoculation of *Glomus intraradices* and *Trichoderma atroviride* acts as a bio-stimulant to promote growth, yield and nutrient uptake of vegetable crops. *Journal of the Science of Food and Agriculture, 95*, 1706–1715.

De Vleesschauwer, D., Xu, J., & Hofte, M. (2014). Making sense of hormone-mediated defense networking: From rice to *Arabidopsis*. *Frontiers in Plant Science, 5*, 1–15.

Du Jardin, P. (2015). Plant bio-stimulants: Definition, concept, main categories and regulation. *Scientia Horticulturae, 196*, 3–14.

Duc, N. H., Csintalan, Z., & Posta, K. (2018). Arbuscular mycorrhizal fungi mitigate negative effects of combined drought and heat stress on tomato plants. *Plant Physiology and Biochemistry, 132*, 297–307.

El-Khallal, S. M. (2007). Induction and modulation of resistance in tomato plants against *Fusarium* wilt disease by bio-agent fungi (arbuscular mycorrhiza) and/or hormonal elicitors (jasmonic acid & salicylic acid): 1 – Changes in growth, some metabolic activities and endogenous hormones related to defence mechanism. *Australian Journal of Basic and Applied Sciences, 1*, 691–705.

El-Kherbawy, M., Angle, J. S., Heggo, A., & Chaney, R. L. (1989). Soil pH, rhizobia, and vesicular arbuscular mycorrhizae inoculation effects on growth and heavy metal uptake of alfalfa (*Medicago sativa* L.). *Biology and Fertility of Soils, 8*, 61–65. https://doi.org/10.1007/BF00260517

Enshasy, H. E. A., Ambehati, K. K., Ashraf, F., Ramchuran, K. K., Sayyed, R. Z., et al. (2020). *Trichoderma*: Biocontrol agents for promoting plant growth and soil health. In A. Yadav, S. Mishra, D. Kour, N. Yadav, & A. Kumar (Eds.), *Agriculturally important fungi for sustainable agriculture: Vol 2: Functional annotation for crop protection* (pp. 239–259). Springer.

Falahian, F., Ardebili, Z. O., Fahimi, F., & Khavarinejad, R. (2007). Effect of mycorrhizal fungi on some defense enzymes against *Gaeumannomyces gaminis* in wheat. *Pakistan Journal of Biological Sciences, 10*, 2418–2422.

Gao, X., Lu, X., Wu, M., Zhang, H., Pan, R., Tian, J., & Liao, H. (2012). Co-inoculation with rhizobia and AMF inhibited soybean red crown rot: From field study to plant defense-related gene expression analysis. *PLoS One, 7*, e33977.

Gianinazzi-Pearson, V., Gollotte, A., Dumas-Gaudot, E., Franken, P., & Gianinazzi, S. (1994). Gene expression and molecular modifications associated with plant responses to infection by arbuscular mycorrhizal fungi. In M. Daniels, J. A. Downic, & A. E. Osbourn (Eds.), *Advances in molecular genetics of plant-microbe interactions* (pp. 179–186). Kluwer Academic Publishers.

Giocoechea, N., Dolezal, K., Antolin, M. C., Strnad, M., & Sanchez-Diaz, M. (1995). Influence of mycorrhizae and *Rhizobium* on cytokinin content in drought-stressed alfalfa. *Journal of Experimental Botany, 46*, 1543–1549. https://doi.org/10.1023/A:1004216225159

Gowtham, H. G., Murali, M., Shilpa, N., Amruthesh, K. N., Gafur, A., et al. (2024). Harnessing abiotic elicitors to bolster plant's resistance against bacterial pathogens. *Plant Stress, 11*, 100371. https://doi.org/10.1016/j.stress.2024.100371

Haikal, N. Z. (2008). Control of *Rhizoctonia solani* in soybean (*Glycine max* L.) by seed-coating with *Trichoderma viride* and *Gliocladium virens* spores. *Journal of Applied Biosciences, 1*, 34–39.

Hamid, B., Zaman, M., Farooq, S., Fatima, S., Sayyed, R. Z., et al. (2021). Bacterial plant biostimulants: A sustainable way towards improving growth, productivity, and health of crops. *Sustainability, 21*(13), 2856. https://www.mdpi.com/2071-1050/13/5/2856.

Hamidian, M., Movahhedi-Dehnavi, M., Sayyed, R. Z., Almalki, W. H., Gafur, A., & Fazeli-Nasab, B. (2023). Co-inoculation of mycorrhiza and methyl jasmonate regulates morpho-physiological & antioxidant responses of *Crocus sativus* (saffron) under salinity stress conditions. *Scientific Reports, 13*, 7378.

Hirrel, M. C., & Gerdemann, J. W. (1980). Improved growth of onion and bell pepper in saline soils by two vesicular-arbuscular mycorrhizal fungi. *Soil Science Society of America Journal, 44*, 654–655.

Ianson, D. C., & Linderman, R. G. (1993). Variation in the response of nodulating pigeonpea (*Cajanus cajan*) to different isolates of mycorrhizal fungi. *Symbiosis, 15*, 105–119.

Ibijbijen, J., Urquiaga, S., Ismail, M., Alves, B. J. R., & Boddey, R. M. (1996). Effect of arbuscular mycorrhizal fungi on growth, mineral nutrition and nitrogen fixation of three varieties of common beans (*Phaseolus vulgaris*). *The New Phytologist, 1*, 353–360.

Ismaiel, G., & Abdelaziz, S. (2019). Effect of arbuscular mycorrhizal fungi and some plant growth promoting rhizobacteria in controlling root-knot nematode (*Meloidogyne incognita*) on tomato under greenhouse conditions. *Example of Design Research, 69*, 131–150.

Jabborova, D., Annapurna, K., Azimov, A., Tyagi, S., Pengani, K. R., et al. (2022). Co-inoculation of biochar and arbuscular mycorrhizae for growth promotion and nutrient fortification in soybean under drought conditions. *Frontiers in Plant Science, 13*, 947547.

Jaizme-Vega, M. C., Sosa-Hernandez, B., Hernandez, J. M., & Galan, S. V. (1998). Interaction of arbuscular mycorrhizal fungi and the soil pathogen *Fusarium oxysporum* f. sp. *cubense* on the first stages of micropropagated Grande Naine banana. *Acta Horticulturae, 490*, 285–295.

Johnson, D., & Gilbert, L. (2015). Interplant signalling through hyphal networks. *The New Phytologist, 205*, 1448–1453.

Johnson, N. C., & Graham, J. H. (2013). The continuum concept remains a useful framework for studying mycorrhizal functioning. *Plant and Soil, 363*, 411–419. https://www.jstor.org/stable/42952872

Kauffman, G. L., Kneivel, D. P., & Watschke, T. L. (2007). Effects of a bio-stimulant on the heat tolerance associated with photosynthetic capacity, membrane thermo-stability, and polyphenol production of perennial ryegrass. *Crop Science, 47*, 261–267.

Kim, K. Y., Jordan, D., & McDonald, G. A. (1998). Effect of phosphate-solubilizing bacteria and vesicular arbuscular mycorrhizae on tomato growth and soil microbial activity. *Biology and Fertility of Soils, 26*, 79–87.

Kohl, J., Kolnaar, R., & Ravensberg, W. J. (2019). Mode of action of microbial biological control agents against plant diseases: Relevance beyond efficacy. *Frontiers in Plant Science, 10*, 845.

Kucey, R. M. N. (1987). Increased phosphorus uptake by wheat and field beans inoculated with a phosphorus solubilizing *Penicillium bilaii* strain and with vesicular-arbuscular mycorrhizal fungi. *Applied and Environmental Microbiology, 53*, 2699–2703.

Kulkarni, S. A., Kulkarni, S., Sreenivas, M. N., & Kulkarni, S. (1997). Interaction between vesicular-arbuscular (VA) mycorrhizae and *Sclerotium rolfsii* Sacc. In groundnut. *Karnataka Journal of Agricultural Sciences, 10*, 919–921.

Lambert, L. (2006). Decouvrez les biostimulants, Quebec Vert. *Medias Transcontinental, 28*(2), 67.

Leelavathi, M. S., Vani, L., & Reena, P. (2014). Antimicrobial activity of *Trichoderma harzianum* against bacteria and fungi. *International Journal of Current Microbiology and Applied, 3*, 96–103.

Li, S. L., Zhao, S. J., & Zhao, L. Z. (1997). Effects of VA mycorrhizae on the growth of eggplant and cucumber and control of diseases. *Acta Phytophylacica Sinica, 24*, 117–120.

Liu, C. Y., Zhang, F., Zhang, D. J., Srivastava, A., Wu, Q. S., & Zou, Y. N. (2018). Mycorrhiza stimulates root-hair growth and IAA synthesis and transport in trifoliate orange under drought stress. *Scientific Reports, 8*, 1–9. https://doi.org/10.1038/s41598-018-20456-4

Lopez-Bucio, J., Pelagio-Flores, R., & Herrera-Estrella, A. (2015). *Trichoderma* as biostimulant: Exploiting the multilevel properties of a plant beneficial fungus. *Scientia Horticulturae, 196*, 109–123.

Lopez-Moral, A., Agusti-Brisach, C., & Trapero, A. (2021). Plant bio-stimulants: New insights into the biological control of Verticillium wilt of olive. *Frontiers in Plant Science, 12*, 662178.

Meyer, J. R., & Linderman, R. G. (1986). Response of subterranean clover to dual inoculation with vesicular arbuscular mycorrhizal fungi and a plant growth promoting bacterium *pseudomonas putida*. *Soil Biology and Biochemistry, 18*, 185–190. https://doi.org/10.1016/0038-0717(86)90025-8

Mitra, D., Uniyal, N., Panneerselvam, P., Senapati, A., & Ganeshamurthy, A. N. (2020). Role of mycorrhiza and its associated bacteria on plant growth promotion and nutrient management in sustainable agriculture. *International Journal of Life Sciences and Applied Sciences, 1*, 1–10.

Mitra, D., Djebaili, R., Pellegrini, M., Mahakur, B., Sarker, A., et al. (2021a). Arbuscular mycorrhizal symbiosis: Plant growth improvement and induction of resistance under stressful conditions. *Journal of Plant Nutrition, 44*(13), 1993–2028. https://doi.org/10.1080/01904167.2021.1881552

Mitra, D., Guerra Sierra, B. E., Khoshru, B., De Los Santos Villalobos, S., Belz, C., et al. (2021b). Impacts of arbuscular mycorrhizal fungi on rice growth, development, and stress management with a particular emphasis on strigolactone effects on root development. *Communications in Soil Science and Plant Analysis, 52*, 1591–1621. https://doi.org/10.1080/00103624.2021.1892728

Mitra, D., Nayeri, F. D., Sansinenea, E., Ortiz, A., Bhatta, B. B., et al. (2023a). Unraveling arbuscular mycorrhizal fungi interaction in rice for plant growth development and enhancing phosphorus use efficiency through recent development of regulatory genes. *Journal of Plant Nutrition, 46*, 3184–3220. https://doi.org/10.1080/01904167.2023.2191638

Mitra, D., Panneerselvam, P., Senapati, A., Chidambaranathan, P., Nayak, A. K., & Mohapatra, P. K. (2023b). Arbuscular Mycorrhizal fungi response on soil phosphorus utilization and enzymes activities in aerobic rice under phosphorus-deficient conditions. *Life, 13*, 1118. https://doi.org/10.3390/life13051118

Mukherjee, P. K., Horwitz, B. A., Herrera-Estrella, A., Schmoll, M., & Kenerley, C. M. (2012). *Trichoderma* research in the genome era. *Annual Review of Phytopathology, 51*, 105–129.

Nabila, & Kasiamdari, R. S. (2021). Antagonistic activity of siderophore-producing bacteria from black rice rhizosphere against rice blast fungus *Pyricularia oryzae*. *Microbiology and Biotechnology Letters, 49*, 217–224.

Najafi, Z. V., Sohrabi, Y., Sayyed, R. Z., Suriani, N. L., & Datta, R. (2021). Effects of combinations of rhizobacteria, mycorrhizae, and seaweeds on growth and yields in wheat cultivars under the influence of supplementary irrigation. *Plants, 10*, 811. https://www.mdpi.com/2223-7747/10/4/811/htm

Nellemann, C., Mac Devette, M., Manders, T., Eickhout, B., Svihus, B., Prins, A. G., & Kaltenborn, B. P. (2009). *The environmental food crisis–The environment's role in averting future food crises. A UNEP rapid response assessment*. United Nations Environment Programme, GRID-Arendal. Birkeland Trykkeri AS, Norway 1-104. ISBN: 978-82-7701-054-0.

Nicolas, C., Hermosa, R., Rubio, B., Mukherjee, P. K., & Monte, E. (2014). *Trichoderma* genes in plants for stress tolerance-status and prospects. *Plant Science, 228*, 71–78.

Oehl, F., Sieverding, E., Palenzuela, J., Ineichen, K., & Alves da Silva, G. (2011). Advances in *Glomeromycota* taxonomy and classification. *IMA Fungus, 2*, 191–199. https://doi.org/10.5598/imafungus.2011.02.02.10

Panneerselvam, P., Sahoo, S., Senapati, A., Kumar, U., Mitra, D., et al. (2019). Understanding interaction effect of arbuscular mycorrhizal fungi in rice under elevated carbon dioxide conditions. *Journal of Basic Microbiology, 59*, 1217–1228. https://doi.org/10.1002/jobm.201900294

Pastucha, N. A., & Patkowska, E. (2005). Wpływ płynów pohodowlanych grzybów antagonistycznych na zdrowotność i plonowanie soi. *Acta Agrobotanica, 58*, 111–124.

Patkowska, E., Mielniczuk, E., Jamiołkowska, A., Skwaryło-Bednarz, B., & Błazewicz-Wozniak, M. (2020). The influence of *Trichoderma harzianum* Rifai T-22 and other biostimulants on rhizosphere beneficial microorganisms of carrot. *Agronomy, 10*, 1637.

Paula, M. A., Urquiaga, S., Siqueira, J. O., & Dobereiner, J. (1992). Synergistic effects of vesicular arbuscular mycorrhizal fungi and diazotrophic bacteria on nutrition and growth of sweet potato (*Ipomoea batatas*). *Biology and Fertility of Soils, 14*, 61–66. https://doi.org/10.1007/BF00336304

Pozo, M. J., & Azcon-Aguilar, C. (2007). Unravelling mycorrhiza induced resistance. *Current Opinion in Plant Biology, 10*, 393–398.

Prasad, R., Singh Gill, S., & Tuteja, N. (2018). *Trichoderma*: Its multifarious utility in crop improvement. In R. Prasad, S. Singh Gill, & N. Tuteja (Eds.), *New and future developments in microbial biotechnology and bioengineering. Crop improvement through microbial biotechnology* (1st ed., pp. 263–291). Elsevier.

Prashanthi, S. K., Kulkarni, S., Sreenivasa, M. N., & Kulkarni, S. (1997). Integrated management of root rot disease of safflower caused by *Rhizoctonia bataticola*. *Environment and Ecology, 15*, 800–802.

Raja, N. (2013). Biopesticides and biofertilisers: Ecofriendly sources for sustainable agriculture. *Journal of Biofertilizers & Biopesticides, 4*, 112. https://doi.org/10.4172/2155-6202.1000e112

Requena, N., Mann, P., & Franken, P. (2000). A homologue of the cell cycle check point TOR2 from *Saccharomyces cerevisiae* exists in the arbuscular mycorrrhizal fungus *Glomus mosseae*. *Protoplasma, 212*, 89–98.

Rollan, M. C., Monaco, C., Lampugnani, G., Arteta, N., Bayo, D., & Urrutia, M. I. (2007). Effects of post-emergent herbicides on *Trichoderma harzianum*, a potential bio-control agent against *Sclerotinia sclerotiorum* in soybean cropping. *Acta Agronomica Hungarica, 55*, 355–362.

Saboor, A., Muhammad, A. A., Hussain, S., El Enshasy, H. E., Hussain, S., et al. (2021). Zinc nutrition and arbuscular mycorrhizal symbiosis effects on maize (*Zea mays* L.) growth and productivity. *Saudi Journal of Biological Sciences, 28*, 6339–6351. https://doi.org/10.1016/j.sjbs.2021.06.096

Santander, C., Sanhueza, M., Olave, J., Borie, F., Valentine, A., & Cornejo, P. (2019). Arbuscular mycorrhizal colonization promotes the tolerance to salt stress in lettuce plants through an efficient modification of ionic balance. *Journal of Soil Science and Plant Nutrition, 19*, 321–331.

Saxena, A. K., Rathi, S. K., & Tilak, K. V. B. R. (1997). Differential effect of various endomycorrhizal fungi on nodulating ability of green gram by *Bradyrhizobium* sp. (*Vigna*) strain S24. *Biology and Fertility of Soils, 24*, 175–178. https://doi.org/10.1007/s003740050227

Shalaby, A. M., & Hanna, M. M. (1998). Preliminary studies on interactions between VA mycorrhizal fungus *Glomus mosseae*, *Bradyrhizobium japonicum* and *Pseudomonas syringae* in soybean plants. *Acta microbiologica Polonica, 47*, 385–391.

Simard, S. W., Beiler, K. J., Bingham, M. A., Deslippe, J. R., Philip, L. J., & Teste, F. P. (2012). Mycorrhizal networks: Mechanisms, ecology and modelling. *Fungal Biology Reviews, 26*, 39–60.

Song, Y., Chen, D., Lu, K., Sun, Z., & Zeng, R. (2015). Enhanced tomato disease resistance primed by arbuscular mycorrhizal fungus. *Frontiers in Plant Science, 6*, 786.

Sood, M., Kapoor, D., Kumar, V., Sheteiwy, M. S., Ramakrishnan, M., et al. (2020). *Trichoderma*: The "secrets" of a multitalented biocontrol agent. *Plants, 9*, 762.

Sreeramulu, K. R., Onkarappa, T., & Swamy, H. N. (1998). Bio-control of damping off and black shank disease in tobacco nursery. *Tobacco Research, 24*, 1–4.

Srivastava, P., Sahgal, M., Sharma, K., Enshasy, H. E., Gafur, A., et al. (2022). Optimization and identification of siderophores produced by *Pseudomonas monteilli* strain MN759447 and its antagonism towards fungus associated with mortality in *Dalbergia sissoo* plantation forests. *Front Pl Sci, 13*, 984522.

Torre, L. A., Battaglia, V., & Caradonia, F. (2016). An overview of the current plant biostimulant legislations in different European member states. *Journal of the Science of Food and Agriculture, 96*, 727–734.

Tortora, M. L., Diaz-Ricci, J. C., & Pedraza, R. O. (2012). Protection of strawberry plants (*Fragaria ananassa* Duch.) against anthracnose disease induced by *Azospirillum brasilense*. *Plant & Soil, 356*, 279–290.

Tyskiewicz, R., Nowak, A., Ozimek, E., & Jaroszuk-Scisel, J. (2022). *Trichoderma*: The current status of its application in agriculture for the bio-control of fungal phytopathogens and stimulation of plant growth. *International Journal of Molecular Sciences, 23*, 2329.

Yakhin, O. I., Lubyanov, A. A., Yakhin, I. A., & Brown, P. H. (2017). Biostimulants in plant science: A global perspective. *Frontiers in Plant Science, 7*, 2049.

Zeilinger, S., & Omann, M. (2007). *Trichoderma* biocontrol: Signal transduction pathways involved in host sensing and mycoparasitism. *Gene Regulation and Systems Biology, 1*, GRSB-S397.

Chapter 14
Fungi as Biocontrol Agents of Insect and Mite Pests

Aditya Singh Ranout, Rahul Kumar, Rupinder Kaur, and Gireesh Nadda

Abstract Agricultural crops are vulnerable to attacks by various insect and mite pests. Research has demonstrated that these pests significantly impact the nutritional and economic value of agricultural outputs. As a result, the usage of chemical-based insecticides and acaricides emerged, leading to health and environmental concerns. Therefore, there is a need for the development and utilization of sustainable, eco-friendly methods to control insect and mite pests. To address this issue, the scientific and industrial communities have shown interest in developing and using biopesticides to control these pests. Amongst biopesticides a number of entomopathogenic fungi (EPF) emerged as a viable option for managing insect and mite populations, despite some inherent limitations. In this chapter, we review the existing literature on EPF, their effectiveness against insect and mite pests, the advantages over chemical-based formulations, their mode of action, constraints, contemporary research to address these limits, and the fungal-based products already available in the market.

Keywords Biological control · Biopesticides · Entomopathogenic fungi · Mycoacaricides · Mycoinsecticides

Authors Aditya Singh Ranout and Rahul Kumar are sharing first authorship to this chapter.

A. S. Ranout · R. Kumar · R. Kaur · G. Nadda (✉)
Academy of Scientific and Innovative Research (AcSIR), Ghaziabad, India

Entomology Laboratory, Agrotechnology Division, CSIR-Institute of Himalayan Bioresource Technology (IHBT), Palampur, India
e-mail: girish@ihbt.res.in

P. Mathur, S. Roy (eds.), *Plant Microbiome and Biological Control*, Sustainability in Plant and Crop Protection 20,
https://doi.org/10.1007/978-3-031-75845-4_14

Abbreviations

DTX	Destruxin
EPA	United States Environmental Protection Agency
EPF	Entomopathogenic fungi
FAO	Food and Agriculture Organization of the United Nations
IPM	Integrated Pest Management
PPE	Personal protective equipment

14.1 Introduction

Agricultural crops are susceptible to infestation by a variety of insect and mite pests, that have a substantial influence on the nutritional and economic worth of agricultural yields. The projected crop loss in India owing to insect pests varied for different crops, with cotton seeing a loss of approximately 30%, followed by rice 25%, oilseeds 20%, sugarcane 20%, pulses 15%, maize 18%, groundnut 15%, sorghum and millets 8%, and wheat 5% (Dhaliwal et al., 2015). Various insecticides, including cypermethrin, deltamethrin, chlorpyrifos, fipronil, malathion, and imidacloprid, are widely used in agriculture to reduce pest populations. However, these synthetic chemicals have detrimental effects on the agroecosystem, human health, and various beneficial non-target organisms (Khan et al., 2020). Furthermore, the continual use of a certain set of pesticides results in the development of resistance and the resurgence of pest populations (Islam et al., 2017). Various effective strategies have been implemented to control insect and mite pests, aiming at minimizing environment exposure to pesticides.

Consequently, a widely used approach for sustainable pest management is the implementation of alternative and environmentally friendly methods to enhance the nutritional value, quality and yield of agricultural crops (Hesham et al., 2021). Given the negative impacts of chemical insecticides and acaricides, biopesticides are necessary due to their environmentally benign, safe, and rapidly degradable characteristics. The microbiome-derived biopesticides offer a non-toxic method for managing insect and mite pest populations. Various agents, including viruses, fungi, and bacteria have been advocated as viable options and are effectively utilized in pest management strategies (Rastegari et al., 2020). Some of the widely used entomopathogenic fungi (EPF) for managing insect and mite pests include species and isolates of genera *Beauveria*, *Metarhizium*, *Hirsutella*, *Verticillium*, *Lecanicillium*, and *Paecilomyces*.

Approximately 750 fungal species have the ability to infect various insect and mite pests (Mantzoukas & Eliopoulos, 2020), that can be explored to develop myco-insecticides and -acaricides. These fungal isolates have a broad spectrum of hosts, exhibit high toxicity, and may control both chewing and sucking insect and mite pests, effectively. Thus, they are considered as an integral component in Integrated Pest Management (IPM) strategies. Unlike bacteria and viruses, EPF enter their hosts directly through the cuticle, which needs to be digested for survival (Khan et al., 2012). Fungi inflict harm to the host by germinating on the cuticle and subsequently multiply within the hemocoel, leading to an infection. The infection rate is contingent upon the fungal species and the quantity of infectious spores.

Out of 31 insect orders, 20 are notably affected by EPF, and it is worth noting that these fungi infect all phases of their development, including eggs, larvae, pupae, nymphs, and adults (Islam et al., 2021). However, the virulence of EPF against the different developmental stages varies (Ranout et al., 2024). In addition, EPF can also trigger insect-related diseases by producing a range of extracellular enzymes that effectively break down the outer covering of insects (Mondal et al., 2016). Nevertheless, it is essential to have a comprehensive knowledge of fungal physiology to effectively implement fungi-mediated biological control.

This chapter covers the utilization of fungi to manage major insect and mite pests. Besides, it also focuses on fungal species used in biocontrol, their pathogenicity mechanisms, application methods, targeted insect and mite species, and the factors that can influence the EPF efficacy. In addition, essential information is provided concerning the advantages and limitations linked to the utilization of fungal-based insecticides and acaricides, their regulatory and safety aspects and available market products.

14.2 Fungal Species Used in Biocontrol

EPF comprise a vast assemblage of over 750 species capable of parasitizing insect and mite pests. They are included in phyla Zygomycota, Ascomycota, and Deuteromycota (Mantzoukas & Eliopoulos, 2020). Frequently employed EPF consists of readily cultivable fungi classified under the Ascomycota (Mascarin & Jaronski, 2016). Various biopesticides derived from species under the genera *Metarhizium*, *Beauveria*, *Paecilomyces*, *Isaria*, and *Lecanicillium* have been employed globally (Table 14.1). These fungi possess a broad range of activity, enabling them to infect a diverse array of arthropod species irrespective of natural predators of insect and mite pests (Khan et al., 2012; Castro et al., 2016). The species currently employed in the production of myco-insecticides and -acaricides include *Beauveria bassiana*, *B. brongniartii*, *Metarhizium anisopliae*, *Lecanicillium* spp. (formerly known as *Verticillium lecanii*), *Isaria fumosorosea* (formerly *Paecilomyces fumosoroseus*), *Hirsutella thompsonii*, and *Cladosporium oxysporium* (Table 14.1).

Table 14.1 Market products of fungal-based insecticides and acaricides

Target Pest	Fungal species	Trade Name	Formulation[a]	Producer
Coleoptera (Scarab beetle grubs, Weevils, Coffee berry borer, Longhorn beetle, etc.)	*Beauveria bassiana*	Bauveril	WP	Laverlam S.A., Colombia
		Bea-Sin	WP	Agrobiologicos Noroeste (Agrobionsa),Mexico
		Boverin	WP	Biodron, Russia
		Boverol-Spofa	WP	Czechoslovakia
		Conidia	WP	AgroEvo, Germany
	B. brongniartii	Beauveria Schweizer	WP	Lbu (formerly Eric Schweizer Seeds), Switzerland
		Betel	WP	Arysta (formerly NPP, Calliope), France
		Biolisa-Kamikiri	WP	Nitto Denko, Japan
		Engerlingspilz	WP	Andermatt Biocontrol AG, Switzerland
		Melocont	G/WC	Andermatt Biocontrol AG, Switzerland
	Metarhizium anisopliae	Biofight	L	Omega Ecotech products, India
		Biologie Bio 1020	G	Bayer AG, Germany
		Biometaz	WDG	Pioneer Agro Industry, India
		Biostorm	WP	Varsha Bio Science and Teachnology, India
		DeepGreen	SC	Live System Technology S.A., Colombia
		GreenMeta	L	Greenlife Biotech Laboratory, India
		JasMeta	WP	Shri Ram Solvent Extraction, India
	M. flavoviride var. *flavoviride*	BioGreen	G	Biocare Technology Pvt. Ltd., Australia
	Isaria fumosorosea	Ancora	WP	OHP, Inc., U.S.A.
		PFR-97	WDG	Thermo Trilogy, U.S.A.

(continued)

Table 14.1 (continued)

Target Pest	Fungal species	Trade Name	Formulation[a]	Producer
Lepidoptera (Beet Armyworm, Diamondback moth, Cabbage looper, Cutworm, Bollworm, etc.)	*Beauveria bassiana*	Bio-Power	L	Bayer CropScience Limited, U.K.
		BotaniGard ES	WP	Laverlam International Corporation, U.S.A.
		Mycotrol	WP	Laverlam International Corporation, U.S.A.
		Myco-Jaal	SC	Gaiagen Technologies Private Limited, India
	B. brongniartii	Beauveria Brongniartii	WSP	MarkNature By Nature, U.S.A.
		Beauveria-Schweizer	WP	Eric Schweizer Seeds, Switzerland
	Metarhizium anisopliae	Green Meta	L	Greenlife biotech Laboratory, India
		Kalichakra ipl	WP	IPL Biologicals Limited, India
		Katyayani Metarhizium Anisopliae	L	Katyayani Organics, India
	Isaria fumosorosea	Ancora Microbial Insecticide	WDG	OHP, Inc., U.S.A.
		Challenger	L	Koppert do Brasil Holdin Ltda., Brazil
		PFR-97	WDG	Certis Biologicals, U.S.A.
	Nomuraea rileyi	Numoraea 50	WP	Ago Biocontrol, Colombia
Hemiptera (Aphids, Whitefly, Mealybugs, Psyllids, Stinkbugs, Leafhoppers, etc.)	*Beauveria bassiana*	Adral	WP	Bio-Crop, Colombia
		Bassicore	SC	Core Biotechnology, Colombia
		Biosoft	WP	Agri land Biotech, India
		Naturalis L	L	Troy Bioscience, U.S.A.
		Racer BB	WP	SOM Phytopharma, India
	Conidiobolus thromboides	Vektor 25SL	SL	Mycolab, South Africa
	Isaria fumosorosea	Bemisin	WP	Probioagro, Venezuela
		Pae-Sin	WP	Agrobionsa, Mexico
		PFR-97	WDG	Thermo Trilogy, U.S.A.
		Preferal	WDG	SePRO Corporation, U.S.A.

(continued)

Table 14.1 (continued)

Target Pest	Fungal species	Trade Name	Formulation[a]	Producer
	Lecanicillium lecanii	Dr. Bacto's Vertigo	L	Anand Agro Care, India
		Green Verticill	L	Greenlife Biotech Laboratory, India
		Mycotal	WP	Koopert Biological Systems, Netherlands
		Vertactec	WP	Koopert Biological Systems, Netherlands
		WertiSphere	WP	Bharat Agro Chemicals and Fertilizers, India
	L. longisporum	Inovert	WP	Inora, India
		Vertalec	WP	Koopert Biological Systems, Netherlands
		Verticare	WP	Vishwamitra Bio Agro, India
	Metarhizium anisopliae	Bio-Catch-M	WP	T. Stanes, India
		Bio Magic	L	T. Stanes, India
Thysanoptera (Thrips)	*Beauveria bassiana*	BBC	WP	Sri Biotech Laboratories India LTD., India
		Bio Expert	WP	Live Systems Technology, Colombia
	Metarhizium anisopliae	Campaign	WP	Real IPM, Ghana, Uganda
Orthoptera (Grasshoppers and Locusts)	*M. flavoviride* var. *flavoviride*	Green Muscle	WP	Biological Control Products Ltd., South Africa
Acari (Rust mites, Spider Mites, etc.)	*Beauveria bassiana*	Bovebio	WP	Biofungi – Industriae Comerciode defensivos Biologicose Inoculantes, Brazil
		Broadband	WP	BASF, South Africa
	Metarhizium anisopliae	Achieve	WP	Real IPM, Kenya
	Hirsutella thompsonii	AlMite	AS	IPL Biologicals Limited, India
		Mycohit	WP	Plantrich Chemicals & Biofertilizer Ltd., India
		No-Mite	AS/WP	Biotech International Limited, India
	Isaria fumosorosea	Priority	WP	T. Stanes, India
		PFR-97	WDG	Thermo Trilogy, U.S.A.

[a]*WP* wettable powder, *WDG* water-disposable powder, *SC* suspension concentrate, *G* granular, *L* liquid, *WSP* water-soluble powder, *AS* aqueous suspension

14.2.1 Beauveria

The *Beauveria* genus belongs to the Cordycipitaceae family which includes the *Cordyceps* genus, recognized for having the largest number of entomopathogenic fungi (Sharma et al., 2023b). Species of *Beauveria* genus, like many other fungal species, has a white mycelial appearance (Bamisile et al., 2019). This genus comprises valuable species, including *B. bassiana*, *B. brongniartii*, *B. amorpha*, and *B. caledonica*. Out of these, *B. bassiana* has the ability to grow naturally in soil across the globe (Sharma et al., 2024a) and possesses pathogenic characteristics that can cause white muscardine disease in many insect and mite hosts. Several strains of *Beauveria* exhibit a great degree of selectivity when it comes to choosing their hosts. The host exoskeleton becomes completely coated with a dense white mat due to the severe infection caused by the mycelial growth that eventually takes the shape of a synnemata (Rohrlich et al., 2018). Microscopic observations readily enable the study and recognition of *B. bassiana*, revealing the presence of conidiogenous cells. These structures have round bases and elongated serrated segments and produce round spores with a diameter of less than 3.5 μm (Kary & Alizadeh, 2017).

Beauveria bassiana ABNBb 101 is capable of parasitizing a wide range of arthropod hosts and is commonly regarded as a nonselective strain that induces white muscardine disease in various insect species (whiteflies, aphids, thrips, grasshoppers, and beetles) belonging to different orders. Commercial formulations are available in the market as a wettable powder or emulsifiable liquid with insecticidal and acaricidal characteristics (Table 14.1).

14.2.2 Metarhizium

Metarhizium, belonging to the Clavicipitaceae family, includes three significant insect and mite pathogenic species, namely *M. anisopliae*, *M. album*, and *M. flavoviride*. Among them, *M. anisopliae* is recognized as a promising entomopathogen (Brunner-Mendoza et al., 2019). This genus is characterized by its visually striking branched pattern of the conidiophores, which are tightly intertwined and have round conical tips. The conidiogenous cells are grouped in a dense hymenium with a precise alignment. The conidia lack septa, and have a pale, green, or yellow-green tint. The structures are either ovoid or cylindrical, arranged in chains that typically cluster together to create prismatic or cylindrical columns, or occasionally a compact arrangement of parallel chains (Leadmon et al., 2020). *Metarhizium anisopliae* is an extensively researched fungus inhabiting soil, and is found worldwide, from the Arctic to the Tropics (Shahid et al., 2012). It is noted that a wide variety of market products of *Metarhizium* are available and are utilized as a biocontrol agent against root weevils, flies, thrips, two-spotted spider mites, and gnats (Table 14.1).

14.2.3 Lecanicillium

The *Lecanicillium* genus consists of several species that are harmful to insects. These fungi were previously classified as a single species called *Verticillium lecanii*. They belong to the fungal order Hypocreales and are classified as the anamorphic (asexual) form of the *Cordyceps* group, in the Clavicipitaceae family. These genera embrace several hosts, including insects, mites, and nematodes. The genus includes significant EPF such as *Lecanicillium lecanii*, *L. attenuatum*, *L. longisporum*, *L. nodulosum*, and *L. muscarium* (Sharma et al., 2023a). Goettel et al. (2008) documented the ability of *Lecanicillium* spp. to function as a solitary microbial control agent against various insect pests such as aphids, mealy bugs, scale insects, and whiteflies. *Lecanicillium lecanii* demonstrated efficacy against aphids, mealybugs, and spider mites, whereas *L. attenuatum* showed effectiveness against *Aedes albopictus* (asian tiger mosquito) and *Plutella xylostella* (diamondback moth) (Table 14.1).

14.2.4 Isaria

The *Isaria* genus belongs to the Cordycipitaceae family and consists of noteworthy EPF, formerly referred to as *Paecilomyces*. The entomopathogenic species of *Isaria* consists of *I. farinosa* and *I. fumosorosea* exhibiting synnematous growth of conidiophores. They are septate and bear whorls of different branches and conidiogenous cells (phialides). The conidiogenous cells are reported to have an oval or flask-like form, or are sub-globose with a noticeable neck. They are found individually or in clusters. The conidia are unicellular, transparent to light-coloured (Das et al., 2021). *Isaria fumosorosea* is a pathogen that causes a condition known as "sickness" in whiteflies, often called "yellow muscardine". It serves as a parasite of whiteflies, has a broader spectrum of hosts, and mostly targets lepidopteran pests. Nevertheless, other strains of *Isaria* have demonstrated effectiveness in combating whiteflies, tea jassids, thrips, aphids, termites, and also some dipterans (Table 14.1).

14.2.5 Hirsutella

The genus *Hirsutella* (Cordycipitaceae) includes entomopathogenic species such as *H. thompsonii*, *H. sinensis*, *H. citriformis*, and *H. gigantea* (Tigano et al., 2006; Sharma et al., 2024b). Identifying and distinguishing the members of this genus from other fungal genera is challenging due to a large number of species and their high variety. If present, the synnemata is upright, cylindrical, or slightly tapered, ranging from being long and hair-like to short and having wart-like projections. It may have few or no branches at all or numerous short side branches (Mongkolsamrit et al., 2012). The conidiogenous cells are sub-globose. *Hirsutella thompsonii* has the potential to be utilized as a biocontrol agent against a variety of pests, including *Phyllocoptruta oleivora* (citrus rust mite), *Varroa destructor* (varroa mite), and certain high nuisance species within the Lepidoptera and Hemiptera groups (Table 14.1) (Behie et al., 2015).

14.3 Mode of Action

EPF possess an intricate and distinctive mechanism by which they get possession of the interior environment of insects and mite pests, either by breaching the outer protective layer, cuticle or through the mouthparts. The cuticle serves as the first line of defence against fungal parasites. From a general perspective, the life cycle of EPF commences with the germination of infectious spores on the host cuticle, leading to the formation of a germ tube, which then penetrates the cuticle and multiplies within the host to generate toxins that ultimately result in the demise of the insect (Fig. 14.1). Under favourable climatic conditions, the fungus emerges from the insect carcass and disseminates in the vicinity. It is observed that favourable environmental circumstances and a host that is vulnerable to infection are the key factors for the survival, reproduction, and effective infection of EPF.

14.3.1 Adhesion and Germination

The spores adhere to the surface of the host through passive transport by air and water and are the first stage of mycosis. They have the ability to germinate on any part of the cuticle. However, researchers have also found specific locations where germination occurs in many insects (Ortiz-Urquiza et al., 2016). According to Holder et al. (2007), desiccated spores of *B. bassiana* were reported to exhibit hydrophobic rodlets coated with protein hydrophobins on their outermost layer. The presence of this hydrophobic layer is exclusive to the conidial stage and is not found on the mycelial portions of these fungi. The dry spores adhere to the cuticle through hydrophobic forces exerted by the rodlets. This process of surface rodlet formation, which results in surface hydrophobicity and virulence, is facilitated by two hydrophobins (Hyd1 and Hyd2) (Cho et al., 2007).

On the genomic level, Wang and St Leger (2007) characterized two mitotic arrest-deficient (*Mad*) genes in *M. anisopliae*, *Mad1* and *Mad2*, which facilitate the adhesion of spores during pathogenesis. The *Mad1* gene is not specific as regards the adhesion process, but its absence leads to reduced spore adherence and germination, resulting in low pathogenicity. The *Mad2* gene, on the other hand, is mostly associated with the active stage and has a specific target (Ortiz-Urquiza & Keyhani, 2013). Upon attaching to the host surface, the spore undergoes germination, with its subsequent growth being significantly influenced by factors such as pH, temperature, oxygen levels, and nutrient availability (Fig. 14.1). Fungi with a wide range of

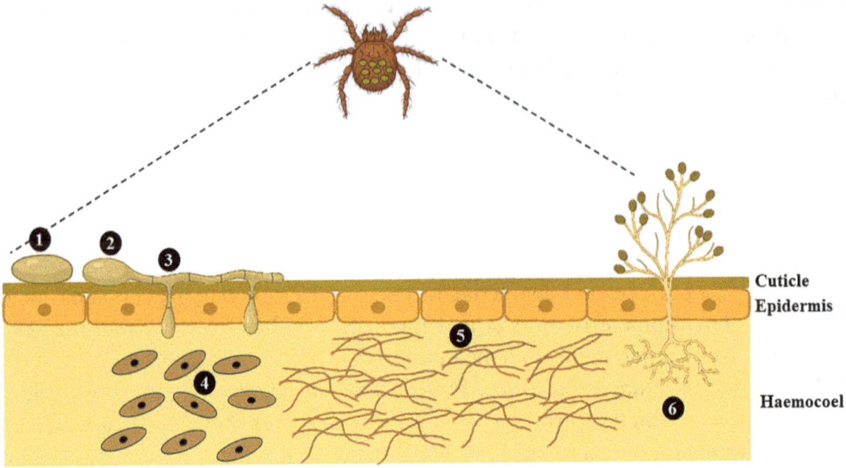

Fig. 14.1 Schematic representation of the mode of action of fungal-based pesticides. (1) Attachment of conidium, (2) Germination, (3) Cuticular and epidermal penetration, (4) Propagation of blastospores in haemocoel, (5) Vegetative fungal growth in haemocoel and (6) Formation of conidiophores bearing conidia

hosts are able to germinate when exposed to a diverse array of carbon and nitrogen sources. Nevertheless, certain EPF necessitate particular chemicals for their growth and development and exhibit a limited host range (Ortiz-Urquiza et al., 2013).

14.3.2 Host Penetration

Following the successful attachment to the cuticle, EPF must next pierce the epidermis and access the internal host body parts in order to absorb nutrients necessary for their vegetative and reproductive development (Fig. 14.1). Both enzymatic breakdown and mechanical stresses are necessary for this process (Cheong et al., 2016). Dark melanotic lesions on the host cuticle are a distinctive trait indicating spore penetration. During penetration, different classes of enzymes, such as proteases, chitinases, and lipases, are secreted in order to degrade the cuticle. Proteases, such as trypsin, chymotrypsin, esterase, chymoelastase, and collagenase, have been identified to have a wide spectrum of functions. The formation of appressoria is associated with the initial production of endoproteases on the host cuticle, particularly subtilisin-like enzymes (Pr1) and trypsin-like enzymes (Pr2), thus aiding the penetration process (St. Leger, 1995). Furthermore, the rate of N-acetylglucosaminidase synthesis is comparatively lower than that of proteolytic enzymes.

Beauveria bassiana requires a total of eight cytochrome *P450* genes (*CYP52-X1*, *CYP655–C1*, *CYP5337-A1*, *CYP52-G11*, *CYP539–B5*, *CYP617–N1*, *CYP53-A26*,

CYP584-Q1) to encode enzymes for the digestion of lipids in the insect cuticle. Additionally, it needs four catalases (CatB, CatP, CatC, CatD), aldehyde dehydrogenases (ADH-2 FALDH), a long-chain alcohol (3-oxoacyl carrier protein reductase), and a recognized hydrocarbon carrier protein (Acyl-CoA oxidase) to be involved in this process (Pedrini et al., 2013). These enzymes collectively cooperate to enhance the penetration process during pathogenesis. After breaching the cuticle, the fungus aims at getting access to the host respiratory system in order to obtain optimal nutrition. The production of blastospores and vegetative hyphae accomplishes this. Host mortality occurs as a consequence of multiple circumstances, including mycotoxins, tissue infiltration, and insufficient nutrition.

14.3.3 Toxin Produced by EPF

Fungi produce a wide array of mycotoxins, which have been found in many tissues of insects and mites, and adversely affect their health. Cytotoxins modify the cellular structure and facilitate hyphae penetration (Sheppard & Filler, 2014). Additionally, neuromuscular toxins have an impact on the well-being of insects, including overall paralysis, reduced responsiveness, and lethargy. Eyal et al. (1994) extracted several toxins from *B. bassiana*, including beauverolides, beauvericin, isarolides, and brassinolide. Moreover, *M. anisopliae* is also known to produce both destruxins (DTXs) and cytochalasins during insect and mite pathogenesis. DTXs have the ability to stimulate Ca channels, leading to the depolarization of the muscle membrane in lepidopteran insects. The subsequent mycotoxins that EPF produce during pest pathogenesis eventually lead to host mortality (Table 14.2).

Table 14.2 Toxins produced by different entomopathogenic fungi (EPF)

Toxin	Fungi	Function	Reference
Destruxins	*Metarhizium anisopliae*	Immunodepressant activity	Haron et al. (2020)
Beauvericin	*Beauveria bassiana, Fusarium semitectum, Fusarium moniliforme, Isaria fumosorosea*	Beauvericin causes cell apoptosis, directs the flux of Ca^{2+}, into the cells and result in cytotoxicity.	Qadri et al. (2011)
Bassianolide	*Verticillium lecani, Beauveria bassiana*	Acts as ionophore with toxic effect on insect	Kanaoka et al. (1978)
Leucinostatins	*Purpureocillium lilacinus, Paecilomyces marquandii, Isaria farinosus*	Cytotoxic effect and insecticidal properties	Mikami et al. (1989)
Efrapeptins	*Tolypocladium inflatum*	Inhibitors of mitochondrial oxidative phosphorylation and ATPase activity	Goettel et al. (2010)

14.3.4 Destruxins

Destruxins (DTXs) are cyclic depsipeptides composed of D-α hydroxyl acid linked to five amino acids and were initially extracted from *M. anisopliae* (Kodaira, 1961). A total of 28 destruxins have been identified, primarily from various *Metarhizium* species, exhibiting varying levels of activity (Peng et al., 2005). The DTX level has been linked to the degree of pathogenicity and the ability to infect specific hosts. Studies on DTXs activity have demonstrated disruption of the insect cellular immune system, the prevention of nodule development, and the interference with defensive processes such as phagocytosis (Pal et al., 2007). DTXs cause an immunosuppressive effect in the host, followed by mycelial development (Table 14.1).

DTX-E, within the DTXs family, exhibited both cytotoxic and cytostatic effects in murine leukaemia cells and also activated calcium channels in insect muscles (Chen et al., 2014). Furthermore, the study also demonstrated that DTX-A and DTX-B caused a fast influx of Ca^{2+} in hemocytes. Nevertheless, the channel inhibitors, specifically 2-APB (2-aminoethoxydiphenyl borane) and U73122, did not have any impact on this fast inflow. Chen et al. (2014) also found that DTX-A at a concentration of 690 μmol/L and DTX-B at a concentration of 20 μmol/L caused a notable reduction in intracellular free H^+. The data elucidate that the primary function of DTXs is to promote the pathogens colonization in host tissues by altering their ion channels. Consequently, these toxins facilitate insect mortality upon infection.

14.3.5 Beauvericin

Beauvericin is sourced from several fungi such as *B. bassiana*, *Fusarium semitectum*, *F. moniliforme*, *I. fumosorosea*, and certain plant pathogenic fungi that belong to the phylum Basidiomycota (Wang & Xu, 2012; Qasim et al., 2018). Beauvericin has a tendency to form complexes with other cations and possesses insecticidal potential against mosquitoes, butterflies, and potato beetles (Luz et al., 2017). Beauvericin induces cell death by elevating oxidative stress through the generation of reactive oxygen species (ROS), disrupting mitochondrial and hormonal signaling pathways (Tedjiotsop Feudjio et al., 2010; Qadri et al., 2011). Like DTX, beauvericin also facilitates the influx of Ca^{2+} into the cells, elevating intracellular Ca^{2+} concentration, eventually leading to cytotoxicity (Qadri et al., 2011).

14.3.6 Bassianolide

Bassianolide is a toxic compound derived from *L. lecanii* and *B. bassiana* strains (Ravindran et al., 2018). Bassianolide is a cyclic polymer composed of d-α-hydroxyisovaleryl L-N-methylleucinol and functions as an ionophore, similar to other depsipeptides and renowned for its toxic impact on insects. Administering a dosage of 2 µg/larva of bassianolide, suspended in water, to larvae of *Bombyx mori* can induce atonic symptoms. Curiously, the toxin exhibited greater harmful effects when administered at a dosage exceeding 5 µg/larva (Kanaoka et al., 1978). A study conducted by Ali et al. (2017) found that bassianolide, produced by *L. muscarium*, can influence the acetylcholine receptors in insect muscles. This biochemical alteration ultimately results in a decrease in acetylcholine-E generation within the whiteflies host body and causes subsequent pest mortality.

14.3.7 Leucinostatins

In the realm of mycotoxins, leucinostatins are gaining a lot of attention due to their pathogenicity against insect and mite pests. They exhibit significant insecticidal properties against the *Leptinotarsa decemlineata* (colorado potato beetle) and are commonly found in the fungal species *Purpureocillium lilacinus*, *Paecilomyces marquandii*, and *I. farinosa* (Gökçe et al., 2006). Chemically, it is described as l-threo-b-hydroxy leucine, 2-amino-6-hydroxy-4 methyl-8-oxodecanoic acid (AHMOD) and cis-4-methyl-L-proline (Mori et al., 1982; Wang et al., 2016).

14.3.8 Efrapeptins

The main sources of efrapeptins, a compound mixture of antibiotic peptides, are the fungal soil hyphomycetes, *B. nivea* and *Tolypocladium niveum* (Niu et al., 2020). It is worth noting that efrapeptins demonstrate toxic effects against the Colorado potato beetle and *Tetranychus urticae* (two-spotted spider mite). In general, all the peptides exhibited inhibitory effects on ATPase activity and mitochondrial oxidative phosphorylation, hence assisting in the control of pest populations. According to Niu et al. (2020), peptides may serve as the catalytic site for competitive inhibitors that bind to the soluble F1 portion of the mitochondrial ATPase. These changes at cellular levels indicate the effectiveness of efrapeptins in managing insect and mite populations (Brückner & Degenkolb, 2020).

14.4 Formulations and Application Methods

Undoubtedly, fungal-based insecticides and acaricides play a crucial role in IPM, and the efficiency of these biopesticides is heavily influenced by the formulation and distribution mechanisms employed. Typical formulations used in practice include wettable powders, suspensions, granules, dusts, and seed coats (Table 14.1). The suspensions are administered by applying a liquid spray on the leaves and by drenching the soil. The foliar spray method of application entails diluting the formulation with water and directly spraying it onto the plant foliage. This method effectively controls a diverse array of foliar pests (Mascarin & Jaronski, 2016). On the other hand, the soil drenching method involves pouring or spraying liquid formulations at the base of plants or directly into the soil. This method specifically targets soil-dwelling pests, such as nematodes and grubs (Jackson & Dunlap, 2003). In addition to suspensions, dust and seed coats are appropriate for soil applications and can be distributed using standard spreaders. Seed coating, on the other hand, provides protection to seedlings from soil inhabiting pests, and commonly, mycorrhizal fungi are employed for this purpose (Jeffries et al., 2003).

Moreover, as a result of technological progress, fungal-based nanoparticles and biopolymer-based formulations are gaining recognition as cutting-edge methods for delivering EPF as insecticides and acaricides. A variety of metals, such as selenium, copper, zinc, iron, and silver are used to develop fungal-based nanoparticles. Silver nanoparticles are the most efficacious and environmentally friendly (Bihal et al., 2023). These nanoparticles are used to control insect populations using spray techniques. Currently, there are no available insecticides and acaricides in the market that specifically pertain to fungal-based nanoparticles. Another technological intervention to overcome fungal-based insecticides and acaricides limitations, such as low germination rate and UV damage, is the introduction of biopolymers for formulation development. Various biopolymers, such as hydroxyethylcellulose, carboxymethylcellulose, chitosan, alginate, and starch, have been utilized for the development of entomopathogenic fungal formulations (Batista et al., 2018). Biopolymers can enhance the efficacy of EPF by prolonging the viability, stability, adherence, and persistence of the fungal spores, thereby increasing the effectiveness of EPF as insecticides and acaricides, under field conditions (Cafarchia et al., 2022).

The development of biopolymer-based formulations involves the utilization of microencapsulation and nanoencapsulation techniques. These techniques entail encapsulating fungi within small polymer spheres or capsules, which serve as protective barriers. The purpose of these barriers is to shield the EPF from various environmental stressors, including UV radiation and desiccation. Encapsulation improves the durability and viability of conidia in outdoor environments, as well as the capacity of fungal spores to attach to the cuticle of the targeted insects. It also extends the product shelf life, enhancing the overall effectiveness of EPF as an insecticide and acaricide (Riseh et al., 2022). Alginate is commonly used for encapsulating and producing granules, nano or microspheres that contain conidia of *M. anisopliae* and *B. bassiana* (Shah et al., 2023). Additionally, researchers have utilized starch and chitosan to create microspheres that encapsulate *B. bassiana* spores (Felizatti et al., 2021). In addition, the delivery of formulations based on biopolymers can pose greater challenges than formulations based on standard or synthetic polymers. This is because the particle size and charge can affect the product effectiveness (Cafarchia et al., 2022). Despite the potential of biopolymer-based formulations of EPF for the production of efficient and eco-friendly insecticides and acaricides, additional research is required to overcome the challenges associated with their development and to enhance their utilization in agricultural systems. This will help achieve a more favourable balance between the cost and benefits of their application. Several ways can be employed to enhance cost efficiency. One possible strategy is to investigate alternate sources of biopolymers that could be more economically viable and easily accessible, such as waste materials and agricultural residues.

14.5 Factors Influencing Efficacy of Fungal-Based Insecticides and Acaricides

Fungal-based insecticides and acaricides have gained popularity due to their eco-friendly nature and relatively low impact on non-target organisms. However, their efficacy can be influenced by a range of factors. Understanding these factors is crucial for maximizing their effectiveness in pest management. Here, we will delve into the key factors influencing the efficacy of fungal-based biopesticides.

14.5.1 Fungal Strain Selection

The selection of the fungal strain is a crucial component that significantly impacts effectiveness. The process entails meticulously evaluating and choosing the optimal fungal strain, considering multiple parameters. Fungal species and strains differ in their host ranges, pathogenicity, and capacity to adapt to different environments (Gul et al., 2014). Choosing the suitable strain for the specific pest is crucial for achieving success. Certain strains have greater efficacy against particular pests, whereas others possess a wider range of effectiveness (Jaronski, 2010).

14.5.2 Environmental Conditions

Fungal-based biopesticides can be greatly influenced by environmental conditions, including temperature, humidity, and UV radiation, which can considerably impact their effectiveness. Different abiotic environmental conditions influence the effectiveness of EPFs at the strain level. Therefore, the most effective approach to enhance their utilization is to assess the tolerance of individual strains to specific environmental constraints and select strains that are well-suited to the target environment or location based on their ability to withstand major climatic stresses (Couceiro et al., 2021). EPFs exhibit optimal growth under particular temperature and humidity conditions and their effectiveness can be diminished by elevated temperatures or decreased humidity levels. It is reported that UV radiation has the ability to eliminate fungus spores (Faria et al., 2017).

14.5.3 Application Methods

The method of application plays a critical role in efficacy. Diverse application techniques are utilized based on the specific pest being targeted, prevailing environmental circumstances, and the particular composition of the EPF. Optimal spray droplet size, pressure, and volume must be achieved to ensure the delivery of an efficient dosage. The primary techniques of administration encompass foliage spray, seed treatment, soil integration, and dusting (Lacey et al., 2015). Inadequate utilization may lead to unequal dispersion and diminished effectiveness (Shah & Pell, 2003).

14.5.4 Application Timing

The timing of the application is critical for any pesticide to achieve the desired results. Optimal efficacy of fungal-derived biopesticides is achieved by applying them at particular developmental phases of the pest, usually during the initial instars or when the pests are most vulnerable. Their effectiveness can be diminished by using EPF either too late or too early (Inglis et al., 1997). For optimal pest control, certain EPF products may necessitate several applications. It is essential to comprehend the endurance of the EPF and the necessity for reapplication in order to determine the appropriate timing for subsequent treatments (Ugine et al., 2007). Additionally, it is important to consider the projected weather conditions, such as intense precipitation or severe temperatures, as these factors might promptly diminish the effectiveness of EPF. Therefore, it is crucial to select a time when the weather is favourable for EPF activity.

14.5.5 Population Density of Target Pests

The density of the target insect population also influences the effectiveness of fungal-based biopesticides. Gaining insight into the dynamics of pest populations is essential for implementing efficient pest management and control techniques, as it enables the precise and timely deployment of control measures. Applying EPF during the height of pest populations can enhance the efficacy of pest management and effectively suppress the pest. Increased pest populations may necessitate greater quantities of the biopesticide for efficient management. It is recommended that the application schedule be modified according to the fluctuations in pest population dynamics (Meyling & Eilenberg, 2007).

14.5.6 Resistance Management

Pests can gradually acquire resistance to fungal biopesticides. In order to reduce resistance, it is crucial to alternate between different strains or modes of action and refrain from using the same fungal strain continuously (Lecuona et al., 2019). To reduce the selective pressure on pests and prevent them from developing resistance to a certain control agent, it is important to use the correct dosages of EPF and avoid overreliance on high dosages (Ambethgar, 2009). Consistent surveillance of pest populations and prompt identification of any indications of resistance are essential for adopting timely tactics for control.

14.6 Advantages of EPF

The utilization of EPF for the purpose of managing insect and mite pests has numerous benefits within the framework of Integrated Pest Management (IPM) (Khan et al., 2012). Fungi are natural organisms that possess the ability to infect and exterminate insects, rendering them a sustainable and ecologically sound alternative for pest management. They avoid or lessen the introduction of synthetic chemicals into the ecosystem, hence mitigating the hazards associated with chemical pesticides. These fungi have a high degree of specificity towards their intended pests, hence reducing the risk of harm to non-target organisms such as humans, pets and wildlife, and beneficial insects. Compared to chemical pesticides, they exert little influence on the environment since they do not accumulate in the food chain or water sources (Islam et al., 2017). Certain EPF exhibit enduring pest control capabilities due to their ability to stay in the environment and consistently infect successive generations of pests.

EPF are permitted in organic agricultural systems due to their classification as natural and sustainable means of pest management. The fungi undergo natural degradation once the pest population has been effectively managed, resulting in the absence of any detrimental leftovers in the environment. Most of these EPF are often regarded as non-hazardous to humans, animals, and beneficial insects, thereby making them a preferable choice for managing pests in locations where people and animals reside (Haase et al., 2015). EPF have a lower likelihood of developing resistance in pests compared to chemical pesticides, because these fungi use many ways and diverse strategies to infect and eliminate pests, including attachment, penetration, and synthesis of mycotoxins, thereby impeding the pest ability to acquire resistance and effectively kill them. These are advocated for their inclusion as a major component of an IPM initiative that employs several tactics to manage pests effectively, in a comprehensive and environment-friendly manner. Although EPF may have a greater initial cost compared to chemical pesticides, they offer a cost-effective and sustainable solution for pest control by minimizing the requirement for frequent pesticide applications. The utilization of these fungi aids in fostering a healthier and more harmonized ecosystem through the regulation of insect populations, the conservation of biodiversity, and the mitigation of adverse effects caused by chemical pesticides (Neale et al., 2024). Certain strains of EPF can be extracted and utilized from nearby surroundings, resulting in enhanced efficacy in managing pests that have adapted to the specific area (Baverstock et al., 2010).

Thus, integrating EPF into pest management strategies provides a sustainable and efficient substitute for chemical pesticides, diminishing the ecological and health hazards of conventional pest control approaches. Nevertheless, it is crucial to take into account the particular environment and pest species when selecting the suitable EPF for management.

14.7 Limitations of EPF

Fungal-derived insecticides and acaricides have garnered interest as ecologically sound substitutes for chemical pesticides. They have numerous benefits, including precise targeting, biodegradability, and minimal harm to non-target organisms. Nevertheless, they also entail a distinct array of difficulties and constraints. Fungal-derived biopesticides are susceptible to environmental variables such as temperature, humidity, and UV radiation. The effectiveness of biological pesticides varies considerably depending on the specific geographic locations (Shah & Pell, 2014). Fungal-derived biopesticides typically exhibit a slower onset of action than chemical alternatives. This delay can impede their efficacy in conditions that demand prompt pest management (Khan et al., 2012). The majority of fungal biopesticides exhibit a limited range of effectiveness, selectively targeting particular pests or categories of pests. This limits their effectiveness in controlling a broad spectrum of agricultural and horticultural pests (Meyling & Eilenberg, 2007).

Utilizing fungal biopesticides can be more intricate than using chemical pesticides, as it necessitates precise conditions and approaches. According to Glare and Inwood (2019), fungal-based biopesticides often have a shorter effective duration than chemical pesticides. This can occur if they are not maintained and handled correctly, decreasing their ability to work effectively. The regulatory procedures involved in approving and registering fungal biopesticides can be protracted and expensive, which may restrict their accessibility and acceptance. The effectiveness of fungal biopesticides may change between various fungal strains and pest species. Improving the practicability and effectiveness of fungal-based insecticides and acaricides is crucial for sustainable pest management and their integration into management programs.

Current investigations into mycological insecticides and acaricides seek to tackle the difficulties and constraints linked to these biological pesticides. Researchers are studying methods to improve the ecological robustness of EPFs, hence increasing their efficacy across many environmental situations. This includes carefully choosing and cultivating strains that possess enhanced resistance to UV radiation, variations in temperature, and moisture levels (Resende et al., 2021). Scientists are currently advancing novel techniques and compositions to streamline the use of fungal biopesticides. This encompasses investigations on sprayable formulations and innovative delivery techniques. Attempts are underway to expand the range of fungal biopesticides by carefully choosing or genetically altering fungal strains to effectively combat a broader array of pests, such as mites and different types of insects (Quesada-Moraga et al., 2007). Research is focused on enhancing the longevity of fungal biopesticides by advancing storage methods and preservation procedures and creating more stable formulations (Tavladoraki et al., 2019). Collaboration between researchers and policymakers is underway to simplify the regulatory procedures for fungal biopesticides, facilitating the approval and market accessibility of these environment-friendly options.

Current research aims at identifying and cultivating exceptionally potent strains of EPF by utilizing genetic selection, optimizing growth conditions, and creating innovative delivery techniques (Ortiz-Urquiza et al., 2013). Continuous research endeavours are crucial for surmounting the obstacles and constraints of fungal-based insecticides and acaricides. They have the ability to enhance the reliability and practicality of these biopesticides as tools for IPM while simultaneously decreasing the environmental repercussions of conventional chemical pesticides.

14.8 Regulatory and Safety Considerations

Fungal-derived insecticides and acaricides have become increasingly popular as a more ecologically sound and enduring substitute for chemical pesticides. Various rules, governing organizations, and safety measures have been established to guarantee the secure handling and application of the subject at hand. The United States Environmental Protection Agency (EPA) has a key role in the regulation of pesticides in the US. Fungal biopesticides are subject to a registration process that is analogous to that of chemical pesticides. The EPA standards guarantee the safety of these items for the environment, humans, and non-target organisms (EPA Pesticide Regulation, 2024). According to the EPA Pesticide Regulation, the Federal Insecticide, Fungicide, and Rodenticide Act (FIFRA) and the Federal Food, Drug, and Cosmetic Act (FFDCA) establish pesticide restrictions.

The European Food Safety Authority (EFSA) assesses the safety and effectiveness of fungal biopesticides in Europe. They offer scientific recommendations that direct the process of approving and registering. Prior to being allowed for use inside the European Union (European Food Safety Authority, 2024), fungal biopesticides must satisfy strict requirements on safety, efficacy, and environmental consequences. Furthermore, the World Health Organization (WHO) assumes responsibility for providing international directives for the management and safety of pesticides (World Health Organization, 2014). Although it does not possess regulatory authority, it plays a role in developing safety processes and standards. The WHO advocates for the judicious application of pesticides to safeguard human well-being and the natural surroundings.

The Central Insecticides Board (CIB) in India is a regulatory entity that oversees the registration and regulation of all pesticides, including biopesticides. Its primary role is to guarantee that these products adhere to safety and efficacy criteria, and it grants permits for their production, importation, and sale. The Registration Committee (RC), which is part of the CIB, is in charge of reviewing and approving the registration of different types of pesticides, including fungal-based biopesticides such as insecticides and acaricides. Prior to awarding registration, they assess safety, efficacy and other relevant data required for registration of pesticides in India (CIB&RC, 2023).

While working with fungal-based insecticides and acaricides, pesticide applicators and handlers must adhere to safety guidelines and regulations, including proper personal protective equipment (PPE). These items may include gloves, coveralls, masks, and protective eyewear in order to reduce skin contact, inhalation, and eye exposure. The efficacy and safety of fungal biopesticides heavily rely on the application method. Farmers, applicators and handlers need to adhere to the guidelines provided by the manufacturer to ensure correct implementation methods, such as using sprays, dusting, or integrating into the soil. It is crucial to calibrate equipment and strictly follow label instructions. Refrain from using fungal biopesticides in close proximity to water bodies or during windy conditions that may result in the accidental dispersion of the substances.

To safeguard non-target organisms and ecosystems, it is imperative to take into account the establishment of buffer zones and strictly follow the guidelines provided on labels. Ensuring the appropriate storage and disposal of fungal biopesticides is of utmost importance. Store them in a cold and dry location, away from food and feed storage. Properly discard containers and any extra product associated with them in accordance with local legislation. Strict compliance with these standards and adherence to safety protocols are essential for the responsible and efficient utilization of fungal-based insecticides and acaricides. By adhering to these principles and optimal methodologies, we may effectively utilize the capabilities of biopesticides while mitigating possible hazards to both human health and the environment.

14.9 Market Status

Regardless of biological or chemical origin, pesticides are used globally to reduce crop damage and increase agricultural productivity. The global pesticide market was valued at $45.35 billion in 2022 and is expected to grow to $53.44 billion by 2035 (The Business Research Company, 2022). The global biopesticides market was worth $1.8 billion in 2021 and is expected to reach $2.3 billion by 2027 (Statista, 2023). FAO (2021) reported that China had the largest pesticide consumption (13.1 kg/ha), whilst India stood at 13th position with a consumption of 0.3 kg/ha and Comoros had the lowest consumption (0.0006 kg/ha) (Fig. 14.2a).

The worldwide and Indian market size for bio-insecticides and -acaricides are increasing, as chemical-based insecticides and acaricides are reported to possess negative impact on human well-being. In India, chemical-based insecticides account for 79% of the total registered insecticides, with the remaining fraction being occupied by bioinsecticides (Fig. 14.2b). Within the bioinsecticide category, the predominantly registered insecticides are derived from fungi, accounting for 43% of the total followed by bacterial and virus-derived bioinsecticides (Fig. 14.2c). It is observed that the genus *Beaveria* had the highest number of registered fungal-based insecticides followed by *Verticillium* and *Metarhizium* (Fig. 14.2d). It is worth mentioning that over the past decade, there has been a noticeable trend towards the increased use and advancement of biopesticides. Furthermore, it is anticipated that upcoming technological progress and the investigation of novel genera for the production of fungal-based insecticides and acaricides could have advantageous outcomes for the agricultural sector. Additionally, this research may aid in mitigating the detrimental impact of chemical pesticides on the environment.

14.10 Conclusions

This chapter examined the various aspects of EPF, including their efficacy, advantages and limitations over chemical pesticides for managing insect and mite pests. In India, the bulk of bioinsecticides are derived from fungi, with *B. bassiana* being predominant and widely explored (Fig. 14.2c, d). With technological advancement,

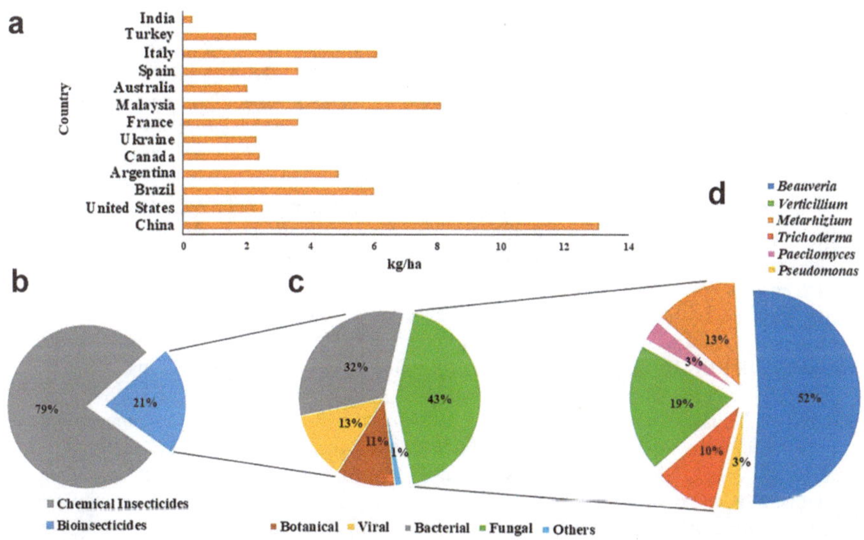

Fig. 14.2 Usage and market share of pesticides. (**a**) Global pesticide usage (FAO, 2021); (**b**), (**c**) and (**d**) Registered insecticides in India along with their origin

researchers have successfully overcome the constraints associated with fungal-based insecticides and acaricides. However, additional research is required to develop more efficacious and stable formulations for field applications. Furthermore, exploration studies are necessary to uncover other efficacious native fungal species that have greater potential than *B. bassiana*. Considering the insect and mite pest management in India, it is advisable to follow the guidelines of CIB&RC regarding the appropriate dosages of fungal-based insecticides and acaricides.

Generally, farmers are influenced by the quick effectiveness of chemical pesticides, which poses a challenge for them in embracing biopesticides. Therefore, it is very important to acquaint them with the negative consequences of chemical pesticides and the advantages of using biopesticides. Over the past few decades, there has been a consumer trend towards organic agricultural products, which limits the use of chemical pesticides. Therefore, it is widely predicted that using EPF as biocontrol agents will see significant and rapid growth. The achievement of chemical-free agricultural output depends on the active participation of agricultural communities. Hence, the State and Central government officials must coordinate campaigns, demonstrations, and hands-on training to educate farmers about the benefits of using biopesticides.

Acknowledgements The authors are thankful to Director of CSIR-IHBT, for supporting this work. Aditya Singh Ranout (HCP-0037), Rahul Kumar (NWP-0100), and Rupinder Kaur (HCP-0101) are thankful to CSIR-IHBT for funding. CSIR-IHBT communication number of this manuscript is 5436.

References

Ali, S., Zhang, C., Wang, Z., Wang, X. M., Wu, J. H., et al. (2017). Toxicological and biochemical basis of synergism between the entomopathogenic fungus *Lecanicillium muscarium* and the insecticide matrine against *Bemisia tabaci* (Gennadius). *Scientific Reports, 7*, 46558.

Ambethgar, V. (2009). Potential of entomopathogenic fungi in insecticide resistance management (IRM): A review. *Journal of Biopesticides, 2*, 177–193.

Bamisile, B. S., Dash, C. K., Akutse, K. S., Qasim, M., Ramos Aguila, L. C., et al. (2019). Endophytic *Beauveria bassiana* in foliar-treated Citrus Limon plants acting as a growth suppressor to three successive generations of *Diaphorina citri* Kuwayama (Hemiptera: Liviidae). *Insects, 10*, 176.

Batista, A. C. D. L., Souza Neto, F. E. D., & Paiva, W. D. S. (2018). Review of fungal chitosan: Past, present and perspectives in Brazil. *Polímeros, 28*, 275–283.

Baverstock, J., Roy, H. E., & Pell, J. K. (2010). Entomopathogenic fungi and insect behaviour: From unsuspecting hosts to targeted vectors. *Biological Control, 55*, 89–102.

Behie, S. W., Jones, S. J., & Bidochka, M. J. (2015). Plant tissue localization of the endophytic insect pathogenic fungi *Metarhizium* and *Beauveria*. *Fungal Ecology, 13*, 112–119.

Bihal, R., Al-Khayri, J. M., Banu, A. N., Kudesia, N., Ahmed, F. K., et al. (2023). Entomopathogenic fungi: An eco-friendly synthesis of sustainable nanoparticles and their nanopesticide properties. *Microorganisms, 11*, 1617.

Brückner, H., & Degenkolb, T. (2020). Sequences of tolypins, insecticidal efrapeptin-type peptaibiotics from species of the fungal genus *Tolypocladium*. *Chemistry & Biodiversity, 17*, e2000276.

Brunner-Mendoza, C., Reyes-Montes, M. D. R., Moonjely, S., Bidochka, M. J., & Toriello, C. (2019). A review on the genus *Metarhizium* as an entomopathogenic microbial biocontrol agent with emphasis on its use and utility in Mexico. *Biocontrol Science and Technology, 29*, 83–102.

Cafarchia, C., Pellegrino, R., Romano, V., Friuli, M., Demitri, C., et al. (2022). Delivery and effectiveness of entomopathogenic fungi for mosquito and tick control: Current knowledge and research challenges. *Acta Tropica, 106627*, 106627.

Castro, T., Mayerhofer, J., Enkerli, J., Eilenberg, J., Meyling, N. V., et al. (2016). Persistence of Brazilian isolates of the entomopathogenic fungi *Metarhizium anisopliae* and *M. Robertsii* in strawberry crop soil after soil drench application. *Agriculture, Ecosystems & Environment, 233*, 361–369.

Chen, X. R., Hu, Q. B., Yu, X. Q., & Ren, S. X. (2014). Effects of destruxins on free calcium and hydrogen ions in insect hemocytes. *Insect Science, 21*(1), 31–38.

Cheong, P., Glare, T. R., Rostás, M., & Haines, S. R. (2016). Measuring chitinase and protease activity in cultures of fungal entomopathogens. *Methods in Molecular Biology, 1477*, 177–189.

Cho, E. M., Kirkland, B. H., Holder, D. J., & Keyhani, N. O. (2007). Phage display cDNA cloning and expression analysis of hydrophobins from the entomopathogenic fungus *Beauveria* (Cordyceps) *bassiana. Microbiology, 153*, 3438–3447.

CIB&RC (Central Insecticides Board & Registration Committee). (2023). https://ppqs.gov.in/divisions/cibrc/registered-products

Couceiro, J., Fatoretto, M., Demétrio, C., Meyling, N., & Delalibera, I. (2021). UV-B radiation tolerance and temperature-dependent activity within the entomopathogenic fungal genus Metarhizium in Brazil. *Frontiers in Fungal Biology, 2*, 645737.

Das, G., Shin, H. S., Leyva-Gómez, G., Prado-Audelo, M. L. D., Cortes, H., et al. (2021). *Cordyceps* spp.: A review on its immune-stimulatory and other biological potentials. *Frontiers in Pharmacology, 11*, 2250.

Dhaliwal, G. S., Jindal, V., & Mohindru, B. (2015). Crop losses due to insect pests: Global and Indian scenario. *Indian Journal Of Entomology, 77*, 165–168.

EFSA (European Food Safety Authority). (2024). https://www.efsa.europa.eu/en

EPA (US Environmental Protection Agency) Pesticide Regulation. (2024). https://www.epa.gov/regulatoryinformation-topic/regulatory-and-guidance-information-topic-pesticides

Eyal, J. M. D. A., Mabud, M. A., Fischbein, K. L., Walter, J. F., Osborne, L. S., & Landa, Z. (1994). Assessment of *Beauveria bassiana* Nov. EO-1 strain, which produces a red pigment for microbial control. *Applied Biochemistry and Biotechnology, 44*, 65–80.

FAO (Food and Agriculture Organization of the United Nations). (2021). *Food and agriculture data*. https://www.fao.org/faostat/en/#home.

Faria, M. R., Wraight, S. P., & Myron Smith, J. E. (2017). Factors affecting the development of fungal bioinsecticides for field and greenhouse use. *Biological Control, 110*, 23–32.

Felizatti, A. P., Manzano, R. M., Rodrigues, I. M. W., da Silva, M. F. D. G. F., Fernandes, J. B., & Forim, M. R. (2021). Encapsulation of *B. bassiana* in biopolymers: Improving microbiology of insect pest control. *Frontiers in Microbiology, 12*, 704812.

Glare, T. R., & Inwood, A. J. (2019). Morbidity of biopesticides to non-target insects: Strains of *Agrotis segetum* infected with *Metarhizium anisopliae* and *Beauveria bassiana*. *Biological Control, 64*, 283–291.

Goettel, M. S., Koike, M., Kim, J. J., Aiuchi, D., Shinya, R., & Brodeur, J. (2008). Potential of *Lecanicillium* spp. for management of insects, nematodes and plant diseases. *Journal of Invertebrate Pathology, 98*, 256–261.

Goettel, M. S., Ellenberg, J., & Glare, T. (2010). 11 Entomopathogenic fungi and their role in regulation of insect populations. In L. I. Gilbert & S. S. Gill (Eds.), *Insect control: Biological and synthetic agents* (pp. 387–430). Academic Press.

Gökçe, A., Isaacs, R., & Whalon, M. E. (2006). Behavioural response of Colorado potato beetle (*Leptinotarsa decemlineata*) larvae to selected plant extracts. *Pest Management Science (formerly Pesticide Science), 62*, 1052–1057.

Gul, H. T., Saeed, S., & Khan, F. A. (2014). Entomopathogenic fungi as effective insect pest management tactic: A review. *Applied Sciences and Business Economics, 1*, 10–18.

Haase, J., Castagneyrol, B., Cornelissen, J. H. C., Ghazoul, J., Kattge, J., et al. (2015). Contrasting effects of tree diversity on young tree growth and resistance to insect herbivores across three biodiversity experiments. *Oikos, 124*, 1674–1685.

Haron, E. N., Ahmed, M., Ali, S. S., Abas, A. A., & Elshaier, M. E. (2020). Evaluate the effects of entomopathogenic fungi isolates on wheat aphid, *Schizaphis graminum* (Rondani) (Hemiptera: Aphididae). *Egyptian Academic Journal of Biological Sciences. A, Entomology, 13*, 149–159.

Hesham, A. E. L., Kaur, T., Devi, R., Kour, D., Prasad, S., et al. (2021). Current trends in microbial biotechnology for agricultural sustainability: Conclusion and future challenges. In A. N. Yadav, J. Singh, C. Singh, & N. Yadav (Eds.), *Current trends in microbial biotechnology for sustainable agriculture* (pp. 555–572). Springer.

Holder, D. J., Kirkland, B. H., Lewis, M. W., & Keyhani, N. O. (2007). Surface characteristics of the entomopathogenic fungus *Beauveria (Cordyceps) bassiana*. *Microbiology, 153*, 3448–3457.

Inglis, G. D., Goettel, M. S., & Johnson, D. L. (1997). Use of hyphomycetous fungi for managing insects in field crops. In L. A. Lacey & H. K. Kaya (Eds.), *Field manual of techniques in invertebrate pathology* (pp. 651–682). Springer.

Islam, T., Biswas, M. J. H., Howlader, M. T. H., & Ullah, M. S. (2017). Laboratory evaluation of *Beauveria bassiana*, some plant oils and insect growth regulators against two-spotted spider mite, *Tetranychus urticae* Koch (Acari: Tetranychidae). *Persian Journal of Acarology, 6*(3).

Islam, W., Adnan, M., Shabbir, A., Naveed, H., Abubakar, Y. S., et al. (2021). Insect-fungal-interactions: A detailed review on entomopathogenic fungi pathogenicity to combat insect pests. *Microbial Pathogenesis, 159*, 105122.

Jackson, M. A., & Dunlap, C. A. (2003). Effects of inorganic and organic fertilizers on turfgrass macrophagous and soil microbes. *Plant & Soil, 256*, 303–313.

Jaronski, S. T. (2010). Ecological factors in the inundative use of fungal entomopathogens. *Biological Control, 55*, 159–185.

Jeffries, P., Gianinazzi, S., Perotto, S., Turnau, K., & Barea, J. M. (2003). The contribution of arbuscular mycorrhizal fungi in sustainable maintenance of plant health and soil fertility. *Biology and Fertility of Soils, 37*, 1–16.

Kanaoka, M., Isogai, A., Murakoshi, S., Ichinoe, M., Suzuki, A., & Tamura, S. (1978). Bassianolide, a new insecticidal cyclodepsipeptide from *Beauveria bassiana* and *Verticillium lecanii*. *Agricultural and Biological Chemistry, 42*, 629–635.

Kary, N. E., & Alizadeh, Z. (2017). Effects of sub-culturing on genetic and physiological parameters in different *Beauveria bassiana* isolates. *Journal of Invertebrate Pathology, 145*, 62–67.

Khan, S., Guo, L., Maimaiti, Y., Mijit, M., & Qiu, D. (2012). Entomopathogenic fungi as microbial biocontrol agent. *Molecular Plant Breeding, 3*, 63–79.

Khan, S., Uddin, M. N., Rizwan, M., Khan, W., Farooq, M., et al. (2020). Mechanism of insecticide resistance in insects/pests. *Polish Journal of Environmental Studies, 29*, 2023–2030.

Kodaira, Y. (1961). Toxic substances to insects, produced by *Aspergillus ochraceus* and *Oopsra destructor*. *Agricultural and Biological Chemistry, 25*, 261–262.

Lacey, L. A., Grzywacz, D., Shapiro-Ilan, D. I., Frutos, R., Brownbridge, M., & Goettel, M. S. (2015). Insect pathogens as biological control agents: Back to the future. *Journal of Invertebrate Pathology, 132*, 1–41.

Leadmon, C. E., Sampson, J. K., Maust, M. D., Macias, A. M., Rehner, S. A., et al. (2020). Several Metarhizium species produce ergot alkaloids in a condition-specific manner. *Applied and Environmental Microbiology, 86*, e00373-20.

Lecuona, R. E., Erazo, M. F., Iannone, L. J., López Lastra, C. C., & Cabello, M. N. (2019). Entomopathogenic fungi against pests and their relevance in integrated pest management in Latin America. In J. S. Tkacz & L. Lange (Eds.), *Advances in fungal biotechnology for industry, agriculture, and medicine* (pp. 111–129). Springer.

Luz, C., Saladino, F., Luciano, F. B., Mañes, J., & Meca, G. (2017). Occurrence, toxicity, bioaccessibility and mitigation strategies of beauvericin, a minor Fusarium mycotoxin. *Food and Chemical Toxicology, 107*, 430–439.

Mantzoukas, S., & Eliopoulos, P. A. (2020). Endophytic entomopathogenic fungi: A valuable biological control tool against plant pests. *Applied Sciences, 10*, 360.

Mascarin, G. M., & Jaronski, S. T. (2016). The production and uses of *Beauveria bassiana* as a microbial insecticide. *World Journal of Microbiology and Biotechnology, 32*, 177.

Meyling, N. V., & Eilenberg, J. (2007). Ecology of the entomopathogenic fungi *Beauveria bassiana* and *Metarhizium anisopliae* in temperate agroecosystems: Potential for conservation biological control. *Biological control, 43*, 145–155.

Mikami, Y., Yazawa, K., Fukushima, K., Arai, T., Udagawa, S. I., & Samson, R. A. (1989). Paecilotoxin production in clinical or terrestrial isolates of *Paecilomyces lilacinus* strains. *Mycopathologia, 108*, 195–199.

Mondal, S., Baksi, S., Koris, A., & Vatai, G. (2016). Journey of enzymes in entomopathogenic fungi. *Pacific Science Review A: Natural Science and Engineering, 18*(2), 85–99.

Mongkolsamrit, S., Kobmoo, N., Tasanathai, K., Khonsanit, A., Noisripoom, W., et al. (2012). Life cycle, host range and temporal variation of *Ophiocordyceps unilateralis/Hirsutella formicarum* on Formicine ants. *Journal of Invertebrate Pathology, 111*(3), 217–224.

Mori, Y., Tsuboi, M., Suzuki, M., Fukushima, K., & Arai, T. (1982). Structure of leucinostatin a, new peptide antibiotic from *Paecilomyces lilacinus* A-267. *Journal of the Chemical Society, Chemical Communications, 2*, 94–96.

Neale, M. D., Cullen, L., & Ranout, A. S. (2024). Improving soil health in the UK: why a microbial approach is indispensable in attaining sustainable soils. Sustainable Microbiology, qvae026.

Niu, X., Thaochan, N., & Hu, Q. (2020). Diversity of linear non-ribosomal peptide in biocontrol fungi. *Journal of Fungi, 6*, 61.

Ortiz-Urquiza, A., & Keyhani, N. O. (2013). Action on the surface: Entomopathogenic fungi versus the insect cuticle. *Insects, 4*, 357–374.

Ortiz-Urquiza, A., Keyhani, N. O., & Quesada-Moraga, E. (2013). Culture conditions affect virulence and production of insect toxic proteins in the entomopathogenic fungus *Metarhizium anisopliae. Biocontrol Science and Technology, 23*, 1199–1212.

Ortiz-Urquiza, A., Fan, Y., Garrett, T., & Keyhani, N. O. (2016). Growth substrates and caleosin-mediated functions affect conidial virulence in the insect pathogenic fungus *Beauveria bassiana. Microbiology, 162*, 1913–1921.

Pal, S., Leger, R. J. S., & Wu, L. P. (2007). Fungal peptide Destruxin a plays a specific role in suppressing the innate immune response in *Drosophila melanogaster. The Journal of Biological Chemistry, 282*, 8969–8977.

Pedrini, N., Ortiz Urquiza, A., Huarte Bonnet, C., Zhang, S., & Keyhani, N. O. (2013). Targeting of insect epicuticular lipids by entomopathogenic fungi: Hydrocarbon oxidation within the context of a host-pathogen interaction. *Frontiers in Microbiology, 4*, 24.

Peng, K. C., Huang, H. S., Tzeng, Y. M., & Liu, S. Y. (2005). Circular dichroism analysis of destruxins from *Metarhizium anisopliae. Journal of Biochemical and Biophysical Methods, 62*(1), 41–50.

Qadri, S. M., Kucherenko, Y., & Lang, F. (2011). Beauvericin induced erythrocyte cell membrane scrambling. *Toxicology, 283*, 24–31.

Qasim, M., Lin, Y., Dash, C. K., Bamisile, B. S., Ravindran, K., et al. (2018). Temperature-dependent development of Asian citrus psyllid on various hosts, and mortality by two strains of Isaria. *Microbial Pathogenesis, 119*, 109–118.

Quesada-Moraga, E., Landa, B. B., Muñoz-Ledesma, F. J., & Santiago-Álvarez, C. (2007). Endophytic ccolonizationof opium poppy, *Papaver somniferum*, by an entomopathogenic *Beauveria bassiana* strain. *Mycological Research, 111*, 1199–1205.

Ranout, A. S., Kaur, R., Kumar, R., & Nadda, G. (2024). Pathogenicity and compatibility studies of native. *Tolypocladium inflatum and Clonostachys krabiensis* against Tetranychus urticae. *Journal of Applied Entomology.*

Rastegari, A. A., Yadav, A. N., & Yadav, N. (Eds.). (2020). *New and future developments in microbial biotechnology and bioengineering* (p. 351). Elsevier.

Ravindran, K., Sivaramakrishnan, S., Hussain, M., Dash, C. K., Bamisile, B. S., Qasim, M., & Liande, W. (2018). Investigation and molecular docking studies of Bassianolide from *Lecanicillium lecanii* against *Plutella xylostella* (Lepidoptera: Plutellidae). *Comparative Biochemistry and Physiology Part C: Toxicology & Pharmacology, 206*, 65–72.

Resende, J. T., Strasser, H., & Rehner, S. A. (2021). Environmental factors influencing the effectiveness of *Beauveria bassiana* against the pecan weevil, *Curculio caryae*. *Journal of Pest Science, 94*, 1003–1016.

Riseh, R. S., Hassanisaadi, M., Vatankhah, M., Soroush, F., & Varma, R. S. (2022). Nano/microencapsulation of plant biocontrol agents by chitosan, alginate, and other important biopolymers as a novel strategy for alleviating plant biotic stresses. *International Journal of Biological Macromolecules, 222*, 1589–1604.

Rohrlich, C., Merle, I., Mze Hassani, I., Verger, M., Zuin, M., et al. (2018). Variation in physiological host range in three strains of two species of the entomopathogenic fungus *Beauveria*. *PLoS One, 13*, e0199199.

Shah, P. A., & Pell, J. K. (2003). Entomopathogenic fungi as biological control agents. *Applied Microbiology and Biotechnology, 61*, 413–423.

Shah, P. A., & Pell, J. K. (2014). Entomopathogenic fungi as biological control agents. *Applied Microbiology and Biotechnology, 98*, 3515–3523.

Shah, S., Ash, G. J., & Wilson, B. A. (2023). Resporulation of *Metarhizium anisopliae* granules on soil and mortality of *Tenebrio molitor*: Implications for wireworm management in sweetpotato. *The Annals of Applied Biology, 182*, 65–76.

Shahid, A. A., Rao, Q. A., Bakhsh, A., & Husnain, T. (2012). Entomopathogenic fungi as biological controllers: New insights into their virulence and pathogenicity. *Archives of Biological Sciences, 64*, 21–42.

Sharma, A., Sharma, S., & Yadav, P. K. (2023a). Entomopathogenic fungi and their relevance in sustainable agriculture: A review. *Cogent Food & Agriculture, 9*(1), 2180857.

Sharma, A., Kaur, E., Joshi, R., Kumari, P., Khatri, A., Swarnkar, M. K.,... & Nadda, G. (2023b). Systematic analyses with genomic and metabolomic insights reveal a new species, *Ophiocordyceps indica* sp. nov. from treeline area of Indian Western Himalayan region. *Frontiers in Microbiology*, 14, 1188649.

Sharma, A., Ranout, A. S., Kaur, R., Kumari, P., & Nadda, G. (2024a). Unearthing diverse culturable fungal communities associated with *Ophiocordyceps indica* sp. nov. from Indian Western Himalaya. *Journal of Basic Microbiology, 64*, 2300461.

Sharma, A., Ranout, A. S., & Nadda, G. (2024b). Insights into cultivation strategies, bioactive components, therapeutic potential, patents, and market products of *Ophiocordyceps sinensis*: A comprehensive review. *South African Journal of Botany, 171*, 546–570.

Sheppard, D. C., & Filler, S. G. (2014). Host cell invasion by medically important fungi. *Cold Spring Harbor Perspectives in Medicine, a019687*.

St. Leger, R. J. (1995). The role of cuticle-degrading proteases in fungal pathogenesis of insects. *Canadian Journal of Botany, 73*(S1), 1119–1125.

Statista. (2023, March 24). Bio-pesticides market size worldwide in 2016 and 2021, with a forecast for 2027. https://www.statista.com/statistics/1350343/bio-pesticides-market-size-globally/#:~:text=The%20global%20bio%2Dpesticides%20market,the%20global%20bio%2Dpesticides%20market

Tavladoraki, P., Rossi, M., Martinelli, F., Pintus, M., & Ruiu, L. (2019). A new approach to the preservation of *Beauveria bassiana* conidia for long-term storage at low temperatures. *Microorganisms, 7*, 651.

Tedjiotsop Feudjio, F., Dornetshuber, R., Lemmens, M., Hoffmann, O., Lemmens-Gruber, R., & Berger, W. (2010). Beauvericin and enniatin: Emerging toxins and/or remedies? *World Mycotoxin Journal, 3*, 415–430.

The business research company. (2022). Pesticide global market report. https://www.thebusinessresearchcompany.com/report/pesticides-global-market-report#:~:text=The%20global%20pesticides%20market%20size,(CAGR)%20of%205.5%25

Tigano, M. S., Adams, B., Maimala, S., & Boucias, D. (2006). Genetic diversity of *Hirsutella thompsonii* isolates from Thailand based on AFLP analysis and partial beta-tubulin gene sequences. *Genetics and Molecular Biology, 29*, 715–721.

Ugine, T. A., Wraight, S. P., & Sanderson, J. P. (2007). Effects of manipulating spray-application parameters on efficacy of the entomopathogenic fungus *Beauveria bassiana* against western flower thrips, *Frankliniella occidentalis*, infesting greenhouse impatiens crops. *Biocontrol Science and Technology, 17*, 193–219.

Wang, C., & St Leger, R. J. (2007). The MAD1 adhesin of *Metarhizium anisopliae* links adhesion with blastospore production and virulence to insects, and the MAD2 adhesin enables attachment to plants. *Eukaryotic Cells, 6*, 808–816.

Wang, Q., & Xu, L. (2012). Beauvericin, a bioactive compound produced by fungi: A short review. *Molecules, 17*, 2367–2377.

Wang, G., Liu, Z., Lin, R., Li, E., Mao, Z., Ling, J., et al. (2016). Biosynthesis of antibiotic leucinostatins in bio-control fungus *Purpureocillium lilacinum* and their inhibition on Phytophthora revealed by genome mining. *PLOS Pathogens, 12*, e1005685.

World Health Organization. (2014). *The International Code of Conduct on Pesticide Management*. https://www.who.int/publications-detail-redirect/9789251085493

Chapter 15
Pathogenic Algae in Host Microbiota: Invasion Mechanism and Management Strategies

Nilamjyoti Kalita, Pratisha Das, Shaswatee Bhattacharjee, Trideep Chetia, and Partha Pratim Baruah

Abstract Microbiomes networks play a crucial role in plant development and underpin many ecological function. Plants microbiota comprises microorganisms that live inside or in their vicinity with interactions that influence the host physiology. A large number of bacteria, fungi, actinomycetes and algae are known to make up the complex network of microbial communities. Within their complex interactions, a few species show pathogenicity towards the host plant. Other species may act as significant antagonists, thereby positively influencing the host. These interactions elicit a cascade of signaling events, such as hormonal and induced systemic responses. An enormous number of pathogenic microbes, including algae, develop diverse strategies over time to infect a wide range of host plants. Upon successful infection, they can cause various impacts on the host at morphological, physiological and biochemical levels. However, understanding the complex connections between plant microbiome and algal pathogens is critical towards developing effective algal disease management techniques and mitigating their pathogenicity. In this chapter, therefore, we review the advancements in studies regarding algae as pathogens capable of causing various plant diseases. We also reviewed present efforts carried out to build long-term management strategies to prevent disease and increase ecosystem health by applying microbes present in the plant microbiota.

Keywords Disease management · Host response · Invasion mechanism · Plant microbiome · Parasitic algae

---

N. Kalita · P. Das · S. Bhattacharjee · T. Chetia · P. P. Baruah (✉)
Plant Ecology Laboratory, Department of Botany, Gauhati University, Guwahati, Assam, India
e-mail: ppbaruah@gauhati.ac.in

P. Mathur, S. Roy (eds.), *Plant Microbiome and Biological Control*, Sustainability in Plant and Crop Protection 20,
https://doi.org/10.1007/978-3-031-75845-4_15

Abbreviations

ISR	Induced Systemic Resistance
MoP	Murate of Potash
BGA	Blue Green Algae

15.1 Introduction

Algae are a group of diverse and highly specialized chlorophyll bearing organisms. The members of the group are either microscopic or macroscopic and can be unicellular or multicellular (Munir et al., 2015). They have been recorded from wide habitats ranging from fresh (Baruah et al., 2020, 2022) to marine water including brackish water as well, in a varied range of temperatures, pH and nutrient levels (Falkowski & Raven, 1997). Some algae are epilithic, epiphytic, or, on some occasions, cryophytic, growing on ice beds (Blazencic, 2007). Being chlorophyll-bearing autotrophic producers (Kalita & Baruah, 2024; Boruah & Baruah, 2024), algae not only provide food to consumers of higher trophic levels but also take part in the productivity of any ecosystem, besides performing a major role in carbon sequestration.

Despite the advantages of phototrophy, many algae have reverted to heterotrophy and shifted towards parasitism as a heterotrophic tactic, losing the ability to photosynthesize due to some hitherto unknown evolutionary adaptive mechanism (Oborník, 2020). Within this diverse realm of parasitic algae, many algal and plant species have been found to be hosts, vulnerable to parasitism (Skovgaard et al., 2012; Preuss et al., 2017). During their evolutionary process, these algal invaders have acquired the ability to enter the host body and not only proliferate but also reproduce to complete their life cycles in and around tissues, and sometimes at cellular levels, until available food and moisture conditions occur (Thomas, 1913; Abdulkhair & Alghuthaymi, 2016). This process leads to reduced productivity and vitality of the host plants. The impact of such algal infections on the host species extends beyond the symptoms that are readily apparent. These parasites not only cause morphological alterations in their host species but are also associated with disturbances in physiology and biochemistry, which ultimately leads to consequences like stunted growth, changed metabolism, poor nutrition absorption, and reduced yield (Ponmurugan et al., 2010; Suto et al., 2014).

Since the role of algae as a pathogen is very intriguing and relatively unexplored in the field of host-pathogens interactions, this chapter reviews available literature on the role of algae as pathogens. We also highlight their interactions with various host plants, including species-specific invasion mechanisms and long-term management strategies to prevent disease.

15.2 Range of Host Plants

To date, algal parasites have been exclusively reported within the class Chlorophyceae, Rhodophyceae and Dinophyceae. Furthermore, the occurrence of parasitic algae in land plants has been solely reported from the Chlorophyceae (Rodríguez et al., 2008; Brooks, 2004; Joubert & Rijkenberg, 1971; Vázquez et al., 2010; Skovgaard et al., 2012). These parasitic algae represent a diverse range of families. They were also found to be capable of infecting a diverse array of host species belonging to various families. Moreover,, while a subset of these parasites exhibited the remarkable capability to infect multiple hosts, some were found to be highly specific, and only capable of infecting one or two hosts.

Within the class Rhodophyceae, as reviewed by Preuss et al. (2017), a total of 125 red algal species were reported as common parasites on other red algal members (Table 15.1). It was observed that 69 out of these parasites, which is approx. 55% of the total red algal parasites, are exclusively host-specific (Table 15.1). On the other hand, 19 parasitic species (15%) were found to be able to infect two hosts. The remaining 35 members (30%) were reported to infect three or more plant species, demonstrating a broader range and signifying their ability to exploit a wider spectrum of hosts.

Four genera, namely, *Gracilaria* (Gracilariaceae), *Gelidium* (Gelidiaceae), *Laurencia* (Rhodomelaceae) and *Polysiphonia* (Rhodomelaceae), were found to be the most common hosts for parasitic red algal species. Among these, members of genera *Polysiphonia* belonging to Rhodomelaceae viz. *P. atlantica*, *P. caespitosa* and *P. devoniensis* were found to be common hosts for all other red algal pathogens. Findings suggest that the vulnerability of these genera to parasitic invasions may be particularly due to ecological factors such as humidity and temperature, as well as the biology of the hosts (Preuss et al., 2017).

Every member identified as a green parasitic alga in class Chlorophyceae can infect higher terrestrial plants. A total of 17 species parasitic in higher terrestrial plants have been reported so far from three families, i.e., Trentepohliaceae, Watanabeaceae and Chlorochytriaceae (Joubert & Rijkenberg, 1971; Pitaloka et al., 2014; Suto et al., 2014; Wonglom et al., 2018). Among them, 11 species were found to be pathogenic over a single host, while only 3 species (*Cephaleuros karstenii*, *C. parasiticus* and *C. virescens*) can exploit multiple host species like *Citrus* spp., *Syzygium* spp., *Camellia sinensis*, *Piper nigrum* etc. (Table 15.1). *Cephaleuros diffusus* was observed to be highly specialized, infecting the leaves of *Acacia auriculiformis*, whereas *C. expansus* could infect multiple hosts, such as *Piper longum* and *Psidium guajava* (Wonglom et al., 2018). *Cephaleuros virescens* also exhibited a wide range of hosts, infecting several plants belonging to various families. In addition, most of the parasitic green algal species of genus *Phyllosiphon*, such as *P. arisari*, *P. deformans*, *P. philodendra* and *P. ari* parasitize specific hosts under the family Araceae (Joubert & Rijkenberg, 1971; Guiry & Guiry, 2023).

The most common host plants of green parasitic algae are either tropical or subtropical shrubs or trees. Broadleaf trees are more vulnerable to these pathogens

Table 15.1 List of parasitic algae infecting various host species in tropics and sub-tropics

Parasitic algae	Family	Host	Reference
Aiolocolax pulchellus Pocock 1956	Rhodomelaceae	*Polysiphonia atlantica* Kapraun et Norris, *Polysiphonia caespitosa* (Pocock) Hollenberg, *Polysiphonia devoniensis* Maggs et Hommersand	Preuss et al. (2017)
Antarctocolax lambii Skottsberg	Rhodomelaceae	*Picconiella plumosa* (Kylin) de Toni	Preuss et al. (2017)
Apoglossocolax pusillus Maggs & Hommersand	Delesseriaceae	*Apoglossum ruscifolium* (Turner) J. Agardh	Preuss et al. (2017)
Asterocolax denticulatus (Tokida) Feldmann & G. Feldmann	Delesseriaceae	*Phycodrys fimbriata* (Kuntze) Kylin	Preuss et al. (2017)
Asterocolax erythroglossi Feldmann & G.Feldmann 1951	Delesseriaceae	*Erythroglossum laciniatum, Erythroglossum sandrianum* (Kützing) Kylin	Preuss et al. (2017)
Asterocolax gardneri (Setchell) Feldmann & G.Feldmann	Delesseriaceae	*Anisocladella pacifica, Phycodrys isabelliae, Phycodrys setchelli, Nienburgia andersoniana, Polyneura latissimi*	Preuss et al. (2017)
Asterocolax hypophyllophilus M.J.Wynne	Delesseriaceae	*Mikamiella ruprechtiana* (Zinova) Wynne	Preuss et al. (2017)
Benzaitenia yenoshimensis Yendo	Rhodomelaceae	*Chondria crassicaulis* Harvey, *Chondria* spp., *Laurencia* spp., *Palisada peniculata* (Kützing) Cassano, Sentíes, Gil-Rodriquez & M. T. Fujii	Preuss et al. (2017)
Bostrychiocolax australis Zuccarello & J.A.West	Rhodomelaceae	*Bostrychia radicans*	Preuss et al. (2017)
Callocolax acicularis M.J.Wynne & J.N.Heine	Kallymeniaceae	*Callophyllis rhynchocarpa* Ruprecht	Preuss et al. (2017)
Callocolax fungiformis Kylin	Kallymeniaceae	*Callophyllis edentata* Kylin, *Callophyllis heanophylla* Setchell, *Callophyllis flabellulata* Harvey, *Callophyllis pinnata* Setchell et Swezy	Preuss et al. (2017)
Callocolax japonica Tsugi (nom. inval.)	Kallymeniaceae	*Callophyllis* spp.	Preuss et al. (2017)
Callocolax neglectus F.Schmitz ex Batters	Kallymeniaceae	*Callophyllis hombroniana* (Montagne) Kützing, *Callophyllis laciniata* (Hudson) Kützing	Preuss et al. (2017)

(continued)

Table 15.1 (continued)

Parasitic algae	Family	Host	Reference
Centrocerocolax ubatubensis A.B. Joly	Ceramiaceae	*Centroceras clavulatum* (C. Agardh) Montagne, *Ceramium tenerrimum* (Martens) Okamura, *Ceramium* spp.	Guiry and Guiry (2023) and Preuss et al. (2017)
Chamaethamnion pocockiae R.E.Norris	Rhodomelaceae	*Kuetzingia natalensis*	Guiry and Guiry (2023) and Preuss et al. (2017)
Chamaethamnion schizandra Falkenberg	Rhodomelaceae	*Polysiphonia decipiens* Montagne, *Micropeuce feredayae* (Harvey) Kylin ex Silva	Preuss et al. (2017)
Champiocolax lobatus Womersley	Champiaceae	*Champia viridis* C. Agardh	Preuss et al. (2017)
Champiocolax sarae Bula-Meyer	Champiaceae	*Champia* compressa Harvey, *Champia parvula* (C. Agardh) Harvey, *Champia salicornoides* Harvey	Preuss et al. (2017)
Choreocolax americanus Reinsch	Rhodomelaceae	*Lophura* spp.	Preuss et al. (2017)
Choreocolax destructor Reinsch	Rhodomelaceae	*Chondracanthus teedei* (Mertens) Kützing	Preuss et al. (2017)
Choreocolax polysiphoniae Reinsch	Rhodomelaceae	*Polysiphonia fastigiata*, *Cystoclonium purpureum* (Hudson) Batters, *Neosiphonia confusa* (Hollenberg) Norris, *Vertebrata lanosa* (L.) Christensen	Guiry and Guiry (2023) and Preuss et al. (2017)
Choreocolax rabenhorstii Reinsch	Rhodomelaceae	*Phycodrys rubens* (Goodenough *et* Woodward) Kützing	Preuss et al. (2017)
Choreocolax rhodymeniae Reinsch	Rhodomelaceae	*Rhodymenia* spp., *Palmaria decipiens* (Reinsch) Ricker, *Palmaria georgica* (Reinsch) Ricker	Guiry and Guiry (2023); Preuss et al. (2017)
Choreocolax tumidus Reinsch		*Ceramiun* spp., *Ceramium virgatum* Roth, *Cystoclonium purpureum*	Preuss et al. (2017)
Choreonema thuretii (Bornet) F.Schmitz	Hapalidiaceae	*Jania micrarthrodia* Lamouroux, *Jania rosea* (Lamarck) Decaisne, *Jania rubens* (L.) Lamouroux, *Jania tenella* (Kützing) Grunow, *Jania verrucosa* Lamouroux	Guiry and Guiry (2023) and Preuss et al. (2017)
Coccotylus hartzii (Rosenvinge) Le Gall & G.W. Saunders	Phyllophoraceae	*Coccotylus truncatus* (Pallas) Wynne & Heine	Preuss et al. (2017)

(continued)

Table 15.1 (continued)

Parasitic algae	Family	Host	Reference
Colacodasya australica Womersley	Delesseriaceae	*Dasya clavigera* (Womersley) Parson	Preuss et al. (2017)
Colacodasya californica Hollenberg	Delesseriaceae	*Heterosiphonia erecta* Gardner	Guiry and Guiry (2023); Preuss et al. (2017)
Colacodasya inconspicua (Reinsch) F.Schmitz	Delesseriaceae	*Merenia microcladioides*, *Heterosiphonia berkeleyi* Montagne, *Heterosiphonia* spp., *Polysiphonia anisogona* Hooker et Harvey, *Polysiphonia* spp.	Preuss et al. (2017)
Colacopsis lophurellae Kylin	Rhodomelaceae	*Lophurella Hookeriana*	Preuss et al. (2017)
Colacopsis pulvinata (F.Schmitz) De Toni	Rhodomelaceae	*Osmundaria serrata* (Suhr) J. Agardh	Preuss et al. (2017)
Colacopsis smitheniae R.E.Norris	Rhodomelaceae	*Aneurianna nozawae* (R.E.Norris) L.E. Philipps	Preuss et al. (2017)
Colacopsis velutina M.T.Martin & Pocock	Rhodomelaceae	*Rhodomelopsis africana* Pocock	Preuss et al. (2017)
Dawsoniocolax bostrychiae (A.B.Joly & Yamaguishi-Tomita) A.B.Joly & Yamaguishi-Tomita	Rhodomelaceae	*Bostrychia calliptera* (Montagne) Montagne, *Bostrychia montagnei* Harvey, *Bostrychia radicans* (Montagne) Montagne	Preuss et al. (2017)
Dipterocolax fernandezianus J.Morrill	Rhodomelaceae	*Dipterosiphonia parva*	Preuss et al. (2017)
Episporium centroceratis Möbius	Ceramiaceae	*Centroceras clavulatum*	Preuss et al. (2017)
Epulo multipedes R.A.Townsend & Huisman	Hapalidiaceae	*Jania verrucosa*	Preuss et al. (2017)
Ezo epiyessoense Adey, Masaki & Akioka	Lithophyllaceae	*Lithophyllum crouaniorium* Foslie, *Lithophyllum yessoense* Foslie, *Titanoderma pustulatum* (Lamouroux) Nägeli	Preuss et al. (2017)
Faucheocolax attenuatus Setchell	Faucheaceae	*Gloiocladia laciniata*, G. fryeana	Preuss et al. (2017)
Gardneriella tuberifera Kylin	Solieriaceae	*Sarcodiotheca gaudichaudii*, *Agardhiella coulteri* (Harvey) Setchell	Preuss et al. (2017)
Gelidiocolax christianae Feldmann & G.Feldmann	Pterocladiophilaceae	*Gelidium spathulatum* (Kützing) Bornet	Preuss et al. (2017)

(continued)

Table 15.1 (continued)

Parasitic algae	Family	Host	Reference
Gelidiocolax deformans Seoane-Camba (nom. inval.)	Pterocladiophilaceae	*Gelidium cantabricum, G. sesquipedale, Gelidium corneum*	Preuss et al. (2017)
Gelidiocolax desikacharyi E.K. Ganesan	Pterocladiophilaceae	*Gelidium floridanum* Taylor	Preuss et al. (2017)
Gelidiocolax lyndae R.E. Norris	Pterocladiophilaceae	*Kuetzingia natalensis*	Preuss et al. (2017)
Gelidiocolax mammillatus K.-C. Fan & Papenfuss	Pterocladiophilaceae	*Pterocladiella capillacea* (Gmelin) Santelices et Hommersand	Preuss et al. (2017)
Gelidiocolax margaritoides (M.T.Martin & Pocock) K.-C.Fan & Papenfuss	Pterocladiophilaceae	*Gelidium pulchellum* (Turner) Kützing, *Ptilophora pinnatifida* J.Agardh	Preuss et al. (2017)
Gelidiocolax microsphaericus N.L.Gardner	Pterocladiophilaceae	*Gelidium coulteri* Harvey, *Gelidium nudifrons* Gardner, *Gelidium pulchrum* Gardner, *Gelidium pusillum* (Stackhouse) Le Jolis, *Gelidium robustum* (Gardner) Hollenberg et Abbott,	Preuss et al. (2017)
Gelidiocolax pustulatus E.C.Oliveira & Yoneshigue	Pterocladiophilaceae	*Pterocladiella capillacea*	Preuss et al. (2017)
Gelidiocolax suhriae (M.T.Martin & Pocock) K.-C. Fan & Papenfuss	Pterocladiophilaceae	*Gelidium vittatum* (L.) Kützing	Preuss et al. (2017)
Gelidiocolax verruculatus Ouahi & Najiim (nom. inval.)	Pterocladiophilaceae	*Gelidum attenuatum* (Turner) Thuret, *Gelidium pulchellum* (Turner) Kützing), *Gelidium pusillum, Pterocladiella capillacea*	Preuss et al. (2017)
Gloiocolax novae-zelandiae Sparling	Faucheaceae	*Gloioderma saccata* (J. Agardh) Norris	Preuss et al. (2017)
Gonimocolax australis (Skottsberg) Kylin	Delesseriaceae	*Myriogramme livida* (J. D. Hooker & Harvey) Kylin, *Schizoseris* spp.	Preuss et al. (2017)
Gonimocolax corymbosus Baardseth	Delesseriaceae	*Myriogramme livida* (J. D. Hooker & Harvey) Kylin, *Schizoseris* spp.	Preuss et al. (2017)
Gonimocolax roscoffensis Feldmann & G.Feldmann	Delesseriaceae	*Myriogramme livida* (J. D. Hooker & Harvey) Kylin, *Schizoseris* spp.	Preuss et al. (2017)

(continued)

Table 15.1 (continued)

Parasitic algae	Family	Host	Reference
Gonimophyllum africanum M.T.Martin & Pocock	Delesseriaceae	*Acrosorium maculatum* (Sonder ex Kützing) Papenfuss, *Acrosorium* spp., *Botryoglossum platycarpus* (Turner) Kützing, *Botryocarpa prolifera* Greville, *Delesseriaceae* spp., *Neuroglossum binderianium* Kützin	Preuss et al. (2017)
Gonimophyllum buffhamii Batters	Delesseriaceae	*Acrosorium ciliolatum* (Harvey) Kylin, *Cryptopleura ramosa* (Hudson) Newton	Preuss et al. (2017)
Gonimophyllum insulare F.S.Wagner	Delesseriaceae	*Hymenena semicostata* (J.Agardh) Kylin	Preuss et al. (2017)
Gonimophyllum skottsbergii Setchell	Delesseriaceae	*Cryptopleura crispa* Kylin, *Hymenena flabelligera* (J. Agardh) Kylin, *Cryptopleura ruprechtianum* (J. Agardh) Kylin	Preuss et al. (2017)
Gracilaria babae (H.Yamamoto) P.-K. Ng, P.-E. Lim & S.-M. Phang	Gracilariaceae	*Hydropuntia* sp., *Gracilaria salicornia* (C.Agardh) Dawson	Preuss et al. (2017)
Gracilariocolax deformans (Weber van Bosse) Gerung & Yamamoto	incertae sedis	*Gracilaria canaliculata* Sonder, *Gracilaria salicornia*	Preuss et al. (2017)
Gracilariocolax henriettae Weber Bosse	incertae sedis	*Gracilaria hauckii* Silva	Preuss et al. (2017)
Gracilariocolax infidelis (Weber-van Bosse) Gerung & Yamamoto	incertae sedis	*Gracilaria canaliculata*, *Gracilaria minor* (Sonder) Durairatnam, *Gracilaria salicornia*	Preuss et al. (2017)
Gracilariocolax setchellii (Weber van Bosse) Gerung & Yamamoto	incertae sedis	*Gracilaria canaliculata*, *Gracilaria salicornia*	Preuss et al. (2017)
Gracilariocolax setchellii var. *aggregatus* (Weber van Bosse) Gerung & Yamamoto	incertae sedis	*Gracilaria minor (Sonder)* Durairatnam	Preuss et al. (2017)
Gracilariocolax sibogae (Weber Bosse) Gerung & Yamamoto 2002	incertae sedis	*Gracilaria arcuata Zanardini*, *Gracilaria canaliculata*, *Gracilaria dura* (C. Agardh) J. Agardh	Preuss et al. (2017)
Gracilariophila oryzoides Setchell & H.L.Wilson	Gracilariaceae	*Gracilaria confervoides*, *Gracillariopsis lamaneiformis*	Preuss et al. (2017)

(continued)

Table 15.1 (continued)

Parasitic algae	Family	Host	Reference
Gracilariophila gardneri Setchell	Gracilariaceae	*Gracilaria cunninghamii* Farlow ex J.Agardh	Preuss et al. (2017)
Grateloupiocolax colombianus Schnetter & Bula-Meyer	Halymeniaceae	*Grateloupia filicina* (Lamouroux) C. Agardh	Preuss et al. (2017)
Harveyella mirabilis (Reinsch) F.Schmitz & Reinke	Rhodomelaceae	*Gonimophyllum skottsbergii, Odonthalia floccosa, O. washingtoniensis* and *R. confervoides.*	Preuss et al. (2017)
Holmsella pachyderma (Reinsch) Sturch	Pterocladiophilaceae	*Gracilaria gracilis* and *Gracilariopsis longissima*	Preuss et al. (2017)
Holmsella australis Noble & Kraft	Pterocladiophilaceae	*Gracilaria ramulosa, G. flagelliformis, Gracilaria cliftonii*	Guiry and Guiry (2023) and Preuss et al. (2017)
Hypneocolax stellaris Børgesen	Cystocloniaceae	*Hypnea cornuta* (Kützing) J. Agardh, *Hypnea musciformis* (Wulfen) Lamouroux, *Hypnea ramentacea* (C. Agardh) J. Agardh, *Hypnea valentiae* (Turner) Montagne, *Hypnea variabilis* Okamura	Preuss et al. (2017)
Hypneocolax stellaris f. *orientalis* Weber Bosse	Cystocloniaceae	*Hypnea* spp., *Hypnea filiformis* (Harvey) Womersley	Preuss et al. (2017)
Janczewskia gardneri Setchell & Guernsey	Rhodomelaceae	*Laurencia gardneri* Hollenberg, *Osmundea spectabilis* (Postels et Ruprecht) Nam, *Osmundea pinnatifida* (Hudson) Stackhouse	Preuss et al. (2017)
Janczewskia hawaiiana Apt	Rhodomelaceae	*Laurencia nidifica* J. Agardh, *Laurencia mcdermidiae* Abbott	Preuss et al. (2017)
Janczewskia lappacea Setchell	Rhodomelaceae	*Chondria nidifica* Harvey (syn. *Laurencia nidifica*)	Preuss et al. (2017)
Janczewskia meridionalis Martin & Pocock	Rhodomelaceae	*Laurencia flexuosa* Kützing, *Laurencia natalensis* Kylin	Preuss et al. (2017)
Janczewskia moriformis Setchell	Rhodomelaceae	*Chondria atropurpurea* Harvey, *Laurencia translucida* Fujii & Cordeiro-Marina, *Palisada flagellifera* (J. Agardh) Nam	Preuss et al. (2017)
Janczewskia morimotoi Tokida	Rhodomelaceae	*Laurencia nipponica* Yamada	Preuss et al. (2017)
Janczewskia ramiformis C.F.Chang & B.M.Xia	Rhodomelaceae	*Laurencia okamurae* Yamada	Preuss et al. (2017)
Janczewskia solmsii Setchell & Guernsey	Rhodomelaceae	*Laurencia subopposita* (J.Agardh) Setchell	Preuss et al. (2017)

(continued)

Table 15.1 (continued)

Parasitic algae	Family	Host	Reference
Janczewskia tasmanica Falkenberg	Rhodomelaceae	*Laurencia forsteri* (Mertens) Greville, *Laurencia heteroclada* (Harvey) Saito & Womersley, *Laurencia* spp.	Preuss et al. (2017)
Janczewskia teysmannii Weber Bosse	Rhodomelaceae	*Acanthophora spicifera* (Vahl) Børgesen	Preuss et al. (2017)
Janczewskia verruciformis Solms-Laubach	Rhodomelaceae	*Laurencia obtusa* (Hudson) Lamouroux	Preuss et al. (2017)
Jantinella sinicola (Setchell & Gardner) Kylin	Rhodomelaceae	*Chondria acrorhizophora* Setchell & Gardner, *Chondria clarionensis* Setchell & Gardner	Preuss et al. (2017)
Jantinella verruciformis (Setchell & McFadden) Kylin	Rhodomelaceae	*Chondria* spp., *Mychodea episcopalis* J. Agardh	Preuss et al. (2017)
Kintokiocolax aggregatoceranthus Tak.Tanaka & Y. Nozawa	Grateloupiaceae	*Carpopeltis angusta*	Preuss et al. (2017)
Kvaleya epilaeve Adey & Sperapani	Mesophyllumaceae	*Phymatolithon laeve* (Rosenvinge) Düwel & Wegeberg	Preuss et al. (2017)
Laurenciocolax polysporus A.D.Zinova & Perestenko	Rhodomelaceae	*Laurencia caspica* Zinova & Zaberzhinskaya	Preuss et al. (2017)
Leachiella pacifica Kugrens	Rhodomelaceae	*Neosiphonia paniculata* (Montagne) Norris, *Polysiphonia hendryi* Gardner, *Pterosiphonia bipinnata* (Postels & Ruprecht) Falkenberg, *Pterosiphonia dendroidea* (Montagne) Falkenberg, *Pterosiphonia* spp.	Preuss et al. (2017)
Levringiella gardneri (Setchell) Kylin	Rhodomelaceae	*Pterosiphonia baileyi*	Preuss et al. (2017)
Levringiella microscopica (Levring) Kylin	Rhodomelaceae	*Polysiphonia virgata*	Preuss et al. (2017)
Masakiella bossiellae (Kloczkova) Guiry & Selivanova 2007	Corallinaceae	*Bossiella* sp.	Preuss et al. (2017)
Meridiocolax bracteatus J.Noble & Kraft	Rhodomelaceae	*Polysiphonia sparsa*	Preuss et al. (2017)
Meridiocolax narcissus J.Morrill	Rhodomelaceae	*Polysiphonia ferulacea* Suhr	Preuss et al. (2017)
Meridiocolax polysiphoniae (E.C.Oliveira & Ugadim) J.Morrill	Rhodomelaceae	*Polysiphonia denudata* (Dillwyn) Greville ex Harvey	Preuss et al. (2017)

(continued)

Table 15.1 (continued)

Parasitic algae	Family	Host	Reference
Microcolax africanus M.T.Martin & Pocock	Rhodomelaceae	*Streblocladia tenuissima* M. A. Pocock	Preuss et al. (2017)
Microcolax botryocarpus (Hooker f. & Harvey) F. Schmitz	Rhodomelaceae	*Streblocladia glomerulata* (Montagne) Papenfuss, *Streblocladia neglecta* Schmitz	Preuss et al. (2017)
Neohalosacciocolax aleuticus I.K.Lee & Kurogi	Palmariaceae	*Halosaccion minjaii*	Preuss et al. (2017)
Neotenophycus ichthyosteus Kraft & I.A.Abbott	Rhodomelaceae	*Neosiphonia poko*	Preuss et al. (2017)
Onychocolax polysiphoniae Pocock	Rhodomelaceae	*Polysiphonia incompta*	Preuss et al. (2017)
Phitycolax inconspicuus M.J.Wynne & F.J.Scott	Delesseriaceae	*Phitymophora amansioides*	Preuss et al. (2017)
Plocamiocolax pulvinatus Setchell	Plocamiaceae	*Plocamium cartilagineum* (L.) Dixon	Preuss et al. (2017)
Plocamiocolax papenfussianus M.F.Martin & Pocock	Plocamiaceae	*Plocamium corallorhiza* (Turner) Hooker *et* Harvey	Preuss et al. (2017)
Polycoryne compacta A.D.Zinova	Delesseriaceae	*Myriogramme kerguelensis* Levring	Preuss et al. (2017)
Polycoryne radiata Skottsberg 1919	Delesseriaceae	*Nitophyllum* sp.	Preuss et al. (2017)
Pterocladiophila hemisphaerica K.-C. Fan & Papenfuss	Pterocladiophilaceae	*Pterocladiella bartlettii (Taylor) Santelices, Pterocladia lucida (R. Brown ex Turner) J. Agardh*	Preuss et al. (2017)
Rhodophyllis parasitica M.Preuss & G.Zuccarello	Cystocloniaceae	*Rhodophyllis membranacea*	Preuss et al. (2017)
Rhodophysema kjellmanii G.W.Saunders & Clayden, (nom. illeg.)	Rhodophysemataceae	*Devaleraea ramentacea* (L.) Guiry, *Palmaria palmata* (L.) F. Weber et Mohr	Preuss et al. (2017)
Rhodymeniocolax botryoideus Setchell	Rhodymeniaceae	*Rhodymenia palmetta, R. pacifica, Rhodymenia* spp.	Preuss et al. (2017)
Rhodymeniocolax mediterraneus Vergés, Izquierdo & Verlaque	Rhodymeniaceae	*Rhodymenia ardissonei*	Preuss et al. (2017)
Scagelonema parasiticum R.E.Norris & M.J.Wynne	Wrangeliaceae	*Antithamnion defectum* Kylin	Preuss et al. (2017)
Sorellocolax stellaris Yoshida & Mikami	Delesseriaceae	*Sorella repens* (Okamura) Hollenberg	Preuss et al. (2017)

(continued)

Table 15.1 (continued)

Parasitic algae	Family	Host	Reference
Sporoglossum lophurellae Kylin	Rhodomelaceae	*Lophurella hookeriana* (J. Agardh) Falkenberg	Preuss et al. (2017)
Spyridiocolax capixabus A.B.Joly & E.C.Oliveira	Callithamniaceae	*Spyridia aculeata var. disticha* Børgesen	Preuss et al. (2017)
Stromatocarpus parasiticus Falkenberg	Rhodomelaceae	*Placophora monocarpa* (Montagne) Papenfuss, *Polysiphonia virgata* (C. Agardh) Sprengel	Preuss et al. (2017)
Symphyocolax koreanus M.-S.Kim	Rhodomelaceae	*Symphyocladia latiuscula*	Preuss et al. (2017)
Syringocolax macroblepharis Reinsch	Ceramiaceae	*Gelidium amansii* (Lamouroux) Lamouroux, *Plocamium cartilagineum* (L.) Gaillon	Preuss et al. (2017)
Tikvahiella candida Kraft & P.W.Gabrielson	Solieriaceae	*Solieria robusta*	Preuss et al. (2017)
Trichidium pedicellatum J.M.Noble & Kraft	Rhodomelaceae	*Lophocladia kuetzingii*;	Preuss et al. (2017)
Tylocolax microcarpus F.Schmitz	Rhodomelaceae	*Lenormandia spectabilis*	Preuss et al. (2017)
Ululania stellata K.E.Apt & K.E.Schlech	Rhodomelaceae	*Acanthophora pacifica* (Setchell) Kraft, *Acanthophora spicifera*	Preuss et al. (2017)
Dawsoniocolax bostrychiae (A.B.Joly & Yamaguishi-Tomita) A.B.Joly & Yamaguishi-Tomita	Rhodomelaceae	*Bostrychia* spp.	Preuss et al. (2017)
Holmsella australis Noble & Kraft	Pterocladiophilaceae	*Gracilaria furcellata, Gracilaria ramulosa, G. flagelliformis*	Preuss et al. (2017)
Meridiocolax bracteatus Noble & Kraft	Rhodomelaceae	*Polysiphonia sparsa*	Preuss et al. (2017)
Chlorophyceae			
Cephaleuros diffusus Thompson & Wujek	Trentepohliaceae	*Acacia auriculiformis*	Wonglom et al. (2018)
Cephaleuros expansus Thompson & Wujek	Trentepohliaceae	*Piper longum* and *Psidium guajava*	Wonglom et al. (2018)
Cephaleuros drouetii Thompson & Wujek	Trentepohliaceae	*Saraca declinata*	Wonglom et al. (2018)
Cephaleuros henningsii Schmidle	Trentepohliaceae	*Largestremia speciosa*	Wonglom et al. (2018)

(continued)

Table 15.1 (continued)

Parasitic algae	Family	Host	Reference
Cephaleuros karstenii Schmidle	Trentepohliaceae	*Citrus aurantiifolia, C. maxima, Ficus benjamina, Syzygium malaccense,* and *Theobroma cacao*	Wonglom et al. (2018)
Cephaleuros lagerheimii Schmidle	Trentepohliaceae	*Barringtonnia acutangula*	Wonglom et al. (2018)
Cephaleuros parasiticus Karsten	Trentepohliaceae	*Psidium guajava, camelia sinensis, Syzygium aromaticum.*	Suto et al. (2014)
Cephaleuros pilosa Thompson & Wujek	Trentepohliaceae	*Garcinia mangostana*	Wonglom et al. (2018)
Cephaleuros solutus Karsten	Trentepohliaceae	*Durio zibethinus*	Wonglom et al. (2018) and Pitaloka et al. (2014)
Cephaleuros tumidae-setae Thompson & Wujek	Trentepohliaceae	*Syzyium austral*	Wonglom et al. (2018)
Cephaleuros virescens Kunze ex E.M.Fries	Trentepohliaceae	*Alstonia scholaris, Anacardium occidentale, Annona muricata, A. squamosa, Artabotrys siamensis, Artocarpusheterophyllus, Baccaurea ramiflora, Bauhinia aureifolia, Bouea macrophylla, Calliandra haematocephala, Callistemon lanceolatus, Camellia sinensis, Carmona retusa, Casearia grewiifolia, Chrysophyllum cainito, Cinnamomum iners, Citrus aurantiifolia, C. maxima, Crescentia cujete, Dimocarpus longan, Diospyros rhodocalyx, Durio zibethinus, Ficus maciellandii, F. microcarpa, Garcinia mangostana, Glycosmis pentaphylla, Hevea brasiliensis, Ixora coccinea, Lagerstroemia floribunda, L. indica, Lansium domesticum, L. parasiticum, Magnolia champaca, Mammea siamensis, Mangifera indica, Manilkara zapota, Michelia alba, Nephelium lappacium, Parinari anamensis, Persea americana, Piper nigrum, Plumeria obtusa, Pterocarpus macrocarpus, Pterospermum littorale, Rauwenhoffia siamensis, Saraca indica, Sauropus androgynus, Shorea obtusa, Spondias mombin*	Wonglom et al. (2018)

(continued)

Table 15.1 (continued)

Parasitic algae	Family	Host	Reference
Rhodochytrium spilanthidis	Chlorochytriaceae	*Ambrosia artemisiifolia*	Atkinson (1908)
Phyllosiphon arisari	Watanabeaceae	*Arisarum* spp.	Joubert and Rijkenberg (1971)
Phyllosiphon deformans	Watanabeaceae	*Anchomanes difformis*	Joubert and Rijkenberg (1971)
Phyllosiphon philodendra	Watanabeaceae	*Philodendron* sp.	Joubert and Rijkenberg (1971)
Phyllosiphon ari Procházková, Němková & Neustupa	Watanabeaceae	*Arum italicum*	Guiry and Guiry (2023)
Phyllosiphon yunnanensis Shuyin Li, Benwen Liu, Huan Zhu, Zhengyu Hu & Guoxiang Liu	Watanabeaceae	*Dracaena draco*	Guiry and Guiry (2023)
Dinophyceae			
Stylodinium sphaera Pascher	Phytodiniaceae	*Oedogonium* sp.	Pfiester and Popovsky (1979)
Cystodinedria inermis (Geitler) Pascher	Phytodiniaceae	*Oedogonium* sp., *Spirogyra* sp.	Pfiester and Popovsky (1979)
Amoebophyra ceratii (Koeppen) Cachon	Amoebophryaceae	*Akashiwa sanguinea* (Hirasaka) G. Hansen, *Gymnodinium instriatum, karlodinium micrum*	Coats and Park (2002)

(Chakravarty & Mishra, 1983; Chapman & Good, 1983). Furthermore, certain host plants, such as citrus (*Citrus aurantiifolia* and *C. maxima*), tea and guava exhibit susceptibility to multiple green parasitic algal species (Holcomb, 1986; Holcomb et al., 1998; Sinclair & Lyon, 2005; Holcomb, 1986; Went, 1895). As these plants are economically important and are extensively cultivated for their fruits and leaves, the infections due to parasitic algae can have negative impacts on their overall productivity and the farmers' income (Holcomb, 1986).

Dinophyceaen parasites, a fascinating subgroup among the parasitic algae, exhibit a striking diversity in host preference. As listed in Table 15.1, these parasitic dinoflagellates can infect species ranging from other dinoflagellates, such as *Akashiwa sanguinea* and *Gymnodinium instriatum*, to filamentous green algae such as *Oedogonium* sp. and *Spirogyra* sp. (Pfiester & Popovsky, 1979; Coats & Park, 2002). Similarly, *Amoebophyra ceratii* exhibits a widespread host range, while *Stylodinium sphaera* demonstrates a relatively narrow range of hosts by infecting only *Oedogonium* sp. This showcases Dinophycean members being similar to the red and green algal parasitic species, capable of infecting numerous hosts (Pfiester & Popovsky, 1979; Coats & Park, 2002).

15.3 Invasion Mechanisms

Understanding the processes behind the invasion of algal parasites in hosts is crucial, as it holds significant implications for both ecological and economic dynamics. Researchers have recently started paying attention to the nature of these invasions, including the diverse interactions between algal parasites and their hosts.

15.3.1 Pre-penetration

Successful infections occur under moist conditions when propagules such as zoospores, sporangia or fragmented pathogenic thalli, released from infected host plants, get deposited on susceptible host surfaces (Thomas, 1913; Browne, 2017). Amongst those, the zoospores are often considered to be the principal inoculant for algae-borne diseases (Chapman & Good, 1983). In the case of *Trentepohlia* parasites, generally 8–16 zoospores with single keeled flagella are released when sporangiophores become wet. Upon landing on a suitable surface they begin to form a thallus for subsequent host penetration (Graham & McBride, 1975; Chapman & Good, 1983; Chapman & Henk, 1985; Thompson & Wujek, 1997). Reproduction in unfavourable circumstances, such as dry environments, leads to developing akinetes, a resistant reproductive cell that grows larger, producing thicker-walled sporangiophore initials (Chapman & Good, 1983). On reaching favourable environmental conditions, these sporangiophore initials form either sporangiophore laterals or quadriflagellate zoospores (Browne, 2017).

The invasion of the host by pathogenic algae exhibits intricate dispersal mechanisms that significantly impact their ability to infect plants. Dispersion of spores in species like *C. parasiticus* occurs from the lower leaf surface, where the zoosporangia become mature. In the case of *C. virescens* instead, the release of spores occurs from the upper surface of the host leaf (Suto et al., 2014). Various factors, such as insects, arachnids, rain, wind etc., have been reported to transport and disperse these spores to the surface of a suitable host (Joubert & Rijkenberg, 1971; Thompson & Wujek, 1997). These factors may either open up new wounds or help the algal spores infect already present wounds, facilitating penetration into the host body.

15.3.2 Penetration

In order to start infection, the algal zoospores must find a suitable substratum immediately after their release from the zoosporangium (Chapman & Good, 1983). A short-lived zoospore might struggle to penetrate the host protective cuticle effectively, and juvenile thalli have not been shown to break down the cuticle (Chapman & Good, 1983). As a result, the exact mechanism of penetrating the host is not clear to date. However, the penetration mechanisms employed by these algae are multifaceted and encompass several crucial aspects. Commonly, they tend to penetrate the host leaves, establishing at sub-cuticular, sub-epidermal or through intramatrix areas, although they have also been observed to invade through stem and developing fruits (Fig. 15.1) (Abdulkhair & Alghuthaymi, 2016; Chanthapatchot & Satjarak, 2019).

15.3.2.1 Leaf Infection

Leaves, as the principal organ performing photosynthesis, serve as a primary conduit for invasion in plants. The implications of leaf infections are particularly significant in plants where leaves hold economic importance (Ponmurugan et al., 2007). To gain insight into the dynamics of leaf infection, it is essential to comprehend the selective nature of pathogenic algae in choosing their hosts and the types of leaves they target. Pathogenic algae have shown greater affinity towards broad and long-lived leaves, as they provide an enormous surface area and available time for infection and invasion (Chapman & Good, 1983). Factors such as thickness and surface texture play pivotal roles in host susceptibility, and during infection, various portions of the leaves, such as lamina, petiole, and veins, might become affected. The preference of pathogenic algae for the upper leaf surface is noteworthy, although

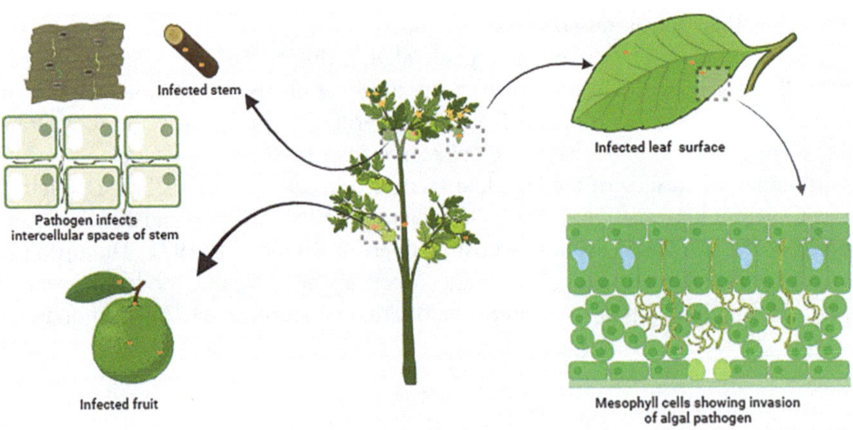

Fig. 15.1 A schematic illustration of the infection process caused by pathogenic algae on plants.

instances of lower surface infections have been observed (Wolf, 1930). An intercellular study on the infection behaviour of pathogenic algae like *Cephaleuros* on tea revealed that primarily algal cells invade the upper surface of the host leaf and eventually migrate to the lower side, resulting in the formation of fructification on both surfaces.

15.3.2.2 Stem and Fruit Infection

Notably, some algal pathogens, commonly exemplified by *Cephaleuros*, are also known to inflict infections on stems. A report by Holcomb (1986) on *Cephaleuros* infection indicated that, among the 287 host plants, 80 were known to have infection on their stems. Stem infections are frequently thought to be more harmful to plants because they may restrict nutrient transport and produce partial or entire girdling (Chapman & Good, 1983). These stem infections, hence, could be considered detrimental due to substantial growth impairment and, in severe cases, may cause death of the host plant.

Studies have shown that fruits of guava (Ruehle, 1941), citrus (Suit & Du, 1946), coffee (Went, 1895), tea (Mann & Hutchinson, 1907) and blackberry (Browne, 2017) have been infected by various algal pathogens. The symptoms of fruit infections are distinct and discernible, characterized by smaller spots and blotches with colour changes ranging from greenish to brown or black. Members of the genus *Cephaleuros* are also capable of infecting the fruits of their hosts (Sunpapao et al., 2016). Such fruit infections are, in most cases, superficial (Joubert & Rijkenberg, 1971).

15.3.3 Post-penetration Events

After a successful invasion, the algal pathogen must try to establish itself in the host tissue, thus affecting various metabolic responses of the host. Such responses contribute towards the formation of ecesis in the host. The impact of parasitic algae on the host plants results from interactions between the two entities (Browne, 2017). These interactions lead to changes in the host structure, anatomy and overall well-being, by significantly affecting health and vitality of the infected plants (Wolf, 1930). This mechanism thus highlights the critical interplay between the alga and host responses as shown below.

15.3.3.1 Morphology

The effects of pathogenic algae on host morphology can be conveniently grouped into three categories, by presenting symptoms on leaves, branches, and fruits (Joubert & Rijkenberg, 1971). At an early stage, generally the infection occurs on the upper leaf surfaces in the form of yellow-green coloured pinpoint lesions. Although it occurs primarily in the foliage, this symptom may also appear on twigs and fruits, forming reddish-brown to orange-coloured patches. Other forms of lesions may be greyish-green to yellowish-green in appearance (Joubert & Rijkenberg, 1971). The sterile hair-like outgrowths of the pathogen that project perpendicular to the vegetative thallus give a velvety appearance to these symptomatic patches (Browne, 2017). These patches are cushion-like, circular-irregular, and dendritically lobed in outline, which can be even seen through an unaided eyes (Fig. 15.2) (Wolf, 1930).

Fig. 15.2 Leaves of citrus (**a**), guava (**b**), mango (**c**) and tea (**d**) infected by *Cephaleuros* sp.

In plants like magnolia, avocado and loquat, whose lower leaf surfaces are bushy, the algal patches are confined only to the upper leaf side. Eventually, the lesions on the leaf surface become dry and form dead necrotic areas. In tea plants, the leaf surfaces from the foliage distal to the girdle turn yellowish, which gives the appearance of "light". These changes in the leaf colour may be due to the chlorosis and leads to a variegation on the surface (Mann & Hutchinson, 1904; Mann & Hutchinson, 1907). During *Cephaleuros* infection on tea, a small purple translucent spot appears on the upper surface of the leaves, which gradually extends to the lower surface, from which the lesions migrate to other parts of the plant such as twigs, stems and fruits (Wolf, 1930; Tunstall & Sarmah, 1947).

The lesions on twigs are often elongated and more extended, due to the fusion of numerous tiny lesions which girdle the twigs (Joubert & Rijkenberg, 1971). After the profuse growth of the parasite on the host twig and the limb, a number of symptoms can be observed. Reports suggest that lesions on young stems or twigs are normally darkish grey or purple in colour but gradually become typically rusty red. After a considerable amount of time, the host turns stunted and bears only a few leaves (Thompson & Wujek, 1997). Similarly, the bark becomes scaly with cracks, and localized swelling can be frequently observed at the site of infection (Chanthapatchot & Satjarak, 2019).

Generally, fruit lesions are smaller than leaf spots and are darkish-green to brown or black, with irregular outlines. Such colouration was also observed in infected blackberry (*Rubus fruticosus*) canes during infection by *Cephaleuros virescens*. Prolonged infection can lead to desiccation of infected areas, may result in the cracking of tissue, and may facilitate secondary infections by other cane pathogens (Brooks, 2004; Browne, 2017). Reduction in the number of fruits per cane with increasing disease intensity were also noted for blackberry plants, but no significant observation was recorded in the cane diameter and fruit size (Browne, 2017).

15.3.3.2 Anatomy

Recently, researchers have also paid ample attention to the study of the effects caused by algal pathogens on the host anatomy. It was reported that many host palisade cells recover their meristematic activity in response to any pathogen attack, generating compact layers of developing cells (Wolf, 1930). The newly created cell layers closer to the algal filament usually die and convert to a cork-like appearance, presumably acting as a barrier against further algal pathogenic attacks. Similarly, Went (1895) reported that *C. coffeae* infected host cells become isolated from the infected cells by a thick barrier. Pathogen-caused lesions generate tissue hyperplasia, with or without suberin, occurring at the edges or below the lesions (Joubert & Rijkenberg, 1971). Moreover, the active division and re-division of the spongy mesophyll cells of the host thicken their cell wall, thus acting as a protective barrier against further invasion of algal pathogens (Thompson & Wujek, 1997).

15.3.3.3 Physiology and Biochemistry

Some important physiological effects on host plants like tea, mango, guava, citrus, vanilla, sapota etc., caused by members of Trentepohliales include reduction in glucose levels and transpiration (Dhillon et al., 1992). Reduction in photosynthetic rate is another negative impact observed on the host, upon infection (Nelson, 2008). In the leaves of the firecracker plant *Cuphea* sp., sporangiophores of the alga *Stomatochroon* spp. were reported to be responsible for a blockage of stomata that inhibited gaseous exchange, leading to a water-soaking condition (Brooks, 2004; Chapman & Good, 1983; Timpano & Pearlmutter, 1983). In a major study, Dasgupta and Bindra (1992) showed a decrease in the transpiration rate and a lowering of sucrose, proteins, glucose, nitrite and amino acids in mango, citrus, vanilla, sapota and guava plant tissues affected by algal pathogens.

Leaves of *Arisarum vulgare* commonly infected by the green alga *Phyllosiphon arisari*, showed a loss of chlorophyll and starch, leading to slow necrosis (Tubeuf, 1897). Moreover, infected leaves of timber-yielding plants showed decreased amounts of phenol and nitrogen content as compared to healthy leaves, whereas amounts of amino acid, sugars and dry weights of leaves were found to be higher in the infected leaves (Chakravarty & Mishra, 1983). *Cephaleuros* spp. have also been observed to cause depletion in minerals and water absorption as well as loss of green photosynthetic regions and necrosis of host tissues (Wolf, 1930; Safeeulla & Govindu, 1948; Muthukumar et al., 2015). Ponmurugan et al. (2009) further documented reduction of various biochemical constituents including sugars, amino acids, proteins, polyphenols and catechin in tea leaves infected by *Cephaleuros parasiticus*.

15.4 Role of Plant Microbiome and Management of Algal Diseases

The term microbiome indicates a distinctive microbial population occupying a relatively well-defined environment, including genetic and physio-chemical features (Berg et al., 2020). Microorganisms such as bacteria and fungi in and around the root system comprise the plant microbiome (Babalola et al., 2020). Within this complex microbial community, these microorganisms interact with one another symbiotically, along with the host plant. Such microbiome has a dynamic impact on the plant nutrient cycling and resistivity against various pathogens, including algae (growth and severity) with both direct (Chen et al., 2012) and indirect effects (Afi et al., 1996; Matsumoto et al., 2003; Mu et al., 2007; Cho, 2012; Jiang et al., 2014). Numerous components of the plant microbiome, such as the rhizosphere microbiome, the endophytic microbiome, the phyllosphere microbiome, the rhizoplane microbiome etc., may operate to contrast algal pathogens, either by competing for resources or creating antimicrobial chemicals that impede the pathogens growth (Demuez et al., 2015).

The plant microbiota has, therefore, been found to contribute to the host plant defense against algal infections by influencing several physiological pathways. The organisms present in the microbiota can improve the defence mechanism of the plants by enhancing their immune systems (Yan et al., 2019; Babalola et al., 2020). The beneficial endophytic microbes contribute to the host defence (Mengistu, 2020; Grabka et al., 2022) by increasing their resistance towards algal diseases through induced systemic resistance (ISR) mechanisms. Bio-molecules such as quinolones, indole alkaloids, pyrroles and cyclic enzymes are produced by bacteria that can act as algaecides (Demuez et al., 2015). A few members of cyanobacteria or blue-green algae (BGA), such as *Nostoc spongiaeforme* (Hirata et al., 2003), *Anabaena flos-aquae* (Kearns & Hunter, 2001), *Microcystis aeruginosa* (Ikawa et al., 1997) and green algae like *Chlorococcum ellipsoideum* (Kroes, 1972) produce peptides, fatty acids and lipid derivatives which can also affect the growth of other algal species present in their vicinity. Microorganisms have also been found to enhance plant hormone levels, making them resilient to biotic stresses (Dreher & Callis, 2007). Furthermore, beneficial microorganisms sometimes compete for resources with algal pathogens, limiting their growth and proliferation (Sasse et al., 2018). Although the appropriate mechanisms have not yet been reported, studies have revealed that rhizobiome communities can promote the host growth and disease-resistance capacity (Omomowo & Babalola, 2019).

As the actual understanding of the complex connections linking the plant microbiome and algal pathogens is limited, an effective disease management technique is needed to mitigate algal pathogenicity. Researchers and practitioners should cooperate to build a sustainable, long-term strategy to prevent algae-borne diseases and enhance ecosystem health by applying potential microbes present in the plant microbiota. Therefore, the management of algal infections in their respective host plants needs an all-inclusive approach. Despite the limitation on the availability of literature on this topic, Ramya et al. (2013) noteworthy research on the control of *Cephaleuros* infection in tea plantations highlighted the effectiveness of various strategies in controlling the disease. They include treatments with different types of chemicals (algicides, fungicides, fertilizers) and biocontrol agents as well as practising various harvesting methods (Fig. 15.3).

Algicide To mitigate algal growth, researchers and industries frequently depend on algicides, which are chemicals engineered to manage or eradicate algae populations. However, Ramya et al. (2013) uncovered that *C. parasiticus* remained unaffected by algicides like algaenozol and algaenozine. In tea fields, the effectiveness of algicides was compromised due to inadequate binding with tea leaves. Factors such as continuous tea plucking practices further reduce the effectiveness of algicides. The specificity of selected algicides can also be another factor which may render them ineffective against host plants such as tea.

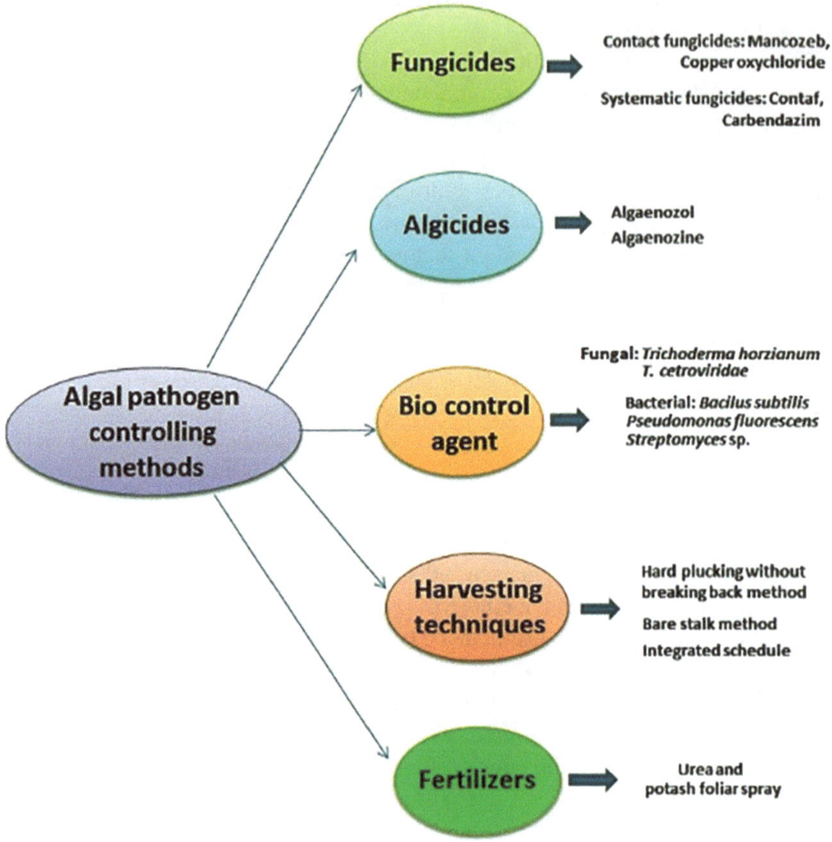

Fig. 15.3 Different strategies for management of algal pathogens

Fungicide Fungicides play a crucial role by protecting crops from fungal diseases that can devastate yields and compromise food security. These chemical agents are specifically formulated to combat fungi, and microorganisms that can cause various plant diseases. While algal pathogens are taxonomically different from fungal pathogens, fungicides with a broad-spectrum activity can effectively suppress their development. They may offer a viable solution to manage algal infections on the host plants by targeting the algal growth and proliferation. During field treatments, it was noticed that the incidence of *Cephaleuros parasiticus* was reduced, with systemic fungicides outperforming contact fungicides Ramya et al., 2013. Among the systemic fungicides, carbendazim showed an effective disease protection, likely due to its immediate action inhibiting spindle formation during mitosis (Kalim et al., 2000).

Fertilizers Foliar spray of fertilizers such as urea (CH_4N_2O) and Muriate of Potash (MoP) solutions have also been found to result in a significant reduction of red rust incidence. Potassium fertilizer enhances bush frame development, provides rigidity, strengthens stalks, and increases disease resistance by thickening cuticles, controlling turgor pressure, and preventing wilting and curling.

Biocontrol Agents Among the non-chemical methods of pathogen control, biocontrol seems to be the most suited for organic cultivation. It is environmentally safe, sustainable, economically viable, and highly specific (Pandit et al., 2022). Various biocontrol agents, including bacteria and fungi, may effectively mitigate pathogenic algal infections by acting as effective barriers to the entry of the pathogen in the host plants. Ramya et al. (2013) noticed a moderate reduction in red rust disease incidence caused by *C. parasiticus* through treatments based on bacterial biocontrol agents such as *Pseudomonus fluorescens*, *Bacillus subtilis*, *Streptomyces* sp. or fungal biocontrol agents such as *Trichoderma atroviride* and *T. harzianum*. These agents inhibited the growth and reproduction of *C. parasiticus*, with antagonists completely colonizing the filaments of *C. parasiticus* and causing wall lysis and rupture of the zoosporangia containing zoospores. Microscopic observations confirmed the production of antibiotic substances by *Trichoderma* sp., suppressing the pathogen development. Bushes treated with biocontrol agents exhibited the highest yield and productivity index with *Streptomyces sannanensis* resulting in the highest green leaf yield. Increased plucking surface, plucking points, leaf moisture, and dry matter content also increased with aplicatin of biocontrol agents. An impact was also observed on the host plant physiological and biochemical parameters, by producing growth-stimulating hormones such as auxins, cytokinin and gibberellins, thereby enhancing the plant capacity to withstand algal infections.

Harvesting Methods Different harvesting methods were found to be effective against the spread and infection of algal pathogens. Harvesting methods such as hard plucking without breaking back method, bare stalk plucking, shear harvesting, integrated schedule method etc., were tested to control red rust disease in tea plantations. The hard plucking without breaking back method showed the lowest disease incidence, while the integrated scheduling yielding the highest production. Continuous harvesting resulted in the highest disease incidence, likely because this method can weaken plant bushes and increase susceptibility to pathogens. It was found that dipping the shears in detergent/fungicide or biocontrol solutions before shearing failed to protect plants, likely due to the endophytic nature of the pathogen.

15.5 Conclusions

The concept of algae as a parasite is very intriguing, yet it has received surprisingly little attention among researchers to date. Our study highlights 145 plant parasitic algal species from the class Rhodophyceae, Chlorophyceae and Dinophyceae, with a special emphasis on their host range and invasion processes. The parasitic algae exhibit varying host specificity: while some of them infect only a single host, other members demonstrate a broader host range with the ability to infect multiple species. The invasion process varies from the dispersal of spores to their penetration into host organs for successful parasitic invasion. Finally, the effects on host plants comprise a variety of symptoms, such as alterations in morphology, anatomy, and physiological and biochemical characteristics. These findings emphasize the complicated relationships between parasitic algae and their host plants, which have substantial ecological and economic repercussions.

Acknowledgements The authors are grateful to the Head, Department of Botany, Gauhati University, for providing facilities developed under DST-FIST, UGC-SAP, OIL, MoEF & CC, ASTEC for rendering help to carry out the research work. The third author (SB) is also thankfully acknowledge fellowship support recieved from SJSGC fellowship granted by UGC, New Delhi.

References

Abdulkhair, W. M., & Alghuthaymi, M. A. (2016). Plant pathogens. *Plant Growth, 49*.

Afi, L., Metzger, P., Largeau, C., Connan, J., Berkaloff, C., & Rousseau, B. (1996). Bacterial degradation of green microalgae: Incubation of *Chlorella emersonii* and *Chlorella vulgaris* with *pseudomonas oleovorans* and *Flavobacterium aquatile*. *Organic Geochemistry, 25*(1–2), 117–130.

Atkinson, G. F. (1908). Note on the occurrence of *Rhodochytrium spilanthidis* Lagerheim in North America. *Science, 28*, 691–692.

Babalola, O. O., Fadiji, A. E., Enagbonma, B. J., Alori, E. T., Ayilara, M. S., & Ayangbenro, A. S. (2020). The nexus between plant and plant microbiome: Revelation of the networking strategies. *Frontiers in Microbiology, 11*, 548037.

Baruah, P. P., Boruah, B., Nath, S., Kalita, H., & Bhattacharjee, S. (2020). Algal diversity in Deepor Beel of Assam: A Ramsar site of north East India. *Nelumbo, 62*, 221–252.

Baruah, P. P., Bhattacharjee, S., Nath, S., Boruah, B., & Kalita, H. (2022). Algal diversity in riparian lentic habitats of Brahmaputra valley. *Nelumbo, 64*, 81–118.

Berg, G., Rybakova, D., Fischer, D., Cernava, T., Vergès, M. C. C., Charles, T., & Schloter, M. (2020). Microbiome definition re-visited: Old concepts and new challenges. *Microbiome, 8*, 1–22.

Blazencic, J. (2007). Sistematikaalgi. Beograd: Brennan L and owned author vitae biofuels from microalgae-a review of technologies for production, processing and extractions of biofuels and co-products. *Renewable and Sustainable Energy Reviews, 14*, 557–577.

Boruah, B., & Baruah, P. P. (2024). Freshwater desmid diversity along Naga thrust: A hydrocarbon province of South-East Asia. *Journal of Asia-Pacific Biodiversity, 17*(2), 375–395.

Brooks, F. E. (2004). Plant-parasitic algae (Chlorophyta: Trentepohliales) in American Samoa. *Pacific Science, 58*(3), 419–428.

Browne, F. B. (2017). *Orange cane blotch of blackberry caused by Cephaleuros virescens: Chemical control and yield losses associated with the disease*. (Doctoral dissertation,. University of Georgia.

Chakravarty, P., & Mishra, R. R. (1983). Studies in forest pathology. V. Host range of *Cephaleuros virescens* Kunze and the biochemical changes of the infected leaves. *European Journal of Forest Pathology, 13*(2), 109–115.

Chanthapatchot, W., & Satjarak, A. (2019). Distribution, diversity, and specificity of a parasitic algal genus *Cephaleuros* in Thailand. *Sains Malaysiana, 48*(8), 1609–1618.

Chapman, R. L., & Good, B. H. (1983). Subaerial symbiotic green algae: Interactions with vascular plant hosts [*Phycopeltis, Stomatochroon, Trentepohlia*]. In L. J. Goff (Ed.), *Algal symbiosis: A continuum of interaction strategies* (pp. 173–204). Cambridge University Press.
Chapman, R. L., & Henk, M. C. (1985). Observations on the habitat, morphology and ultrastructure of *Cephaleuros parasiticus* (Chlorophyta) and a comparison with *C. virescens* 1. *Journal of Phycology, 21*, 513–522.
Chen, W. M., Sheu, F. S., & Sheu, S. Y. (2012). *Aquimarin asalinaria* sp.nov., a novel algicidal bacterium isolated from a saltpan. *Archives of Microbiology, 194*, 103–112.
Cho, J. Y. (2012). Algicidal activity of marine Alteromonas sp.KNS-16 and isolation of active compounds. *Bioscience, Biotechnology, and Biochemistry, 76*, 1452–1458.
Chowdary, Y. B. K. (1969). Induction of reproductive organs in *Cephaleuros virescens*. *Indian Journal of Microbioiology, 8*(1), 53–58.
Coats, D. W., & Park, M. G. (2002). Parasitism of photosynthetic dinoflagellates by three strains of Amoebophrya (dinophyta): Parasite survival, infectivity, generation time, and host Specificity1. *Journal of Phycology, 38*(3), 520–528.
Dasgupta, W. S., & Bindra, A. S. (1992). Some biochemical changes induced in powdery mildew infected grape wine leaves. *Plant Disease Research, 7*, 248–258.
Demuez, M., González-Fernández, C., & Ballesteros, M. (2015). Algicidal microorganisms and secreted algicides: new tools to induce microalgal cell disruption. *Biotechnology Advances, 33*(8), 1615–1625.
Dhillon, W. S., Bindra, A. S., & Kapoor, S. P. (1992). Some biochemical changes are induced in powdery mildew-infected grapevine leaves. *Plant Disease Research, 7*(2), 248–250.
Dreher, K., & Callis, J. (2007). Ubiquitin, hormones and biotic stress in plants. *Annals of Botany, 99*(5), 787–822.
Falkowski, P. G., & Raven, J. A. (1997). *Aquatic photosynthesis*. Blackwell Science.
Grabka, R., d'Entremont, T. W., Adams, S. J., Walker, A. K., Tanney, J. B., Abbasi, P. A., & Ali, S. (2022). Fungal endophytes and their role in agricultural plant protection against pests and pathogens. *Plants, 11*(3), 384.
Graham, L. E., & McBride, G. E. (1975). The ultrastructure of multilayered structures associated with flagellar bases in motile cells of *Trentepohlia aurea* 1. *Journal of Phycology, 11*(1), 86–96.
Guiry, M. D. & Guiry, G. M. (2023). *AlgaeBase. World-wide electronic publication*. National University of Ireland. https://www.algaebase.org; searched on October 29, 2023.
Hirata, K., Yoshitomi, S., Dwi, S., Iwabe, O., Mahakhant, A., & Polchai, J. (2003). Bioactivities of nostocine a produced by a freshwater cyanobacterium *Nostoc spongiaeforme* TISTR 8169. *Journal of Bioscience and Bioengineering, 95*, 512–517.
Holcomb, G. E. (1986). Hosts of the parasitic alga *Cephaleuros virescens* in Louisiana and new host records for the continental United States. *Plant Disease, 70*(11), 1080–1083.
Holcomb, G. E., Vann, S. R., & Buckley, J. B. (1998). First report of Cephaleuros virescens in Arkansas and its occurrence on cultivated blackberry in Arkansas and Louisiana. *Plant Disease, 82*(2), 263.
Ikawa, M., Sasner, J. J., & Haney, J. F. (1997). Inhibition of Chlorella growth by degradation and related products of linoleic and linolenic acids and the possible significance of polyunsaturated fatty acids in phytoplankton ecology. *Hydrobiologia, 356*, 143–148.
Janczewski, E. (1872). Le parasitisme du *Nostoc lichenoides*. *Annales des Science Naturelles; Botanique, 5*(16), 306–316.
Jiang, X., Ren, C., Hu, C., & Zhao, Z. (2014). Isolation and algicidal characterization of Bowmanella denitrificans S088 against *Chlorella vulgaris*. *Journal of Microbiology and Biotechnology, 30*, 621–629.
Joubert, J. J., & Rijkenberg, F. H. J. (1971). Parasitic green algae. *Annual Review of Phytopathology, 9*(1), 45–64.
Kalim, S., Luthra, K. P., & Gandhi, S. K. (2000). Influence of bavistin seed treatment on morphological and biochemical parameters of cow pea roots susceptible to *Rhizoctonia* species. *Indian Journal of Mycology and Plant Pathology, 30*, 375–379.

Kalita, N., & Baruah, P. P. (2024). Copper removal efficacy and stress tolerance potential of Leptolyngbya sp. GUEco1015. *Heliyon, 10*(8).
Kearns, K. D., & Hunter, M. D. (2001). Toxin-producing *Anabaena flos-aquae* induces settling of *Chlamydomonas reinhardtii*, a competing motile alga. *Microbial Ecology, 42*, 80–86.
King, J. W. (1954). An investigation of hematochrome accumulation in the alga *Phycopeitis hawaiiensis* n. sp. *Pacific Science Review, 8*, 205–208.
King, J. W. (1955). Further investigation on hematochrome accumulation in the alga *Phycopeltis hawaiiensis*. *Biological Bulletin, 109*, 361–362.
Kroes, H. W. (1972). Extracellular products from *Chlorococcum ellipsoideum* and *Chlamydomonas globosa*. *Archives of Microbiology, 84*, 270–274.
Mann, H. H., & Hutchinson, C. M. (1904). Red rust; a serious blight of the tea plant. *Indian Tea Association Bulletin, 4*, 1–26.
Mann, H. H., & Hutchinson, C. M. (1907). *Cephaleuros virescens* Kunze the "red rust" of tea. *Department of Agriculture, India Botany, 1*(6), 1–33.
Matsumoto, M., Yokouchi, H., Suzuki, N., Ohata, H., & Matsunaga, T. (2003). Saccharification of marine microalgae using marine bacteria for ethanol production. *Applied Biochemy Biotechnol, 105*, 247–254.
Mengistu, A. A. (2020). Endophytes: Colonization, behaviour, and their role in defense mechanism. *Int JMicrobiol, 2020*, 6927219–6927218.
Merker P. (1888). *Gunnera macrophylla* Blum. Inaugural-Dissertation, G. Schirling.
Shave, C. D., Millyard, L., & May, R. C. (2021). Now for something completely different: Prototheca, pathogenic algae. *PLoS Pathogens, 17*, e1009362.
Mu, R. M., Fan, Z. Q., Pei, H. Y., Yuan, X. L., Liu, S. X., & Wang, X. R. (2007). Isolation and algae-lysing characteristics of the algicidal bacterium B5. *Journal of Environmental Sciences, 19*, 1336–1340.
Munir, N., Imtiaz, A., Sharif, N., & Naz, S. (2015). Optimization of growth conditions of different algal strains and determination of their lipid contents. *The Journal of Animal and Plant Science, 25*(2), 546–553.
Muthukumar, T., Uma, E., & Priyadharsini, P. (2015). Occurrence of Foliicolous parasitic alga *Cephaleuros virescens* on cultivated ornamental plants in southern India. *Botanica Lithuanica, 20*, 87–98.
Nelson, S. C. (2008). *Cephaleuros* species, the plant-parasitic green algae. *Plant Disease, 43*, 1–6.
Oborník, M. (2020). Photoparasitism as an intermediate state in the evolution of apicomplexan parasites. *Trends in Parasitology, 36*(9), 727–734.
Omomowo, O. I., & Babalola, O. O. (2019). Bacterial and fungal endophytes: Tiny giants with immense beneficial potential for plant growth and sustainable agricultural productivity. *Microorganisms, 7*(11), 481.
Pandit, M. A., Kumar, J., Gulati, S., Bhandari, N., Mehta, P., Katyal, R., Rawat, C. D., Mishra, V., & Kaur, J. (2022). Major biological control strategies for plant pathogens. *Pathogens, 11*(2), 273.
Pfiester, L. A., & Popovsky, J. (1979). Parasitic, amoeboid dinoflagellates. *Nature, 279*(5712), 421–424.
Pitaloka, M. K., Petcharat, V., & Sunpapao, A. (2014). *Cephaleuros solutus* Karsten, as a causal agent of durian (*Durio zibethinus* Murray) algal spot disease in Thailand. *Khon Kaen Agriculture Journal, 42*(Suppl 3), 644–648.
Ponmurugan, P., Baby, U. I., & Rajkumar, R. (2007). Growth, photosynthetic and biochemical responses of tea cultivars infected with various diseases. *Photosynthetica, 45*, 143–146.
Ponmurugan, P., Saravanan, D., Ramya, M., Srinivasan, T. R., Baby, U. I., & Ajay, D. (2009). Studies on *Cephaleuros parasiticus* Karst, a pathogenic alga causing red rust disease in tea plantations. *Journal of Plantation Crops, 37*, 70–73.
Ponmurugan, P., Saravanan, D., & Ramya, M. (2010). Culture and biochemical analysis of a tea Algal pathogen, *Cephaleuros parasiticus*. *Journal of Phycology, 46*, 1017–1023.

Preuss, M., Nelson, W., & Zuccarello, G. (2017). Red algal parasites: A synopsis of described species, their hosts, distinguishing characters and areas for continued research. *Botanica Marina, 60*, 13–25.

Printz, H. (1939). Vorarbeiten zu einer Monographie der Trentepohliaceen. *Nyt Magazin for Naturevidenskaberne, 80*, 137–210.

Ramya, M., Ponmurugan, P., & Saravanan, D. (2013). Management of Cephaleuros parasiticaus Karst (Trentepohliales: Trentepohliaceae), an algal pathogen of tea plant, *Camellia sinsensis* (L)(O. Kuntze). *Crop Protection, 44*, 66–74.

Rodríguez, F., Feist, S. W., Guillou, L., Harkestad, L. S., Bateman, K., Renault, T., & Mortensen, S. (2008). Phylogenetic and morphological characterization of the green algae infesting blue mussel *Mytilus edulis* in the North and South Atlantic oceans. *Diseases of Aquatic Organisms, 81*, 231–240.

Ruehle, G. D. (1941). Algal leaf and fruit spot of guava. *Phytopathology, 3*(1), 95–96.

Safeeulla, K. M., & Govindu, H. C. (1948). Some new hosts for *Cephaleuros*. *Journal of the Mysore University. Section B, 11*, 47–49.

Sasse, J., Martinoia, E., & Northen, T. (2018). Feed your friends: Do plant exudates shape the root microbiome? *Trends in Plant Science, 23*, 25–41.

Sinclair, W. A., & Lyon, H. H. (2005). *Diseases of trees and shrubs* (p. 575). Comstock Publishing Associates.

Skovgaard, A., Karpov, S. A., & Guillou, L. (2012). The parasitic dinoflagellates *Blastodinium* spp. inhabiting the gut of marine, planktonic copepods: Morphology, ecology, and unrecognized species diversity. *Frontiers in Microbiology, 3*, 305.

Suit, R. F., & Du, C. E. P. (1946). The *Cephaleuros* disease of citrus. *Citrus India, 27*(3).

Sunpapao, A., Thithuan, N., Bunjongsiri, P., & Arikit, S. (2016). *Cephaleuros parasiticus*, associated with algal spot disease on *Psidium guajava* in Thailand. *Australasian Plant Disease Notes, 11*, 1–4.

Suto, Y., Ganesan, E. K., & West, J. A. (2014). Comparative observations on *Cephaleuros parasiticus* and *C. virescens* (Trentepohliaceae, Chlorophyta) from India. *Algae, 29*, 121–126.

Thomas, N. (1913). Notes on Cephaleuros. *Annals of Botany, 27*(108), 781–792.

Thompson, R. H., & Wujek, D. E. (1997). *Trentepohliales: Cephaleuros, phycopeltis, and stomatochroon: Morphology, taxonomy, and ecology*. Science Publishers.

Timpano P & Pearlmutter NL (1983). Algal invasion of angiosperm mesophyll. SEM Inc., AMF O'Hare, .

Tubeuf, K. V. (1897). *Diseases of plants induced by cryptogamic parasites: An introduction to the study of pathogenic fungi, slime-fungi, bacteria and algae*. Longmans.

Tunstall, A. C., & Sarmah, K. C. (1947). Notes on stem diseases of tea in north East India. *Tocklai Experimental Station: Indian Tea Association Mem, 16*, 1–77.

Vázquez, N., Rodríguez, F., Ituarte, C., Klaich, J., & Cremonte, F. (2010). Host–parasite relationship of the geoduck Panopea abbreviata and the green alga *Coccomyxa parasitica* in the Argentinean Patagonian coast. *Journal of Invertebrate Pathology, 105*, 254–260.

Ward, H. M. (1884). On the structure, development, and lifehistory of a tropical epiphyllous lichen (*Strigula complanata* Fee). *Trans Linn Soc Lond, 2nd Series Botany, 2*, 87–119.

Went, F. A. F. C. (1895). *Cephaleuros coifeae*, eine neue pal'asitische Chroolepidee. *Center Bacteriol, 1*, 681–687.

Wolf, F. A. (1913). Another host for *Rhodochytrium*. *Phytopathology, 3*, 311.

Wolf, F. A. (1930). A parasitic alga, *Cephaleuros virescens* Kunze, on citrus and certain other plants. *Journal of the Elisha Mitchell Scientific Society, 45*, 187–205.

Wonglom, P., Thithuan, N., Bunjongsiri, P., & Sunpapao, A. (2018). Plant-parasitic algae (*Cephaleuros* spp.) in Thailand, including four new records. *Pacific Science, 72*, 363–371.

Yan, L., Zhu, J., Zhao, X., Shi, J., Jiang, C., & Shao, D. (2019). Beneficial effects of endophytic fungi colonization on plants. *Applied Microbiology and Biotechnology, 103*, 3327–3340.

Chapter 16
Disease Resistance in Mycorrhiza-Colonised Vegetable Crop: Understanding and Opportunity

Debajyoti Saha, Nabanita Sarkar, and Sanjoy Sadhukhan

Abstract Mycorrhizae, the symbiotic associations between fungi and plant roots, are fundamental to plant biology, ecology, and agricultural practice. This chapter delves into the mutually beneficial relationships between plants and fungi, emphasizing their critical role in nutrient uptake, disease resistance, and environmental sustainability. Through the lens of sustainable agriculture, it critically assesses the potential for leveraging mycorrhizal fungi in reducing dependency on chemical pesticides, thereby mitigating environmental impact and enhancing food safety. Beyond nutrient acquisition, mycorrhizae are pivotal in bolstering plant resilience against abiotic and biotic stresses. This symbiosis is crucial for plant health and productivity, soil health, and ecosystem dynamics, influencing nutrient cycling and carbon storage. The practical applications of mycorrhizal associations in agriculture are vast, offering sustainable solutions to some of the most pressing challenges in food production, including disease management in vegetable crops. Through a combination of historical insights, current research findings, and case studies, this chapter illustrates how mycorrhizae contribute to disease resistance and the potential for these associations in sustainable agricultural practices. Looking forward, the chapter explores innovative strategies to harness mycorrhizal benefits fully, including genetic approaches and precision agriculture, underscoring the importance of inter-

Debajyoti Saha and Nabanita Sarkar were contributed equally.

D. Saha
Department of Botany, Siksha Bhavana, Visva-Bharati, Santiniketan, Birbhum, West Bengal, India

N. Sarkar
Department of Botany, Ballygunge Science College, University of Calcutta, Kolkata, West Bengal, India

S. Sadhukhan (✉)
Plant Molecular Biology Laboratory, Department of Botany, Raiganj University, Raiganj, Uttar Dinajpur, West Bengal, India

P. Mathur, S. Roy (eds.), *Plant Microbiome and Biological Control*, Sustainability in Plant and Crop Protection 20, https://doi.org/10.1007/978-3-031-75845-4_16

disciplinary research and collaboration in advancing our understanding and application of mycorrhizal symbioses, for a resilient and sustainable agriculture.

Keywords Disease resistance · Mycorrhiza · Nutrient uptake · Stress tolerance · Sustainable agriculture · Vegetable crops

Abbreviations

ABA	Abscisic Acid
AM	Arbuscular Mycorrhiza
AMF	Arbuscular Mycorrhizal Fungi
BR	Brassinosteroid
CK	Cytokinin
ECM	Ectomycorrhiza
ET	Ethylene
GA	Gibberellic Acid
GPS	Global Positioning System
IAA	Indole Acetic Acid (Auxin)
IPM	Integrated Pest Management
ISR	Induced Systemic Resistance
JA	Jasmonic Acid
MAMP	Microbe-Associated Molecular Pattern
MIR	Mycorrhizal-Induced Resistance
PAMP	Pathogen-Associated Molecular Pattern
PPA	Pre-Penetration Apparatus
PR	Pathogenesis-Related
ROS	Reactive Oxygen Species
SA	Salicylic Acid

16.1 Introduction

Mycorrhiza, a fascinating and mutually beneficial symbiotic relationship between fungi and plants, lies beneath the soil, often hidden from view but profoundly influential in plant biology and ecology. This remarkable association has been the subject of scientific inquiry and practical interest for centuries, and its importance continues to unfold in contemporary research and agricultural practices (Chen et al., 2018; Nogueira De Sousa, 2023).

Mycorrhiza represents a partnership evolved over millions of years in which plants and fungi collaborate for mutual advantage. The word "mycorrhiza" is derived from the Greek word, with "mýkēs" meaning fungus and "rhiza" signifying root. This aptly describes the nature of this association as it involves the close interaction between a roots and specific types of fungi (Friberg, 1999; Dhiman et al., 2022).

The relationship begins when fungal hyphae and delicate thread-like structures infiltrate a root system. This infiltration is not invasive but somewhat cooperative, as the fungi establish themselves within the root cells or form a network around them. In return for carbohydrates and other organic compounds provided by the plant through photosynthesis, mycorrhizal fungi offer an array of services that significantly benefit the host (Begum et al., 2019; Huey et al., 2020; Wilkes, 2021).

One of the most prominent advantages of mycorrhizae is the enhancement of nutrient uptake. These fungi possess a vast network of hyphae that extend far beyond the reach of the root system, effectively increasing the surface area for nutrient absorption. This expanded reach allows fungi to access nutrients, such as phosphorus, nitrogen, and minerals, which might otherwise remain inaccessible to the plant (Huey et al., 2020; Etesami et al., 2021; Sheikh-Assadi et al., 2023).

Furthermore, mycorrhizae contribute to the resilience of plants under challenging environmental conditions. These fungi can improve host tolerance to drought, salinity, and heavy metals by secreting various substances. They also play a crucial role in disease resistance, as mycorrhizal associations may suppress pathogenic microorganisms and promote natural plant defence mechanisms (Begum et al., 2019; Zhu et al., 2022; Wahab et al., 2023).

The implications of mycorrhiza extend well beyond the plant-fungus relationship. They ripple through ecosystems, influencing plant community dynamics, nutrient cycling, and soil carbon storage. Researchers are continually uncovering the complex mechanisms and ecological significance of mycorrhizal associations, shedding light on their potential in agriculture, forestry, and environmental restoration (Dell, 2002; Begum et al., 2019; Huey et al., 2020; Wahab et al., 2023).

16.2 Types of Mycorrhizal Associations

There are several types of mycorrhizal associations, but the two most prevalent and well-studied are arbuscular mycorrhiza (AM) and ectomycorrhiza (ECM).

16.2.1 Mycorrhizal Symbiosis

Mycorrhizal symbiosis is a remarkable and intricate partnership between certain fungi types and root systems of most terrestrial plants. This cooperative association, which evolved over hundred million years, plays a fundamental role in the ecology of ecosystems and the vitality of plants (Van Der Heijden et al., 2015). Arbuscular Mycorrhiza (AM) fungi are ancient and widespread, forming mutualistic relationships with most plant species, including many crops. They penetrate the root cells of host plants, creating structures called arbuscules within cells. Arbuscules serve as sites for nutrient exchange, primarily facilitating the uptake of phosphorus and other essential minerals from soil. AM fungi play a vital role in enhancing nutrient acquisition, water absorption, and overall plant health (Chen et al., 2018; Wang et al., 2019; Begum et al., 2019; Wilkes, 2021; Singh et al., 2022; Sheikh-Assadi et al., 2023).

16.2.2 Ectomycorrhiza (ECM)

ECM fungi are typically associated with woody plants like trees and shrubs. In ECM symbiosis, the fungal hyphae surround the root cells without penetrating them. This external sheathing of the root tips forms a specialised structure known as the Hartig net, facilitating the exchange of nutrients and water (Massicotte et al., 1989; Henke et al., 2015). ECM fungi are significant in forest ecosystems, contributing to tree growth, soil structure, and carbon cycling.

16.3 Benefits of Mycorrhizal Symbiosis

The mycorrhizal symbiosis offers several advantages to both the host plant and the fungus.

16.3.1 Enhanced Nutrient Uptake

Mycorrhizal fungi extend the reach of a root system, increasing its access to essential nutrients, especially phosphorus, nitrogen, and minerals. Phosphorus is essential for plant growth and is often found in limited quantities in soil. These fungi can release enzymes that break down organic matter or minerals, releasing bound phosphorus and making it available for the plant. This increased access to phosphorus promotes healthier root development, better flowering, and overall plant growth. Mycorrhizal associations are also beneficial for nitrogen uptake. These fungi can capture and transport soil nitrogen to the plant, particularly in the form of ammonium and amino acids. This nitrogen supply is crucial for various plant processes, such as protein synthesis, chlorophyll production, and overall metabolic activities. Mycorrhizal fungi thus help plants maintain their health and vigour by facilitating nitrogen uptake. In addition to phosphorus and nitrogen, mycorrhizal fungi aid in absorbing various minerals, including essential micronutrients like iron, zinc, and copper. These minerals are vital for numerous biochemical processes within plants, such as photosynthesis, enzyme activation, and development of healthy cell structures. By extending the root system reach to access these minerals, mycorrhizal fungi contribute to the plant overall nutrient balance. Mycorrhizal fungi not only increase nutrient uptake but also improve its use efficiency. They facilitate the transfer of nutrients to the plant in a controlled manner, reducing the risk of excess nutrient loss through leaching or runoff. This efficiency contributes to more sustainable agriculture and minimizes the environmental impact of fertilizers (Wang et al., 2019; Sheikh-Assadi et al., 2023).

16.3.2 Improved Stress Tolerance

Mycorrhizal associations enhance plants resilience to environmental stressors, including drought, salinity, and heavy metals. Fungi can help mitigate the adverse effects of these conditions on plant health. The fungal hyphal networks act as extensions of the root system, reaching deeper into the soil and accessing water sources that are beyond the reach of roots. This improved access to moisture enables the plant to maintain adequate hydration, even during dry periods. Furthermore, mycorrhizal fungi can stimulate the production of stress-resistance compounds within the plant, such as osmolytes and antioxidants, which help the plant endure water scarcity. Soil salinity, a common stressor in many regions, can hinder plant growth by disrupting the balance of ions and water uptake. Mycorrhizal fungi can alleviate salt stress by regulating ion absorption and improving water management in the root zone.

Additionally, these fungi can enhance the plant ability to exclude or compartmentalize toxic ions, reducing the damaging effects of salts on cellular processes. In soils contaminated with heavy metals, which are toxic to most plants, mycorrhizal associations provide a protective shield. The fungi can form barriers that prevent the uptake of heavy metals into the plant, thus reducing metal toxicity. They can also facilitate the sequestration of heavy metals in fungal structures, preventing these contaminants from accumulating in the root tissues. This detoxification process helps the plant thrive in metal-contaminated environments. Mycorrhizal associations can also influence the plant hormonal balance. They can stimulate the production of stress-responsive hormones such as abscisic acid and cytokinins, which help plants adapt to environmental challenges. These hormonal changes can trigger various stress-mitigating responses, such as altered stomatal closure, improved water use efficiency, and enhanced root growth (Sa et al., 2019; Hanaka et al., 2021; Singh et al., 2022; Samal et al., 2023; Shi et al., 2023).

16.3.3 Disease Resistance

Mycorrhizal symbiosis has been linked to enhanced disease resistance in plants. Some mycorrhizal fungi release secondary metabolites and antimicrobial compounds that can directly inhibit the growth and activity of pathogenic microorganisms. These compounds act as a protective shield for the plant, preventing the invasion and proliferation of pathogens in the root zone. Mycorrhizal fungi help to reduce the risk of disease development by creating an unfavourable environment for pathogens. Mycorrhizal associations can trigger the plant natural defence mechanisms, a phenomenon known as induced systemic resistance (ISR). When mycorrhizal fungi colonize the roots, they can induce the plant to produce defence-related compounds, such as phytoalexins, pathogenesis-related (PR) proteins, and reactive oxygen species (ROS). These compounds directly inhibit pathogen growth and prime the plant immune system to respond more effectively to pathogen attacks in other parts or tissues of the same plant. Improved nutrient uptake facilitated by mycorrhizal fungi indirectly contributes to disease resistance. When plants receive a balanced and ample supply of essential nutrients, they can allocate more resources to their immune responses, making them better equipped to defend against pathogens. Mycorrhizal fungi can indirectly suppress pathogens by outcompeting them for nutrients and by producing antibiotics or other inhibitory substances. By depriving pathogens of essential resources and releasing toxic compounds, these fungi create an environment less conducive to pathogen growth and establishment (Hao et al., 2019; Enebe & Erasmus, 2023).

16.3.4 Carbon Cycling

Through the extensive mycorrhizal network, carbon compounds, primarily sugars and other organic compounds produced during photosynthesis, are transported from the host to the surrounding soil. The carbon compounds transferred by mycorrhizal fungi contribute to organic matter formation in soil. This organic matter includes many substances, such as dead plant material, fungal hyphae, and various microbial residues. It is a crucial component of healthy soils and is vital in maintaining soil structure, moisture retention, and nutrient availability. Mycorrhizal fungi also play a significant role in forming humus, a stable and long-lasting component of soil organic matter. Humus is rich in carbon and serves as a reservoir for nutrients, particularly nitrogen and phosphorus. It helps to improve soil structure, making it more stable and resistant to erosion and enhancing its water-holding capacity. The carbon-rich substances introduced into the soil through mycorrhizal associations can act as a glue-like material that binds soil particles together, promoting the formation of soil aggregates. These aggregates improve soil structure by creating pore spaces for better air and water circulation. Improved soil structure is essential for root growth, microbial activity, and overall soil fertility. The carbon transferred to the soil via mycorrhizal networks also contributes to carbon sequestration, a process that stores carbon in the soil for extended periods. This is important for soil health and mitigating climate change, as it helps reduce atmospheric carbon dioxide levels (Hawkins et al., 2023).

16.3.5 Ecological Significance

Beyond their immediate benefits to plants, mycorrhizal associations have profound ecological implications, influencing the composition and dynamics of plant communities. Different plant species form specific types of mycorrhizal relationships, and the presence (or absence) of mycorrhizal partners can impact the success of plant species within a community. Some plants may establish stronger mutualistic associations with mycorrhizal fungi, giving them a competitive advantage in nutrient-poor soils. This, in turn, can influence a plant relative abundance and distribution in an ecosystem, where mycorrhizal interactions are closely linked to species diversity. The diverse array of mycorrhizal fungi and their specific associations with various plant species contribute in fact to the overall diversity of plant communities. These fungi create niche differentiation, enabling different plant species to coexist within the same habitat by accessing different nutrient sources and responding to varying environmental conditions.

Mycorrhizal networks significantly impact nutrient cycling within ecosystems. Ectomycorrhizal (ECM) fungi in forests create a vast underground network connecting individual trees. This phenomenon is often called the "wood wide web" and allows for the spatial redistribution of nutrients. This resource-sharing system enhances overall ecosystem nutrient cycling and helps maintaining the health of the forest community (Read & Perez-Moreno, 2003; Peay, 2016). These relationships showcase how different organisms collaborate to maximize resource use, and they are key components of the intricate web of ecological interactions that sustain biodiversity and ecosystem health (Garg & Chandel, 2010).

Mycorrhizal symbiosis exemplifies the intricate interdependence of life forms in nature. It's a testament to the power of cooperation between seemingly disparate organisms—plants and fungi—to thrive in diverse environments and sustain life on Earth. Understanding and leveraging this symbiosis holds great promise for addressing challenges in agriculture and environmental conservation, in an era of changing climates and growing resource constraints (Chen et al., 2018; Averill et al., 2019).

16.4 Mechanisms of Disease Resistance in Mycorrhizal-Colonized Plants

In the modern world scenario, the demand to develop high-yielding, disease-resistant crops and decrease use of chemical fertilisers is growing. Improved symbiotic associations with microbes help to combat biotic and abiotic stress, combined with efficient nutrient assimilation (Jacott et al., 2017). AM can form symbionts with 80–90% of terrestrial plants and are found to have roles in stress tolerance, nutrient uptake, plant growth, crop productivity, and disease resistance in return for photosynthetic carbon from their host (Zhu et al., 2022). Reports showed increased vulnerability of mycorrhizal-colonised plants *vs* the attacks of viruses, powdery mildew, and fungal rust (caused by i.e., *Blumeria*, *Oidium*, *Uromyces*, etc.) along with increased plant biomass and crop yield (Pozo De La Hoz et al., 2021). Mycorrhizal-associated plants have a variety of defence mechanisms to perform disease resistance.

16.4.1 Mycorrhizal-Induced Resistance (MIR)

Defences in plants can be innate, regardless of herbivory, or induced, expressed in response to external stimuli, preserving the resources available to plants for development and reproduction. Through induced resistance, plants can express several defensive mechanisms to combat the biotic and abiotic stresses they encounter. Mycorrhizal-Induced Resistance (MIR) is a systemic response, meaning that resistance is not limited to the site of infection but can extend to other parts of the plant (Bhavanam & Stout, 2021).

It is widely acknowledged that to produce resistance, an AM symbiosis must be well-established (Pozo De La Hoz et al., 2021). Roots release the hormone strigolactone to establish AM colonisation. It stimulates fungal spore germination, metabolism, hyphal branching and the exudation of a lipochitooligosaccharide, known as the mycorrhizal (Myc) factor, which the host plant receptors can recognise. Following several transcriptional pathways, the pre-penetration apparatus (PPA), a sub-cellular structure of plant epidermal cells, is produced in response to some chemical stimulation from AMF hyphopodia. The PPA directs the production of hyphae that grow through root epidermal cells and start branching inside the inner cortical cells to form arbuscules, the nutrient-exchanging structure (Jacott et al., 2017).

16.4.2 Hormonal Signalling

Following the successful colonisation of mycorrhiza in roots, the host recognises pathogen-associated molecular patterns (PAMPs) or microbe-associated molecular patterns (MAMPs) from AMF, resulting in the initial introduction of immunity against pathogens. A cascade of signalling pathways is elicited in response to recognising PAMPs/MAMPs, producing several plant defence hormones (Jacott et al., 2017). AMFs can directly induce the production of hormones such as Salicylic acid (SA), Jasmonic acid (JA), Auxin (IAA), Cytokinin (CK), Gibberellic acid (GA), Brassinosteroids (BR), Ethylene (ET), and Abscisic acid (ABA). AMFs stimulate the production of these hormones, which can activate many genes to impede stress conditions (Weng et al., 2022). Research has found that SA can stimulate defence mechanisms against biotrophic infection. It was also found that AMFs increase endogenous SA levels, although the initial increase of SA level is downregulated to allow the symbiosis to form (Fig. 16.1). JA and ET pathways act against necrotrophic diseases (Hohmann & Messmer, 2017). Also ABA has a critical role in defence mechanisms, encouraging early-responding defence mechanisms like MAMP-dependent stomatal closure, ROS induction, cell wall reinforcements, etc. Thus, ABA is mobile in both the xylem and phloem. It can act as a long-range MIR signal to the shoot (Cameron et al., 2013).

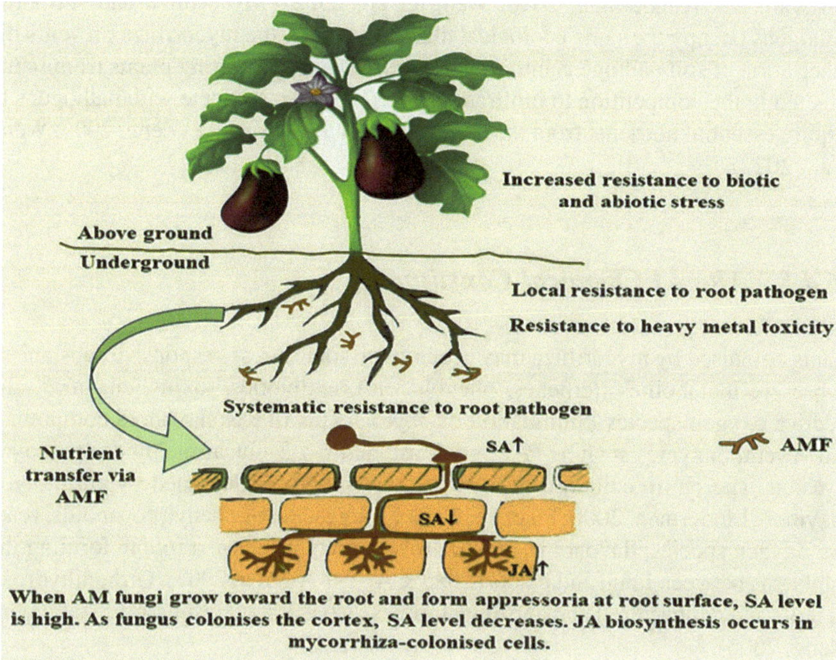

Fig. 16.1 Mycorrhizal association with plant roots, highlighting various functions and hormonal signalling (SA and JA) in the root cortical cells

16.4.3 Improved Nutrient-Uptake

Plants with better nutritional status can stand up against both biotic and abiotic stresses (Fig. 16.1). The AMF hyphal network increases phosphorus acquisition efficiency, which boosts plant output and enhances nutrient uptake (mostly of phosphorus) (Hohmann & Messmer, 2017). The phosphate supply in most soil types is constrained because free phosphate cations are immobilised. The mycorrhiza hyphae can get around the mobilisation restriction, increasing the amount of soil that is accessible for phosphate uptake (Jacott et al., 2017). Some alterations in root exudate caused by phosphate may impact pathogens spores ability to germinate (Linderman, 2000).

16.4.4 Improved Physical Barrier and Pathogen Suppression

As a result of AMF colonisation, plants release chemicals that change their morphology and anatomy to protect themselves from harmful soil pathogens such as bacteria. These changes include increased lignification in the root cells epidermis, the formation of wound-barrier in root cells, increased branching, thickening of the cell wall, etc. (Linderman, 2000; Weng et al., 2022). Mycorrhiza and soilborne pathogens occupy the same ecological niche. However, the mycorrhiza prevents the penetration of soil pathogens into root epidermal cells, protecting plants from being infected in the competition to infiltrate cells. AMFs also compete with pathogens to acquire essential nutrients from the rhizosphere and root surface (Zeng, 2006; Weng et al., 2022).

16.4.5 Altered Chemical Constituents

Plants colonised by mycorrhiza may accumulate some stress-responsive substances. Those are metabolites (terpenes, phenols, etc), antibiotics, toxins, enzymes, and reactive oxygen species. Stimulation of phytoalexins like isoflavonoid compounds and several enzymes such as PAL, chitinase and β-1,3-glucanase hinder pathogen invasion. The chitin cell wall of a pathogenic fungus is degraded by chitinolytic enzymes (Linderman, 2000; Pu et al., 2022). Expression of phenylpropanoids, reactive oxygen species, flavonoids, and hydrolytic enzymes have a role in forming the symbiosis between fungi and plants (Pozo & Azcón-Aguilar, 2007). Orthodihydroxy phenols and antibacterial and antifungal toxics also work against harmful pathogens (Zeng, 2006).

16.4.6 Beneficial Microbial Communities in Mycorrhizosphere

Combining mycorrhiza with other beneficial microbes can provide increased protection against harmful pathogens. Some examples of these microbes include *Trichoderma harzianum*, non pathogenic strains of *F. oxysporum, Glomus mosseae, Acaulospora laevis,* etc. (Weng et al., 2022). By altering the chemical components in the mycorrhizosphere, plants can limit infection by other harmful microbes. After the formation of the mycorrhizosphere, plants can recognise MAMPs present in mycorrhiza, which may trigger induced systematic responses. The contribution of the Jasmonate pathway in this induced response has also been reported (Cameron et al., 2013).

16.5 Case Studies

Case studies provide valuable insights into the practical application of mycorrhizal associations to enhance disease resistance in vegetable crops. A few notable cases of the potential of mycorrhiza-mediated disease management are given in Table 16.1.

Table 16.1 Mycorrhizal roles in disease resistance and its applications

Host	Pathogen/Pest	Disease	Mechanism	Application	References
Solanum lycopersicum	*Fusarium oxysporum*	Fusarium Wilt	Induced systemic resistance. Stimulated the production of resistance-related substances Competition with other pathogens to acquire nutrients.	Mycorrhizal inoculation can be applied in tomato cultivation to reduce the incidence of Fusarium wilt. This application is particularly beneficial in soils where Fusarium wilt is a known problem. Mycorrhizal fungi form a symbiotic relationship with tomato roots, promoting the production of pathogenesis-related proteins and activating systemic resistance pathways. This response helps the plant resist Fusarium wilt by strengthening its natural defences. Additionally, mycorrhizal fungi can compete with the *Fusarium* pathogen for resources and root colonization sites, reducing disease incidence.	Song et al. (2015), Sinno et al. (2020), Ma et al. (2023), and Meddad-Hamza et al. (2023)
Cucumis sativus	*Pythium* spp. (*P. aphanidermatum, P. debaryanum, P. ultimum* etc.)	Pythium root rot	Protective barrier around roots. Improved nutrient uptake.	Mycorrhizal associations are applied to protect cucumber plants from Pythium root rot, especially in hydroponic and greenhouse systems where the disease is prevalent. Mycorrhizal fungi create a protective shield around cucumber roots, forming a physical and chemical barrier that prevents *Pythium* spp. pathogens from infecting the root tissues. In addition to this direct defence, mycorrhizal-colonized cucumber plants exhibit improved nutrient and water uptake, enhancing overall plant health and resilience against root rot.	Moulin et al. (1994), Siddiqui et al. (2008) and Raaijmakers et al. (2009)
Capsicum annum	*Phytophthora capsici*	Phytophthora Blight	Systematic resistance mechanism Production of antimicrobial compounds and pathogenesis-related proteins Competition with pathogens for resources.	Mycorrhizal inoculation is employed in pepper crops to enhance resistance against Phytophthora blight, a common and destructive disease. Mycorrhizal associations trigger the plant systemic resistance mechanisms, including producing antimicrobial compounds and pathogenesis-related proteins. It enhances the plants ability to ward off *Phytophthora* infections. The mycorrhizal fungi can also compete with the pathogen for resources, making it a valuable tool in regions where *Phytophthora* is a significant threat.	Cardarelli et al. (2020)

(continued)

Table 16.1 (continued)

Host	Pathogen/Pest	Disease	Mechanism	Application	References
Lactuca sativa	*Rhizoctonia solani*	Bottom rot	Enhanced physical and chemical structure of plant Improved nutrient status.	Mycorrhiza-mediated disease resistance can be employed in lettuce cultivation to combat bottom rot, especially in fields with a history of the disease. Mycorrhizal fungi enhance lettuce root health and produce antimicrobial compounds, strengthening the roots and making them less susceptible to bottom rot pathogens. This application is beneficial in regions where bottom rot is known to cause crop losses.	Erlacher et al. (2014) and Wallon et al. (2020)
Cucurbita pepo	*Golovinomyces cichoracearum, Podosphaera xanthii, Leveillula taurica* etc	Powdery mildew	Promoting jasmonic acid and other defense-related compounds. Altering hormonal balance.	Mycorrhizal inoculation reduced powdery mildew infections in zucchini plants, particularly in regions with high humidity where this disease thrives. Mycorrhizal colonization can alter the hormonal balance in zucchini, promoting the production of jasmonic acid and other defense-related compounds. This hormonal response helps the plant resist powdery mildew infections. Additionally, improved nutrient uptake contributes to overall disease resistance.	Beslemes et al. (2023)
Solanum tuberosum	*Phytophthora infestans*	Late blight	Inducing the production of pathogenesis-related proteins and antimicrobial compounds. Extended root network.	Mycorrhizal treatments are used in potato farming to enhance resistance against late blight, a devastating disease for potato crops. Potato plants are prepared to mount a better and quicker defence response against late blight by mycorrhizal fungus. They induce the production of pathogenesis-related proteins and antimicrobial compounds. The extended root network and enhanced nutrient uptake contribute to plant vigour and disease resistance.	Tsedaley (2014)
Daucus carota	*Meloidogyne* sp.	Root Knot Nematodes	Reducing susceptibility to nematode infestation. Improving root health and nutrient uptake.	Mycorrhizal associations can be applied in carrot cultivation to reduce the damage caused by root-knot nematodes in infested soils. Mycorrhizal fungi enhance root health and improve nutrient acquisition, making carrot roots less attractive to nematodes. This application is particularly beneficial in regions with a history of nematode infestations.	Seo et al. (2015) and Kim et al. (2017)

Brassica oleracea	*Sclerospora* spp.	Downy Mildew	Improving nutrient and water uptake. Altering chemical constituents against pathogen.	Mycorrhizal treatments applied to broccoli crops enhanced resistance to downy mildew infections, especially in regions prone to this disease. Mycorrhizal fungi induced the plant defence responses, including the production of pathogenesis-related proteins, in response to downy mildew infections. It enhanced resistance against the pathogen—Improved nutrient and water uptake, further supporting disease resistance.	Hong et al. (2008)
Spinacia oleracea	*Fusarium oxysporum*	Fusarium infection	Improving systemic resistance. Improving root health and nutrient uptake.	Mycorrhiza-mediated disease resistance can be employed in spinach cultivation to reduce Fusarium infections, improving crop health and yield by triggering systemic resistance. Enhanced nutrient uptake and improved root health contributed to overall disease resistance. This application is precious in regions where Fusarium is a significant threat.	Gatch and Du Toit (2017)
Raphanus sativus	*Rhizoctonia solani*	Rhizoctonia root rot	Inhibiting growth and activity of pathogen. Production of antimicrobial-compounds.	Mycorrhizal inoculation is applied in radish farming to control Rhizoctonia root rot, particularly in soils with a history of the disease. Mycorrhizal fungi directly inhibit the growth and activity of *Rhizoctonia* pathogens in the root zone. Additionally, they may induce the production of antimicrobial compounds in radish plants, protecting against infection. This application enhances the overall health and quality of radish crops.	Elsharkawy et al. (2022)
Brassica oleracea	*Xanthomonas campestris*	Black rot	Enhancing plant ability to resist pathogen. Induction of several defense-related pathways.	Mycorrhizal associations are employed in cabbage crops to enhance resistance against black rot, especially in areas where this disease is prevalent. Mycorrhizal inoculation enhances the plant ability to resist infection by black rot-causing pathogens like *Xanthomonas campestris*. It involves the activation of defence-related pathways and the production of antimicrobial compounds. Cabbage crops benefit from increased disease resistance, leading to higher yields and quality.	Rubel et al. (2020) and Liu et al. (2022)

(continued)

Table 16.1 (continued)

Host	Pathogen/Pest	Disease	Mechanism	Application	References
Phaseolus vulgaris	More than 7000 species of fungi of phylum Basidiomycota	Rust pathogen	Induction of systemic resistance pathways. Production of Phytoalexins and other compounds.	Mycorrhiza-mediated disease resistance in beans involves the induction of systemic resistance pathways, including the production of phytoalexins and other antimicrobial compounds. This application is vital in regions prone to rust outbreaks, as it helps reduce the impact of rust pathogens, ensuring healthier bean crops and higher yields.	Santoyo et al. (2021) and Enebe and Erasmus (2023)
Brassica oleracea	*Plasmodiophora brassicae*	Clubroot disease	Enhancing physical resistance. Improving nutrient uptake.	Mycorrhizal inoculation can be used in cauliflower farming to manage clubroot disease, particularly in soils with a history of infection. Mycorrhizal associations enhance cauliflower root health and suppress pathogen growth, improving resistance to clubroot. This application contributes to healthier cauliflower crops and higher-quality produce.	Auer and Ludwig-Müller (2023)
Cucurbita pepo	*Tobacco mosaic virus* (TMV)	Mosaic virus	Production of antiviral proteins.	Mycorrhizal treatments are applied to squash crops to enhance resistance to mosaic viruses, which can significantly impact yield. Mycorrhizal colonization stimulates the production of antiviral proteins in squash plants, reducing their susceptibility to mosaic virus infections. This application contributes to improved crop health and higher yields of squash.	Abdelkhalek et al. (2022)
Solanum melongena	*Verticillium dahliae, V. albo-atrum*	Verticillium Wilt	Enhancing resistance by priming defence responses. Production of disease-resistance compounds.	Mycorrhiza used in eggplant cultivation reduced the impact of Verticillium wilt, a soilborne disease. Mycorrhizal inoculation enhances eggplant resistance to Verticillium wilt by priming the plant defence responses. It includes the production of pathogenesis-related proteins and antimicrobial compounds. This application contributes to healthier eggplant crops and improved yield.	Singh et al. (2014)
Abelmoschus esculentus	*Sclerotinia sclerotiorum*	Sclerotinia Rot	Suppressing pathogens by production of antimicrobial compounds.	Mycorrhizal inoculation is employed in okra farming to suppress Sclerotinia rot, particularly in areas with favourable conditions for the pathogen. Mycorrhizal associations can directly inhibit the growth of *Sclerotinia* pathogens in okra roots and induce the production of antimicrobial compounds, reducing disease incidence. This application enhances overall plant health and reduces disease-related losses in okra crops.	Prova et al. (2018)

Brassica oleracea	*Alternaria alternata*, *A. brassicae*, *A. brassicicola*	Alternaria Leaf spot	Production of antifungal compounds. Enhancing physical barrier.	Mycorrhiza-mediated resistance can be applied in kale cultivation to reduce Alternaria leaf spot infections, improving crop quality. Mycorrhizal associations stimulate the plant defence mechanisms against Alternaria leaf spot. It includes the production of antifungal compounds and the strengthening of leaf physical barriers. This application helps ensure healthier kale crops with improved visual and nutritional quality.	Poveda et al. (2022)
Brassica rapa	*Golovinomyces cichoracearum*, *Podosphaera xanthii*, *Leveillula taurica*, etc.	Powdery Mildew	Induction of hormonal pathways. Enhancing nutrient uptake.	Mycorrhizal treatments are used in turnip farming to enhance resistance to powdery mildew, particularly in regions with high humidity where this disease is prevalent. Mycorrhizal colonization can alter the hormonal balance in turnip plants, promoting the production of jasmonic acid and other defence-related compounds. This hormonal response helps the plant resist powdery mildew infections. Additionally, improved nutrient uptake contributes to disease resistance, leading to healthier turnip crops.	Widawati (2020)
Pisum sativum	*Pratylenchus* spp.	Root lesion Nematodes	Improving root health and nutrient acquisition.	Mycorrhizal associations can be applied in pea cultivation to reduce root lesion nematode damage in infested soils. Mycorrhizal fungi enhance the plant resistance towards root lesion nematodes by improving root health and nutrient acquisition. They can also reduce nematode attraction to pea roots. This application helps to maintain healthier pea crops and prevents yield losses caused by nematode damage.	Parihar et al. (2020)
Allium cepa	*Stromatinia cepivora*	White Rot	Suppressing the growth of pathogens. Production of some antimicrobial compounds.	Mycorrhizal inoculation is employed in onion farming to suppress white rot infections, especially in soils with a history of the disease. Mycorrhizal fungi can suppress the growth and activity of white rot pathogens in onion roots, protecting against infection. They may also induce the production of antimicrobial compounds in onion tissues. This application is critical for regions where white rot is a common and destructive disease, ensuring healthier onion crops and reducing post-harvest losses.	Pakarinen et al. (2021)

16.6 Applications of Mycorrhiza-Mediated Disease Resistance in Vegetable Crops

Understanding mycorrhizal associations and their role in disease resistance has led to several practical applications in agriculture. These examples showcase the effectiveness of mycorrhiza-mediated disease management in vegetable crops.

16.6.1 Organic Farming Practices

The symbiotic partnership between plant roots and mycorrhizal fungi is highly beneficial in organic farming. Mycorrhizal fungi effectively expand a plant's root system, increasing its capacity to absorb water and nutrients, which is crucial in nutrient-poor organic soils. One of the primary goals of organic farming is to minimise the use of synthetic pesticides. By enhancing the disease resistance of vegetable crops, mycorrhizal associations help to reduce the reliance on chemical pesticides. It is a vital aspect of organic farming, as it promotes a healthier and more environmentally sustainable agricultural ecosystem (Meemken & Qaim, 2018; Fernández et al., 2022; Gamage, 2023).

16.6.2 Soil Health Improvement

Mycorrhizal symbiotic relationships have a long-lasting impact on soil health and can persist in soil for multiple growing seasons. As a result, they contribute to long-term soil health improvement, reducing the presence of soilborne pathogens. Mycorrhizal fungi can actively suppress soilborne pathogens through various mechanisms. They compete with pathogens for root space and nutrients, produce antifungal compounds, and even induce plant systemic resistance. Reducing soilborne diseases is essential for maintaining crop productivity (Corato, 2020; Harman et al., 2021; Chen et al., 2023).

16.6.3 Crop Rotation and Disease Suppression

Mycorrhizal associations persist in the soil even when different crops are rotated. It is a significant advantage, as crop rotations are commonly used to disrupt the life cycles of soilborne pathogens. When strategically integrated into rotation schemes, mycorrhiza and crops continue to provide disease suppression benefits. Crop rotations are a vital strategy for reducing the buildup of soilborne pathogens. By including mycorrhiza and crops in these rotations, farmers can further mitigate disease issues and maintain a healthy soil environment for their crops (Ryan, 2018; Page, 2020).

16.6.4 Disease-Prone Crop Protection

Vulnerable crops like tomato and pepper are often susceptible to soilborne pathogens. Inoculating these crops with mycorrhizal fungi before planting enhances their disease resistance. Mycorrhizal fungi create a biological barrier that hinders the entry and establishment of pathogens in the root system. By reducing the risk of infections in disease-prone areas, mycorrhizal associations offer a preventive measure that can lead to higher crop yields and quality, particularly in regions where these crops are economically significant (Van Bruggen & Finckh, 2016; Panth et al., 2020; Santoyo et al., 2021).

16.6.5 Hydroponics and Controlled Environment Agriculture

In hydroponic and indoor farming systems, where natural soil microbiota are absent, mycorrhizal associations may enhance root health by improving nutrient and water uptake. It is especially critical in soilless growing environments where plants depend entirely on nutrient solutions. Mycorrhizal fungi also protect against root pathogens in hydroponic systems, by reducing the risk of root diseases, which can be particularly problematic in closed indoor environments (Berruti et al., 2016; Thomas et al., 2023).

16.6.6 Rehabilitation of Degraded Soils

Mycorrhizal fungi have the remarkable ability to remediate contaminated soils. They can help to break down and neutralise toxins, facilitating the restoration of soil health, with a valuable dual role where soil contamination is a concern. Simultaneously, introducing mycorrhizal associations enhances disease resistance in newly planted crops in rehabilitated soils. This is especially beneficial when restoring land for agricultural use (Ma et al., 2023; Wahab et al., 2023).

16.6.7 Integrated Pest Management (IPM)

Mycorrhizal associations are a biological component of IPM. They provide a natural means of controlling soilborne diseases, complementing other IPM practices. They reduce the reliance on chemical pesticides and promote a holistic and environment-friendly approach to pest and disease management. Mycorrhizal associations work well in tandem with other IPM strategies, such as the use of beneficial insects, cultural practices, and resistant crop varieties. This integrated approach minimises the impact of pests and diseases on crops, ecologically and sustainably (Mukerji & Ciancio, 2007; Alyokhin et al., 2020).

Field applications illustrate the versatility and effectiveness of mycorrhiza-mediated disease resistance in vegetable crop production. By harnessing the power of symbiotic relationships between plants and fungi, farmers can adopt more sustainable and environmentally friendly approaches for disease management, ultimately contributing to healthier crops, increased yields, and reduced environmental impacts. Mycorrhizal associations offer numerous benefits in organic farming and sustainable agriculture, including reduced pesticide use, improved soil health, disease suppression, and environmental sustainability. Their versatility and ability to persist in various agricultural contexts make them a valuable tool for farmers aiming to adopt more eco-friendly and sustainable practices (Begum et al., 2019; Chen et al., 2023; Wahab et al., 2023).

16.7 Challenges and Limitations of Mycorrhiza-Mediated Disease Resistance in Vegetable Crops

While mycorrhizal associations offer promising solutions for disease resistance in vegetable crops, several challenges and limitations must be considered in their practical application.

16.7.1 Host Specificity

Not all plant species form mycorrhizal associations, and the effectiveness of these associations can vary among species. Some plants depend highly on mycorrhizal fungi for nutrient uptake, while others may benefit less. Selecting the suitable mycorrhizal species for a specific crop is essential. It requires understanding the preferences of different fungal species and their compatibility with different hosts. Researchers may conduct experiments to identify which mycorrhizal species provides the most significant disease-resistance benefits for specific vegetable crops (Campos et al., 2018; Li et al., 2021; Kuyper & Jansa, 2023).

16.7.2 Environmental Conditions

Mycorrhizal associations are sensitive to environmental factors. Soil type, pH, temperature, and moisture levels can all influence the establishment and effectiveness of these symbiotic relationships. For example, arbuscular mycorrhizal fungi may perform better in sandy soils, while ectomycorrhizal fungi are often associated with forest ecosystems. Researchers and farmers must consider adapting mycorrhizal strategies to various environmental conditions, such as adjusting inoculation methods or selecting mycorrhizal species suited to specific soil types. Developing strategies for maintaining mycorrhizal associations under suboptimal conditions is a research challenge that can enhance the reliability of this approach (Begum et al., 2019; Huey et al., 2020; Etesami et al., 2021).

16.7.3 Pathogen Specificity

Mycorrhizal associations are not universally effective against all pathogens. The effectiveness of mycorrhiza-mediated resistance can vary depending on the pathogen biology. Mycorrhiza does not significantly suppress some diseases, and the resistance level can vary based on the pathogen biology and mode of infection. Mycorrhizal associations may be more successful in suppressing soilborne pathogens than foliar diseases. Research is required to understand the mechanisms by which mycorrhizal associations confer disease resistance and identify the most effectively managed diseases through this symbiosis. Scientists may explore how mycorrhizal fungi influence the plant immune response and how they compete with or inhibit pathogenic organisms (Huey et al., 2020; Pozo De La Hoz et al., 2021).

16.7.4 Crop Variability

Different crop varieties within a species can respond differently to mycorrhizal inoculation regarding disease resistance. This variability is often linked to genetic differences. Understanding the genetic basis of these differences and selecting crop varieties more responsive to mycorrhizal fungi can optimise disease resistance strategies. Plant breeders can work on developing crop varieties with enhanced compatibility with mycorrhizal fungi, potentially through traditional breeding or genetic engineering. Plant breeding efforts can identify and select crop varieties with genetic traits that promote stronger mycorrhizal associations and disease resistance. It may involve traditional breeding methods or genetic modification (Kokkoris et al., 2019; Orrù et al., 2021).

16.7.5 Inoculum Availability

Access to viable mycorrhizal inocula can be limited and costly in some regions, making this approach less accessible to small-scale farmers. Researchers and agricultural supply chains need to focus on expanding the production and distribution of mycorrhizal products to ensure their availability to a broader range of farmers. Developing cost-effective production methods and packaging strategies is essential to address this challenge. Collaboration between research institutions and agricultural supply chains can expand the distribution of affordable and high-quality inoculants (Mukhongo et al., 2016; Chen et al., 2018; Basiru et al., 2020; Wilkes, 2021).

16.7.6 Symbiotic Establishment

The establishment of mycorrhizal associations can take several weeks to months, depending on the plant and fungal species involved. During this establishment period, crops may remain vulnerable to diseases, as mycorrhizal fungi need time to colonise roots and establish symbiosis. During this establishment phase, crops may be vulnerable to diseases. Research is needed to develop methods for expediting the establishment of mycorrhizal symbiosis to enhance early disease resistance. Techniques may include priming plants with mycorrhizal inoculants or implementing strategies to facilitate rapid colonisation. Research efforts may explore methods to accelerate the process of mycorrhizal colonisation, potentially through the use of pre-colonised plants or mycorrhizal spore coatings (Begum et al., 2019; Huey et al., 2020; Santoyo et al., 2021; Wahab et al., 2023).

16.7.7 Crop Rotation and Soil Fumigation

In cases of severe soilborne diseases, agricultural practices such as soil fumigation or crop rotation may be necessary. However, these practices can disrupt mycorrhizal associations. After such disruptions, additional efforts are needed to reintroduce mycorrhizal fungi to the soil and re-establish the symbiosis. Research can focus on developing methods to mitigate the adverse effects of these practices on mycorrhizal associations. For instance, finding ways to reintroduce mycorrhizal fungi to the soil promptly after soil treatments can help maintain disease resistance (Chen et al., 2018; Panth et al., 2020).

16.7.8 Management Practices

Conventional farming practices, such as excessive fertiliser use and pesticide application, can negatively impact mycorrhizal associations. Achieving a balance between conventional practices and mycorrhiza-mediated disease resistance can be challenging. Research is necessary to develop sustainable farming practices promoting plant health and symbiotic relationships with fungi. Education and outreach efforts can help farmers understand the potential benefits of these practices and encourage more sustainable management approaches. Reducing excessive chemical inputs and promoting organic or agroecological farming methods may help protect mycorrhizal symbiosis (Hage-Ahmed et al., 2019; Kuila & Ghosh, 2022; Pagano et al., 2023).

16.7.9 Research Gaps

While there is a significant body of research on mycorrhiza and disease resistance, there are still gaps in our understanding. Further research is needed to optimise mycorrhizal applications, including selecting the most appropriate fungal species for specific crops, refining management strategies, and improving the overall efficacy of mycorrhizal strategies. Researchers may focus on elucidating the specific biochemical and molecular mechanisms underlying disease resistance in mycorrhizal associations. Moreover, studies can explore the ecological dynamics of these interactions in diverse agricultural ecosystems (Jacott et al., 2017; Kuyper & Jansa, 2023).

16.7.10 Cost-Benefit Analysis

Implementing mycorrhiza-mediated disease resistance may involve costs, including purchasing inoculum, application expenses, and potential adjustments to farming practices. Farmers must conduct cost-benefit analyses to determine whether disease control and crop yield benefits outweigh the costs. Research can contribute by providing data on the economic viability of mycorrhizal strategies and helping farmers make informed decisions regarding their adoption (Fasusi et al., 2021; Díaz-Urbano et al., 2023).

Despite these challenges and limitations, mycorrhiza-mediated disease resistance remains a promising and sustainable approach for managing diseases in vegetable crops. Addressing these limitations requires ongoing research, developing practical solutions, and integrating mycorrhizal strategies into broader agricultural practices to maximise their effectiveness while minimising potential drawbacks.

16.8 Future Directions

As research in the field of mycorrhiza-mediated disease resistance continues to advance, several exciting directions and opportunities are emerging that hold promise for the future of sustainable agriculture. These directions include:

16.8.1 Microbial Consortia

Understanding the dynamics of microbial consortia involving mycorrhizal fungi, rhizobacteria, and endophytes requires studying the interactions between these microorganisms and their impact on plant health and disease resistance. Researchers may investigate how specific combinations of microorganisms influence plant immune responses, nutrient uptake, and growth. The goal is to optimise consortia for different crops and environmental conditions, potentially through microbiome engineering (Backer et al., 2018; Santoyo et al., 2021; Sagar et al., 2021).

16.8.2 Genetic Approaches

Breeding crop varieties with enhanced compatibility with mycorrhizal fungi involves identifying the genetic traits associated with successful mycorrhizal associations and selecting these traits. Traditional breeding methods and cutting-edge genetic engineering techniques, such as CRISPR-Cas9, can be used to develop mycorrhiza-responsive crop genotypes. Research efforts must also address potential regulatory and ethical concerns associated with genetically modified crops (Porter & Sachs, 2020; Ramírez-Flores et al., 2020; Wang et al., 2021; Sandrini et al., 2022).

16.8.3 Precision Agriculture

In precision agriculture, technologies like GPS, remote sensing, and soil sensors can assist in accurately applying mycorrhizal inoculants to specific field areas. Researchers must investigate the optimal timing, placement, and dosage of inoculants for different crops and soil types. Real-time data analysis can further refine these applications, making them more efficient and cost-effective for farmers (Rillig et al., 2016; Brito et al., 2021; Basiru & Hijri, 2022; George & Ray, 2023).

16.8.4 Bioinformatics and -Omics

High-throughput sequencing and omics technologies enable scientists to examine plants and mycorrhizal fungi genetic, transcriptomic, and proteomic responses during their interactions. This research can be highly detailed, involving identifying specific genes, proteins, and metabolic pathways involved in the symbiotic relationship. A collaborative effort of molecular biologists, bioinformaticians, and plant scientists is needed to unravel the intricacies of this interaction (Crandall et al., 2020; Yang et al., 2021; Shaw et al., 2022; Wang et al., 2023; Wijayawardene et al., 2023).

16.8.5 Climate Change Adaptation

Investigating how mycorrhizal associations help crops adapt to climate change involves multifaceted studies. Researchers need to expose plants to varying environmental conditions in controlled environments to assess their responses and measure the impacts on drought resistance, heat tolerance, and disease susceptibility. This research aims at developing strategies for selecting or engineering crop genotypes that are more resilient in the face of changing climate conditions (Pons & Müller, 2022).

16.8.6 Scaling Up Production

Increasing production of mycorrhizal inoculants requires optimising mycorrhizal fungi cultivation and formulation. It involves research into bioreactor systems, growing media, and quality control methods to ensure consistent and effective inoculant production. Packaging and distribution strategies may include developing cost-effective, user-friendly formats that cater to different agricultural contexts (Berruti et al., 2016; Rosa et al., 2020; Oviatt & Rillig, 2021; Wahab et al., 2023).

16.8.7 Ecosystem Services

Researching the ecosystem services provided by mycorrhizal associations encompasses ecological and soil science. It involves quantifying the carbon sequestration potential of mycorrhizal networks, studying their role in nutrient cycling, and assessing how they improve soil structure and fertility. This research provides insights into the broader environmental benefits of mycorrhizal strategies (M. Tahat et al., 2020; Fall et al., 2022; Wahab et al., 2023).

16.8.8 Innovative Farming Systems

Integrating mycorrhiza-mediated disease resistance into innovative farming systems is a multidisciplinary effort. Researchers must design and test agroforestry, intercropping, and vertical farming systems that effectively leverage mycorrhizal associations. It includes selecting crop combinations, optimising planting densities, and considering spatial and temporal dynamics of mycorrhizal colonisation (De Carvalho et al., 2010; Oruru & Njeru, 2016; Maitra et al., 2021).

16.8.9 Education and Outreach

Educational efforts should be designed to reach farmers, extension agents, and policymakers. These programs may include workshops, informational materials, and online resources to raise awareness and understanding of mycorrhiza-mediated disease resistance. Collaboration with agricultural extension services and partnerships with agricultural organisations can facilitate outreach to the farming community (Dhiman et al., 2022).

16.8.10 Global Collaboration

International collaboration in mycorrhizal research requires knowledge sharing, data exchange, and collaborative projects. Researchers from different regions can benefit from each other's experiences and adapt strategies to suit local agricultural challenges. This collaborative effort can foster innovation and improve the global understanding of mycorrhizal practices (Johnson & Gehring, 2007; Ferlian et al., 2018).

16.8.11 Regulatory Considerations

Developing regulatory frameworks for mycorrhizal products involves collaboration between researchers, policymakers, and industry stakeholders. This process may include defining inoculant quality standards, establishing safety guidelines, and implementing labelling requirements to ensure safe and effective use of mycorrhizal products in agriculture (Salomon et al., 2022).

16.8.12 Long-Term Monitoring

Evaluating the long-term impacts of mycorrhiza-mediated disease resistance on soil health, crop productivity, and disease dynamics requires extended research efforts. This longitudinal research can involve continuous monitoring of fields, assessing changes in soil properties, and tracking crop performance over multiple growing seasons. It is essential for assessing sustainability and long-term benefits of mycorrhiza-based approaches (Fall et al., 2022; George & Ray, 2023). The future of mycorrhiza-mediated disease resistance in vegetable crops is promising, with numerous opportunities for research, innovation, and practical applications. By harnessing the potential of mycorrhizal associations, agriculture can move closer to more sustainable and resilient crop production systems that benefit both farmers and the environment.

16.9 Implications for Sustainable Agriculture

The integration of mycorrhiza-mediated disease resistance into sustainable agriculture practices carries significant implications for the future of food production, environmental stewardship, and global food security. Here are the critical implications:

16.9.1 Reduced Chemical Dependency

Integrating mycorrhiza-mediated disease resistance into agriculture reduces chemical dependency, reducing reliance on synthetic pesticides and fertilisers. This reduces environmental impacts and lowers chemical residues in food products. Mycorrhizal associations preserve beneficial organisms, contribute to soil health, and are cost-effective for farmers. It aligns with organic farming principles and contributes to long-term sustainability, making farming more sustainable and environmentally friendly (Tudi et al., 2021; Fasusi et al., 2021; Kumar et al., 2022).

16.9.2 Enhanced Soil Health

Mycorrhizal associations improve soil structure, nutrient cycling, and organic matter accumulation. They promote stable aggregates, reduce soil compaction, and facilitate root growth and penetration. Mycorrhizal networks also help plants access and transfer nutrients, and suppress soilborne pathogens. They stabilise soil, reduce erosion, and influence soil pH levels. Improved soil structure and organic matter content lead to better water retention, ensuring a consistent water supply. Mycorrhizal-mediated disease resistance contributes to long-term soil health and sustainability (Khangura et al., 2023; Wahab et al., 2023).

16.9.3 Resource Efficiency

Mycorrhizal associations improve plant nutrient uptake, reducing the need for synthetic fertilisers and minimising environmental impact. It leads to increased crop yields and reduced costs for farmers. Mycorrhiza-mediated disease resistance also improves water use efficiency, reducing water stress and irrigation needs. Resource-efficient agriculture benefits farmers economically and reduces the environmental footprint, reducing chemical residues and nutrient runoff. It also contributes to long-term sustainability, promoting sustainable farming practices and reduced carbon footprint (Brito et al., 2021; Kuila & Ghosh, 2022; George & Ray, 2023).

16.9.4 Climate Resilience

By extending the root system reach, mycorrhizal associations enhance plant drought resilience, improving water uptake, and reducing water wastage. They also help plants to cope with heat stress by improving nutrient uptake and stimulating heat shock proteins. Mycorrhizal associations also promote disease resistance, reducing susceptibility to diseases and promoting water retention in soil. They also help to maintain consistent crop productivity by minimising disease impact and reducing climate-related crop losses. Overall, mycorrhizal associations contribute to climate change mitigation (Usman et al., 2021; Loo et al., 2022; Eckardt et al., 2023).

16.9.5 Biodiversity Conservation

Mycorrhizal associations involving soil microorganisms foster microbial diversity, supporting soil ecosystems and biodiversity conservation. They protect beneficial soil organisms like earthworms and predatory mites, maintaining habitat and well-being. Promoting mycorrhizal associations in agriculture indirectly supports native plant growth, supports pollinators, and promotes disease resistance, preserving rare or endangered plant species. It raises awareness about symbiotic relationships and biodiversity conservation, offering educational opportunities for farmers and the broader community (Asmelash et al., 2016).

16.9.6 Reduced Soil Erosion

By forming stable aggregates, mycorrhizal associations stabilise soil, preventing soil erosion and promoting long-term soil fertility and agricultural productivity. They prevent topsoil erosion, sedimentation, and soil disturbance, preserving beneficial soil microorganisms and promoting soil health. Mycorrhizal associations persist in soil, preventing erosion and sediment runoff and promoting sustainable agriculture practices. They also prevent soil degradation, reducing arable land loss and agricultural productivity, promote responsible land management practices and sustainable land use (Bot & Benites, 2005; Simard & Austi, 2010; Fall et al., 2022).

16.9.7 Improved Food Security

Mycorrhiza-mediated disease resistance is a sustainable and environmentally friendly practice that helps maintain stable crop yields by reducing disease impact. It is crucial for food security, reducing crop losses and increasing disease resilience.

Mycorrhiza-mediated disease resistance can be applied to various vegetable crops, promoting crop diversity and reducing reliance on single crops. It also mitigates climate change impacts, improves food availability and access, and enhances economic stability for farmers. This practice can be applied globally to address food security challenges (Thirkell et al., 2017; Loo et al., 2022).

16.9.8 Sustainable Crop Rotation

Sustainable crop rotation is crucial for maintaining soil health and reducing disease impact. Mycorrhiza-mediated disease resistance can help to achieve this by reducing disease pressure in successive crops. It promotes soil microbial diversity, nutrient cycling, and soil erosion prevention. Mycorrhizal associations also help to prevent soil erosion, especially in vulnerable soils. Both sustainable crop rotation and mycorrhiza-mediated disease resistance align with organic farming principles, promoting natural processes and reducing synthetic inputs. This long-term sustainability ensures fertile soils, reduced environmental impact, and continued farmland productivity (Cappelli et al., 2022; Wahab et al., 2023).

16.9.9 Organic Farming Integration

Organic farming emphasises the reduction of synthetic pesticides, herbicides, and fertilisers. Mycorrhiza-mediated disease resistance, a natural method of disease management, reduces the need for chemical pesticides and enhances plant immunity. It also contributes to soil health, environmental stewardship, and nutrient cycling. This approach can be easily integrated into organic farming practices, improving crop yields and reducing input costs. It also aligns with consumer preferences for low-chemical farming methods and supports organic farming principles (Siddiqui et al., 2008; Gamage, 2023) (Fig. 16.2).

16.9.10 Economic Benefits

Mycorrhiza-mediated disease resistance leads to higher crop yields, reduced input costs, improved crop quality, and increased farmer profitability. It also serves as a risk mitigation strategy, reducing crop losses and optimising resource use. This sustainable agricultural practice also reduces environmental liabilities, making it more economically viable for small-scale farmers and attracting premium consumers. The positive effects of mycorrhizal associations persist over time (Maitra et al., 2021; Yang et al., 2021).

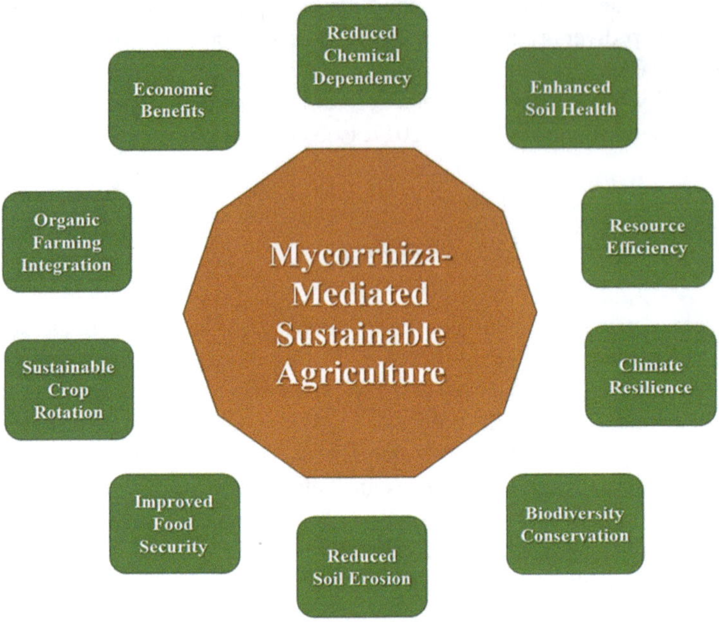

Fig. 16.2 Role of mycorrhiza in supporting sustainable agriculture by interlinking ecological benefits and agricultural productivity for long-term environmental resilience

16.9.11 Research and Education

Mycorrhiza-mediated disease resistance in agriculture offers new research opportunities, enhancing understanding of plant-fungi relationships and soil microbiome interactions. This approach encourages sustainable practices and soil biology, leading to the discovery of beneficial microorganisms. Education about its benefits encourages wider adoption, environmental literacy, and consumer awareness. Research and education can also lead to policy support, international collaboration, and community formation, ultimately improving crop health, yields, and related economic benefits (Sosa-Hernández et al., 2019; Kuila & Ghosh, 2022).

16.9.12 Policy Support

Governments and agricultural agencies can promote the adoption of mycorrhiza-based strategies by incorporating them into agricultural policies and programs that encourage sustainable and environmental friendly practices (M. Tahat et al., 2020; Gamage, 2023).

In summary, the implications of mycorrhiza-mediated disease resistance for sustainable agriculture are profound. This approach aligns with reducing environmental impact, improving soil health, and ensuring long-term food security. By harnessing the power of mycorrhizal associations, agriculture can move closer to a more resilient, efficient, and environmentally responsible future.

16.10 Policy and Industry Perspectives

Adopting and integrating mycorrhiza-mediated disease resistance into agriculture have policy and industry implications. An overview follows, showing how policies and industry perspectives are shaping the use of mycorrhizal associations in vegetable crop production:

16.10.1 Policy Perspectives

16.10.1.1 Regulatory Frameworks

Governments and regulatory bodies are crucial in setting guidelines and standards for mycorrhizal products used in agriculture. They assess these products safety, potential risks to human health, non-target species, and the environment. Mycorrhizal products must meet stringent safety standards before approval for agricultural use. Regulatory agencies require rigorous testing to determine their effectiveness in crop health, yield, and disease resistance. They monitor and enforce compliance with established standards through inspections, product testing, and audits. They may also work towards harmonisation with international standards to facilitate trade and avoid undue restrictions (Chen et al., 2018; Huey et al., 2020; Salomon et al., 2022).

16.10.1.2 Support for Sustainable Agriculture

Policymakers are recognising the potential of mycorrhiza-mediated disease resistance in promoting sustainable agricultural practices. Financial incentives, such as grants, subsidies, and tax benefits, can help farmers integrate mycorrhizal associations into crop production. These ations should aim at improving soil health, reducing soilborne diseases, and promoting organic matter accumulation. Supporting sustainable agriculture, especially in the context of mycorrhiza-mediated disease resistance, addresses environmental and societal concerns, encouraging innovative techniques for agricultural productivity and ecological resilience (Sosa-Hernández et al., 2019; Oviatt & Rillig, 2021; George & Ray, 2023).

16.10.1.3 Environmental Conservation

Use of mycorrhizal associations reduces the need for synthetic fertilisers and pesticides, facilitates reduced tillage, supports plant diversity, and plays a vital role in erosion control. Mycorrhizal fungi create a symbiotic environment for soil microorganisms, fostering soil biodiversity and ecosystem health. Environmental conservation is closely linked to agroecological farming practices, with mycorrhizal technologies playing a key role in regenerative agriculture. Integrating mycorrhizal technologies into agriculture supports environmental objectives and sustainable food production (Bach et al., 2020; Cappelli et al., 2022).

16.10.1.4 Pesticide Reduction

Mycorrhizal associations are essential in pest management by creating healthier plants that naturally resist pests and diseases, reducing the need for chemical insecticides and fungicides. This approach promotes sustainable pest management and agricultural ecosystem health by allowing beneficial insects to thrive. Mycorrhizal approaches also contribute to sustainable agricultural practices by reducing environmental impact on water quality, soil health, and non-target organisms. Incorporating mycorrhizal fungi into agriculture supports crop health and minimises chemical interventions. Pesticide reduction can increase farmers profitability and drive demand for crops grown with mycorrhizal technologies (Pretty & Bharucha, 2015; Alyokhin et al., 2020; Helepciuc & Todor, 2022).

16.10.1.5 Research and Extension

Scientific research is crucial for understanding mycorrhizal associations and their benefits in agriculture. It investigates mycorrhizal fungi biology, interactions with plants, and effects on soil health and crop performance. The goal is to uncover mechanisms underlying disease resistance and develop improved inoculants. Extension services provide educational materials, technical assistance, and best management practices for mycorrhiza-mediated disease resistance. They also monitor the impact of mycorrhizal technologies on crop yields, soil health, and environmental sustainability (Chave et al., 2019; Tahat et al., 2020; Di Bene et al., 2022).

16.10.1.6 Certification and Labelling

Certification programs set rigorous standards for using mycorrhizal technologies in agriculture, including inoculant type, quality, application methods, and sustainable farming practices. Farmers must undergo verification processes to confirm adherence. Certified products display a label or logo, signalling their use of mycorrhiza-mediated disease resistance and sustainable practices. Certification programs advocate for adopting these technologies at the industry level, working with government agencies and industry associations. Educational initiatives raise awareness and provide incentives, creating a market for environmentally friendly practices (Rillig et al., 2016; Salomon et al., 2022).

16.10.2 Industry Perspectives

16.10.2.1 Product Development

Mycorrhizal inoculants are developed to enhance nutrient uptake, disease resistance, and plant growth in crops and soils. They come in various forms and undergo compatibility tests to avoid negative interactions with other microorganisms or agricultural inputs. Considering soil conditions and pH levels, the inoculants are designed for specific crops. Quality control processes ensure consistency and reliability. Collaboration between developers, research institutions, and agricultural organisations accelerates research and dissemination. Mycorrhizal products must meet regulatory requirements and safety standards (Batista & Singh, 2021; O'Callaghan et al., 2022).

16.10.2.2 Market Expansion

Market expansion strategies are essential for promoting the use of mycorrhiza-mediated disease resistance in agriculture. These strategies emphasise the benefits of mycorrhizal products, their impact on crop yields, disease resistance, health, sustainability, and economic benefits like cost savings and reduced synthetic fertiliser and pesticide use. Real-world examples and testimonials from successful farmers are used as persuasive evidence. Collaborations with distributors, retailers, research institutions, consumers, and government agencies can further expand the market (Jacquet et al., 2022).

16.10.2.3 Collaboration

Mycorrhiza-mediated disease resistance in agriculture requires collaboration between industry players, research institutions, and agricultural organisations. Research partnerships contribute scientific knowledge about mycorrhizal associations, their effects on crops, and application methods. Industry stakeholders work with research institutions to develop inoculants, while agricultural extension services disseminate information and offer training programs. Collaborative trial programs provide data on mycorrhizal technology performance. Collaboration between industrial associations and government agencies promotes policy advocacy and regulatory environment. International collaborations promote global trends and adoption of best practices (Vosátka et al., 2012).

16.10.2.4 Education and Training

Education and training require awareness campaigns, workshops, and dedicated sessions to inform farmers and consumers about the benefits of these technologies. Educational materials, on-farm demonstrations, and expert consultations provide knowledge. Public awareness campaigns influence consumers' preferences and encourage youth involvement. The education process is ongoing, with continuous learning opportunities and research updates. Support systems like helplines and forums help farmers improve crop productivity, sustainability, and resilience (Berruti et al., 2016; Chave et al., 2019; Basiru & Hijri, 2022).

16.10.2.5 Supply Chain Management

Effective supply chain management, including planning production, quality control, and storage facilities are required for mycorrhiza exploitations. Efficient distribution networks, collaboration with retailers, and sustainable practices like eco-friendly packaging and transportation routes are crucial. Accurate demand forecasting and contingency plans are essential for maintaining supply. Global expansion necessitates adapting supply chain management practices to serve international markets. The agricultural community can access sustainable and resilient crop production by optimising production, storage, distribution, and partnerships (Mukhongo et al., 2016; Oviatt & Rillig, 2021; Dhiman et al., 2022).

16.10.2.6 Global Expansion

The global expansion of mycorrhiza-mediated disease resistance in agriculture requires analysing international markets, adapting technologies to local conditions, and complying with regulations. It involves managing import and export logistics, understanding cultural and social aspects, and partnering with local distributors, agricultural supply stores, and cooperatives. Building relationships with local government agencies can help navigate regulatory requirements and support global expansion efforts (Vosátka et al., 2012; Dhiman et al., 2022).

16.10.2.7 Research and Innovation

Research and innovation are crucial for understanding mycorrhizal associations, selecting and improving strains, and exploring optimal application techniques. Research examines the benefits of different crops, while economic studies evaluate the cost-effectiveness of adopting mycorrhizal technologies. Advanced bioinformatics and molecular biology techniques enable targeted strain development and improved product design. Collaboration of research institutions, industry stakeholders, and agricultural organisations is encouraged to share knowledge and accelerate research and development (Kuila & Ghosh, 2022; Dhiman et al., 2022; Wahab et al., 2023).

In summary, the adoption of mycorrhiza-mediated disease resistance in vegetable crop production is influenced by both policy and industry perspectives. Policies that support sustainable agriculture and environmental conservation can encourage the adoption of mycorrhizal technologies, while the agricultural industry plays a crucial role in product development, marketing, and education. Collaboration between these stakeholders can drive the integration of mycorrhizal associations into mainstream agricultural practices, promoting sustainable and resilient food production systems.

16.11 Conclusions

Mycorrhiza-mediated disease resistance in vegetable crops represents a dynamic and promising agricultural frontier with great potential for a sustainable and resilient food production system. Mycorrhizal associations offer modern solutions to age-old challenges in agriculture. These symbiotic relationships are pivotal in enhancing disease resistance while improving nutrient uptake and environmental sustainability. Researchers uncovered the intricate mechanisms by which mycorrhiza-colonised plants bolster their defence mechanisms. This understanding has paved the way for practical applications in disease management. Real-world case studies demonstrate effectiveness of mycorrhiza-mediated disease resistance in various vegetable crops. These successes underscore the versatility and applicability of this approach. While challenges and limitations exist, including host specificity and environmental factors, research is actively addressing these issues.

Genetic approaches, microbial consortia, and precision agriculture are among the innovative strategies on the horizon. The implications for sustainable agriculture are profound. Mycorrhiza-mediated disease resistance reduces chemical dependency, enhances soil health, promotes resource efficiency, and contributes to climate resilience. It aligns with the broader goals of environmental conservation and food security. Policymakers recognise the benefits of mycorrhiza-mediated disease resistance and support its adoption through regulatory frameworks and incentives. The agricultural industry is investing in research, product development, and market expansion. Collaboration among stakeholders, including researchers, policymakers, industry players, and farmers, is essential to realize the full potential of mycorrhizal associations in vegetable crop production. Global knowledge sharing can accelerate progress and innovation.

In conclusion, mycorrhiza-mediated disease resistance represents a robust and sustainable tool in the toolkit of modern agriculture. Its capacity to enhance crop health, reduce environmental impact, and promote long-term food security positions it as a cornerstone of future agricultural practices. As we move forward, the continued exploration of mycorrhizal associations and their integration into sustainable agriculture is a testament to our ability to harness the wisdom of nature, to address the challenges of tomorrow.

References

Abdelkhalek, A., Király, L., Al-Mansori, A.-N. A., Younes, H. A., Zeid, A., Elsharkawy, M. M., & Behiry, S. I. (2022). Defense responses and metabolic changes involving phenylpropanoid pathway and PR genes in squash (*Cucurbita pepo* L.) following Cucumber mosaic virus infection. *Plants, 11*, 1908. https://doi.org/10.3390/plants11151908

Alyokhin, A., Nault, B., & Brown, B. (2020). Soil conservation practices for insect pest management in highly disturbed agroecosystems—A review. *Entomologia Experimentalis et Applicata, 168*, 7–27.

Asmelash, F., Bekele, T., & Birhane, E. (2016). The potential role of arbuscular mycorrhizal fungi in the restoration of degraded lands. *Frontiers in Microbiology, 7*, 1095.

Auer, S., & Ludwig-Müller, J. (2023). Biocontrol of clubroot disease: How successful are endophytic fungi and bacteria? *European Journal of Plant Pathology, 433-451*, 433–451.

Averill, C., Bhatnagar, J. M., Dietze, M. C., Pearse, W. D., & Kivlin, S. N. (2019). Global imprint of mycorrhizal fungi on whole-plant nutrient economics. *Proceedings of the National Academy of Sciences of the United States of America, 116*, 23163–23168. https://doi.org/10.1073/pnas.1906655116

Bach, E. M., Ramirez, K. S., Fraser, T. D., & Wall, D. H. (2020). Soil biodiversity integrates solutions for a sustainable future. *Sustainability, 12*, 2662. https://doi.org/10.3390/su12072662

Backer, R., Rokem, J. S., Ilangumaran, G., Lamont, J., Praslickova, D., Ricci, E., Subramanian, S., & Smith, D. L. (2018). Plant growth-promoting Rhizobacteria: Context, mechanisms of action, and roadmap to commercialization of biostimulants for sustainable agriculture. *Frontiers in Plant Science, 9*, 1473. https://doi.org/10.3389/fpls.2018.01473

Basiru, S., & Hijri, M. (2022). The potential applications of commercial arbuscular mycorrhizal fungal inoculants and their ecological consequences. *Microorganisms, 10*, 1897. https://doi.org/10.3390/microorganisms10101897

Basiru, S., Mwanza, H. P., & Hijri, M. (2020). Analysis of Arbuscular Mycorrhizal fungal inoculant benchmarks. *Microorganisms, 9*, 81. https://doi.org/10.3390/microorganisms9010081

Batista, B. D., & Singh, B. K. (2021). Realities and hopes in the application of microbial tools in agriculture. *Microbial Biotechnology, 14*, 1258–1268. https://doi.org/10.1111/1751-7915.13866

Begum, N., Qin, C., Ahanger, M. A., Raza, S., Khan, M. I., Ashraf, M., Ahmed, N., & Zhang, L. (2019). Role of Arbuscular Mycorrhizal fungi in plant growth regulation: Implications in abiotic stress tolerance. *Frontiers in Plant Science, 10*, 1068. https://doi.org/10.3389/fpls.2019.01068

Berruti, A., Lumini, E., Balestrini, R., & Bianciotto, V. (2016). Arbuscular Mycorrhizal fungi as natural biofertilizers: let's benefit from past successes. *Frontiers in Microbiology, 6*, 1559. https://doi.org/10.3389/fmicb.2015.01559

Beslemes, D., Tigka, E., Roussis, I., Kakabouki, I., Mavroeidis, A., & Vlachostergios, D. (2023). Effect of Arbuscular Mycorrhizal fungi on nitrogen and phosphorus uptake efficiency and crop productivity of two-rowed barley under different crop production systems. *Plants, 12*, 1908. https://doi.org/10.3390/plants12091908

Bhavanam, S., & Stout, M. J. (2021). Assessment of Silicon- and Mycorrhizae- mediated constitutive and induced systemic resistance in rice, *Oryza sativa* L., against the Fall Armyworm, *Spodoptera frugiperda* Smith. *Plants, 10*, 2126. https://doi.org/10.3390/plants10102126

Bot, A., & Benites, J. (2005). *The importance of soil organic matter: Key to drought-resistant soil and sustained food and production*. Food and Agriculture Organization of the United Nations.

Brito, I., Carvalho, M., & Goss, M. J. (2021). Managing the functional diversity of arbuscular mycorrhizal fungi for the sustainable intensification of crop production. *Plants People Planet, 3*, 491–505. https://doi.org/10.1002/ppp3.10212

Cameron, D. D., Neal, A. L., Van Wees, S. C. M., & Ton, J. (2013). Mycorrhiza-induced resistance: More than the sum of its parts? *Trends in Plant Science, 18*, 539–545.

Campos, C., Carvalho, M., Brígido, C., Goss, M. J., & Nobre, T. (2018). Symbiosis specificity of the preceding host plant can dominate but not obliterate the association between wheat and its arbuscular mycorrhizal fungal partners. *Frontiers in Microbiology, 9*, 2920.

Cappelli, S. L., Domeignoz-Horta, L. A., Loaiza, V., & Laine, A.-L. (2022). Plant biodiversity promotes sustainable agriculture directly and via belowground effects. *Trends in Plant Science, 27*, 674–687. https://doi.org/10.1016/j.tplants.2022.02.003

Cardarelli, M., Rouphael, Y., Kyriacou, M. C., Colla, G., & Pane, C. (2020). Augmenting the sustainability of vegetable cropping systems by configuring rootstock-dependent rhizomicrobiomes that support plant protection. *Agronomy, 10*, 1185. https://doi.org/10.3390/agronomy10081185

Chave, M., Angeon, V., Paut, R., Collombet, R., & Tchamitchian, M. (2019). Codesigning biodiversity-based agrosystems promotes alternatives to mycorrhizal inoculants. *Agronomy for Sustainable Development, 39*, 48. https://doi.org/10.1007/s13593-019-0594-y

Chen, M., Arato, M., Borghi, L., Nouri, E., & Reinhardt, D. (2018). Beneficial services of Arbuscular Mycorrhizal fungi—From ecology to application. *Frontiers in Plant Science, 9*, 1270. https://doi.org/10.3389/fpls.2018.01270

Chen, W., Modi, D., & Picot, A. (2023). Soil and phytomicrobiome for plant disease suppression and management under climate change: A review. *Plants, 12*, 2736.

Corato, U. D. (2020). Soil microbiota manipulation and its role in suppressing soil-borne plant pathogens in organic farming systems under the light of microbiome-assisted strategies. *Chemical and Biological Technologies in Agriculture, 7*, 1–26.

Crandall, S. G., Gold, K. M., Jiménez-Gasco, M. D. M., Filgueiras, C. C., & Willett, D. S. (2020). A multi-omics approach to solving problems in plant disease ecology. *PLoS One, 15*, e0237975. https://doi.org/10.1371/journal.pone.0237975

De Carvalho, A. M. X., De Castro, T. R., Cardoso, I. M., & Kuyper, T. W. (2010). Mycorrhizal associations in agroforestry systems. In P. Dion (Ed.), *Soil biology and agriculture in the tropics* (pp. 185–208). Springer.

Dell, B. (2002). Role of Mycorrhizal fungi in ecosystems. *CMU Journal, 1*, 47–60.

Dhiman, M., Sharma, L., Kaushik, P., Singh, A., & Sharma, M. M. (2022). Mycorrhiza: An eco-friendly bio-tool for better survival of plants in nature. *Sustainability, 14*, 10220.

Di Bene, C., Dolores Gómez-López, M., Francaviglia, R., Farina, R., Blasi, E., Martínez-Granados, D., & Calatrava, J. (2022). Barriers and opportunities for sustainable farming practices and crop diversification strategies in mediterranean cereal-based systems. *Frontiers in Environmental Science, 10*, 861225.

Díaz-Urbano, M., Goicoechea, N., Velasco, P., & Poveda, J. (2023). Development of agricultural bio-inoculants based on mycorrhizal fungi and endophytic filamentous fungi: Co-inoculants for improve plant-physiological responses in sustainable agriculture. *Biological Control, 182*, 105223.

Eckardt, N. A., Ainsworth, E. A., Bahuguna, R. N., Broadley, M. R., Busch, W., et al. (2023). Climate change challenges, plant science solutions. *Plant Cell, 35*, 24–66.

Elsharkawy, M. M., Kuno, S., Hyakumachi, M., Mostafa, Y. S., Alamri, S. A., & Alrumman, S. A. (2022). PCR-DGGE analysis proves the suppression of *Rhizoctonia* and *Sclerotium* root rot due to successive inoculations. *Journal of Fungi, 8*, 133.

Enebe, M. C., & Erasmus, M. (2023). Susceptibility and plant immune control–A case of mycorrhizal strategy for plant colonization, symbiosis, and plant immune suppression. *Frontiers in Microbiology, 14*, 1178258.

Erlacher, A., Cardinale, M., Grosch, R., Grube, M., & Berg, G. (2014). The impact of the pathogen *Rhizoctonia solani* and its beneficial counterpart *Bacillus amyloliquefaciens* on the indigenous lettuce microbiome. *Frontiers in Microbiology, 5*, 175.

Etesami, H., Jeong, B. R., & Glick, B. R. (2021). Contribution of Arbuscular Mycorrhizal fungi, phosphate–solubilizing bacteria, and silicon to P uptake by plant. *Frontiers in Plant Science, 12*, 699618.

Fall, A. F., Nakabonge, G., Ssekandi, J., Founoune-Mboup, H., Apori, S. O., et al. (2022). Roles of Arbuscular Mycorrhizal fungi on soil fertility: Contribution in the improvement of physical, chemical, and biological properties of the soil. *Frontiers in Fungal Biology, 3*, 723892.

Fasusi, O. A., Cruz, C., & Babalola, O. O. (2021). Agricultural sustainability: Microbial biofertilizers in rhizosphere management. *Agriculture, 11*, 163.

Ferlian, O., Biere, A., Bonfante, P., Buscot, F., Eisenhauer, N., et al. (2018). Growing research networks on mycorrhizae for mutual benefits. *Trends in Plant Science, 23*, 975–984.

Fernández, J. A., Ayastuy, M. E., Belladonna, D. P., Comezaña, M. M., & Contreras, J. (2022). Current trends in organic vegetable crop production: Practices and techniques. *Horticulturae, 8*(10), 893. https://doi.org/10.3390/horticulturae8100893

Friberg, S. (1999). Distribution and diversity of arbuscular mycorrhizal fungi in traditional agriculture on The Niger inland delta, Mali, West Africa. *CBM:s Skriftserie, **3***, 53–80.

Gamage, A. (2023). Role of organic farming for achieving sustainability in agriculture. *Farming System, 1*, 100005. https://doi.org/10.1016/j.farsys.2023.100005

Garg, N., & Chandel, S. (2010). Arbuscular mycorrhizal networks: Process and functions. A review. *Agronomy for Sustainable Development, 30*, 581–599.

Gatch, E. W., & Du Toit, L. J. (2017). Limestone-mediated suppression of Fusarium wilt in spinach seed crops. *Plant Disease, 101*, 81–94.

George, N. P., & Ray, J. G. (2023). The inevitability of arbuscular mycorrhiza for sustainability in organic agriculture—A critical review. *Frontiers in Sustainable Food Systems, 7*, 1124688.

Hage-Ahmed, K., Rosner, K., & Steinkellner, S. (2019). Arbuscular mycorrhizal fungi and their response to pesticides. *Pest Management Science, 75*, 583–590.

Hanaka, A., Ozimek, E., Reszczyńska, E., Jaroszuk-Ściseł, J., & Stolarz, M. (2021). Plant tolerance to drought stress in the presence of supporting bacteria and fungi: An efficient strategy in horticulture. *Horticulturae, 7*, 390. https://doi.org/10.3390/horticulturae7100390

Hao, Z., Xie, W., & Chen, B. (2019). Arbuscular mycorrhizal symbiosis affects plant immunity to viral infection and accumulation. *Viruses, 11*, 534.

Harman, G., Khadka, R., Doni, F., & Uphoff, N. (2021). Benefits to plant health and productivity from enhancing plant microbial symbionts. *Frontiers in Plant Science, 11*, 610065.

Hawkins, H.-J., Cargill, R. I. M., Van Nuland, M. E., Hagen, S. C., Field, K. J., Sheldrake, M., et al. (2023). Mycorrhizal mycelium as a global carbon pool. *Current Biology, 33*, R560–R573.

Helepciuc, F.-E., & Todor, A. (2022). Greener European agriculture? Evaluating EU member states' transition efforts to integrated pest management through their national action plans. *Agronomy, 12*, 2438.

Henke, C., Jung, E.-M., & Kothe, E. (2015). Hartig' net formation of *Tricholoma vaccinum*-spruce ectomycorrhiza in hydroponic cultures. *Environmental Science and Pollution Research, 22*, 19394–19399.

Hohmann, P., & Messmer, M. M. (2017). Breeding for mycorrhizal symbiosis: Focus on disease resistance. *Euphytica, 213*, 113. https://doi.org/10.1007/s10681-017-1900-x

Hong, S. Y., Jee, H. J., Choi, Y. J., & Shin, H. D. (2008). First confirmed report of downy mildew caused by Hyaloperonospora parasitica on broccoli in Korea. *Plant Pathology, 57*, 777–777.

Huey, C. J., Gopinath, S. C. B., Uda, M. N. A., Zulhaimi, H. I., Jaafar, M. N., Kasim, F. H., & Yaakub, A. R. W. (2020). Mycorrhiza: a natural resource assists plant growth under varied soil conditions. *3 Biotech, 10*, 204.

Jacott, C., Murray, J., & Ridout, C. (2017). Trade-offs in arbuscular mycorrhizal symbiosis: Disease resistance, growth responses and perspectives for crop breeding. *Agronomy, 7*, 75.

Jacquet, F., Jeuffroy, M.-H., Jouan, J., Le Cadre, E., Litrico, I., Malausa, T., Reboud, X., & Huyghe, C. (2022). Pesticide-free agriculture as a new paradigm for research. *Agronomy for Sustainable Development, 42*, 8.

Johnson, N. C., & Gehring, C. A. (2007). Mycorrhizas: Symbiotic mediators of rhizosphere and ecosystem processes. In Z. G. Cardon & J. L. Whitbeck (Eds.), *The rhizosphere* (pp. 73–100). Elsevier.

Khangura, R., Ferris, D., Wagg, C., & Bowyer, J. (2023). Regenerative agriculture–A literature review on the practices and mechanisms used to improve soil health. *Sustainability, 15*, 2338.

Kim, E., Seo, Y., Kim, Y. S., Park, Y., & Kim, Y. H. (2017). Effects of soil textures on infectivity of root-knot nematodes on carrot. *The Plant Pathology Journal, 33*, 66–74.

Kokkoris, V., Hamel, C., & Hart, M. M. (2019). Mycorrhizal response in crop versus wild plants. *PLoS One, 14*, e0221037.

Kuila, D., & Ghosh, S. (2022). Aspects, problems and utilization of Arbuscular Mycorrhizal (AM) application as bio-fertilizer in sustainable agriculture. *Current Research in Microbial Sciences, 3*, 100107.

Kumar, S., Diksha, S. S. S., & Kumar, R. (2022). Biofertilizers: An ecofriendly technology for nutrient recycling and environmental sustainability. *Current Research in Microbial Sciences, 3*, 100094.

Kuyper, T. W., & Jansa, J. (2023). Arbuscular mycorrhiza: Advances and retreats in our understanding of the ecological functioning of the mother of all root symbioses. *Plant & Soil, 489*, 41–88.

Li, T., Yang, W., Wu, S., Selosse, M.-A., & Gao, J. (2021). Progress and prospects of mycorrhizal fungal diversity in orchids. *Frontiers in Plant Science, 12*, 646325.

Linderman, R. G. (2000). Effects of mycorrhizas on plant tolerance to diseases: Mycorrhiza-disease interactions. In Y. Kapulnik & D. D. Douds (Eds.), *Arbuscular mycorrhizas: Physiology and function* (pp. 345–365). Springer.

Liu, Z., Wang, H., Wang, J., Lv, J., Xie, B., Luo, S., et al. (2022). Physical, chemical, and biological control of black rot of brassicaceae vegetables: A review. *Frontiers in Microbiology, 13*, 1023826.

Loo, W. T., Chua, K.-O., Mazumdar, P., Cheng, A., Osman, N., & Harikrishna, J. A. (2022). Arbuscular mycorrhizal symbiosis: A strategy for mitigating the impacts of climate change on tropical legume crops. *Plants, 11*, 2875.

Ma, M., Taylor, P. W. J., Chen, D., Vaghefi, N., & He, J.-Z. (2023). Major soilborne pathogens of field processing tomatoes and management strategies. *Microorganisms, 11*, 263.

Maitra, S., Hossain, A., Brestic, M., Skalicky, M., Ondrisik, P., et al. (2021). Intercropping—A low input agricultural strategy for food and environmental security. *Agronomy, 11*, 343.

Massicotte, H. B., Peterson, R. L., & Melville, L. H. (1989). Hartig net structure of ectomycorrhizae synthesized between *Laccaria bicolor* (Tricholomataceae) and two hosts: *Betula alleghaniensis* (Betulaceae) and *Pinus resinosa* (Pinaceae). *American Journal of Botany, 76*, 1654–1667. https://doi.org/10.1002/j.1537-2197.1989.tb15150.x

Meddad-Hamza, A., Benzina, F., Meddad, C., Hamza, N., Reghmit, A., Ziane, H., & Ksentini, H. (2023). Biological control of arbuscular mycorrhizal fungi and *Trichoderma harzianum* against *Fusarium oxysporum* and *Verticillium dahliae* induced wilt in tomato plants. *Egyptian Journal Of Biological Pest Control, 33*, 91. https://doi.org/10.1186/s41938-023-00737-5

Meemken, E.-M., & Qaim, M. (2018). Organic agriculture, food security, and the environment. *Annual Review of Resource Economics, 10*, 39–63.

Moulin, F., Lemanceau, P., & Alabouvette, C. (1994). Pathogenicity of Pythium species on cucumber in peat-sand, rockwool and hydroponics. *European Journal of Plant Pathology, 100*, 3–17.

Mukerji, K. G., & Ciancio, A. (2007). Mycorrhizae in the integrated pest and disease management. In A. Ciancio & K. G. Mukerji (Eds.), *General concepts in integrated Pest and disease management* (pp. 245–266). Springer.

Mukhongo, R. W., Tumuhairwe, J. B., Ebanyat, P., AbdelGadir, A. H., Thuita, M., & Masso, C. (2016). Production and use of arbuscular mycorrhizal fungi inoculum in sub-Saharan Africa: Challenges and ways of improving. *International Journal of Soil Science, 11*, 108–122.

Nogueira De Sousa, R. (2023). Mycorrhizal fungi–A current overview on agricultural productivity and soil health. In R. Nogueira De Sousa (Ed.), *Arbuscular Mycorrhizal fungi in agriculture – New insights* (p. 272). IntechOpen.

O'Callaghan, M., Ballard, R. A., & Wright, D. (2022). Soil microbial inoculants for sustainable agriculture: Limitations and opportunities. *Soil Use and Management, 38*, 1340–1369.

Orrù, L., Canfora, L., Trinchera, A., Migliore, M., Pennelli, B., Marcucci, A., et al. (2021). How tillage and crop rotation change the distribution pattern of fungi. *Frontiers in Microbiology, 12*, 634325.

Oruru, M. B., & Njeru, E. M. (2016). Upscaling arbuscular mycorrhizal symbiosis and related agroecosystems services in smallholder farming systems. *BioMed Research International, 2016*, 1–12.

Oviatt, P., & Rillig, M. C. (2021). Mycorrhizal technologies for an agriculture of the middle. *Plants People Planet, 3*, 454–461.

Pagano, M. C., Kyriakides, M., & Kuyper, T. W. (2023). Effects of pesticides on the arbuscular mycorrhizal symbiosis. *Agrochemicals, 2*, 337–354.

Page, K. L. (2020). The ability of conservation agriculture to conserve soil organic carbon and the subsequent impact on soil physical, chemical, and biological properties and yield. *Frontiers in Sustainable Food Systems, 4*, 31.

Pakarinen, A., Fritze, H., Timonen, S., Kivijärvi, P., & Velmala, S. (2021). Boreal soil microbial diversity and seed onion mycorrhizal colonization is unaffected by preceding one season crop cultivation. *European Journal of Soil Biology, 105*, 103335.

Panth, M., Hassler, S. C., & Baysal-Gurel, F. (2020). Methods for management of soilborne diseases in crop production. *Agriculture, 10*, 16.

Parihar, M., Rakshit, A., Rana, K., Tiwari, G., & Jatav, S. S. (2020). The effect of arbuscular mycorrhizal fungi inoculation in mitigating salt stress of pea (*Pisum Sativum* L.). *Communications in Soil Science and Plant Analysis, 51*, 1545–1559.

Peay, K. G. (2016). The mutualistic niche: Mycorrhizal symbiosis and community dynamics. *Annual Review of Ecology, Evolution, and Systematics, 47*, 143–164.

Pons, C., & Müller, C. (2022). Impacts of drought stress and mycorrhizal inoculation on the performance of two spring wheat cultivars. *Plants, 11*, 2187.

Porter, S. S., & Sachs, J. L. (2020). Agriculture and the disruption of plant–microbial symbiosis. *Trends in Ecology & Evolution, 35*, 426–439.

Poveda, J., Díaz-González, S., Díaz-Urbano, M., Velasco, P., & Sacristán, S. (2022). Fungal endophytes of Brassicaceae: Molecular interactions and crop benefits. *Frontiers in Plant Science, 13*, 932288.

Pozo, M. J., & Azcón-Aguilar, C. (2007). Unraveling mycorrhiza-induced resistance. *Current Opinion in Plant Biology, 10*, 393–398.

Pozo De La Hoz, J., Rivero, J., Azcón-Aguilar, C., Urrestarazu, M., & Pozo, M. J. (2021). Mycorrhiza-induced resistance against foliar pathogens is uncoupled of nutritional effects under different light intensities. *Journal of Fungi, 7*, 402.

Pretty, J., & Bharucha, Z. (2015). Integrated pest management for sustainable intensification of agriculture in Asia and Africa. *Insects, 6*, 152–182.

Prova, A., Akanda, A. M., Islam, S., & Hossain, M. M. (2018). Characterization of *Sclerotinia sclerotiorum*, an emerging fungal pathogen causing blight in hyacinth bean (*Lablab purpureus*). *The Plant Pathology Journal, 34*, 367–380.

Pu, C., Ge, Y., Yang, G., Zheng, H., Guan, W., Chao, Z., Shen, Y., Liu, S., Chen, M., & Huang, L. (2022). Arbuscular mycorrhizal fungi enhance disease resistance of *Salvia miltiorrhiza* to Fusarium wilt. *Frontiers in Plant Science, 13*, 975558.

Raaijmakers, J. M., Paulitz, T. C., Steinberg, C., Alabouvette, C., & Moënne-Loccoz, Y. (2009). The rhizosphere: A playground and battlefield for soilborne pathogens and beneficial microorganisms. *Plant & Soil, 321*, 341–361.

Ramírez-Flores, M. R., Perez-Limon, S., Li, M., Barrales-Gamez, B., Albinsky, D., Paszkowski, U., et al. (2020). The genetic architecture of host response reveals the importance of arbuscular mycorrhizae to maize cultivation. *eLife, 9*, e61701.

Read, D. J., & Perez-Moreno, J. (2003). Mycorrhizas and nutrient cycling in ecosystems–A journey towards relevance? *The New Phytologist, 157*, 475–492.

Rillig, M. C., Sosa-Hernández, M. A., Roy, J., Aguilar-Trigueros, C. A., Vályi, K., & Lehmann, A. (2016). Towards an integrated mycorrhizal technology: Harnessing mycorrhiza for sustainable intensification in agriculture. *Frontiers in Plant Science, 7*, 1625.

Rosa, D., Pogiatzis, A., Bowen, P., Kokkoris, V., Richards, A., Holland, T., & Hart, M. (2020). Performance and establishment of a commercial mycorrhizal inoculant in viticulture. *Agriculture, 10*, 539.

Rubel, M. H., Abuyusuf, M., Nath, U. K., Robin, A. H. K., Jung, H. J., Kim, H. T., et al. (2020). Glucosinolate profile and glucosinolate biosynthesis and breakdown gene expression manifested by black rot disease infection in cabbage. *Plants, 9*, 1121.

Ryan, M. H. (2018). Little evidence that farmers should consider abundance or diversity of arbuscular mycorrhizal fungi when managing crops. *The New Phytologist, 220*, 1092–1107.

Sa, G., Yao, J., Deng, C., Liu, J., Zhang, Y., Zhu, Z., Zhang, Y., Ma, X., Zhao, R., Lin, S., Lu, C., Polle, A., & Chen, S. (2019). Amelioration of nitrate uptake under salt stress by ectomycorrhiza with and without a Hartig net. *The New Phytologist, 222*, 1951–1964.

Sagar, A., Rathore, P., Ramteke, P. W., Ramakrishna, W., Reddy, M. S., & Pecoraro, L. (2021). Plant growth promoting rhizobacteria, arbuscular mycorrhizal fungi and their synergistic interactions to counteract the negative effects of saline soil on agriculture: Key macromolecules and mechanisms. *Microorganisms, 9*, 1491.

Salomon, M. J., Watts-Williams, S. J., McLaughlin, M. J., Bücking, H., Singh, B. K., et al. (2022). Establishing a quality management framework for commercial inoculants containing arbuscular mycorrhizal fungi. *iScience, 25*, 104636.

Samal, I., Bhoi, T. K., Majhi, P. K., Murmu, S., Pradhan, A. K., et al. (2023). Combatting insects mediated biotic stress through plant associated endophytic entomopathogenic fungi in horticultural crops. *Frontiers in Plant Science, 13*, 1098673. https://doi.org/10.3389/fpls.2022.1098673

Sandrini, M., Moffa, L., Velasco, R., Balestrini, R., Chitarra, W., & Nerva, L. (2022). Microbe-assisted crop improvement: A sustainable weapon to restore holobiont functionality and resilience. *Horticulture Research, 9*, uhac160.

Santoyo, G., Gamalero, E., & Glick, B. R. (2021). Mycorrhizal-bacterial amelioration of plant abiotic and biotic stress. *Frontiers in Sustainable Food Systems, 5*, 672881.

Seo, Y., Kim, Y. S., Park, Y., & Kim, Y. H. (2015). Comparisons of pathological responses in carrot to root-knot nematodes. *The Plant Pathology Journal, 31*, 441–445.

Shaw, R. K., Shen, Y., Yu, H., Sheng, X., Wang, J., & Gu, H. (2022). Multi-omics approaches to improve clubroot resistance in *Brassica* with a special focus on *Brassica oleracea* L. *International Journal of Molecular Sciences, 23*, 9280. https://doi.org/10.3390/ijms23169280

Sheikh-Assadi, M., Khandan-Mirkohi, A., Taheri, M. R., Babalar, M., Sheikhi, H., & Nicola, S. (2023). Arbuscular mycorrhizae contribute to growth, nutrient uptake, and ornamental characteristics of statice (*Limonium sinuatum* [L.] Mill.) subject to appropriate inoculum and optimal phosphorus. *Horticulturae, 9*, 564.

Shi, J., Wang, X., & Wang, E. (2023). Mycorrhizal symbiosis in plant growth and stress adaptation: From genes to ecosystems. *Annual Review of Plant Biology, 74*, 569–607.

Siddiqui, Z. A., Akhtar, M. S., & Futai, K. (Eds.). (2008). *Mycorrhizae: Sustainable agriculture and forestry*. Springer. https://doi.org/10.1007/978-1-4020-8770-7

Simard, S., & Austi, M. (2010). The role of mycorrhizas in forest soil stability with climate change. In S. Simard (Ed.), *Climate change and variability* (p. 502). IntechOpen. https://doi.org/10.5772/9813

Singh, D., Ambroise, A., Haicour, R., Sihachakr, D., & Rajam, M. V. (2014). Increased resistance to fungal wilts in transgenic eggplant expressing alfalfa glucanase gene. *Physiology and Molecular Biology of Plants, 20*, 143–150.

Singh, M., Bisht, S., Singh, S., & Sharma, J. G. (2022). Implications of abiotic stress tolerance in arbuscular mycorrhiza colonized plants: Importance in plant growth and regulation. *Journal of Applied Biology & Biotechnology*, 1–11. https://doi.org/10.7324/JABB.2022.100601

Sinno, M., Ranesi, M., Gioia, L., d'Errico, G., & Woo, S. L. (2020). Endophytic fungi of tomato and their potential applications for crop improvement. *Agriculture, 10*, 587.

Song, Y., Chen, D., Lu, K., Sun, Z., & Zeng, R. (2015). Enhanced tomato disease resistance primed by arbuscular mycorrhizal fungus. *Frontiers in Plant Science, 6*, 786.

Sosa-Hernández, M. A., Leifheit, E. F., Ingraffia, R., & Rillig, M. C. (2019). Subsoil arbuscular mycorrhizal fungi for sustainability and climate-smart agriculture: A solution right under our feet? *Frontiers in Microbiology, 10*, 744.

Tahat, M. M., Alananbeh, K. M., Othman, Y. A., & Leskovar, D. I. (2020). Soil health and sustainable agriculture. *Sustainability, 12*, 4859. https://doi.org/10.3390/su12124859

Thirkell, T. J., Charters, M. D., Elliott, A. J., Sait, S. M., & Field, K. J. (2017). Are mycorrhizal fungi our sustainable saviours? Considerations for achieving food security. *Journal of Ecology, 105*, 921–929.

Thomas, P., Knox, O. G. G., Powell, J. R., Sindel, B., & Winter, G. (2023). The hydroponic rockwool root microbiome: Under control or underutilised? *Microorganisms, 11*, 835.

Tsedaley, B. (2014). Late blight of potato (*Phytophthora infestans*) biology, economic importance and its management approaches. *Journal of Biology, Agriculture and Healthcare, 4*, 215–225.

Tudi, M., Daniel Ruan, H., Wang, L., Lyu, J., Sadler, R., Connell, D., et al. (2021). Agriculture development, pesticide application and its impact on the environment. *International Journal of Environmental Research and Public Health, 18*, 1112.

Usman, M., Ho-Plágaro, T., Frank, H. E. R., Calvo-Polanco, M., Gaillard, I., Garcia, K., & Zimmermann, S. D. (2021). Mycorrhizal symbiosis for better adaptation of trees to abiotic stress caused by climate change in temperate and boreal forests. *Frontiers in Forests and Global Change, 4*, 742392.

Van Bruggen, A. H. C., & Finckh, M. R. (2016). Plant diseases and management approaches in organic farming systems. *Annual Review of Phytopathology, 54*, 25–54.

Van Der Heijden, M. G. A., Martin, F. M., Selosse, M., & Sanders, I. R. (2015). Mycorrhizal ecology and evolution: The past, the present, and the future. *The New Phytologist, 205*, 1406–1423.

Vosátka, M., Látr, A., Gianinazzi, S., & Albrechtová, J. (2012). Development of arbuscular mycorrhizal biotechnology and industry: Current achievements and bottlenecks. *Symbiosis, 58*, 29–37.

Wahab, A., Muhammad, M., Munir, A., Abdi, G., Zaman, W., Ayaz, A., et al. (2023). Role of Arbuscular Mycorrhizal fungi in regulating growth, enhancing productivity, and potentially influencing ecosystems under abiotic and biotic stresses. *Plants, 12*, 3102.

Wallon, T., Sauvageau, A., & Van Der Heyden, H. (2020). Detection and quantification of *Rhizoctonia solani* and *Rhizoctonia solani* AG1-IB causing the bottom rot of lettuce in tissues and soils by multiplex qPCR. *Plants, 10*, 57.

Wang, J., Fu, Z., Ren, Q., Zhu, L., Lin, J., Zhang, J., et al. (2019). Effects of Arbuscular Mycorrhizal fungi on growth, photosynthesis, and nutrient uptake of *Zelkova serrata* (Thunb.) Makino seedlings under salt stress. *Forests, 10*, 186.

Wang, L., Wang, X., Maimaitiaili, B., Kafle, A., Khan, K. S., & Feng, G. (2021). Breeding practice improves the mycorrhizal responsiveness of cotton (*Gossypium* spp. L.). *Frontier in Plant Science, 12*, 780454.

Wang, M., Li, R., & Zhao, Q. (2023). Multi-omics techniques in genetic studies and breeding of forest plants. *Forests, 14*, 1196.

Weng, W., Yan, J., Zhou, M., Yao, X., Gao, A., Ma, C., et al. (2022). Roles of Arbuscular Mycorrhizal fungi as a biocontrol agent in the control of plant diseases. *Microorganisms, 10*, 1266.

Widawati, S., & Suliasih. (2020). Comprehensive test of rhizobacteria as biostimulant, vesicular arbuscular mycorhizza (VAM) and graded dose of NPK fertilizer on the growth of bok choy (*Brassica rapa* L.). *IOP Conference Series: Earth and Environmental Science, 572*, 012023. https://doi.org/10.1088/1755-1315/572/1/012023

Wijayawardene, N. N., Boonyuen, N., Ranaweera, C. B., De Zoysa, H. K. S., Padmathilake, R. E., et al. (2023). Omics and other advanced technologies in mycological applications. *Journal of Fungi, 9*, 688.

Wilkes, T. I. (2021). Arbuscular Mycorrhizal Fungi in agriculture. *Encyclopedia, 1*, 1132–1154. https://doi.org/10.3390/encyclopedia1040085

Yang, Y., Saand, M. A., Huang, L., Abdelaal, W. B., Zhang, J., Wu, Y., et al. (2021). Applications of multi-omics technologies for crop improvement. *Frontiers in Plant Science, 12*, 563953. https://doi.org/10.3389/fpls.2021.563953

Zeng, R.-S. (2006). Disease resistance in plants through mycorrhizal fungi induced allelochemicals. In Inderjit & K. G. Mukerji (Eds.), *Allelochemicals: Biological control of plant pathogens and diseases* (pp. 181–192). Kluwer Academic Publishers.

Zhu, B., Gao, T., Zhang, D., Ding, K., Li, C., & Ma, F. (2022). Functions of arbuscular mycorrhizal fungi in horticultural crops. *Scientia Horticulturae, 303*, 111219.

Chapter 17
Optimizing Post-harvest Disease Management Through Plant Microbiome Intervention Strategies

Prasann Kumar and Debjani Choudhury

Abstract The plant microbiome plays a significant role in post-harvest disease management, influencing disease development and severity after harvest. Understanding the dynamics and functions of the plant microbiome can provide valuable insights into post-harvest disease management strategies. The microbiome can influence post-harvest diseases by competing with and antagonizing pathogens, producing antimicrobial compounds, and modulating plant defence responses. Manipulating the plant microbiome through pre- and post-harvest interventions, such as biocontrol agents, sanitisers, and storage conditions, can help suppress post-harvest diseases, extend produce shelf life, and maintain its quality. Harnessing the potential of the plant microbiome in post-harvest disease management offers promising avenues for sustainable and eco-friendly approaches to enhance food safety and reduce post-harvest losses.

Keywords Antimicrobial · Disease · No poverty · Plant microbes · Post harvest · Zero hunger

Abbreviations

PGRs	Plant growth regulators
SA	Salicylic acid
JA	Jasmonic acid
SAR	Systemic Acquired Resistance

P. Kumar (✉)
Department of Agronomy, School of Agriculture, Lovely Professional University, Phagwara, Punjab, India

D. Choudhury
Department of Plant Pathology, School of Agriculture, Lovely Professional University, Phagwara, Punjab, India

P. Mathur, S. Roy (eds.), *Plant Microbiome and Biological Control*, Sustainability in Plant and Crop Protection 20,
https://doi.org/10.1007/978-3-031-75845-4_17

BTH	Benzothiadiazole
ASM	Acibenzolar-S-methyl
Gas	Gibberellins
ABA	Abscisic acid
ISR	Induced Systemic Resistance
1-MCP	1-methylcyclopropene

17.1 Introduction

Post-harvest diseases represent a significant challenge in the agricultural industry, causing substantial economic losses and compromising food safety and quality. The plant microbiome, which includes bacteria, fungi, archaea, viruses, and other microorganisms, by interacting with the plant and the surrounding environment, also influences the development and severity of post-harvest diseases (Bhandari et al., 2022; Kumar et al., 2022b; Mukhtar et al., 2023).

The plant microbiome begins to establish during seed germination and continues to evolve throughout the plant life cycle. It encompasses various plant surfaces, such as leaves, stems, flowers, and fruits, as well as internal tissues, including the endosphere (internal tissues) and phyllosphere (above-ground plant surfaces). The microorganisms in the plant microbiome may have beneficial and pathogenic characteristics, and their interactions shape the plant overall health and resilience (Alam et al., 2022; Mangalam & Giri, 2022; Otero-Jiménez et al., 2021).

Post-harvest diseases occur after crops have been harvested, during transportation, storage, and processing. These diseases are predominantly caused by pathogenic microorganisms, including fungi, bacteria, and viruses, that infect and colonize harvested produce. Controlling post-harvest diseases is essential for preserving the quality and safety of agricultural commodities, extending their shelf life, and reducing related economic losses (Arshad et al., 2021; Elolimy et al., 2021; Mazzola et al., 2018; Zhimo et al., 2021).

The plant microbiome influences post-harvest disease management through several mechanisms:

1. *Competition and Antagonism.* Beneficial microorganisms within the plant microbiome can compete with and antagonize post-harvest pathogens, inhibiting their growth and activity. Competition for nutrients and space limits the availability of resources for pathogens, impeding their colonization. Moreover, antagonistic microorganisms produce antimicrobial compounds, such as antibiotics, enzymes, and volatile organic compounds, directly inhibiting the growth of pathogens.
2. *Induced Resistance.* The plant microbiome can induce systemic resistance in plants, priming them for enhanced defence against post-harvest pathogens. Microorganisms within the microbiome can trigger the plant immune response, producing defence-related compounds and activating defence pathways. This

induced resistance provides long-lasting protection against a range of post-harvest pathogens.

3. *Biocontrol Agents.* Certain microorganisms within the plant microbiome act as biocontrol agents that can actively suppress post-harvest pathogens. These biocontrol agents may colonize the plant surfaces or reside within its tissues, exerting their antagonistic effects on pathogens through various mechanisms. Examples include yeasts, bacteria, and fungi that produce antimicrobial compounds or compete with pathogens for nutrients and space.
4. *Environmental Interactions.* The plant microbiome interacts with the post-harvest environment, influencing the survival and activity of post-harvest pathogens. Environmental factors, such as temperature, humidity, and storage conditions, can impact the composition and dynamics of the plant microbiome. Optimizing post-harvest conditions, such as temperature-controlled storage or modified atmospheres, can enhance the abundance and activity of beneficial microorganisms, thereby suppressing post-harvest diseases.

To understand the composition, dynamics, and functions of the plant microbiome and its interaction with post-harvest pathogens it is crucial to manage post-harvest diseases effectively. Strategies for post-harvest disease management can involve manipulating the plant microbiome through various interventions. These may include applying biocontrol agents, sanitisers, and post-harvest treatments to enhance the abundance of beneficial microorganisms, suppress pathogen growth, and maintain post-harvest product quality (Garello et al., 2023; Reveles et al., 2021).

Using the plant microbiome in post-harvest disease management offers promising avenues for sustainable and eco-friendly approaches. By harnessing the beneficial interactions between plants and their associated microorganisms, it is possible to develop effective strategies that reduce reliance on chemical treatments, extend shelf life, and enhance food safety and quality. Understanding the complex dynamics of the plant microbiome in post-harvest disease management is essential for advancing agricultural practices and ensuring a sustainable and resilient food supply chain (Sangiorgio et al., 2022; Shi et al., 2022; Taparia et al., 2021; Wassermann et al., 2022).

17.2 Research Gap in Plant Microbiome and Post-harvest Disease Management

While significant progress has been made in understanding the role of the plant microbiome in post-harvest disease management, several research gaps warrant further investigation. They include:

1. *Dynamics of the Post-Harvest Microbiome.* There is a need to explore the dynamics of the plant microbiome during post-harvest handling and storage, Understanding how the microbiome composition and structure change during

different stages of post-harvest processing can provide insights into the microbial factors influencing disease development, and the potential for manipulating the microbiome to mitigate post-harvest diseases.

2. *Functional Characterization of Microbial Interactions.* Further research is required to comprehensively characterize the functional interactions between microorganisms within the plant microbiome and post-harvest pathogens. Investigating the mechanisms by which beneficial microorganisms antagonize pathogens, compete for resources, and induce plant resistance will enhance our understanding of the complex microbial dynamics involved in post-harvest disease management.
3. *Identification of Key Microbial Players.* Identifying specific microbial taxa and strains that contribute most significantly to disease suppression in the post-harvest environment is crucial. Targeted studies focusing on the functional and ecological roles of specific microbial species or strains will provide valuable insights into their potential as biocontrol agents or microbial inoculants for post-harvest disease management.
4. *Impact of Environmental Factors.* The influence of environmental factors, such as temperature, humidity, and storage conditions, on the composition and activity of the post-harvest microbiome, requires further investigation. Understanding how these factors shape the dynamics of beneficial microorganisms and their interactions with pathogens will aid in developing optimized storage and handling practices for disease management.
5. *Microbial Community Succession.* Research is needed to elucidate the temporal succession patterns of the post-harvest microbiome and its impact on disease development. Longitudinal studies that track microbial community dynamics over time will provide insights into the stability, resilience, and potential shifts in the microbiome profiles, that can affect disease outcomes.
6. *Integration of Microbiome-Based Approaches.* The development and implementation of microbiome-based approaches for post-harvest disease management are still in their infancy. More research is needed to explore the practical application and effectiveness of interventions such as biocontrol agents, microbial inoculants, and post-harvest treatments in real-world scenarios, including commercial settings.
7. *Safety and Regulatory Considerations.* There is a need to assess the safety and regulatory aspects associated with microbial-based interventions in post-harvest disease management. Investigating the potential impacts of these interventions on food safety, microbial ecology, and consumer acceptance will facilitate their adoption in a manner that aligns with regulatory guidelines and consumer expectations.

Addressing these research gaps will contribute to a more comprehensive understanding of the plant microbiome role in post-harvest disease management. It will enable the development of effective and sustainable strategies to mitigate post-harvest diseases, enhance food safety, and optimize quality and shelf life of agricultural commodities (Ali et al., 2021; Castillo-de la Cruz et al., 2019; Fournier et al., 2022; Hanson et al., 2022).

17.3 Causes of Post-harvest Diseases

Several factors contribute to the occurrence and development of diseases during the post-harvest stage. Post-harvest diseases can result in significant economic losses and negatively impact food quality and safety (Brandl et al., 2023; Islam et al., 2023; Wei et al., 2018). Understanding their causes is crucial for implementing effective management strategies. Here are some fundamental causes of post-harvest diseases:

1. *Pathogenic Inocula.* Pathogens, including fungi, bacteria, and viruses, contribute significantly to post-harvest diseases. They may already be present in the field during the crop growth or can be introduced during harvest, handling, and storage. Pathogenic inocula can proceed from infected plant debris, contaminated soil, equipment, or other sources. These pathogens can infect harvested crops, developing disease during storage or transportation.
2. *Wounds and Mechanical Damage.* Wounds and mechanical damage to harvested crops provide entry points for pathogens, facilitating their invasion and colonization. Mechanical damage can occur during harvesting, handling, sorting, or packaging. Bruising, cuts, abrasions, or punctures create vulnerable sites for pathogen penetration, which can lead to secondary infections and disease progression.
3. *Environmental Factors.* Environmental conditions are crucial in post-harvest disease development. Temperature, humidity, and air circulation influence the growth and activity of pathogens and their interaction with harvested crops. High humidity levels can create a favourable environment for fungal growth and spore germination, while inadequate air circulation can lead to moisture buildup, promoting disease development. Temperature fluctuations during storage or transportation can also affect the rate of disease progression.
4. *Storage Conditions.* Improper storage conditions can contribute to disease development in harvested crops. Inadequate temperature and humidity control, poor ventilation, and inappropriate packaging materials can create conditions that favour pathogen growth and survival. Improper storage practices may increase respiration rates, moisture buildup, or ethylene production, accelerating disease progression and produce spoilage.
5. *Secondary Infections.* Harvested crops can become susceptible to secondary infections by opportunistic pathogens or saprotrophic microorganisms in the environment. These organisms take advantage of pre-existing injuries, senescing tissues, or the release of volatile compounds from damaged crops. Secondary infections can further deteriorate the quality and marketability of harvested crops, leading to increased disease incidence.

6. *Post-Harvest Treatments.* Post-harvest treatments, such as applications of chemicals or physical interventions, may inadvertently contribute to disease development. Improper application of chemicals, using contaminated water sources, or inadequate rinsing procedures can introduce pathogens or create conditions that favour their growth. Similarly, improper handling of harvested crops during washing, waxing, or sanitization processes can lead to cross-contamination or provide favourable conditions for disease development.
7. *Storage Pests and Vectors.* Storage pests, such as insects, mites, or rodents, can damage harvested crops and facilitate pathogen invasion. These pests may physically wound the crops or act as carriers for pathogens, transferring them from infected to healthy tissues.

Addressing the causes of post-harvest diseases requires implementing appropriate management strategies. These strategies may include effective sanitation practices, proper handling techniques, improved storage conditions, implementation of preventive measures, and post-harvest treatments that minimize the risk of disease development (Bulleri et al., 2022; Liu et al., 2020b; Lupwayi et al., 2022). Understanding and mitigating these causes makes it possible to reduce post-harvest losses, maintain crop quality, and ensure the safety of harvested crops for consumers (Table 17.1).

17.4 Post-harvest Disease Causal Factors

Post-harvest diseases in crops are influenced by various factors that contribute to their development and progression (Alberoni et al., 2021; Bui-Xuan et al., 2022; Chatterjee et al., 2022; Farag et al., 2021; Liu, Brettell, et al., 2020a; Morales-Cedeño et al., 2021; Shen et al., 2023). Understanding these factors is crucial for effective disease management. The critical factors involved in post-harvest diseases in crops include pathogens such as fungi, bacteria, viruses, and other microorganisms. Different pathogens have specific host preferences and infection strategies. They can be present on the harvested crops or introduced during handling, storage, or transportation processes. Other factors include the susceptibility of crops to post-harvest diseases, that varies depending on the crop species, cultivar, and physiological state. Some crops have natural defence mechanisms that make them more resistant to specific pathogens, while others may be more susceptible. The stage of maturity at harvest also influences the produce susceptibility to diseases. The residues left on or near harvested crops can also serve as a source of inoculum for pathogens. Infected or diseased plant debris, such as fallen leaves, fruits, or stems, can harbour pathogens that can infect freshly harvested crops. Proper removal and disposal of crop residues can help reduce the inoculum load and minimize disease incidence. As previously indicated, wounds and mechanical damage incurred during harvesting, handling, or processing provide entry points for pathogens. Environmental conditions, including temperature, humidity, and air circulation, are

Table 17.1 Most common post-harvest diseases and related causal agents

Produce	Post-harvest disease	Causal agent
Apple	Blue Mold	*Penicillium expansum*
Apple	Gray Mold	*Botrytis cinerea*
Apple	Black Rot	*Botryosphaeria* spp.
Banana	Anthracnose	*Colletotrichum musae*
Banana	Crown Rot	*Phytophthora* spp.
Banana	Stem-End Rot	*Lasiodiplodia theobromae*
Banana	Black Sigatoka	*Pseudocercospora fijiensis*
Tomato	Fruit Rot	*Rhizopus* spp., *Alternaria* spp.
Tomato	Bacterial Canker	*Clavibacter michiganensis* subsp. *michiganensis*
Tomato	Early Blight	*Alternaria solani*
Tomato	Gray Mold	*Botrytis cinerea*
Potato	Late Blight	*Phytophthora infestans*
Potato	Pink Rot	*Phytophthora erythroseptica*
Potato	Soft Rot	*Pectobacterium* spp.
Potato	Black Scurf	*Rhizoctonia solani*
Citrus	Green Mold	*Penicillium digitatum*
Citrus	Sour Rot	*Geotrichum citri-aurantii*
Citrus	Alternaria Rot	*Alternaria* spp.
Citrus	Citrus Canker	*Xanthomonas citri* subsp. *citri*
Grapes	Botrytis Bunch Rot	*Botrytis cinerea*
Grapes	Powdery Mildew	*Erysiphe necator*
Grapes	Black Rot	*Guignardia bidwellii*
Grapes	Downy Mildew	*Plasmopara viticola*
Mango	Anthracnose	*Colletotrichum* spp.
Mango	Stem-End Rot	*Lasiodiplodia* spp.
Mango	Post-Harvest Deterioration	Various microorganisms
Pineapple	Pink Disease	*Acetobacteraceae bacteria*
Pineapple	Black Rot	*Ceratocystis paradoxa*
Pineapple	Fruit Collapse	*Phytophthora cinnamomi*
Avocado	Anthracnose	*Colletotrichum* spp.
Avocado	Stem-End Rot	*Colletotrichum gloeosporioides*
Avocado	Post-Harvest Rots	Various pathogens
Carrot	Sclerotinia Rot	*Sclerotinia* spp.
Carrot	Bacterial Soft Rot	*Pectobacterium* spp.
Carrot	Alternaria Leaf Blight	*Alternaria* spp.
Carrot	Powdery Mildew	*Erysiphe* spp.
Cabbage	Black Rot	*Xanthomonas campestris* pv. *campestris*
Cabbage	Soft Rot	*Pectobacterium* spp.
Cabbage	Clubroot	*Plasmodiophora brassicae*
Onion	Neck Rot	*Botrytis allii*

(continued)

Table 17.1 (continued)

Produce	Post-harvest disease	Causal agent
Onion	Bacterial Soft Rot	*Pectobacterium* spp.
Onion	Botrytis Leaf Blight	*Botrytis* spp.
Onion	Pink Root	*Phoma terrestris*
Lettuce	Downy Mildew	*Bremia lactucae*
Lettuce	Gray Mold	*Botrytis cinerea*
Lettuce	Bottom Rot	*Rhizoctonia solani*
Lettuce	Bacterial Soft Rot	*Pectobacterium carotovorum*
Strawberry	Gray Mold	*Botrytis cinerea*
Strawberry	Anthracnose	*Colletotrichum* spp.
Strawberry	Leather Rot	*Phytophthora* spp.
Strawberry	Rhizopus Rot	*Rhizopus* spp.
Blueberry	Anthracnose	*Colletotrichum* spp.
Blueberry	Botrytis Fruit Rot	*Botrytis cinerea*
Blueberry	Mummy Berry	*Monilinia vaccinii-corymbosi*
Blueberry	Phytophthora Root Rot	*Phytophthora* spp.
Peach	Brown Rot	*Monilinia* spp.
Peach	Rhizopus Rot	*Rhizopus* spp.
Peach	Bacterial Spot	*Xanthomonas arboricola* pv. *pruni*
Pear	Soft Rot	*Erwinia* spp.
Pear	Black Spot	*Alternaria alternata*
Pear	Blue Mold	*Penicillium* spp.
Pear	Fire Blight	*Erwinia amylovora*
Cherry	Brown Rot	*Monilinia* spp.
Cherry	Powdery Mildew	*Podosphaera clandestina*
Cherry	Bacterial Canker	*Pseudomonas syringae*
Cherry	Cherry Leaf Spot	*Blumeriella jaapii*
Olive	Olive Knot	*Pseudomonas savastanoi*
Olive	Peacock Spot	*Spilocaea oleagina*
Olive	Anthracnose	*Colletotrichum* spp.
Olive	Olive Fly	*Bactrocera oleae*
Peanut	Aspergillus Rot	*Aspergillus* spp.
Peanut	Sclerotinia Rot	*Sclerotinia* spp.
Peanut	Web Blotch	*Phoma* spp.
Peanut	Leaf Spot	*Cercospora* spp.
Walnut	Walnut Blight	*Xanthomonas arboricola* pv. *juglandis*
Walnut	Anthracnose	*Colletotrichum* spp.
Walnut	Walnut Husk Fly	*Rhagoletis completa*
Walnut	Crown Gall	*Agrobacterium tumefaciens*
Almond	Brown Rot	*Monilinia* spp.
Almond	Shot Hole	*Wilsonomyces carpophilus*
Almond	Almond Scab	*Cladosporium* spp.
Almond	Almond Leaf Spot	*Stigmina carpophila*

(continued)

Table 17.1 (continued)

Produce	Post-harvest disease	Causal agent
Hazelnut	Eastern Filbert Blight	*Anisogramma anomala*
Hazelnut	Bacterial Blight	*Xanthomonas arboricola* pv. *corylina*
Hazelnut	Filbertworm	*Cydia* spp.
Hazelnut	Phytophthora Root Rot	*Phytophthora* spp.
Coffee	Coffee Berry Disease	*Colletotrichum* spp.
Coffee	Coffee Leaf Rust	*Hemileia vastatrix*
Coffee	Coffee Wilt Disease	*Fusarium xylarioides*
Coffee	Coffee Bacterial Blight	*Pseudomonas syringae*
Rice	Grain Discoloration	*Aspergillus* spp., *Penicillium* spp.
Rice	Rice Blast	*Pyricularia oryzae*
Rice	Bacterial Leaf Blight	*Xanthomonas oryzae* pv. *oryzae*
Rice	Sheath Blight	*Rhizoctonia solani*
Wheat	Fusarium Head Blight	*Fusarium* spp.
Wheat	Loose Smut	*Ustilago tritici*
Wheat	Tan Spot	*Pyrenophora tritici-repentis*
Wheat	Septoria Leaf Blotch	*Septoria* spp.
Maize	Fusarium Ear Rot	*Fusarium* spp.
Maize	Aspergillus Ear Rot	*Aspergillus* spp.
Maize	Diplodia Ear Rot	*Diplodia* spp.
Maize	Common Rust	*Puccinia sorghi*
Soybean	Soybean Rust	*Phakopsora pachyrhizi*
Soybean	Sudden Death Syndrome	*Fusarium virguliforme*
Soybean	Pod and Stem Blight	*Diaporthe* spp.
Soybean	Brown Stem Rot	*Phialophora aggregata*

crucial in post-harvest disease development. Also the storage conditions greatly influence the development of post-harvest diseases. Improper storage practices, such as high humidity, inadequate ventilation, or incorrect temperature control, can create conditions that favour pathogen growth and spread. Storage pests like insects or rodents can also cause physical damage and facilitate pathogen entry. Other factors include:

1. *Interactions with Beneficial Microorganisms.* The presence and activity of beneficial microorganisms within the crop microbiome can impact post-harvest diseases. Beneficial microorganisms, such as antagonistic bacteria or fungi, can suppress the growth of a pathogen and protect produce from diseases. Imbalances or disruptions in the crop microbiome can influence disease development.
2. *Harvesting and Handling Practices.* The methods and practices employed during harvesting and handling can influence disease incidence. Rough handling, improper hygiene, and contaminated equipment can introduce pathogens to the harvested crops. Sanitary practices, including regular cleaning and disinfection, can minimize the risk of pathogen contamination.

3. *Transportation and Distribution.* Transportation and distribution processes can further contribute to post-harvest diseases. Prolonged transportation, inadequate temperature control, and rough handling during transit can exacerbate disease development. Cross-contamination between batches or crops can occur without proper sanitation measures.

17.5 Plant Growth Regulator in Post-harvest Disease Management

Plant growth regulators (PGRs) are chemical compounds that can be applied to crops to regulate and modify their growth and development. While PGRs are primarily known for their effects on plant growth and yield, they also have potential applications in post-harvest disease management (Asiri et al., 2022; Bellini et al., 2023; Bose et al., 2023; Cazzaniga et al., 2023a; Chaisiri et al., 2023; Chen et al., 2021; Hong et al., 2022; Kumar et al., 2022a). Here are some ways in which PGRs can be applied for post-harvest disease management:

1. *Delayed Senescence.* PGRs can delay the senescence process in harvested crops, extending their shelf life. Senescence is a natural physiological process that leads to tissue deterioration and susceptibility to pathogens. By applying PGRs that inhibit ethylene production or action, such as 1-methylcyclopropene (1-MCP), the ripening and senescence process can be delayed, reducing the risk of post-harvest diseases.
2. *Enhancement of Defense Mechanisms.* PGRs can stimulate the plant natural defence mechanisms, making it more resistant to post-harvest pathogens. For example, salicylic acid (SA) and jasmonic acid (JA) are PGRs known for their involvement in plant defence responses. These PGRs can enhance the production of defence-related compounds, activate defense signalling pathways, and strengthen the plant ability to resist pathogen attacks.
3. *Induction of Systemic Acquired Resistance (SAR).* SAR is a plant defence response that provides long-lasting protection against various pathogens. PGRs such as benzothiadiazole (BTH) or acibenzolar-S-methyl (ASM) can induce SAR in harvested crops. SAR induction primes the plant immune system, enhancing resistance to subsequent infections, including post-harvest pathogens.
4. *Reduction of Pathogen Growth.* Certain PGRs possess antimicrobial properties and can directly inhibit the growth and development of post-harvest pathogens. For instance, chitosan, a natural biopolymer derived from chitin, has been shown to have antifungal activity against various fungal pathogens. Treating harvested crops with chitosan-based formulations can suppress the growth and colonization of post-harvest pathogens, reducing disease incidence.
5. *Regulation of Wound Healing.* PGRs can promote wound healing and tissue repair in harvested crops. Rapid wound closure minimizes the entry points for

pathogens and reduces the risk of infection. PGRs like auxins or cytokinins can promote cell division and tissue regeneration, aiding healing and preventing further damage and disease development.

6. *Preservation of Fruit Quality.* PGRs can also help to maintain the quality attributes of harvested fruits and vegetables, indirectly contributing to disease management. PGRs like gibberellins (GAs) or abscisic acid (ABA) can influence fruit size, colour development, texture, and flavour. By optimizing these quality parameters, PGRs can enhance harvested produce marketability and consumer appeal, reducing post-harvest losses and the likelihood of disease occurrence, due to compromised fruit quality.

It is important to note that PGRs for post-harvest disease management should follow recommended application rates and comply with regulations. Proper timing, dosage, and application techniques are crucial to achieve desired effects without negatively impacting crop quality or human health (Chung et al., 2021; Ma et al., 2020; Ramakrishnan et al., 2021; Villa-Rodriguez et al., 2022). Additionally, the effectiveness of PGRs may vary depending on crop species, variety, environmental conditions, and specific pathogens involved. Therefore, it is advisable to conduct thorough research and consult with experts in the field before implementing PGR-based strategies for post-harvest disease management.

17.6 Post-harvest Residue Retention and Fungal Attack

Retention of post-harvest residues refers to leaving crop residues, such as stalks, leaves, and other plant materials, in the field after harvest instead of removing or burning them (where this action is possible). Retention has been recognized for positively impacting soil health and biodiversity, including enhancing soil fungal biodiversity. Here is an overview of how the retention of post-harvest residues can contribute to the enhancement of soil fungal biodiversity:

1. *Organic Matter Input.* Post-harvest residues are rich in organic matter, providing a valuable source of nutrients and energy for soil microorganisms, including fungi that play a vital role in decomposing organic matter, breaking it down into simpler compounds that other organisms can utilize. A continuous supply of organic matter is introduced into the soil by retaining post-harvest residues, promoting fungal growth and diversity.
2. *Substrate Availability.* Post-harvest residues are a substrate for fungal colonization and growth. Different fungal species have specific substrate preferences, and the diversity of post-harvest residues can support a wide range of fungal species. By providing diverse substrates, including residues with varying chemical composition and physical structure, the retention of post-harvest residues can support the growth of different microbial functional groups, contributing to overall biodiversity.

3. *Habitat Creation.* Post-harvest residues produce a favourable habitat for soil fungi. The residues offer protection from harsh environmental conditions, such as temperature extremes and UV radiation, which can harm fungal survival. The residues also provide physical structure and moisture retention, creating micro-environments conducive to fungal growth and colonization. The presence of diverse habitats within the soil supports the establishment and persistence of various fungal species.
4. *Nutrient Cycling.* Fungi are essential drivers of nutrient cycling in the soil ecosystem. They play a crucial role in decomposing organic matter and releasing essential nutrients for plant growth. By retaining post-harvest residues, organic matter input promotes the activity of fungi involved in nutrient-cycling processes, such as decomposition, mineralization, and immobilization. This enhances nutrient availability and cycling within the soil, supporting a diverse fungal community.
5. *Mutualistic Associations.* Some fungi, such as mycorrhizal fungi, form mutualistic associations with plants. Mycorrhizal fungi establish symbiotic relationships with plant roots, enhancing nutrient uptake and improving plant health. Retaining post-harvest residues can provide a substrate for mycorrhizal fungi, promoting their establishment and facilitating beneficial plant interactions. These mutualistic associations contribute to the overall fungal biodiversity in soil.
6. *Disease Suppression.* Certain fungi have antagonistic properties and can suppress soil-borne plant pathogens. The abundance and diversity of antagonistic fungi can be increased by retaining post-harvest residues. These fungi compete with pathogens for resources, produce antimicrobial compounds, and stimulate plant defence responses, reducing the incidence and severity of soil-borne diseases. The presence of a diverse fungal community enhances the soil natural disease suppression potential.

Overall, the retention of post-harvest residues provides various benefits for soil fungal biodiversity. It supports the growth of diverse species, promotes nutrient cycling, facilitates mutualistic associations with plants, and contributes to disease suppression (X. Liu et al., 2021; Mącik et al., 2023; Mishra et al., 2023; Rizvi et al., 2022; Yang et al., 2023; Zeng et al., 2022). By implementing practices that encourage residue retention, farmers and land managers can enhance soil health and promote a thriving fungal community, which benefits overall ecosystem functioning and crop productivity.

17.7 PGPR and Disease Control

Plant growth-promoting bacterial endophytes (PGBEs) are beneficial bacteria that inhabit the internal tissues of plants without causing any harm (Annapure & Bagul, 2022; Callewaert et al., 2021; Culp & Goodman, 2023; Mazzola et al., 2020;

Valentino et al., 2022; Wani et al., 2016; Zeng et al., 2021). These endophytes have been increasingly recognized for their potential as biocontrol agents against pre- and post-harvest diseases. Here follows an overview of how PGBEs can serve as biocontrol agents for pre- and post-harvest diseases.

17.7.1 Pre-harvest Disease Biocontrol

1. *Induced Systemic Resistance (ISR).* PGBEs can elicit the plant's natural defence mechanisms, leading to induced systemic resistance. They stimulate the production of defence-related compounds and activate signalling pathways that enhance the plant ability to resist pathogen attacks. This pre-harvest protection reduces the incidence and severity of diseases, ultimately improving crop health and yield.
2. *Antagonistic Activity.* Many PGBEs can produce antimicrobial compounds, including antibiotics and enzymes, that inhibit the growth and development of plant pathogens. These bioactive metabolites can directly suppress the pathogens responsible for pre-harvest diseases, limiting colonization and disease symptoms.
3. *Competition and Exclusion.* PGBEs compete with pathogens for resources, such as nutrients and space within the plant tissues. By colonizing the same niche as the pathogens, PGBEs can outcompete them, preventing their establishment and reducing disease development. This competitive exclusion mechanism helps protect crops from pre-harvest diseases.

17.7.2 Post-harvest Disease Biocontrol

1. *Biofilm Formation.* PGBEs can form biofilms on the surfaces of harvested crops, creating a protective barrier against post-harvest pathogens. The biofilm acts as a physical barrier and impedes pathogen attachment, reducing the risk of infection. Additionally, biofilm-associated PGBEs can produce antimicrobial compounds that inhibit pathogen growth, further enhancing the biocontrol effect.
2. *Antifungal Activity.* Several PGBEs have been shown to produce antifungal compounds effective against post-harvest pathogens. These compounds can inhibit spore germination, mycelial growth, or the production of virulence factors, leading to reduced disease incidence. PGBEs with antifungal properties offer a sustainable alternative to synthetic fungicides in post-harvest disease management.
3. *Indirect Defense Mechanisms.* PGBEs can enhance post-harvest disease management indirectly by promoting the plant overall health and quality. By stimulating plant growth and improving nutrient uptake, PGBEs enhance crop vigour and resistance to stress, making the harvested crops less susceptible to post-harvest diseases.

4. *Disease Suppression in Storage.* PGBEs can be applied to storage environments to prevent the proliferation of post-harvest pathogens during storage. The presence of beneficial endophytes can limit the growth of pathogens in storage facilities, reducing the risk of disease outbreaks and extending the shelf life of harvested crops.

PGBEs offer a sustainable and environmentally friendly pre- and post-harvest disease management approach. Their multifaceted mechanisms provide effective biocontrol against various pathogens. By harnessing the potential of PGBEs, farmers and growers can reduce reliance on chemical pesticides, promote plant health, and improve the quality and safety of harvested crops throughout the production and post-harvest stages (Cazzaniga et al., 2023b; Dong & Gupta, 2019; Droby & Wisniewski, 2018; Fruscella et al., 2023; Li et al., 2019; Selvan & Mukhopadhyay, 2022; Yan et al., 2023; Zeng et al., 2021).

17.8 Storage Condition and Disease Management

Storage affects post-harvest disease management, as proper conditions are essential to minimize the development and spread of diseases in harvested produce. Storage conditions involve:

Temperature: maintaining appropriate storage temperatures is crucial in preventing the growth and activity of pathogens. Most pathogens have temperature ranges where they thrive; deviations from these ranges can inhibit or promote their growth. In general, lower temperatures are preferred for most crops to slow down metabolic processes and reduce pathogen activity. However, different crops have specific temperature requirements, and it is essential to adhere to the recommended storage temperature for each crop to prevent disease development.

Humidity: controlling humidity levels in storage facilities is vital for disease management. High humidity can create a favourable environment for fungal growth and encourage the development of post-harvest diseases such as mould, rot, and mildew. To minimize humidity, adequate ventilation and air circulation should be maintained in storage areas. Additionally, certain crops may require specific humidity levels to prevent moisture-related diseases or desiccation.

Air Circulation: proper air circulation is essential to prevent the buildup of moisture and the formation of microclimates that can promote the growth of pathogens. Adequate airflow helps to maintain uniform temperature and humidity throughout the storage area, reducing the risk of localized areas becoming conducive to pathogen development. Good ventilation also helps remove ethylene gas, which can accelerate ripening and senescence processes, leading to increased susceptibility to diseases.

Ethylene Control: ethylene is a natural plant hormone involved in the ripening and senescence of fruits and vegetables. However, excessive ethylene accumulation can accelerate the deterioration of harvested crops and increase susceptibility to diseases. Monitoring and controlling ethylene levels in storage facilities can help extend the shelf life of produce and reduce the risk of post-harvest diseases. Ethylene absorbers or inhibitors can minimize ethylene production and maintain production quality.

Sanitation: maintaining cleanliness and sanitation in storage areas is crucial for post-harvest disease management. Any residual plant debris, soil, or decaying produce can be a source of pathogens, providing an ideal environment for disease development. Regular cleaning and disinfection of storage facilities, containers, and equipment can help reduce the presence of pathogens and minimize the risk of contamination.

Integrated Pest Management (IPM): implementing an integrated pest management approach is essential for effective disease management during storage. This includes proper packaging, pest monitoring, and pest control measures. Rodents, insects, and other pests can damage stored crops and introduce pathogens. The risk of disease transmission can be reduced by implementing measures to prevent pest infestations, such as proper sealing, pest-resistant containers, and targeted pest control methods.

Periodic Inspections: regular inspections of stored crops are necessary to identify any sign of disease or deterioration. Prompt removal of infected or damaged produce can prevent the spread of diseases to healthy crops. It is essential to conduct thorough visual inspections, checking for visible symptoms, mould growth, or any indications of decay. Regular monitoring allows for timely intervention and appropriate disease management strategies.

The risk of post-harvest diseases can be minimized by implementing and maintaining appropriate storage conditions, including temperature control, humidity management, proper air circulation, sanitation, and integrated pest management practices. Creating an environment that discourages growth and spread of pathogens helps to preserve the quality and marketability of harvested produce, reducing economic losses and ensuring food safety for consumers (Almeida et al., 2022; Chilakala et al., 2023; Feye et al., 2020; Przemieniecki et al., 2021; Si et al., 2021; Smith, 2020; Wicaksono et al., 2022; Yuan et al., 2015).

17.9 Modelling Plant Microbiome and Post-harvest Disease Interactions

Developing a comprehensive model for plant microbiome and post-harvest disease management requires a detailed understanding of microbial dynamics, host-pathogen interactions, and environmental factors (Ogundeji et al., 2022; Prestt & Roberts, 2023). Here we outline the key components that such a model should include:

1. *Microbial Dynamics.* The model should incorporate the dynamics of the different microbial populations within the plant microbiome, including beneficial and pathogenic microorganisms. This would involve equations describing microbial species growth, death, and interaction rates.
2. *Host-Pathogen Interactions.* The interactions between the host plant and pathogens should be considered, including the host susceptibility to different pathogens, the pathogens ability to infect and colonize the host, and the host defence mechanisms. These interactions should be represented by equations capturing infection rates, colonization dynamics, and host defence responses.
3. *Environmental Factors.* The model would account for environmental factors that influence post-harvest disease development, such as temperature, humidity, and nutrient availability, based on equations describing their effects on microbial growth, host physiology, and pathogen activity.
4. *Disease Progression.* To simulate the progression of post-harvest diseases the dynamics of pathogens growth and spread within the harvested crop should be incorporated, based on factors such as pathogen reproduction rates, dispersal mechanisms, and the impact of environmental conditions on disease development.
5. *Microbial Interactions.* The model should also account for the interactions among different microbial species within the plant microbiome. This should include positive (e.g., synergistic effects) and negative interactions (e.g., competition or inhibition), to capture their influence on disease suppression or promotion.
6. *Management Strategies.* The effectiveness of various disease management strategies, such as applying biocontrol agents, fungicides, or changes in storage conditions, should be integrated into the model through parameters that represent their effects on microbial populations, pathogen activity, or host resistance.
7. *Validation and Calibration.* The model must be calibrated using experimental data from field studies or controlled environments. This would involve fitting the model functions and variables to observed data to ensure its accuracy and reliability in predicting disease dynamics, and the effectiveness of management strategies.

Aa plant microbiome and post-harvest disease management model requires in-depth knowledge of the underlying biological processes and extensive data collection, to set up equations, numerical simulations, and parameter estimation techniques. Collaborations between microbiologists, plant pathologists, mathematicians, and data scientists would be essential to develop and refine such a model. Developing a comprehensive model for plant microbiome and post-harvest disease management is, however, a very complex task, due to the many factors involved (Almeida et al., 2022; Dong & Gupta, 2019; Selvan & Mukhopadhyay, 2022; Smith, 2020; Yuan et al., 2015). in spite of such complexity, we provide here a simplified equation that captures the fundamental dynamics of microbial populations and disease progression. Please note that this equation is conceptual and does not encompass all system intricacies:

$$\mathrm{dM}/\mathrm{dt} = \gamma M\left(1-\mathrm{M}/\mathrm{K}\right)-\beta M\left(\mathrm{P}/\mathrm{H}\right)$$

where

- dM/dt = the rate of microbial population (M) change over time.
- γ = the growth rate constant, representing the intrinsic growth potential of the microbial population.
- K = the carrying capacity, representing the maximum size of the microbial population that the environment can support.
- β = the disease transmission rate, representing the likelihood of a pathogen to transfer from the microbial population to the host plant.
- P = the population of pathogens present in the environment.
- H = a factor representing the susceptibility or health status of the host plant.

This simplified equation assumes a logistic growth model for the microbial population, dampening the growth rate as the population approaches the carrying capacity. The term $\beta M(P/H)$ represents the transmission of pathogens from the microbial population to the host plant, taking into account the microbial population density and the host plant susceptibility to disease.

It is important to note that this equation does not consider specific microbial species, environmental factors, or the influence of management strategies. As previously indicated, a more complex model should incorporate additional parameters and equations for specific interactions, environmental conditions, and management interventions. Furthermore, calibration and validation with experimental data would be necessary to refine and validate the model accuracy.

Modelling plant microbiome and post-harvest disease management requires a multidisciplinary approach. The model should be tailored to specific crops, pathogens, and management practices to accurately capture the system dynamics, to guide or evaluate effective disease management strategies.

17.10 Microbiome-Based Approaches to Prevent Post-harvest Food Losses

Microbiome-based biotechnology offers promising solutions for reducing food loss in post-harvest. The plant microbiome, consisting of diverse microbial communities associated with plants, plays a crucial role in plant health and disease resistance (Asiri et al., 2022; Bellini et al., 2023; Liu et al., 2021; Mącik et al., 2023; Mishra et al., 2023; Rizvi et al., 2022). Microbiome-based biotechnologies aim at enhancing crop quality, extending produce shelf life, to reduce post-harvest losses. Critical aspects of microbiome-based biotechnologies for reducing food losses in post-harvest include:

1. *Biological Control of Pathogens.* A number of microorganisms within the plant microbiome often show antagonistic properties and may suppress the growth of post-harvest pathogens. By harnessing these beneficial microbes, biotechnology offers a natural and sustainable approach to controlling post-harvest diseases.

These microbes can produce antimicrobial compounds, compete with pathogens for resources, or induce systemic resistance in plants, thus reducing disease incidence and severity.

2. *Spoilage Prevention.* Microbiome-based biotechnology can help to prevent spoilage and extend the shelf life of harvested produce. Beneficial microbes can inhibit the growth of spoilage-causing microorganisms by competing for nutrients or producing antimicrobial substances. Additionally, they can modulate the production of enzymes involved in the breakdown of fruits and vegetables, slowing down the ageing and senescence processes and preserving the quality of the produce, for more extended periods.
3. *Biocontrol Agents and Biofertilizers.* Specific microbial taxa from the plant microbiome can be isolated, characterized, and formulated into biocontrol agents or biofertilizers. These products can be applied directly to harvested crops to enhance their quality, nutritional value, and resistance to post-harvest diseases. Biocontrol agents can colonize the crop surfaces, out-compete pathogens, and secrete antimicrobial compounds, while biofertilizers can enhance nutrient availability and uptake, promoting healthier and more resilient crops.
4. *Microbial Consortia and Synthetic Communities.* Microbiome-based biotechnology concepts include the use of microbial consortia or synthetic communities to optimize desired functions. Combining beneficial microbes with complementary traits, such as pathogen suppression, nutrient cycling, or biofilm formation, improves overall effectiveness and stability of microbiome-based interventions. The consortia may be tailored to specific crops, post-harvest conditions, and disease challenges to achieve the desired outcomes.
5. *Precision Microbiome Engineering* Advances in genetic engineering and synthetic biology techniques enable the targeted modification of the plant microbiome for desired traits. Engineering or modulating the microbiome to enhance post-harvest characteristics require the identification of vital microbial players and their functional genes. Main goal is introducing or enhancing beneficial traits, optimizing microbial interactions, and fine-tuning the microbiome composition to maximize the desired outcomes.

Microbiome-based biotechnology holds immense potential for reducing post-harvest food losses by harnessing the power of beneficial microbes associated with plants (Asiri et al., 2022; Bellini et al., 2023; Cazzaniga et al., 2023b; Chaisiri et al., 2023; Chen et al., 2021; Kumar et al., 2022a). By leveraging the natural functions of the plant microbiome, it offers sustainable and eco-friendly alternatives to conventional post-harvest treatments. However, further research, development, and field trials are needed to optimize microbiome-based interventions, tailoring them to specific crops and environments, and ensuring their safety and efficacy. With continued advancements in this field, microbiome-based biotechnology has the potential to revolutionize post-harvest management and contribute to global food security by reducing food losses and waste (Table 17.2).

Table 17.2 Microbiome-based applications

Application	Description
Biological control	Utilizing beneficial microbes to suppress post-harvest pathogens by producing antimicrobial compounds or competing for resources
Spoilage prevention	Using beneficial microbes to inhibit the growth of spoilage-causing Microorganisms, prolonging the shelf life of harvested produce
Biocontrol agents	Formulating beneficial microbes into biocontrol agents for direct application on harvested produce to enhance quality and disease resistance
Biofertilizers	Formulating beneficial microbes into biofertilizers to improve nutrient availability and uptake, promoting healthier and more resilient crops
Microbial consortia and synthetic communities	Developing mixed populations of beneficial microbes with complimentary traits to optimize post-harvest outcomes and stability
Precision microbiome	Using genetic engineering and synthetic biology techniques to modify and optimize the plant microbiome for desired post-harvest traits
Engineering microbial formulations	Developing formulations and delivery methods for practical application of beneficial microbes on harvested crops
Quality enhancement	Harnessing the plant microbiome to improve the quality attributes Of harvested produce, such as flavour, aroma, colour, and texture
Disease resistance	Enhancing natural defence mechanisms through the manipulation of the produce microbiome to increase resistance to diseases
Nutrient management	Optimizing nutrient cycling and availability through the the activity of beneficial microbes, improving the post-harvest nutrient content
Shelf life extension	Using microbiome-based interventions to delay ageing and senescence processes, extending the shelf life of harvested crops
Genetic manipulation	Applying genetic engineering techniques to modify the genetics of beneficial microbes for improved post-harvest performance
Environmental factors	Studying and manipulating environmental factors such as temperature, humidity, and atmosphere to optimize microbiome-based interventions
Integrated approaches	Combining microbiome-based interventions with other post-harvest management strategies, such as temperature control or packaging

17.11 A Network Model for Post-harvest Microbiome Biotechnology

A network model may describe interconnected components representing different aspects of microbiome-based biotechnology and their influence on reducing food losses in post-harvest (Chung et al., 2021; Ramakrishnan et al., 2021; Villa-Rodriguez et al., 2022). Basic components are:

1. Microbial communities associated with plants, comprising bacteria, fungi, and other microorganisms. They are involved in plant health, disease resistance, and post-harvest quality.
2. Beneficial microorganisms that exhibit beneficial traits for reducing post-harvest diseases and spoilage within the microbial communities. They include biocontrol agents, growth-promoting bacteria, or fungi with antimicrobial properties.
3. Disease Suppression: the network model should include the interactions between beneficial microbes and post-harvest pathogens, highlighting how beneficial species can inhibit the growth of pathogens, by competing for resources or producing antimicrobial compounds, thus reducing disease incidence and severity.
4. Spoilage Prevention: The model should account for the role of beneficial microbes in preventing spoilage of harvested produce, showcasing how they inhibit the growth of spoilage-causing microorganisms, modulate enzyme activity, and extend the produce shelf life.
5. Microbial Consortia, wherein multiple beneficial species with complementary functions work together to enhance post-harvest outcomes. The interactions and cooperation between microbial species should be represented, highlighting their combined impact on reducing food losses.
6. Plant-Microbe Interactions: the model should emphasize the dynamic interactions between the plant and its associated microbial communities, showcasing how the plant microbiome influences plant health, nutrient uptake, and defence mechanisms, ultimately contributing to post-harvest diseases and spoilage reduction.
7. Environmental Factors such as temperature, humidity, and atmosphere should be integrated as influential nodes. These factors can shape the dynamics of microbial communities, their impact on post-harvest quality. Their interactions should be represented in the model.
8. Disease Management Strategies connecting the microbiome-based biotechnology components with other disease management strategies such as temperature control, packaging techniques, or fungicide treatments. This approach should illustrate how these strategies can complement and enhance the effectiveness of microbiome-based interventions.

A network modelling approach should represent the interconnectedness and complexity of microbiome-based biotechnology to reduce food losses in post-harvest. It appears helpful to highlight and visualize the complex relationships and dependencies linking microbial communities, beneficial microbes, plant-microbe interactions, environmental factors, and disease management strategies. Such an approach will result informative about the intricate dynamics of potential microbiome interventions, required to improve post-harvest outcomes and minimize food losses.

17.12 Future Prospects

The prospects for plant microbiome and post-harvest disease management are promising, with ongoing advancements and opportunities for innovation. Here are some critical prospects in this field:

1. *Enhanced understanding of microbial interactions.* Further research will deepen our understanding of the intricate interactions within the plant microbiome, to gain insights into microbial community dynamics, beneficial microbe-host interactions, and mechanisms of disease suppression. This understanding will enable us to design more targeted and effective microbiome-based interventions.
2. *Microbiome engineering and synthetic communities.* Advances in genetic engineering and synthetic biology will facilitate the design and engineering of microbial communities with desired traits. By manipulating the composition and functionality of the plant microbiome, we can enhance disease resistance, nutrient uptake, and post-harvest quality. Synthetic communities of microorganisms with optimized interactions and functions will provide new tools for managing post-harvest diseases.
3. *Precision microbiome management.* Technologies such as high-throughput sequencing and metabolomics will enable more precise characterization of the plant microbiome. This information can be used to develop personalized management strategies tailored to specific crops, environments, and disease challenges. Precision microbiome management will enhance the efficacy and efficiency of post-harvest disease control.
4. *Integration of multi-omics data.* Integrating multi-omics data, including genomics, transcriptomics, and metabolomics, will provide comprehensive insights into the plant-microbe interactions. This integrative approach will uncover key pathways, genes, and metabolites involved in disease management, allowing for targeted interventions and the development of biomarkers for disease prediction and monitoring.
5. *Application of machine learning and artificial intelligence.* Machine learning and artificial intelligence techniques will enable the analysis of large-scale microbiome data and the identification of patterns and predictive models. These tools will enhance our ability to predict disease outbreaks, assess disease risk, and optimize management strategies. Machine learning algorithms can also assist in selecting and designing effective biocontrol agents and formulations.
6. *Commercialization of microbiome-based products.* As research progresses, the commercialization of microbiome-based products for post-harvest disease management will expand. This includes developing biocontrol agents, probiotics, and biostimulants designed for post-harvest applications. Improved formulations, application methods, and delivery systems will increase the availability and accessibility of these products to growers and food producers.
7. *Integration of microbiome-based approaches in IPM.* Microbiome-based interventions will be integrated into holistic and sustainable IPM strategies. Combining biological control agents, cultural practices, and targeted chemical

interventions will maximize disease management while minimizing environmental impact. The integration of multiple approaches will yield synergistic effects and promote long-term sustainability.

8. *Regulatory framework and standardization.* Establishing regulatory frameworks and standards will ensure the safety, efficacy, and quality control of microbiome-based products. Robust guidelines will facilitate the development and commercialization of these products, fostering confidence among stakeholders and promoting their adoption in post-harvest disease management practices.

In summary, the future of plant microbiome and post-harvest disease management holds immense potential. Advances in microbiome research, genetic engineering, data analysis, and IPM will drive innovation and develop sustainable, effective, and personalized approaches for reducing post-harvest losses and ensuring food security. Continued collaboration among scientists, industry, and policymakers will be crucial in translating research findings into practical solutions, realizing the full potential of microbiome-based interventions in agriculture.

17.13 Conclusions

The plant microbiome plays a significant role in post-harvest disease management. Harnessing the power of beneficial microbes associated with plants offers promising avenues for reducing post-harvest losses and ensuring food security. Through microbiome-based interventions, we can enhance crop quality, extend shelf life, and minimize the incidence and severity of post-harvest diseases. The interactions within the plant microbiome are complex and diverse, involving a dynamic network of microorganisms. Some beneficial microbes have a potential to suppress pathogens, prevent spoilage, and promote plant health through competition, antimicrobial compound production, and systemic resistance induction. By leveraging these interactions, we can develop sustainable and eco-friendly approaches to disease management, moving away from conventional chemical treatments.

Microbiome-based biotechnology provides a range of strategies for post-harvest disease management. This includes developing biocontrol agents and biofertilizers, manipulation of microbial consortia, precision microbiome engineering, production and use of microbial formulations. These approaches offer targeted and tailored solutions to specific crops, environments, and disease challenges. Furthermore, advancements in genetic engineering, synthetic biology, and data analytics enable us to delve deeper into the complexities of the plant microbiome and develop more sophisticated interventions. By understanding the functional genomics and metabolic pathways of beneficial microbes, we can optimize their performance and enhance their potential to reduce post-harvest diseases. While the field of plant microbiome and post-harvest disease management is rapidly evolving, there are still challenges to address. Standardizing microbiome-based interventions, optimizing formulations and application methods, and scaling up the technology for commercial use remain essential focus areas. Additionally, regulatory frameworks and safety assessments must be established to ensure the efficacy and safety of microbiome-based products.

In conclusion, the plant microbiome presents a promising avenue for sustainable post-harvest disease management. By harnessing the interactions and functions of beneficial microbes, we can reduce food losses, improve food quality, and contribute to global food security. Continued research, innovation, and collaboration among scientists, industry stakeholders, and policymakers will drive the advancement and application of microbiome-based approaches in post-harvest disease management.

Acknowledgements We express our sincere gratitude to the Department of Agronomy for their support and assistance provided throughout the writing. The Department's commitment to academic excellence and research has been instrumental in completing this endeavour.

Author's Contribution The authors of this work have made significant contributions to the research project/study. Each author has participated sufficiently in the research project/study, made intellectual contributions, and is responsible for the work accuracy and integrity. The authors have collaborated closely, ensuring the completion of this work through collective effort, expertise, and dedication.

References

Alam, M. J., Puppala, V., Uppulapu, S. K., Das, B., & Banerjee, S. K. (2022). Chapter nine – Human microbiome and cardiovascular diseases. In B. Das & V. Singh (Eds.), *Human microbiome in health and disease – Part B* (pp. 231–279). Academic. https://doi.org/10.1016/bs.pmbts.2022.07.012

Alberoni, D., Favaro, R., Baffoni, L., Angeli, S., & Di Gioia, D. (2021). Neonicotinoids in the agroecosystem: In-field long-term assessment on honeybee colony strength and microbiome. *Science of the Total Environment, 762*, 144116.

Ali, A., Ghani, M. I., Ding, H., Iqbal, M., Cheng, Z., & Cai, Z. (2021). Arbuscular mycorrhizal inoculum coupled with organic substrate induces synergistic effects for soil quality changes, and rhizosphere microbiome structure in long-term monocropped cucumber planted soil. *Rhizosphere, 20*, 100428.

Almeida, J. I., Tenreiro, M. F., Martinez-Santamaria, L., Guerrero-Aspizua, S., Gisbert, J. P., Alves, P. M., et al. (2022). Hallmarks of the human intestinal microbiome on liver maturation and function. *Journal of Hepatology, 76*, 694–725.

Annapure, U. S., & Bagul, A. (2022). Relationship between gut microbiome and diabetes. In G. Goel, T. Requena, & S. Bansal (Eds.), *Developments in microbiology* (pp. 107–125). Academic.

Arshad, M. A., Hassan, F., Rehman, M. S., Huws, S. A., Cheng, Y., & Din, A. U. (2021). Gut microbiome colonization and development in neonatal ruminants: Strategies, prospects, and opportunities. *Annals of Nutrition, 7*, 883–895.

Asiri, F., Kim, J., & Chu, K.-H. (2022). Chitosan-harvested polyhydroxybutyrate-rich Zobellella denitrificans ZD1 as a multifunction feed for aquaculture. *Aquaculture, 561*, 738655. https://doi.org/10.1016/j.aquaculture.2022.738655

Bellini, A., Gilardi, G., Idbella, M., Zotti, M., Pugliese, M., Bonanomi, G., & Gullino, M. L. (2023). *Trichoderma* enriched compost, BCAs and potassium phosphite control Fusarium wilt of lettuce without affecting soil microbiome at genus level. *Applied Soil Ecology, 182*, 104678. https://doi.org/10.1016/j.apsoil.2022.104678

Bhandari, S., Kukreja, S., Kumar, V., Dey, A., & Goutam, U. (2022). Chapter 9 – Meicrobiome stimulants and their applications in crop plants. In J. Samuel, A. Kumar, & J. Singh (Eds.), *Relationship between microbes and the environment for sustainable ecosystem services, Vol. 1* (pp. 177–195). Elsevier.

Bose, T., Vivas, M., Slippers, B., Roux, J., Kemler, M., et al. (2023). Retention of post-harvest residues enhances soil fungal biodiversity in eucalyptus plantations. *Forest Ecology and Management, 532*, 120806. https://doi.org/10.1016/j.foreco.2023.120806

Brandl, M. T., Mammel, M. K., Simko, I., Richter, T. K. S., Gebru, S. T., & Leonard, S. R. (2023). Weather factors, soil microbiome, and bacteria-fungi interactions as drivers of the epiphytic phyllosphere communities of romaine lettuce. *Food Microbiology, 113*, 104260. https://doi.org/10.1016/j.fm.2023.104260

Bui-Xuan, D., Tang, D. Y. Y., Chew, K. W., Nguyen, T. D. P., Le Ho, H., et al. (2022). Green biorefinery: Microalgae-bacteria microbiome on tolerance investigations in plants. *Journal of Biotechnology, 343*, 120–127.

Bulleri, F., Pretti, C., Bertolino, M., Magri, M., Pittaluga, G. B., et al. (2022). Adding functions to marine infrastructure: Pollutant accumulation, physiological and microbiome changes in sponges attached to floating pontoons inside marinas. *Science of the Total Environment, 848*, 157773. https://doi.org/10.1016/j.scitotenv.2022.157773

Callewaert, C., Knödlseder, N., Karoglan, A., Güell, M., & Paetzold, B. (2021). Skin microbiome transplantation and manipulation: Current state of the art. *Computational and Structural Biotechnology Journal, 19*, 624–631. https://doi.org/10.1016/j.csbj.2021.01.001

Castillo-de la Cruz, P., Wanek, A. G., Kumar, P., An, X., Elsegeiny, W., et al. (2019). Intestinal IL-17R signaling constrains Il-18-driven liver inflammation by the regulation of microbiome-derived products. *Cell Reports, 29*(8), 2270–2283.e7. https://doi.org/10.1016/j.celrep.2019.10.042

Cazzaniga, S. G., Braat, L., van den Elsen, S., Lombaers, C., Visser, J., et al. (2023a). Pinpointing the distinctive impacts of ten cover crop species on the resident and active fractions of the soil microbiome. *Applied Soil Ecology, 190*, 105012. https://doi.org/10.1016/j.apsoil.2023.105012

Cazzaniga, S. G., van den Elsen, S., Lombaers, C., Kroonen, M., Visser, J., et al. (2023b). On the legacy of cover crop-specific microbial footprints. *Soil Biology and Biochemistry, 184*, 109080. https://doi.org/10.1016/j.soilbio.2023.109080

Chaisiri, K., Linsuwanon, P., & Makepeace, B. L. (2023). The chigger microbiome: Big questions in a tiny world. *Trends in Parasitology, 39*(8), 696–707. https://doi.org/10.1016/j.pt.2023.05.002

Chatterjee, G., Negi, S., Basu, S., Faintuch, J., O'Donovan, A., & Shukla, P. (2022). Microbiome systems biology advancements for natural Well-being. *Science of the Total Environment, 838*, 155915. https://doi.org/10.1016/j.scitotenv.2022.155915

Chen, O., Sudakaran, S., Blonquist, T., Mah, E., Durkee, S., & Bellamine, A. (2021). Effect of arabinogalactan on the gut microbiome: A randomized, double-blind, placebo-controlled, crossover trial in healthy adults. *Nutrition, 90*, 111273. https://doi.org/10.1016/j.nut.2021.111273

Chilakala, A. R., Pandey, P., Durgadevi, A., Kandpal, M., Patil, B. S., et al. (2023). Drought attenuates plant responses to multiple rhizospheric pathogens: A study on a dry root rot-associated disease complex in chickpea fields. *Field Crops Research, 298*, 108965. https://doi.org/10.1016/j.fcr.2023.108965

Chung, E., Elmassry, M. M., Cao, J. J., Kaur, G., Dufour, J. M., et al. (2021). Beneficial effect of dietary geranylgeraniol on glucose homeostasis and bone microstructure in obese mice is associated with suppression of proinflammation and modification of gut microbiome. *Nutrition Research, 93*, 27–37. https://doi.org/10.1016/j.nutres.2021.07.001

Culp, E. J., & Goodman, A. L. (2023). Cross-feeding in the gut microbiome: Ecology and mechanisms. *Cell Host & Microbe, 31*(4), 485–499. https://doi.org/10.1016/j.chom.2023.03.016

Dong, T. S., & Gupta, A. (2019). Influence of early life, diet, and the environment on the microbiome. *Clinical Gastroenterology and Hepatology, 17*(2), 231–242. https://doi.org/10.1016/j.cgh.2018.08.067

Droby, S., & Wisniewski, M. (2018). The fruit microbiome: A new frontier for postharvest biocontrol and postharvest biology. *Postharvest Biology and Technology, 140*, 107–112. https://doi.org/10.1016/j.postharvbio.2018.03.004

Elolimy, A. A., Liang, Y., Lopes, M. G., & Loor, J. J. (2021). Antioxidant networks and the microbiome as components of efficiency in dairy cattle. *Livestock Science, 251*, 104656. https://doi.org/10.1016/j.livsci.2021.104656

Farag, M. A., von Bergen, M., Saleh, B. M., Homsi, M. N., & Abd El-Al, M. S. (2021). How do green and black coffee brews and bioactive interaction with gut microbiome affect its health outcomes? Mining evidence from mechanistic studies, metagenomics and clinical trials. *Trends in Food Science & Technology, 118*, 920–937. https://doi.org/10.1016/j.tifs.2021.11.004

Feye, K. M., Thompson, D. R., Rothrock, M. J., Kogut, M. H., & Ricke, S. C. (2020). Poultry processing and the application of microbiome mapping. *Poultry Science, 99*(2), 678–688. https://doi.org/10.1016/j.psj.2019.12.019

Fournier, P., Pellan, L., Barroso-Bergadà, D., Bohan, D. A., Candresse, T., et al. (2022). The functional microbiome of grapevine throughout plant evolutionary history and lifetime. In D. A. Bohan & A. Dumbrell (Eds.), *Functional microbiomes* (Vol. 67, pp. 27–99). Academic.

Fruscella, L., Kotzen, B., Paradelo, M., & Milliken, S. (2023). Investigating the effects of fish effluents as organic fertilisers on onion (*Allium cepa*) yield, soil nutrients, and soil microbiome. *Scientia Horticulturae, 321*, 112297. https://doi.org/10.1016/j.scienta.2023.112297

Garello, M., Piombo, E., Prencipe, S., Schiavon, G., Berra, L., Wisniewski, M., Droby, S., & Spadaro, D. (2023). Fruit microbiome: A powerful tool to study the epidemiology of dry lenticel rot and white haze–Emerging postharvest diseases of apple. *Postharvest Biology and Technology, 196*, 112163.

Hanson, M. C., Petch, G. M., Ottosen, T.-B., & Skjøth, C. A. (2022). Climate change impact on fungi in the atmospheric microbiome. *Science of the Total Environment, 830*, 154491. https://doi.org/10.1016/j.scitotenv.2022.154491

Hong, S., Sun, Y., Sun, D., & Wang, C. (2022). Microbiome assembly on *Drosophila* body surfaces benefits the flies to combat fungal infections. *IScience, 25*(6), 104408. https://doi.org/10.1016/j.isci.2022.104408

Islam, S. U., Mangral, Z. A., Hussain, K., Tariq, L., Bhat, B. A., Khuroo, A. A., & Hassan Dar, T. U. (2023). Unravelling diversity, drivers, and indicators of soil microbiome of *Trillium govanianum*, an endangered plant species of the Himalaya. *Environmental Research, 227*, 115819. https://doi.org/10.1016/j.envres.2023.115819

Kumar, D., Saraf, M., Joshi, C. G., & Joshi, M. (2022a). Rhizosphere microbiome analysis of healthy and infected cumin (*Cuminum cyminum* L.) varieties from Gujarat, India. *Current Research in Microbial Sciences, 3*, 100163. https://doi.org/10.1016/j.crmicr.2022.100163

Kumar, V., Umrao, P. D., & Kaistha, S. D. (2022b). Beneficial plant microbiome assisted chromium *phytoremediation*. In K. Bauddh & Y. Ma (Eds.), *Advances in microbe-assisted phytoremediation of polluted sites* (pp. 301–346). Elsevier.

Li, K., DiLegge, M. J., Minas, I. S., Hamm, A., Manter, D., & Vivanco, J. M. (2019). Soil sterilization leads to re-colonization of a healthier rhizosphere microbiome. *Rhizosphere, 12*, 100176. https://doi.org/10.1016/j.rhisph.2019.100176

Liu, H., Brettell, L. E., Qiu, Z., & Singh, B. K. (2020a). Microbiome-mediated stress resistance in plants. *Trends in Plant Science, 25*, 733–743.

Liu, J., Song, Y., Tang, M., Lu, Q., & Zhong, G. (2020b). Enhanced dissipation of xenobiotic agrochemicals harnessing soil microbiome in the tillage-reduced rice-dominated agroecosystem. *Journal of Hazardous Materials, 398*, 122954. https://doi.org/10.1016/j.jhazmat.2020.122954

Liu, X., Hannula, S. E., Li, X., Hundscheid, M. P. J., Klein Gunnewiek, P. J. A., Clocchiatti, A., Ding, W., & de Boer, W. (2021). Decomposing cover crops modify root-associated microbiome composition and disease tolerance of cash crop seedlings. *Soil Biology and Biochemistry, 160*, 108343. https://doi.org/10.1016/j.soilbio.2021.108343

Lupwayi, N. Z., Blackshaw, R. E., Geddes, C. M., Dunn, R., & Petri, R. M. (2022). Multi-year and multi-site effects of recurrent glyphosate applications on the wheat rhizosphere microbiome. *Environmental Research, 215*, 114363. https://doi.org/10.1016/j.envres.2022.114363

Ma, H., Pineda, A., Hannula, S. E., Kielak, A. M., Setyarini, S. N., & Bezemer, T. M. (2020). Steering root microbiomes of a commercial horticultural crop with plant-soil feedbacks. *Applied Soil Ecology, 150*, 103468. https://doi.org/10.1016/j.apsoil.2019.103468

Mącik, M., Gryta, A., Sas-Paszt, L., & Frąc, M. (2023). New insight into the soil bacterial and fungal microbiome after phosphorus biofertilizer application as an important driver of regenerative agriculture including biodiversity loss reversal and soil health restoration. *Applied Soil Ecology, 189*, 104941. https://doi.org/10.1016/j.apsoil.2023.104941

Mangalam, A. K., & Giri, S. (2022). Role of microbiome and metabolome in the pathobiology of MS. *Clinical Immunology, 235*, 108934. https://doi.org/10.1016/j.clim.2022.108934

Mazzola, M., Muramoto, J., & Shennan, C. (2018). Anaerobic disinfestation induced changes to the soil microbiome, disease incidence and strawberry fruit yields in California field trials. *Applied Soil Ecology, 127*, 74–86. https://doi.org/10.1016/j.apsoil.2018.03.009

Mazzola, M., Graham, D., Wang, L., Leisso, R., & Hewavitharana, S. S. (2020). Application sequence modulates microbiome composition, plant growth and apple replant disease control efficiency upon integration of anaerobic soil disinfestation and mustard seed meal amendment. *Crop Protection, 132*, 105125. https://doi.org/10.1016/j.cropro.2020.105125

Mishra, A., Singh, D., Hathi, Z., Purohit, H. J., Jessy, M. D., et al. (2023). Soil microbiome dynamics associated with conversion of tropical forests to different rubber based land use management systems. *Applied Soil Ecology, 188*, 104933. https://doi.org/10.1016/j.apsoil.2023.104933

Morales-Cedeño, L. R., del Orozco-Mosqueda, M. C., Loeza-Lara, P. D., Parra-Cota, F. I., de los Santos-Villalobos, S., & Santoyo, G. (2021). Plant growth-promoting bacterial endophytes as biocontrol agents of pre- and post-harvest diseases: Fundamentals, methods of application and future perspectives. *Microbiological Research, 242*, 126612. https://doi.org/10.1016/j.micres.2020.126612

Mukhtar, H., Wunderlich, R. F., Muzaffar, A., Ansari, A., Shipin, O. V., Cao, T. N.-D., & Lin, Y.-P. (2023). Soil microbiome feedback to climate change and options for mitigation. *Science of the Total Environment, 882*, 163412. https://doi.org/10.1016/j.scitotenv.2023.163412

Ogundeji, A. O., Meng, L., Cheng, Z., Hou, J., Yin, T., Zhang, S., Liu, X., Liu, X., & Li, S. (2022). Integrated crop practices management stimulates soil microbiome for Verticillium wilt suppression. *European Journal of Agronomy, 140*, 126594. https://doi.org/10.1016/j.eja.2022.126594

Otero-Jiménez, V., del Carreño-Carreño, J. P., Barreto-Hernandez, E., van Elsas, J. D., & Uribe-Vélez, D. (2021). Impact of rice straw management strategies on rice rhizosphere microbiomes. *Applied Soil Ecology, 167*, 104036. https://doi.org/10.1016/j.apsoil.2021.104036

Prestt, A. J., & Roberts, M. R. (2023). Specific suppression of vascular wilt disease of onion is linked with fungal soil community structure. *Rhizosphere, 25*, 100671. https://doi.org/10.1016/j.rhisph.2023.100671

Przemieniecki, S. W., Damszel, M., Ciesielski, S., Kubiak, K., Mastalerz, J., Sierota, Z., & Gorczyca, A. (2021). Bacterial microbiome in *Armillaria ostoyae* rhizomorphs inhabiting the root zone during progressively dying Scots pine. *Applied Soil Ecology, 164*, 103929. https://doi.org/10.1016/j.apsoil.2021.103929

Ramakrishnan, B., Maddela, N. R., Venkateswarlu, K., & Megharaj, M. (2021). Linkages between plant rhizosphere and animal gut environments: Interaction effects of pesticides with their microbiomes. *Environmental Advances, 5*, 100091. https://doi.org/10.1016/j.envadv.2021.100091

Reveles, K. R., Young, E. H., Zeidan, A. R. A. L., & Dong, Q. (2021). Chapter 17- Microbiome changes in aging. In N. Musi & P. Hornsby (Eds.), *Handbooks of aging* (pp. 367–389). Academic. https://doi.org/10.1016/B978-0-12-815962-0.00017-2

Rizvi, A., Ahmed, B., Saghir Khan, M., Rajput, V. D., Umar, S., Minkina, T., & Lee, J. (2022). Maize associated bacterial microbiome linked mitigation of heavy metal stress:

A multidimensional detoxification approach. *Environmental and Experimental Botany, 200*, 104911. https://doi.org/10.1016/j.envexpbot.2022.104911

Sangiorgio, D., Cellini, A., Spinelli, F., Pastore, C., Farneti, B., et al. (2022). Contribution of fruit microbiome to raspberry volatile organic compounds emission. *Postharvest Biology and Technology, 183*, 111742. https://doi.org/10.1016/j.postharvbio.2021.111742

Selvan, C., & Mukhopadhyay, S. (2022). Chapter 13 – Harnessing gut friendly microbiomes to combat metabolic syndrome. In D. Bagchi & B. Downs (Eds.), *Immunity, digestive health and nutrition* (pp. 185–191). Academic. https://doi.org/10.1016/B978-0-12-822238-6.00033-9

Shen, Y., Li, X., Xiong, R., Ni, Y., Tian, S., & Li, B. (2023). Effect of peach trichome removal on post-harvest brown rot and on the fruit surface microbiome. *International Journal of Food Microbiology, 402*, 110299. https://doi.org/10.1016/j.ijfoodmicro.2023.110299

Shi, Y., Yang, Q., Zhao, Q., Dhanasekaran, S., Ahima, J., Zhang, X., Zhou, S., Droby, S., & Zhang, H. (2022). *Aureobasidium pullulans* S-2 reduced the disease incidence of tomato by influencing the postharvest microbiome during storage. *Postharvest Biology and Technology, 185*, 111809. https://doi.org/10.1016/j.postharvbio.2021.111809

Si, J., Froussart, E., Viaene, T., Vázquez-Castellanos, J. F., Hamonts, K., et al. (2021). Interactions between soil compositions and the wheat root microbiome under drought stress: From an *in silico* to *in planta* perspective. *Computational and Structural Biotechnology Journal, 19*, 4235–4247. https://doi.org/10.1016/j.csbj.2021.07.027

Smith, M. (2020). Microorganisms and microbiome. In M. Smith (Ed.), *Gene environment interactions* (pp. 177–203). Academic.

Taparia, T., Hendrix, E., Hendriks, M., Nijhuis, E., de Boer, W., & van der Wolf, J. (2021). Casing soil microbiome mediates suppression of bacterial blotch of mushrooms during consecutive cultivation cycles. *Soil Biology and Biochemistry, 155*, 108161. https://doi.org/10.1016/j.soilbio.2021.108161

Valentino, V., Sequino, G., Cobo-Díaz, J. F., Álvarez-Ordóñez, A., De Filippis, F., & Ercolini, D. (2022). Evidence of virulence and antibiotic resistance genes from the microbiome mapping in minimally processed vegetables producing facilities. *Food Research International, 162*, 112202. https://doi.org/10.1016/j.foodres.2022.112202

Villa-Rodriguez, E., Lugo-Enríquez, C., Ferguson, S., Parra-Cota, F. I., Cira-Chávez, L. A., & de los Santos-Villalobos, S. (2022). *Trichoderma harzianum sensu lato* TSM39: A wheat microbiome fungus that mitigates spot blotch disease of wheat (*Triticum turgidum* L. subsp. *durum*) caused by *Bipolaris sorokiniana*. *Biological Control, 175*, 105055. https://doi.org/10.1016/j.biocontrol.2022.105055

Wani, Z. A., Mirza, D. N., Arora, P., & Riyaz-Ul-Hassan, S. (2016). Molecular phylogeny, diversity, community structure, and plant growth promoting properties of fungal endophytes associated with the corms of saffron plant: An insight into the microbiome of *Crocus sativus* Linn. *Fungal Biology, 120*(12), 1509–1524. https://doi.org/10.1016/j.funbio.2016.07.011

Wassermann, B., Abdelfattah, A., Cernava, T., Wicaksono, W., & Berg, G. (2022). Microbiome-based biotechnology for reducing food loss post harvest. *Current Opinion in Biotechnology, 78*, 102808. https://doi.org/10.1016/j.copbio.2022.102808

Wei, Z., Hu, J., Gu, Y., Yin, S., Xu, Y., Jousset, A., Shen, Q., & Friman, V. P. (2018). Ralstonia solanacearum pathogen disrupts bacterial rhizosphere microbiome during an invasion. *Soil Biology and Biochemistry, 118*, 8–17. https://doi.org/10.1016/j.soilbio.2017.11.012

Wicaksono, W. A., Buko, A., Kusstatscher, P., Sinkkonen, A., Laitinen, O. H., et al. (2022). Modulation of the food microbiome by apple fruit processing. *Food Microbiology, 108*, 104103. https://doi.org/10.1016/j.fm.2022.104103

Yan, M., Tian, H., Song, S., Tan, H. T. W., Lee, J. T. E., et al. (2023). Effects of digestate-encapsulated biochar on plant growth, soil microbiome and nitrogen leaching. *Journal of Environmental Management, 334*, 117481. https://doi.org/10.1016/j.jenvman.2023.117481

Yang, Q., Lanhuang, B., Zhu, M., Solairaj, D., Ngea, G. L. N., & Zhang, H. (2023). Biocontrol efficacy of *Wickerhamomyces anomalus* on tomato field diseases and study of the relevant mechanisms via microbiome analysis. *Biological Control, 184*, 105285. https://doi.org/10.1016/j.biocontrol.2023.105285

Yuan, J., Chaparro, J. M., Manter, D. K., Zhang, R., Vivanco, J. M., & Shen, Q. (2015). Roots from distinct plant developmental stages are capable of rapidly selecting their own microbiome without the influence of environmental and soil edaphic factors. *Soil Biology and Biochemistry, 89*, 206–209. https://doi.org/10.1016/j.soilbio.2015.07.009

Zeng, H., Xu, H., Liu, G., Wei, Y., Zhang, J., & Shi, H. (2021). Physiological and metagenomic strategies uncover the rhizosphere bacterial microbiome succession underlying three common environmental stresses in cassava. *Journal of Hazardous Materials, 411*, 125143. https://doi.org/10.1016/j.jhazmat.2021.125143

Zeng, H., Hu, W., Liu, G., Xu, H., Wei, Y., et al. (2022). Microbiome-wide association studies between phyllosphere microbiota and ionome highlight the beneficial symbiosis of *Lactococcus lactis* in alleviating aluminium in cassava. *Plant Physiology and Biochemistry, 171*, 66–74. https://doi.org/10.1016/j.plaphy.2021.12.029

Zhimo, V. Y., Kumar, A., Biasi, A., Salim, S., Feygenberg, O., et al. (2021). Compositional shifts in the strawberry fruit microbiome in response to near-harvest application of *Metschnikowia fructicola*, a yeast biocontrol agent. *Postharvest Biology and Technology, 175*, 111469. https://doi.org/10.1016/j.postharvbio.2021.111469

Chapter 18
Exploring the Role of Plant Microbiome in Post-harvest Tomato Disease Management (*Lycopersicon esculentum* Mill)

Pratima Vishwakarma, Ragini Vishwakarma, and Smriti Mall

Abstract In tomato fruit, post-harvest diseases are the chief problem affecting fruit quality, quantity, and shelf life. They result in noteworthy economic losses in the agricultural industry. To tackle this problem, this paper reviews the emergent field of plant microbiome research and explores how it can help to lessen tomato post-harvest diseases. Plant microbiome denotes the microorganism community living in and on the plant, such as bacteria, fungi, and viruses. It refers to the important role played by microorganisms in sustaining the plants health and defense mechanisms. We also review the complex relationship between an important microbiome and the development of significant strategies for tomato post-harvest disease management. Recent advancements in this field concern the use of important beneficial microorganisms, such as bacteria and fungi, as biocontrol agents of post-harvest diseases. These microorganisms suppress growth, development and spread of plant pathogens, offering promising substitutes of chemical pesticides. They show no negative impact on the environment or human health. This chapter highlights the possibility of harnessing the plant microbiome by means of eco-friendly and sustainable disease management strategies. Farmers can use beneficial microbes as biocontrol agents to reduce chemical pesticides and promote healthy and disease-resistant plants. This tactic can improve tomato quality, reduce post-harvest losses in tomato, and positively impact agriculture and the environment.

Keywords Antagonistic bacteria · Microbiome · Post-harvest pathogens · Rhizosphere · Siderophore

---

P. Vishwakarma
Department of Botany, St. Joseph's College for Women, Gorakhpur, UP, India

R. Vishwakarma · S. Mall (✉)
Molecular Plant Pathology Laboratory, Department of Botany, DDU Gorakhpur University, Gorakhpur, UP, India
e-mail:smriti.mall@rediffmail.com

P. Mathur, S. Roy (eds.), *Plant Microbiome and Biological Control*, Sustainability in Plant and Crop Protection 20,
https://doi.org/10.1007/978-3-031-75845-4_18

Abbreviations

ISR Induced systemic resistance
PGPR Plant growth promoting rhizobacteria
SAR Systemic acquired resistance

18.1 Introduction

Pathogenic bacteria and fungi infect fresh fruits and vegetables during various phases of crop growth, viz. harvesting, post-harvest, storage, and consumption, leading to severe economic losses due to the increased cost of commodities. Post-harvest losses significantly affect the productivity of fruit and vegetable crops. They may even lead to one-third loss of produce, every year. Tomato is the world most important vegetable crop, but it faces over 200 insect and pest challenges (including bacterial, fungal, and viral infections), ultimately affecting the nutritive significance and value of this crop (Sharma et al., 2023). The current tomato production of is around 100 million tons per year, from about 3.7 million ha, worldwide (FAOSTAT, 2001). Its production was about 180.8 million tons in 2019, with a value of 93.9 billion USD in 2018 (FAOSTAT, 2020).

During the last 20 years (2011–2020), numerous new or re-emerging tomato diseases (mainly caused by viruses) have been reported globally. From the various reports, it has been clear that about 312 viruses, viroid, and satellite viruses belonging to 22 families and 39 genera cause numerous diseases in tomato. This is undoubtedly one of the highest number of disease-causing agents ever discovered for any plant (Rivarez et al., 2021).

In recent years, the agricultural crop output has expanded very much, but the gloomiest part is that till now, the development and implementation of post-harvest technologies lag significantly, which leads to excess losses. Fruit crops experience post-harvest losses due to mechanical, microbiological, and physiological factors ranging between 16–36%, per year. Horticultural products are examples of a perishable commodity which needs great attention and strategies to ensure proper processing and value addition. In order to enhance national food security and uphold food safety, it is essential to take care of proper processing to prevent post-harvest diseases in fruits and vegetables.

Many pre-harvest elements, viz. harvesting, market preparation (sorting, pre-cooling, grading, on-farm storage and packaging), storage, transportation, value addition and by-product management, may lead to improvement in post-harvest management of horticultural commodities. However, tomato crops are prone to numerous post-harvest diseases, viz. Rhizopus rot (*Rhizopus stolonifer*), gray mould (*Botrytis cinerea*), anthracnose (*Colletotrichum coccoides, C. gloeosporoides, C. dematium*), early blight (*Alternaria solani*), phoma rot (*Phoma destructiva*), bacterial soft rot (*Erwinia carotovora* subsp. *carotovora*), southern blight (*Sclerotium rolfsii*), and *Fusarium* rot (Devappa et al., 2021).

18.2 Plant Microbiome Role in Plant Health

Growth and output of fruit and vegetables benefit from the presence of bacteria in the roots. In this respect, PGPR (plant growth promoting rhizobacteria) are bacterial species that influence the development of plants and, indirectly, also benefit human health. PGPR reside in the soil near the roots and stimulate plant growth and fruitfulness. They act as natural resources for agriculture, boosting plant resistance towards numerous phytopathogens occurring in a broad range of stages such as fruits, vegetables and diverse forest trees (Zia et al., 2021).

Different rhizosphere bacteria are used as biological indicators to evaluate the soil quality and fertility. They are considered as biofertilizers, maintaining the soil composition and helping to sustain an ecological balance. These bacteria produce phytohormones that help endorse sustainable agriculture. They also perform many critical services, viz., atmospheric nitrogen fixation, proper distribution of essential nutrients, improving the organic carbon content, increasing water retention, regulating local soil pH, also regulating rhizosphere soil porosity.

Soil rhizosphere is a unique biological niche that favours the growth of many important beneficial microorganisms associated with roots. The bacteria present in rhizosphere soil mostly belong to diverse groups, viz., *Alphaproteobacteria*, *Bacteroidetes* (e.g., *Burkholderia*, *Pseudomonas*, *Rhizobia*), *Firmicutes* (e.g., *Bacillus*), *Betaproteobacteria* and *Gammaproteobacteria*. They are widely used in biotechnological applications, which help improve soil fertility, upsurge crop production, being environment-friendly (Santoyo et al., 2021.

18.3 Pathogens Responsible for Tomato Post-harvest Diseases

Tomato diseases are mainly caused by bacteria and fungi, with other pathogens such as viruses contributing to post-harvest losses. Some of the common diseases caused by these microbes are briefly reviewed in the following paragraphs.

18.3.1 Bacterial Diseases

18.3.1.1 Bacterial Spot

This disease is caused by various bacterial species which belong to the genus *Xanthomonas*. The most commonly involved species is *X. euvesicatoria*, which was formerly known as *X. axonopodis* pv. *vesicatoria* or *X. campestris* pv. *vesicatoria* (Hernández-Huerta et al., 2021). *Xanthomonas* is gram-negative, motile, and strictly aerobic. Symptoms of the disease usually include the appearance of circular to irregular spots on the stems and leaves, which feel slightly greasy. The spots on the leaves show a yellow halo as they grow and merge, which causes the leaves to wither and turn brown. Symptoms on fruits are more conspicuous than those of leaves or stems. The main symptoms on fruits are black spots, with raised pimple-like dots surrounded by water-soaked areas. At maturity, the spots expand and are grey-brown in colour with scabby, sunken, and pitted centres (Biswas et al., 2013).

18.3.1.2 Bacterial Canker

The causal agent of Bacterial Canker is *Clavibacter michiganensis* sub sp. *michiganensis*, a gram-positive bacterium. This is a highly destructive plant disease known to affect also pepper and tomato crops. This disease usually expresses itself in young transplants, leading to complete leaf defoliation. In older leaves, a browning at the edges occurs (Nandi et al., 2018). The disease development occurs inwards towards the midrib foliage, resulting in dark-coloured necrotic lesions on the fruits, with white halos measuring from 16 to 32 mm. The lesions appear on the fruit surface as a slightly raised white spot, which soon turns brown and dark-coloured after maturity. The disease leads to significant losses in affected crops and becomes difficult to control once onset (Wang et al., 2022).

18.3.2 Fungal Diseases

Fungi, also known as moulds, cause various diseases in crop plants, including significant agents of tomato fruit deterioration. Eradication of fungi is tougher than for bacteria because the fungal cells are larger, and the spores they produce are highly resistant to drying and various environmental stresses (Etebu et al., 2013). The chief post-harvest diseases caused by fungi are *Rhizopus* rot, Sour rot, Black mold rot, and Buckeye rot (Paul et al., 2021).

18.3.2.1 Sour Rot

Geotrichum candidum, a type of yeast, causes this disease. Its main symptoms are the formation of lesions, which gradually develop thicker, gelatinous masses and resemble cottage cheese. Firstly, the lesions are watery, harden, and become highly coated with fungal growth. The disease is so named because of the characteristic odour of sour rot produced by lactic acid bacteria (Bartz et al., 2015).

18.3.2.2 Rhizopus Rot

This disease is caused by the fungus *Rhizopus stolonifer.* Its preliminary symptoms are watery lesions progressively producing a clear liquid with time. A cotton-like structure is present over the affected area. The main route of entry of fungi in plants is through natural openings or wounds due to mechanical injury (Chuku et al., 2008).

18.3.2.3 Buckeye Rot

Buckeye rot is a post-harvest disease of tomato caused by the pathogen *Phytophthora parasitica*. This pathogen affects both unripe and ripe fruits. The initial symptoms of this disease are the presence of circular water-soaked spots. As the disease progresses, dark spots are covered with the fungus thin white mycelia (Ma et al., 2023).

18.3.2.4 Black Mold

Several fungal pathogens, including *Alternaria arborescens*, *Stemphyllinm botryosum* and *S. consortiale* cause Black mold rot. Rots characterize the main symptoms of this disease observed on the stem or the bud end of the tomato fruit. Initially, the lesions are sunken and quickly covered with a dark brown to black mould. If the vascular strand connected with the scar is infected, it leads to the development of internal lesions.

18.4 Current Management Practices and Limitations

Extensive chemical pesticides are used for the tomato disease management, negatively affecting soil health and reducing nutritional quality of fruits. Continuous use of chemicals leads to disease resistance, which in turn leads to uncontrolled disease spread and pathogens growth. Proper handling practices and treatment processes such as sorting, pre-cooling, packaging, cleaning, and storage are important measures that can be used to manage post-harvest diseases of tomato. These methods help to maintain the quality of the tomato fruits and also increase their shelf-life. However, post-harvest disease management also has some limitations, especially in developing nations where advanced methods of post-harvest disease management are not accessible. If ideal practices are not followed, this situation may lead to a high rate of post-harvest produce loss (Singh et al., 2017; Arah et al., 2016). Many current practices rely on the continuous use of chemical pesticides, which can negatively impact the nutritional content of tomatoes and soil productivity. Moreover, chemical pesticides can pose health risks to humans and hurt the environment (Arah et al., 2016).

The use of chemical derivatives in agriculture, *viz.*, fertilizers, fungicides, herbicides, and insecticides, to control diseases may impart several health concerns. Many chemicals have the potential to accumulate in human tissues, plants, and animals, with various noxious effects. Chemicals that preserve or protect fruits and vegetables from infections accumulate over them and cannot be removed easily after washing (Gallo et al., 2020). Prolonged exposure to these compounds leads to various health problems in humans. Their residues can accumulate in humans through various modes, viz., inhalation, skin diffusion, oral ingestion, etc., leading to different allergies and hypersensitivity reactions in many individuals.. So, it becomes necessary to minimize the use of chemical pesticides and opt for biological products that help minimize the harm to humans and the environment.

Microorganisms causing plant disease negatively impact crop yield by drastically lowering crop performance and quality (Dilantha Fernando et al., 2005). Traditionally, these phytopathogens are controlled by pesticides that, apart of the cited health and environmental issues, may become ineffective due to development of resistance in pathogens, *vs* specific chemicals. The development of natural pesticides appears hence as a sustainbale way to overcome these problems. In this respect, rhizobacteria represent an effective substitute of chemicals, due to their potential in the biocontrol of many diseases or insect pests.

Many studies are focusing worldwide on efforts to develop chemical-free, healthy crops. Microbes are one of the best options to reduce harsh chemicals from agriculture, as they are safer and effective. These microbes act as biopesticides, biostimulants or biofertilizers and play an important role in maintaining or even increasing soil fertility and health. They promote crop yields and also act as a barrier to prevent the entry of pathogens in plants. Most of these microbes are present in the rhizosphere of healthy plants and help maintaining their ecologic and natural balance.

18.5 Beneficial Biocontrol Agents *vs* Tomato Post-harvest Diseases

The rhizosphere is one of the best areas to identify and select different types of microbes and plant interactions. This space is the best site containing various root exudates, including various chemical signals and numerous metabolites, including carbon compounds, *viz.*, organic acids, sugars, and various amino acids, present in the soil matrix (Oku et al., 2012). Although most rhizospheric bacteria have few effects on plant health, a minority of taxa i.e., PGPR and mycorrhizal fungi, appears associated to highly beneficial effects, observable on the plant growth and health (Glick, 2012) (Table 18.1). This beneficial effect is achieved by facilitating high nutrients uptake, stimulating plant growth, increasing yield, and mitigating the effect of abiotic and biotic stress factors. The intrinsic balance in nature can be best explained by the symbiotic relationship between these microorganism and their host

Table 18.1 List of tomato diseases, causal agents, antagonistic organisms and effects on plants

Disease	Causal organism	Antagonist	Effect	Reference
Early blight	*Alternaria alternata*	*Trichoderma asperellum*	Plant growth-promoting traits viz., IAA production, phosphate solubilization	Sehim et al. (2023)
Early blight	*Alternaria solani*	*Bacillus* and *Pantoea*	AHL and IAA production activities, alpha-tomatine resistance	
Early blight	*Alternaria tenuissima*	*Enterobacter cowanii*	Suppression of T34 siderophore synthesis	Shi and Sun (2017)
Bacterial speck	*Botrytis cinerea*	*Exiguobacterium aurantiacum*, *Exiguobacterium* spp., *Staphylococcus xylosus*, *Pantoea eucalypti*, *Bacillus methylotrophicus*, *Pseudomonas veronii*, *P. rhodesiae* and *P. cichorii*	Production of siderophores, auxin indole-3-acetic and solubilization of inorganic phosphate	Romero et al. (2016)
Gray mold	*Botrytis cinerea*	*Trichoderma asperellum*	Triggers pathways of system acquired resistance (SAR) and induced systemic resistance (ISR), thereby enhancing disease resistance	Wang et al. (2021)
Gray mold	*Botrytis cinerea*	*Trichoderma harzianum*	Induced changes related to jasmonic acid	Harel et al. (2014)
Gray mold	*Botrytis cinerea*	*Enterobacter cowanii*	Suppression of T34 siderophore synthesis	Shi and Sun (2017)
Gray mold	*Botrytis cinerea*	*Crytococcus laurentii*	–	
Gray mold	*Botrytis cinerea*	*Bacillus velezebsis*, *B. subtilis* subsp. *subtilis* and *B. amyloliquefaciens*	Decrease fruits water loss rate in post-harvest and form biofilm on their surface, effectively reducing the disease occurrence	Cheng et al. (2023)
Gray mold	*Botrytis cinerea*	*Bacillus* and *Pantoea*	AHL and IAA production activities, alpha-tomatine resistance	
Gray mold	*Botrytis cinerea*	*T. harzianum*, *Aureobasidium pullulans*, *Cryptococcus albidus*	Reduced stem lesions and plants death by 40–100%	Dik and Elad (1999)

(continued)

Table 18.1 (continued)

Disease	Causal organism	Antagonist	Effect	Reference
Gray mold	*Botrytis cinerea*	*Rhodotorula glutinis*	Reduced infected wounds by 50%	Kalogiannis et al. (2006)
Gray mold	*Botrytis cinerea*	*Bacillus mojavensis*	Enhanced the activities of peroxidase (POD), catalase (CAT), polyphenol oxidase (PPO), superoxide dismutase (SOD), and phenyl alanine lyase (PAL) and increased the content of malondialdehyde (MDA). Enhanced the transcript levels of systemic acquired resistance-related marker genes (*SlLoxD*, *SlMyc2*, *SlPR1-a*, *SlPAL5*, *SlEIN2*, and *SlEIN3*)	Zheng et al. (2023)
Gray mold	*Botrytis cinerea*	*Bacillus subtilis*	–	Bu et al. (2021)
Leaf mold	*Fulvia fulva*	*Bacillus* and *Pantoea*	AHL and IAA production activities, alpha-tomatine resistance	
Fusarium wilt	*Fusarium gramineraru*m	*T. asperellum*	Plant growth-promoting traits viz., IAA production, phosphate solubilization	Sehim et al. (2023)
Fusarium wilt	*Fusarium oxysporum* f. sp. *lycopersici*	*T. asperellum*	Increased growth of roots and shoots of tomato plants and total lycopene, sugar, K, N, Ca, P and Mg contents, after 120 days	Hasan et al. (2020)
Fusarium wilt	*F. oxysporum* f. sp. *lycopersici*	*T.asperellum*	Suppression of T34 siderophore synthesis	Segarra et al. (2010)
Fusarium wilt	*F.oxysporum, F. solani*	*T. asperellum*	Plant growth-promoting traits viz., IAA production, phosphate solubilization	Sehim et al. (2023)
Fusarium wilt	*Fusarium verticillioides*	*Enterobacter cowanii*	Suppression of T34 siderophore synthesis	Shi and Sun (2017)

(continued)

Table 18.1 (continued)

Disease	Causal organism	Antagonist	Effect	Reference
Bacterial speck	*Pseudomonas syringae*	*Exiguobacterium aurantiacum*, *Exiguobacterium* spp., *Staphylococcus xylosus*, *Pantoea eucalypti*, *B. methylotrophicus*, *Pseudomonas veronii*, *P.rhodesiae*, *P. cichorii*	Produced siderophores, the auxin indole-3-acetic and solubilize inorganic phosphate	Romero et al. (2016)
Damping off	*Pythium aphanidermatum*	*Crytococcus laurentii*	–	
Tomato bacterial wilt	*Ralstonia solanacearum*	*B. amyloliquefacien* CM-2 and T-5	Positive reactions for ammonia, indole acetic acid and siderophores production, and phosphate solubilizing activity	Tan et al. (2013)
Tomato bacterial wilt	*R. solanacearum*	*Pseudomonas putida*	Biofilm formation	Sun et al. (2017)
Damping-off of seedlings	*Rhizoctonia solani*	*Streptomyces mutabilis* and *S. cyaneofuscatus*	Plant-growth promoting activities	Goudjal et al. (2014)
Damping-off of seedlings	*R. solani*	*T. asperellum*	Plant growth-promoting traits viz., IAA production, Phosphate solubilization	Sehim et al. (2023)
Bacterial spot	*Xanthomonas citri* subsp. *citri*, *X. oryzae* pv. *oryzae*, *X. oryzae* pv. *oryzicola*	*Pseudomonas putida*	Biofilm formation	Sun et al. (2017)
Bacterial spot	*Xanthomonas vesicatoria*	*Bacillus velezensis* and *Pseudomonas soli*	Produce various novel metabolic products	Felipe et al. (2021)

plants, where each member plays a unique role in maintaining the ecosystem equilibrium (Smith & Smith, 2011).

Rhizosphere bacteria various positive roles in the rhizosphere, are achieved both directly (*viz*., via modulation of phytohormones, solubilization of phosphate and production of ammonia) and indirectly (*viz*., through antibiotics, siderophores and HCN production) (Fig. 18.1, Table 18.1).

Several bacterial genera, viz., *Alcaligenes*, *Arthrobacter*, *Azospirillum*, *Azotobacter*, *Bacillus*, *Burkholderia*, *Enterobacter*, *Klebsiella*, *Pseudomonas*, and *Serratia* include species that enhance the growth of plants and hence act as PGPR (Jasim et al., 2014; Kumar et al., 2016). Various strains show a positive effect on the growth and also help in the disease management of tomato. These valuable taxa

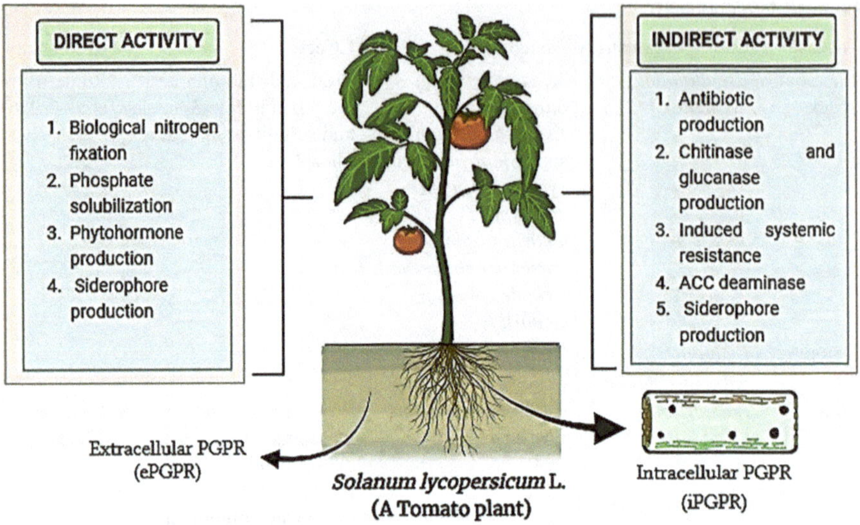

Fig. 18.1 PGPR type and activity

include species of *Azotobacter, Bacillus, Micromonospora, Pseudomonas* (i.e., *P. fluorescens*) and *Serratia*. These strains play an important role in plant growth by assisting the growth of tomato plants and showed protection against various diseases (Pastor et al., 2012; Narendra Babu et al., 2015; Martínez-Hidalgo et al., 2015).

18.6 Potential of Bacteria as Biocontrol Agents

Botrytis cinerea Pers. ex Fr. is a global fungal disease that causes gray mould disease in plants. This pathogen can significantly reduce the quality and quantity of tomato fruits, resulting in a devastating effect on their production. Wang et al. (2020) evaluated the biocontrol bacterium CQ-4, extracted and tested from the rhizosphere soil of tomato plants, for its morphological, physiological, and biochemical properties. The 16S rDNA sequence showed that the isolated belonged to the species *Pseudomonas aeruginosa.* It showed a potent antagonistic effect on *Botrytis cinerea*. This strain also showed a strong antibacterial spectrum and inhibited 12 other plant diseases.

A new *Pseudomonas* strain, QBA5, was isolated from healthy plant of tomato and was identified through morphology, biochemical and molecular methods. It showed a strong antifungal activity *vs B. cinerea*, inhibiting the conidial germination, elongation of the germ tube and mycelial growth. QBA5 effectively prevented gray mold disease development on tomato plants and fruit. QBA5 was found to damage the plasma membrane of *B. cinerea* (Gao et al., 2018). Other bacterial strains *viz.*, BL1, BT5, BR8, and BF11 were isolated from tomato healthy tissues proceeding from leaves, roots, fruits and stems, and were tested for their antagonistic activity against *B. cinerea*. The strains exhibited strong inhibitory activity on the fungal growth ranging from 27 to 53%. Belonging to the genus *Bacillus* they were also known for producing lipopeptides that inhibit the pathogen growth (Kefi et al., 2015).

Various researchers isolated strains of bacteria *viz.*, *Bacillus methylotrophicus* (isolate MT3), *Exiguobacterium aurantiacum* (isolates BT3 and MT8), *Exiguobacterium* spp. (isolate GT4), *Staphylococcus xylosus* (isolate BT5), *Pantoea eucalypti* (isolate NT6), *Pseudomonas veronii* (isolates BT4 and NT2), *Pseudomonas rhodesiae* (isolate BT2) and *Pseudomonas cichorii* (isolate NT3) from tomato leaves. In other studies, scientists isolated strains *viz.*, NT2, BT2, MT3, and NT6 and performed the seed inoculation. They found that the isolates inhibited the growth of *B. cinerea* and *Pseudomonas syringae* and produced various antimicrobial compounds. The isolates promoted growth of tomato plants, produced iron-binding siderophores, synthesized auxins (plant hormones), also helping in the solubilization of inorganic phosphate. These properties made them a valuable agent to improve plant health and control post-harvest diseases (Fig. 18.2) (Romero et al., 2016).

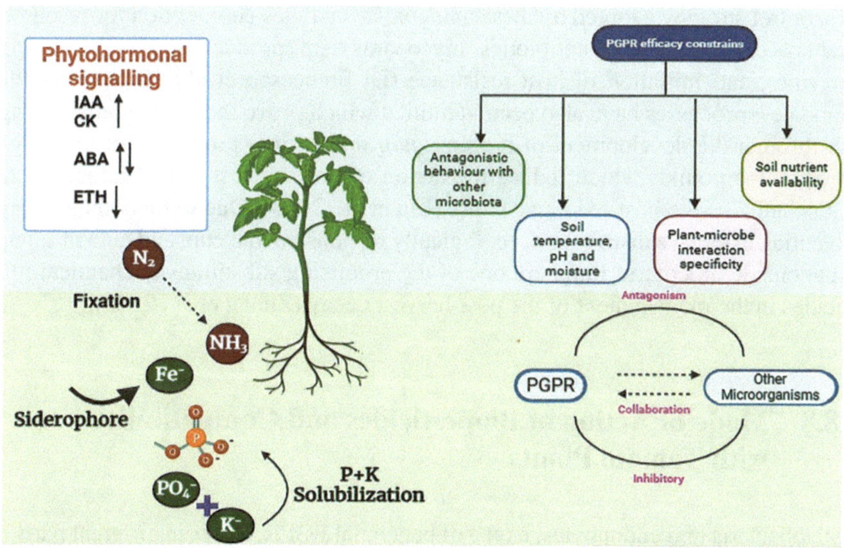

Fig. 18.2 Outline of PGPR traits affecting plant growth

18.7 Fungal Antagonists in Post-harvest Disease Management

Botrytis cinerea, the organism responsible for causing gray mold disease in plants, also grows on tomato fruits and leaves. Its growth is hindered by the yeast *Rhodotorula glutinis*, which is acquired from the tomato phyllosphere (Kalogiannis et al., 2006). Several *Trichoderma* spp., (asexually reproducing fungi) act in soil as powerful opportunistic invaders, able to grow quickly, being potent spore and antibiotic producers and strong competitors for space, nutrition and light (Montero-Barrientos et al., 2011). Because of these traits, *Trichoderma* strains are dominant and widespread, flourishing well on native grasslands, marsh, agriculture, salt and desert soils, in all climatic zones (Gal-Hemed et al., 2011).

Marine *Trichoderma* isolates were tested to assess their potential as halotolerant biocontrol agents. The isolates were very effective against *Rhizoctonia solani,* as they elicited a systemic defense in plants (Gal-Hemed et al., 2011). Few fungi secrete mycotoxins like aflatoxins, during the post-harvest phase and help in the spread of causative agents to colonize seeds of crop plants. *Trichoderma* strains include important mycoparasites and as such appear as effective biocontrol agents..

Trichoderma are a well-known, versatile soil antagonists of fungi that inhibit various soil fungal infections and plant pathogen incursions. They are commonly used as biofungicides against various disease-causing pathogens *viz*., *B. cinerea*, *Fusarium* spp., *Pythium* spp., *Rhizoctonia solani*, *Sclerotium rolfsii* etc., which affect a wide range of high value economic crops (Mohiddin et al., 2010). The main biocontrol strategy adopted by these antagonists includes competition for resources and space, production of antibiotics, mycoparasitism, production of cell wall lytic enzymes, and initiation of host resistance (Di Francesco et al., 2016). Recently, some new processes have also been identified which prove these antagonistic fungi act through the development of biofilm, *quorum* sensing, producing of antifungal volatile compounds which reduce the action of post-harvest fungal infections on fruits and also the host oxidative stress (Liu et al., 2013). Due to their antagonistic potential, extensive distribution, ecologically responsive makeup, and human safety, antagonistic biocontrol fungi are one of the promising substitutes to chemical fungicides in the management of the post-harvest decay (Zhang et al., 2020).

18.8 Mode of Action of Biopesticides and Compatibility with Tomato Plants

Rhizobacteria and endophytes, a type of beneficial PGPR, thrive in different parts of tomato plants (Jambon et al., 2018). They act as biofertilizers that promote proper nutrient assimilation and enhance the overall quality of the crops. PGPR significantly improve the development of plants, protect against various infections, and help plants tolerate various environmental stresses such as drought and salinity

(Numan et al., 2018). Their capability to support plant growth and health mark them as a vital component of current agriculture. PGPR enhance the growth and development of plants through various direct and indirect mechanisms (Berger et al., 2017). They promote the growth of tomato plants by helping them properly assimilate essential nutrients and minerals essential for proper growth. For example, tomato seedlings inoculated with an actinomycetes PGPR strain increased the final fruit content of glucose, fructose, maleate, zinc, nitrate and phosphorus (Gouda et al., 2018).

Biopesticides include a set of microbes which exhibit antibacterial and antifungal activity in plants, which may have an industrial potential. In agricultural processes, microorganic bioinoculants have a specific role in biological control technologies. They produce extracellular hydrolytic enzymes and secondary metabolites that help eradicate tomato pathogens at a minimum inhibitory concentration. In most cases, competition for nutrients by these bioinoculants could be described as the potential mechanism deployed by the bioinoculant agents (Lau et al., 2020). Other biocontrol measures which affect the host plant defense machinery include systemic acquired resistance, elicited by microbes that influence the production of hormones in plants, *viz.*, auxin, cytokine and ethylene, lowering the effects of harmful pathogens (Alori & Babalola, 2018).

In plants, biocontrol includes a wide range of mechanisms (Fig. 18.2) *viz.*, direct antagonism (by production of antibiotics, HCN, siderophores, hydrolytic enzymes like chitinases, proteases, lipases etc.), or indirectly when the biocontrol agents act as probiotics and compete with the pathogens for a niche, especially for nutrients (van Loon, 2007). Plants have two types of responses: induced systemic resistance (ISR) and systemic acquired resistance (SAR) (see also Chaps. 3, 4, 5, 6, 7, 8, 9 and 10). Apart from that, variations in the level of hormones in the plants tissues also function as the primary facilitator for some biocontrol agents (Martínez-Viveros et al., 2010).

There is an additional type of interaction linking two or more organisms in which one inhibits the other from performing life-sustaining functions *viz.*, feeding and growth. This type of interaction is known as "antagonistic association", in which one organisms inhibits the other (Karthika et al., 2020). This type of organisms produce unique compounds in the rhizosphere of tomato plants that inhibit or kill other species. They tend to produce various enzymes i.e., protease, cellulase, chitinases, etc., which act as natural catalysts that can annihilate or dissolve the cell wall of the fungi attacking food.

Rhizobacteria also control the physiological activity of spoilage organisms producing, as mentioned above, various enzymes or volatiles and inhibiting their activity in the rhizosphere soil. For example, soil bacteria belonging to *Actinobacteria*, *Firmicutes*, and *Proteobacteria*, which reside in the rhizosphere, inhibited *Rhizoctonia solani*, the causal agent of soft rot infection in tomato (Guerrieri et al., 2020; Piechulla et al., 2017). This clearly shows how these interactions are very specific and help prevent diseases.

Rhizobacteria like *Rhizobium* and *Bacillus* secret siderophores that inhibit iron absorption by the spoilage microbes present in their rhizosphere environment, competing with and limiting pathogens survival (Lurthy et al., 2020). Siderophores are secondary metabolites with low molecular weight and high affinity towards iron-chelating molecules. In response to iron deficiency, both Gram-positive and Gram-negative bacteria synthesize siderophores, which are also important components of the metabolism of pathogenic bacteria.

According to various studies, a compromised iron acquisition machinery and/or the addition of manufactured molecules lead in bacteria to a decreased ability to cause diseases (Khasheii et al., 2021). Siderophores have recently drawn attention because of their potential involvement in several metabolic pathways. In microbial ecology, siderophores are used to endorse the development of a number of uncultivable microorganisms and can modify microbial communities. Treatments with siderophores sustained the growth of various plant species and raised their output by improving the uptake of Fe by the plants. Siderophores may substitute hazardous pesticides and function against various harmful phytopathogens (Saha et al., 2016). They may also act as bioremediation agents, as their application helps detoxify samples contaminated by heavy metals.

Production of phytohormones by PGPR is highly recognized and is of considerable significance in sustainable agriculture. Certain PGPR are well-known to produce phytohormones (such as auxins, gibberellic acid, cytokinins, ethylene and abscisic acid), thus sustaining plant growth (Stefen et al., 2022). Natural, semi-synthetic, and synthetic hormones grouped into six major classes of phytohormones are characterized by their significant role in breaking seed dormancy in various plants, leading to germination. On the contrary, hormones influence plants growth sustainably, activate their immune response, promoting seedling growth and fruiting.

Analytical methods such as mass spectrometry indicated that certain strains of *Bacillus* produce varied and complex antibiotics. In general, strains *viz.*, BL1 and BT5 are known to produce diverse combinations of fengycin and surfactin antibiotics, whereas BR8 is characterized for both iturin and surfactin antibiotics. However, a complex mixture of heterogeneous antibiotics such as bacillomycin D, fengycin, and surfactin wass produced by BF11 (Kefi et al., 2015).

18.9 Conclusions

Post-harvest diseases are a major constraint in the production of tomato, resulting in poor fruit quality and financial losses. Chemical pesticides, known for their harmful effects on the environment, depleting soil quality and human health, are used in traditional methods of disease management. The impact of plant microbiomes and beneficial bacteria as biocontrol agents is, however, gaining considerable attention in the enhancement of sustainable agriculture.

Besides other advantages, many bacteria also enhance plant growth and development, soil fertility, nitrogen uptake by roots, etc. As a result of systemic resistance, host plants become more disease-resistant. Therefore, biocontrol agents obtained from plant microbiomes provide a pathway towards sustainable agriculture. Overall, there is a need to understand and utilize beneficial plant microbiomes to develop efficient methods to manage post-harvest diseases of tomato. The application of biocontrol agents enhances the quality of the plant, reducing the dependency on chemical pesticides and decreasing post-harvest losses. This generates a pathway to sustainable development in the agricultural sector. However, these solutions demand additional research and further studies to boost tomato production and reduce the agriculture ecological footprint.

Acknowledgement Author (Smriti Mall) Research and Development Department Higher education UP government, Lucknow.

References

Alori, E. T., & Babalola, O. O. (2018). Microbial inoculants for improving crop quality and human health in Africa. *Frontiers in Microbiology, 9*, 2213.

Arah, I. K., Ahorbo, G. K., Anku, E. K., Kumah, E. K., & Amaglo, H. (2016). Postharvest handling practices and treatment methods for tomato handlers in developing countries: A mini review. *Advances in Agriculture, 2016*, 1–8. https://doi.org/10.1155/2016/6436945

Bartz, J. A., Yuk, H. G., Mahovic, M. J., Warren, B. R., Sreedharan, A., & Schneider, K. R. (2015). Internalization of Salmonella enterica by tomato fruit. *Food Control, 55*, 141–150.

Berger, B., Baldermann, S., & Ruppel, S. (2017). The plant growth-promoting bacterium *Kosakonia radicincitans* improves fruit yield and quality of *Solanum lycopersicum*. *Journal of the Science of Food and Agriculture, 97*(14), 4865–4871. https://doi.org/10.1002/jsfa.8357

Biswas, C., Dey, P., Bera, A., Satpathy, S., & Mahapatra, B. S. (2013). First report of bacterial leaf spot caused by *Xanthomonas campestris* pv. *olitorii* on jute grown for seed in India. *Plant Disease, 97*(8), 1109–1109. https://doi.org/10.1094/PDIS-12-12-1196-PDN

Bu, S., Munir, S., He, P., Li, Y., Wu, Y., Li, X., et al. (2021). *Bacillus subtilis* L1-21 as a biocontrol agent for postharvest gray mold of tomato caused by *Botrytis cinerea*. *Biological Control, 157*, 104568.

Cheng, M., Meng, F., Lv, R., Wang, P., Wang, Y., Chen, X., & Wang, A. (2023). Postharvest preservation effect of composite biocontrol agent on tomatoes. *Scientia Horticulturae, 321*, 112344.

Chuku, E. C., Ogbonna, D. N., Oneegbu, B. A., & Adeleke, M. T. V. (2008). Comparative studies on fungi and bio-chemical characteristics of snake gourd (Trichosanthes curcumerina Linn) and tomato (Lycopersicon esculentum Mill). *Journal of Applied Science 8*(1), 168–172.

Devappa, V., Sangeetha, C. G., Vinay, M. R., Allada, S. N., & Jhansirani, P. S. (2021). Postharvest handling and diseases of horticultural produce. In *Postharvest Handling and Diseases of Horticultural Produce*. CRC Press. https://doi.org/10.1201/9781003045502

Di Francesco, A., Martini, C., & Mari, M. (2016). Biological control of postharvest diseases by microbial antagonists: How many mechanisms of action? *European Journal of Plant Pathology, 145*, 711–717.

Dik, A., & Elad, Y. (1999). Comparison of antagonists of *Botrytis cinerea* in greenhouse-grown cucumber and tomato under different climatic conditions. *European Journal of Plant Pathology, 105*, 123–137.

Dilantha Fernando, W. G., Nakkeeran, S., & Zhang, Y. (2005). Biosynthesis of antibiotics by PGPR and its relation in biocontrol of plant diseases. In Z. A. Siddiqui (Ed.), *PGPR: Biocontrol and biofertilization* (pp. 67–109). Springer.

Etebu, E., Nwauzoma, A. B., & Bawo, D. D. (2013). Postharvest spoilage of tomato (*Lycopersicon esculentum* Mill.) and control strategies in Nigeria. *Journal of Biology, Agriculture and Healthcare, 3*, 51–61.

FAOSTAT (2020). FAOSTAT Database. https://www.fao.org/faostat/en/#home

FAOSTAT (2001). FAOSTAT Dtaabase. https://www.fao.org/faostat/en/#data

Felipe, V., Bianco, M. I., Terrestre, M., Mielnichuk, N., Romero, A. M., & Yaryura, P. M. (2021). Biocontrol of tomato bacterial spot by novel *Bacillus* and *Pseudomonas* strains. *European Journal of Plant Pathology, 160*, 935–948.

Gal-Hemed, I., Atanasova, L., Komon-Zelazowska, M., Druzhinina, I. S., Viterbo, A., & Yarden, O. (2011). Marine isolates of *Trichoderma* spp. as potential halotolerant agents of biological control for arid-zone agriculture. *Applied and Environmental Microbiology, 77*, 5100–5109.

Gallo, M., Ferrara, L., Calogero, A., Montesano, D., & Naviglio, D. (2020). Relationships between food and diseases: What to know to ensure food safety. *Food Research International, 137*, 109414.

Gao, P., Qin, J., Li, D., & Zhou, S. (2018). Inhibitory effect and possible mechanism of a *Pseudomonas* strain QBA5 against gray mold on tomato leaves and fruits caused by *Botrytis cinerea*. *PLoS One, 13*, e0190932.

Glick, B. R. (2012). Plant growth-promoting bacteria: Mechanisms and applications. *Scientifica, 2012*, 1–15. https://doi.org/10.6064/2012/963401

Gouda, S., Kerry, R. G., Das, G., Paramithiotis, S., Shin, H. S., & Patra, J. K. (2018). Revitalization of plant growth promoting rhizobacteria for sustainable development in agriculture. *Microbiological Research, 206*, 131–140.

Goudjal, Y., Toumatia, O., Yekkour, A., Sabaou, N., Mathieu, F., & Zitouni, A. (2014). Biocontrol of *Rhizoctonia solani* damping-off and promotion of tomato plant growth by endophytic actinomycetes isolated from native plants of Algerian Sahara. *Microbiological Research, 169*(1), 59–65. https://doi.org/10.1016/j.micres.2013.06.014

Guerrieri, M. C., Fanfoni, E., Fiorini, A., Trevisan, M., & Puglisi, E. (2020). Isolation and screening of extracellular PGPR from the rhizosphere of tomato plants after long-term reduced tillage and cover crops. *Plants, 9*, 668.

Harel, Y. M., Mehari, Z. H., Rav-David, D., & Elad, Y. (2014). Systemic resistance to gray mold induced in tomato by benzothiadiazole and *Trichoderma harzianum* T39. *Phytopathology, 104*, 150–157.

Hasan, Z. A. E., Mohd Zainudin, N. A. I., Aris, A., Ibrahim, M. H., & Yusof, M. T. (2020). Biocontrol efficacy of *Trichoderma asperellum* -enriched coconut fibre against Fusarium wilts of cherry tomato. *Journal of Applied Microbiology, 129*, 991–1003.

Hernández-Huerta, J., Tamez-Guerra, P., Gomez-Flores, R., Delgado-Gardea, M. C. E., García-Madrid, M. S., Robles-Hernández, L., & Infante-Ramirez, R. (2021). Prevalence of *Xanthomonas euvesicatoria* (formally *X. perforans*) associated with bacterial spot severity in *Capsicum annuum* crops in South Central Chihuahua, Mexico. *PeerJ, 9*, e10913.

Jambon, I., Thijs, S., Weyens, N., & Vangronsveld, J. (2018). Harnessing plant-bacteria-fungi interactions to improve plant growth and degradation of organic pollutants. *Journal of Plant Interactions, 13*, 119–130.

Jasim, B., Joseph, A. A., John, C. J., Mathew, J., & Radhakrishnan, E. K. (2014). Isolation and characterization of plant growth promoting endophytic bacteria from the rhizome of *Zingiber officinale*. *3 Biotech, 4*, 197–204.

Kalogiannis, S., Tjamos, S. E., Stergiou, A., Antoniou, P. P., Ziogas, B. N., & Tjamos, E. C. (2006). Selection and evaluation of phyllosphere yeasts as biocontrol agents against grey mould of tomato. *European Journal of Plant Pathology, 116*, 69–76.

Karthika, S., Varghese, S., & Jisha, M. S. (2020). Exploring the efficacy of antagonistic rhizobacteria as native biocontrol agents against tomato plant diseases. *3Biotech, 10*, 320.

Kefi, A., Ben Slimene, I., Karkouch, I., Rihouey, C., Azaeiz, S., Bejaoui, M., et al. (2015). Characterization of endophytic *Bacillus* strains from tomato plants (*Lycopersicon esculentum*) displaying antifungal activity against *Botrytis cinerea* Pers. *World Journal of Microbiology and Biotechnology, 31*, 1967–1976.

Khasheii, B., Mahmoodi, P., & Mohammadzadeh, A. (2021). Siderophores: Importance in bacterial pathogenesis and applications in medicine and industry. *Microbiological Research, 250*, 126790.

Kumar, A., Singh, R., Yadav, A., Giri, D. D., Singh, P. K., & Pandey, K. D. (2016). Isolation and characterization of bacterial endophytes of *Curcuma longa* L. *3 Biotech, 6*, 60.

Lau, E. T., Tani, A., Khew, C. Y., Chua, Y. Q., & Hwang, S. S. (2020). Plant growth-promoting bacteria as potential bio-inoculants and biocontrol agents to promote black pepper plant cultivation. *Microbiological Research, 240*, 126549. https://doi.org/10.1016/j.micres.2020.126549

Liu, J., Sui, Y., Wisniewski, M., Droby, S., & Liu, Y. (2013). Review: Utilization of antagonistic yeasts to manage postharvest fungal diseases of fruit. *International Journal of Food Microbiology, 167*, 153–160.

Lurthy, T., Cantat, C., Jeudy, C., Declerck, P., Gallardo, K., Barraud, C., et al. (2020). Impact of bacterial siderophores on iron status and ionome in pea. *Frontiers in Plant Science, 11*, 730.

Ma, M., Taylor, P. W. J., Chen, D., Vaghefi, N., & He, J.-Z. (2023). Major soilborne pathogens of field processing tomatoes and management strategies. *Microorganisms, 11*(2), 263.

Martínez-Hidalgo, P., García, J. M., & Pozo, M. J. (2015). Induced systemic resistance against *Botrytis cinerea* by *Micromonospora* strains isolated from root nodules. *Frontiers in Microbiology, 6*.

Martínez-Viveros, O., Jorquera, M., Crowley, D., Gajardo, G., & Mora, M. (2010). Mechanisms and practical considerations involved in plant growth promotion by rhizobacteria. *Journal of Soil Science and Plant Nutrition, 10*.

Mohiddin, F. A., Khan, M. R., Khan, S. M., & Bhat, B. H. (2010). Why *Trichoderma* is considered super hero (super fungus) against the evil parasites? *The Plant Pathology Journal, 9*, 92–102.

Montero-Barrientos, M., Hermosa, R., Cardoza, R. E., Gutiérrez, S., & Monte, E. (2011). Functional analysis of the *Trichoderma harzianum nox1* gene, encoding an NADPH oxidase, relates production of reactive oxygen species to specific biocontrol activity against *Pythium ultimum*. *Applied and Environmental Microbiology, 77*, 3009–3016.

Nandi, M., Macdonald, J., Liu, P., Weselowski, B., & Yuan, Z. (2018). Clavibacter michiganensis ssp. michiganensis: Bacterial canker of tomato, molecular interactions and disease management. *Molecular Plant Pathology, 19*, 2036–2050.

Narendra Babu, A., Jogaiah, S., Ito, S., Kestur Nagaraj, A., & Tran, L. S. P. (2015). Improvement of growth, fruit weight and early blight disease protection of tomato plants by rhizosphere bacteria is correlated with their beneficial traits and induced biosynthesis of antioxidant peroxidase and polyphenol oxidase. *Plant & Science, 231*, 62–73.

Numan, M., Bashir, S., Khan, Y., Mumtaz, R., Shinwari, Z. K., Khan, A. L., et al. (2018). Plant growth promoting bacteria as an alternative strategy for salt tolerance in plants: A review. *Microbiological Research, 209*, 21–32.

Oku, S., Komatsu, A., Tajima, T., Nakashimada, Y., & Kato, J. (2012). Identification of chemotaxis sensory proteins for amino acids in *Pseudomonas fluorescens* Pf0-1 and their involvement in chemotaxis to tomato root exudate and root colonization. *Micro Environment, 27*, 462–469.

Pastor, N., Carlier, E., Andrés, J., Rosas, S. B., & Rovera, M. (2012). Characterization of rhizosphere bacteria for control of phytopathogenic fungi of tomato. *Journal of Environmental Management, 95*, S332–S337.

Paul, N. C., Park, S., Liu, H., Lee, J. G., Han, G. H., Kim, H., & Sang, H. (2021). Fungi associated with postharvest diseases of sweet potato storage roots and in vitro antagonistic assay of *Trichoderma harzianum* against the diseases. *Journal of Fungi, 7*, 927.

Piechulla, B., Lemfack, M. C., & Kai, M. (2017). Effects of discrete bioactive microbial volatiles on plants and fungi. *Plant, Cell & Environment, 40*, 2042–2067.

Rivarez, M. P. S., Vučurović, A., Mehle, N., Ravnikar, M., & Kutnjak, D. (2021). Global advances in tomato virome research: Current status and the impact of high-throughput sequencing. *Frontiers in Microbiology, 12*, 671925.

Romero, F. M., Marina, M., & Pieckenstain, F. L. (2016). Novel components of leaf bacterial communities of field-grown tomato plants and their potential for plant growth promotion and biocontrol of tomato diseases. *Research in Microbiology, 167*, 222–233.

Saha, M., Sarkar, S., Sarkar, B., Sharma, B. K., Bhattacharjee, S., & Tribedi, P. (2016). Microbial siderophores and their potential applications: A review. *Environmental Science and Pollution Research, 23*, 3984–3999.

Santoyo, G., Urtis-Flores, C. A., Loeza-Lara, P. D., del Orozco-Mosqueda, M. C., & Glick, B. R. (2021). Rhizosphere colonization determinants by plant growth-promoting Rhizobacteria (PGPR). *Biology, 10*, 475. https://doi.org/10.3390/biology10060475

Segarra, G., Casanova, E., Avilés, M., & Trillas, I. (2010). *Trichoderma asperellum* strain T34 controls Fusarium wilt disease in tomato plants in soilless culture through competition for iron. *Microbial Ecology, 59*, 141–149. https://doi.org/10.1007/s00248-009-9545-5

Sehim, A. E., Hewedy, O. A., Altammar, K. A., Alhumaidi, M. S., & Abd Elghaffar, R. Y. (2023). *Trichoderma asperellum* empowers tomato plants and suppresses *Fusarium oxysporum* through priming responses. *Frontiers in Microbiology, 14*, 1140378.

Sharma, V., Kaushik, M., Agnihotri, C., Agnihotri, S., & Singh, B. P. (2023). Postharvest disease management of tomato (*Solanum lycopersicum* L.) using endophytic actinobacteria as natural biocontrol agent. In M. K. Solanki, M. K. Yadav, B. P. Singh, & V. Gupta (Eds.), *Microbial endophytes and plant growth* (pp. 137–150). Elsevier. https://doi.org/10.1016/B978-0-323-90620-3.00005-2

Shi, J. F., & Sun, C. Q. (2017). Isolation, identification, and biocontrol of antagonistic bacterium against *Botrytis cinerea* after tomato harvest. *Brazilian Journal of Microbiology, 48*(4), 706–714.

Singh, V. K., Singh, A. K., & Kumar, A. (2017). Disease management of tomato through PGPB: Current trends and future perspective. *3 Biotech, 7*(4), 255. https://doi.org/10.1007/s13205-017-0896-1

Smith, S. E., & Smith, F. A. (2011). Roles of arbuscular mycorrhizas in plant nutrition and growth: New paradigms from cellular to ecosystem scales. *Annual Review of Plant Biology, 62*, 227–250.

Stefen, D. L. V., Nunes, F. R., Rodolfo, G. R., Segatto, C., Anastácio, T. C., & Lajus, C. R. (2022). How phytohormones synthesized by PGPR affect plant growth? In R. Z. Sayyed & V. G. Uarrota (Eds.), *Secondary metabolites and volatiles of PGPR in plant-growth promotion* (pp. 119–131). Springer. https://doi.org/10.1007/978-3-031-07559-9_7

Sun, D., Zhuo, T., Hu, X., Fan, X., & Zou, H. (2017). Identification of a *Pseudomonas putida* as biocontrol agent for tomato bacterial wilt disease. *Biological Control, 114*, 45–50.

Tan, S., Jiang, Y., Song, S., Huang, J., Ling, N., Xu, Y., & Shen, Q. (2013). Two *Bacillus amyloliquefaciens* strains isolated using the competitive tomato root enrichment method and their effects on suppressing *Ralstonia solanacearum* and promoting tomato plant growth. *Crop Protection, 43*, 134–140.

van Loon, L. C. (2007). Plant responses to plant growth-promoting rhizobacteria. *European Journal of Plant Pathology, 119*, 243–254.

Wang, X., Zhou, X., Cai, Z., Guo, L., Chen, X., Chen, X., et al. (2020). A biocontrol strain of *Pseudomonas aeruginosa* CQ-40 promote growth and control *Botrytis cinerea* in tomato. *Pathogens, 10*, 22. https://doi.org/10.3390/pathogens10010022

Wang, R., Chen, D., Khan, R. A. A., Cui, J., Hou, J., & Liu, T. (2021). A novel *Trichoderma asperellum* strain DQ-1 promotes tomato growth and induces resistance to gray mold caused by *Botrytis cinerea*. *FEMS Microbiology Letters, 368*(20), fnab140. https://doi.org/10.1093/femsle/fnab140

Wang, Y., Deng, S., Li, Z., & Yang, W. (2022). Advances in the characterization of the mechanism underlying bacterial canker development and tomato plant resistance. *Horticulturae, 8*, 209.

Zhang, X., Li, B., Zhang, Z., Chen, Y., & Tian, S. (2020). Antagonistic yeasts: A promising alternative to chemical fungicides for controlling postharvest decay of fruit. *Journal of Fungi, 6*, 158.

Zheng, L., Gu, X., Xiao, Y., Wang, S., Liu, L., Pan, H., & Zhang, H. (2023). Antifungal activity of *Bacillus mojavensis* D50 against *Botrytis cinerea* causing postharvest gray mold of tomato. *Scientia Horticulturae, 312*, 111841.

Zia, R., Nawaz, M. S., Siddique, M. J., Hakim, S., & Imran, A. (2021). Plant survival under drought stress: Implications, adaptive responses, and integrated rhizosphere management strategy for stress mitigation. *Microbiological Research, 242*, 126626.

Chapter 19
Counteracting Post-harvest Diseases Through Horticultural Plants Microbiome

Ankan Das

Abstract Horticulture diseases in the post-harvest are worrying as they are leading to aggravated losses. Different chemical methods for controlling these diseases have already been tried, and many are still in use. However, in their residual forms, these chemicals are entering the food chain causing all sorts of medical concerns to higher beings. Therefore, some non-chemical but effective methods are needed. Several microorganisms, unlike disease-causing pathogens, are also present in association with plants and have some capabilities and action mechanisms for combating the harmful species. If studied, identified, and utilized effectively against problematic pathogens, they would help eliminating post-harvest diseases and, in turn, reduce post-harvest losses. Therefore, the present chapter is based on the above concept to discuss interesting microbial agents and their antagonistic properties against post-harvest horticultural pathogens. The review also highlights some recent developments in research, market-based biocontrol products, and future requirements in the sector.

Keywords Bio-control · Beneficial · Exploration · Harmful · Microbes

Abbreviations

BCA Biological Control Agents
ISR Initiation of Systemic Resistance

A. Das (✉)
Department of Horticulture, Institute of Agricultural Science, University of Calcutta, Kolkata, West Bengal, India

P. Mathur, S. Roy (eds.), *Plant Microbiome and Biological Control*, Sustainability in Plant and Crop Protection 20,
https://doi.org/10.1007/978-3-031-75845-4_19

19.1 Introduction

Plants microbiomes, comprising hundreds of species from various microbial lineages such as bacteria, fungi, protists, and archaea, are very specialized and complex systems (Wassermann et al., 2022; Berg et al., 2020). Plants evolved over million years in an ecosystem dominated by microbes (Kusstatscher et al., 2020; Cordovez et al., 2019). Several microorganisms develop in association with plants, in different parts and establish themselves accordingly. Microbes effectively invade the rhizosphere, phyllosphere, and carposphere, starting from the seed (Kusstatscher et al., 2020; Adam et al., 2016; Berg & Raaijmakers, 2018; Mitter et al., 2017). Microbial colonization, however, varies with respect to the plant type and biotic and abiotic features as well. According to recent research, seeds, and fruits are the primary means by which the microbiota is transferred to the next generation of plants (Kusstatscher et al., 2020; Berg, Grube, et al., 2014b; Bergna et al., 2018; Sánchez-Cañizares et al., 2017). Due to the close and strong relationship that exists between the resident microbe and the plant during the course of their life cycle, this relationship is known as the second genome and is considered as a single entity known by the term 'holobiont' (Verma et al., 2022).

Plants and associated microorganism can lead together to unpredictable outputs. The microbial–plant associations work consistently in providing stability during plant growth and development under living and nonliving stress circumstances (Verma et al., 2022; Hassani et al., 2018). Many studies identified the connection linking plants and their microbial communities. Direct and indirect mechanisms involved in the promotion of plant growth and phytopathogen management include nutrient acquisition, phytohormone modulation, the production of antioxidant enzymes, lowering and control of diseases, incorporation of systemic resistance and productin and role of volatile compounds (Verma et al., 2022; Kumari et al., 2022; Pathak et al., 2022).

Wastage and loss of food-based commodities are global problems that require significant attention. The Food and Agriculture Organization of the United Nations estimates that, between production and consumption, over half of farmed fruits and vegetables are lost (Kusstatscher et al., 2020; Gustafsson et al., 2013). In addition to post-harvest food losses resulting from technological limitations in harvesting and handling, undesirable development of sprouts or depletion of sugar levels and moisture content, the spoilage and storage rot caused by pathogens continue as a significant global issue (Wassermann et al., 2022). In poor nations, post-harvest losses of fruit and vegetable account for 40–50% of total yield (Liu et al., 2013a, 2020). When fruit and vegetables are stored after harvest, fungal pathogens cause decays which typically result in significant degradation of qualitative attributes and nutritional parameters, including spoilage by mycotoxins, and decreased market value (Liu et al., 2020; El Ghaouth et al., 2004). Food losses that occur after harvesting can be either direct or quantitative, such as during harvesting, or proceed from food quality losses, such as low water and sugar content, physiological damage, or unwanted sprouting initiations (Kusstatscher et al., 2020; Aulakh & Regmi, 2013). According to Kusstatscher et al. (2020), significant damage to horticultural commodities happens during storage as well, notwithstanding the significant waste generated during preparation and transportation.

Huge volumes of chemicals used in the field for horticultural substances, such as growth regulating agents and pesticides, might have detrimental effects on the environment and public health in addition to the limited effectiveness of present management methods (Wassermann et al., 2022). However, over the past 10 years, the following factors have contributed to a decrease in the use of synthetic fungicides for the prevention of post-harvest diseases: (i) pathogen resistance to numerous important fungicides hasemerged, leading to new pathogenic biotypes, (ii) ineffective substitute fungicides, (iii) rising of fungicide residue levels in agricultural products, iv) toxicological issues pertaining to human health and (v) other adverse environmental effects (Dukare et al., 2019; Droby, 2005). Moreover, the public became more and more aware of the true reality that many toxic substances are entering the human food chain, and that this threat is a main reason for amplified health-related problems. Therefore, the hunt for safer and more environmentally friendly alternatives to prevent post-harvest infections and decay is becoming more popular, worldwide (Dukare et al., 2019).

Microorganisms are extremely varied and represent a huge radiatin of microbial lineages. They can be found in plant tissues and organs as endophytes or as epiphytes on their surface as well (Kusstatscher et al., 2020). There are countless microorganisms active in plant associations and their role on the plant system is variable. A high number of these microorganisms are thought to be important for fruit health, quality, and disease resistance, but others are known to be plant pathogens (Kusstatscher et al., 2020; Berg et al., 2014a; Droby & Wisniewski, 2018; Leff & Fierer, 2013). Beneficial microbe-plant relationship thus can be utilized to get serviceable outputs, valuable in the horticultural sector for post-harvest disease management. We now have more opportunities to gain a deeper understanding of how microbes interact with their environment, different plant hosts, or how they change as concerns their diversity in relation to diseases. The interactions and role of biocontrol agents in the microbiome can be studies in more detail, thanks to recent technological advancements, which include high-throughput sequencing methods and metagenomic technologies (Kusstatscher et al., 2020; Abdelfattah et al., 2018; Berg et al., 2016). Thus, the present chapter mainly focuses on how the microbial community can help us to manage various post-harvest diseases of horticultural crops. Several problem-causing pathogens play indeed a vital role in magnifying horticultural crops losses, leading to economic negative consequences, not in India alone but worldwide as well.

19.2 Mechanisms of Post-harvest Disease Development

Numerous fungal infections cause vegetables, fruits, and other horticultural items to deteriorate and develop post-harvest diseases. The main disease-causing organisms in the post-harvest stage of fruit and vegetables include members of fungal genera such as *Fusarium*, *Gloeosporium*, *Aspergillus*, *Penicillium*, *Geotrichum*, *Mucor*, *Alternaria*, *Monilinia*, *Botrytis*, and *Rhizopus* (Dukare et al., 2019; Barkai-Golan, 2001). Plants may be attacked at various points during their life cycle, and post-harvest diseases can emerge in a complex way (Wassermann et al., 2022).

A number of biotic and abiotic stressors, including harvesting, ripening, and mechanical damage, frequently trigger the process by which a disease expansion occurs. Fungal infections processes start from conidia that germinate and enter the host tissues through wounds and injuries (Dukare et al., 2019; Alkan & Fortes, 2015). In earlier reports, it has been documented that a huge number of microorganism-based pathogens can cause field diseases, such as bacterial soft rots and fungal grey molds, which manifest symptoms on infected fruits quickly, after harvest. However, the problem can also occur through damaged tissues (Wassermann et al., 2022). The pathogens also enter the host cuticle during the fruit growing cycle. At the original location of early, unripe fruit, a number of fungal species, including *Lasiodiplodia*, *Colletotrichum*, *Alternaria*, *Monilinia*, *Botrytis*, *Phomopsis* and *Botryosphaeria* may live inactively (Dukare et al., 2019; Prusky et al., 2013). This indolent state where the fungus is present but no symptoms are evident is called "latent infection". During storage, these pathogens stay dormant and cannot be seen or detected until the fruits develop. At this stage the fungal infection inocula multiply rapidly as the fruits ripen. The pathogenic fungus consumes resources from the host and kills host cells in a necrotrophic manner during the early development phase. This causes the tissues to decompose and decay begins (Dukare et al., 2019). This pathogenic activity severely hampers the resilience of the fruit, thus pushing itself in a zone of susceptibility. Therefore, throughout the ripening process, the innate disease resistance mechanism that shields the fruits from fungal attack weakens or becomes ineffective, and the fruits begin to exhibit the traits of the disease (Dukare et al., 2019; Prusky et al., 2013).

19.3 Grouping Post-harvest Diseases

Christensen and Kaufmann (1965) split pathogens into two groups based on their origin, as field or post-harvest species (Sakib et al., 2020). Both of these two types of disease-causing organisms are uniformly accountable for losses in the horticultural sector.

19.3.1 Field Pathogens

According to Sakib et al. (2020), these microorganisms are amongst those that infect plants or their products throughout the growing season, prior to harvest. Environmental conditions, faulty cultivation aspects, improper management practices, etc., are culprits behind the pathogens accelerative and augmented effects on plants.

19.3.2 Pathogen in Storage

The microorganisms responsible for infection while produce is being stored are known as storage pathogens. During the period of harvest, field pathogen-induced illness symptoms might be quite subtle. Even after harvest, field pathogen infection can still occur in juicy or meaty fruits and vegetables (Sakib et al., 2020). As mentioned above, innumerable pathogenic genera occur, causing rot and decay in the post-harvest life of vegetables and fruits. During storage, vegetables and fruits may get an infection from another pathogen or from the very same field pathogen(s). Field pathogen-caused illness in seeds and grains stops progressing quickly after harvest. However, during storage, seeds and grains can become infected again by additional diseases (Sakib et al., 2020).

19.3.3 Post-Harvest Diseases

Nowadays, post-harvest diseases are an important issue in the horticultural industry that suffers enormous economic losses. Tables 19.1 and 19.2 show a description of a few of the major post-harvest diseases that affect most common horticulture commodities (Rawal & Sharma, 2015).

19.4 Microbial Antagonism

Antagonism occurs when nearby phytopathogens experience inhibition or disruption in their normal growth, development, and activity due to the actions of antagonistic organisms known as Biological Control Agents(BCA) able to manage infections and pests that affect horticulture crops (Dukare et al., 2019; Heydari & Pessarakli, 2010). A good deal of research has been carried out in this vast sector to find and identify best antagonistic microorganisms, acting against harmful pathogens. Numerous microbial species exhibit an antagonistic action against pre and post-harvest infections. These microbes create metabolites and antifungal

Table 19.1 Post-harvest diseases of major fruits

Fruit & diseases	Causal organisms	Remarks
Mango		
Stem end rot	*Botryadiplodia theobromeae*	The affected fruit has reduced non-reducing sugar levels in the pulp and quickly loses ascorbic acid.
Anthracnose	*Colletotrichum gloeosporiodies*	Causes a significant loss of product both prior to and following harvest
Citrus		
Aspergillus rot	*Aspergillus niger*	Within 8–10 days, the rot completely kills the fruit and releases a fermented stench.
Alternaria fruit rot	*Alternaria citri*	In the impacted fruits, the fungus lowers soluble solids, lowers sugars, and raises the percentage of acidity.
Brown rot	*Phytophthora nicotianae*	When stored, healthy fruits that harbour a latent illness eventually become ill.
Anthracnose	*Colletotrichum gloeosporiodies*	This disease is more common in Coorg mandarin throughout the main growing season and less common in monsoon crops.
Apple		
Pink Mould rot	*Trichothecium roseum*	Typically, the decay is connected to apple scab.
Bitter rot	*Glomerella cingulata*	The Golden delicious variety suffers greatly
Soft rot	*Penicillium expansum*	The fungus's invasion may cause the fruits to accumulate patulin, a mycotoxin, which when consumed might have major negative effects on health.
Core rot	*Alternaria alternate*	Frequently observed in apples with an open hole near the calyx tip
Grape		
Cladosporium rot	*Cladosporium herbarum*	A dark, hard, shallow decay is caused by the pathogen and is typically limited to a single side or flower end.
Bitter rot	*Greenaria uvicola*	In transportation and keeping, grapes with minor blemishes and scars lose their quality.
Grey Mould rot	*Botrytis cinerera*	Because the disease may spread by touch, even a single contaminated berry might result in "nest rot" while being transported or stored. After rain, bunches are more likely to deteriorate.
Blue Mould rot	*Penicillium digitatum*	The fruits that are affected, acquires a decaying flavour.
Papaya		
Marcrophomia fruit rot	*Marcrophomina phaseolina*	Deep within the fruit, the infection promotes decaying and breakdown.
Ascochyta rot	*Ascochyta caricae*	In extreme temperatures and excessive relative humidity, the illness worsens.

(continued)

Table 19.1 (continued)

Fruit & diseases	Causal organisms	Remarks
Rhizopus rot	*Rhizopus stolonifer*	When cold-stored fruit is subjected to outside heat while being sold, rot develops.
Banana		
Cigar end rot	*Verticillium theobromae*	Fruits can become contaminated while they are still developing. Certain banana types from the Cavendish group are particularly susceptible to this infection.
Pink Mould rot	*Trichothecium roseum*	Only the peel is affected by the decay.
Botriyodiplodia fruit rot	*Botryadiplodia theobromeae*	The fungal infection is a parasite of wounds.
Fusarium fruit rot	*Fusarium monaliformae*	This was initially reported from Allahabad, India, and is currently widespread across the nation.
Anthracnose	*Colletoetrichum musae*	On mature bananas, a relative humidity of more than 85% promotes the occurrence of the illness as well as conidial germination.
Pomegranate		
Soft rot	*Rhizophus arrhizus*	A significant obstacle to pomegranate preservation
Black blotch	*Botryadiplodia theobromeae*	A major issue that ruins the fruit entirely and is especially harmful in the months after the monsoon
Dry rot	*Botryadiplodia theobromeae Colletotrichum gloeosporiodies, Coniella granati*	Most of these diseases are preharvest ones, and they might manifest as firm dry brown rot in storage.
Guava		
Fruit canker	*Pestalotiopsis psidii, Colletotrichum gloeosporiodies*	The flesh remains unaffected, but its price value drops significantly. Safeda is a cultivar that is exceptionally resistant to this problem.
Colletotrichum fruit rot	*Colletotrichum gloeosporiodies*	A dangerous post-harvest illness in the northeast of India alongside other wet regions.
Macrophoma fruit rot	*Macrophoma allahabadensis*	Following infection, the fruit's sugar content is fully digested.

Source: Rawal & Sharma, 2015

chemicals active vs specific diseases, which stop the development and metabolism of their causal agents (Dukare et al., 2019).

The activity of antifungal metabolites is also variable, as they may restrict the growth or completely eradicate the pathogen. Thus, these organisms reduce the potential for additional fruit spoiliage during storage by preventing, inhibiting or killing the propagules of pathogens developing on fruit. To reduce the growth of fungal pathogens, antifungal microbial agents elicit such inhibition by means of both direct and indirect mechanisms. The precise identification of microbes which show antagonistic effects against trouble-causing pathogens helped in fighting against this crucial problem. Currently, filamentous fungus, bacteria, and yeast

Table 19.2 Post-harvest diseases of vegetables

Vegetables & diseases	Causal organisms	Remarks
Tomato		
Phoma rot	*Phoma destructiva*	Causing significant losses during storage and transportation.
Bacterial spot	*Xanthomonas vasicatoria*	The illness spreads quickly when the mean temperature is 20 °C and the relative humidity is >85%. Usually affecting the green fruits, although occasionally seen in the older ones as well
Sour rot	*Geotrichum candidum*	Most dangerous disease when tomato fruits are produced during the monsoon season
Buck eye rot	*Phytophthora nicotianae*	Seed and soil borne disease, encouraged by high humidity and temperatures
Malustela rot	*Malustella seria*	The disease outward signs resemble those of Alternaria rot.
Cabbage		
Soft rot	*Erwinia carotovora*	The illness could be undiagnosed for an extended amount of time.
Bacterial zonate rot	*Pseudomonas cichorii*	A significant market disease, occasionally mistaken for Alternaria rot because of its zonate impact.
Black rot	*Xanthomonas campestris*	A major issue with cabbage and a vascular condition
Potato		
Dry rot	*Fusarium solani* var. *coerulum*	The illness, which is prevalent around the world, appears in stores a few months after harvest.
Brown rot	*Pseudomonas solanacearum*	The pathogen is capable of surviving for months on the tuber epidermis.
Rhizopus rot	*Rhizopus oryzae*	Losses during transportation and storage might achieve 100% when there is a significant infection prevalence
Charcoal rot	*Macrophomina phaseolina*	The rot primarily occurs during warm, dry seasons, resulting in significant losses at various times and locations
Bean		
Pod rot	*Pythium* sp.	Difficult to forecast when the illness may manifest, unless there has been a previous history of *Pythium* issues in the field.
Grey mould	*Botrytis cinerea*	The mould quickly ruins produce by spreading to nearby beans
Cottony leak	*Phytopthora capsici* and other oomycetes	Separating healthy pod tissue from diseased pod tissue is clearly crimson one.
Onion		
Neck rot	*Botrytis allii*	The illness becomes more severe with high moisture present right before and during harvest, as well as during the field curing of onions

(continued)

Table 19.2 (continued)

Vegetables & diseases	Causal organisms	Remarks
Purple blotch	*Alternaria porri*	Onion thrips development is the first sign of the illness. The virus worsens during storage
Bulb rot	*Fusarium solani*	The bulb is totally consumed by decaying after 7–8 days. Below 15 °C the rate of degradation is much slower.
Cephalosporium rot	*Cephalosporium curtipes*	When the illness reaches bulbs, more than 26% of them show a decay

Source: Rawal & Sharma, 2015

represent the most important taxonomic groups of antagonistic species utilized to control most fungal diseases (Dukare et al., 2019).

19.4.1 Types of Antagonists

Fungi and bacteria are mostly employed to biologically manage bacterial and fungal plant diseases. According to Hajek and Eilenberg (2018), bacteria and fungi may both be utilized to manage other fungi and bacteria.

19.4.1.1 Fungal Antagonists

Under this category, the ability of microbial antagonism was recognized not prior to the early part of the nineteenth century. Alexander Fleming discovered in 1928 that the ascomycete fungus *Penicillium notatum* hindered the development of some bacteria, which was the first significant finding of a microorganism hostility towards other microorganisms. This finding paved the way for the creation of penicillin which is still an extremely effective medication that treats bacterial infections in humans. Fleming received the 1945 most prestigious Nobel Prize in the category of medicine in recognition of such discovery (Hajek & Eilenberg, 2018). There are multifarious categories and types of fungi, but only a few of them shows antagonistic effects against field and post-harvest disease-causing pathogens. The green-colored mould *Trichoderma* is arguably the most often utilized fungus employed as an antagonist. *Trichoderma* antagonists are very potent because of their rapid growth, aggression, and adaptability as they employ multiple, different mechanisms of activity (Hajek & Eilenberg, 2018). However, *Trichoderma* is not the only type of fungal antagonist used in fungal bio control. Although they never cause plant diseases, certain fungi that are employed as antagonists are, in fact, close relatives of fungal disease-causing microorganism. *Aspergillus flavus* is one example of this (Hajek & Eilenberg, 2018). The fungal antagonistic approach has been utilized *vs* several plant-related ailments, on and off the field. Antagonistic fungi have mostly been used to reduce plant diseases that are transmitted through soil. Fungal antagonists have also been produced to combat infections that target injuries in trees, leaves, and harvested products (Hajek & Eilenberg, 2018).

19.4.1.2 Bacterial Antagonists

These microorganisms show antagonistic effects against many detrimental species. Although not closely connected to insect pathogens like i.e., *Bacillus thuringiensis* many species showing disease antagonistic properties are members of the family Bacillaceae. The spore-forming bacteria from this family are antagonists of plant diseases, and few of them are known to generate antibiotics (Hajek & Eilenberg, 2018). Thus, as per fungi, also the microorganisms under this group have more than one type mechanism to contrast deleterious pathogens. Because of their environmental persistence capacity, these spore-forming organisms persist for a long time after treatments. Bacillaceae members are used to combat both soil-borne and foliar diseases (Hajek & Eilenberg, 2018).

19.4.1.3 Oomycete Antagonists

This group of microorganisms today does not fall under the major divisions of microbial antagonists but still appears relevant enough for discussion. Because they share many traits with fungi, water moulds were originally considered to be fungi, members of the oomycetes group. But according to Hajek and Eilenberg (2018), these species are actually protists and are currently categorized under the Chromista Kingdom, falling the oomycota class in the broader division of eukaryotes. This group has its own capability, which might not be at par with the fungal and the bacterial antagonistic types, but still is apt and considered. For instance, through competitive interactions, *Pythium oligandrum* which is an oomycete, acts as an antagonist to decrease illnesses brought on by oomycetes and phyto-pathogenic fungi (Hajek & Eilenberg, 2018).

19.5 Mode of Action

The term "Biological Control Agent" (BCA) refers to organisms that inhibits pests/pathogens and regulate their densities (Kusstatscher et al., 2020). BCA and other beneficial microorganisms deploy their antagonistic effects *vs* the disease-causing pathogens through distinct mechanisms. Some of them are described below.

19.5.1 Antibiosis

Antibiotics are low molecular weight, secreted secondary metabolites that, in little amounts, are dangerous to other bacteria (Alizadeh et al., 2020; Fravel, 1988). One of the primary mechanisms shown by the antibiotics generated by biocontrol agents is the lowering of illness symptoms (Alizadeh et al., 2020; Haas & Défago, 2005). For instance, *Pseudomonas cepacia* has been shown to reduce various disease indicators by restricting the growth of post-harvest pathogens such as *Pencillium expansum* and *Botrytis cinerea* in apples, by creating an antibiotic called pyrrolnitrin (Janisiewicz & Roitman, 1988; Sakib et al., 2020). Additionally, at sub-inhibitory doses, the antibiotics may prevent bacteria from releasing exogenous factors related to virulence and adhesion (Alizadeh et al., 2020; Kumar et al., 2009). Iturin, an antibiotic, is developed by the bacteria *Bacillus subtilis* and *Pseudomonas cepacia* is known to destroy pathogens (Sakib et al., 2020; Pusey & Wilson, 1984). It has been established, therefore, that the synthesis of antibiotics is a significant strategy used by microbes to control a variety of plant diseases (Alizadeh et al., 2020; Gardener & Fravel, 2002). Antibiotics can alter an organism physiology even at sub-inhibitory doses (Alizadeh et al., 2020). By generating antibiotics, *P. cepacia* also proved successful in suppressing green mould *Penicillium digitatum* on lemons (Sakib et al., 2020).

Other studies showed that quinolone and macrolide antibiotics prevented *Pseudomonas aeruginosa* from producing virulence factors and cell signaling (Alizadeh et al., 2020). Additionally, *Bacillus* spp. synthesize exotoxin metabolites and enzymes with nematicidal properties (Alizadeh et al., 2020; Engelbrecht et al., 2018).

19.5.2 Resource Competition

Scramble and contest are the two wide divisions into which Nicholson separated rivalry over a limited resource. According to Hibbing et al. (2010), scramble competition, also known as exploitation competition, entails the quick use of limited resources or resources without any direct competitor engagement. In exploitative competition, one strain exaustes a limited resource, preventing the opponent from using it. Enhanced nutrition absorption or extracellular release of nutrient-harvesting molecules underpins this mechanism (Ghoul & Mitri, 2016). In this process, a direct accomplishment of antagonism by competition is not shown, however competition is manifested indirectly. The yeast *Saccharomyces cerevisiae* and the bacterium *Escherichia coli* provide two examples of this kind of competition: in presence of oxygen, they can both metabolically switch from fermentation to respiration, producing high growth rates but low yields that enable them to take up nutrients more quickly than their rivals (Ghoul & Mitri, 2016; MacLean & Gudelj, 2006; Pfeiffer et al., 2001; Vulić & Kolter, 2001).

The "champion" of a contest competition, also known as an interference competition, appropriates the resource(s) through direct, hostile encounters among rivals (Hibbing et al., 2010). The synthesis of digestive enzymes, helping hydrolyze complex molecules present in food, and siderophores, metal-scavenging compounds that access insoluble iron, are two instances of this paradigm of rivalry planning (Ghoul & Mitri, 2016). To safeguard crops against after-harvesting decays, antagonistic bacteria can also be employed to compete with fungal pathogens for nutrients and space (Wang et al., 2020b; Lastochkina et al., 2019). Therefore, acquiring accessible resource sites may be the initial mechanism in competition even if there may be already an adversarial relationship (such as antibiosis) between two distinct species (Alizadeh et al., 2020; Maloy, 1993). For example, Scuderi et al. (2009) discovered antifungal mediators which acted through competition for resources and space as means of action, through which *Burkholderia gladioli* suppressed green mold of citrus produce (Wang et al., 2020b). Likewise, Wang et al. (2020a) reported that *Metschnikowia citriensis*, a biocontrol yeast, quickly colonised citrus wound tissue and adhered strongly to its surface. This resulted in competition with the infectious agent *Geotrichum citri-aurantii* for nutrients and space (Verma et al., 2022). The outbreak of *Penicillium expansum* on apples was effectively controlled by *Candida saitona* at a dosage of 10^7 CFU/ml, where the strategy again was aggressive competition of resources (Sakib et al., 2020; McLaughlin et al., 1990). Wang et al. (2018) suggested that the ability of *Pseudomonas fluorescens* to suppress green mould on orange fruit was mostly dependent on struggle for habitat and nutrients (Wang et al., 2020c). El Ghaouth et al. (2004) found in a different study that a level of 10^8 CFU/ml of *Candida saitona* was more effective in reducing apple blue mould (*Penicillium expansum*) by providing competition (Sakib et al., 2020). *Aureobasidium pullulans* is an oppositional fungal-like yeast that stops *Penicillium expansum* from developing in juice from fruit (Verma et al., 2022). Through iron exhaustion, *Metschnikowia pulcherrima* was found to compete towards *Pencillium expansum* and *Botrytis cinerea* in apples (Sakib et al., 2020; Saravanakumar et al., 2008).

19.5.3 Parasitism

The capacity of fungal hostile agents to parasitize other fungi using them as a nutrient source is called mycoparasitism, direct parasitism, or hyperparasitism. According to Alizadeh et al. (2020) and Maloy (1993), mycoparasitism either completely destroys fungal propagules or causes their structure to be destroyed and lysed. Therefore, we can understand that the phenomenon of parasitism is not solely targeted towards higher macro-organisms but that micro-life forms also rely on it as a tool for survival and dominance. The following conditions must be accomplished in a chronological order for mycoparasitism to occur: intimate contact with a fungal pathogen, bilateral pathogen and antagonist identification, lytic enzyme release by the antagonist, infiltration inside the target host, ongoing development of the antagonist inside the host, and final escape (Talibi et al., 2014; Alizadeh et al., 2020).

The parasitic action of microorganisms has been found to be useful in providing an antagonistic approach toward several plant or post-harvest pathogens. Therefore, increasing the effectiveness of microbial agents in regulating post-harvest illnesses of fruits and vegetables may depend on selection of microbial antagonists with increased activity of factors or molecules i.e., cell wall breakdown enzymes (Sakib et al., 2020; Wisniewski et al., 1989). The molecular interplay, however, appears complex. The extrinsic β-1,3-glucanase enzyme released by biocontrol yeasts, for examples, also causes lysis of their cell wall (Alizadeh et al., 2020). According to Bonaterra et al. (2003), one of the main mechanisms that *Pantoea agglomerans* deployed to manage *Monilinia laxa* was a direct parasitism (Sakib et al., 2020). According to Wisniewski et al. (1991), who investigated the biological inhibition of *B. cinerea* by the fungal antagonist *Meyerozyma guilliermondii*, the antagonist cell adhered firmly to *B. cinerea* as a result of lectin-like contact (Alizadeh et al., 2020).

19.5.4 Induced Resistance

This is an ancillary or indirect way in which the plants contrast disease-causing pathogens due to the amplification of their resistive power, bestowed by some beneficial microorganisms. Therefore, rhizobacteria enhancing the growth and development of plants can elicit the initiation of systemic resistance (ISR) that protects plants from pathogens (Alizadeh et al., 2020; Sikora, 1992). Additionally, antagonistic yeasts that directly cause induced resistance, as shown for *Candida oleophila* (Wisniewski et al., 1995), *Candida sake* (Teixido et al., 1999; Usall et al., 2001) and *Candida saitona* (El-Ghaouth et al., 1998) were able to reduce apple decay (Wenneker & Thomma, 2020). Cruz et al. (2018) investigated the impact on the strawberry microflora of the biocontrol agents *Bacillus amyloliquefaciens, Beauveria bassiana* and *Trichoderma harzianum* alone and in combination and found them useful by providing resistance. In addition to preventing the propagation of the after-harvest pathogen *B. cinerea* the combined administration of biocontrol agents increased their efficacy (Verma et al., 2022). Fusarium wilt in cucumbers produced by the fungus *F. oxysporum* f.sp. *cucumerinum* was prevented by ISR stimulation induced by the bacteria *Pseudomonas putida* and *Serratia marcescens* (Alizadeh et al., 2020).

Inducing host defense is one of the ways used to antagonize a variety of fruit post-harvest degradation processes (Wang et al., 2020b; Dukare et al., 2019; Moraes Bazioli et al., 2019; Talibi et al., 2014). In order to protect apple host cell walls against the growth of *Penicillium expansum*, *Cyptococcus saitona* deployed a chitinase activity also forming some structural barriers (papillae) (Sakib et al., 2020; El-Ghaouth et al., 1998). Citrus fruit has been shown to express induced resistance by a variety of antagonistic yeasts and bacteria, involving *Debaryomyces hansenii* (Arras & Arru, 1999), *Pichia membranaefaciens* (Luo et al., 2012; Wang et al.,

2020b), *Yarrowia lipolytica* (Zhu et al., 2019) and *Candida oleophila* (Droby et al., 2002). These antagonists also include *P. fluorescens* (Wang et al., 2018) and *B. subtilis* (Waewthongrak et al., 2015). According to Sakib et al. (2020) and Ippolito et al. (2000), *Aureobasidium pullulans* temporarily increased the expression of the enzymes 1, 3-gluconase, peroxidase and chitinase in apple wounds, which accelerated the healing process providing a defense mechanism against *Penicillium expansum*. *Methylobacterium* spp. strains were inoculated against several diseases by Ardanov et al. (2012), who found that the resistance to disease depended on the proportion of the bacterium inoculum, cultivar, and pathogens (Verma et al., 2022).

19.5.5 Cell Wall Degradation Enzymes

Enzyme-producing microorganisms can break down chitin, cellulose, amino acids, and hemicellulose. They may also contribute to the inhibition of plant diseases. Major components of fungal cell walls include chitin and β-1,3-glucans (Alizadeh et al., 2020; Lam & Gaffney, 1993). This is an important feature of antagonistic microorganisms by which they kill the pathogen through a cellular breakdown due to the digestion of the cell wall. The capacity to produce enzymes such as chitinases, glucanases, and proteases that dissolve pathogen cell walls is a primary characteristic of *Trichoderma* strains with antagonistic potential (Alizadeh et al., 2020; López-Mondéjar et al., 2011). Some other microorganisms have shown this antagonism ability, but further research is still required in this field to identify more species with this mode of action. Another study showed that the chitinase-producing *Serratia marcescens* suppresses the development of *F. oxysporum*, *Rhizoctonia solani* and *Botrytis* spp. through this mode of action (Alizadeh et al., 2020; Ningaraju, 2006).

19.5.6 Siderophores

This is a peculiar compound secreted by some antagonistic species. It is different from antibiotics and enzymes responsible for the degradation of the cellular wall, and acts through a fascinating action mechanism. Siderophores are low-molecular-weight chelators that have an extremely high and precise attraction for Fe (III) (Alizadeh et al., 2020; Barbeau et al., 2002). Microorganisms that may produce siderophores, both facultatively, anaerobic and aerobically, may play a significant role in the interactions with pathogens (Alizadeh et al., 2020; Haggag & Mohamed, 2007). The way of siderophore action is different but is equally effective in controlling the disease development. Many bacteria, including BCAs, inhibit the development, proliferation, and metabolism activity of phytopathogens by siderophore-mediated metal chelation. They have been shown to rely on siderophores as a major defense mechanism against phytopathogens (Alizadeh et al., 2020; Haggag & Abo-Sadera, 2000).This is a relatively new field of study, however some examples show a potential in horticulture post-harvest disease control. *Rahnella aquatilis* produces siderophores that have the ability to suppress post-harvest infections by *B. cinerea* and *P. expansum* (Alizadeh et al., 2020; Calvo et al., 2007). The apple-related illnesses caused by *B. cinerea*, *Alternaria alternata* and *P. expansum* were controlled by the siderophore pulcherrimin, produced by *Monilinia fructicola* and *Metschnikowia pulcherrima* (Alizadeh et al., 2020; Saravanakumar et al., 2008).

19.6 Commercial Microbial Antagonists

In an effort to reduce the damages caused by post-harvest rot, biocontrol agents have been developed as viable alternatives to synthetic fungicides, in recent years (Qing, 2000). The solutions found to shield the harvested produce, result from a significant effort deployed over the last 30 years to identify specific organisms within horticultural microbiomes (Wisniewski & Droby, 2019; Droby et al., 2016; Wisniewski et al., 2016). Since they have no detrimental effect on the environment and the well-being of humans, combative yeasts are considered among the finest biological antagonistic agents among the microorganisms identified thus far (Liu et al., 2013a, 2020).

A number of business ventures that rely on the usage of a single antagonist have flourished because of this strategy (Wisniewski & Droby, 2019; Droby et al., 2016; Wisniewski et al., 2016). Yeast biocontrol agents that show resistance to hypoxia, high levels of carbon dioxide, and hypothermia, as well as suitability for circumstances of storage, have been produced and commercialized, over the past few years (Liu et al., 2013b, 2020; Calvo-Garrido et al., 2014; Spadaro & Droby, 2016). The global marketplace for biological goods is currently tiny, and only a few items established a foothold while others were terminated after a short period, despite their promise to control a number of post-harvest pests (Kusstatscher et al., 2020; Wisniewski et al., 2016). Furthermore, due to their lower efficacy rate and stability, in comparison to chemical antimicrobials, fewer antagonistic yeast agents have been used in clinical settings (Liu et al., 2020; Spadaro & Droby, 2016).

Biological control variants may be often obtained from the naturally occurring epiphytic microbiota of horticultural produce (Droby et al., 2009; Liu et al., 2013a). According to Kusstatscher et al. (2020) *B. amyloliquefaciens* (Arrebola et al., 2010), *Pseudomonas syringae* (Bull et al., 1997), *Bacillus megaterium* (Kong et al., 2010), *Erwinia herbicola* (Bryk et al., 1998), and *Pantoea agglomerans* (Usall et al., 2008) are a few of the most well-researched bio-control bacteria. *Pseudomonas syringae* is the basis for BioSave® (JetHarvest Solutions, Florida, USA), one of the first and longest-running post-harvest, bio-control product (Wisniewski et al., 2015; Janisiewicz & Jeffers, 1997). A number of products more widely available on the market include Shemer® (Koppert, The Netherlands) that contains *Metschnikowia fructicola*, Nexy® (BioNext, France) based on *Candida oleophilia*, and Candifruit® developed by IRTA (Spain) that uses *Candida sake* (Kusstatscher et al., 2020).

Other potent, pre- and post-harvest biocontrol agents are derived from bacteria, mostly from the fermentation byproducts of *B. subtilis*, known to produce the antibiotic iturin (Wisniewski et al., 2016). As previously indicated, BioSave was effective in managing post-harvest infections of potatoes and sweet potatoes (Kusstatscher et al., 2020). Despite being produced concurrently with BioSave, the *Candida oleophilia* -based product Aspire® (Ecogen, Inc.) was eventually discontinued from the market due to inadequate marketing strategies and management (Wisniewski et al., 2016). The *B. subtilis* based Serenade® (Bayer, Germany) protects tomato, strawberry, and stone fruit from many diseases. Pantovital® (Domca, Spain) contains *Pantoea agglomerans* and is used on a variety of crops. Several fruit cultivars are treated with Amylo-X® (Biogard CBC, Italy) in conjunction with *B. amyloliquefaciens* (Kusstatscher et al., 2020).

19.7 Potential in Post-harvest Disease Management

A new era in the exploration of post-harvest technology and biological processes has begun with the development of microbiome-based studies. This approach also presents new possible chances for innovative concepts and methods in biocontrol, based on a deeper comprehension of the interactions among the microbes that live on harvested fruits and vegetables (Wang et al., 2020a; Kusstatscher et al., 2020; Wisniewski & Droby, 2019). From an ecological and financial standpoint, using the microbiome as a resource is emerging as a viable area for biotechnological applications (Wassermann et al., 2022; Janisiewicz & Korsten, 2002). The products of microbial applications are biodegradable, do not require harsh temperatures or high energy amounts, and have practically a wide potential for improving plants health, quality, and tolerance to a variety of stressors (Wassermann et al., 2022). In order to increase accessibility to food in a secure, effective, and sustainable manner, Droby et al. (2016) highlighted the significance of biological control in post-harvest protection of food and its nutritional value (Wisniewski & Droby, 2019).

Crop-associated bacteria are relevant to post-harvest treatments because they support plant health throughout growth and maintain the the quality of fruit and plant parts following harvest (Kusstatscher et al., 2020; Droby & Wisniewski, 2018). The previous three decades have seen an increasing effor in scientific research on post-harvest biocontrol, leading to the production of several natural control products (Wisniewski & Droby, 2019; Droby et al., 2016; Wisniewski et al., 2016),. However, the advent of microbiome-based research has opened up a whole new field of study that will significantly advance our understanding of post-harvest pathology and biological interactions, presenting new avenues for microbials usage (Wisniewski & Droby, 2019; Droby & Wisniewski, 2018). As noted by Kusstatscher et al. (2020) and Smalla et al. (2001), the microbiome associated with plants is particularly, and individually, distinctive. According to Kusstatscher et al. (2020) and Berg, Grube, et al. (2014b), plant root exudates are essential chemo-attractants and repellents that draw specific bacteria from the surrounding soil. They can also be utilized to manage diseases that arise in the post-harvest life by direct or indirect means (Kusstatscher et al., 2020).

An increasing body of research indicates that microbiomes and vegetation coevolved, or at least combined to form an amalgamated metaorganism (Berg et al., 2016). This view provides a fresh viewpoint on how the microbiome could be used to create novel food preservation techniques (Wisniewski & Droby, 2019). In their findings, Kusstatscher et al. (2020) stated that strains that have been the focus of recent investigation — i.e., *C. oleophilia* (Lima et al., 1997; Liu et al., 2012), *Metschnikowia fructicola* (Kurtzman & Droby, 2001; Piombo et al., 2018), *C. sake* (Nunes et al., 2001), *P. guilliermondii* (Wilson et al., 1993), *Pichia anomala* (Haïssam, 2011), *Cryptococcus laurentii* (Filonow, 1998), and *Aureobasidium pullulans* (Leibinger et al., 1997; Lima et al., 1997; Mari et al., 2012) consistently demonstrated ensuring bio-control possibilities.

19.8 Future Prospects

The horticulture sector suffers from huge losses, and one of the principal reasons behind them is their high internal moisture content of produce and their storage environments, which elicit and favor the insurgence of microorganism-based pathologies. Thus, attacking pathogens result in the infestation of several diseases, in and outside the field. The challenge in horticulture today concern yield sustainability more than production, where the post-harvest diseases act as a major challenge whose control would consequently lead to manage food losses.

The problem of inorganic chemical approaches resulting in residual toxicity, biomagnifications, etc., has engulfed us in the present situation. The microbial approach is a non-chemical method, which is eco-friendly in nature and, therefore, is the need of the hour. Countless microorganisms may be found in every square inch of soil. Knowing more about them by a rigorous study and research work can open many new possibilities.

Research works is needed in the sector of microbial antagonism as the answers for many of the questions are still unknown. A focus is required in this field by scientists to shed sufficient light on the fascinating world of microorganisms.

Biocontrol products with antagonistic effect *vs* harmful pathogens have been launched in the market. However, they are not yet used in bulk or as a prominent and promising strategy for disease control, because of many reasons. Therefore we need efforts to identify the gaps related to the observed market falls, and design our research accordingly, to expand biological control as a major tool to restrict post-harvest pathogens and the subsequent losses they cause.

19.9 Conclusions

The nutritional value received of horticultural crops is immense, acting as a backbone for sustaining the world population. Hence, it is our duty to take care of our horticultural yield so that their produce can properly reach consumers. As we know, innumerable challenges occur in crop maintenance, both at the field and the storage chamber levels, out of which post-harvest disease prevention is a major one. Thus, we can see how one group of microorganisms can be used to antagonize the other group of pathogens, responsible for disease manifestation in post-harvest life. Therefore, we can state that microbial antagonism is an excellent technique for chemical-free disease control. However, as the microbial world is very diverse, we are required to undertake more extensive efforts to explore the sector and make microbial biocontrol a promising strategy to combat plant and produce diseases in the close future.

References

Abdelfattah, A., Malacrinò, A., Wisniewski, M., Cacciola, S. O., & Schena, L. (2018). Metabarcoding: A powerful tool to investigate microbial communities and shape future plant protection strategies. *Biological Control, 120*, 1–10.

Adam, E., Groenenboo, A. E., Kurm, V., Rajewska, M., Schmidt, R., Ty, O., et al. (2016). Controlling the microbiome: Microhabitat adjustments for successful biocontrol strategies in soil and human gut. *Frontiers in Microbiology, 7*, 1079.

Alizadeh, M., Vasebi, Y., & Safaie, N. (2020). Microbial antagonists against plant pathogens in Iran: A review. *Open Agriculture, 5*, 404–440.

Alkan, N., & Fortes, A. M. (2015). Insights into molecular and metabolic events associated with fruit response to post-harvest fungal pathogens. *Frontiers in Plant Science, 6*, 889.

Ardanov, P., Sessitsch, A., Häggman, H., Kozyrovska, N., & Pirttilä, A. M. (2012). *Methylobacterium*-induced endophyte community changes correspond with protection of plants against pathogen attack. *PLos One, 7*(10), e46802.

Arras, G., & Arru, S. (1999). Integrated control of postharvest citrus decay and induction of phytoalexins by *Debaryomyces hansenii*. *Advances in Horticultural Science*, 76–81.

Arrebola, E., Jacobs, R., & Korsten, L. (2010). Iturin A is the principal inhibitor in the biocontrol activity of *bacillus amyloliquefaciens* PPCB004 against postharvest fungal pathogens. *Journal of Applied Microbiology, 108*(2), 386–395.

Aulakh, J., & Regmi, A. (2013). *Post-harvest food losses estimation: Development of consistent methodology*. Food and Agriculture Organization of the United Nations.

Barbeau, K., Zhang, G., Live, D. H., & Butler, A. (2002). Petrobactin, a photoreactive siderophore produced by the oil-degrading marine bacterium *Marinobacter hydrocarbonoclasticus*. *Journal of the American Chemical Society, 124*, 378–379.

Barkai-Golan, R. (2001). *Postharvest diseases of fruits and vegetables: Development and control*. Elsevier.

Berg, G., & Raaijmakers, J. M. (2018). Saving seed microbiomes. *The ISME Journal, 12*(5), 1167–1170.

Berg, G., Erlacher, A., Smalla, K., & Krause, R. (2014a). Vegetable microbiomes: Is there a connection among opportunistic infections, human health and our 'gut feeling'? *Microbial Biotechnology, 7*(6), 487–495.

Berg, G., Grube, M., Schloter, M., & Smalla, K. (2014b). Unraveling the plant microbiome: Looking back and future perspectives. *Frontiers in Microbiology, 5*.

Berg, G., Rybakova, D., Grube, M., & Köberl, M. (2016). The plant microbiome explored: Implications for experimental botany. *Journal of Experimental Botany, 67*(4), 995–1002.

Berg, G., Rybakova, D., Fischer, D., Cernava, T., Vergès, M. C. C., Charles, T., et al. (2020). Microbiome definition re-visited: Old concepts and new challenges. *Microbiome, 8*(1), 1–22.

Bergna, A., Cernava, T., Rändler, M., Grosch, R., Zachow, C., & Berg, G. (2018). Tomato seeds preferably transmit plant beneficial endophytes. *Phytobiomes Journal, 2*(4), 183–193.

Bonaterra, A., Mari, M., Casalini, L., & Montesinos, E. (2003). Biological control of *Monilinia laxa* and *Rhizopus stolonifer* in postharvest of stone fruit by Pantoea agglomerans EPS125 and putative mechanisms of antagonism. *International Journal of Food Microbiology, 84*(1), 93–104.

Bryk, H., Sobiczewski, P., & Dyki, B. (1998). Antagonistic effect of *Erwinia herbicola* on in vitro spore germination and germ tube elongation of *Botrytis cinerea* and *Penicillium expansum. Biological Control, 43*, 97–106.

Bull, C. T., Stack, J. P., & Smilanick, J. L. (1997). *Pseudomonas syringae* strains ESC-10 and ESC-11 survive in wounds on citrus and control green and blue molds of citrus. *Biological Control, 8*(1), 81–88.

Calvo, J., Calvente, V., de Orellano, M. E., Benuzzi, D., & de Tosetti, M. I. S. (2007). Biological control of postharvest spoilage caused by *Penicillium expansum* and *Botrytis cinerea* in apple by using the bacterium *Rahnella aquatilis. International Journal of Food Microbiology, 113*(3), 251–257.

Calvo-Garrido, C., Teixidó, N., Roudet, J., Viñas, I., Usall, J., & Fermaud, M. (2014). Biological control of botrytis bunch rot in Atlantic climate vineyards with Candida sake CPA-1 and its survival under limiting conditions of temperature and humidity. *Biological Control, 79*, 24–35.

Christensen, C. M., & Kaufmann, H. H. (1965). Deterioration of stored grains by fungi. *Annual Review of Phytopathology, 3*(1), 69–84.

Cordovez, V., Dini-Andreote, F., Carrión, V. J., & Raaijmakers, J. M. (2019). Ecology and evolution of plant microbiomes. *Annual Review of Microbiology, 73*, 69–88.

Cruz, A. F., Barka, G. D., Sylla, J., & Reineke, A. (2018). Biocontrol of strawberry fruit infected by *Botrytis cinerea*: Effects on the microbial communities on fruit assessed by next-generation sequencing. *Journal of Phytopathology, 166*(6), 403–411.

Droby, S. (2005, March). Improving quality and safety of fresh fruits and vegetables after harvest by the use of biocontrol agents and natural materials. In: *International symposium on natural preservatives in food systems 709* (pp. 45–52).

Droby, S., & Wisniewski, M. (2018). The fruit microbiome: A new frontier for postharvest biocontrol and postharvest biology. *Postharvest Biology and Technology, 140*, 107–112.

Droby, S., Vinokur, V., Weiss, B., Cohen, L., Daus, A., Goldschmidt, E. E., & Porat, R. (2002). Induction of resistance to *Penicillium digitatum* in grapefruit by the yeast biocontrol agent *Candida oleophila. Phytopathology, 92*(4), 393–399.

Droby, S., Wisniewski, M., Macarisin, D., & Wilson, C. (2009). Twenty years of postharvest biocontrol research: Is it time for a new paradigm? *Postharvest Biology and Technology, 52*(2), 137–145.

Droby, S., Wisniewski, M., Teixidó, N., Spadaro, D., & Jijakli, M. H. (2016). The science, development, and commercialization of postharvest biocontrol products. *Postharvest Biology and Technology, 122*, 22–29.

Dukare, A. S., Paul, S., Nambi, V. E., Gupta, R. K., Singh, R., Sharma, K., & Vishwakarma, R. K. (2019). Exploitation of microbial antagonists for the control of postharvest diseases of fruits: A review. *Critical Reviews in Food Science and Nutrition, 59*(9), 1498–1513.

El Ghaouth, A., Wilson, C., & Wisniewski, M. (2004). Biologically-based alternatives to synthetic fungicides for the control of postharvest diseases of fruit and vegetables. In S. A. M. H. Naqvi (Ed.), *Diseases of Fruits and Vegetables: Volume II: Diagnosis and Management* (pp. 511–535). Springer.

El-Ghaouth, A., Wilson, C. L., & Wisniewski, M. (1998). Ultrastructural and cytochemical aspects of the biological control of *Botrytis cinerea* by *Candida saitoana* in apple fruit. *Phytopathology, 88*(4), 282–291.

Engelbrecht, G., Horak, I., Jansen van Rensburg, P. J., & Claassens, S. (2018). *Bacillus*-based bionematicides: Development, modes of action and commercialisation. *Biocontrol Science and Technology, 28*(7), 629–653.

Filonow, A. B. (1998). Role of competition for sugars by yeasts in the biocontrol of gray mold of apple. *Biocontrol Science and Technology, 8*(2), 243–256.

Fravel, D. R. (1988). Role of antibiosis in the biocontrol of plant diseases. *Annual Review of Phytopathology, 26*(1), 75–91.

Gardener, B. B. M., & Fravel, D. R. (2002). Biological control of plant pathogens: Research, commercialization, and application in the USA. *Plant Health Progress, 3*(1), 17.

Ghoul, M., & Mitri, S. (2016). The ecology and evolution of microbial competition. *Trends in Microbiology, 24*(10), 833–845.

Gustafsson J, Cederberg, C., Sonesson, U., & Emanuelsson, A. (2013). The methodology of the FAO study: Global food losses and food waste-extent, causes and prevention"-FAO, 2011.

Haas, D., & Défago, G. (2005). Biological control of soil-borne pathogens by fluorescent pseudomonads. *Nature Reviews Microbiology, 3*(4), 307–319.

Haggag, W. M., & Mohamed, H. A. A. (2007). Biotechnological aspects of microorganisms used in plant biological control. *American-Eurasian Journal of Sustainable Agriculture, 1*(1), 7–12.

Haggag, W. M., & Abo-Sadera, S. A. (2000). Influence of iron sources and siderophore producing pseudomonas on crown-rot disease incidence and seed contamination of peanut with pathogenic aspergilli. *Annals of Agricultural Science, Moshtohor, 38*(4), 1933–1948.

Haïssam, J. M. (2011). *Pichia anomala* in biocontrol for apples: 20 years of fundamental research and pratical applications. *Antonie Van Leeuwenhoek, 99*, 93–105.

Hajek, A., & Eilenberg, J. (2018). Microbial antagonists combating plant pathogens and plant parasitic nematodes. In A. E. Hajek & J. Eilenberg (Eds.), *Natural enemies: An introduction to biological control* (pp. 308–324). Cambridge University Press.

Hassani, M. A., Durán, P., & Hacquard, S. (2018). Microbial interactions within the plant holobiont. *Microbiome, 6*, 1–17.

Heydari, A., & Pessarakli, M. (2010). A review on biological control of fungal plant pathogens using microbial antagonists. *Journal of Biological Sciences, 10*, 273–290.

Hibbing, M. E., Fuqua, C., Parsek, M. R., & Peterson, S. B. (2010). Bacterial competition: Surviving and thriving in the microbial jungle. *Nature Reviews Microbiology, 8*, 15–25.

Ippolito, A., El Ghaouth, A., Wilson, C. L., & Wisniewski, M. (2000). Control of postharvest decay of apple fruit by *Aureobasidium pullulans* and induction of defense responses. *Postharvest Biology and Technology, 19*, 265–272.

Janisiewicz, W. J., & Jeffers, S. N. (1997). Efficacy of commercial formulation of two biofungicides for control of blue mold and gray mold of apples in cold storage. *Crop Protection, 16*, 629–633.

Janisiewicz, W. J., & Korsten, L. (2002). Biological control of postharvest diseases of fruits. *Annual Review of Phytopathology, 40*(1), 411–441.

Janisiewicz, W. J., & Roitman, J. (1988). Biological control of blue mold and gray mold on apple and pear with pseudomonas cepacia. *Phytopathology, 78*(12), 1697–1700.

Kong, Q., Shan, S., Liu, Q., Wang, X., & Yu, F. (2010). Biocontrol of aspergillus flavus on peanut kernels by use of a strain of marine bacillus megaterium. *International Journal of Food Microbiology, 139*, 31–35.

Kumar, A., Saini, P., & Shrivastava, J. (2009). Production of peptide antifungal antibiotic and biocontrol activity of *Bacillus subtilis*. *Indian Journal of Experimental Biology, 47*, 57–62.

Kumari, M., Qureshi, K. A., Jaremko, M., White, J., Singh, S. K., Sharma, V. K., et al. (2022). Deciphering the role of endophytic microbiome in postharvest diseases management of fruits: Opportunity areas in commercial up-scale production. *Frontiers in Plant Science, 13*, 4390.

Kurtzman, C. P., & Droby, S. (2001). Metschnikowia fructicola, a new ascosporic yeast with potential for biocontrol of postharvest fruit rots. *Systematic and Applied Microbiology, 24*, 395–399.
Kusstatscher, P., Cernava, T., Abdelfattah, A., Gokul, J., Korsten, L., & Berg, G. (2020). Microbiome approaches provide the key to biologically control postharvest pathogens and storability of fruits and vegetables. *FEMS Microbiology Ecology, 96*, fiaa119.
Lam, S. T., & Gaffney, T. D. (1993). Biological activities of bacteria used in plant pathogen control. In I. Chet (Ed.), *Biotechnology in plant disease control* (pp. 291–320). Wiley-Liss Press.
Lastochkina, O., Seifikalhor, M., Aliniaeifard, S., Baymiev, A., Pusenkova, L., Garipova, S., et al. (2019). *Bacillus* spp.: Efficient biotic strategy to control postharvest diseases of fruits and vegetables. *Plants, 8*, 97.
Leff, J. W., & Fierer, N. (2013). Bacterial communities associated with the surfaces of fresh fruits and vegetables. *PLoS One, 8*(3), e59310.
Leibinger, W., Breuker, B., Hahn, M., & Mendgen, K. (1997). Control of postharvest pathogens and colonization of the apple surface by antagonistic microorganisms in the field. *Phytopathology, 87*, 1103–1110.
Lima, G., Ippolito, A., Nigro, F., & Salerno, M. (1997). Effectiveness of *Aureobasidium pullulans* and *Candida oleophila* against postharvest strawberry rots. *Postharvest Biology and Technology, 10*, 169–178.
Liu, J., Wisniewski, M., Droby, S., Norelli, J., Hershkovitz, V., Tian, S., & Farrell, R. (2012). Increase in antioxidant gene transcripts, stress tolerance and biocontrol efficacy of *Candida oleophila* following sublethal oxidative stress exposure. *FEMS Microbiology Ecology, 80*(3), 578–590.
Liu, J., Sui, Y., Wisniewski, M., Droby, S., & Liu, Y. (2013a). Utilization of antagonistic yeasts to manage postharvest fungal diseases of fruit. *International Journal of Food Microbiology, 167*(2), 153–160.
Liu, J., Wisniewski, M., Artlip, T., Sui, Y., Droby, S., & Norelli, J. (2013b). The potential role of PR-8 gene of apple fruit in the mode of action of the yeast antagonist, *Candida oleophila*, in postharvest biocontrol of *Botrytis cinerea*. *Postharvest Biology and Technology, 85*, 203–209.
Liu, X., Gao, Y., Yang, H., Li, L., Jiang, Y., Li, Y., & Zheng, J. (2020). *Pichia kudriavzevii* retards fungal decay by influencing the fungal community succession during cherry tomato fruit storage. *Food Microbiology, 88*, 103404.
López-Mondéjar, R., Ros, M., & Pascual, J. A. (2011). Mycoparasitism-related genes expression of *Trichoderma harzianum* isolates to evaluate their efficacy as biological control agent. *Biological Control, 56*(1), 59–66.
Luo, Y., Zeng, K., & Ming, J. (2012). Control of blue and green mold decay of citrus fruit by *Pichia membranefaciens* and induction of defense responses. *Scientia Horticulturae, 135*, 120–127.
MacLean, R. C., & Gudelj, I. (2006). Resource competition and social conflict in experimental populations of yeast. *Nature, 441*(7092), 498–501.
Maloy, O. C. (1993). *Plant disease control: Principles and practice*. Wiley.
Mari, M., Martini, C., Spadoni, A., Rouissi, W., & Bertolini, P. (2012). Biocontrol of apple postharvest decay by *Aureobasidium pullulans*. *Postharvest Biology and Technology, 73*, 56–62.
McLaughlin, R. J., Wilson, C. L., Chalutz, E., Kurtzman, C. P., Fett, W. F., & Osman, S. F. (1990). Characterization and reclassification of yeasts used for biological control of postharvest diseases of fruits and vegetables. *Applied and Environmental Microbiology, 56*(11), 3583–3586.
Mitter, B., Pfaffenbichler, N., Flavell, R., Compant, S., Antonielli, L., Petric, A., et al. (2017). A new approach to modify plant microbiomes and traits by introducing beneficial bacteria at flowering into progeny seeds. *Frontiers in Microbiology, 8*, 11.
Moraes Bazioli, J., Belinato, J. R., Costa, J. H., Akiyama, D. Y., Pontes, J. G. D. M., Kupper, K. C., et al. (2019). Biological control of citrus postharvest phytopathogens. *Toxins, 11*(8), 460.
Ningaraju, T. M. (2006). Cloning and characterization of chitinase gene/s from native isolates of *Serratia marcescens*. *Indian National Agricultural Research System, 1748*, 1–20.

Nunes, C., Usall, J., Teixidó, N., Miró, M., & Viñas, I. (2001). Nutritional enhancement of biocontrol activity of *Candida sake* (CPA-1) against *Penicillium expansum* on apples and pears. *European Journal of Plant Pathology, 107*, 543–551.

Pathak, P., Rai, V. K., Can, H., Singh, S. K., Kumar, D., Bhardwaj, N., et al. (2022). Plant-endophyte interaction during biotic stress management. *Plants, 11*(17), 2203.

Pfeiffer, T., Schuster, S., & Bonhoeffer, S. (2001). Cooperation and competition in the evolution of ATP-producing pathways. *Science, 292*(5516), 504–507.

Piombo, E., Sela, N., Wisniewski, M., Hoffmann, M., Gullino, M. L., Allard, M. W., et al. (2018). Genome sequence, assembly and characterization of two *Metschnikowia fructicola* strains used as biocontrol agents of postharvest diseases. *Frontiers in Microbiology, 9*, 593.

Prusky, D., Alkan, N., Mengiste, T., & Fluhr, R. (2013). Quiescent and necrotrophic lifestyle choice during postharvest disease development. *Annual Review of Phytopathology, 51*, 155–176.

Pusey, P. L., & Wilson, C. (1984). Postharvest biological control of stone fruit brown rot by *Bacillus subtilis*. *Plant Disease, 68*(9), 753–756.

Qing, T. S. F. (2000). Biological Technologies for the Control of postharvest diseases of fruits and vegetables. *Botany Letters, 17*, 211–217.

Rawal, R. D., & Sharma, J. (2015). *Postharvest diseases. Managing postharvest quality and losses in horticultural crops. 1*. Daya Publishing House. 110002.

Sakib, N., Bhat, M. A., Dar, J. A., & Sultan, A. (2020). Biological control and its role in management of post-harvest diseases. *International Journal of Current Microbiology and Applied Sciences, 11*, 2251–2264.

Sánchez-Cañizares, C., Jorrín, B., Poole, P. S., & Tkacz, A. (2017). Understanding the holobiont: The interdependence of plants and their microbiome. *Current Opinion in Microbiology, 38*, 188–196.

Saravanakumar, D., Ciavorella, A., Spadaro, D., Garibaldi, A., & Gullino, M. L. (2008). *Metschnikowia pulcherrima* strain MACH1 outcompetes *Botrytis cinerea*, *Alternaria alternata* and *Penicillium expansum* in apples through iron depletion. *Postharvest Biology and Technology, 49*(1), 121–128.

Scuderi, G., Bonaccorsi, A., Panebianco, S., Vitale, A., Polizzi, G., & Cirvilleri, G. (2009). Some strains of *Burkholderia gladioli* are potential candidates for postharvest biocontrol of fungal rots in citrus and apple fruits. *Journal of Plant Pathology*, 207–213.

Sikora, R. A. (1992). Management of the antagonistic potential in agricultural ecosystems for the biological control of plant parasitic nematodes. *Annual Review of Phytopathology, 30*(1), 245–270.

Smalla, K., Wieland, G., Buchner, A., Zock, A., Parzy, J., Kaiser, S., et al. (2001). Bulk and rhizosphere soil bacterial communities studied by denaturing gradient gel electrophoresis: Plant-dependent enrichment and seasonal shifts revealed. *Applied and Environmental Microbiology, 67*(10), 4742–4751.

Spadaro, D., & Droby, S. (2016). Development of biocontrol products for postharvest diseases of fruit: The importance of elucidating the mechanisms of action of yeast antagonists. *Trends in Food Science & Technology, 47*, 39–49.

Talibi, I., Boubaker, H., Boudyach, E. H., & Ait Ben Aoumar, A. (2014). Alternative methods for the control of postharvest citrus diseases. *Journal of Applied Microbiology, 117*(1), 1–17.

Teixido, N., Usall, J., & Vinas, I. (1999). Efficacy of preharvest and postharvest *Candida sake* biocontrol treatments to prevent blue mould on apples during cold storage. *International Journal of Food Microbiology, 50*(3), 203–210.

Usall, J., Teixidó, N., Torres, R., de Eribe, X. O., & Viñas, I. (2001). Pilot tests of *Candida sake* (CPA-1) applications to control postharvest blue mold on apple fruit. *Postharvest Biology and Technology, 21*(2), 147–156.

Usall, J., Smilanick, J., Palou, L., Denis-Arrue, N., Teixidó, N., Torres, R., & Vinas, I. (2008). Preventive and curative activity of combined treatments of sodium carbonates and *Pantoea agglomerans* CPA-2 to control postharvest green mold of citrus fruit. *Postharvest Biology and Technology, 50*(1), 1–7.

Verma, S., Azevedo, L. C. B., Pandey, J., Khusharia, S., Kumari, M., Kumar, D., et al. (2022). Microbial intervention: An approach to combat the postharvest pathogens of fruits. *Plants, 11*(24), 3452.

Vulić, M., & Kolter, R. (2001). Evolutionary cheating in *Escherichia coli* stationary phase cultures. *Genetics, 158*(2), 519–526.

Waewthongrak, W., Pisuchpen, S., & Leelasuphakul, W. (2015). Effect of *Bacillus subtilis* and chitosan applications on green mold (*Penicillium digitatum* Sacc.) decay in citrus fruit. *Postharvest Biology and Technology, 99*, 44–49. https://doi.org/10.1016/j.postharvbio.2014.07.016

Wang, Z., Jiang, M., Chen, K., Wang, K., Du, M., Zalán, Z., et al. (2018). Biocontrol of *Penicillium digitatum* on postharvest citrus fruits by *Pseudomonas fluorescens*. *Journal of Food Quality, 2018*, 1–10.

Wang, S., Ruan, C., Yi, L., Deng, L., Yao, S., & Zeng, K. (2020a). Biocontrol ability and action mechanism of *Metschnikowia citriensis* against *Geotrichum citri-aurantii* causing sour rot of postharvest citrus fruit. *Food Microbiology, 87*, 103375.

Wang, S., Zhou, Y., Luo, W., Deng, L., Yao, S., & Zeng, K. (2020b). Primary metabolites analysis of induced citrus fruit disease resistance upon treatment with oligochitosan, salicylic acid and *Pichia membranaefaciens*. *Biological Control, 148*, 104289.

Wang, Z., Mei, X., Du, M., Chen, K., Jiang, M., Wang, K., et al. (2020c). Potential modes of action of *Pseudomonas fluorescens* ZX during biocontrol of blue mold decay on postharvest citrus. *Journal of the Science of Food and Agriculture, 100*(2), 744–754.

Wassermann, B., Abdelfattah, A., Cernava, T., Wicaksono, W., & Berg, G. (2022). Microbiome-based biotechnology for reducing food loss post harvest. *Current Opinion in Biotechnology, 78*, 102808.

Wenneker, M., & Thomma, B. P. (2020). Latent postharvest pathogens of pome fruit and their management: From single measures to a systems intervention approach. *European Journal of Plant Pathology, 156*, 663–681.

Wilson, C. L., Wisniewski, M. E., Droby, S., & Chalutz, E. (1993). A selection strategy for microbial antagonists to control postharvest diseases of fruits and vegetables. *Scientia Horticulturae, 53*(3), 183–189.

Wisniewski, M., & Droby, S. (2019). The postharvest microbiome: The other half of sustainability. *Biological Control, 137*, 104025.

Wisniewski, M., Wilson, C., & Hershberger, W. (1989). Characterization of inhibition of *Rhizopus stolonifer* germination and growth by *Enterobacter cloacae*. *Canadian Journal of Botany, 67*(8), 2317–2323.

Wisniewski, M., Biles, C., Droby, S., McLaughlin, R., Wilson, C., & Chalutz, E. (1991). Mode of action of the postharvest biocontrol yeast, *Pichia guilliermondii*. I. Characterization of attachment to *Botrytis cinerea*. *Physiological and Molecular Plant Pathology, 39*(4), 245–258.

Wisniewski, M., Droby, S., Chalutz, E., & Eilam, Y. (1995). Effects of Ca^{2+} and Mg^{2+} on *Botrytis cinerea* and *Penicillium expansum* in vitro and on the biocontrol activity of *Candida oleophila*. *Plant Pathology, 44*(6), 1016–1024.

Wisniewski, M., Norelli, J., Droby, S., Ballester, A. R., Abdelfattah, A., & Levin, E. (2015). Genomic tools for developing markers for postharvest disease resistance in Rosaceae fruit crops. In *III International symposium on postharvest pathology: Using science to increase food availability, 1144*, 7–16.

Wisniewski, M., Droby, S., Norelli, J., Liu, J., & Schena, L. (2016). Alternative management technologies for postharvest disease control: The journey from simplicity to complexity. *Postharvest Biology and Technology, 122*, 3–10.

Zhu, H., Zhao, L., Zhang, X., Foku, J. M., Li, J., Hu, W., & Zhang, H. (2019). Efficacy of *Yarrowia lipolytica* in the biocontrol of green mold and blue mold in *Citrus reticulata* and the mechanisms involved. *Biological Control, 139*, 104096.

Chapter 20
Management of Mycotoxin Problems: Prospects and Limitations

S. Kiran, M. Surekha, and S. M. Reddy

Abstract Mycotoxins are secondary metabolites of storage moulds toxic to man and animals when consumed through infested foods and feeds causing a variety of health hazards. They are not only thermostable but also do not degrade easily and act at threshold concentrations causing irreversible health disorder. Mycotoxins are hard to remove from food grains and remain in the food chain. The substantial financial losses linked to mycotoxins and the effects on animal productivity, human health and both domestic and international trades have kept them in the public eye. Mycotoxins are unavoidable, more toxic than pesticides and can only be contained at food safety permissible levels. They have a potential to cause cancer apart from effects such as mutagenicity, teratogenicity, neurotoxicity, immunotoxicity or sudden death. By implementing different pre- and post-harvest strategies based on good field agricultural and manufacturing practices during food handling, storage and distribution after harvest, negative effects of food grains mycotoxins can be prevented. Hazard analysis and critical control points aid in their prevention in addition to management methods. Even with thorough and in-depth study, a sizable fraction of farmers, merchants, processors and consumers still only partially adhere to the recommendations made for the mycotoxins management in foods and feeds. Various measures must be taken to contain health hazardous mycotoxins while foods are in the production stage in the field, during harvest, post-harvest stage, storage, processing and consumption are herein discussed. Measures which can be taken after they enter in to food chain such as decontamination, detoxification by different physical, chemical and biological methods are also discussed. An integrated approach to this problem appears of immense help to protect the health of man and animals, and the global economy.

S. Kiran (✉)
Department of Botany, Satavahana University, Karimnagar, Telangana, India

M. Surekha · S. M. Reddy
Department of Botany, Kakatiya University, Warangal, Telangana, India

P. Mathur, S. Roy (eds.), *Plant Microbiome and Biological Control*, Sustainability in Plant and Crop Protection 20,
https://doi.org/10.1007/978-3-031-75845-4_20

Keywords Biocontrol detoxification · Biotransformation · Food and feed · Prevention of moulds · Mycotoxin contamination · Pre-and post-harvest · Quality regulating measures

Abbreviations

2,4-D	2,4-Dichlorophenoxyacetic acid
AFs	aflatoxins
AOAC	Association of Official Agricultural Chemists
BHA	Butylated hydroxy anisole
BHT	Butylated hydroxytoluene
CAPP	Cold atmospheric pressure plasma
CAS	Controlled atmosphere storage
CCP's	Codex code principles
BCHA	Branchedchain higher acids
DON	Deoxynivalenol
EFSA	European Food Safety Authority
ELISA	Enzymelinked immunosorbent assay
EST	Expressed sequence tages
FASSAI	Food safety and standards Authority of India
FB	Fumonisin B
FCMAP	Food Contamination Monitoring and Assessment Programme
GAP	Good Agricultural Practice
GEMS	Global Environment Monitoring System
GHP	Good Hygienic Practice
GRAS	Generally recognized as safe
GSP	Good storage practices
HACCP	Hazard Analysis and Critical Control Point
HIGS	Host induced gene silencing
LAB	Lactic acid bacteria
MAF	Modified atmosphere packing
MGBG	Methylglyoxal bisguanylhydrazone
OTA	Ochratoxin A
PAT	Patulin
PG	Propyl gallate
RH	Relative humidity
RNAi	RNA interference
TCT	Trichothecenes
TMTD	Trimethyl thiuram disulphide
UNEP	United Nations Environment Programme
WHO	World Health Organization
ZEN	Zearalenone

20.1 Introduction

Mycotoxins, which are commonly recognized as the secondary metabolites of storage moulds, are said to play a critical role in the global economy of any given country. Food containing mycotoxins beyond a predetermined threshold is either rejected outright or sold for a reduced price and used for a different purpose in regional or even international markets, incurring significant financial losses (Bhat & Vasanthi, 1999). Cereals, nuts, corn, beans, spices, fruits, vegetables, oil seeds and farm animal feed are naturally contaminated by mycotoxins. The majority of mycotoxins are difficult to break down chemically. They remain in the plough to plate and constitute a major problem because they are difficult to remove before and after harvest (Awad et al., 2010).

Gbashi et al. (2019) examined the socioeconomic consequences of mycotoxin exposure. The majority of countries experience a significant problem with mycotoxin contamination of their food sources. According to data from the Global Environment Monitoring System (GEMS), Food Contamination Monitoring and Assessment Programme (FCMAP), the risk to the health of humans and animals can be significantly reduced by strict regulation of food and feed as well as the implementation of suitable public health initiatives. As a result, several nations have passed laws to regulate the amount of mycotoxins in both domestic food supplies and food that is traded internationally. Guidelines or laws that set maximum acceptable limits for mycotoxin levels in foods and feeds are in practice in 18 countries worldwide.

In India, food commodities may contain up to 30 parts per billion of AFB, but there are no restrictions on the amount of other mycotoxins that can be present. Therefore, through a network of knowledgeable teams of the Save Grain Campaign, government agencies should purchase food grains that confirm to prescribe requirements in order to minimize the possibility of contamination and guarantee the supply of highquality food grains to consumers. Even though more than 100 fungal species produce about 400 different types of mycotoxins, only a few are most prevalent, enter the food chain and infest crops both before and after harvest, posing the greatest threat to human and livestock health and causing losses to the agricultural, animal husbandry and public health sector. The first line of defense against food contaminants is the human gut and mycotoxin targets the gastrointestinal tract first. Contamination of interior environments with mycotoxin is a global issue, particularly in less technologically advanced nations. Mycotoxin exposure occurs to human and animals through food, inhalation or touch and can cause a variety of buildingrelated ailments, such as cancer, neurological disorders and kidney disease.

There is an urgent need to frame a National Guidelines for prevention and control of mycotoxins contamination specially aflatoxins in agricultural commodities, which should include: (1) harvesting of fully matured grains/cobs/ears and (2) drying to safe moisture level (less than 10%), (3) separating the healthy, sound grains/cobs/ears from damaged or rotted or infected, (4) drying the grains/cobs/ears on raised bamboo floors along with other appropriate drying methods as required for the merchants, (5) using a water proof sheet (plastic not advisable) to cover dried produce from rain, if required mechanical drying can be done, (6) storage at proper dry conditions, along with good ventilation, efficient sanitation practices to destroy or discard infested ears/cobs/etc., (7) and effective fumigation for control of insect problem.

Inspite of intensive researches on mycotoxins during last 30 years, we do not have a distinct picture of the impact of mycotoxins exposure and assessment of the risk. There is an urgent need for coordination and demarcation of work on the basis of regional frontiers and expertise available. Various physical (moisture, temperature, mechanical damage and length of storage etc.,), chemical (level of CO_2 and chemical nature of substrate) and biological (plant varietal differences, availability of inoculum spore load, strain variation (lectins) and microbial ecosystem) which are reported to influence the mycotoxin contamination of foods and feeds are to be studied in detail. Therefore, ecology of moulds which produce mycotoxins should be better understood and appropriate measures should be taken to control the mycotoxin contamination. Kabak et al. (2006) and Agriopoulou et al. (2020) have discussed various aspects and strategies to be developed for prevention of mycotoxin contamination and have also suggested different molecular approaches for eliminating mycotoxins from foods. Nada et al. (2022) discussed different strategies to be adopted in the management of mycotoxin problem of wheat and maize. Figure 20.1 illustrates the possibilities of postharvest problem of mycotoxins and their management.

In most of the regions of India the relative humidity remains fairly high during monsoon, when the temperature shows considerable fluctuations, favours the growth of mycotoxigenic moulds. Further, unseasonal rains and flash floods are frequent

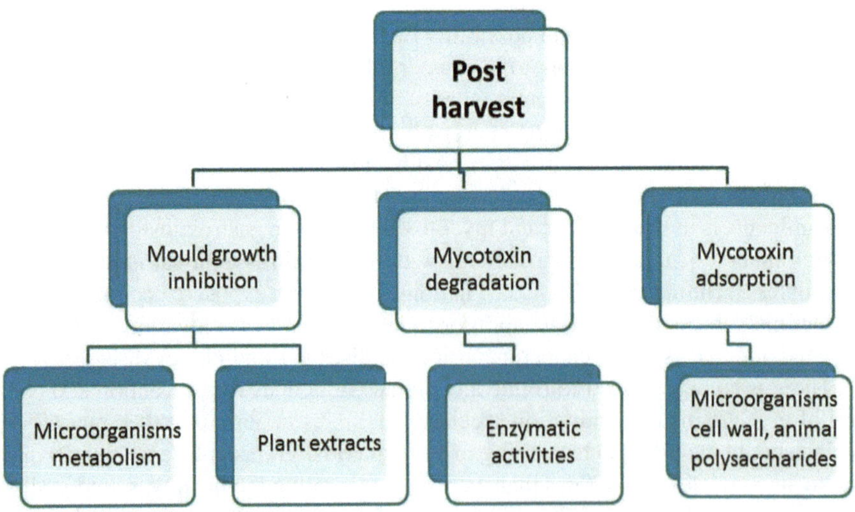

Fig. 20.1 The possibilities of post-harvest control of mold growth for mycotoxin elimination

which also results in increased moisture level of the grains during the most of time in a year. Critical perusal of reports indicate that river basins of Ganga, Brahmapurtra, Godavari, Krishna as well as coastal areas of Andhra Pradesh, Tamil Nadu and Karnataka have high incidence of mycotoxigenic fungi both on standing crop as well as in stored agricultural produce. Similarly, almost every year coastal belt of Telangana State is affected by floods which promote mould infestation and damage both in stored and standing crops (Kiran et al., 2015). Practically all food and feed commodities are prone to mould infestation. On the basis of reports of natural occurrence and levels of mycotoxins contamination from different parts of India, maize, groundnut, rice, cotton seed and some millets can be categorized as high risk commodities. Insects and birds are reported to promote mould infestation due to the damage caused by them to grains on standing crop (Sinha & Ranjan, 1989).

20.2 Prevention of Mould Growth and Mycotoxin Contamination

It is now realized that contamination of food stuffs by moulds and mycotoxins is worldwide phenomenon and the problem is more acute in tropical and subtropical countries like India. Further, people of developing countries are likely to be victims of first order due to increasing incidence of these mycotoxins over a wide range of foods and feeds. Hence, there is an urgent need to prevent infestation of food products by mycotoxin producing moulds by manipulation of the environment. The goal of various control strategies should be to either turn the toxin into a nontoxic, biodegradable product or lower the mycotoxin content to a safe level without compromising the nutritional value of the commodities that have been treated.

Liu et al. (2022) discussed merits and demerits of different methods in the management of mycotoxin in foods and feeds. Awuchi et al. (2021) and Nada et al. (2022) have also discussed different strategies to be adapted in the control of mycotoxins contamination of wheat and maize. They are of the opinion that while choosing the methods to control mycotoxin contamination of the foods, the following measures are to be taken up:

1. Technically and financially feasible as well as relevant
2. Avoid significant altering the nutritional value of food and producing additional harmful substances.
3. Avoid releasing harmful and poisonous residues.
4. Bioacceptable, friendly environment to crop producer and consumer.

Two primary categories can be used to group mycotoxin control techniques:

1. Avoiding the growth of mould and the contamination of mycotoxin.
2. Detoxification of toxic product.

Understanding the fungus that produce mycotoxins is necessary to develop mitigation or removal plans. The primary source of mycotoxin generation is the development of fungus in crops and agricultural goods. Mycotoxin production is enhanced by a multitude of environmental conditions such as: (1) Plant sensitivity to fungal infestation, (2) Suitability of substrate for fungal growth, (3) Temperature and moisture content, (4) Insect and pest related physical damage to seeds.

20.2.1 Preharvest Measures

Preharvest methods for controlling mycotoxin include breeding, timing of sowing, irrigation, optimum crop rotation, tillage, application of fungicide to control crop infestations in fields and weed control.

Selection
The growth of fungi and the formation of mycotoxins are inhibited by the selection of resistant or less susceptible varieties to fungal contamination. It is the most effective method of managing fungi that produce mycotoxin (Luo et al., 2018).

Sowing Time
Plant blossoming is influenced by the timing of sowing. Cereals enter their blossoming phase depending on sowing date. The correlation observed between the time of planting and the attainment of several growth phases suggests that the timing of planting and crop maturity can have a substantial impact on the proliferation of mycotoxigenic fungi and mycotoxin contamination. The crop is most heavily contaminated with mycotoxin during the flowering period or spore formation (Munkvold, 2003). For example, early planting dates for maize tend to minimize mycotoxin contamination levels (Munkvold, 2003). Winter wheat and barley cultivars have a higher risk of *Fusarium* infestation because they develop and mature earlier than spring varieties (Jouany, 2007). Unfavourable environmental conditions favour increases fungal danger of contamination.

Irrigation At some growth stages, irrigation helps plants to become less stressed. However, during excessive rainfall, particularly flowering encourages the spread of *Fusarium* sp. and crop infestation. So irrigation should not be used during the flowering stage and maturation of crops, particularly wheat, barley and rye (Awad et al., 2010).

Correct Crop Rotation Preventing grain damage is crucial because it reduces the likelihood of fungal infection and other damages caused by insects, birds or mechanical means. Fungal spores infect plants during their first year of life and they have the ability to colonise other plants and spread through the soil. After maize is harvested, stubble is left in the ground where mould can proliferate and if it is cultivated in the same land the next year (Jouany, 2007).

Crop disease spreads more readily depending on cropping systems. When wheat is sown in the same field every year fungal infestation is more when compared to or when maize and wheat are rotated. There is a greater chance of mycotoxin contamination in the absence of crop rotation. The incidence of mycotoxin contamination can be decreased by using a multifield crop rotation such as alternate cultivation of rapeseed, sugar beet root, sunflower or soybeans (Jouany, 2007; Awad et al., 2010). Schaafsm et al. (2001) noted that planting crops other than wheat 2 years prior to growing wheat in the study field significantly reduced deoxynivalenol (DON) contamination of wheat.

Tillage Appropriate tillage and management of postharvest plant waste in the field are critical to lowering the danger of mycotoxin contamination. For instance, minimal tillage (to a depth of 10–20 cm) and no tillage are less successful than ploughing the soil (to a depth of 10–30 cm) in order to counteract mycotoxin contamination. Ploughing after harvest slows down the growth of *F. graminearum,* which is responsible for mycotoxin DON contamination and redrot on maize cobs (Gibberella stage). Furthermore, crop waste can be burned, removed or ploughed to reduce the likelihood of fungal infestation in upcoming harvests.

Proper Fertilization No information exists regarding the impact of fertilizers on the degree of fungal and mycotoxin contamination in crops. This can be achieved by changing the crop residues, influencing plant development and modifying the soil's microbial activity and structure.

Weed, Insect and Pest Control A vast variety of moulds can infest weeds. Any crop like wheat that has a lot of weeds in it, will be highly contaminated. Pests and insects can harm kernels, making it easier for moulds and mycotoxin contamination to spread widely. Toxigenic fungi can be categorized into three classes based on the stage of their infestation on grains: (a) field fungi (b) storage fungi and (c) advanced degradation fungi. Plant pathogenic fungi belonging to the genus *Fusarium* (*F. moniliforme, F. roseus, F. tricinctum* and *F. nivale*) are included in the first group. The majority of the storage fungi are species belonging to *Penicillium* and *Aspergillus* genera. The fungi responsible for advanced deterioration often avoid infecting healthy grains, but they can readily target damaged ones and need high moisture content. Species of *Aspergillus, Chaetomium, Scopulariopsis, Rhizopus, Mucor* and *Absidia* are a few fungal genera of the third category. It is a difficult task to prevent mycotoxin contamination in a varied environment conditions. Mycotoxins elaborated during the stage of food and feed production can be effectively prevented using postharvest prevention measures such as applying processing procedures that will minimize subsequent entry into the food and feed chain by conceivable ways of inhibition, detoxification or destruction of mycotoxins and fungal contamination that occurred at preharvest can be minimized at postharvest. There are 5 main keys to prevent and control mycotoxins in grains which can be taken up at three steps as presented in Table 20.1.

Table 20.1 Five keys to prevent and control mycotoxins in grains

Recommendations	Actionable steps to achieve recommendations	Reasoning behind the recommendations
Sustain plant's vigour and health	1.Timely plant agroecologically adapted varieties, 2.Plant sound seeds, 3.Maintain soil fertility by applying fertilizers and manure, and/or following other integrated approaches, 4.Maintain adequate soil, maintain adequate soil moisture during plant growth, 5.Use correct plant spacing, 6.Use pest and disease resistant and tolerant varieties,7.Control pests and diseases,8.Timely weed (remove weeds)	A vigourous and healthy plant is capable of fighting pests and diseases including fungi. A diseased/wounded crop is prone to fungal attack.
Reduce toxigenic fungal population in growing plants and in storage	1. Follow crop rotation, 2.Use biocontrol if available, 3. Remove and burn all diseased plants, 4. Do not let produce come into contact with soil, 5.Optimally apply fungicides to growing plants and in storage whenever necessary, 6.Clean and disinfect storage bags and facilities	While fungi are omnipresent, some environments have higher population than others and consequently present higher mycotoxin contamination risk. Strategies that reduce the chance of grains coming in contact with high fungal population or reduce the fungi themselves, lower the risk of mycotoxin contamination.
Rapidly reduce moisture content of grains and avoid rehydration	1. Timely harvest, 2. Dry crop rapidly and completely, as soon as possible after harvest, 3.Keep and transport produce under dry conditions (well aerated, no leakage, no contact with walls and floors),4.Do not keep water in a commodity store, 5.Avoid pest (rodents, insect etc) attack 6.Avoid combining stored grains and domesticated animals and people (sleeping), 7.Use desiccants such as calcium chloride ($CaCl_2$) to maintain low grain moisture	High grain moisture content in harvested grains and high relative humidity of the surrounding air are critical factors for fungal proliferation and mycotoxin contamination. Rapidly drying grains immediately after harvest and keeping them under dry environment reduces the risk of fungal proliferation and mycotoxin contamination.
Safeguard husks/hulls or pericarp/testa	1. Avoid mechanical damage during harvesting and threshing, 2. A void rodent and insect attack.	The outer structures of seeds/grains such as the hull/pods and seed coat/ testa, provide first line of defence against microorganisms threatening the seed integrity and fungal penetration. Conversely, damaged grains are more prone to fungal penetration and mycotoxin contamination
Clean and remove mycotoxin high risk components (mouldy/ diseased, immature and damaged fraction, dirt and debris)	1. Scour, clean grain surfaces, sieve and winnow off fines/ screenings such as dirt and debris, 2.Electronically or mechanically or manually sort and discard diseased, immature, shriveled and broken grains, 3. Separate grains based on density and discard buoyant fractions.	Mycotoxins are concentrated in fine material such as dirt and debris and in few grains particularly those with extensive mold growth, broken, insect damaged, shriveled and immature. Cleaning and removal of these risky components potentially reduce mycotoxin concentration

Source: Matumba et al. (2021)

20.2.2 Prevention

Primary

The preventative measures ought to be implemented prior to the appearance of fungal infestation and mycotoxin contamination. The most crucial and successful strategy for decreasing fungal growth and mycotoxin generation is at this level of prevention. Some of the practices recommended to create unfavorable conditions for the growth of mycotoxin producing fungi are:

1. Selecting crop varieties which are resistant to mycotoxigenic fungi
2. Management of field infection of planting crops by fungi
3. Formulation of schedule for appropriate preharvest, harvest, and postharvest activities
4. Reducing the moisture content of seeds during postharvest and storage
5. When feasible, store goods at a low temperature.
6. To prevent mould growth use food preservatives and fungicides.
7. Using authorized insecticides to control insect infestation in bulk stored grains.

Secondary

In order to stop further degradation and mycotoxin contamination, toxic fungi that invade commodities at an early stage should either be removed or their growth should be halted. Some of the methods recommended are:

1. Re-dry the products to prevent further growth of mycotoxigenic fungi.
2. Elimination of infested seeds
3. Detoxification or inactivation of food grains contaminated with mycotoxins
4. Keep stored goods away from environment that encourages the growth of mycotoxigenic fungi.

Tertiary

Both the primary and secondary preventions would be ineffective once the products were heavily infested by mycotoxigenic fungi. Some of the methods that can check further increase of myctoxigenic fungal population.

1. Total discard of the infested products.
2. Protect products from further increase of mycortoxigenic fungal growth by storing at unfavorable conditions.

Seeds rich in antifungal substances are likely to be resistant to fungal infestation and mycotoxin contamination. Fungal-resistant varieties of seeds should be used for sowing. The quest for the "super" plant to combat mould infestation is beset with other technical difficulties as well. It is still feasible and encouraged to generate fungal-resistant plant cultivars that use genetic resistance to mycotoxin contamination. Following the harvest, it is crucial to dry the produce, store it properly and transport it in an appropriate manner. Mould growth and mycotoxin production are influenced by a number of factors including high moisture content, humid climate, warm temperature (25–40 °C), insect infestation, and pest damage. Therefore, appropriate measures should be taken up to prevent fungal contamination. The Food and Agriculture Organization (FAO) also offers advice services to its member nations to help them to realize the significance of the mycotoxin contamination and to prepare plans and tactics that will effectively control mycotoxin contamination. FAO has organized two international conferences and suggested the following recommendations

1. The development of workable and realistic strategies to stop food crops get contaminated after harvest and to improve postharvest storage at the local and warehouse levels to prevent fungal infestation
2. The expansion and synchronisation of studies on mycotoxin toxicity, prevalence, avoidance and removal of mycotoxins.
3. The creation of practical, affordable and secure methods to prevent the foods from mycotoxin contamination.
4. The strengthening of existing research centers of mycotoxin control with modern facilities and training of laboratory personnel and extension staff.
5. Developing and standardizing the application and harmonization of dependable, affordable and globally recognized techniques for sampling, sample preparation and analysis, standardization of laws especially those establishing allowable threshold of mycotoxin levels in food.

20.3 Control Strategies for Mycotoxin Contamination

According to FAO (Scott, 1998), the mycotoxin decontamination process must:

1. Eliminate, deactivate or eradicate the mycotoxin
2. Ensure that no harmful, cancer causing or mutagenic substances are left behind infinished goods or food items derived from animals fed with decontaminated feed.
3. Preserve the product's acceptability and nutritional value.
4. Not to change the crucial technology characteristics
5. Eliminate fungal spores and mycelia, which have the potential to multiply and produce new mycotoxins under ideal conditions.

Additional information on the method's environmental impact is required by the US Food and Drug Administration. Scott (1998) has given an overview of methods used in industrial decontamination of foods and feeds (Fig. 20.2).

According to Umesha et al. (2017) it is challenging to prevent mycotoxin contamination in the field throughout the harvesting, storage and transportation of food and feed commodities. Mycotoxin contamination is caused by different reasons, including soil wetness, insect invasion and mineral shortages (Murphy et al., 2006). Preventing fungal contamination of food and feed commodities is the best approach to safeguard consumers and assure their safety as it ultimately results in the production of mycotoxin. For the sake of public health, it is imperative to manage mycotoxins in food production. This can be achieved in a number of ways including by

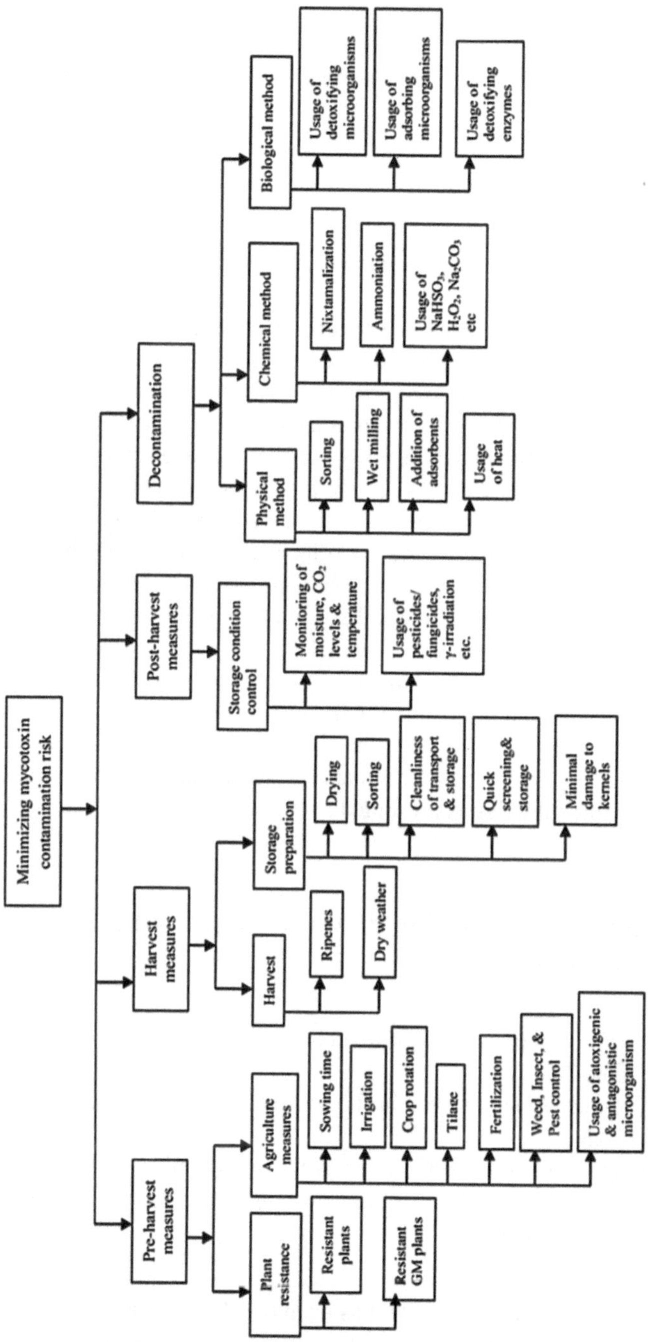

Fig. 20.2 Different techniques for reducing mycotoxins in industrial food and feeds

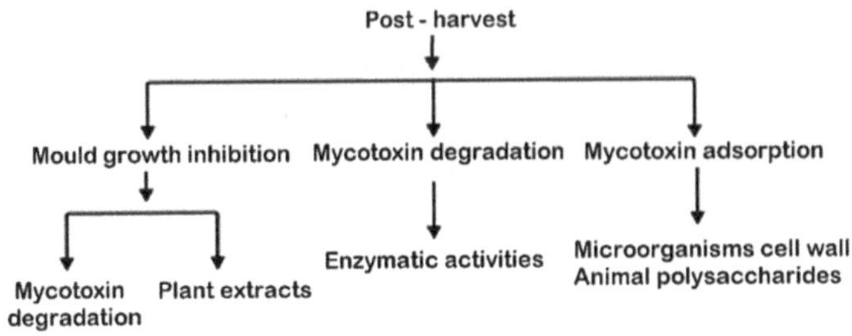

Fig. 20.3 Mycotoxin elimination in food and feeds at post-harvest stages

preventing fungal contamination of food and feed, decontaminating mycotoxins and maintaining ongoing surveillance (Chein et al., 2019). Decontamination is crucial since mycotoxin contamination cannot always be prevented (Aiko & Mehta, 2015). Both chemical and physical therapies are effective in decontaminating mycotoxins. Physical treatments such as heating, boiling, extrusion and radiation are used to decontaminate mycotoxins. Food processing involves applying heat to food, which may cause mycotoxins to become inactivated. However, according to Bullerman and Bianchini (2007), the majority of mycotoxins are heat stable and may be difficult to inactivate with heat processing. The inactivation of mycotoxins through heat depends on several critical parameters including temperature, water content and length of heat exposure. Chemical treatment is one of the methods of decontaminating mycotoxins. Nonetheless, the degradation products that follow may have an impact on the safety and quality of food (Aiko & Mehta, 2015). Fungi that produce toxins can infest seeds during the preharvest phase in the field, harvest, postharvest handling and storage. Schematic presentation of control mould growth and mycotoxin contamination is presented in Fig. 20.3.

20.4 Methods of Prevention of Mycotoxin Contamination at Post-harvest

Different methods of management of mycotoxin contaminations common in India are shown in Table 20.2.

Some simple methods employed in the decontamination of foods and feeds are given in Table 20.3.

Mould damage of stored grain may be prevented mainly by four types of treatments viz. (i) drying of grain (ii) controlled storage atmosphere (iii) use of natural products and (iv) chemical treatment.

Drying of Grain

It is a well-known truth that grain that has been dried can be stored for a long time, protecting it from mould growth and insect infestation because they need moisture to grow and develop. Grain drying techniques that work well include solar energy drying, inbin drying, infrared, microwave or sonic drying.

Table 20.2 Mycotoxins management practices in India

Item	Area of research	Reference
Aqueous extract of tea and coffee	Decreased AFB_1 toxicity	Mo et al. (2013)
Influence of cropping pattern	AFB_1 contamination	Bilgrami et al. (1992)
Plant extracts	AFB_1, Penitrem, Satrotoxin Citrinin	Panda et al. (2015), Iram et al. (2016) and Surekha and Reddy (1992)
Food preservatives	AFB_1 inhibition	Tian et al. (2012)
Effect of vitamins and alcohol	Aflatoxin B1 production	Verma et al. (2001)
Ferulic acid, Ovanillin	Aflatoxin B1 production	Silvac et al. (2020)
Growth regulators influence	Citrinin production	Giridhar and Reddy (1997)
Food preservatives Black pepper seeds Phenolic antioxidants	CPA and other mycotoxins	Laxma Reddy and Reddy (1992) and Giridhar and Reddy (1996, 2001)
Interaction between various fungi	Mycotoxins production	Choudhary and Sinha (1993)
Elicitation influence of phytoalexins on plants	AFB1, CPA and satratoxin	Giridhar et al. (2001)
Volatile compounds	Patulin and terreic acid	Girisham and Reddy (1986)
H_2O_2 and irradiation	AFB_1 inactivation	Sreenivasa et al. (1967) and Patel et al. (1989)
Spice oils	Mycotoxin production inhibition	Juglal and Rand (2002)
Pressure parboiling	AFB_1, mycotoxins	Dors et al. (2009)

Controlled Atmosphere Storage

Most of the mycotoxigenic fungi are favoured by warm humid conditions. Hence, strategies to modify these conditions may help to fight against the mycotoxin problem. It may be safe and to avoid mould growth and bug infestation. To continue the longstanding practice of storing grain underground, which causes the grain to cool and deplete oxygen until it reaches the appropriate level.

Use of Natural Products

The constituents of numerous plants have been found to exhibit antifungal activity against storage fungi. Oregano, clove, cinnamon, garlic and thyme are some of such plants which are reported to inhibit different mycotoxins production should be employed.

Chemical Treatment

A combination of propionic, acetic and benzoic acids has been shown to be moderately effective in preventing *Aspergillus* and *Penicillium* species from degrading stored grains. Phenol and 2,4D inhibited the cyclopiazonic acid production by *Penicillium griseofulvum* (Krishna Reddy & Reddy, 1989). Similarly, Laxma Reddy and Reddy (1992), have reported that pnitrophenol and αnaphthol proved to be effective inhibitors of satratoxin production by *Stachybotrys atra*. Clavero et al. (1993) reported to destroy aflatoxin with the help of hydrogen peroxide.

Table 20.3 Mycotoxin decontamination methods

Decontamination strategy/process	Processing methods	References
Cleaning of grains		
(a) Cleaning	1. Screening to remove under or oversized grain particles	Schofield (2005)
	2. Removal of dust, crushed and fungus infested kernels	Bullerman and Bianchini (2007)
(b) Milling process	1. Processing of edible parts and inedible byproducts separately	Vanara et al. (2018)
	2. Processing of whole grains using roller mill or hammer mill or disk mill or pulverizer	Thomas et al. (2018)
(c) Separation system (based on screening, gravity and airseparation)	1. Using several classifiers to remove from fine to coarse dust particles	Vidosavljevic et al. (2018)
	2. Removal of broken kernels	
(d) Dry cleaning	Polishing grain surface using brushes (having polypropylene bristles)	Vidosavljevic et al. (2018)
(e) Sorting (manual or optic)	To separate same sized healthy from contaminated kernels	Colovic et al. (2019), and Karlovsky et al. (2016)
Thermal processing at 150 °C or more	1. Roasting	Kabak (2009)
	2. Extrusion processing	
Feed additives	1. Use of microorganisms, yeast cell wall, mineral clay for detoxification	Kolosova and Stroka (2011)
	2. Reduction in mycotoxin absorbance using absorbents or binders like zeolite, bentonite, phyllosilicate (kaolinites, illite, and smectite clay), tectosilicates, activated charcoal (nonspecific sequester), cellulose, peptidoglycans, glucomannans, synthetic polymers (polyvinylpyrrolidone and cholestryramine)	EFSA (2009)
	3. Carbon-based nanoparticles	Horky et al. (2018)
	4. Sequesters like wheat straw, alfalfa hay, grape pomace	Stangroom and Smith (1984)
	5.Humic acid derivatives and live bacteria (lactic acid) and yeast (*Saccharomyces cerevisiae*)	Jard et al. (2009)
Dietary supplements and antioxidants	Provide protection against mycotoxins	Sorrenti et al. (2013)

Wheat grain was completely protected against *Aspergillus* and *Penicillium* species for a year of storage when seeds treated with Bavistin and TMTD (Trimethyl thiuram disulphide) at a 0.25 percent concentration (Srivastava, 1986). Aflatoxin production in soybean seeds has also been reported to be inhibited by a few phenolics, including phydroxybenzoic, pcoumaric acid, and Ovanillin (Silvac et al., 2020). Phenol and 2,4dichlorophenol inhibited the cyclopiazonic acid production by

P. griseofulvum (Krishna Reddy & Reddy, 1992), while onitrophenol, pnitrophenol, pchloronitrtophenol and pentachlorophenol were effective in checking the growth of *P. griseofulvum. Stachybotrys atra* was sensitive to pnitrophenol, βnaphthol, pcresol and αnaphthol and failed to produce satratoxin (Laxma Reddy & Reddy, 1992). By decreasing level of oxygen and in its place CO_2 or N_2 will increase anaerobic conditions and discourages mould growth and mycotoxins contamination. Some of the organic acids such as acetic acid (Buchanan & Ayres, 1979), propionic acid and butyric acid (Gosh & Haggblom, 1985), malic acid (Megalla & Hafez, 1982), benzoic acid (Uraih & Offonre, 1981), sorbic acid (Youssef & Marth, 1984), lactic acid (Reiss, 1979), citric acid and their sodium salts are reported to be effective in the control of mycotoxigenic fungal infestation and mycotoxin contamination. Benzoic acid derivatives such as onitrobenzoate, oaminobenzoate, benzocain (ethyl aminobenzoate), enthyl benzoate, methyl benzoate and aspirin (Oacetoxy benzoic acid) are reported to be useful in control of mycotoxin contamination (Davis & Diener, 1967).

It has been observed that potassium fluoride and potassium sulfite are useful in managing the mycotoxin problem (Davis & Diener, 1967). According to Graham and Graham (1986), allicin and similar substances from extracts of onions and garlic are useful in the management of mycotoxin. Cuero et al. (1988) have reported chitosan or chitin derivatives extracted from crustacean shells were also useful in control of mycotoxin contamination. Cinnamon extract containing transcinnamaldehyde and ferulic acid and Phydroxy3 methyl cinnamic acid are reported to be useful in controlling mycotoxin contamination (Bullerman et al., 1977).

Vandergraft et al. (1975) advocated the use of propionic acid to protect the high moisture corn without affecting its nutritive value. Richard and Hamilton (1981) have suggested that the increase in acidity by feed ingredients can enhance the antifungal activity of propionic acid. Youssef and Marth (1984) have reported the efficacy of sorbic acid in the control of aflatoxin production by *A. parasiticus.* Similarly Garza et al. (1993) reported that potassium sulphate could inhibit the production of penicillic acid by *A. ochraceus* and *P. aurantiogriseum.* Dichlorvos was reported to reduce the aflatoxin production by *A. flavus* without affecting its growth (Hsieh, 1973). Our investigations also revealed that propionic acid could check the production of cyclopiazonic acid, patulin, penitrems, roridin and satratoxin production by different fungi (Krishna Reddy et al., 1997). Though mycotoxins can be chemically removed, the degradation products left behind could have an impact on the safety and quality of food.

20.4.1 Improved Farm Management

Recent researches reveal that fungal attacks happen in the field, which makes it crucial to prevent mycotoxin contamination of developing plants. This was confirmed by Bilgrami and Choudhary (1990) on the standing crop of maize. Further, field contamination of mycotoxins is reported to be more under stress conditions.

Bilgrami et al. (1991) have reported that agronomic practices such as planting date, plant density and irrigation levels of crops are reported to influence the mycotoxin contamination in general and aflatoxin in particular. Therefore, it is important to use effective farm management techniques such as enhanced irrigation techniques, fungusfree planting seeds and insect and plant disease prevention. Moreover, the crop should be harvested when it is mature and the right amount of seed moisture. Proper harvesting equipment should also be used to prevent seed damage. It is possible to stop the development of mould and mycotoxin elaboration by providing suitable storage facilities and environmental conditions. Reducing the hazards to human and animal health greatly depends on strict regulation of food and feed as well as adequate public health interventions. Munkvold (2003) discussed in detail the various facets of genetic and cultural approaches to regulate mycotoxin contamination in maize.

20.4.2 Separation of Infested Grains

Minimizing mycotoxin contamination can be achieved effectively through physical separation of contaminated grains either through a manual process or an electronic sorter. Certain mycotoxins, such as aflatoxins and fumonisins, have been demonstrated to be effectively reduced by physical screening followed by removal of damaged kernels by air blowing, washing with water or by physial gravity methods (Sinha, 2001).

20.4.3 Volatile Compounds

Volatile compounds which move fast, leave no residues and highly active at low concentrations are being advocated in the protection of food grains at postharvest stage (Girisham & Reddy, 1987). Vapours of cyclohexane, ethyl acetate inhibited the patulin production by *A. terreus,* while terreic acid production was inhibited by vapours of nhutanol (Girisham & Reddy, 1988). Production of cyclopiazonic acid by *Penicillium griseofulvum* was inhibited by vapours of acetic anhydride, nbutanol, formic acid and formaldehyde (Krishna Reddy & Reddy, 1989). *P. crustosum* was sensitive to vapours, of aniline, acetone, ethyl acetate and isoamyl alcohol and inhibited the peniterm B production (Surekha & Reddy, 1992). Satratoxin production by *S. atra* was totally inhibited by vapours of acetic acid, formaldehyde and propionic acid (Laxma Reddy & Reddy, 1992).

20.4.4 Food Preservatives

Use of food preservative against mycotoxigenic fungi has also been advocated. Lal and Kapoor (1980) reported the efficiency of boric acid in protecting maize seeds against mycotoxin contamination. Similarly, Krishna Reddy and Reddy (1990) have suggested the use of food preservatives in the protection of different food grains from mycotoxin contamination.

20.4.5 Antioxidants

Antioxidants like nicotinic acid, methylxanthine, kaempferol, flavonoids, polyphenols, para benzoquinone, vitamin A,C,E,B, carotene, butylated hydroxy anisole (BHA) and propyl gallate (PG) which are common additives to human food and animal feed to protect from accumulation of unsaturated lipids and other materials from oxidative degradation spoilage are reported to check the growth of mycotoxigenic fungi (Thompson, 1992). Garlic and cabbage seed extracts which have antioxidants, had a significant protective effect against fumonisins. This effect is most likely caused by scavenging free radicals, enhancing antioxidantdriven detoxification mechanism or by inhibiting microsomal lipid peroxidation. BHA, BHT and PG were effective in checking the growth and mycotoxin production by many mycotoxigenic fungi (Thompson, 1992). Carvacol could inhibit the fusarial toxins production but it had no effect on the species of *Aspergillus* and *Penicillium.* Passone et al. (2007) reported the antiaflatoxigenic property of food grade antioxidants on peanut grains.

20.4.6 Polyamine Biosynthetic Inhibitors

Polyamines which regulate the growth and development of different living organisms are key metabolites for the very existence of an organism. When these polyamines are reported to be synthesized in plants at least in two biosynthetic pathways, fungi are reported to have only one pathway. Taking the advantage of this difference in polyamine biosynthesis, Rajam et al. (1985) and Rajam and Galston (1985) have suggested exploiting this difference, for protecting the plants from fungal attack. Wallace and Fraser (2004) and Khurana et al. (1996) have successfully checked the intensity of diseases studied by them with the help of polyamine biosynthetic inhibitors. BCHA could effectively check the elaboration of aflatoxins, ochratoxin A, patulin, terreic acid, citrinin, penitrem B and cyclopiazonic acid by respective fungi (Krishna Reddy ct al., 1997). MGBG showed specificity towards fungi in degree of inhibition.

20.4.7 Antagonists

According to Arseneault and Filion (2017) antagonistic interactions such as competition, hyperparasitism and antibiosis can result in biological control. Biological control agents employ diverse mechanisms of action to counteract the infection.

Since phylloplane and grains is natural source of fungal contamination of grains, it is generally favoured area for antagonist screening (Janisiewicz & Korsten, 2002). Various techniques have been developed to address the need for selective isolation of distinct antagonist groups (Pal & Gardener, 2006; Lahlali et al., 2022).

Aflatoxin B1 may be broken down by *Flavobacterium aurantiacum* (NRRL B184) variety of products, including peanut cream, soybean oil, corn and corn oil (Chen et al., 2016). Aflatoxin is metabolised by bacteria and produce variety of products that can be dissolved in CO_2, chloroform and water. Research by Smiley and Draughon (2000) revealed that dead bacterial cells can bind aflatoxin but failed to break it down completely. *Lactobacillus rhamnosus* GG (LGG) was most efficient microbe for removing AFB1 and zearalenone in liquid medium. Additionally, it has been reported that in a study conducted with live, heat-treated bacteria, AFB_1 bonding took place physically outside the cell. Scientists, as well as the biological and chemical industries have been driven to develop a massproduction of biological control agents protect crops from fungal infections and mycotoxin contamination. Sharma et al. (2009) reported variety of bioproducts designed to regulate mycotoxins (Table 20.4).

The effectiveness of using bioagents to inhibit toxigenic fungi in raw or processed food commodities has been the focus of research. Additionally, the use of bioagents to target the biosynthetic genes involved in mycotoxin biosynthesis has reduced the mycotoxin contamination significantly. (Medina et al., 2017). Strategies to be adapted for the development of bioagents are illustrated in Fig. 20.4. Droby et al. (2009) outlined the scheme of development of biocontrol agent (Fig. 20.5).

Abd El Fadeel et al. (2022) have reported probiotic bacteria and yeast strains could inhibit the growth and subsequent aflatoxins and ochratoxin A production by *A. flavus* and *P. nordicum* respectively. Moller et al. (2021) have also reported the efficacy of lactic acid bacteria on the growth and aflatoxin production by *A. parasiticus*. Gomaa et al. (2018) have reported that the antifungal substances from *Lactobacillus paracasei* and *L. brevis* isolated from dairy products detoxified aflatoxin B_1. An organism intended for use in mycotoxin biocontrol should concentrate on its effects, growth or application to food including both raw materials and processed goods. Use of more than one organism instated of single source can give better control.

Bacillus subtilis, an endophytic bacterium, has shown promise in inhibiting fungal hyphae and the subsequent synthesis of fumonisin (Bacon et al., 2001). Fumonisin B_1 and B_2 in maize grains can be decreased by applying *Bacillus amyloliquefaciens* and *Microbacterium oleovorans* as seed coatings at a concentration of 107 colony forming units per millimetre (Pereira et al., 2007). Wagacha and Muthomi (2008) have excellently discussed on the handling of mycotoxins in

Table 20.4 Bioproducts developed for control of mycotoxin contamination

Product	Microbial agent	Food commodity	Manufacturer/distributor
AF36	*Aspergillus flavus* AF36	Corn and cotton	Arizona Cotton Research and Protection Council, USA
Aflaguad	*A. flavus* strain NRRL21882	Peanuts and corn	Syngenta Crop Protection, USA
AQ10 biofungicide	*Ampelomyces quisquolis*	Apples, grapes, strawberries, tomato, cucurbits	Ecogen, Inc., USA
Aspire	*Candida oleophila* strain 1182	Apple, pear, citrus, cherries, and potato	Ecogen, Inc., USA
Biosave 10LP, 110	*Pseudomonas syringae* (strain 10 LP, 110)	Apple, pear, strawberries, potato	Eco Science Corporation, USA
Blight Ban A 506	*Pseudomonas fluorescence* A506	Onion	Nu Farm, Inc., USA
Contains WG, intercept WG	*Coniothyrium minitans*	Vegetables	Prophyta Biologischer, Germany
Messenger	*Erwinia amylovora*	Vegetables	EDEN Bioscience Corporation, USA
Rhioplus	*Bacillus subtilis* FZB 24	Potato and other vegetables	KFZB Biotechnik, Germany
Serenade	*B. subtilis*	Apple, pear, grapes, and vegetables	Agro Quess, Inc., USA

Source: Fapohunda et al., 2017

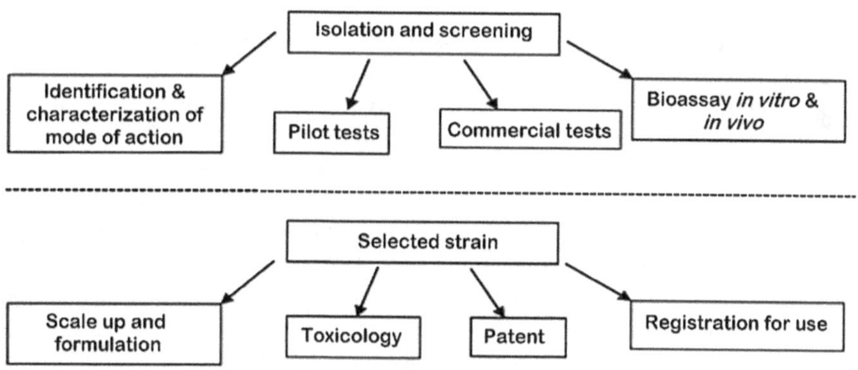

Fig. 20.4 Steps in the development of antagonists (bio agents)

African food. Deoxynivalenol (DON) concentrations in grain were reduced by 89–97% with the help of *Bacillus* strain AS 43.4 (Pirgozliev et al., 2003). Rhizobacteria such as *Azotobacter*, *Bacillus*, *Pseudomonas* and *Arthrobacter* reduced *Fusarium* population and consequently its mycotoxin production in maize (Cavaglieri et al., 2005). Kagot et al. (2019) discussed the merits and demerits of control of mycotoxins through bioagents. Lactic acid bacteria (LAB), which have

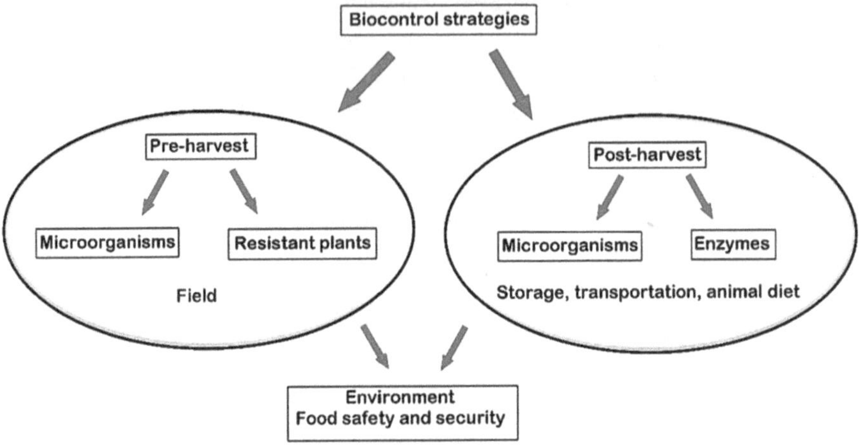

Fig. 20.5 A schematic presentation of the biological strategies in the development of biocontrol agents

been granted safe status and are GRAS, may be used commercially as biological control agents in food to stop the growth of mould, extend the shelf life of fermented goods and lessen the health risks related to mycotoxins (Bianchini & Bullerman, 2010). However, the majority of research on biocontrol of mycotoxin is conducted either in laboratories or on basic systems. Thus, it is important to critical assessment of the applications of biocontrol agents or their enzymes in industrial, field and storage conditions.

Several microorganisms, including bacteria, actinomycetes and fungi, prevent the growth of mycotoxigenic fungi and production of mycotoxin. Still some organisms are reported to degrade mycotoxin already produced. *Flavobacterium aurantiacum* is known to metabolize aflatoxins (Smiley & Draughon, 2000). *Absidia repens* can convert toxic aflatoxin to nontoxic aflatoxicol within 34 days (Detroy & Hesseltine, 1969). Similarly *Lactobacillus acidophilus, L. bulgaricus* and *L. plantarum* could prevent the mould growth which in turn may pave the way for inhibition of mycotoxin production (Karunaratne et al., 1990). Further, moulds producing mycotoxins may also be responsible for degrading the mycotoxin depending on multitude factors (Wilson et al., 2002). The aflatoxin degrading activity of *A. parasiticus* was attributed to the production of peroxidase (Doyle & Marth, 1979). Similarly, species of *Fusarium* deacetylate or acetylate trichothecene mycotoxins and convert them to nontoxic. *Citrobacterium* was able to assimilate T2 toxin as a carbon and energy source. *Morganella morganil,* an intestinal organism, can degrade aflatoxin B_1 (Itoh & Aibara, 1992).

A variety of fungi, particularly yeasts and bacteria, have been found to be useful in managing postharvest illnesses in fresh foods. Nevertheless, little research has been done on the fungi that grow on agricultural products and their biological regulation. It has been observed that competition for nutrients and space may be one of the processes underlying the limitation of fungal growth and mycotoxin formation. Apart from that induction of antibiosis, hyperparasitic interactions and defense mechanisms also plays a substantial role in prevention of growth of mycotoxic fungi. Girisham and Reddy (1986) reported fungi *A. parasiticus, A. niger, C. brachyspora, F. solani* and *M. roridum* that inhibit the patulin production by *A. terreus*. Similarly, a study by Reddy and Surekha (1994) demonstrated the complete inhibition of the production of penetrim B by *P. aurantiogriseum* under the effect of *P. griseofulvum* and *A. niger.*

Several microorganisms such as *Trichoderma* sp. (Ren et al., 2022), *Lactobacillus* sp. (Ahlberg et al., 2017), atoxigenic strain of *Aspergillus flavus* (Silvac et al., 2020) and *Flavobacterium aurantiacum* (Line et al., 1994) are reported to inhibit aflatoxin production by *A. flavus*. Similarly, *Streptococcus salvarius* (Skrinjar et al., 1996) could eliminate ochratoxin A from yoghurt. Yates et al. (1999) reported that *Trichoderma viride* and *Saccharomyces cervisiae* inhibited growth and fumonisin production by *F. moniliforme*. Etcheverry et al. (1998) reported that *Aspergillus parasiticus* and *B. subtilis* could suppress alternariol and alternariol monomethylether and tenuazoic acid produced by *A. alternata*. Skrinjar et al. (1996) reduced aflatoxin with the help of *Lactobacillus delbrukii*. Peteri et al. (2007) reported the degradation of ochratoxin A by astaxanthin producing yeasts whereas Yabe et al. (2008) observed that *Achromobacter xylosoxidans* reduced aflatoxin production by *A. parasiticus* by producing an unknown inhibitory substance. Microorganisms capable of degrading different mycotoxins are listed in Table 20.5.

The discovery of substances that are inhibitory to storage fungi could be a significant advantage from these antagonists even though the feasibility of biological control in the management of mycotoxin in foods and feeds is still unclear. Dorner (2009) suggested the control of mycotoxins problem by employing nontoxigenic strains against toxigenic strains.

In brief use of an integrated strategy to reduce mycotoxin levels in food and feed is advised. It is increasingly understood that, particularly in areas with limited food supplies, this strategy should involve initiatives to both prevent contamination and detoxify contaminated food. Analytical quality assurance in agricultural export produce is required (Sashidhar et al., 1992). Certain mycotoxins can be eliminated using biological detoxification techniques like fermentation. Cell free extracts of *Phoma* sp. were able to degrade AFB_1.

20.4.8 Genetic Engineering Approach

In order to prevent mould infestation and of mycotoxin producton, efforts have been made to raise moldresistant cultivars that resist mould growth not only in the field on standing crops but also during storage. Out of 15 commonly cultivated maize varieties screened for aflatoxin production, four varieties (Him123, Ganga2, Ganga 5 and EH2420) proved to be resistant to aflatoxin production (Bilgrami et al., 1980). Brown et al. (1999) reported the varying resistance of corn to mold growth and aflatoxin production and possible exploitation of these in the management of aflatoxin problem. Similarly Krishna Reddy and Reddy (1992), during screening of

Table 20.5 Microbial biodegradation of mycotoxins

Mycotoxin	Microbes	Biodegradable products	Mechanism	Reference
Aflatoxin	*Nocardia corynebacteroides*, *Corynebacterium rubrum*, *Pseudomonas putida*, *Rhodococcus* spp., *Saccharomyces cerevisiae*	Aflatoxicol, Aflatoxin M1, Aflatoxin B2a	Act on lactone ring	Adebo et al. (2017) and Du et al. (2017)
Fumonisin B1	*Exophiala spinifera*, *Sphingopyxis* spp., *Sphingomonas* spp.	Fumonisin, hydrolyzed B1 (HFB1)	Removes stricarballylate groups with carboxylesterases in C14 and C15	Du et al. (2017) and Vanhoutte et al. (2016)
Ochratoxin A	*Bacillus licheniformis*, *Bacillus* spp., *Brevibacterium iodinum*, *Acinetobacter calcoaceticus*, *Brevibacterium epidermidis*, *Lactobacillus acidophilus*, *Aspergillus versicolor*, *A. niger*, *A. ochraceus*, *Rhodoturula* spp., *Saccharomyces* spp.	Phenylalanine, Ochratoxin A	Hydrolyzes the amide bond	Du et al. (2017) and Vanhoutte et al. (2016)
Trichothecenes	*Blastobotrys capitulate*, *Trichomonascus*, *Aspergillus*, *Curtobacterium* spp., *Anaerovibrio lipolytica*, *Selenomonas* and *Saccharomyces* spp.	3-acetyl T2 toxin, T2 toxin 3-glucoside, neosolaniol	Acetylation deacetylation, deep oxidation, oxygenation, epimerization, and glucosylation	Du et al. (2017) and Vanhoutte et al. (2016)
Zearalenone	*Mucor bainieri*, *Rhizopus* spp., *Cunninghamella bainieri*, *Alternaria alternate*, *Thamidium elegans*, *Aspergillus ochraceous*, *Rhodococcus* spp., *Streptomyces rimosus*, *Trichosporon mycotoxinivorans*, *Pseudomonas* spp., *Aspergillus niger*, *Acinetobacter* spp.	(αZEL) and β-zearalenol (βZEL), ZOM1, 2, 4 dimethoxyl zearalenone, zearalenone 4-β-D -glucoside	Cleavage of the lactone ring and change of the hydroxyl group C4	Du et al. (2017) and Vanhoutte et al. (2016)

Source: Akbar et al., 2022

maize cultivars for prevention of cyclopiazonic acid contamination, expressed its feasibility. Girisham and Reddy (1987), have reported the resistance of some of the cultivars of bajra to *Aspergillus terreus* infestation and patulin contamination. However, evaluation and selection for resistance to mycotoxins at postharvest stage may be more difficult than it has been for plant resistance to field diseases.

However, evaluation and selection for resistance to mycotoxins may be more challenging in the postharvest stage.

Bekalu et al. (2020) reported that excessive expression of antifungal HvNEP1 gene in the endosperm reduces barley susceptibility and prevents grains from absorbing fumonisin B_1. Targeted gene silencing is a crucial method for managing *Fusarium* infestation in cereal crops. One of the methods that controls gene expression is called RNA interference or RNAi. Selecting a resistant cultivar and reducing mycotoxin accumulation and production are two ways to lower the levels of mycotoxins in cereals (Majumdar et al., 2017). Boutigny et al. (2008) and Kluger et al. (2015) reported the development of grains that resistant to the buildup and breakdown of mycotoxins by converting them into nontoxic derivatives or preventing the production of toxic metabolites.

Technological developments in genetic engineering, genomics and marker development may enhance food safety by reducing the risk of mycotoxin contamination. The identification of several genes involved in the interaction between mycotoxigenic fungi and host plants as well as the development of mycotoxin has been made possible by advancements in microarray, fungal expressed sequence tags (EST) and whole genome sequencing research. Beginning with the discovery of candidate genes for silencing (HIGS), it moves on to mutagenesis, molecular breeding techniques, and genetic engineering. Lastly, a process known as hostinduced gene silencing (HIGS) is used to instruct the mycotoxigenic fungus to suppress the expression of one of its own genes by the host plant (Nakayashiki, 2005).

Ehrlich (2014) believes that in addition to offering a reliable and costeffective method of indirect evaluation of fungal mycotoxin control, gene manipulation approaches may be able to find resistance mechanisms that prevent mycotoxin elaboration and fungal growth.

In mycotoxin management proteomics is also being used as a novel locus to find RAPs and potential resistance genes linked to the resistance mechanisms in resistant lines relative to susceptible lines. Breeders may find it helpful to identify appropriate techniques to increase the plant resistance and stress tolerance of host plants against mycotoxigenic fungal contamination, as a result of the finding of storage and stressrelated proteins as biomarkers for mycotoxin (Wang et al., 2010; Assefa & Geremew, 2018).

Rapid Screening Techniques

Infected seeds typically contain mycotoxins in low percentage of the kernels. It is ideal to have a quick screening test for mycotoxins in order to guarantee the quality of food systems. Manual, mechanical and electronic devices are suggested to minimize mycotoxin problem in processed foods. Microscopic observations and BGY fluorescence are known to help in the detection of *A. flavus* growth and aflatoxin production (Sinha & Ranjan, 1989). Segregation of healthy corn on the basis of buoyancy reported to reduce the aflatoxin contamination to a minimum level (Huff & Hagler, 1985). Kos et al. (2016), Maslowska et al. (2018) and Taniwaki et al. (2003) have suggested ELISA technique for detection of aflatoxin and ochratoxin A respectively.

Control of Environmental Condition
Environment of storage place is of paramount importance in keeping the food commodities free from mould infestation and mycotoxin contamination. Among various factors, moisture, temperature, presence of contaminating microorganisms and the nature and physiological state of the commodity are important in minimizing the mycotoxin contamination (Fountain et al., 2014).

Moisture
Water content is a critical factor for keeping quality of stored product. Drying of seed and other agricultural commodities to the safe moisture levels is a critical factor (9% kernels and 13.5% for corn) in the control of mycotoxin contamination. When analyzing moisture and fungal development, water activity rather than water content is the crucial factor. The fraction of a material's water that is "free" is measured by its water activity. Given that water separates into "bound" and "free" states in any natural material. Because "bound" water is chemically or physically bonded to the substance, it cannot sustain the growth of fungi. The remaining material is free water (aw), which is ideal for the growth of fungi. For the majority of xerophilic fungi, the growth limit is approximately aw 0.7. It is crucial to remember that water activity depends on temperature, meaning that moisture sorption isotherms are always created at a particular temperature. In actuality, water activity rises with temperature for a given water content. A temperature increase may cause water activity to rise to dangerous level and permits the prevention of fungus development, which has practical implications for commodity storage. Thus, controlling the temperature in the storage environment is crucial. The amount of safe water varies depending on the product. It is approximately 14% for maize, 15% for wheat, and 7% for groundnuts. This reflects both the proportional amount of "free" water and the commodity's chemical composition. Generally moisture and temperature are interrelated and are likely to regulate mould growth and mycotoxin contamination. Generally storage moulds are known to be xerophils and can grow at 13% and above. The moisture of 13% is equivalent to 70% RH and *aw* 0.7. The available water (aw) is more important in permitting the growth of moulds. An increase in moisture content results in the ten-fold increase of mould growth. During storage even though major portion of seed lot is dry, certain pockets may have seeds with high moisture content which persists for long time in view of lack of aeration. Thus, those pockets become foci for mycotoxin elaboration and spread to other regions. Aeration of stored products is important to prevent mould growth, mycotoxin elaboration and biodeterioration. Keeping the warehouse or container at a low temperature and humidity, is another way of protecting food grains from mould infestation and mycotoxin contamination.

Temperature
The growth of mould and the production of mycotoxin are both significantly influenced by temperature. *A flavus* produces aflatoxin over a wide range of temperature (7.5 °C to 35 °C). If commodities are stored at 4 °C, one can prevent *A. flavus* growth and aflatoxin production. Air tight storage is not desirable which may

provide ideal moisture, encouraging mould growth. Dry and oil roasting resulted in reduction to about 50–70% of aflatoxin B production (Martins et al., 2017). Baking reduced a_w levels up to two thirds only (ValleAlgarra et al., 2009).

Modified Gaseous Atmosphere
If normal composition of gases of the atmosphere is changed the moulds get discouraged. Generally oxygen level is reduced and in its place CO_2 and N_2 are increased resulting in anaerobic condition. This atmosphere is called modified atmosphere packing (MAF) and controlled atmosphere storage (CAS). CAS is generally used for bulk storage for a period of 6 to 9 months and by this method it was possible to control the growth of *A. flavus* and aflatoxin elaboration in peanuts. Further, it can check mould growth indirectly by controlling insect infestation in CAS. MAP is the enclosure of food products in gas barrier materials which results in reduced microbial growth.

20.5 Detoxification of Mycotoxins

Detoxification process involves in the destruction, degradation or inactivation of mycotoxins by physical, chemical or biological methods. Pamela and Wim (2019) reported the detoxification of mycotoxins by some insects. However, Detoxification has its own set of restrictions because the food items that are treated must be free of harmful substances and maintain their vital nutritional content. The ideal detoxification procedure should include:

1. Totally deactivate, or eliminate the mycotoxin, or lower its concentration to a manageable level.
2. To avoid leaving harmful residues in food or feed
3. Not to change the produce's technological qualities or acceptance
4. Preserve the food or feed's nutritional worth.
5. Eliminate fungal spores and mycelia to stop resuscitation and produce mycotoxin.
6. If feasible adopt the standard procedures for food preparation and processing.
7. Use money wisely
8. It should be simple to operate
9. It should not harm or put employees' health in danger
10. Receive regulatory agencies approval

The detoxification of different mycotoxin contaminated foods and feeds has been reviewed from time to time by several workers including Zhu et al. (2017), Adebiyi et al. (2019) and Nazhand et al. (2020). ElSharkawy and AbulHajj (1988) have reported the detoxification of zearalenone by*Gliocladium roseum*. Varga et al. (2005) reported that an isolate of *Rhizopus stolonifer* was able to metabolize ochratoxin A in moistened wheat. Awuchi et al. (2021) have suggested use of microorganisms will be effective in removal of mycotoxins from foods and feeds. Microorganisms may bind, modify or degrade the mycotoxins to a lesser toxic compounds through decarboxylation, hydrolysis, deamination, glucosylation and/or acetylation (Hojnik et al., 2017). According to Colovic et al. (2019) certain bacteria and yeasts have the ability to convert ochratoxin A into phenylalanine. By attaching to their cell surface or changing them into a less harmful form, yeasts and lactic acid bacteria proved to lower the amount of mycotoxins (Markov et al., 2019).

While several treatments have been shown to lower the concentration of particular mycotoxin, no single technique has been proved to be equally safe and efficient against the large range of mycotoxins that can coexist in different commodities. List of microorganisms capable of detoxifying different mycotoxins are précised in Table 20.6.

Recently, Assaf et al. (2018) have developed a novel machine or detoxification of mycotoxins from liquid foods like beverages (Fig. 20.6). The device uses a specialized cartridge with probiotic LAB biofilm adhered to it, allowing liquid food to flow through the adsorbents and produce a detoxified beverage.

Table 20.6 Microorganisms capable of detoxifying mycotoxins

Microorganisms	Mycotoxins	Reference
Phenylobacterium immobile	AFB, OTA, ZEA	Schatzmayr et al. (2006)
Gliocladium roseum, Eubacterium	BBSH 797 DON	Foroud et al. (2019)
Devosia genus	DON	Hassan and Zhou (2018)
Bacillus, Brevibacterium	AFB, ZEA, DON, OTA	Luo et al. (2018)
Candida krusei AUMC 8161	AFB, OTA, ZEN, alternariol	Zohri and Abdel Kareem (2018)
Saccharomyces cerevisiae	DON	Liu et al. (2019)
Candida utilis	AFB, OTA, ZEA	Jakopovic et al. (2018)
Clonostachys rosea	ZEA	Hackbart et al. (2014)
Trichosporon mycotoxinivorans	ZEA	Utermark and Karlovsky (2006)

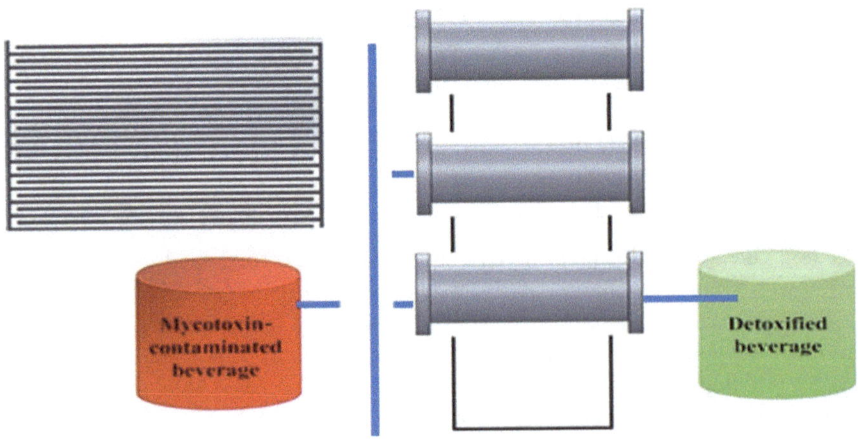

Fig. 20.6 Prototype machine for detoxification of mycotoxins from beverage

For the purpose of enhancing detoxification, fumigation and immersion with the chemical utilised, specific conditions have been mixed with the chemicals at different treatment intervals including moisture content, heat, ultraviolet gamma irradiation, sunshine and pressure. Detoxification can be achieved by (1) Physical methods, (2) Chemical methods and (3) Biological methods.

20.5.1 Physical Methods

In order to lower the mycotoxin level using physical measures to remove mycotoxins mostly involve sorting contaminated and damaged foods and feeds using techniques like sorting, steeping, dehulling, washing, density segregation, sieve cleaning, etc. Physical techniques that help to lower mycotoxins levels during postharvest applications include floating, quickdrying, and UV treatment (Marshall et al., 2020). When rice is heated and cooked under pressure as opposed to at atmospheric pressure, around 70% of the aflatoxin can be eliminated. Nevertheless, its use as a detoxification technique revealed numerous shortcomings, including minimal potential as a detoxification strategy when applied to meals, limited implementation, and negligible efficacy (Assaf et al., 2019). On the other hand, it has also been noted that mycotoxins can be eliminated from food by chemical reactions such as peroxidation, ozonation, ammoniation (Norred et al., 1991). Food spoilage species, such as bacteria, moulds, and yeasts, can be stopped from growing by ionising radiations like gamma rays. Additionally, it renders harmful organisms inactive, such as mosquitoes and parasitic worms. Reduction has been linked to gamma irradiation (5–10 Mred). They eliminate the mycotoxin's mutagenic properties. For the purpose of decontaminating aflatoxin, gamma irradiation and ammoniation should be combined. Nevertheless, these approaches have shown to be less successful since they do not meet the requirements for a good detoxification process and have detrimental effects on food's nutritional content, efficacy, and safety. Furthermore, the process of detoxification via chemical approaches is costly and necessitates intricate specifications (Li et al., 2020). Therefore, it is believed that some physical procedures can effectively remove mycotoxins from food or feed. Some of the simple and cheap physical methods are précised in Table 20.7.

Mycotoxin decontamination of food and feeds can be achieved by following physical methods such as: (i) heat inactivation, (ii) irradiation, (iii) adsorption, (iv) use of ozone, (v) nanoparticles, (vi) cold plasma and (vii) photocatalytic treatments.

(i) *Heat inactivation.* Heating at 25 °C may degrade aflatoxins. Increase of moisture and pressure may improve the efficiency of aflatoxin removal (Pankaj et al., 2018). Approximately half of the mycotoxins can be destroyed by cooking at atmospheric pressure. Reduction of aflatoxin from dry roasting and oil roasting of groundnuts is substantial. When rice was cooked under 15 pounds of pressure for 5 min, the most of the aflatoxins were destroyed, as opposed to when it was cooked normally or with too much water (Sani et al., 2014).

Table 20.7 Physical methods of mycotoxin decontamination employed in food industry

Procedure	Mycotoxin	Product	Efficiency
Automated removal of damaged kernels	Aflatoxins	Peanuts, pistachio	+++ +++
Fluorescence sorting	Aflatoxins	Maize, cotton seeds, dried figs	+++
Sieving	Fumonisins	Maize	++
Flotation	Aflatoxins	Maize, peanuts	+++
Rinsing	DON, ZEA, Fumonisins	Wheat, maize	++
Wetmilling	Aflatoxins, ZEA	Maize	+/–
Roasting	Aflatoxins OTA OTA	Coffee, maizepeanutsCoffee	+ or ++
Heat processing	OTA Trichothecenes DON FB1	Flour All food All food Maize-based food	++
Γradition	T2 toxin, ZEA DON, Aflatoxins	Wheat	++
Sunlight	Aflatoxins	Wheat	+3

Source: Peraica et al., 2002

Another effective method for eliminating aflatoxin from crude groundnut oil is the use of light. According to studies, visible light works better than infrared or ultraviolet light (Shen & Singh, 2021). Similarly oven drying and autoclaving detoxified AFB1 by more than 72–80% (Qun Ying Su et al., 2019). However, the above said drying treatment efficiency varied with mycotoxins. Some of the major cooked products such as roti and gutka (Poridge) prepared from contaminated maize and sorghum flour are detoxified to safe level.

(ii) *Irradiation*. UV irradiation is of limited use in the control of mycotoxin. However, gamma irradiation on large scale may be considered (Karlovsky et al., 2016; Mao et al., 2016). However, because people are hesitant of eating irradiated food, irradiation is not used extensively. As radiation may enter cells, it can damage of DNA lead to mutations. Zachetti et al. (2019) showed a decrease in DON and FB levels in irradiated maize and wheat grains after low molecular weight chitosan with deacetylation was applied. Gamma radiation is useful in reduction of AFB1 in processed rice (Gonçalves et al., 2019). In naturally infested maize, electron beam radiation led to the destruction of ZEN and OTA (Shi et al., 2018). However, radiation can result in chemical, biological and physical reactions that are hazardous to man and animals. Therefore, wider use of radiation techniques in the food sector is still a dubious practice.

(iii) *Adsorption*. Cosme et al. (2020) have suggested the use of bentonite for removal of aflatoxins from the white and red wine. Prandini et al. (2009) reported the removal of aflatoxin M_1 from milk with alumina silica, alumino silicates which adsorb mycotoxins from solution. Hydrated sodium and calcium, aluminosilicates have high affinity to aflatoxin B_1. Di Gregorio et al. (2014) have reported the adsorption of T2 toxin by intestine wall by addition of bentonite to feed. Huwig et al. (2001) have discussed elegantly use of different adsorbents in detoxification of animal feed.

(iv) *Use of ozone.* Use of ozone appears as the safest decontamination technique for the following reasons:

1. Ozone has been used by the food industry for surface disinfection and sanitation.
2. It breaks down quickly, creates oxygen and leaves little residue on food due to its short halflife.
3. It is a broadspectrum decontaminator that eliminates microorganisms like fungus that produce mycotoxin without leaving any residues.
4. It is high oxidizing property makes it a safe antibacterial and antifungal agent in food industry.
5. Since ozone reacts with the functional groups of mycotoxin molecules to modify their molecular structures, forming end products with low molecular weight and few, less hazardous double bonds.
6. Ozone completely incapacitates fungi that produce mycotoxin. Spore germination and mycotoxin generation are reduced by ozone fumigation.
7. It breaks down rapidly and has a limited ability to penetrate.

Ozone treatment is a wellknown environmentally friendly and safe method of detoxification. Further, ozone precursors are easily available and can be applied both in gaseous form and aqueous form. Piemontese et al. (2018), Santos Alexandre et al. (2018) and Porto et al. (2019) have reported the effectiveness of ozone in controlling different mycotoxin contaminants. According to Dwarakanath et al. (1968), aflatoxin B1 and G1 were completely destroyed in an hour when cotton seed and peanut meal were stored in ozonized air, but B2 remained unaffected. Further, this process decreased the amino acid lysine and protein efficiency ratio of the meal (Dollear et al., 1968).

(v) *Nanoparticles* Chitosan coated Fe_3O_4 nanoparticles were utilized for detoxification of Patulin. According to Luo et al. (2018) and Tarazona et al. (2019) silver nanoparticles shown growthrestraining qualities against *Fusarium* sp. and even demonstrated efficacy in lowering mycotoxin levels. Excellent mycotoxin detoxifying capabilities were demonstrated by a nano composite containing aluminium oxide, benzonite and activated carbon (Gonzalez Jartin et al., 2019). According to Zhou et al. (2020) increased research on nanoparticles revealed that a novel type of photocatalysis nanoparticle called UCNO @ TiO2 (up conversion nanoparticles) breaks down deoxynivalenol molecules in cereals.

(vi) *Cold plasma.* Also referred to as the "fourth state of matter," cold plasma is mostly composed of photons, ions, and free radicals such as reactive oxygen and nitrogen. It is utilised in food processing and has special physical and chemical characteristics that are said to have antibacterial effects. Cold atmospheric pressure plasma (CAPP) is a lowcost and environmentally friendly technology that could be used for mycotoxin decontamination (Wielogorska et al., 2019). Hojnik et al. (2017) reported the reduction of aflatoxin, trichothecens, zearalenone and fumonisins with the exposure to cold atmospheric plasma. Similarly, Basaran et al. (2008) reported the elimination of DON and FB in irradiated maize and wheat grains following the application of low molecularweight chitosan with deacetylation (Zachetti et al., 2019). A recent revolutionary nonthermal physical approach for the elimination of fungi and mycotoxin uses ozone, with partially ionised cold plasma destroyed mycotoxins either partially or totally (Misra et al., 2019). The application of cold plasma results in the breakdown of fungal DNA and the release of internal components through the cell wall.

(vii) *Photocatalytic:* A new nonthermal method for removing mycotoxin from food is called photocatalytic detoxification. During this procedure, photons are absorbed by the solid photocatalyst surface, which starts a chemical reaction that produces free radicals. These free radicals interact with mycotoxins and aid in their degradation or oxidation into less harmful substances (Murugesan et al., 2021).

20.5.2 Chemical Methods

Different substances have been tested as reagents to break down mycotoxins. Chemical approaches involve the use of synthetic or in organic oxidising, reducing, alkalizing and acidic chemicals. According to Karlovsky et al. (2016) there are four ways to introduce chemicals in to foods such as packing, fumigation, mixing, and immersion. Jiao et al. (2019) have reported the formation of hemiacetals from Aflatoxins B_2 and G_2 by some acids through incorporation of water. Inorganic or organic bases are inexpensive and can degrade aflatoxins more effectively (Pankaj et al., 2018). Calvin and Cameron (1986) evolved one step procedure for the selective removal of the 12,13 epoxide ring of diacetoxyscirpenol and deoxynivalenol. Janisiewicz and Conway (2010), suggested chemical treatment to prevent mycotoxin contamination. Hydrated NaCa aluminosilicate can degrade ochratoxin A and diacetoxyscirpenol (Santin et al., 2002). Aflatoxin levels in groundnuts have also been found to be effectively reduced by aldehydes like formaldehyde. Ammonia gas appears to be the most practical and costeffective among others for lowering aflatoxin levels in wide range of foods. Nowadays, Senegal, France, and the United States employ ammonia on a commercial basis to remove aflatoxins from food products. Aflatoxin B_1 is formed at high temperatures and pressures as a result of this process decreased protein decarboxylation of α-keto acid, which is a drawback. Though Park et al. (1988) suggested the use of methylamine and ethylamine in the

Table 20.8 Chemicals used in food industry for mycotoxin decontamination

Chemicals	Mycotoxin	Product	Efficient
Calcium hydroxide preparation	FB_1, ZEA DON	Maize	++
Hydrogen peroxide	Aflatoxins	Peanut	+++
Hydrogen peroxide/sodium bicarbonate	FB_1	Maize	+++
Sodium bisulphite	Aflatoxins	Peanuts	+++
Sodium chloride	Aflatoxins	Peanuts	+++
Ammonia	Aflatoxins Aflatoxins Fumonisins	Maize Peanuts meal Maize	+++ +++ +
Ammonia with calcium hydroxide (at 96 °C)	OTA	Swine feed	+++
Ammonia with increased pressure and ambient temperature	Aflatoxins Aflatoxins Fumonisins	Cotton seed Maize, peanut meal Maize	+++ +++ +
Ammonia with atmospheric pressure and ambient temperature	Aflatoxins	Peanut meal	+++

Source: Peraica et al., 2002

management of mycotoxin contamination, this practice did not gain acceptance due to formation of toxic products. Treatment of ground nut with hydrogen peroxide resulted in the removal of aflatoxin upto 97% (Clavero et al., 1993). Singh et al. (2003) reported the total removal of aflatoxins from food by treating with sodium hypochlorite. Partial reduction of aflatoxins during fermentation of yogurt was reported by Megalla and Hafez (1982). Some of the chemical methods employed in mycotoxin decontamination of foods and feeds are shown in Table 20.8.

The best methods reported for removing mycotoxins from contaminated goods are chemical treatments. The process should guarantee that the toxin may be transformed into a harmless derivative or derivatives without negatively altering the original product. It is important to evaluate the treated products for their mutagenicity. Some of these chemicals like acetic acid (Aiko et al., 2016), formaldehyde (Karlovsky et al., 2016), hydrogen peroxide (Sreenivasa et al., 1967), methylamine (Park et al., 1988), sodium hypochlorite (Singh et al., 2003), sodium bicarbonate and sodium bisulfate (Moerck et al., 1980) have been successfully employed in the decontamination of mycotoxins different food commodities. The following chemical methods are in use: (i) solvent extraction, (ii) oxidizing agents, (iii) bisulfate.

(i) *Solvent extraction*: aflatoxins from agricultural commodities can he removed without leaving any traces of aflatoxin by 95% ethyl alcohol, 90% aqueous acetone, 80% isopropyl alcohol, hexanolethanol and hexanolmethanol (Anonymous, 1968). However, it is not only costly but also pose problems in removal of solvents after treatment, loss of nutrients and problem in disposal of extract.

(ii) *Oxidizing agents*: Hydrogen peroxide is an oxidizing agent which has a potential to destroy aflatoxin (Clavero et al., 1993). Aflatoxin B1 and G1 can be broken down by lactoperoxidase when it is exposed to hydrogen peroxide and chloride ions, as demonstrated by Doyle and Marth (1979). Treating the corn with 0.5% H_2O_2 resulted in the decrease of aflatoxin to a significant level. Doyle et al. (1982) reduced the aflatoxin in milk by treating with 0.04 g/10 ml o hydrogen peroxide.

(iii) *Bisulfite*: Common uses for bisulfite as a food preservative include inhibiting browning as an antioxidant, reducing agent and preventing microbial development in fruits, vegetables, and drinks. Aflatoxin B1 and G1 have also been known to be degraded by bisulfate (Doyle & Marth, 1979). According to Moerck et al. (1980) sodium bisulfite can destroy aflatoxin in maize. It breaks down aflatoxin more quickly than ammonia.

20.5.3 Biological Methods

While maintaining the nutritional value and organoleptic quality of food commodities and drinks, microorganisms or their enzymes can be utilised as food and feed additives or as a detoxifying agents during food processing (Nahle et al., 2022). Awuchi et al. (2021) have suggested use of microorganisms will be effective decontaminants of mycotoxins from foods and feeds by using decarboxylation, hydrolysis, deamination, glucolysation, and/or acetylation, microorganisms have the ability to bind, alter, or degrade mycotoxins into less hazardous molecules (Hojnik et al., 2017). According to Colovic et al. (2019) some bacteria and yeast can convert ochratoxin A into phenyalanise. Yeast and lactic acid bacteria have been shown to lower the level of mycotoxin by attaching to their cell surface and control changing them into a less harmful (Pleadin et al., 2019).

When selecting a method to control mycotoxin contamination, it is important to keep in mind that the method(s) should:

(i) be technically and economically appropriate and applicable;
(ii) not significantly alter the nutritional value of food material or produce additional toxic compounds;
(iii) not release harmful and health threatening residues.

The process for selecting microorganisms for commercial usage in the biological control of mycotoxigenic fungus has been superbly evaluated by Kohl et al. (2011). The following factors must be taken into account while developing biological control of mycotoxins:

1. Mycotoxigenic species life cycle
2. The changes in the surrounding circumstances
3. The agronomy of plants

Table 20.9 Microorganisms used in the management of common mycotoxins

Mycotoxins	Microorganisms
Aflatoxins	*Lactobacillus plantarum* LOCK 0945, *L. brevis* LOCK 0944, *L. paracasei* LOCK 0920, *L. kefiri*, *Bacillus pumilus*, *B. subtilis* ANSB060, *Kazachstania servazzii*, *Acetobacter syzygii*, *Rhodococcus erythropolis*, *Pseudomonas putida*, *Mycobacterium fluoranthenivorans* DSM 44556 T, *Streptomyces lividans* TK 24, *Saccharomyces cerevisiae*, *Pichia anomala*, *Fusarium aurantiacum* strain NRRLB 184, *Pseudomonas putida*, *Flavobacterium aurantiacum*
Ochratoxin A	*L. acidophilus* VM 20, *L. bulgaricus*, *L. helveticus*, *L. rhamnosus GG*, *B. lichniformis*, *B. subtilis*, *Bifidobacterium animalis* VM 12, *Brevibacterium*, *Cupriavidus basilensis* OR16, *Pediococcus parvulus*, *S. cerevisiae* O11, *S. bayanus*, *Yarrowia lipolytica*
Zearalenone	*B. licheniformis* CK1, *B. pumilus* ES21, *B. subtilis*, *L. mucosae* lm4208, *L. rhamnosus*, *P. otitidis*THN1, *Rhodococcus*, *Lysinibacillus* sp., *Geobacillus* and *Tepidimicrobium* sp.
Trichothecenes (DON, T2/HT2)	*Nocardioides* and *Devosia*, *Lactobacillus sakei* KTU056, *Pediococcus acidilactici* KTU057, *Pediococcus pentosaceus* KTU058, KTU0509 and KTU0510, *Eggerthella* sp. DII9

Source: Nesic et al., 2021

4. Interactions with pests that frequently serve as ports of entry for mycotoxigenic fungus, particularly when cereals are maturing and postharvest processing (coffee, chocolate and some nuts).

Medina et al. (2017) have discussed practical aspects of production and fermentation hurdles to develop effective biocontrol of mycotoxigenic fungi. Researchers have to look for effective measures of safer and better alternatives for mycotoxin detoxification. The use of microorganisms, such as fungi, yeast and lactic acid bacteria (LAB) are subject of many researches recently to eliminate mycotoxins from foods and feeds (Abrunhosa et al., 2010).

The European Food Safety Authority (EFSA, 2009) indicated that certain yeasts, bacteria or fungal enzymes can breakdown mycotoxins into nontoxic chemicals and can be achieved mycotoxin decontamination. For instance ochratoxin A and zearalenone can be detoxified by the yeast *Trichosporon mycotoxinivorans* (Vekiru et al., 2010). Fadia et al. (2020) reviewed different strategies for detoxification of mycotoxins using microorganisms. Some of the microorganisms used to control some common mycotoxins are précised in Table 20.9.

(i) *Microbial enzymes.* In the processes of degrading mycotoxin, microbial enzymes and their metabolites also demonstrated efficiency (Li et al., 2020). One of the most promising approaches in controlling mycotoxins contamination is enzymatic detoxification which have serious drawbacks compared to chemical techniques like nutrient loss, chemical contamination of raw materials or time and moneyconsuming procedures (Wang et al., 2019). The enzymatic detoxification of mycotoxins involves a broad spectrum of microorganisms such as bacteria, moulds and yeasts (Hathout & Aly, 2014). Mycotoxins can typically be broken down into less harmful or inactive metabolites by microbial enzymes (Lyagin & Efremenko, 2019). Most bacteria employ the process of degradation in their daily life activity to break down particular mycotoxin and turn them into less or even nontoxic products. Enzymes are essential active molecules released by bacteria throughout the breakdown process (Guan et al., 2021). Enzymatic degradation has been the focus of the majority of scientific research to detoxify most prominent mycotoxins such as aflatoxins, ochratoxin A, zearalenone, trichothecenes and patulin (Loi et al., 2018; Zhang et al., 2020). According to Wang et al. (2011) manganese peroxidase is a promising option for mycotoxin detoxification. AFB, ZEN and other mycotoxins can be broken down by this peroxidase isolated from different lignocellulose degrading fungi, including *Irpex lacteus*, *Phanerochaete chrysosporium*, *Ceriporiopsis subvermispora* and *Nematoloma frowardii* (Tang et al., 2013). According to Loi et al. (2018) laccase that has been isolated from *Pleurotus eryngii* has the capacity to degrade ZEN and AFB concurrently by 86% and 100%, respectively. Some of the microbial enzymes used in the control of mycotoxins are précised in Table 20.10.

The enzyme of *Trichosporon mycotoxinivorans* retains its deactivating ability even after fermentation, concentration, freezedrying and stabilisation. A viable alternative for ochratoxin A and/or zearalenone detoxification appears to be its use in postharvest feed supplement (Chaytor et al., 2011). Moreover, certain compoundsproducing actinomycetes have the fascinating ability to disrupt the mycotoxin biosynthesis pathway which may also be used to regulate mycotoxins in postharvest circumstances. According to Sakuda et al. (1996) this new study line has shown promising results in controlling aflatoxin contamination. The recommended actinomycete is *Streptomyces* and several substances produced by this organism including blastatin A and B. Blasticidin A and dioctatin A have stopped the formation of aflatoxin without affecting the growth of the fungus (Yoshinari et al., 2007). Since these compounds do not interfere in the fungal development and the treated fungus may confer an extra protection by niche competition against further infestations.

Another strategy for controlling mycotoxin is the use of antagonistic moulds as well as atoxigenic moulds. One of the most effective strategies is to lower the aflatoxin burden in maize through the use of *Aspergillus flavus* spores from atoxigenic strains (Mauro et al., 2015). Following processing with the strain *Streptomyces* sp. and artificial inoculation of wheat with *F. graminearum*, a notable reduction in the development of *Fusarium* toxin production and an increase in the yield of wheat (Colombo et al., 2019).

(ii) *Biotransformation.* This techniques that decontaminate mycotoxins has been identified as a viable solution since it has a high specificity, yield harmless products and even result in full detoxification in mild and ecofriendly conditions (Ji et al., 2016).

Microorganisms use an enzymatic process called biotransformation to convert mycotoxins into less harmful chemicals. It has been observed that a broad variety of bacteria, fungi, and yeasts are capable of converting mycotoxins into a less hazardous or nontoxic form. These microorganisms function in the animal digestive

Table 20.10 Microbial enzymes involved in the degradation of mycotoxins

Source	Mycotoxin	Enzyme	Degradation (%)	Reference
Armillariella tabescens	Aflatoxin	Aflatoxin oxidase enzyme (AFO)	Not reported (n.r.)	Loi et al. (2017)
Aspergillus niger	OTA	Carboxypeptidase	99	Abrunhosa et al. (2006)
	OTA	Ochratoxinase	n. r.	Dobritzsch et al. (2014)
	FAB1	n. r.	23.6	Zhang et al. (2014)
	Zearalenone Ochratoxin	Lipase, Protease A and Amidase 2	n. r.	Loi et al. (2017)
Aspergillus tubingensis MO36	OTA	n. r.	97.5	Cho et al. (2016)
Bacillus amyloliuefaciencs	OTA	Carboxypeptidase	72	Cho et al. (2016)
Bacillus velezenis	AFB1	n. r.	94.7	Alberts et al. (2009)
Bacillus pumilus DY 3108	AFB1	n. r.	88	Alberts et al. (2009)
Bacillus pumilus ES21	ZEN	n. r.	95.7	Wang et al. (2017)
E. coli	ZEN	Lactonohydrolase Zhd518	n. r.	Wang et al. (2018)
Gliocladium roseum	ZEN	ZENG	60	Zhang et al. (2020)
Lignocellulose degrading fungi	ZEN	Manganese Peroxidase	34.0	Tang et al. (2013)
	AFB1	Manganese Peroxidase	84.9	Tang et al. (2013)
Mycobacterium smegmatis	Aflatoxin	F420H2dependent reductases	n. r.	Loi et al. (2017)
Myxococcus fulvus ANSM068	Aflatoxin	Myxobacteria aflatoxin degrading enzyme (MADE)	n. r.	Loi et al. (2017) and Wang et al. (2017)
Peniophora sp.	AFB1	n. r.	40.45	Alberts et al. (2009)
Phanerochaete sorfida	AFB1	Manganese peroxidase	86	Wang et al. (2011)
Pleurotus eryngii	AFB1	Laccase	86	Loi et al. (2018)
Pleurotus eryngii	ZEN	Laccase	100	Loi et al. (2018)
Pleurotus pulmonarius (ITEM 17144)	Aflatoxin	Laccase (lac2)	n. r.	Loi et al. (2017)

(continued)

Table 20.10 (continued)

Source	Mycotoxin	Enzyme	Degradation (%)	Reference
Pleurotus eryngii (PS419)	Aflatoxin	Ery4	n. r.	Loi et al. (2017)
Pleurotus ostreatus	Aflatoxin	Mn peroxidase	n. r.	Loi et al. (2017)
Pleurotus ostreatus	Aflatoxin	Aflatoxin degradation enzyme	n. r.	Loi et al. (2017)
Rhodococcus erythropolis	AFB1	n. r.	90	Hackbart et al. (2014)
Trametes versicolor	AFB1	Laccase	67	Zeinvand et al. (2015)
Trametes versicolor	Zearalenone Ochratoxin	Laccase	n. r.	Loi et al. (2017)
Streptomyces coelicolor	Do	Laccase	n. r.	Loi et al. (2017)
Clonostachys rosea	Do	Lactono hydrolase	n. r.	Loi et al. (2017)
Acinetobacter sp.	Do	2cys peroxiredoxin	n. r.	Loi et al. (2017)
Bovine pancreas	Do	Carboxypeptidase A: CPA Bovine pancreas	n. r.	Loi et al. (2017)
Saccharomyces cerevisiae	Do	Carboxypeptidase Y: CPY	n. r.	Loi et al. (2017)
Trametes versicolor (Sigma Aldrich, USA)	Trichothecenes	Cytochrome P450 system (Ddna + Kdx + KdR)	n. r.	Loi et al. (2017)
Streptomyces coelicolor		UDP glycosyltransferase	n. r.	Loi et al. (2017)
Sphingomonas sp. ATCC55552	Fumonisin	Carboxylesterase and aminotransferase	n. r.	Loi et al. (2017)
Sphingopyxis sp. MTA144	Fumonisin	Carboxylesterase B and aminotransferase	n. r.	Loi et al. (2017)
Sphingopyxis sp. MTA144	Fumonisin	Fumonisin esterase	n. r.	Loi et al. (2017)
Trametes versicolor	AFB1	Laccase	67	Zeinvand et al. (2015)
Yarrowia lipolytica Y2	OTA	Carboxypeptidases	n. r.	Zhang et al. (2018)

system before the mycotoxins are reabsorbed. Acetylation, glycosylation, ring cleavage, hydrolysis, deamination and decarboxylation are examples of biological transformation reactions. Lactic acid bacteria and yeast are two examples of microorganisms that are effective in binding and/or removing mycotoxins. They also have an impact on the stability of mycotoxin sequestrations, which helps mycotoxin be biochemically broken down and transformed into substances that are physiologically safe for the intestines of humans and animals. Hassan and Zhou (2018) reported that the microbial transformation of deoxynivalenol into nontoxic compounds. However, the process must be economical and it must remove all traces of toxin and at the same time, the nutritional quality of the commodity must not be impaired.

(iii) *Plant Products*. Juglal and Rand (2002) and Oliveira et al. (2020) reported the inhibition of aflatoxin production by *A. flavus* by essential oil, eugenol of *Cinnamornum tamala* and eucalyptus oil respectively. Hussain et al. (2012) recorded the inhibition of aflatoxin by herbal drugs screened by them. Garlic extracts (Negera & Washe, 2019) and *Ocimum basalicum* (Kumar et al., 2011) and Cardamom (Mo et al., 2013) inhibited aflatoxin production by *A. flavus*. Sinha (1990) reported the inhibition of aflatoxin production by *A. flavus* in the presence of tannic acid, caffeic acid and phloroglucinol and some other plant extracts, whereas Maryamma et al. (1990) recorded amelioration of aflatoxicosis of ducks with extracts of *Tinospora cordifolia.*

Other herbs such as thyme, star anise seeds (Hitokoto et al., 1978), black and white pepper (Nguegwouo et al., 2018), *Plumbago indica* are reported to be effective in the control of mycotoxin contamination. Flavonoids from *Vingilia orobiodes* and chlorophorin from *Chlorophora excela,* a drimane type of susquiterpinoids from tree, *Warburgia salutaris* offer safe strategies for reducing mycotoxin levels (Samie & Mashau, 2013). According to Noreen and Russell (2004), the components of walnut seed coat, particularly from the cultivar Tulare, are strong inhibitors of aflatoxin production and have the ability to almost completely transform aflatoxigenic *A. flavus* atoxigenic. Some of the plant products used to control the mycotoxicity of fungi are shown in Table 20.11.

According to Pundir and Jain (2010) investigation on the effectiveness of 22 plant extracts against mycotoxigenic fungi associated with food. Ginger and clove extracts were more effective than other plant extracts. Hema et al. (2009) in their study of antifungal and antibacterial properties of a few South Indian spices and herbs against *A. flavus* proved to be effective. Plant products, especially essential oils are recognized as most promising group of natural compounds for the development of safer antifungal agents (Varma & Dubey, 2001). Neem oil has been reported to be effective in controlling mycotoxigenic fungi. Essential oils of *Morinda lucida* and *Azadirachta indica* inhibited considerably the growth of a toxigenic *A. flavus* and aflatoxin production in inoculated maize grains (Bankole et al., 2006).

Combination of biological control with other postharvest techniques including altered storage conditions followed by chemical and physical treatments proved to be more effective in the control of mycotoxin contamination (Mari et al., 2007; Janisiewicz & Conway, 2010).

An advantage of this strategy is presumably its decreased usage of pesticides, which is a desirable to human and environmental health (Janisiewicz & Korsten, 2002). Unlike chemical pesticides, the biological methods usually offer many mycotoxin management alternatives which are safe and ecofriendly (Fravel, 2005).

Table 20.11 Efficacy of plant products and their components in mycotoxin detoxification

Plant	Mechanism	Reference
Adhatoda vasica	Degraded AFB_1	Aiko and Mehta (2015)
Curcuma longa L.	Hepatoprotective effects against AFB_1 toxicity *in vitro* and *in vivo*	Limaye et al. (2018)
Curcuma amada	Inhibited AFB_1, STE and PAT toxicity	Vipin et al. (2017)
Corymbia citriodora	Detoxify AFB_1 and AFB_2	Iram et al. (2015)
Lycopene	Against ochratoxin A induced renal oxidative stress and apoptosis	Palabiyik et al. (2013)
Ocimum tenuiflorum	Detoxify aflatoxins	Panda and Mehta (2013)
Piper nigrum L.	Disruption of cell wall and plasma membrane, coagulation of the cytoplasm, damage cellular organelles and ergosterol biosynthesis.	Castellanos et al. (2020)
Rosmarinus officinalis	Antifungal and antiaflatoxigenic	Da Silva et al. (2020)
Syzygium aromaticum	Inhibited AF, STE & CIT toxicity	Hussain et al. (2012)
Thymus vulgaris	Ameliorated oxidative stress and genotoxicity effects of AFB_1	Abdel Fattah et al. (2010)

20.6 Quality Control and Regulatory Measures

Methods and mechanisms of biocontrol of mycotoxins are:

1. Growth inhibition of mycotoxigenic fungi
2. Mycotoxin binding
3. Mycotoxin decontamination
4. Mycotoxin adsorption
5. Enzymatic degradation of mycotoxin

Governments, through their various departments and agencies i.e., the Ministry of Health, Agriculture, the Food and Drug Administration, the National Environment Committee Board or the Consumer Protection Committee Board, oversee the careful control of mycotoxins. A special administrative committee and the legislative body oversees the national policy of food safety and fixes maximum tolerance levels for mycotoxins will help in the mycotoxin problem in foods and feeds. Mycotoxins contamination should be wellunderstood by farmers, middlemen, food and feed manufacturers and exporters. They should also be urged to prevent and limit the establishment of mycoflora and mycotoxin elaboration to that pose health risk less to consumers in food commodities. Some of the major organizations should actively participate in formulating worldwide and country oriented regulations.

It is also necessary for international collaboration in regulation of mycotoxin in commodities or product trade. Limits on quality control should be set by the nations for specific goods meant for import or export. The producing nations would be encouraged to be conscious of the possibility of mycotoxin contamination in the commodities to be exported. The European Union, for instance, already set maximum tolerance limits for aflatoxins in animal feeds. These limits are 20 parts per billion (ppb) for complete feed for figs, poultry, and dairy animals and 50 parts per billion for each produce that is processed into mixed feed and complete feed for cattle, sheep, and goats. The establishment of regulation concerning the limits of mycotoxin contamination of food/feed is based on:

1. Toxicity related data
2. Mechanism of toxicity
3. Information regarding the distribution of mycotoxin concentration
4. Code of conduct and traceability procedures which will help in gathering knowledge about the trade contact that existed before the outbreak mycotoxin poisoning
5. Food supply sufficiency

Tolerance levels for different mycotoxins are fixed by commissions of different countries and indicated in related official documents.

20.6.1 HACCP in Mycotoxins Management

Although it is a relatively new method, the application of HACCP to mycotoxin control is based on sound principles and it is desirable to prevent the production of mycotoxin in the first place. They consider the following aspects:

1. Mycotoxins are stable once they formed and can be challenging to eradicate using current decontamination
2. Getting representative samples for mycotoxin research from bulk stores retailers posses a serious issues.
3. Accurate mycotoxin analysis is too expensive and frequently too slow to be useful in many supply chains.
4. Keep in mind that in the end, a well run HACCP plan generally eliminates the requirement for comprehensive quality control testing.

Critical control points identified in the management of mycotoxins are illustrated in Fig. 20.7.

1. *Preharvest and harvest (field) situation.* Contamination of certain commodities with mycotoxin types is essentially a field event. Trichothecenes produced by *Fusarium* species on developing cereal crops, such as T2 toxin and deoxynivalenol (DON) are important examples of preharvest mycotoxins. Thus, the domain of agronomists, mycologists and plant pathologists in understanding of mycotoxin contamination and experts in these fields will need to be a part of the

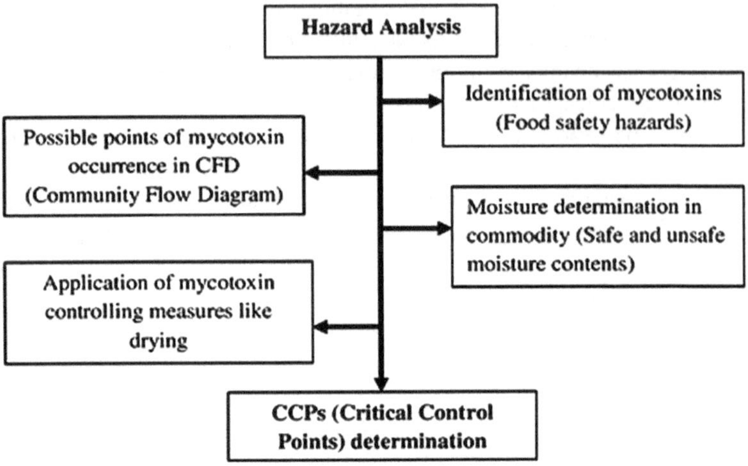

Fig. 20.7 Application of HACCP in mycotoxin hazard analysis

HACCP team. Good quality management principles such as 1. Good agricultural practice, 2. Good manufacturing practice, 3. Good hygienic practice and 4. Good storage practice are essential for effective control of mycotoxins.

The system protects consumers against mycotoxin contaminated of food. The following are the different contamination and preventive approaches to check mycotoxin contamination during different stages of agricultural commodities.

I. **Food grains and feeds**

i. **Preharvest stage**

(a) Prevention techniques at farm level and
(b) Selecting fungal and pestresistant seed. Selection of seeds with high tolerance, resistance and integrity and correct application of fungicides are important to prevent mycotoxin contamination.

ii. **Postharvest stages**

(a) Procedures for safe harvesting, processing, drying and stocking
(b) Dehydration of crops and storage in a lowmoisture environment are vital important to produce a mycotoxin free commodity.

HACCP systems are sometimes thought to be "difficult" and unduly complex. This has to be recognised as a genuine risk in the current situation when we are thinking holistically. We need to have certain other systems in place before we start if we are to avoid having a system that is too complicated to be useful. Good Agricultural Practice (GAP), a loose set of rules of practice intended to promote safe primary food production is the most significant quality system in the field level. The Codex Code of General Principles on Food Hygiene (1997) includes GAP characteristics. OEFRA and the Health and Safety

Commission have released some GAPrelated guidelines and codes of practice in the UK. The majority of crop production in the UK is governed by crop proficiency schemes, which typically include precise guidelines for cultivating crops that closely resemble GAPs. Scottish Quality Cereals (Scotland) and the Assured Combinable Crops Scheme (England and Wales) are two examples of such schemes. The application of GAP in a HACCP scheme has the practical consequence of reducing the likelihood of codex code principles (CCPs) emerging at the field stage. Some examples of controls that fall beyond GAP's purview and are therefore not necessary, CCP's while creating a HACCP plan are as follows:

1. The application of cultivars resistant to pests and fungi
2. Proper land preparation including fertilizer application, crop rotation, waste clearance and proper ploughing
3. Reducing the damage caused by insects and other pests (since this can promote fungal infection)
4. Fungal infection management.
5. The mitigation of weed stress and drought
6. Harvesting at the proper maturity stage and moisture content
7. Proper handling and storage right after harvest.

It makes sense economically to produce wholesome crops. If only for the corresponding rise in yields and value; crops free of mycotoxin contamination. It is important to keep in mind that controlling mycotoxin frequently associated visible fungal growth. Nonetheless, there are certain issues related to mycotoxin contamination that are not covered by GAP. For instance, using fungicides improperly can occasionally result in higher mycotoxin contamination. This might occur if the fungus is agitated and produces more mycotoxins. Nonetheless, there will be situations where it is appropriate to identify CCPs in the field. This usually occurs when measures are taken to avoid the generation of mycotoxin through particular control measures. There are several HACCP plan examples available where CCPs have been determined at the field stage.

II. **Codes of Practice.** It has been observed that during the preharvest stage, HACCP plans can be partially streamlined by utilising Good Agricultural Practices (GAP) and the associated Good Hygienic Practices (GHP), which pertain to the handling of crop material on the farm after harvest. Furthermore, good storage practices (GSP) must be adhered to. Some of the best storage procedures are maintenance of hygienic conditions, preventing rodent and bird access, preventing moisture ingress (e.g., by using moisture barriers on floors), controlling temperature and aeration and preventing invertebrate infestation.

i. **Postharvest control.** Maintenance of proper moisture dominates the control of mycotoxin contamination at the postharvest stage. In preventing fungal contamination in industry is aware of significance of drying and controlling moisture during storage. Similar to preharvest controls, the issue is not simply presence of mycotoxins in the much the commodity but may become obviously mouldy add advances effect on pricing and acceptance.

Maintaining the commodity at a safe level is impossible where fungal growth and mycotoxin generation are prevalent. Remember that certain growers, particularly in underdeveloped nations like India do not have access to cuttingedge drying equipment. The most popular approach in many areas is outdoor sun drying, which might be problematic in bad weather. The relevant fungi in the storage environment are often thrive in xerophilic that are conditions and are called xerotolerant. The fact that the metabolic water produced by the developing fungus can prolong and intensify the process of growth and even facilitate the emergence of other moisturedependent mycotoxigenic fungal species and causes complicated. Preventing pest harvest damage is important for controlling mycotoxin production because, as a result of their metabolic activities, insects and vermin that devour stored material can produce moisture and focal heating. These warmer, more humid regions offer the perfect environment for the start and progression of local fungal growth and mycotoxin synthesis. Conditions in bulk storages become complicate due to localised influences like these.

ii. **Postharvest CCPs.** At the postharvest stage, the designation of CCPs will often come after a system's assessment and the outcomes of any further surveillance research and controlled tests. After that, it is possible to pinpoint the stages that are essential for mycotoxin management and put the right controls in place.

In South East Asia, yellow maize is produced for animal feed composition. This sample was produced as a result of a project that was started in the middle of the 1980s when aflatoxin levels consistently exceeded legal restrictions, such as 20 mg kg1 in the EU, endangering the commodity's exports. In South East Asia, yellow maize is grown as a wet and dry season crop. Aflatoxin contamination in preharvest dry season maize is more likely effected because of the higher danger of drought damage, but postharvest contamination in the longer wet season crop is more likely because of issues with sun drying. The harvested crop was usually kept on the farm for a while before being transferred to silos for export or huge feed mills via a local primary trader and a regional secondary trader. These are all places where things are stored. After a period of harvesting and storing the cobs on the farm, the maize was "shelled" or the kernels were removed from the cob.

Moisture control and drying are of major concern. It is also evident that until the later phases of bulk storage, mycotoxin analysis itself has a relatively minor role in the supply chain. For many commodity categories in underdeveloped nations, this is extremely typical which may be due to lack of funding and necessary expertise. Even in the West, the time needed to complete analysis reduces their value as a monitoring tool. However, particular assays can be helpful in confirming a HACCP plan.

Since the majority of these control measures are ineffective, strict programmes have been formed to prevent animals and humans being exposed to contaminated meals and feeds efficient techniques for tracking the amount of toxins in food have been devised. While certain mycotoxins have had risk assessments completed, additional epidemiological data regarding human exposure is required to define safe and toxicological limits. We may have to deal with mycotoxin issues for a long time and more research is needed on a variety of mycotoxicoses. Through complete eradication of mycotoxin contamination is unachievable the issue can be resolved by using a variety of strategies and sound farm management.

20.7 The Integrated Approach

Mycotoxin management may be achieved by the synergy of combination of different control methods at a lower level than it is used alone. Integrated approach includes creation of modified atmosphere and gamma irradiation, fungistats, natural products and bacteriocins. Although this approach could become important because each component is employed at only a relative low level, it still has limitations. Biotechnological approaches such as replacing toxigenic strains with nontoxigenic ones in the field, incorporation of antifungal genes into crop plants, breeding resistant varieties and prevention of mycotoxin biosynthesis are some of the strategies which may provide long lasting solutions to mycotoxin problems.

Worldwide organizations under the UN system, such as FAO, WHO and UNEP are tasked with giving vital information on a range of topics related to the prevention and management of mycotoxin issues. The following are guidelines for international trade: (a) sampling and analysis procedures; (b) surveillance and food control inspection systems; (c) feeding various animals contaminated produce; (d) strict formulation and implementation of protocols for detoxification and product quality control. It is important to encourage conferences, symposiums, training sessions, and seminars on the most recent research on mycotoxins. Developing nations should receive the development and transfer of lowcost technology for the assessment, prevention and control of environmental mycotoxins from developed nations.

Recent developments in biotechnology led to the isolation of genes involved in biosynthesis of mycotoxins. Information gained through enzymology and molecular biology of these mycotoxigenic fungi can be used in several ways in the management of these obnoxious fungi and mycotoxin elaboration in one of the following ways: (1) To engineer a biological control agent, and (2) To identify the compounds which block mycotoxin formation and to identify the compound that inhibits the mycotoxigenic fungus.

Maryamma et al. (1990) recorded amelioration of aflatoxin of ducks with extracts of *Tinospora cardifolia*. Control of mycotoxin contamination likely to involve a combination of methods and a search of all possible avenues are needed. The approach should be an integrated one involving genetics, biochemistry, ecology and plant breeding. Though factors affecting mycotoxin production are known and biochemical basis of these effects is unknown. The genes responsible for inhibition of mycotoxins in some of the resistant varieties have been identified but plant breeders do not have markers for resistance. Mycotoxin problem can be managed by engineering a biological control agent and identifying the compounds that block mycotoxin formation and compounds in the resistant variety that inhibit the fungus.

Finding substances that selectively block a gene's transcription pathway or the route's enzymes is another control tactic. This would make it possible for plant breeders to choose plants with chemicals that either impede or induce growth.

Mycotoxins are unavoidable which causes natural contamination of foods and feeds but further research is needed to minimize mycotoxin levels which are essential for public health and economy of the nation. A comprehensive approach starting from the field before the planting, continued throughout the entire food chain is required to control mycotoxin contamination. Despite the fact that international organizations such as the FAO and EFSA have developed harmonized limits and rules for mycotoxins, their implementation has been difficult because of resource constraints and the compromised production of food in poor nations through subsistence farming.

In India, the problem of mycotoxin is increasingly realized recently and intensive surveys at geographical and climatic region needs to be undertaken. A strategy to control mycotoxins of that region has to be taken up accordingly. There is an urgent need for intensive and extensive work on the natural incidence of mycotoxins. Intricacies involved in the elaboration of these mycotoxins needs to be taken up at the molecular level. An integrated approach in the management of this problem needs special attention.

The current state of global warming, climatic alterations and fungal shifts in maize and cereals will necessitate ongoing food safety monitoring with regard to various secondary metabolites that are harmful to human beings and animals that are to be both regulated and unregulated. Higher environmental consciousness is likely to relegate the ecological aspects of food production. A greater demand for cereals and food will also increase the need for bioacceptable solutions for larger cropping fields and a reduction in ecological damage through fungicidal treatment. Bioacceptable practices should benefit not only the crop but also the environment, producers, and consumers. Introducing and utilizing natural and biological defenses against fungus contamination is undoubtedly one of the most important initiatives that will contribute to maintaining environmentally friendly food production while ensuring human health. Important area for future studies is to include evaluation of potentials of lactic acid bacteria (LAB) for binding mycotoxins and the factors that affect the process. Another area of research is to generate information on metagenomic DNA obtained from the environment shared by toxigenic moulds can be used for the construction of genomic libraries that could be searched for the genes of specific detoxification enzymes.

The highly favourable climate conditions for growth of mycotoxigenic fungi are prevailing in tropical countries like India and socioeconomic and climate changes pose greater challenge to food safety and security and in turn the health of human beings and animals. Its effects on sustainable development of India damage agricultural export and food security needs immediate attention of policy makers, all stakeholders along with food supply chain, to identify critical areas of collaboration and strengthen alliances in order to ameliorate the effects of these mycotoxins in India in particular and the world at large.

20.8 Future Perspectives

Future researches should focus on generating data on epidemiological effects and longterm toxicity especially in human beings. The development of inexpensive mycotoxin detection instruments that are portable, reliable and easy to use in the field is an important aspect of research. Another important aspect to be investigated is the development of new, genetically modified plant that can resist the fungal invasion. Research should be taken up for detail characterization of molecular mechanism of the action of mycotoxins and its effect on animal and human health which may help for the creation of strategy for prevention and therapy after poisoning.

References

Abd El Fadeel, R. H., Shalaby, S. O., Arafa, M. M., & Khater, A. A. (2022). Biological control of some types of mycotoxins by probiotic bacteria and yeast strains. *Egyptian Journal of Dairy Science, 8896.*

Abdel Fattah, S. M., Abosrea, Y. H., Shehata, F. E., Flourage, M. R., & Helal, A. D. (2010). The efficacy of thyme oil as antitoxicant of aflatoxin(s) toxicity in sheep. *Journal of American Science, 6*, 948–960.

Abrunhosa, L., Santos, L., & Venancio, A. (2006). Degradation of Ochratoxin a by proteases and by a crude enzyme of *aspergillus Niger. Food Biotechnology, 20*, 231–242.

Abrunhosa, L., Paterson, R. R. M., & Venancio, A. (2010). Biodegradation of ochratoxin a for food and feed decontamination. *Toxins, 2*, 1078–1099.

Adebiyi, J. A., Kayitesi, E., Adebo, O. A., Changwa, R., & Njobeh, P. B. (2019). Food fermentation and mycotoxin detoxification: An African perspective. *Food Control, 106*, 10631.

Adebo, O. A., Njobeh, P. B., Gbashi, S., Nwinyi, O. C., & Mavumengwana, V. (2017). Review on microbial degradation of aflatoxins. *Critical Reviews in Food Science and Nutrition, 57*, 3208–3217.

Agriopoulou, S., Stamatelopoulou, E., & Varzakas, T. (2020). Advances in occurrence, importance, and mycotoxin control strategies: Prevention and detoxification in foods. *Foods, 9*, 137.

Ahlberg, S., Joutsjoki, V., Laurikkala, S., Varmanen, P., & Korhonen, H. (2017). *Aspergillus flavus* growth inhibition by *Lactobacillus* strains isolated from traditional fermented Kenyan milk and maize products. *Archives of Microbiology, 199*, 457–464.

Aiko, V., & Mehta, A. (2015). Occurrence, detection and detoxification of mycotoxins. *Journal of Biosciences, 40*, 943–954.

Aiko, V., Edamana, P., & Mehta, A. (2016). Decomposition and detoxification of aflatoxin B1 by lactic acid. *Journal of the Science of Food and Agriculture, 96*, 1959–1966.

Akbar, A., Majeed, F. A., Sadiq, M. B., Khan, S. A., & Rabaan, A. A. (2022). Mycotoxins occurrence in food commodities, their associated hazards and control strategies. *Journal of the Hellenic Veterinary Medical Society, 73*(1), 3853–3866. https://doi.org/10.12681/jhvms.26023

Alberts, J. F., Gelderblom, W. C. A., Botha, A., & van Zyl, W. H. (2009). Degradation of aflatoxin B1 by fungal lactase enzymes. *International Journal of Food Microbiology, 135*, 47–52.

Anonymous. (1968). Indian standard specification for edible groundnut flour expeller pressed and solvent extracts. *IS, 4684*, 4875.

Arseneault, T., & Filion, M. (2017). Biocontrol through antibiosis: Exploring the role played by subinhibitory concentrations of antibiotics in soil and their impact on plant pathogens. *Canadian Journal of Plant Pathology, 39*, 267–274.

Assaf, J. C., El Khoury, A., Atoui, A., Louka, N., & Chokr, A. (2018). A novel technique for aflatoxin M1 detoxification using chitin or treated shrimp shells: *in vitro* effect of physical and kinetic parameters on the binding stability. *Applied Microbiology and Biotechnology, 102*, 6687–6697.

Assaf, J. C., El Khoury, A., Chokr, A., Louka, N., & Atoui, A. (2019). A novel method for elimination of aflatoxin M1 in milk using lactobacillus rhamnosus GG biofilm. *International Journal of Dairy Technology, 72*, 248–256.

Assefa, T., & Geremew, T. (2018). Major mycotoxins occurrence, prevention and control approaches. *Biotechnology and Molecular Biology Reviews, 12*, 111.

Awad, W. A., Ghareeb, K., Bohm, J., & Zentek, J. (2010). Decontamination and detoxification strategies for the Fusarium mycotoxin deoxynivalenol in animal feed and the effectiveness of microbial biodegradation. *Food Additives and Contaminants Part A Chemistry, Analysis, Control, Exposure and Risk Assessment, 27*, 510–520.

Awuchi, C. G., Ondari, E. N., Eseoghene, I. J., Twinomuhwezi, H., Amagwula, I. O., & Morya, S. (2021). *Fungal growth and mycotoxins production: Types, toxicities, control strategies, and detoxification. Fungal reproduction and growth* (p. 121). Intech Open.

Bacon, C. W., Yates, I. E., Hinton, D. M., & Meredith, F. (2001). Biological control of *Fusarium moniliforme* in maize. *Environmental Health Perspectives, 109*, 325–332.

Bankole, S., Aogunsanwo, B., Mosho, A., & Adewuyi, G. O. (2006). Fungal contamination and aflatoxin B_1 of 'egusi' melon seeds in Nigeria. *Food Control, 17*, 814–818.

Basaran, P., Basaran Akgul, N., & Oksuz, L. (2008). Elimination of aspergillus parasiticus from nut surface with low pressure cold plasma (LPCP) treatment. *Food Microbiology, 25*, 626–632.

Bekalu, Z. E., Krogh Madsen, C., Dionisio, G., Bæksted Holme, I., Jorgensen, L. N., Fomsgaard, I., & Brinch Pedersen, H. (2020). Over expression of Nepenthesin *HvNEP1* in barley endosperm reduces *Fusarium* head blight and mycotoxin accumulation. *Agronomy, 10*(2), 203.

Bhat, R., & Vasanthi, S. (1999). Occurrence of aflatoxin and its economic impact on human nutrition and animal feed natural occurrence. *Agriculture at Developpement, 23*, 3 p.

Bianchini, A., & Bullerman, L. B. (2010). Biological control of molds and mycotoxins in foods. Chapter 1. In Appell et al. (Eds.), *Mycotoxin prevention and control in agriculture – ACS symposium series*. American Chemical Society.

Bilgrami, K. S., & Choudhary, A. (1990). Incidence of *Aspergillus flavus* in the aerosphere of maize fields of Bhagalpur. *Indian Phytopathology, 43*(11), 38–42.

Bilgrami, K. S., Prasad, T., Misra, R. S., & Sinha, K. K. (1980). *Survey and study o f mycotoxin producing fungi associated with grains in the standing maize crop*. Final Technical Report. ICAR. Project,. Bhagalpure University.

Bilgrami, K. S., Choudhary, A., & Masood, A. (1991). Aflatoxin contamination in mustard in relation to agronomic practices. *Journal of the Science of Food and Agriculture, 54*(2), 221–228.

Bilgrami, K. S., Kumar, S. R., & Ahmad, M. (1992). Influence of the cropping pattern on aflatoxin contamination in preharvest Kharif (monsoon) maize crop (*Zea mays*). *Journal of the Science of Food and Agriculture, 58*(1), 101–106.

Boutigny, A. L., Richard Forget, F., & Barreau, C. (2008). Natural mechanisms for cereal resistance to the accumulation of *Fusarium* trichothecenes. *European Journal of Plant Pathology, 121*, 411–423.

Brown, R. L., Chen, Z. Y., Cleveland, T. E., & Russin, J. S. (1999). Advances in the development of host resistance in corn to aflatoxin contamination by *Aspergillus flavus*. *Phytopathology, 89*(2), 113–117.

Buchanan, R. L., & Ayres, J. G. (1979). Effect of sodium acetate on growth and aflatoxin production by *Aspergillus parasiticus* NRRL 2999. *Journal of Food Science*, 128–132.

Bullerman, L. B., & Bianchini, A. (2007). Stability of mycotoxins during food processing. *International Journal of Food Microbiology, 119*, 140–146.

Bullerman, L. B., Lieu, F. Y., & Seier, A. S. (1977). Inhibition of growth and aflatoxin production by cinnamon and clove oils, cinnamic aldehyde and eugenol. *Journal of Food Science, 42*, 1107–1108.

Calvin, E. W., & Cameron, S. (1986). Chemical deoxygenation of the trichothecenes, diacetoxy-scirpenol and deoxyn ivalenol. *Journal of the Chemical Society, Chemical Communications, 46*, 1084–1085.

Castellanos, L. M., Olivas, N. A., AyalaSoto, J., Contreras, C. M. D., Ortega, M. Z., Salas, F. S., & Ochoa, L. H. (2020). *In vitro and in vivo antifungal activity of clove (Eugenia caryophyllata)* and pepper (*Piper nigrum* L.) essential oils and functional extracts against *Fusarium oxysporum* and *Aspergillus Niger* in tomato (*Solanum lycopersicum* L.). *International Journal of Microbiology, 18*.

Cavaglieri, L., Orlando, J., Rodríguez, M. I., Chulze, S., & Etcheverry, M. (2005). Biocontrol of *Bacillus subtilis* against *Fusarium verticillioides* in vitro and at the maize root level. *Research in Microbiology, 156*, 748–754.

Chaytor, A. C., Jeff, A. H., Eric van Heugten, M., Todd, S., & Sung, W. K. (2011). Occurrence and decontamination of mycotoxins in swine feed. *Asian-Australasian Journal of Animal Science Seoul, 24*, 723–738.

Chein, S. H., Sadiq, M. B., & Anal, A. K. (2019). Antifungal effects of chitosan films incorporated with essential oils and control of fungal contamination in peanut kernels. *Journal of Food Processing and Preservation, 43*, e14235.

Chen, A., Frisvad, J. C., Sun, B., Varga, J., Kocsube, S., Dijksterhuis, J., Kim, D. H., Hong, S. B., Houbraken, J., & Samson, R. A. (2016). *Aspergillus* section Nidulantes (formerly *Emericella*): Polyphasic taxonomy, chemistry and biology. *Studies in Mycology, 84*, 1–118.

Cho, S. M., Jeong, S. E., Lee, K. R., Sudhani, H. P. K., Kim, M., Hong, S. Y., & Chung, S. H. (2016). Biodegradation of ochratoxin a by *Aspergillus tubingensis* isolated from meju. *Journal of Microbiology and Biotechnology, 26*, 1687–1695.

Choudhary, A. K., & Sinha, K. K. (1993). Competition between a toxigenic *Aspergillus flavus* strain and other fungi on stored maize kernels. *Journal of Stored Products Research., 29*(75), 80.

Clavero, M. R. S., Huno, Y. C., Beuchat, L. R., & Nakayama, T. (1993). Separation of aflatoxin-contaminated kernels from sound kernels by hydrogen peroxide treatment. *Journal of Food Protection, 56*(2), 130–133.

Colombo, E. M., Kunova, A., Pizzatti, C., Saracchi, M., Cortesi, P., & Pasquali, M. (2019). Selection of an Endophytic *Streptomyces* sp. strain DEF09 from wheat roots as a biocontrol agent against *Fusarium graminearum*. *Frontiers in Microbiology, 10*, 2356.

Colovic, R., Nikola, P., Federica, C., Giuseppina, A., Donato, G., Olivera, D., Jovana, K., & Luciano, P. (2019). Decontamination of mycotoxin contaminated feedstuffs and compound feed. *Toxins (Basel), 25*(11), 617.

Cosme, F., Ines, A., Ferreira, B., Silva, D., Filipe Ribeiro, L., Abrunhosa, L., & Nunes, F. M. (2020). Elimination of aflatoxins B1 and B2 in white and red wines by bentonite fining. Efficiency and impact on wine quality. *Foods, 9*, 1789.

Cuero, R. G., Lillehoj, E. B., Cleveland, T. E., & Reine, A. H. (1988). Chitosan as a control agent of toxigenic fungal growth and aflatoxin production. *Proceedings of the Japanese Association of Mycotoxicology*. Supplement No. 1, 194–198 Japanese Association of Mycotoxicology.

Da Silva, B. N., Kohiyama, C. Y., Nakasugi, L. P., Nerilo, S. B., Mossini, S. A. G., Romoli, J. C. Z., Mikcha, J. M. G., de Abreu Filho, B. A., & Machinski, M., Jr. (2020). Antifungal and antiaflatoxigenic activity of rosemary essential oil (*Rosmarinus officinalis* L.) against *Aspergillus flavus*. *Food Additives and Contaminants, 37*(1), 153–161.

Davis, N. D., & Diener, U. L. (1967). Inhibition aflatoxin synthesis by paminobenzoic acid, potassium sulfite and potassium fluoride. *Applied Microbiology, 15*, 1517–1518.

Detroy, R. W., & Hesseltine, C. W. (1969). Transformation of aflatoxin B1 by steroidhydroxylating fungi. *Canadian Journal of Microbiology, 15*, 67.

Di Gregorio, M. C., Neeff, D. V., De Jager, A. V., Corassin, C. H., Carao, A. C. D. P., & de Albuquerque, R. (2014). Mineral adsorbents for prevention of mycotoxins in animal feeds. *Toxin Reviews, 33*, 125–135.

Dobritzsch, D., Wang, H., Schneider, G., & Yu, S. (2014). Structural and functional characterization of ochratoxinase, a novel mycotoxindegrading enzyme. *The Biochemical Journal, 462*, 441–452.

Dollear, F. G., Mann, G. E., Codifer, L. P., Gardner, H. K., Koltun, S. P., & Vix, H. L. E. (1968). Elimination of aflatoxins from peanut meal. *Journal of the American Oil Chemists Society, 45*(12), 862–865.

Dorner, J. W. (2009). Biological control of aflatoxin contamination in corn using a nontoxigenic strain of *Aspergillus flavus*. *Journal of Food Protection, 72*, 801–804.

Dors, G. C., Pinto, L. A. A., & Badiale Furlong, E. (2009). Migration of mycotoxins into rice starchy endosperm during the parboiling process. *LWT–Food Science and Technology, 42*, 433–437.

Doyle, M. P., & Marth, E. H. (1979). Peroxidase activity in mycelia of *Aspergillus parasiticus* that degrade aflatoxin. *European Journal of Applied Microbiology, 7*, 211–217.

Doyle, M. P., Applebaum, R. S., Brackett, R. E., & Marth, E. H. (1982). Physical, chemical and biological degradation of mycotoxins in foods and agricultural commodities. *Journal of Food Protection, 45*, 964–971.

Droby, S., Wisniewski, M., Macarisin, D., & Wilson, C. (2009). Review: Twenty years of postharvest biocontrol research: Is it time for a new paradigm. *Postharvest Biology and Technology, 52*, 137–145.

Du, K., Wang, C., Liu, P., Li, Y., & Ma, X. (2017). Effects of dietary mycotoxins on gut microbiome. *Protein and Peptide Letters, 24*(5), 397–405.

Dwarakanath, C. T., Rayner, E. T., Mann, G. E., & Dollear, F. G. (1968). Reduction of aflatoxin levels in cottonseed and peanut meals by ozonization. *Journal of the American Oil Chemists Society, 45*(2), 93–95.

Ehrlich, K. C. (2014). Nonaflatoxigenic *Aspergillus flavus* to prevent aflatoxin contamination in crops, advantages and limitations. *Frontiers in Microbiology, 5*, 19.

ElSharkawy, S. H., & AbulHajj, Y. J. (1988). Microbial transformation of zearalenone, reduction, hydroxylation, and methylation products. *The Journal of Organic Chemistry, 53*, 515–519.

Etcheverry, M., Cavaglieri, L., & Chulze, S. (1998). Microbial interaction of *Aspergillus parasiticus* and *Bacillus subtilis* with *Alternaria alternata*. Production of alternariol, alternariol monomethylether and tenuazonic acid on sunflower seeds. *Mycotoxin Research, 14*(1), 28.

European Food Safety Authority EFSA. (2009). Review of mycotoxindetoxifying agents used as feed additives, mode of action, efficacy and feed/food safety. *Science Report*. Available in, <http,//www.efsa.europa.eu/en/scdocs/doc/22e.pdf>. Accessed on 26 Aug 2012.

Fadia, B. T., Bochra, K., Yasir, M., Al, Q., Jalila Ben, S. A., & Kamel, C. (2020). Review: Biotechnology of mycotoxins detoxification using microorganisms and enzymes. *Toxicon, 160*, 12–22.

Fapohunda, S. O., Adetoun, O. E., & Toba, S. A. (2017). Biological control of mycotoxins, An update. *World's Veterinary Journal, 7*(4), 117–127.

Foroud, N. A., Baines, D., Gagkaeva, T. Y., Thakor, N., Badea, A., Steiner, B., Bürstmayr, M., & Burstmayr, H. (2019). Trichothecenes in cereal grains an update. *Toxins, 11*, 634.

Fountain, J. C., Scully, B. T., Ni, X., Kemerait, R. C., Lee, R. D., & Chen, Z. Y. (2014). Environmental influences on maize aspergillus flavus interactions and aflatoxin production. *Frontiers in Microbiology, 5*, 40.

Fravel, D. R. (2005). Commercialization and implementation of biocontrol. *Annual Review of Phytopathology, 43*, 337–359.

Garza, S. R., Canela, I. V., & Sanchis, V. (1993). Effects of potassium sorbate on growth and penicillic acid production by *aspergillus ochraceus* and *Penicillium aurantiogriseum*. *Zentralblatt für Mikrobiologie, 148*(5), 343–350.

Gbashi, S., Ntakadzeni. E. M., Sarah De, S., Marthe De, B., Ifeoluwa, A., Oluwafemi, A. A., & Patrick, B.N. (2019). The socio economic impact of mycotoxin contamination in Africa. *Mycotoxins impact and management strategies*.

Giridhar, P., & Reddy, S. M. (1996). Effect of some plant extracts on citrinin production by *P.citrinum in vitro*. *Indian Botanical Society, 75*, 153–154.

Giridhar, P., & Reddy, S. M. (1997). Effect of growth regulators on Citrinin production by Penicillium *citrinum*: Study II. *National Academy Science Letters, 20*, 62–63.

Giridhar, P., & Reddy, S. M. (2001). Phenolic antioxidants for the control of some mycotoxigenic fungi. *Journal of Food Science and Technology, 38*, 397–399.

Giridhar, P., Orlich, G., Komor, E., & Reddy, S. M. (2001). Elicitation of phytoalexins accumulation and phloem characteristics by some fungi and mycotoxins in *Ricinus communis* Linn. *Advances in Plant Sciences, 14*, 211–214.

Girisham, S., & Reddy, S. M. (1986). Efficacy of volatile compounds and food preservatives in tree control of growth and patulin production by *Aspergillus terreus*. *National Academy Science Letters, 9*, 373–374.

Girisham, S., & Reddy, S. M. (1987). Influence of volatile compounds on the seed mycoflora of pearl millet (*Pennisetum typhoides*). *Journal of Indian Botanical Society*, 176–178.

Girisham, S., & Reddy, S. M. (1988). Efficacy of volatile compounds and food preservatives in the control of terreic acid production by *aspergillus terreus*. *Journal of Indian Botanical Society, 67*.

Gomaa, E. Z., Abdelall, M. F., & ElMahdy, O. M. (2018). Detoxification of aflatoxin B1 by antifungal compounds from lactobacillus brevis and lactobacillus paracasei, isolated from dairy products. *Probiotics and Antimicrobial Proteins, 10*(2), 201–209.

Gonçalves, A., Gkrillas, A., Dorne, J. L., Dall Asta, C., Palumbo, R., Lima, N., Battilani, P., Venancio, A., & Giorni, P. (2019). Pre and postharvest strategies to minimize mycotoxin contamination in the rice food chain. *Reviews in Food Science and Food Safety, 18*, 441–454.

Gonzalez Jartin, J. M., de Castro, A. L., Alfonso, A., Pineirob, Y., Vilar, S. Y., Gomez, M. G., Osorio, Z. V., Sainz, M. J., Vieytes, M. R., & Rivas, J. (2019). Detoxification agents based on magnetic nanostructured particles as a novel strategy for mycotoxin mitigation in food. *Food Chemistry, 294*, 60–66.

Gosh, J., & Haggblom, P. (1985). Effect of sublethal concentration of propionic or butyric acid on growth and aflatoxin production by *Aspergillus flavus*. *International Journal of Food Microbiology, 2*, 323–330.

Graham, H. D., & Graham, E. J. F. (1986). Inhibition of *Aspergillus parasiticus* growth and toxin production by garlic. *Journal of Food Safety, 8*, 101–108.

Guan, Y., Chen, J., Nepovimova, E., Long, M., Wu, W., & Kuca, K. (2021). Aflatoxin detoxification using microorganisms and enzymes. *Toxins (Basel), 13*, 1–17.

Hackbart, H. C. S., Machado, A. R., Christ Ribeiro, A., Prietto, L., & Badiale, F. E. (2014). Reduction of aflatoxins by *Rhizopus oryzae* and *Trichoderma reesei*. *Mycotoxin Research, 30*, 141–149.

Hassan, Y. I., & Zhou, T. (2018). Addressing the mycotoxin deoxynivalenol contamination with soilderive bacterialand enzymatic transformations targeting the C3 carbon. *World Mycotoxin Journal, 11*, 101–111.

Hathout, A. S., & Aly, S. E. (2014). Biological detoxification of mycotoxins, a review. *Annales de Microbiologie, 64*, 905–919.

Hema, R., Kumaravel, S., & Elanchezhiyan, N. (2009). Antimicrobial activity of some of the south Indian spices and herbals against food pathogens. *Global Journal of Pharmacology, 3*, 38–40.

Hitokoto, H., Morozumi, S., Wauke, T., Sakai, S., & Ueno. (1978). Inhibitory effects of condiments and herbal drugs on the growth and toxin production of toxigenic fungi. *Mycopathologia, 66*, 161–165.

Hojnik, N., Cvelbar, U., Tavčar Kalcher, G., Walsh, J. L., & Krizaj, I. (2017). Mycotoxin decontamination of food, cold atmospheric pressure plasma versus "classic" decontamination. *Toxins, 9*, 151.

Horky, P., Skalickova, S., Baholet, D., & Skladanka, J. (2018). Nanoparticles as a solution for eliminating the risk of mycotoxins. *Nanomaterials, 8*, 727.

Hsieh, D. P. H. (1973). Inhibition of aflatoxin biosynthesis by dichlorvos. *Journal of Food Science, 21*, 468–470.

Huff, W. E., & Hagler, W. M. (1985). Density segregation of corn and wheat naturally contaminated with aflatoxin, deoxynivalenol and zearalenone. *Journal of Food Protection, 48*, 416–420.

Hussain, A., Shafqatullah Ali, J., & UrRehman, Z. (2012). Inhibition of aflatoxin producing fungus growth using chemical, herbal compounds/spices and plants. *Pure and Applied Biology, 1*, 8–13.
Huwig, A., Freimund, S., Kappeli, O., & Dutler, H. (2001). Mycotoxin detoxification of animal feed by different adsorbents. *Toxicology Letters, 122*, 179–188.
Iram, W., Anjum, T., Iqbal, M., Ghaffar, A., & Abbas, M. (2015). Mass spectrometric identification and toxicity assessment of degraded products of aflatoxin B_1 and B_2 by *Corymbia citriodora* aqueous extracts. *Scientific Reports, 5*(1).
Iram, W., Anjum, T., Iqbal, M., Ghaffar, A., Abbas, M., & Khan, A. M. (2016). Structural analysis and biological toxicity of aflatoxins B1 and B2 degradation products following detoxification by *Ocimum basilicum* and *Cassia fistula* aqueous extracts. *Frontiers in Microbiology, 7*, 1105.
Itoh, Y., & Aibara, K. (1992). Production of substance that degrades aflatoxin B1 by *Morganella morganii* A1731, degradation of aflatoxins by intestinal microorganisms, Part IV. *Nippon Nogeikagaku Kaishi, 66*(9), 1335–1343.
Jakopovic, Z., Karla, H. C., Jasna, M., Irina, P., Iva, C., Jadranka, F., Jelka, P., Damir, S., Slaven, Z., & Ksenija, M. (2018). Properties and fermentation activity of industrial yeasts *Saccharomyces cerevisiae, S. uvarum, Candida Utilis* and *Kluyveromyces marxianus* exposed to AFB1, OTA and ZEA. *Food Technology and Biotechnology, 56*, 208–217.
Janisiewicz, W. J., & Conway, W. S. (2010). Combining biological control with physical and chemical treatments to control fruit decay after harvest. *Stewart Postharvest Review, Redhill, 6*(1), 1–16.
Janisiewicz, W. J., & Korsten, L. (2002). Biological control of postharvest diseases of fruits. *Annual Review of Phytopathology, 40*, 411–441.
Jard, G., Liboz, T., Mathieu, F., Guyonvarch, A., & Lebrihi, A. (2009). Adsorption of zearalenone by *Aspergillus japonicus* conidia, new trends for biological decontamination in animal feed. *World Mycotoxin Journal, 2*, 391–397.
Ji, C., Fan, Y., & Zhao, L. (2016). Review on biological degradation of mycotoxins. *Animal Nutrition, 2*, 127–133.
Jiao, Z., Yifang, Z., Yi, Z., Baoling, W., & Dandan, G. (2019). Rapid and effective sample cleanup based on polyvinylpyrrolidonecoated magnetite nanoparticles coupled with new precolumn derivatization for determination of aflatoxins B1, B2, G1 and G2 in animal feeds by high performance liquid chromatography with fluorescence detection. *Journal of the Brazilian Chemical Society, 30*(9), 1814–1824.
Jouany, J. P. (2007). Methods for preventing, decontaminating and minimizing the toxicity of mycotoxins in feeds. *Animal Feed Science and Technology, 137*, 342–362.
Juglal, S. G., & Rand, O. B. (2002). Spice oils for the control of cooccurring mycotoxin producing fungi. *Journal of Food Protection, 65*, 683–687.
Kabak, B. (2009). The fate of mycotoxins during thermal food processing. *Journal of the Science of Food and Agriculture, 89*(4), 549–554.
Kabak, B., Dobson, A. D. W., & Var, I. (2006). Strategies to prevent mycotoxin contamination of food and animal feed, a review. *Critical Reviews in Food Science and Nutrition, 46*, 593–619.
Kagot, V., Okoth, S., De Boevre, M., & De Saeger, S. (2019). Biocontrol of aspergillus and Fusarium mycotoxins in Africa, benefits and limitations. *Toxins, 11*, 109.
Karlovsky, P., Suman, M., Berthiller, F., Meester, J., Eisenbrand, G., Perrin, I., Oswald, I. P., Speijers, G., Chiodini, A., & Recker, T. (2016). Impact of food processing and detoxification treatments on mycotoxin contamination. *Mycotoxin Research, 32*, 179–205.
Karunaratne, A., Wezenberg, E., & Bullerman, L. B. (1990). Inhibition of mold growth and aflatoxin production by *lactobacillus* spp. *Journal of Food Protection, 53*(3), 230–236.
Khurana, N., Saxena, R. K., Gupta, R., & Rajam, M. V. (1996). Polyamines as modulators of microcycle conidiation in *Aspergillus flavus*. *Microbiology, 142*(3), 517–523.
Kiran, S., Surekha, M., & Reddy, S. M. (2015). Fungal infestation and mycotoxins contamination of paddy of flooded areas of Godavari belt region of Telangana State, India. *Research Science Reporter, 5*, 30–35.

Kluger, B., Bueschl, C., Lemmens, M., Michlmayr, H., Malachova, A., Koutnik, A., Maloku, I., Berthiller, F., Adam, G., & Krska, R. (2015). Biotransformation of the mycotoxin deoxynivalenol in *Fusarium* resistant and susceptible near isogenic wheat lines. *PLoS One, 10*, e0119656. https://doi.org/10.1371/journal.pone.0119656

Kohl, J., Joeke, P., Philippe, N., Michelina, R., & Bernard, B. (2011). Stepwise screening of microorganisms for commercial use in biological control of plantpathogenic fungi and bacteria. *Biological Control, 57*, 1–12.

Kolosova, A., & Stroka, J. (2011). Substances for reduction of the contamination of feed by mycotoxins, Areview. *World Mycotoxin Journal, 4*, 225–256.

Kos, J., Hajnal, E., Jajic, I., & Jovanov, P. (2016). Comparison of ELISA, HPLCFLD and HPLCMS/MS methods for determination of aflatoxin M1 in natural contaminated milk samples. *Acta Chimica Slovenica, 63*, 747–756.

Krishna Reddy, V., & Reddy, S. M. (1989). Efficacy of certain volatile compounds in the control of cyclopiazonic acid production by *Penicillium griseofulvum. National Academy Science Letters, 12*, 67–68.

Krishna Reddy, V., & Reddy, S. M. (1990). Efficacy of certain food preservatives in the control of cyclopiazonic acid production by *Penicillium griseofulvum. Journal of Food Science and Technology, 27*, 180–181.

Krishna Reddy, V., & Reddy, S. M. (1992). Efficacy of phenols in the control of cyclopiazonic acid production by *Penicilium griseofulum. National Academy Science Letters, 15*, 141–143.

Krishna Reddy, V., Giridhar, P., Surekha, M., Reddy, S. M., & Rajam, M. V. (1997). Polamines biosynthetic inhibitors in the control of mycotooxins. In A. Pakash (Ed.), *Fungi in Biotechnology* (pp. 205–208). CBS Publishers and Distributors.

Kumar, A., Ravindra, S., Priyanka, S., Bhanu, P., & Nawal, K. D. (2011). Chemical composition of *Ocimum basilicum* L. essential oil and its efficacy as a preservative against fungal and aflatoxin contamination of dry fruits. *International Journal of Food Science & Technology, 46*, 1840–1846.

Lahlali, R., Ezrari, S., Radouane, N., Kenfaoui, J., Esmaeel, Q., El Hamss, H., Belabess, Z., & Barka, E. A. (2022). Biological control of plant pathogens, A global perspective. *Microorganisms, 10*, 596.

Lal, S. P., & Kapoor, J. N. (1980). Efficacy of certain food preservatives in the control of *Aspergillus* in stored maize. *Indian Phytopathology, 33*, 336.

Laxma Reddy, G., & Reddy, S. M. (1992). Chemical control of satratoxin production by *Stachybotrys atra. Indian Journal of Microbiology, 32*, 401–403.

Li, P., Su, R., Yin, R., Lai, D., Wang, M., Liu, Y., & Zhou, L. (2020). Detoxification of mycotoxins through biotransformation. *Toxins (Basel), 12*, 1–37.

Limaye, A., Yu, R. C., Chou, C. C., & Liu, J. R. (2018). Protective and detoxifying effects conferred by dietary selenium and curcumin against AFB_1mediated toxicity in livestock, a review. *Toxins., 10*, 25.

Line, J. E., Brackett, R. E., & Wilkinson, R. E. (1994). Evidence for degradation of aflatoxin B1 by *Flavobacterium aurantiacum. Journal of Food Protection, 57*, 788–791.

Liu, Y., Chang, J., Wang, P., Yin, Q., Huang, W., Liu, C., et al. (2019). Effects of *Saccharomyces cerevisiae* on alleviating cytotoxicity of porcine jejunal epithelia cells induced by deoxynivalenol. *AMB Express, 9*, 137.

Liu, M., Ling, Z., Guoxin, G., Lei, Z., Lei, S., Jiefan, D., Yanming, H., Yuanyuan, W., Mahmoud, M. K., & Lvhui, S. (2022). Remediation strategies for mycotoxin control in feed. *Journal of Animal Science and Biotechnology, 13*(19), 116.

Loi, M., Fanelli, F., Liuzzi, V. C., Logrieco, A. F., & Mulé, G. (2017). Mycotoxin biotransformation by native and commercial enzymes, present and future perspectives. *Toxins, 9*, 111.

Loi, M., Fanelli, F., Cimmarusti, M. T., Mirabelli, V., Haidukowski, M., Logrieco, A. F., Caliandro, R., & Mule, G. (2018). In vitro single and combined mycotoxins degradation by Ery4 laccase from *Pleurotus eryngii* and redox mediators. *Food Control, 90*, 401–406.

Luo, Y., Liu, X., & Li, J. (2018). Updating techniques on controlling mycotoxins. A review. *Food Control, 89*, 123–132.

Lyagin, I., & Efremenko, E. (2019). Enzymes for detoxification of various mycotoxins: Origins and mechanisms of catalytic action. *Molecules, 24*, 2362.

Majumdar, R., Rajasekaran, K., & Cary, J. W. (2017). RNA interference (RNAi) as a potential tool for control of mycotoxin contamination in crop plants, Concepts and considerations. *Frontiers in Plant Science, 8*, 200.

Mao, J., He, B., Zhang, L. X., Li, P. W., Zhang, Q., Ding, X. X., & Zhang, W. (2016). A structure identification and toxicity assessment of the degradation products of aflatoxin B1 in peanut oil under UV irradiation. *Toxins, 8*, 332.

Mari, M., Neri, F., & Bertolini, P. (2007). Novel approaches to prevent and control postharvest diseases of fruit. *Stewart Postharvest Review, 3*(6), 17.

Markov, K., Frece, J., Pleadin, J., Bevardi, M., Barisic, L., Kljusuric, J. G., Vulic, A., Jakopovic, Z., & Mrvcic, J. (2019). *Gluconobacter oxydans*potential biological agent for binding or biotransformation of mycotoxins. *World Mycotoxin Journal, 12*, 153–161.

Marshall, H., Julie, P. M., Brian, Q., Yueju, Z., Paula, B., Brendan, F. G., Guangtao, Z., & Christopher, T. E. (2020). Novel decontamination approaches and their potential application for postharvest aflatoxin control. *Trends in Food Science & Technology, 106*, 489–496.

Martins, L. M., Santana, A. S., Iamanaka, B. T., Berto, M. I., Pitt, J. I., & Taniwaki, M. H. (2017). Kinetics of aflatoxin degradation during peanut roasting. *Food Research International, 97*(24), 178–183.

Maryamma, K., Ismail, P., Manomohan, C., & Rajan, A. (1990). Ameliorating effect of amruthu (*Tinospora cordifolia*) in aflatoxicosis of ducks. *Journal of Veterinary and Animal Sciences, 21*, 93–96.

Maslowska, J., Owczarek, A., & Kucharska, U. (2018). Determination of aflatoxins in food products by the ELISA method. *Czech Journal of Food Sciences, 19*, 8–12.

Matumba, L., Sydney, N., Theresa, N., Nyadani, M., Marthe De, B., Antonio, F. L., & Sarah De, S. (2021). Five keys to prevention and control of mycotoxins in grains, A proposal. *Global Food Security, 30*, 17.

Mauro, A., Battilani, P., & Cotty, P. J. (2015). Atoxigenic *Aspergillus flavus* endemic to Italy for biocontrol of aflatoxins in maize. *Biological Control, 60*, 125–134.

Medina, A., Sejakhosi, M., Nik, I., Putra, S., Alicia, R. S., Alicia, R., & Naresh, M. (2017). Biocontrol of mycotoxins, dynamics and mechanisms of action. *Current Opinion in Food Science, 17*, 41–48.

Megalla, S. E., & Hafez, A. H. (1982). Detoxification of aflatoxin B1 by acidogenous yoghurt. *Mycopathologia, 77*, 89–91.

Misra, N. N., Yadav, B., Roopesh, M. S., & Jo, C. (2019). Cold plasma for effective fungal and mycotoxin control in foods, mechanisms, inactivation effects, and applications. *Comprehensive Reviews in Food Science and Food Safety, 18*(1), 106–120.

Mo, H. Z., Zhang, H., Wu, Q. H., & Hu, L. B. (2013). Inhibitory effects of tea extract on aflatoxin production by *Aspergillus flavus*. *Letters in Applied Microbiology, 56*, 462–466.

Moerck, K. E., Lfresh, P. M., Alan, W., & Barney, W. H. (1980). Aflatoxin destruction in corn using sodium bisulfite, sodium hydroxide and *Aqueous Ammonia. Journal of Food Protection, 43*(7), 571–574.

Moller, C. O. D. A., Freire, L., Rosim, R. E., Margalho, L. P., Balthazar, C. F., Franco, L. T., & Oliveira, C. A. F. D. (2021). Effect of lactic acid bacteria strains on the growth and aflatoxin production potential of *Aspergillus parasiticus*, and their ability to bind aflatoxin B1, ochratoxin A, and zearalenone *in vitro*. *Frontiers in Microbiology, 12*, 65586.

Munkvold, G. P. (2003). Cultural and genetic approaches to managing mycotoxins in maize. *Annual Review of Phytopathology, 41*, 99–116.

Murphy, P. A., Hendrich, S., Landgren, C., & Bryant, C. M. (2006). Food mycotoxins, an update. *Journal of Food Science, 71*(5), 51–65.

Murugesan, P., Brunda, D. K., Moses, J. A., & Anandharamakrishnan, C. (2021). Photolytic and photocatalytic detoxification of mycotoxins in foods. *Food Control, 123*(3), 10748.

Nada, S., Tomic, N., Udovicki, B., Djekic, I., & Rajkovic, A. (2022). Prevention and practical strategies to control mycotoxins in the wheat and maize chain. *Food Control, 136*, 108855.

Nahle, S., Andre El, K., Ioannis, S., Ali, C., Nicolas, L., & Ali, A. (2022). Detoxification approaches of mycotoxins, by microorganisms, biofilms and enzymes. *International Journal of Food Contamination., 9*, 3. https://doi.org/10.1186/s40550022000892

Nakayashiki, H. (2005). RNA silencing in fungi, mechanisms and applications. *FEBS Letters, 579*, 5950–5957.

Nazhand, A., Durazzo, A., Lucarini, M., Souto, E. B., & Santini, A. (2020). Characteristics, occurrence, detection and detoxification of aflatoxins in foods and feeds. *Food, 9*, 644. https://doi.org/10.3390/foods9050644

Negera, M., & Washe, A. P. (2019). Use of natural dietary spices for reclamation of food quality impairment by aflatoxin. *Hindawi Journal of Food Quality, 2*, 1–10.

Nesic, K., Habschied, K., & Mastanjevic, K. (2021). Possibilities for the biological control of mycotoxins in food and feed. *Toxins, 13*, 198. https://doi.org/10.3390/toxins13030198

Nguegwouo, E., Sone, L. E., Tchuenchieu, A., Tene, H. M., Mounchigam, E., Njayou, N. F., & Nama, G. M. (2018). Ochratoxin a in black pepper, white pepper and clove sold in Yaounde (Cameroon) markets, contamination levels and consumers' practices increasing health risk. *International Journal of Food Contamination, 5*, 1.

Noreen, M., & Russell, J. M. (2004). Phytochemical inhibition of aflatoxigenicity in *Aspergillus flavus* by constituents of walnut (*Juglans regia*). *Journal of Agricultural and Food Chemistry, 52*(7), 1882–1889.

Norred, W. P., Voss, K. A., Bacon, C. W., & Riley, R. T. (1991). Effectiveness of ammonia treatment in detoxification of fumonisincontaminated corn. *Food and Chemical Toxicology, 29*, 815–819.

Oliveira, R. C., Carvajal Moreno, M., Mercado Ruaro, P., Rojo Callejas, F., & Correa, B. (2020). Essential oils trigger an antifungal and antiaflatoxigenic effect on *Aspergillus flavus* via the induction of apoptosislike cell death and gene regulation. *Food Control, 110*, 107038.

Pal, K. K., & Gardener, M. S. B. (2006). Biological control of plant pathogens. *The Plant Health Instructor, 10*, 125.

Palabiyik, S. S., Erkekoglu, P., Zeybek, N. D., Kizilgun, M., Baydar, D. E., Sahin, G., & Giray, B. K. (2013). Protective effect of lycopene against ochratoxin a induced renal oxidative stress and apoptosis in rats. *Experimental and Toxicologic Pathology, 65*, 853–861.

Pamela, S., & Wim, W. (2019). Insects for food and feedsafety aspects related to mycotoxins and metals. *Food, 8*, 288.

Panda, P., & Mehta, A. (2013). Aflatoxin detoxification potential of *Ocimum tenuiflorum*. *Journal of Food Safety, 33*, 265–272.

Panda, P., Aiko, V., & Alka, M. (2015). Effect of aqueous extracts of *Menthaarvensis* (mint) and Piper betle (betel) on growth and citrinin production from toxigenic *Penicilliumcitrinum*. *Journal of Food Science and Technology, 52*(6), 3466–3474.

Pankaj, S. K., Shi, H., & Keener, K. M. (2018). A review of novel physical and chemical decontamination technologies for aflatoxin in food. *Trends in Food Science and Technology, 71*, 73–83.

Park, D. L., Lee, L. S., Price, R. L., & Pohland, A. E. (1988). Review of the decontamination of aflatoxins by ammoniation, current status and regulation. *Journal of the Association of Official Analytical Chemists, 71*(4), 685–703.

Passone, M. A., Silvia, R., & Miriam, G. E. (2007). Antiaflatoxigenic property of food grade antioxidants under different conditions of water activity in peanut grains. *International Journal of Food Microbiology, 118*(1), 8–14.

Patel, U. D., Govindarajan, P., & Dave, P. J. (1989). Inactivation of aflatoxin B1 by using the synergistic effect of hydrogen peroxide and gamma radiation. *Applied and Environmental Microbiology, 55*(2), 465–467.

Peraica, M., Domijan, A. M., Jurjevic, Z., & Cvjetkovic, B. (2002). Prevention of exposure to mycotoxins from food and feed. *Arhiv za Higijenu Rada i Toksikologiju, 53*, 229–237.

Pereira, P., Nesci, A., & Etcheverry, M. (2007). Effects of biocontrol agents on *Fusarium verticillioides* count and fumonisin content in the maize agroecosystem, impact on rhizospheric bacterial and fungal groups. *Biological Control, 42*, 281–287.

Peteri, Z., Teren, J., Vagvolgyi, C., & Varga, J. (2007). Ochratoxin degradation and adsorption caused by astaxanthinproducing yeasts. *Food Microbiology, 24*, 205–210.

Piemontese, L., Messia, M. C., Marconi, E., Falasca, L., Zivoli, R., Gambacorta, L., Perrone, G., & Solfrizzo, M. (2018). Effect of gaseous ozone treatments on DON, microbial contaminants and technological parameters of wheat and semolina. *Food Additives and Contaminants: Part A, 35*, 760–771.

Pirgozliev, S. R., Edwards, S. G., Hare, M. C., & Jenkinson, P. (2003). Strategies for the control of *Fusarium* head blight in cereals. *European Journal of Plant Pathology, 109*, 731–742.

Pleadin, J., Frece, J., & Markov, K. (2019). Mycotoxins in food and feed. *Advances in Food and Nutrition Research, 89*, 297–345.

Porto, Y. D., Trombete, F. M., Freitas Silva, O., de Castro, I. M., Direit, G. M., & Ascheri, J. L. R. (2019). Gaseous ozonation to reduce aflatoxins levels and microbial contamination in corn grits. *Microorganisms, 7*, 220.

Prandini, A., Tansini, G., Sigolo, S., Filippi, L., Laporta, M., & Piva, G. (2009). On the occurrence of aflatoxin M1 in milk and dairy products. *Food and Chemical Toxicology, 47*, 984–991.

Pundir, R. K., & Jain, P. (2010). Antifungal activity of twentytwo ethanolic plant extracts against foodassociated fungi. *Journal of Pharmacy Research, 3*, 506–510.

Qun Ying Su. (2019). The Toxification and detoxification mechanisms of aflatoxin B1 in human, An Update. https://doi.org/10.5772/intechopen.89221.

Rajam, M. V., & Galston, A. W. (1985). The effects of some polyamine biosynthetic inhibitors on growth and morphology of phytopathogenic fungi. *Plant Cell Physiology, 26*, 683–692.

Rajam, M. V., Wilnstein, L. H., & Galston, A. W. (1985). Prevention of a plant disease by specific inhibition of fungal polyamine biosynthesis. *Proeeedings of the National Aeademy of Scienee USA, 82*, 6874–6878.

Reddy, S. M., & Surekha, M. (1994). Incidence and production of mycotoxin by fungi in a competitive environment. ISCA Symposium, 1991. In G. P. Agarwal & S. K. Hasija (Eds.), *ed* (pp. 133–144). Narendra Publishing House.

Reiss, J. (1979). Prevention of the formation of mycotoxins in whole wheat bread by citric acid and lactic acid. *Experientia, 32*, 168–169.

Ren, X., Brana, M. T., Miriam, H., Antonia, G., Qi, Z., Antonio, F. L., Peiwu, L., Shancang, Z., & Claudio, A. (2022). Potential of *Trichoderma* spp. for biocontrol of aflatoxin producing *Aspergillus flavus*. *Toxins (Basel), 14*(2), 86.

Richard, C. D., & Hamilton, P. B. (1981). Effect of feed ingredients on the antifungal activity of propionic acid. *Poultry Science, 60*(11), 2407–2411.

Sakuda, S., Makoto, O., Kazuo, F., Jiro, N., Akinori, S., & Akira, I. (1996). Aflastatin A, a novel inhibitor of aflatoxin production of *Aspergillus parasiticus*, from Streptomyces. *Journal of the American Chemical Society Washington., 118*(33), 7855–7856.

Samie, A., & Mashau, F. (2013). Antifungal activities of 15 south African medicinal plants against 5 *Fusarium spp*. *Journal of Medicinal Plants Research, 7*(28), 1839–1843.

Sani, A. M., Azizi, E. G., Salehi, E. A., & Rahimi, K. (2014). Reduction of aflatoxin in rice by different cooking methods. *Toxicology and Industrial Health, 30*, 46–550.

Santin, E., Maiorka, A., Krabbe, E. L., Paulillo, A. C., & Alessi, A. C. (2002). Effect of hydrated sodium calcium aluminosilicate on the prevention of the toxic effects of ochratoxin. *Journal of Applied Poultry Research, 11*, 22–28.

Santos Alexandre, A. P., Vela Paredes, R. S., Santos, A. S., Costa, N. S., Canniatti, B. S. G., Calori Domingues, M. A., & Augusto, P. E. D. (2018). Ozone treatment to reduce deoxynivalenol (DON) and zearalenone (ZEN) contamination in wheat bran and its impact on nutritional quality. *Food Additives and Contaminants: Part A, 35*, 1189–1199.

Sashidhar, R. B., Murthy, H. V. V., & Bhat, R. V. (1992). Analytical quality assurance for analysis of aflatoxins in agricultural export produce. *Food Chemistry, 43*, 403–405.
Schaafsm, A. W., Tamburic Ilinic, L., Miller, J. D., & Hooker, D. C. (2001). Agronomic considerations for reducing deoxynivalenol in wheat grain. *Canadian Journal of Plant Pathology, 23*, 279–285.
Schatzmayr, G., Zehner, F., Taubel, M., Schatzmayr, D., Klimitsch, A., Loibner, A. P., & Binder, E. M. (2006). Microbiologicals for deactivating mycotoxins. *Molecular Nutrition & Food Research, 50*, 543–551.
Schofield, E. K. (2005). *Feed manufacturing technology*. American Feed Industry Association Inc..
Scott, P. M. (1998). Industrial and farm detoxification processes for mycotoxins. *Revista de Medicina Veterinaria, 149*, 543–548.
Sharma, R. R., Singh, D., & Singh, R. (2009). Biological control of postharvest diseases of fruits and vegetables by microbial antagonists, A review. *Biological Control, 50*, 205–221.
Shen, M. H., & Singh, R. (2021). Effect of rotating peanuts on aflatoxin detoxification by ultraviolet C light and irradiation uniformity evaluated by AgClbased dosimeter. *Food Control, 120*, 107533.
Shi, H., Li, S., Bai, Y., Prates, L. L., Lei, Y., & Yu, P. (2018). Mycotoxin contamination of food and feed in China, occurrence, detection techniques, toxicological effects and advances in mitigation technologies. *Food Control, 91*, 202–215.
Silvac, B., Manuel, M. S., & Eliana, B. F. (2020). Antioxidant and antifungal activity of phenolic compounds and their relation to aflatoxin B1 occurrence in soybeans (*Glycine max* L.). *Journal of the Science of Food and Agriculture, 100*(3), 1256–1264.
Singh, N., Jand, S. K., & Baxi, K. K. (2003). Chemical detoxification of aflatoxins in contaminated poultry poultry and fish feed detoxification 157 feed. *Indian Journal of Animal Sciences, 73*(197), 199.
Sinha, K. K. (1990). Prevention of aflatoxin production in some cereals and oil seeds by natural plant constituents. *Irish Journal of Food Science and Technology, 14*(2), 109–120.
Sinha, K. K. (2001). Decontamination of mycotoxins and food safety. *Toxicology Letters, 122*, 179–188.
Sinha, A. K., & Ranjan, K. S. (1989). BGY fluorescence and aflatoxin contamination in insect damaged maize crop. *Indian Phytopathology, 42*, 514–518.
Skrinjar, M., Rasic, J. L., & Stojicic, V. (1996). Lowering of ochratoxin a level in milk by yoghurt bacteria and bifidobacteria. *Folia Microbiologica, 1*, 26–28.
Smiley, R. D., & Draughon, F. A. (2000). Preliminary evidence that degradation of aflatoxin B1 by *Flavobacterium aurantiacum* is enzymatic. *Journal of Food Protection, 63*, 415–418.
Sorrenti, V., Di Giacomo, C., Acquaviva, R., Barbagallo, I., Bognanno, M., & Galvano, F. (2013). Toxicity of Ochratoxin a and its modulation by antioxidants, a review. *Toxins, 5*, 1742–1766.
Sreenivasa, M. J., Parpia, H. A. B., Srikanta, S., & Murti, A. S. (1967). Detoxification of aflatoxin in peanut meal by hydrogen peroxide. *Journal of the Association of Official Analytical Chemists, 50*, 350–354.
Srivastava, P. K. (1986). *Bull Grain Tech, 24*(1), 64–69.
Stangroom, K. E., & Smith, T. K. (1984). Effect of whole and fractionated dietary alfalfa meal on zearalenone toxicosis and metabolism in rats and swine. *Canadian Journal of Physiology and Pharmacology, 62*, 1219–1224.
Surekha, M., & Reddy, S. M. (1992). Efficacy of some plant extracts in the control of Penitem A production by *Penicillium aurantiogriseum*. *National Academy Science Letters, 15*(3), 69–71.
Tang, Y., Xiao, J., Chen, Y., Yu, Y., Xiao, X., Yu, Y., & Wu, H. (2013). Secretory expression and characterization of a novel peroxiredoxin for zearalenone detoxification in *Saccharomyces cerevisiae*. *Microbiological Research, 168*, 6–11.
Taniwaki, M., Pitt, J. I., Teixeira, A., & Iamanaka, B. T. (2003). The source of ochratoxin A in Brazilian coffee and its formation in relation to processing methods. *International Journal of Food Microbiology, 82*, 173–179.

Tarazona, A., Gomez, J. V., Mateo, E. M., Jiménez, M., & Mateo, F. (2019). Antifungal effect of engineered silver nanoparticles on phytopathogenic and toxigenic *Fusarium* spp. and their impact on mycotoxin accumulation. *International Journal of Food Microbiology, 306*, 108–259.

Thomas, M., Hendriks, W. H., & van der Poel, A. F. B. (2018). Size distribution analysis of wheat, maize and soybeans and energy efficiency using different methods for coarse grinding. *Animal Feed Science and Technology, 240*, 11–21.

Thompson, D. P. (1992). Inhibition of mycelial growth of mycotoxigenic fungi by phenolic antioxidants. *Mycologia, 84*(5), 791–793.

Tian, J., Huang, B., Luo, X., Zeng, H., Ban, X., He, J., & Wang, Y. (2012). The control of *Aspergillus flavus* with *Cinnamomum jensenianum* hand. Mazz essential oil and its potential use as a food preservative. *Food Chemistry, 130*(3), 520–527.

Umesha, S., Manukumar, H. M. G., Chandrasekhar, B., Shivakumara, P., Shiva Kumar, J., Raghava, S., & Nandhini, M. (2017). Aflatoxins and food pathogens, impact of biologically active aflatoxins and their control strategies. *Journal of Science and Food Agriculture, 97*(6), 1698–1707.

Uraih, N., & Offonre, S. (1981). Inhibition of aflatoxin production in groundnut with benzoic acid derivatives and possible toxic effect of their aromatic residues. *Microbios, 31*, 93–102.

Utermark, J., & Karlovsky, P. (2006). Role of zearalenone lactonase in protection of *Gliocladium roseum* from fungitoxic effects of the Mycotoxin Zearalenone. *Applied and Environmental Microbiology, 73*, 637–642.

ValleAlgarra, F. M., Mateo, R., Medina, A., Mateo, F., Mateo, E. M., Gimeno, A. J. V., & Jimenez, M. (2009). Effect of the baking process on the reduction of ochratoxin A in wheat flour. *Current Research Topics in Applied Microbiology and Microbial Biotechnology*, 382–385.

Vanara, F., Scarpino, V., & Blandino, M. (2018). Fumonisin distribution in maize drymilling products and byproducts, impact of two industrial degermination systems. *Toxins, 10*, 357.

Vandergraft, E. E., Hesseltine, C. W., & Shotwell, O. L. (1975). Grain preservatives, effect on aflatoxin and ochratoxin production. *Cereal Chemistry, 52*, 79–84.

Vanhoutte, I., Audenaert, K., & De Gelder, L. (2016). Biodegradation of mycotoxins, Tales from known and unexplored worlds. *Frontiers in Microbiology, 7*, 561.

Varga, J., Peteri, Z., Tabori, K., & Teren, J. (2005). Degradation of ochratoxin a and othermycotoxins by Rhizopus isolates. *International Journal of Food Microbiology, 99*, 321–328.

Varma, J., & Dubey, N. K. (2001). Efficacy of essential oils of *Caesulia axillaris* and *Mentha arvensis* against some storage pests causing biodeterioration of food commodities. *International Journal Food Microbiology, 68*, 207–210.

Vekiru, E., Christian, H., Rudolf, M., Justyna, R., Gerhard, A., Gerd, S., Rudolf, K., & Rainer, S. (2010). Cleavage of Zearalenone by *Trichosporon mycotoxinivorans* to a novel nonestrogenic metabolite. *Applied and Environment Microbiology Washington, 76*(7), 2353–2359.

Verma, R. J., Shukla, R. S., & Mehta, D. N. (2001). Amelioration of cytotoxic effects of aflatoxin by vitamin a, an in vitro study on erythrocytes. *Toxicology In Vitro, 15*(1), 39–42.

Vidosavljevic, S., Kos, J., Banjac, V., Janic Hajnal, E., Dragojlovic, D., Duragic, O., & Colovic, R. (2018). *Proceedings of the feed technology. Volume 18* (pp. 81–88). University of Novi Sad, Institute of Food Technology.; Novi Sad, Serbia, Simple technologies for removal of mycotoxins in feed.

Vipin, A. V., Raksha, R. K., Nawneet, K. K., Anu, A. K., & Venkateswaran, G. (2017). Protective effects of phenolics rich extract of ginger against aflatoxin B_1induced. *Biomedicine & Pharmacotherapy, 91*, 415–424.

Wagacha, J., & Muthomi, J. (2008). Mycotoxin problem in Africa, current status, implications to food safety and health and possible management strategies. *International Journal of Food Microbiology, 124*, 1–12.

Wallace, H. M., & Fraser, A. V. (2004). Inhibitors of polyamine metabolism, review article. *Amino Acids, 26*, 353–365.

Wang, T., Zhang, E., Chen, X., Li, L., & Liang, X. (2010). Identification of seed proteins associated with resistance to preharvested aflatoxin contamination in peanut (*Arachis hypogaea* L.). *BMC Plant Biology, 10*, 267.

Wang, J., Ogata, M., Hirai, H., & Kawagishi, H. (2011). Detoxification of aflatoxin B1 by manganese peroxidase from the whiterot fungus *Phanerochaete Sordida* YK624. *FEMS Microbiology Letters, 314*, 164–169.

Wang, G., Yu, M., Dong, F., Shi, J., & Xu, J. (2017). Esterase activity inspired selection and characterization of zearalenone degrading bacteria *Bacillus pumilus* ES21. *Food Control, 77*, 57–64.

Wang, M., Yin, L., Hu, H., Selvaraj, J. N., Zhou, Y., & Zhang, G. (2018). Expression, functional analysis and mutation of a novel neutral zearalenonedegrading enzyme. *International Journal of Biological Macromolecules, 118*, 1284–1292.

Wang, N., Wu, W., Pan, J., & Long, M. (2019). Detoxification strategies for zearalenone using microorganisms, A review. *Microorganisms, 7*(7), 208.

Wielogorska, E., Ahmed, Y., Meneely, J., Graham, W. G., Elliott, C. T., & Gilmore, B. F. (2019). A holistic study to understand the detoxification of mycotoxins in maize and impact on its molecular integrity using cold atmospheric plasma treatment. *Food Chemistry, 301*, 125–281.

Wilson, D. M., Mubatanhema, W., & Jurjevic, Z. (2002). Biology and ecology of mycotoxigenic *aspergillus* species as related to economic and health concerns, p. 317. In J. W. de Vries, M. W. Trucksess, & L. S. Jackson (Eds.), *Mycotoxins and food safety*. Kluwer Academic Plenum Publications.

Yabe, K. S., Yan, Y. S., Ichinomiya, M., Nakagawa, H., Shima, Y., Ando, Y., Sakuno, E., & Nakajima, H. (2008). Isolation of microorganisms and substances inhibitory to aflatoxin production. *Food Additives & Contaminants, 25*, 1111–1117.

Yates, I. E., Meredith, F., Smart, W., Bacon, C. W., & Jaworski, A. J. (1999). *Trichoderma viride* suppresses fumonisin B1 production by *Fusarium moniliforme*. *Journal of Food Protection, 66*, 1326–1332.

Yoshinari, T., Tetsuo, A., Keita, N., Tatsuhiko, K., Yoshikazu, T., Yasuhiko, M., Yoshiaki, N., Hiromichi, N., & Shohei, S. (2007). Dioctatin A is a strong inhibitor of aflatoxin production by *Aspergillus parasiticus*. *Microbiology Spencers Wood, 153*, 2774–2780.

Youssef, A. E., & Marth, E. H. (1984). Growth and synthesis of aflatoxin by *Aspergillus parasiticus* in the presence of sorbic acid. *Journal of Food Protection, 44*, 736–741.

Zachetti, V. G. L., Cendoya, E., Nichea, M. J., Chulze, S. N., & Ramirez, M. L. (2019). Preliminary study on the use of chitosan as an ecofriendly alternative to control *Fusarium* growth and mycotoxin production on maize and wheat. *Pathogens, 8*, 29.

Zeinvand, L. H., Sabzevari, O., Setayesh, N., Amini, M., Faramarzib, A., Ahmadabadi, A. N., & Faramarzi, A. M. (2015). Comparative study of in vitro prooxidative properties and genotoxicity induced by aflatoxin B1 and its laccasemediated detoxification products. *Chemosphere, 135*, 1–6.

Zhang, W., Xue, B., Li, M., Mu, Y., Chen, Z., Li, J., & Shan, A. (2014). Screening a strain of *Aspergillus Niger* and optimization of fermentation conditions for degradation of aflatoxin B1. *Toxins (Basel), 6*, 3157–3172.

Zhang, X., Yang, H., Apaliya, M. T., Zhao, L., Gu, X., Zheng, X., Hu, W., & Zhang, H. (2018). The mechanisms involved in ochratoxin A elimination by *Yarrowia lipolytica* Y2. *The Annals of Applied Biology, 173*, 164–174.

Zhang, Z., Xu, W., Wu, H., Zhang, W., & Mu, W. (2020). Identification of a potent enzyme for the detoxification of Zearalenone. *Journal of Agricultural and Food Chemistry, 68*, 376–383.

Zhou, Y., Wu, S., Wang, F., Li, Q., He, C., Duan, N., & Wang, Z. (2020). Assessing the toxicity *in vitro* of degradation products from deoxynivalenol photocatalytic degradation by using up conversion nanoparticles@ TiO_2 composite. *Chemosphere, 238*, 124648.

Zhu, Y., Hassan, Y. I., Lepp, D., Shao, S., & Zhou, T. (2017). Strategies and methodologies for developing microbial detoxification systems to mitigate mycotoxins. *Toxins, 9*, 130. https://doi.org/10.3390/toxins9040130

Zohri, A. A., & Abdel Kareem, M. M. (2018). Four strains of yeasts, as effective biocontrol agents against both growth and mycotoxins formation by selected 11 toxigenic fungi. *Global Advanced Research Journal of Microbiology, 7*, 132–135.

Chapter 21
Genetic Basis Providing Crop Plants Resistance to Phytopathogens

Hausila Prasad Singh, Swati Singh, Sonika Kalia, Vinay Rojaria, Shanvi Nautiyal, Prahlad Masurkar, Jasima Ali, Rishabh Maheshwari, and Smriti Mall

Abstract Increasing human population demanding more food resulted in a major need to control plant pathogens. Sustainable management of most common plant diseased causing pathogens relies on biocontrol agents, genetic engineering and microbial volatilomes, which might be fungal or bacterial in origin. Approaches based on -omics tools are utilized by researchers for controlling plant pathogens under resistant breeding programme, which starts from the identification of desired genes in biocontrol agents such as *Tricoderma virens*, *Ampelomyces quisqualis*, *Coniothyrium minitans* etc. Genome sequencing, protoplast fusion and genetic transformation via biotechnological tools also help in controlling plant pathogens. Actual focus of research is on resistant breeding programmes which are currently most popular and represent cost effective measure for developing resistant cultivars/hybrids. This chapter critically reviews information on various genetic methods providing resistance to plants against their associated pathogens.

H. P. Singh (✉) · V. Rojaria
Department of Genetics and Plant Breeding, BRD PG College, Deoria, India

S. Singh
Department of Phytochemistry, CSIR- Central Institute of Medicinal and Aromatic Plants, Lucknow, India

S. Kalia · S. Nautiyal
Department of Biotechnology, School of Applied and Life Sciences, Uttaranchal University, Dehradun, India

P. Masurkar
Department of Plant Pathology, School of Agriculture, Lovely Professional University, Phagwara, Punjab, India

J. Ali · R. Maheshwari
Department of Plant Pathology, Punjab Agricultural University, Ludhiana, India

S. Mall
Molecular Plant Pathology Laboratory, Department of Botany, Deen Dayal Upadhyaya Gorakhpur University, Gorakhpur, UP, India

P. Mathur, S. Roy (eds.), *Plant Microbiome and Biological Control*, Sustainability in Plant and Crop Protection 20,
https://doi.org/10.1007/978-3-031-75845-4_21

Keywords Biotechnology · Biotic stress · Breeding · Crop improvement · Genome · Pathogen

Abbreviations

AMV	Alfalfa mosaic virus
BCA	Biological control agents
CWDEs	Cell wall-degrading enzymes
CRISPR	Clustered Regularly Interspaced Short Palindromic Repeats
GMOs	Genetically modified crops
MLO	Mycoplasma-like organisms
NLR	Nucleotide-binding leucine-rich repeats
TALEN	Transcription Activator-like Effector Nuclease
TMV	Tobacco mosaic virus (TMV)
VOCs	Volatile Organic Compounds

21.1 Introduction

The survival of the human and animal populations is directly dependent on crop availability and productivity. Unavailability of food crops in adequate amounts often results in economic and financial food crises in developing countries, and leads to deadly famine-like conditions. One of the most potent factors determining the quality, quantity, and production of plants is their resistance to various disease-causing pathogens, in particular in a climate changing environment (Singh et al., 2023). Although most plants recognize pathogens and generate appropriate defense mechanisms in responses, certain pathogens being able to suppress their defense mechanisms or interfere in their recognition by the plant. An infected plant is recognized by stunted growth, disruption of functions and disruption of structure caused by the prevalence of biotic and abiotic environmental stress conditions (Nazarov et al., 2020). Fungi, bacteria, viruses, and protozoa are among the most significant infectious disease-causing pathogens.

The probability of the occurrence or onset of a disease is determined by the presence of a susceptible host, a virulent pathogen, and favourable environmental conditions. The disease severity and effectiveness of the infection depend on the genetic profile, potency, and susceptibility of the organisms involved.

Plant diseases contribute worldwide to about 16% of crop yield losses and directly threaten food security and safety. The five majorly grown crops like wheat, rice, soybean, maize, and potato, recorded a reduction of about 10 to 30% in the yield due to pathogenic infections (Savary et al., 2012). Since ancient times, classical seed breeding techniques for crop selection have been utilized to suppress most pathogens. It was in the late 1940s that genetic interactions and their involvement gained momentum in understanding their role in controlling disease resistance in plants. Agriculture-based global economies witnessed a significant loss and reduction in gross domestic product as agriculture exports and trade contribute to a major portion of their economies, leading to a socio-economic crisis (Ghabrial & Suzuki, 2009). Therefore, herein we critically reviewed the information about various approaches of plant disease management, emerging biocontrol strategies, genes involved in biological control, development of genetically resistant plants and plant -omics role in biological control agent *vs* different pathogens.

21.2 Microbial Volatiles Role in Biological Control

Biological control agents (BCA) are eco-friendly and sustainable tools for plant disease management (Saxena et al., 2019). They act as antagonists and/or suppressors of pathogens. Direct antagonism deals with hyperparasitisın and predation, whereas indirect means seek at competition or host resistance induction. BCA may also trigger the development of an unfavourable condition for microbial growth, leading to the control of a target species (Köhl et al., 2019). They also deal indirectly by exudation of volatile organic compounds (VOCs), and directly by parasitism and induction of, either induced systemic resistance (ISR) or systemic acquired resistance (SAR).

In a pathogen suppression process, BCA may also produce volatiles, antibiotics, and other secondary metabolites. Volatile organic compounds are a mixture of volatile metabolites produced by living microorganisms, including BCA, that are not hazardous for humans/animals but have a specific detrimental effect on plant pathogens (Tilocca et al., 2020).

The volatile compound, which may have both plant and microbial origins, is called "volatilome." The volatilome is gaining steadily increasing interest (Pandit et al., 2022). Volatile molecules can be divided into two forms either inorganic and organic: CO, CO_2, H_2, N_2, O_2, NH_3, H_2S, NO_2^-, SO_2, SO_3, and HCN are the most relevant molecules in the inorganic volatiles. While the volatile organic molecules are small (less than 300 Da), carbon-based molecules feature a low water solubility and a high vapor pressure that make them available in a gaseous state, in normal ambient conditions (Fufurin et al., 2021).

Biosynthetic pathways leading to the production of microbial VOCs are yet poorly understood. However, omics-based studies so far have linked the production of VOCs to either metabolic transformation products of lipids and proteins or as the result of catabolic reactions (Ali et al., 2023).

21.2.1 Bacterial Volatilome

Volatiles secreted by bacteria play an important role in bacterial-plant, bacterial-bacterial, and bacterial-fungal interactions. Alkanes, alkenes, alcohols, esters, ketones, and sulfurs compounds and terpenoids act as bacterial volatiles. Rhizospheric bacteria emit VOCs that may change the root system architecture with plant growth promotion activities (Ana et al., 2020). Bacterial VOCs, i.e., 2-pentanone, 4-heptanone, 2-heptanol, 2-undecanone, 2-tridecanone and 2-pentadecanone, with a compositally compound like sodorifen (a bicyclic oligometus octadiene), were found to be released by the bacterium *Serratia odorifera* and appeared as a very effective volatilome (Cernosek et al., 2020).

One important VOCs is the isoprene diterpenoid (-C_5H_8), which includes the monolactones, oryzalexins, and gibberellins. The capacity to biosynthesize diterpenoids is present in plants, fungi, and selected bacteria (Fischer et al., 2015). These diterpenoids act in pathogen defence, plant growth, signaling, and in the biotic stress response. *Fusarium oxysporum* f. sp. *cubense*, which causes fusarium wilt in banana plants, was found to be inhibited by the metabolite NJN-6 obtained by the bacterium *Bacillus amyloliquefaciens.* Mycotoxic fungi were also found to be inhibited by *Streptomyces* spp. (Khan et al., 2017).

21.2.2 Fungal Volatilome

Apart from bacterial VOCs, fungi also produce volatile compounds, ranging from antagonism to symbiotic relationships with other fungi. Some VOCs the fungi produce are sabinene-I, butanol, 3-methyl benzene ethanol, 1-propanol, 2-methyl, and 2-propanone (Tilocca et al., 2019). An artificial mixture of such volatiles tremendously affects plant pathogenic fungi and bacteria. VOCs such as isobutyl alcohol, isopentyl alcohol, and 3-methyl butanol from *Trichoderma viride* promote plant growth (Elsherbiny et al., 2020). *Talaromyces* sp. produced β-caryophyllene which also induces plant growth, while *Cladosporium* sp. and *Ampelomyces* sp. reduced the pathogen infection and elicited resistance by producing m-cresomethylbenzoate.

An important group of single-celled fungi, i.e., yeasts, also acts as a great source of volatiles and has been found to inhibit some mycotoxin-producing fungi (Oufensou et al., 2023). Yeast-derived volatiles are mainly attributed to ethyl acetate that was especially found to cure *Penicillium roqueforti* infection in airtight storage (Tilocca et al., 2020).

21.2.3 Microbiome-Based Solutions in Plant Protection

Plants cope with stresses with the help of their microbiome, the community of microorganisms (such as bacteria, fungi and viruses) existing in an endophytic, rhizosphere, or phyllosphere condition (Bashir et al., 2022). The rhizospheric condition sustains the development of plants by enhancing their growth and also mitigating negative, environmental effects. Various beneficial taxa are simultaneously active in microbiome-based processes such as the induction of plant hormone secondary metabolites and induced or systemic resistance. Some microbiome species may act as biofertilizers as they improve the uptake of nutrients, nitrogen fixation and solubilization of unavailable soil phosphorous. Another important aspect of the microbiome community is the mitigation of heavy metal stress by transforming these compounds into less toxic forms. The processes of acidity, bioleaching, and detox are also influenced by several microbial mechanisms (Barra Caracciolo & Terenzi, 2021).

21.3 Genetic Engineering and Genome Editing

Genome editing is also known as precision breeding, which focuses on changing genome sequences. This became the basis for developing GMOs (genetically modified crops), obtained by insertion, deletion, and substitution of nucleotides within one or more genes, often causing a loss in the functionality of the target sequence. Genetic editing is made possible by technologies such as CRISPR-Cas9 (Clustered Regularly Interspaced Short Palindromic Repeats) and TALEN (Transcription Activator-Like Effector Nuclease) (Belhaj et al., 2013). CRISPR is a family of sequences of DNA found in prokaryote genomes. These small sequences have been derived from DNA segments of pathogens, which previously affected prokaryotes.

There are a number of successful examples of CRISPR and TALEN for improving defense responses in plants against *Erysiphe necator* in grapes and the development of *Triticum aestivum (*hexaploid wheat) to synthesize *mlo* (mildew locus genes) varieties, by removing three homologous genes (Bai et al., 2013). A recent successful example of CRISPR was seen against citrus canker caused by *Xanthomonas citri*, where the *CsLOB1* resistance gene was identified in Duncan grapefruit (*Citrus* × *paradisi*) and resistant lines were developed. Acevedo-Garcia et al. (2017) focused on the knockdown of three important genes, viz., *TaMLO-A1*, *TaMLO-B1* and *TaMLO-D1* against powdery mildew of wheat caused by *Erysiphe graminis*, using CRISPR-Cas9. Similarly, in the case of apple fire blight disease, three genes (*DIPM1, DIPM2, DIPM4*) were targeted for knockdown using CRISPR-Cas9 technology (Malnoy et al., 2016).

A typical definition of genetic engineering is a molecular procedure that introduces a gene or gene construct typically at random to a recipient organism in a process that would not happen naturally. The resulting strain is referred to as a GMO (genetically modified organism). Regarding the legal regulation of the process, the technology is classified into two categories based on whether the introduced gene originates from the same species (called *cis*-genic technology) or from a different species (called transgenic technology). Despite this, the origin of the gene is irrelevant. Many contend that since cis-genesis may lead to the development of a product by natural crossing, it should be regarded as a non-biotech technology (NBT) and free from GMO laws. This is especially relevant for crops like potato, where it might take decades for a resistant gene to be introduced, or prized cultivars such as grape varieties for wine production.

When it comes to viral infections, r-DNA technology has produced astounding success. Transferring and expressing the coat protein genes of the Tobacco Mosaic Virus (TMV) and Alfalfa Mosaic Virus (AMV) in tobacco represents a notable achievement. This progress prevented or postponed the development of diseases in the transgenic plants (Rajakaruna & Taylor-Robinson, 2016). The theory underlying the addition of coat protein genes to provide resistance against the virus is that the coat protein produced in plant cells in some way prevents the viral RNA from proliferating. A further example is the intoduction of chitinase enzymes, produced by engineered plants, that break down the fungus cell wall to eradicate the soil-borne pathogen *Rhizoctonia solani*.

Comparative genomics technologies, such R (resistance) gene enrichment sequencing (RenSeq), are among the most widely used but least studied methods for genome editing. This method finds the candidate R gene, which recognizes the core effectors, by high throughput screening of germplasms or mutants (Dong & Ronald, 2019). Further investigations revealed that members of the nucleotide-binding leucine-rich repeats (NLR) gene family are involved in distant resistance in RenSeq. In RenSeq, genomic DNA fragments containing NLR from resistant germplasms are captured and enriched using a library of NLR-specific DNA fragments as baits (Monteiro & Nishimura, 2018). According to Jost et al. (2023), the enriched DNA encoding NLR is sequenced and compared with a reference genome to find polymorphisms in the NLR genes that might explain the observed resistance in the resistant germplasm of interest.

Identifying wild relatives of any crop plant is an essential process of any breeding program. At present, many breeding programs are using this technique. In this identification process, a first step is the phenotypic identification of a useful disease resistant wild type of the relative crop. The comparative analysis of presence-absence single nucleotide polymorphism and insertion and deletion polymorphism (InDels) assists in identifying NLRs. Upon identification of the R gene in the resistant wild relative, resistance can be introduced into the crop by hybridization and consecutive backcrosses. RenSeq is a valuable tool to reduce genome complexity and assist in selecting phylogeny (Monteiro & Nishimura, 2018).

In the *Solanum lycopersicum*, *S. pinpinellifolium*, and *S. tuberosum* clones 105, 126, and 317, respectively, unreported NLRs were identified through RenSeq technology (Andolfo et al., 2014). RenSeq techniques also allow us to find wild relatives for the disease resistance whose draft genome is not available. One of the best examples was identifying a wild relative of tomato resistant to *Phytophthora infestans* (Giolai et al., 2017).

21.4 Role of -Omics in Biocontrol Agents of Different Plant Pathogens

Biological control agents (BCAs) have sparked significant enthusiasm as a safe, eco-friendly and sustainable tool for protecting plants from pathogens. Despite this, the practical application of BCAs faces challenges due to their variable effectiveness, when compared with synthetic chemical compounds. BCAs operate mainly via four mechanisms against plant pathogens: competition, antibiosis, hyperparasitism, and stimulation of host defense. However, ideal strains can be selected using standard microbiological techniques, expanding our knowledge of how BCAs might enhance their consistency and efficacy in real-world applications (Puopolo et al., 2014; Massart et al., 2015). Current developments in molecular methods clarified how BCAs work at the molecular level to fight plant diseases (Massart & Jijakli, 2007). The development of novel biocontrol agents is facilitated by advances in high-throughput sequencing techniques, data generated from omics-based technologies, and a better understanding of the functions of the plant microbiome (Berg et al., 2013).

The term "omics" encompasses diverse scientific disciplines such as proteomics, genomics, transcriptomics, phenomics, peptidomics, metabolomics, and secretomics (Sharma et al., 2017). Utilizing these powerful techniques, the metabolome and transcriptome of many diverse BCAs are being identified and analyzed.. Integration of these advanced technologies will yield new generation plant protection products which will sustain plants against different biotic stresses (Berg et al., 2013).

21.4.1 Genomics of Biocontrol Agents

Genomics is the complete genome study and is one of the -omics disciplines. Roderick coined the term "genomics", defined as "the study of the total of all the genes in an organism" (Ram et al., 2018; Majumdar, 2023). Genomics significantly contributed to the understanding and advancement of bio-control. The first biocontrol agent genome fully sequenced through Sanger sequencing was *Pseudomonas protegens* Pf5, followed by *Bacillus amyloliquefaciens* FBZ42 (Paulsen et al., 2005; Chen et al., 2007). Advancements in sequencing the microbial genome have been accelerated through next-generation sequencing (NGS), reducing the cost of NGS analysis for BCAs (Knief, 2014). Out of 27 published prokaryotic genomes of BCAs, only eight have undergone complete genome sequencing. The genomes of some biocontrol agents that are sequenced include *Trichoderma atroviride*, *T. virens*, *T. reesei*, *Bacillus brevis*, *Pantoea vagan Microbacterium barkeriri*, and *Pseudomonas fluorescens* F113 (Ram et al., 2018).

Traditional microbiological methods have improved our comprehension of the mechanisms by which biocontrol agents function. Plant pathogen cell walls are broken down by extracellular cell wall-degrading enzymes secreted by mycoparasites. Genome study of BCAs can reveal genes that code for the synthesis of these enzymes (Massart et al., 2015). According to Oni et al. (2015), genome mining can identify unique orphan gene clusters, which makes it easier to find linked secondary metabolites in biocontrol strains. Also, genome sequencing in combination with annotated-gene-function studies has generated valuable insights related to the BCAs mechanisms. For example, Ma et al. (2018) identified eight gene clusters in *Bacillus atrophaeus* GQJK17 with a potential to produce antimicrobial secondary metabolites. PacBio sequencing showed that the secondary metabolite production in the YB-130 strain of *B. velezensis* is regulated by 12 clusters, and especially by lanthipeptide (Xu et al., 2020). Similarly, Samaras et al. (2021) demonstrated in cucumber that *B. subtilis* (4,076,736 bp genome) has high antagonism and rhizosphere colonization against *Fusarium oxysporum*, *F. oxysporum*. f. sp. *radicis cucumerinum*, and *Pythium aphanidermatum*. Genome sequencing of *Bacillus subtilis* strain Bbv57 (4,302,465 bp) showed that it possesses nine gene clusters that are responsible for the production of antimicrobial secondary metabolites (Thiruvengadam et al., 2022). Kubicel and his co-workers compared the complete genome sequences of *Trichoderma atroviride* and *T. virens* with *T. reesei*, followed by the genome sequencing of *T. hamatum* GD12 (Studholme et al., 2013). Such genomic in depth analyses led to the availability of multiple genome sequences for *Trichoderma* species.

Another crucial aspect of analyzing BCAs genome sequences involves identifying genes and pathways associated with their interaction with plants and their adaptation to the environment, including competition for nutrients and space. Such genes can be identified through various methods, such as comparing them with plant-pathogen interaction pathways outlined in the KEGG database (Kanehisa et al., 2002) or by employing comparative genomics studies with other bacterial stains isolated from the rhizosphere or endosphere of plants, including plant growth-promoting rhizobacteria (PGPR), plant pathogens, or other beneficial microorganisms (Mathimaran et al., 2012; Niu et al., 2011; Brown et al., 2012). Mukherjee et al. (2013) documented that genome sequencing facilitates the identification of high levels of redundancy for functions that are crucial to biocontrol properties, such as the presence of chitinase genes in the genome of *Trichoderma* species.

When metagenomics and meta transcriptomics are combined, they provide a deep understanding of the whole microbial community in a given area. Identifying subcomponents within these microbiomes allows for a possible modulation of the pathogenic activity. The "genomic revolution" — the combination of high-throughput sequencing and declining sequencing costs — has greatly aided in the genome sequencing from plant pathogens, BCAs, and other helpful microbes. The availability of genomes facilitates high-throughput analyses that can accelerate the investigation of biocontrol properties by: (a) identifying strain genome features, (b) identifying and characterizing genes involved in biocontrol properties, (c) characterizing gene clusters with unknown functions via genome mining, and (d) enabling the study of gene transcription (Massart et al., 2015).

21.4.2 Transcriptomics of Biocontrol Agents

Advanced -omics technologies have been critical to improve our understanding of gene expression. Transcriptome investigations are the most effective way of defining candidate transcripts, relevant for various biological processes (Herrera-Estrella,

2014). Transcriptomic studies on biocontrol agents have traditionally used a targeted strategy focused on candidate genes, frequently adopting techniques such as real-time RT-PCR. Genes were often chosen based on microbiological, histochemical, or biochemical data. This technique extensively explored genes linked with pathogen cell wall degrading enzymes, such as glucanases, chitinases, proteases, and others, in the case of fungal biocontrol agents (Massart & Jijakli, 2007). The study of genes generating secondary metabolites has focused chiefly on bacterial biological control agents. This hypothesis-driven method has improved our understanding of gene expression for certain genes but falls short of a whole transcriptome study. 'Discovery-driven' methods were developed to solve this constraint (Massart & Jijakli, 2006; Viterbo et al., 2004). These strategies try to find genes with differential transcription in diverse settings without relying on pre-existing notions. Until a decade ago, the methodologies were distinct, but real-time PCR was routinely employed to confirm the differential expression of genes found using the 'discovery-driven' strategy. With the introduction of microarrays and other high-throughput technologies, targeted and 'discovery-driven' techniques have been integrated. Microarrays may be used to investigate the transcriptome of a variety of host plants, and custom microarrays can be created for any biocontrol agent with a sequenced and annotated genome or cDNA sequences. However, RNA sequencing has actually overtaken microarray technology, which has various advantages (McGettigan, 2013; Nookaew et al., 2012), enabling the investigation of the host plant, the plant pathogen, and the BCAs transcriptomes in their tri-trophic interactions (Massart et al., 2015).

Understanding the mechanisms underlying the BCAs actions and the gene regulation that maintains their harmful effects has been greatly enhanced by transcriptomic investigations. The genes involved in niche adaptation, competition for nutrients and space, intricate interaction and synchronisation of gene transcription (Reithner et al., 2011), and cell physiology necessary for effective mycoparasitism (Seidl et al., 2009) have all been clarified (Adomas et al., 2006; Wu et al., 2011). Moreover, studies on plants transcriptome have shown defence mechanisms *vs* plant diseases activated by biocontrol products. Numerous reactive pathways, including jasmonic acid, abscisic acid and PR proteins, have been discovered through transcriptome research (Okubara et al., 2010; Sun et al., 2011; Feng et al., 2012).

Transcriptome-based studies have been extremely useful as a reference for studying the occurrence of variable gene expression during transcription, focusing entirely on active mRNAs (Meteignier et al., 2017). Numerous transcriptome studies have investigated BCAs, as in the study of *Trichoderma*-plant-pathogen relationship. Reithner et al. (2011) examined the transcriptome of *Trichoderma atrovirides* strain MI206040 in various stages of interaction with *Botrytis cinerea Rhizoctonia solani* and *Phytophthora capsici*. Using quantitative RT-PCR, differential transcription was identified for thirteen genes, most encoding expansin-like proteins, acetylxylanesterases, aspartyl proteases, and trypsin-like proteases, especially when *R. solani* was present. Comparative gene expression showed a synergistic transcription of many genes involved in fungal cell wall disintegration against *B. cinerea* and *P. capsici*, during different stages of parasitism (Okubara et al., 2010). *Pseudomonas fluorescens* Q8r1-96 improved the jasmonic acid pathway in wheat and in the system *Pinus sylvestris-Pingucula gigantea* (Sun et al., 2011). Alternatively, *T. harzianum* strain T34 suppressed this pathway in *Arabidopsis thaliana* (Morán-Diez et al., 2012), while *Ralstonia solanacearum* DhrpB did not affect it in *A. thaliana* (Feng et al., 2012).

The methodologies used to identify differentially expressed genes must be able to confirm high throughput results independently, which will greatly facilitate large-scale transcriptome investigations in the future. Transcriptome analyses of single species, irrespective of biocontrol agents, host plant, or plant pathogens, have dominated gene expression studies. Pioneering studies using traditional real-time PCR or microarray techniques have proved the value of these methodologies in understanding the complex interactions between two or three organisms.

21.4.3 BCAs Proteomics

The proteome is "the entire set of proteins expressed by a genome." Proteomics is the qualitative and quantitative analysis of proteins expressed by a genome in a specific tissue (Wilkins et al., 1996). Understanding the relationships between the host, pathogen, and BCas is critical, as proteomics focuses on the final gene products, i.e., proteins, which are ultimately responsible for the biocontrol properties. A targeted approach to protein quantification entails focusing on proteins that are known to be associated with BCAs, while a discovery-driven strategy is used to explore all final gene products produced throughout the biocontrol process, to identify new proteins and a global view of underlying mechanisms. Specific antibodies or enzymatic assays are widely used to examine the quantity and activity of enzymes related to BCAs, such as chitinases, cellulases, and proteases (Aegerter & Gordon, 2006). Furthermore, gel-based approaches were initially used primarily for proteome studies linked to biocontrol activities, but nowadays improved mass spectrometry-based in gel-free approaches expanded proteomics research on biocontrol agents (Otto et al., 2012).

Proteomics is emerging as a fundamental field that operates alongside genomics. Numerous investigations have emphasized the complexity of the protein complement and demonstrated how it dynamically changes in response to environmental cues. It is anticipated that the cell proteome will provide solutions to important and crucial issues, with consequences for BCAs and the metabolites they produce. Proteomics is useful for identifying proteins that are differentially expressed, have biotechnological applications, and are associated with biocontrol mechanisms (Esposito et al., 2016). Numerous fungi release proteins during their saprotrophic phase, essential for processes including attaching to plant surfaces, penetrating hosts, and controlling virulence factors (Esposito et al., 2016). For example, a proteome investigation of *Trichoderma* spp. discovered the presence of fungal cell wall-degrading enzymes (CWDEs), including three unique proteins such as N-acetyl-b-d-glucosaminidase and endochitinase, generated by *T. atroviride* in response to *R. solani* (Grinyer et al., 2005). Yang et al. (2009) showed that *T. harzianum* increased the manufacture of different cell wall degrading enzymes in culture media during contact with *B. cinerea*. Moreover, proteomics may aid in the identification of immune responses in a BCA-treated plants. Multiple effector proteins produced were found during maize root and *T. virens* interaction, with the effectors inhibiting defense pathways and causing tolerance to the fungus systematic spread (Lamdan et al., 2015).

Furthermore, adding pathogens to a plant-BCA interactive system may provide a better understanding of the BCA-mediated defence responses in hosts. Prabhukarthikeyan et al. (2019) used two-dimensional gel electrophoresis-matrix-assisted laser desorption ionization time-of-flight mass spectrometry (2DE-MALDI-TOF-MS) to to study the three-way interaction among paddy, *Bipolaris oryzae*, and *Bacillus amyloliquefaciens*. Data showed the production of plant defense and metabolism proteins. A further similar system (paddy, *Rhizoctonia solani*, and *B. subtilis* EPB24), showed enhanced expression of putative disease resistance genes (Durgadevi et al., 2021). 2DE-MALDI-TOF analysis, showed that enzymes such as phenylalanine ammonia lyase, serine-threonine protein kinase, and nucleoside diphosphate kinase were among the defense-related proteins upregulated when *Pseudomonas fluorescens* Pf1 was inoculated on tomato roots (Paranthaman et al., 2023). Ultimately, the host defense from *Fusarium oxysporum* f. sp. *lycopersici* attack was aided by these proteins.

Research on proteomics can also shed light on how a disease responds to a particular BCA, which is crucial information for creating efficient management strategies. A proteomics study of *Cytospora mali* (Apple valsa canker) inoculated with *B. velezensis* L-1 showed the activation of the lysine acetylome. Although the BCA decreased the pathogenicity of *C. mali*, the pathogen in turn alleviated BCA suppression by upregulating acetylated proteins linked with cytochrome P450, HSP-70 (Heat Shock Protein) and autophagy (Sun et al., 2022).

The proteomic research of agriculturally important microorganisms has grown in prominence during the last decade (Ram et al., 2018). Numerous methodological advances on tissue and cell fractionation have resulted in successful protein extraction, purification, separation, and subsequent MS-based analyses. Furthermore, efficient protein recognition algorithms and bioinformatic tools for data analysis aided in the hunt for novel BCAs (Esposito et al., 2016).

The proteome responds differently depending on whether a helpful and/or an antagonistic pathogen colonize the host, alone or both simultaneously (Kang et al., 2019). However, when compared to individual inoculations, simultaneous colonization resulted in a more extensive transcriptional and translational response, indicating that the pathogen played a more substantial role in activating the response than the beneficial microbe.

Proteomics research has emerged as a critical tool for thoroughly understanding the metabolic and physiological changes that occur within cells and tissues. The discovery of several differentially expressed proteins provides a comprehensive picture of plant-pathogen interactions. To achieve effective and environment-friendly plant disease management solutions, molecular research into fungal biology and host interactions is required. Integrating proteomics and other molecular approaches provides a tool for evaluating phytopathogen pathogenicity and virulence (Ram et al., 2018).

21.5 Advanced Molecular Strategies to Increase BCAs Efficacy

Biological control offers an optimistic solution for combating plant diseases. Despite the fact that many microorganisms have been identified as potential biocontrol agents, only a limited number have been effectively brought into commercial use. A noteworthy obstacle in the conversion of effective BCAs into commercially viable products is the variable effectiveness in a variety of environmental conditions, ranging from the production stage til the cultivated field (Raguchander et al., 2011). This aspect underscores the necessity of developing more robust BCAs with potential to withstand various adverse conditions, while consistently delivering high performance to effectively reduce the pathogens induced losses. Grasping the mechanisms by which BCAs exert their preventive impacts is crucial for their effective use. The information pertaining to this aspect will help expedite the registration procedures by ensuring that the right agents are chosen, produced, formulated, and applied. Despite several BCAs action mechanisms have been investigated, many still remain largely unclear.

Numerous processes are documented to either jointly or sequentially contribute to the biocontrol capabilities of microbes. Such mechanisms are mostly focused on competition for space and nutrients, mycoparasitism, antibiosis, or the induction of plant defences. Various microscopic, microbiological, and biochemical explorations have focused on biocontrol mechanisms, for many years (Elad, 1996, 2000). However, the advancement in biotechnology research opens a window for numerous opportunities in the field of BCAs research and development. Cutting-edge molecular techniques significantly contributed to the amelioration of innovative, alternative tools, enhancing our understanding of BCAs antagonistic mechanisms. These innovative tools, complementing microbiological, microscopic, molecular and biochemical studies, have been instrumental in unravelling the genetic underpinnings of biocontrol properties (Majumdar, 2023). The manifestation of pertinent genes enables a comprehensive grasp of the mechanisms at play, with gene inactivation and overexpression studies shedding light on gene transcription and regulation.

21.6 Genes Involved in Biocontrol Agents

Numerous genes involved in biocontrol activity have been characterized. Several strategies available—targeted gene sequence from a known database, differential display techniques, and the whole genome sequence of an organism—may provide better insights into mechanisms of action (Haggag & Mohamed, 2007) (Table 21.1).

Table 21.1 Genes identified in biocontrol agents for their presumed role in biocontrol properties

Biocontrol agents	Genes identified	References
Ampelomyces quisqualis	Chitinase class 1	Weiss et al. (1996)
Acremonium ophalosporium	Chitinase	Giczey et al. (2001)
	α-N-acetylgalactosaminidase	Ashida et al. (2000)
Tricoderma virens	Endochitinases	Baek et al. (1999), Carsolio et al. (1994), García et al. (1994), Kim et al. (2002) and Steyaert et al. (2003)
	Serine protease	Pozo et al. (2004)
	Exochitinases	Kim et al. (2002)
	N-acetylglucosaminidases	Kim et al. (2002)
	Exo-β-1,3-glucanases	Kim et al. (2002)
	Peptaibol synthase	Wiest et al. (2002)
	Peptide synthase	Wilhite et al. (2001)
	G-protein alpha subunits	Mukherjee et al. (2004)
	MAPK kinase	Mukherjee et al. (2003)
Tricoderma hamatum	Exo-β-1,3-glucanase	Steyaert et al. (2003)
	Endochitinase	Steyaert et al. (2003)
Tricoderma harzianum, Tricoderma atroviride	Endochitinase	Draborg et al. (1996)
	Exochitinases	Draborg et al. (1995)
	Exo-α-1,3-glucanases	Ait-Lahsen et al. (2001)
Coniothyrium minitans	Exo-β-1,3-glucanase	Giczey et al. (2001)
Pichia anomala	Exo-β-1,3-glucanases	Grevesse (2002)

21.6.1 Gene Identification

Characterizing a gene involves its isolation and sequencing. Gene identification strategies are broadly classified into two types i.e., targeted and open. A targeted strategy requires the prior selection to identify one or a few, close genes precisely. This selection can be informed by existing data or extrapolated from models developed by studying other biocontrol or similar agents. The first step in the targeted technique is to create degenerated primers that amplify certain sequence fragments specific for the studied gene. The careful selection of these primers is based on various considerations, including the amino acid sequence of the studied protein, sequence alignment with related genes/proteins from other microbes, or usage of primers already employed for other BCAs. The DNA fragment is amplified using the degenerate primers in a polymerase chain reaction (PCR), then proceeding to downstream cloning and sequencing. For instance, this method has been successful in isolating the genes for the endo-1,3-glucanase and the exo-1,3-glucanase of *T. harzianum* (Donzelli et al., 2001; Ait-Lahsen et al., 2001). Similarly, Grevesse (2002) used degenerate primers to determine the exoglucanase activity in culture filtrates by targeting conserved regions discovered by several amino acid alignments of known exoglucanase genes. They successfully amplified the genes encoding exo-1,3-glucanase in post-harvest microbial BCAs, including *Pichia anomala*.

The open approach seeks to identify additional genes. Molecular techniques such as cDNA amplified fragment length polymorphism analysis (cDNA-AFLP), differential display, and subtractive hybridization are commonly employed for differential expression studies. cDNA-AFLP was employed in an investigation on the genes involved in the biological regulation of *B. cinerea* by *P. anomala* strain Kh5 (Massart & Jijakli, 2006). This strain was grown in a medium containing either glucose or cell wall components of *B. cinerea*. This method allowed the identification of eleven cDNA segments associated with genes that were overexpressed when *B. cinerea* cell wall components were present. Vasseur et al. (1995) used subtractive hybridization to examine the gene expression in *T. harzianum*, grown in a medium with either glucose or the cell wall of *R. solani* as sole carbon source. Through this method, a permease (*inda1*) and another gene (*indc11*) that exhibited similarities to an oxidoreductase were successfully isolated.

21.6.2 Genome Sequencing

The whole genome sequencing of many species, such as *Bacillus*, *Trichoderma*, or *Pseudomonas*, allowed substantial progress towards our understanding of helpful BCAs. For example, *P. fluorescens* Pf-5 genome data revealed that the manufacture of secondary metabolites, such as antibiotics and siderophores, involves ca. 6% of its 7.07 Mb genome. Using a genomic approach, a novel class of bioactive cyclic lipopeptides known as orfamide A was found inside three orphan gene clusters (Paulsen et al., 2005).

Takeuchi et al. (2014) conducted a thorough genome sequencing of the rhizospheric bacteria *Pseudomonas* sp. Cab57. They used whole-genome shotgun and paired-end sequencing methods, using PCR amplifications to fix gaps between contigs and produce a more cohesive genome representation. This approach allowed the identification of four key gene clusters: *plt*, *hcn*, *prn*, and *phl*. *Phl* is involved in the synthesis of antibiotics; *prn* is involved in the absorption of nitrite and nitrate; whereas *plt* and *hcn* are involved in the Gac/Rsm signal transduction pathway. Surprisingly, it was discovered that these genes improved the bacterium resistance to harmful microbes. The entire genome sequence of the naturally occurring strain of *B. cereus* AR156, showed a strong potential vs a number of soilborne diseases (Jiang et al., 2017). It was effectively established that this BCA protected tomato plants against both *Ralstonia solanacearum*, the cause of bacterial wilt, and the root-knot nematode *Meloidogyne incognita*.

21.6.3 Protoplast Fusion

Protoplast fusion serves as a rapid and straightforward tactic to enhance strains efficacy, by inducing genetic recombination and developing hybrid BCAs (Lalithakumari & Mathivanan, 2003; Abdel-Salam et al., 2007). It is primarly used to amalgamate beneficial characteristics from different promising strains, thereby enhancing the biocontrol capabilities of BCAs. Protoplast isolation, fusion, and regeneration have been effectively achieved with *Trichoderma* strains. Inter-specific protoplast fusion led to enhanced cellulase production in *T. reesei* (Ogawa et al., 1989). Similarly, self-fusion of protoplasts from *T. harzianum* strain PTh18 showed a twofold increase in chitinase and biocontrol activity as compared to the parent strains, against *R. solani* (Prabavathy et al., 2006). Balasubramanian et al. (2012) used this method to manufacture *Trichoderma* HF9 interspecific hybrid (fusant) of *Trichoderma harzianum* × *T. viride*. The gamma ray mutants of *T. harzianum*, *T.h.* M8 and *T.h.* M15, showed more chitinase and endoglucanase synthesis than the wild types, (Ghasemi et al., 2019). This enhanced enzymatic activity was found to effectively suppress the growth of *R. solani*.

To increase the nematicidal potential of *B. amyloliquefaciens* and *Lysinibacillus sphaericus vs Meloidogyne incognita*, protoplast fusion techniques were utilized. In another study, the effectiveness of fusants produced from the parental strains of *Pseudomonas aeruginosa* and *B. cereus* was assessed in terms of impact on root knot nematodes and weed seed germination. Fusants F7, F20, and F35 enhanced antagonistic capability against *M. incognita* juveniles in comparison to the parental strains. Notably, these fusants completely prevented seeds of *Echinochloa crusgalli* and *Portulaca oleracea* from germinating (Mohamed et al., 2021).

21.6.4 Genetic Transformation

Enhancing the biocontrol activity of a biocontrol agents can be achieved by incorporating foreign DNA into their genome, specifically encoding biocontrol metabolites The genetic recombination process entails creating strains that generate elevated quantities of lytic enzymes and antibiotics. Conversely, suppressing or deleting certain genes from BCA may also amplify the enduring biocontrol activity in specific strains. This approach has proven effective in controlling crown gall disease, mediated by *Agrobacterium tumefaciens*. To prevent *A. tumefaciens* strains from becoming resistant to agrocin 84, a modified variant of *A. radiobacter* K84 was utilised, by transferring plasmid genes (de Boer et al., 2003). A *P. fluorescens* strain proceeding from micropropagated apple plantlets that acted as an endophyte received a primary chitinase-encoding gene (*chiA*) from *Serratia marcescens*. Bean seedlings cultivated in a growth chamber, showed an effective control from *R. solani*. The pathogen was successfully suppressed in beans by the *P. fluorescens* Rif1 strain, expressing the *chiA* gene with an increased chitinase activity.

Recombinant *Pseudomonas synxantha* strains, ZHW 15 and ZHW 25, showed improved biocontrol effectiveness against a variety of wheat pathogens (Zhang et al., 2020). By introducing pyrrolnitrin-coding genes from *Pseudomonas protegens* Pf-5 into *P. synxantha* 2–79, the bacterium biocontrol efficiency was enhanced. Comparing the two recombinants to their original parent strains, the latter showed lower biocontrol efficiency.

The antagonistic mechanism known as mycoparasitism entails the breakdown of pathogen cell walls by lytic enzymes such as chitinases, proteases, and glucanases. Successful cloning of chitinase-encoding genes has been achieved in various microorganisms, including *S. marcescens* (Sundheim et al., 1988) and *T. harzianum*. According to Baek et al. (1999), *T. virens* cotton strains that carried two copies of the ech42 gene displayed increased antagonistic activity *vs R. solani*, and strains that had a faulty endochitinase *ech42* gene fragment displayed a lower antagonistic activity. Similar results with *ech42* were obtained from *T. harzianum* and *T. atroviride* which enhanced disease resistance of potato and tobacco towards *Alternaria alternata, A. solani* and *B. cinerea.* Similarly, *Clonostachys rosea* was modified to incorporate the endochitinase-encoding gene *Chi67-1*, which increased the plant biological activity against Sclerotinia rot of soybean (Sun et al., 2017).

21.7 Conclusions

Farmers are still able today to fulfill the food requirements of current growing world population, yet biggest challenges posed by phytopathogens on crops are threatening. In the actual world context, use of resistant varieties is highly preferable due to their environment friendly and sustainable nature. Resistant varieties can be developed in crop improvement programme using traditional to molecular approaches. However, major challenges remain due to the selectin and insurgence of new pathogen strainsand races.

References

Abdel-Salam, M. S., Abd El-Halim, M. M., & El-Hamshary, O. I. M. (2007). Improvement of *pseudomonas* antagonism against *Fusarium oxysporium* through protoplast fusion: I-Fusants induction. *Research Journal of Cell and Molecular Biology, 1*, 37–41.

Acevedo-Garcia, J., Spencer, D., Thieron, H., Reinstädler, A., Hammond-Kosack, K., et al. (2017). Mlobased powdery mildew resistance in hexaploid bread wheat generated by a non-transgenic TILLING approach. *Plant Biotechnology Journal, 15*, 367–378.

Adomas, A., Eklund, M., Johansson, M., & Asiegbu, F. O. (2006). Identification and analysis of differentially expressed cDNAs during nonself-competitive interaction between *Phlebiopsis gigantea* and *Heterobasidion parviporum. FEMS Microbiology Ecology, 57*, 26–39.

Aegerter, B. J., & Gordon, T. R. (2006). Rates of pitch canker induced seedling mortality among *Pinusradiata* families varying in levels of genetic resistance to *Gibberella circinata* (anamorph *Fusariumcircinatum*). *Forest Ecology and Management, 235*, 14–17.

Ait-Lahsen, H., Soler, A., Rey, M., de la Cruz, J., Monte, E., & Llobell, A. (2001). An antifungal exo-α-1, 3-glucanase (AGN13.1) from the biocontrol fungus *Trichoderma harzianum. Applied and Environmental Microbiology, 67*, 5833–5839.

Ali, A., Mughal, A., Ahmad, S., Abbas, S., Mumtaz, A., & Ahmed, I. (2023). Microbial metabolomics: Recent advancements and applications in infectious diseases and drug discovery. In S. De Mandal, X. Xu, F. Jin, P. A. Kumari, & K. S. Ibrahim (Eds.), *Recent advances and future perspectives of microbial metabolites* (pp. 269–335). Academic.

Ana, A. G. S., Carrillo-Cerda, H. A., Rodriguez-Campos, J., Velázquez-Fernández, J. B., Patrón-Soberano, O. A., & Contreras-Ramos, S. M. (2020). Dynamics of volatilomes emitted during cross-talking of plant-growth-promoting bacteria and the phytopathogen, *Fusarium solani*. *World Journal of Microbiology and Biotechnology, 36*, 1–15.

Andolfo, G., Jupe, F., Witek, K., Etherington, G. J., Ercolano, M. R., & Jones, J. D. (2014). Defining the full tomato NB-LRR resistance gene repertoire using genomic and cDNA RenSeq. *BMC Plant Biology, 14*(1), 1–12.

Ashida, H., Tamaki, H., Fujimoto, T., Yamamoto, K., & Kumagai, H. (2000). Molecular cloning of cDNA encoding α-N-acetylgalactosaminidase from *Acremonium sp.* and its expression in yeast. *Archives of Biochemistry and Biophysics, 384*(2), 305–310.

Baek, J. M., Howell, C. R., & Kenerley, C. M. (1999). The role of an extracellular chitinase from *Trichoderma virens* Gv29-8 in the biocontrol of *Rhizoctonia solani*. *Current Genetics, 35*, 41–50.

Bai, Y., Zheng, Z., & Visser, R. G. F. (2013). Lack of susceptibility factors: A novel breeding strategy for durable resistance. *Breakthroughs in the genetics and breeding of capsicum and eggplant*, XVth EUCARPIA. p.97.

Balasubramanian, N., Thamil Priya, V., Gomathinayagam, S., & Lalithakumari, D. (2012). Fusant *Trichoderma* HF9 with enhanced extracellular chitinase and protein content. *Applied Biochemistry and Microbiology, 48*, 409–415.

Barra Caracciolo, A., & Terenzi, V. (2021). Rhizosphere microbial communities and heavy metals. *Microorganisms, 9*, 1462.

Bashir, I., War, A. F., Rafiq, I., Reshi, Z. A., Rashid, I., & Shouche, Y. S. (2022). Phyllosphere microbiome: Diversity and functions. *Microbiological Research, 254*, 126888.

Belhaj, K., Chaparro-Garcia, A., Kamoun, S., & Nekrasov, V. (2013). Plant genome editing made easy: Targeted mutagenesis in model and crop plants using the CRISPR/Cas system. *Plant Methods, 9*(1), 1–10.

Berg, G., Zachow, C., Müller, H., Philipps, J., & Tilcher, R. (2013). Next-generation bio-products sowing the seeds of success for sustainable agriculture. *Agronomy, 3*(4), 648–656.

Brown, S. D., Utturkar, S. M., Klingeman, D. M., Johnson, C. M., Martin, S. L., et al. (2012). Twenty-one genome sequences from *pseudomonas* species and 19 genome sequences from diverse bacteria isolated from the rhizosphere and endosphere of *Populusdeltoides*. *Journal of Bacteriology, 194*, 5991–5993.

Carsolio, C., Gutiérrez, A., Jiménez, B., Van Montagu, M., & Herrera-Estrella, A. (1994). Characterization of ech-42, a *Trichoderma harzianum* endochitinase gene expressed during mycoparasitism. *Proceedings of the National Academy of Sciences, 91*(23), 10903–10907.

Cernosek, T., Eckert, K. E., Carter, D. O., & Perrault, K. A. (2020). Volatile organic compound profiling from postmortem microbes using gas chromatography–mass spectrometry. *Journal of Forensic Sciences, 65*, 134–143.

Chen, X. H., Koumoutsi, A., Scholz, R., Eisenreich, A., Schneider, K., et al. (2007). Comparative analysis of the complete genome sequence of the plant growth–promoting bacterium *Bacillus amyloliquefaciens* FZB42. *Nature Biotechnology, 25*, 1007–1014.

de Boer, M., Bom, P., Kindt, F., Keurentjes, J. J., van der Sluis, I., Van Loon, L. C., & Bakker, P. A. (2003). Control of *Fusarium* wilt of radish by combining *Pseudomonas putida* strains that have different disease-suppressive mechanisms. *Phytopathology, 93*(5), 626–632.

Dong, O. X., & Ronald, P. C. (2019). Genetic engineering for disease resistance in plants: Recent progress and future perspectives. *Plant Physiology, 180*(1), 26–38.

Donzelli, B. G. G., Lorito, M., Scala, F., & Harman, G. E. (2001). Cloning, sequence and structure of a gene encoding an antifungal glucan 1, 3-β-glucosidase from *Trichoderma atroviride* (*T. harzianum*). *Gene, 277*(1–2), 199–208.

Draborg, H., Kauppinen, S., Dalbøge, H., & Christgau, S. (1995). Molecular cloning and expression in S. cerevisiae of two exochitinases from *Trichoderma harzianum*. *Biochemistry and Molecular Biology International, 36*, 781–791.

Draborg, H., Christgau, S., Halkier, T., Rasmussen, G., Dalbøge, H., & Kauminen, S. (1996). Secretion of an enzymatically active *Trichoderma harzianum*endochitinase by *Saccharomyces cerevisiae*. *Current Genetics, 29*, 404–409.

Durgadevi, D., Harish, S., Manikandan, R., Prabhukarthikeyan, S. R., Alice, D., & Raguchander, T. (2021). Proteomic profiling of defense/resistant genes induced during the tripartite interaction of Oryza sativa, *Rhizoctonia solani* AG1-1A, and *Bacillus subtilis* against rice sheath blight. *Physiological and Molecular Plant Pathology, 115*, 101669.

Elad, Y. (1996). Mechanisms involved in the biological control of *Botrytis cinerea* incited diseases. *European Journal of Plant Pathology, 102*, 719–732.

Elad, Y. (2000). Biological control of foliar pathogens by means of *Trichoderma harzianum* and potential modes of action. *Crop Protection, 19*(8–10), 709–714.

Elsherbiny, E. A., Amin, B. H., Aleem, B., Kingsley, K. L., & Bennett, J. W. (2020). *Trichoderma* volatile organic compounds as a biofumigation tool against late blight pathogen *Phytophthorainfestans* in postharvest potato tubers. *Journal of Agricultural and Food Chemistry, 68*, 8163–8171.

Esposito, A., Colantuono, C., Ruggieri, V., & Chiusano, M. L. (2016). Bioinformatics for agriculture in the next-generation sequencing era. *Chemical and Biological Technologies in Agriculture, 3*, 1–12.

Feng, D. X., Tasset, C., Hanemian, M., Barlet, X., Hu, J., et al. (2012). Biological control of bacterial wilt in *Arabidopsis thaliana* involves abscissic acid signalling. *The New Phytologist, 194*, 1035–1045.

Fischer, M. J., Rustenhloz, C., Leh-Louis, V., & Perrière, G. (2015). Molecular and functional evolution of the fungal diterpene synthase genes. *BMC Microbiology, 15*, 1–16.

Fufurin, I. L., Anfimov, D. R., Kareva, E. R., Scherbakova, A. V., Demkin, P. P., Morozov, A. N., & Golyak, I. S. (2021). Numerical techniques for infrared spectra analysis of organic and inorganic volatile compounds for biomedical applications. *Optical Engineering, 60*(8), 082016–082016.

García, I., Lora, J. M., de la Cruz, J., Benítez, T., Llobell, A., & Pintor-Toro, J. A. (1994). Cloning and characterization of a chitinase (CHIT42) cDNA from the mycoparasitic fungus *Trichoderma harzianum*. *Current Genetics, 27*, 83–89.

Ghabrial, S. A., & Suzuki, N. (2009). Viruses of plant pathogenic fungi. *Annual Review of Phytopathology, 47*, 353–384.

Ghasemi, S., Safaie, N., Shahbazi, S., Shams-Bakhsh, M., & Askari, H. (2019). Enhancement of lytic enzymes activity and antagonistic traits of *Trichoderma harzianum* using γ-radiation induced mutation. *Journal of Agricultural Science and Technology, 21*(4), 1035–1048.

Giczey, G., Kerényi, Z., Fülöp, L., & Hornok, L. (2001). Expression of cmg1, an exo-β-1, 3-glucanase gene from *Coniothyrium minitans*, increases during sclerotial parasitism. *Applied and Environmental Microbiology, 67*(2), 865–871.

Giolai, M., Paajanen, P., Verweij, W., Witek, K., Jones, J. D., & Clark, M. D. (2017). Comparative analysis of targeted long read sequencing approaches for characterization of a plant's immune receptor repertoire. *BMC Genomics, 18*(1), 1–15.

Grevesse, C. (2002). Study of the implication of exo-beta-1, 3-glucanase encoding genes in the biocontrol activity of *Pichia anomala* (Hansen) Kurtzman (strain K) against *Botrytis cinerea* on post-harvest apples.

Grinyer, J., Hunt, S., McKay, M., Herbert, B. R., & Nevalainen, H. (2005). Proteomic response of the biological control fungus *Trichoderma atroviride* to growth on the cell walls of *Rhizoctoniasolani*. *Current Genetics, 47*, 381–388.

Haggag, W. M., & Mohamed, H. A. A. (2007). Biotechnological aspects of microorganisms used in plant biological control. *World Journal of Agricultural Sciences, 3*, 771–776.

Herrera-Estrella, A. (2014). Genome-wide approaches toward understanding mycotrophic *Trichoderma* species. In V. K. Gupta, M. Schmoll, A. Herrera-Estrella, R. S. Upadhyay, I. Druzhinina, & M. G. Tuohy (Eds.), *Biotechnology and biology ofTrichoderma* (pp. 455–464). Elsevier.

Jiang, C. H., Chen, Y., Yan, F., Fan, Z. H., & Guo, J. H. (2017). Whole-genome sequence of *Bacillus cereus* AR156, a potential biocontrol agent with high soilborne disease biocontrol efficacy and plant growth promotion. *Genome Announcements, 5*(35), 10–1128.

Jost, M., Outram, M. A., Dibley, K., Zhang, J., Luo, M., & Ayliffe, M. (2023). Plant and pathogen genomics: Essential approaches for stem rust resistance gene stacks in wheat. *Frontiers in Plant Science, 14*.

Kanehisa, M., Goto, S., Kawashima, S., & Nakaya, A. (2002). The KEGG databases at GenomeNet. *Nucleic Acids Research, 30*(1), 42–46.

Kang, X., Wang, L., Guo, Y., Ul Arifeen, M. Z., Cai, X., Xue, Y., et al. (2019). A comparative transcriptomic and proteomic analysis of hexaploid wheat's responses to colonization by *Bacillus velezensis* and *Gaeumannomyces graminis*, both separately and combined. *Molecular Plant-Microbe Interactions, 32*, 1336–1347.

Khan, N., Maymon, M., & Hirsch, A. M. (2017). Combating *Fusarium* infection using *Bacillus*-based antimicrobials. *Microorganisms, 5*(4), 75.

Kim, D. J., Baek, J. M., Uribe, P., Kenerley, C. M., & Cook, D. R. (2002). Cloning and characterization of multiple glycosyl hydrolase genes from *Trichoderma virens*. *Current Genetics, 40*, 374–384.

Knief, C. (2014). Analysis of plant microbe interactions in the era of next generation sequencing technologies. *Frontiers in Plant Science, 5*, 216.

Köhl, J., Kolnaar, R., & Ravensberg, W. J. (2019). Mode of action of microbial biological control agents against plant diseases: Relevance beyond efficacy. *Frontiers in Plant Science, 845*.

Lalithakumari, D., & Mathivanan, N. (2003). Strain improvement in filamentous fungi by protoplast fusion. In N. Mathivanan, V. R. Prabavathy, & S. Gomathinayagam (Eds.), *Innovative methods and techniques for integrated pest and disease management. Centre for advanced studies in botany* (pp. 76–97). University of Madras.

Lamdan, N. L., Shalaby, S., Ziv, T., Kenerley, C. M., & Horwitz, B. A. (2015). Secretome of *Trichoderma* interacting with maize roots: Role in induced systemic resistance*[S]. *Molecular & Cellular Proteomics, 14*(4), 1054–1063.

Ma, J., Wang, C., Wang, H., Liu, K., Zhang, T., et al. (2018). Analysis of the complete genome sequence of *Bacillusatrophaeus* GQJK17 reveals its biocontrol characteristics as a plant growth-promoting rhizobacterium. *BioMed Research International*.

Majumdar, A. (2023). Molecular techniques for the improvement of microbial biocontrol agents against plant pathogens. *Egyptian Journal of Biological Pest Control, 33*(1), 103.

Malnoy, M., Viola, R., Jung, M. H., Koo, O. J., Kim, S., et al. (2016). DNA-free genetically edited grapevine and apple protoplast using CRISPR/Cas9 ribonucleoproteins. *Frontiers in Plant Science, 7*, 1904.

Massart, S., & Jijakli, M. H. (2006). Identification of differentially expressed genes by cDNA-amplified fragment length polymorphism in the biocontrol agent *Pichia anomala* (strain Kh5). *Phytopathology, 96*(1), 80–86.

Massart, S., & Jijakli, H. M. (2007). Use of molecular techniques to elucidate the mechanisms of action of fungal biocontrol agents: A review. *Journal of Microbiological Methods, 69*(2), 229–241.

Massart, S., Perazzolli, M., Höfte, M., Pertot, I., & Jijakli, M. H. (2015). Impact of the omic technologies for understanding the modes of action of biological control agents against plant pathogens. *Biological Control, 60*, 725–746.

Mathimaran, N., Srivastava, R., Wiemken, A., Sharma, A. K., & Boller, T. (2012). Genome sequences of two plant growth-promoting fluorescent *Pseudomonas* strains, R62 and R81. *Journal of Bacteriology, 194*, 3272–3273.

McGettigan, P. A. (2013). Transcriptomics in the RNA-seq era. *Current Opinion in Chemical Biology, 17*(1), 4–11.

Meteignier, L. V., El Oirdi, M., Cohen, M., Barff, T., Matteau, D., et al. (2017). Translatome analysis of an NB-LRR immune response identifies important contributors to plant immunity in *Arabidopsis*. *Journal of Experimental Botany, 68*(9), 2333–2344.

Mohamed, S. A., Ameen, H. H., Elkelany, U. S., El Wakeel, M. A., Hammam, M., & Soliman, G. M. (2021). Genetic improvement of *Pseudomonas aeruginosa* and *Bacillus cereus* for controlling root knot nematode and two weeds under laboratory conditions. *Jordan Journal of Biological Sciences, 14*(4).

Monteiro, F., & Nishimura, M. T. (2018). Structural, functional, and genomic diversity of plant NLR proteins: An evolved resource for rational engineering of plant immunity. *Annual Review of Phytopathology, 56*, 243–267.

Morán-Diez, E., Rubio, B., Domínguez, S., Hermosa, R., Monte, E., & Nicolás, C. (2012). Transcriptomic response of *Arabidopsisthaliana* after 24 h incubation with the biocontrol fungus *Trichodermaharzianum*. *Journal of Plant Physiology, 169*(6), 614–620.

Mukherjee, P. K., Latha, J., Hadar, R., & Horwitz, B. A. (2003). TmkA, a mitogen-activated protein kinase of *Trichoderma virens*, is involved in biocontrol properties and repression of conidiation in the dark. *Eukaryotic Cell, 2*(3), 446–455.

Mukherjee, P. K., Latha, J., Hadar, R., & Horwitz, B. A. (2004). Role of two G-protein alpha subunits, TgaA and TgaB, in the antagonism of plant pathogens by *Trichoderma virens*. *Applied and Environmental Microbiology, 70*(1), 542–549.

Mukherjee, P. K., Horwitz, B. A., Herrera-Estrella, A., Schmoll, M., & Kenerley, C. M. (2013). *Trichoderma* research in the genome era. *Annual Review of Phytopathology, 51*, 105–129.

Nazarov, P. A., Baleev, D. N., Ivanova, M. I., Sokolova, L. M., & Karakozova, M. V. (2020). Infectious plant diseases: Etiology, current status, problems and prospects in plant protection. *Acta Naturae, 12*(3), 46–59.

Niu, B., Rueckert, C., Blom, J., Wang, Q., & Borriss, R. (2011). The genome of the plant growth-promoting rhizobacterium *Paenibacilluspolymyxa* M-1 contains nine sites dedicated to nonribosomal synthesis of lipopeptides and polyketides. *Journal of Bacteriology, 193*, 5862–5863.

Nookaew, I., Papini, M., Pornputtapong, N., Scalcinati, G., Fagerberg, L., Uhlén, M., & Nielsen, J. (2012). A comprehensive comparison of RNA-Seq-based transcriptome analysis from reads to differential gene expression and cross-comparison with microarrays: A case study in *Saccharomyces cerevisiae*. *Nucleic Acids Research, 40*(20), 10084–10097.

Ogawa, K., Ohara, H., Koide, T., & Toyama, N. (1989). Intraspecific hybridization of *Trichoderma reesei* by protoplast fusion. *Journal of Fermentation and Bioengineering, 67*, 207–209.

Okubara, P. A., Call, D. R., Kwak, Y. S., & Skinner, D. Z. (2010). Induction of defense gene homologues in wheat roots during interactions with *Pseudomonasfluorescens*. *Biological Control, 55*, 118–125.

Oni, F. E., Kieu Phuong, N., & Höfte, M. (2015). Recent advances in *Pseudomonas* biocontrol. In J. Murillo, B. A. Vinatzer, R. W. Jackson, & D. L. Arnold (Eds.), *Bacteria-plant interactions: Advanced research and future trends* (pp. 167–198). Caister Academic Press..

Otto, A., Bernhardt, J., Hecker, M., & Becher, D. (2012). Global relative and absolute quantitation in microbial proteomics. *Current Opinion in Microbiology, 15*(3), 364–372.

Oufensou, S., Ul Hassan, Z., Balmas, V., Jaoua, S., & Migheli, Q. (2023). Perfume guns: Potential of yeast volatile organic compounds in the biological control of mycotoxin-producing fungi. *Toxins, 15*, 45.

Pandit, M. A., Kumar, J., Gulati, S., Bhandari, N., Mehta, P., Katyal, R., Rawat, C. D., Mishra, V., & Kaur, J. (2022). Major biological control strategies for plant pathogens. *Pathogens, 11*(2), 273.

Paranthaman, L., Seethapathy, P., Pandita, D., Gopalakrishnan, C., Sankaralingam, S., et al. (2023). Levering proteomic analysis of *Pseudomonas fluorescens* mediated resistance responses in tomato during pathogenicity of *Fusarium oxysporum* f. sp. *oxysporum*. *Frontiers in Sustainable Food Systems, 7*, 1157575.

Paulsen, I. T., Press, C. M., Ravel, J., Kobayashi, D. Y., Myers, G. S., Mavrodi, D. V., et al. (2005). Complete genome sequence of the plant commensal *Pseudomonas fluorescens* Pf-5. *Nature Biotechnology, 23*(7), 873–878.

Pozo, M. J., Baek, J. M., Garcıa, J. M., & Kenerley, C. M. (2004). Functional analysis of tvsp1, a serine protease-encoding gene in the biocontrol agent *Trichoderma virens*. *Fungal Genetics and Biology, 41*(3), 336–348.

Prabavathy, V. R., Mathivanan, N., Sagadevan, E., Murugesan, K., & Lalithakumari, D. (2006). Self-fusion of protoplasts enhances chitinase production and biocontrol activity in *Trichoderma harzianum*. *Bioresource Technology, 97*(18), 2330–2334.

Prabhukarthikeyan, S. R., Yadav, M. K., Anandan, A., Aravindan, S., Keerthana, U., et al. (2019). Bio-protection of brown spot disease of rice and insight into the molecular basis of interaction between *Oryza sativa, Bipolaris oryzae* and *Bacillus amyloliquefaciens*. *Biological Control, 137*, 104018.

Puopolo, G., Cimmino, A., Palmieri, M. C., Giovannini, O., Evidente, A., & Pertot, I. (2014). *Lysobacter capsici* AZ78 produces cyclo (LProLTyr), a 2,5-diketopiperazine with toxic activity against sporangia of *Phytophthora infestans* and *Plasmopara viticola*. *Journal of Applied Microbiology, 117*(4), 1168–1180.

Raguchander, T., Saravanakumar, D., & Balasubramanian, P. (2011). Molecular approaches to improvement of biocontrol agents of plant diseases. *Journal of Biological Control, 25*(2), 71–84.

Rajakaruna, A., & Taylor-Robinson, A. W. (2016). Application of recombinant DNA technology (genetically modified organisms) to the advancement of agriculture, medicine, bioremediation and biotechnology industries. *Journal of Applied Biotechnology & Bioengineering, 1*, 78–80.

Ram, R. M., Keswani, C., Bisen, K., Tripathi, R., Singh, S. P., & Singh, H. B. (2018). Biocontrol technology: Eco-friendly approaches for sustainable agriculture. In D. Barh & V. Azevedo (Eds.), *Omics technologies and bio-engineering* (pp. 177–190). Academic.

Reithner, B., Ibarra-Laclette, E., Mach, R. L., & Herrera-Estrella, A. (2011). Identification of mycoparasitism-related genes in *Trichoderma atroviride*. *Applied and Environmental Microbiology, 77*(13), 4361–4370.

Samaras, A., Nikolaidis, M., Antequera-Gómez, M. L., Cámara-Almirón, J., Romero, D., et al. (2021). Whole genome sequencing and root colonization studies reveal novel insights in the biocontrol potential and growth promotion by *Bacillussubtilis* MBI 600 on cucumber. *Frontiers in Microbiology, 11*, 600393.

Savary, S., Ficke, A., Aubertot, J. N., & Hollier, C. (2012). Crop losses due to diseases and their implications for global food production losses and food security. *Food Security, 4*(4), 519–537.

Saxena, P., Srivastava, J., Pandey, S., Srivastava, S., Maurya, N., et al. (2019). Plants for biocontrol and biological control of plant pathogens. In *Plant Biotic Interactions: State of the Art* (pp. 147–179). Springer.

Seidl, V., Song, L., Lindquist, E., Gruber, S., Koptchinskiy, A., et al. (2009). Transcriptomic response of the mycoparasitic fungus *Trichoderma atroviride* to the presence of a fungal prey. *BMC Genomics, 10*(1), 1–13.

Sharma, V., Salwan, R., Sharma, P. N., & Gulati, A. (2017). Integrated translatome and proteome: Approach for accurate portraying of widespread multifunctional aspects of *Trichoderma*. *Frontiers in Microbiology, 8*, 1602.

Singh, B. K., Delgado-Baquerizo, M., Egidi, E., Guirado, E., Leach, J. E., Liu, H., & Trivedi, P. (2023). Climate change impacts on plant pathogens, food security and paths forward. *Nature Reviews Microbiology, 21*, 1–17.

Steyaert, J. M., Ridgway, H. J., Elad, Y., & Stewart, A. (2003). Genetic basis of mycoparasitism: A mechanism of biological control by species of *Trichoderma*. *New Zealand Journal of Crop and Horticultural Science, 31*(4), 281–291.

Studholme, D. J., Harris, B., Le Cocq, K., Winsbury, R., Perera, V., et al. (2013). Investigating the beneficial traits of *Trichodermahamatum* GD12 for sustainable agriculture—Insights from genomics. *Frontiers in Plant Science, 4*, 258.

Sun, H., Paulin, L., Alatalo, E., & Asiegbu, F. O. (2011). Response of living tissues of Pinus sylvestris to the saprotrophic biocontrol fungus *Phlebiopsisgigantea*. *Tree Physiology, 31*(4), 438–451.

Sun, Z. B., Sun, M. H., Zhou, M., & Li, S. D. (2017). Transformation of the endochitinase gene Chi67-1 in *Clonostachys rosea* 67-1 increases its biocontrol activity against *Sclerotinia sclerotiorum*. *AMB Express, 7*(1), 1–9.

Sun, P., Ma, Q., & Zhang, L. (2022). Comprehensive acetyl-proteomic analysis of *Cytosporamali* provides insight into its response to the biocontrol agent *Bacillusvelezensis* L-1. *Frontiers in Sustainable Food Systems, 6*, 999510.

Sundheim, L., Poplawsky, A. R., & Ellingboe, A. H. (1988). Molecular cloning of two chitinase genes from *Serratia marcescens* and their expression in *pseudomonas species*. *Physiological and Molecular Plant Pathology, 33*(3), 483–491.

Takeuchi, K., Noda, N., & Someya, N. (2014). Complete genome sequence of the biocontrol strain *Pseudomonas protegens* Cab57 discovered in Japan reveals strain-specific diversity of this species. *PLoS One, 9*(4), e93683.

Thiruvengadam, R., Gandhi, K., Vaithiyanathan, S., Sankarasubramanian, H., Loganathan, K., Lingan, R., Rajagopalan, V. R., Muthurajan, R., Ebenezer Iyadurai, J., & Kuppusami, P. (2022). Complete genome sequence analysis of *Bacillussubtilis* Bbv57, a promising biocontrol agent against phytopathogens. *International Journal of Molecular Sciences, 23*(17), 9732.

Tilocca, B., Balmas, V., Hassan, Z. U., Jaoua, S., & Migheli, Q. (2019). A proteomic investigation of *Aspergillus carbonarius* exposed to yeast volatilome or to its major component 2-phenylethanol reveals major shifts in fungal metabolism. *International Journal of Food Microbiology, 306*, 108265.

Tilocca, B., Cao, A., & Migheli, Q. (2020). Scent of a killer: Microbial volatilome and its role in the biological control of plant pathogens. *Frontiers in Microbiology, 11*, 41.

Vasseur, V., Van Montagu, M., & Goldman, G. H. (1995). *Trichoderma harzianum* genes induced during growth on *Rhizoctonia solani* cell walls. *Microbiology, 141*(4), 767–774.

Viterbo, A., Harel, M., & Chet, I. (2004). Isolation of two aspartyl proteases from *Trichodermaasperellum* expressed during colonization of cucumber roots. *FEMS Microbiology Letters, 238*(1), 151–158.

Weiss, N., Sztejnberg, A., & Yarden, O. (1996). The chsA gene, encoding a class-I chitin synthase from *Ampelomycesquisqualis*. *Gene, 168*(1), 99–102.

Wiest, A., Grzegorski, D., Xu, B. W., Goulard, C., Rebuffat, S., et al. (2002). Identification of peptaibols from *Trichoderma virens* and cloning of a peptaibol synthetase. *The Journal of Biological Chemistry, 277*(23), 20862–20868.

Wilhite, S. E., Lumsden, R. D., & Straney, D. C. (2001). Peptide synthetase gene in *Trichoderma virens*. *Applied and Environmental Microbiology, 67*(11), 5055–5062.

Wilkins, M. R., Sanchez, J. C., Gooley, A. A., Appel, R. D., Humphery-Smith, I., et al. (1996). Progress with proteome projects: Why all proteins expressed by a genome should be identified and how to do it. *Biotechnology and Genetic Engineering Reviews, 13*(1), 19–50.

Wu, D. Q., Ye, J., Ou, H. Y., Wei, X., Huang, X., et al. (2011). Genomic analysis and temperature-dependent transcriptome profiles of the rhizosphere originating strain *Pseudomonasaeruginosa* M18. *BMC Genomics, 12*, 1–17.

Xu, W., Zhang, L., Goodwin, P. H., Xia, M., Zhang, J., et al. (2020). Isolation, identification, and complete genome assembly of an endophytic *Bacillusvelezensis* YB-130, potential biocontrol agent against *Fusariumgraminearum*. *Frontiers in Microbiology, 11*, 598285.

Yang, H. H., Yang, S. L., Peng, K. C., Lo, C. T., & Liu, S. Y. (2009). Induced proteome of *Trichodermaharzianum* by Botrytis cinerea. *Mycological Research, 113*(9), 924–932.

Zhang, J., Mavrodi, D. V., Yang, M., Thomashow, L. S., Mavrodi, O. V., Kelton, J., & Weller, D. M. (2020). *Pseudomonas synxantha* 2-79 transformed with pyrrolnitrin biosynthesis genes has improved biocontrol activity against soilborne pathogens of wheat and canola. *Phytopathology, 110*(5), 1010–1017.

Chapter 22
Small RNAs—Game Changers of Plant Defense System Against Bacteria

Subashini Jegadeesan and Jebasingh Tennyson

Abstract Biotic stress is one of the major issues plants had faced during their evolution. To cope with a wide range of pathogens they evolved distinct molecular pathways like small RNA-mediated gene silencing. Small RNAs (sRNAs) are non-coding RNAs with a nucleotide (nt) size of 20–30. Based on their biogenesis pathways, they are broadly classified into miRNA, siRNA, circRNA, and tRFs. These small RNAs have the ability to control the expression of certain genes in response to biotic stress *via* post-transcriptional or RNA silencing. This review focuses on small RNAs, their biogenesis pathway, and their regulatory role in plant resistance against bacterial pathogens.

Keywords Biotic stress · CircRNA · miRNA · sRNA

Abbreviations

CircRNA	Circular RNA
DAMPs	Danger-associated molecular patterns
ETI	Effector-triggered immunity
miRNA	Micro RNA
PAMPs	Pathogen-associated molecular patterns
PTI	PAMP triggered immunity
PRRs	Pattern-recognition receptors
RLKs	Receptor-like kinases
RLPs	Receptor-like proteins

S. Jegadeesan · J. Tennyson (✉)
Department of Plant Sciences, School of Biological Sciences, Madurai Kamaraj University, Madurai, Tamil Nadu, India
e-mail: jebasingh.biological@mkuniversity.ac.in

P. Mathur, S. Roy (eds.), *Plant Microbiome and Biological Control*, Sustainability in Plant and Crop Protection 20,
https://doi.org/10.1007/978-3-031-75845-4_22

rRFs	rRNA derived fragments
sRNAs	Small RNAs
tRFs	tRNA-derived fragments

22.1 Introduction

Plant growth and development are greatly influenced by surrounding microbes, in particular by soil species. The largest repository of biological diversity known to exist on earth is represented by soil microbial communities. The rhizosphere, the narrow volume of soil surrounding roots, contains 10^{11} cells per gm of root and may harbor more than 30,000 species, including bacteria, fungi, nematodes, oomycetes and viruses (Berendsen et al., 2012). These communities are called rhizosphere microbiomes, and include taxa that can either be beneficial or pathogenic. The pathogens are responsible for around 30% of global crop production losses, both at pre- and post-harvest levels (Berendsen et al., 2012; Huang et al., 2016). Due to their pathogenicity, these microbes are referred to as biotic stress factors of plants. Bacterial phytopathogens may cause serious crop catastrophes and considerable monetary losses. They are estimated to inflict more than 1 billion USD in annual losses to the food production, worldwide (Buttimer et al., 2017). Pathogenic bacteria are a severe threat to plant defense systems. To combat pathogens manifestations, plants have developed an arsenal of defense responses, activating or repressing a wide variety of genes, thereby altering the synthesis of corresponding proteins and the accumulation of metabolites. During the onslaught of a pathogen, plants recognize an array of pathogen-associated molecular patterns (PAMPs) or host danger-associated molecular patterns (DAMPs) using cell-surface localized pattern-recognition receptors (PRRs). The majority of PRRs are transmembrane receptor-like kinases (RLKs) or receptor-like proteins (RLPs). Pathogen stimulated PAMPs and DAMPs activate PAMP triggered immunity (PTI) and effector-triggered immunity (ETI) in plants (Fei et al., 2016; Huang et al., 2016) (see also Chaps. 1 and 3 in this Volume).

PTI is the primary defense immune response and activates a variety of plant immune responses, such as stomatal closure, a burst of reactive oxygen species (ROS), increased deposition of callose, signal transduction mediated by mitogen-activated protein kinases, and more (Bigeard et al., 2015). Effective pathogens have developed protein effectors to counteract PTI, leading to effector-triggered susceptibility (ETS). Plants have evolved a secondary defense immune response, effector-triggered immunity (ETI). ETI is more powerful in terms of defensive amplitude, and it is typically mediated by a family of nucleotide-binding/leucine-rich repeat (NB-LRR or NLR) proteins encoded by NLR genes. ETI recognizes pathogen-secreted effectors, initiating the hypersensitive response (HR), which results in programmed host cell death (Cui et al., 2015; Liu et al., 2014). In recent decades, many studies have been carried out on NLR genes and small RNA-mediated pathways. This chapter specifically focuses on the role of host small RNAs in developing defense mechanisms against bacterial phytopathogens.

22.2 Small RNAs Types

The non-coding RNAs with 20–30 nucleotide (nt) long are termed small RNAs (sRNAs). They regulate gene expression through RNA silencing or post-transcriptional gene silencing (Axtell, 2013; Fei et al., 2016). In plants, sRNAs are classified into micro (mi) RNAs and small interfering (si) RNAs, based on their biogenesis pathway and functions. miRNAs and siRNAs are predominantly involved in post-transcriptional and transcriptional regulations, respectively. Other sRNAs recently identified in plants include tRNA-derived fragments (tRFs), circular RNA (CircRNA) and rRNA-derived fragments (rRF). Recent research revealed their vital role in resistance to plant diseases (Jin, 2008; Katiyar-Agarwal & Jin, 2010).

22.3 MicroRNAs

MicroRNAs (miRNA) are 20–24 nucleotide long, endogenous small non-coding RNAs discovered in eukaryotes. Instead of encoding proteins, they influence the post-transcriptional expression of their target genes. They regulate their targeted genes through cleavage or by inhibiting the target mRNA (Ali et al., 2020). During biotic and abiotic stress conditions, plant miRNAs regulates physiological and biochemical processes, by influencing the gene functionality, affecting the plant growth and development.

22.3.1 Biogenesis of miRNAs

miRNA biogenesis is a complicated process involving an assortment of proteins. miRNA (miRNA gene) is initially coded from DNA by a RNA polymerase II, which produces a primary transcript (pri-miRNA) with a 5′ cap and 3′ poly-A tail and a self-complementary stem loop structure of 100–120 nucleotides (Fig. 22.1). The RNase III endo ribonuclease DICER-LIKE1 (DCL1) protein, along with other proteins — such as a nuclear cap-binding complex (CBC), dsRNA-binding protein HYPONASTIC LEAVES (HYL1), C2H2-zinc finger structural protein SERRATE (SE), and the G-patch structural protein TGH — sequentially splices pri-miRNAs in the nucleus to form the precursor miRNA. The first cleavage occurs at the base of the stem, resulting in a 70-bp pre-miRNA that is a miRNA/miRNA (guide strand/passenger strand) duplex (Axtell & Meyers, 2018; Brant & Budak, 2018; Gao et al., 2021).

Pre-miRNA also features 2 or 3 nucleotide overhangs at the 3′ end. A second cleavage occurs after 20–24 nucleotides from the initial cleavage point, releasing the miRNA/miRNA duplex from the stem. HUA ENHANCER1 (HEN1) shields the duplex from deterioration by 2′-O-methylation at the 3′ terminal residues. The duplex may be transferred from the nucleus to the cytoplasm with the help of HST

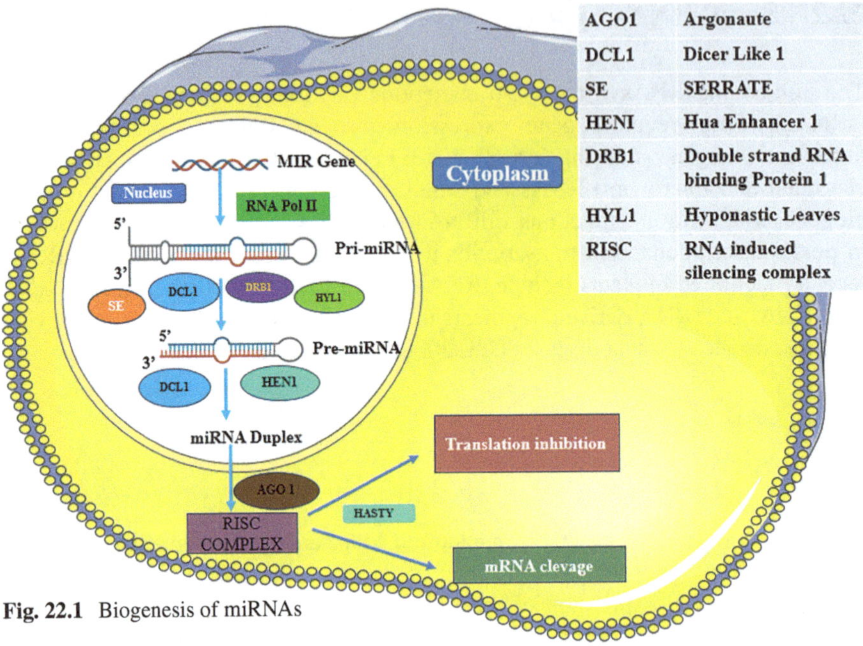

Fig. 22.1 Biogenesis of miRNAs

(HASTY, the plant homolog of exportin-5/Exp5) and other unknown factors. The time of movement of miRNA duplex from the nucleus to the cytoplasm is still uncertain. A few studies demonstrated that miRNA transport into the cytoplasm may occur after integration into the RISC complex (Park et al., 2005). The mature miRNA guide strand is then integrated into an Argonaute 1 (AGO1) protein, which is a component of the RNA-induced silencing complex (RISC). In order to control the expression of the target mRNA by either cleaving the transcript or inhibiting translation, mature miRNA is guided to the target mRNA by the RISC complex with virtually perfect complementarity (Bologna et al., 2018).

22.3.2 *Regulatory Role of Plant miRNAs*

Inflicting numerous diseases on plants, bacteria are significant biotic stress factors for plants. Many miRNAs are implicated during bacterial-plant interactions, modulating the auxin, abscisic acid, and jasmonic acid signalling pathways, influencing the plant defense system, its organogenesis and development (Chen, 2005). Most miRNAs are reported to have conserved sequences and functions across different plant species and are categorized into different types, based on their similarities. Among the group, only a small number of miRNAs show complete species divergence (Dezulian et al., 2005; Fahlgren et al., 2010).

Table 22.1 Plant miRNAs involved in immunity

mi RNA	Origin	Target	Pathogen	Effect on immunity	Reference
miR393	*Arabidopsis thaliana*	TIR1, AFB2 and AFB3	*Pseudomonas syringae*	Positive	Navarro et al. (2006)
miR398	*A. thaliana*	CSD1, CSD2	*P. syringae* pv. tomato	Negative	Jagadeeswaran et al. (2009)
miR399	Citrus plant	PHO2	*Candidatus* Liberibacter asiaticus	Not known	Zhao et al. (2013)
miR160	*A. thaliana*	ARF10, ARF16, ARF17	*P. syringae*	Positive	Li et al. (2010b)
miR167	*A. thaliana*	ARF6 ARF8	*P. syringae*	Positive	Caruana et al. (2020)
miR773	*A. thaliana*	DMT1, DMT2	*P. syringae*	Negative	Li et al. (2010b)
miR400	*A. thaliana*	PPR1, PPR2	*P. syringae* pv. tomato	Negative	Park et al. (2014)
miR472	*A. thaliana*	IR-NB_LRR and CC-NB-LRR resistance genes.	*P. syringae* pv. tomato	Negative	Boccara et al. (2014)
miR482	*Solanum lycopersicum*	Nucleotide Binding Site-Leucine-Rich Repeats (NBS-LRRs)	*P. syringae* pv. tomato	Negative	Canto-Pastor et al. (2019)

In the case of positive regulation, a pathogen-induced miRNA promotes disease resistance, suggesting that the over-expression of that miRNA could be an effective method to overcome infection in plants. However, in the case of negative regulation, suppressing mRNA expression with miRNAs would be an ideal technique for improving plant stress tolerance (Ali et al., 2020). Few miRNAs involved in plant immunity against bacterial pathogens are mentioned in Table 22.1.

22.3.3 *Plant miRNAs* vs *Bacteria*

miR393 The first miRNA discovered to play a function in plant antibacterial PTI is *Arabidopsis* miR393 acting against the bacterial pathogen *Pseudomonas syringae* through the regulation of the auxin signalling pathway (Navarro et al., 2006). The bacterial flagellin (flg22) is an immunogenic peptide (PAMP) recognized by FLAGELLIN SENSING 2 (FLS2), a receptor kinase. This recognition triggers downstream reactions, including the production of ROS and the hormone salicylic acid (SA), which plays a critical role in immune responses (Noshi et al., 2016). miR393 is activated in 20–60 min after an infection when it is exposed to flg22. This activation results in lower levels of expression of auxin receptor TIR1 (Transport Inhibitor Response 1) and its two functional paralogs (AFB2 and AFB3). In contrast to miR393 over-expression lines, constitutive over-expression of another TIR1 paralog, AFB1, which is not a target of miR393, increased disease sensitivity to *P. syringae* pv. tomato. Thus, miR393 contributes to flg22-induced basal defense or PTI in Arabidopsis by targeting auxin receptor mRNAs and inhibiting auxin signalling (Jin, 2008).

miR398 The first miRNA discovered to be down-regulated during a biotic stress is miR398, first identified in Arabidopsis through computational prediction and sequence analysis of a stress-treated *Arabidopsis* small-RNA library. miR398 family is highly conserved in rice, Lotus and *Medicago* (Jones-Rhoades & Bartel, 2004; Sunkar & Zhu, 2004; Zhu et al., 2011). Avirulent strains of *P. syringae* pv. tomato, (Pst) DC3000 (avrRpml or avrRptJ) produce reduced levels of mir398, differing from the virulent strain Pst DC300. When *Pseudomonas* infects plants, it causes an oxidative burst. Experimental data showed that the level of miR398 decreased in reaction to abiotic stress such as ozone fumigation (Dempsey et al., 1999; Jagadeeswaran et al., 2009; Lamb & Dixon, 1997). miR398 target genes of copper superoxide dismutases (*CSD1*, *CSD2*) which aid in the scavenging of ROS by converting O_2 to H_2O_2 (Yamasaki et al., 2007). So, elevated levels of CSDI were expected to detoxify ROS produced by the pathogen infection. These findings support previous findings that miR398 influences pathogen resistance in *Arabidopsis*.

miR399 *Candidatus Liberibacter asiaticus*, the bacterium that causes Huanglongbing (HLB), also known as citrus greening disease, caused miR399 induction. Additional research indicates that the rise in miR399 could result from HLB-induced phosphorus deficit. Its target is a putative ubiquitin-conjugating enzyme (PHO2), which is important for degrading P transporter proteins (PT). Upon infection, miR399 is induced, and upregulation of PHO2 takes place (Huang et al., 2016; Zhao et al., 2013).

miR160 This is a conserved miRNA that was first identified in *Arabidopsis* and has been found as widely present in both model and non-model plants, including rice, poplar, and tomato (Reinhart et al., 2002; Hao et al., 2022). miR160 targets Auxin Response Factors (*ARF10*, *ARF16*, *ARF17*), transcription factors found in the auxin signalling pathways. In *Arabidopsis*, bacterial and fungal pathogens may induce miR160 and down-regulate ARFs, causing variations in auxin-responsive gene expression (AXR3/IAA17, BDL/IAA12, and GH3-like) and callose deposition. This leads to the activation of the primary defense immune response, PAMP-triggered immunity (PTI) (Li et al. 2010a, b; Stork et al., 2010).

miR167 This microRNA has a significant role in controlling auxin-mediated development. Auxin Response Factors 6 (*ARF6*) and 8 (*ARF8*), are two members of a wide family of transcription factors that control the gene expression in response to auxin. Evidence suggests that miR167 targets the mRNAs of these genes. When *Arabidopsis thaliana* is exposed to *P. syringae*, miR167 is over-expressed. Overexpression of miR167 maintains small stomatal apertures, resulting in extremely high levels of resistance against *P. syringae* (Caruana et al., 2020).

miR773 This miRNA is involved in PTI. The mRNA encoding DNA methyltransferase 2 (DMT2) is a target of this miRNA. During an *Agrobacterium* infection, tumor growth is decreased by RNAi-mediated gene silencing of DMT2 and a distinct DNA methyltransferase (DMT1). Remarkably, under biotic and salicylic acid (SA) stress conditions, MET1 is down-regulated (Dowen et al., 2012). Following flg22 treatment, miR773 accumulation was reduced, as shown by deep sequencing analysis of AGO1-immunoprecipitated small RNAs of *A. thaliana*. Accumulation of the miRNA target the DMT2 was also induced in response to flg22 treatment. Transgenic *Arabidopsis* plants over-expressing miR773 reduced the MET2 mRNA levels, decreased callose deposition, and increased susceptibility to the Pst DC3000 and Pst DC3000 hrC− strains. A negative control of PTI can be seen in the reduction of miR773 during biotic stress (Li et al. 2010a, b; Pelaez & Sanchez, 2013).

miR400 Accumulation of miR400 reduces resistance to *P. syringae* pv. tomato *DC3000*. Potential target genes of miR400 include mRNAs encoding pentatricopeptide repeat (PPR) proteins. miR400 cleaves *PPR1* and *PPR2* mRNAs, altering the *Arabidopsis* response to *B. cinerea* and *Pst DC3000*. Due to the PPR protein malfunction caused by miR400, *Arabidopsis* is more vulnerable to harmful bacteria and fungi (Park et al., 2014; Rose et al., 2019).

miR472 It functions as a negative regulator for PTI and ETI, most likely post-transcriptionally, by suppressing a portion of CNLs. miR472 depends on the RNA-dependent RNA polymerase 6 (RDR6), a major actor of RNA silencing that acts at the post-transcriptional level (Boccara et al., 2014).

mir482 It is a homologous counterpart of miR472. It specifically targets nucleotide binding site-leucine-rich repeats (NBS-LRRs), vital elements of disease resistance in dicot plants (Canto-Pastor et al., 2019; Shivaprasad et al., 2012). Based on the type of pathogen infection and the species involved, miR482/2118 can function as a positive or negative regulator (Zhang et al., 2022). miR482 transgenic expression in soybeans significantly increased the number of nodules, indicating that it might sustain the production of nodules in the legume-rhizobial symbiosis (Li et al. 2010a, b). This is supported by the observation that miR482 was up-regulated in response to an infection with the rhizobial bacterium *Bradyrhizobium japonicum* (Subramanian et al., 2008). On the contrary, silencing of miR482 and miR2118b *via* short tandem target mimic (STTM) increased tomato resistance to *P. syringae*, indicating that miR482/2118 has a negative influence on the bacterial infection (Canto-Pastor et al., 2019).

22.3.4 Long miRNAs

Plant miRNAs range in length from 21 to 22 nt. Nevertheless, certain 24 nt long siRNAs in rice and *Arabidopsis* enter the heterochromatic siRNA effector route after their synthesis and cause chromatin changes at their target genes. Remarkably, the majority of long miRNAs identified so far in rice and *Arabidopsis* are also miRNAs distinctive for particular lineages. Because long miRNAs do not appear to make the transition from the proto-miRNA stage to canonical, conserved miRNA suggests that newly evolving "proto-miRNAs" may occasionally go through a phase where they produce 24-nt species (Chellappan et al., 2010; Vazquez et al., 2008). It may also imply that long miRNAs are not favourable for long-term selection. The biological or phenotypic significance of chromatin changes caused by long miRNAs is not yet known (Axtell, 2013).

22.3.5 Small Interfering RNA (siRNA)

The members of this group of RNAs known as siRNA, short siRNA, or silencing RNA, typically have a length of 20–27 base pairs. The four eukaryote kingdoms (protists, fungi, plants, and animals) are known to depend heavily on small interfering RNAs (siRNAs), with the exception of the yeast *Saccharomyces cerevisiae* (Vaucheret, 2006). They are central in the RNA interference (RNAi) pathway (Raja & Tennyson, 2023).

22.4 RNAi and siRNA

RNAi is the collective term for small RNAs (sRNAs) that regulate plant genes by translational suppression of mRNAs or by targeting them for a post-transcriptional destruction (RNA interference) (Pattanayak et al., 2013). RNAi employs a reverse genetic strategy to silence a gene at the post-transcriptional levels. It is a natural plant immune system defensive mechanism acting against biotic stresses. Single-stranded RNAs, such as miRNAs and siRNAs, play important role in RNAi in plants.

The complementary base pairing of siRNAs with target mRNA is caused by the siRNAs incorporation into an AGO-family protein. The RNA-induced silencing complex (RISC) is formed when the AGO protein associates with other proteins following its integration with siRNA (Pratt & MacRae, 2009). The RISC complex can cleave the target mRNA, which serves as a template for another round of dsRNA synthesis, or it can inhibit the mRNA at the translation level, starting the second cycle of RNAi that produces secondary siRNAs. The internal cleavage of mRNAs by endonuclease in the RISC complex determines the mechanism by which particular mRNAs undergo RNAi. The RISC complex directs the siRNA or miRNA cleavage, resulting in sliced mRNA being either degraded by exonucleases or converted into double stranded RNA by RNA-dependent RNA polymerase. This stage decides whether mRNA will be degraded or enter the RNAi feed forward cycle (Hung & Slotkin, 2021).

22.4.1 Biogenesis and Types of siRNA

siRNAs are formed from completely paired double-stranded RNA (dsRNA) precursors, in contrast to miRNAs, which are derived from hairpin loop structures with poor base pairing. These dsRNA precursors are produced *via* antisense transcription or by a cellular RNA-dependent RNA polymerase (RDR) (Katiyar-Agarwal & Jin, 2010). DCL2, DCL3, and DCL4 are principally responsible for processing endogenous siRNAs in plants.

Plants siRNAs are classified into four different types based on their diverse biogenesis. They are *trans*-acting siRNAs (ta-siRNAs), natural antisense transcripts (NATs) derived siRNAs (nat-siRNAs), heterochromatic siRNAs (hc-siRNAs) or repeat-associated siRNAs (ra-siRNAs), and long siRNAs (lsiRNAs) (Katiyar-Agarwal & Jin, 2010).

Non-coding TAS genes are transcribed by RNA Pol II. The lengthy primary transcript products are first cut into a 5′ or 3′ fragment by miRNAs that are loaded with RNA-induced silencing complexes (RISCs). RDR6 and SGS3 then work together to synthesize a complementary strand using these fragments as a template (Fig. 22.2). DCL4 and DRB4 interact with the resulting dsRNA molecule to initiate the subsequent synthesis of ta-siRNAs (Howell et al., 2007).

Nat-siRNAs are produced from the overlap regions of sense and antisense transcripts of *cis*-NATs. A significant proportion of most eukaryotic genomes encode overlapping *cis*-NAT genes, which have the potential to generate siRNAs when base pairing occurs between sense and antisense transcripts. DCL1 and/or DCL2, HYL1, and HEN1 are the cellular components involved in the generation of nat-siRNAs (Borsani et al., 2005; Katiyar-Agarwal et al., 2006).

The main sources of hc-siRNAs and ra-siRNAs, which are typically 24 nt long, include transposons, repeat elements, and heterochromatin areas. They are dependent on the DCL3-RDR2-Pol IV pathway for their biogenesis. They mediate DNA methylation and/or histone modification at the target locations and contribute to the RNA-dependent DNA methylation (RdDM) pathway. The synthesis of lsiRNAs is dependent on DCL1, HYL1, HEN1, AGO7, and HST, and is partially dependent on RDR6 and Pol IV (Katiyar-Agarwal et al., 2007).

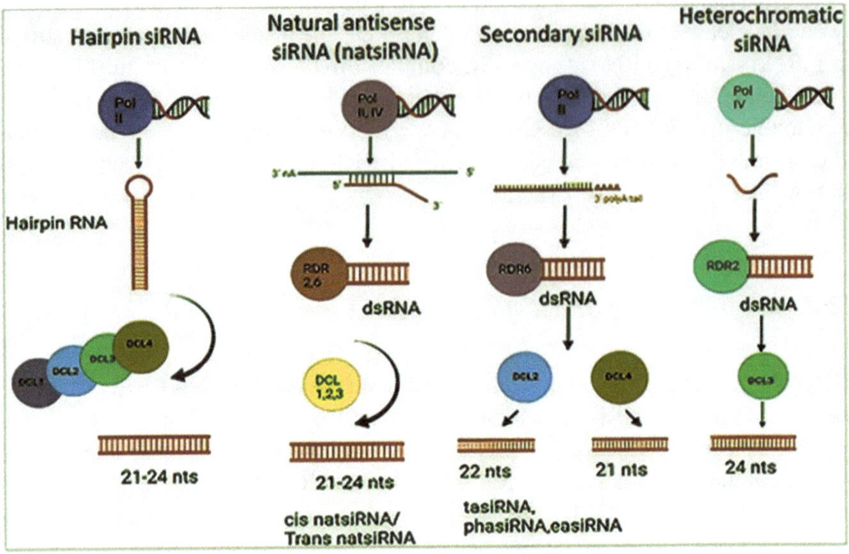

Fig. 22.2 Biogenesis of siRNAs

Table 22.2 Plant siRNAs involved, with a positive effect, in immunity

siRNA	Origin	Target	Pathogen	Reference
nat-siRNAATGB2	*Arabidopsis thaliana*	PPRL, At4g35850	*P. syringae*	Katiyar-Agarwal et al. (2006)
At-lsiRNA-1	*A. thaliana*	AtRAP gene	*P. syringae*	Katiyar-Agarwal et al. (2007)

22.4.2 *Plant Immunity Involving siRNAs* vs *Bacterial Infections*

22.4.2.1 Nat-siRNAATGB2

Pseudomonas syringae carries effector protein Rpt2, notably inducing an endogenous siRNA, nat-siRNAATGB2 (Table 22.2). The analogous host disease resistance gene *RPS2* and the *NDR1* gene, which is necessary for RPS2-specified resistance, are necessary for its induction. Through suppressing PPRL, a suspected negative regulator of the RPS2 resistance pathway, this siRNA supports RPS2-mediated race-specific disease resistance (Fig. 22.3) (Katiyar-Agarwal et al., 2006).

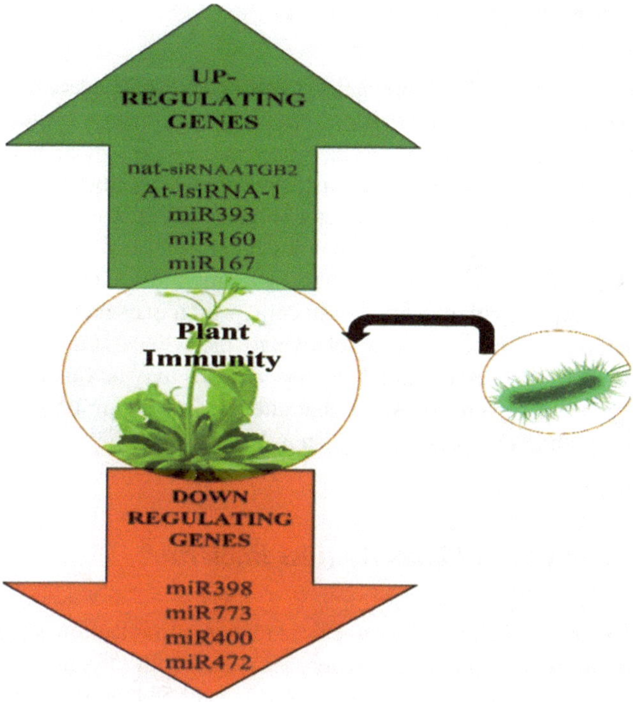

Fig. 22.3 Small RNAs influencing plant immunity

22.4.2.2 At-lsiRNA-1

Pst infection in *Arabidopsis thaliana* causes the induction of AtlsiRNA-1. It inhibits RAP-domain protein involved in disease resistance, encoded by the AtRAP gene. Through XRN4-mediated 5′ to 3′ degradation and decapping, AtlsiRNA-1 may destabilize the target mRNA (Katiyar-Agarwal et al., 2007).

22.5 Circular RNA

Circular RNA (circRNA) is a non-coding RNA resistant to RNase R and has a size range of 100 bp to 4 Kbp. CircRNAs are ubiquitously present, and their prevalence and evolutionary conservation across different species point to some significant, as-yet-unidentified roles (Wang et al., 2017).

22.5.1 Biogenesis of Circular RNAs

CircRNAs are generated from precursor mRNAs (pre-miRNAs) by head-to-tail splicing, or back splicing, in which a downstream 5′ splicing donor site is connected to an upstream 3′ splicing acceptor site (Ye et al., 2015). There have also been reports of alternate methods for producing circRNAs from both introns and intergenic regions (Zhang et al., 2013). Effective back splicing requires both canonical splice signals and canonical spliceosomal machinery (Chen & Yang, 2015). When spliceosomes are adversely formed at back-splicing sites, canonical splicing and back-splicing compete with one another in cells, which may account for the generally lower abundance of circRNAs relative to their linear mRNA counterparts (Chen, 2016). The intricacy of circRNA transcriptions may be further increased by the suggestion made by a recent study that alternative splicing (AS) events play a role in the biosynthesis of circRNAs (Chen & Yang, 2015).

22.5.2 CircRNAs in Plants Against Bacteria

The earliest evidence of circRNA synthesis in response to infection was found in the leaves of *Arabidopsis*. 803 circRNAs were identified with infection of *P. syringae* pv. tomato (Sun et al., 2016). Differential expression of 584 circRNAs was observed in kiwi plant when infected with *P. syringae* pv. *actinidiae* (Psa). These findings suggested, at the very least, that circRNAs contributed to plant sensitivity to pathogen invasion and resistance (Wang et al., 2017). A total of 2098 circRNAs were detected and differentially expressed in potato when infected with *Pectobacterium carotovorum*. The enrichment of parental genes and miRNAs targeted mRNAs in Gene Ontology (GO) revealed that these differentially expressed (DE) circRNAs were engaged in defense response (GO:0006952) (Zhou et al., 2017) (Fig. 22.4).

22.6 Cross-Kingdom RNAi and Role of Pathogen sRNA in Plant Immunity

Mobility of sRNA across and within organisms might significantly impact plant-pathogen interactions. The first evidence of short-distance sRNA cell-to-cell transport in plants was found by Palauqui et al. (1997). Long-distance intra- and interspecies mobility has now been seen for over a decade, including plant-to-pathogen and pathogen-to-plant transfer (Knip et al., 2014). Cross-kingdom RNAi is a remarkable mechanism that induces gene silencing with sRNAs between unrelated species, such as a plant host and its associated bacterium. *Botrytis cinerea*, a fungal pathogen that causes gray mold disease, infects over 200 plant species.

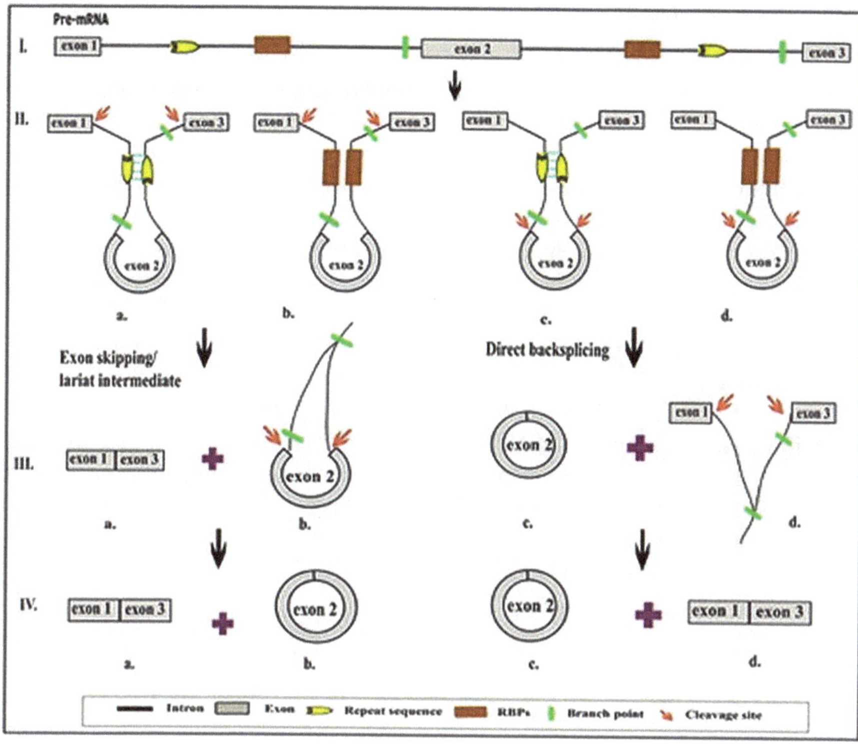

Fig. 22.4 Role of *cis* sequence and *trans* factor in different models of circRNA biogenesis. Normally, the pre-mRNA is enriched with the exonic and intronic regions (I). The non-coding intronic region harbours the highly conserved sequence in the 5′ and 3′, which is essential for splicing by spliceosomal machinery (I). In addition to the conserved sequence, flanking introns consist of a repeat sequence or RBP site, which help to bring the 5′ and 3′ ends of intervening exon closer together due to either the base pairing (IIa, IIc) or by the binding of RBP (IIb, IId). Due to this proximity, circRNAs are generated by exon skipping or direct back splicing (IIIa, IVa). Exon skipping produces linear RNA first, followed by the circularization of an intervening exon along with the formation of a lariat containing flanking introns (IIIb). Subsequent splicing yields a circular exonic RNA (IVb). In contrast, in direct back splicing, exonic circRNA is generated first (IIIc, IVc) and is then followed by an exon–intron lariat (IIId). The latter is further processed to convert into linear RNA (IVd). The pictorial representation is not to scale. (Adapted from: Guria et al., 2020)

Evidences suggest that its small RNA bind to *Arabidopsis* Argonaute (AGO1) and hijacks host RNA interference machinery (RNAi), leading to the selective silencing of host immunity genes (Weiberg et al., 2013).

Another cross-kingdom interaction is transcriptional regulation between *Rhizobium* and soybean. Recent research revealed that bacteria-generated tRFs (tRNA-derived fragments) directed RNAi that can silence certain plant genes (Ren et al., 2019). More experimental evidence on tRFs in this regard can demonstrate that they can play a function in coping with biotic stress.

22.7 Conclusion

Plant sRNA research has made great progress during the last few decades. Subsequent research in the field contributed to a better understanding of their biogenesis, classification and role in biological mechanisms. One of sRNA significant roles is influencing plant disease resistance. An overview of the biosynthesis and regulatory roles of different noncoding RNAs (miRNAs, siRNAs, and circRNAs) that are involved in plant immunity against bacterial infection, is provided in this chapter. Some of them are engaged in the up-regulation of plant immunity, while others are involved in its down-regulation. miRNAs such as miR393, miR398, miR773, miR399, miR160, miR167, miR400, miR472, miR482 and siRNAs including nat-siRNAATGB2, At-lsiRNA-1 were discussed.

Recent advancements like HIGS (Host-induced gene silencing), CRISPR/Cas9 and hairpin RNAs seek use of sRNA to combat biotic stress in an eco-friendly manner. Due to its ability to repress a multitude of important gene targets in pathogens, RNA interference technology is a promising strategy for global food security and economic growth. However, extensive research is needed to overcome the obstacles when approaching sRNA technologies in plants.

References

Ali, M., Javaid, A., Naqvi, S. H., Batcho, A., Kayani, W. K., Lal, A., et al. (2020). Biotic stress triggered small RNA and RNAi defense response in plants. *Molecular Biology Reports, 47*, 5511–5522.

Axtell, M. J. (2013). Classification and comparison of small RNAs from plants. *Annual Review of Plant Biology, 64*, 137–159.

Axtell, M. J., & Meyers, B. C. (2018). Revisiting criteria for plant micro RNA annotation in the era of big data. *The Plant Cell, 30*, 272–284.

Berendsen, R. L., Pieterse, C. M. J., & Bakker, P. A. H. M. (2012). The rhizosphere microbiome and plant health. *Trends in Plant Science, 17*, 478–486.

Bigeard, J., Colcombet, J., & Hirt, H. (2015). Signaling mechanisms in Pattern-Triggered Immunity (PTI). *Molecular Plant, 8*, 521–539.

Boccara, M., Sarazin, A., Thiébeauld, O., Jay, F., Voinnet, O., Navarro, L., & Colot, V. (2014). The *Arabidopsis* miR472-RDR6 silencing pathway modulates PAMP- and effector-triggered immunity through the post-transcriptional control of disease resistance genes. *PLoS Pathogens, 10*, e1003883.

Bologna, N. G., Iselin, R., Abriata, L. A., Sarazin, A., Pumplin, N., et al. (2018). Nucleo-cytosolic shuttling of ARGONAUTE1 prompts a revised model of the plant MicroRNA pathway. *Molecular Cell, 69*, 709–719.

Borsani, O., Zhu, J., Verslues, P. E., Sunkar, R., & Zhu, J.-K. (2005). Endogenous siRNAs derived from a pair of natural cis-antisense transcripts regulate salt tolerance in *Arabidopsis*. *Cell, 123*(7), 1279–1291. https://doi.org/10.1016/j.cell.2005.11.035

Brant, E. J., & Budak, H. (2018). Plant small non-coding RNAs and their roles in biotic stresses. *Frontiers in Plant Science, 9*, 1038. https://www.frontiersin.org/articles/10.3389/fpls.2018.0

Buttimer, C., McAuliffe, O., Ross, R. P., Hill, C., O'Mahony, J., & Coffey, A. (2017). Bacteriophages and bacterial plant diseases. *Frontiers in Microbiology, 8*, 34.

Canto-Pastor, A., Santos, B. A. M. C., Valli, A. A., Summers, W., Schornack, S., & Baulcombe, D. C. (2019). Enhanced resistance to bacterial and oomycete pathogens by short tandem target mimic RNAs in tomato. *Proceedings of the National Academy of Sciences of the United States of America, 116*(7), 2755–2760. https://doi.org/10.1073/pnas.1814380116

Caruana, J. C., Dhar, N., & Raina, R. (2020). Overexpression of *Arabidopsis* microRNA167 induces salicylic acid-dependent defense against *Pseudomonas syringae* through the regulation of its targets ARF6 and ARF8. *Plant Direct, 4*(9), e00270. https://doi.org/10.1002/pld3.270

Chellappan, P., Xia, J., Zhou, X., Gao, S., Zhang, X., et al. (2010). siRNAs from miRNA sites mediate DNA methylation of target genes. *Nucleic Acids Research, 38*(20), 6883–6894. https://doi.org/10.1093/nar/gkq590

Chen, X. (2005). microRNA biogenesis and function in plants. *FEBS Letters, 579*(26), 5923–5931. https://doi.org/10.1016/j.febslet.2005.07.071

Chen, L.-L. (2016). The biogenesis and emerging roles of circular RNAs. *Nature Reviews Molecular Cell Biology, 17*(4), 205–211. https://doi.org/10.1038/nrm.2015.32

Chen, L.-L., & Yang, L. (2015). Regulation of circRNA biogenesis. *RNA Biology, 12*(4), 381–388. https://doi.org/10.1080/15476286.2015.1020271

Cui, H., Tsuda, K., & Parker, J. E. (2015). Effector-triggered immunity: From pathogen perception to robust defense. *Annual Review of Plant Biology, 66*(1), 487–511. https://doi.org/10.1146/annurev-arplant-050213-040012

Dempsey, D. A., Shah, J., & Klessig, D. F. (1999). Salicylic acid and disease resistance in plants. *Critical Reviews in Plant Sciences, 18*(4), 547–575. https://doi.org/10.1080/07352689991309397

Dezulian, T., Palatnik, J. F., Huson, D., & Weigel, D. (2005). Conservation and divergence of microRNA families in plants. *Genome Biology, 6*(11), P13. https://doi.org/10.1186/gb-2005-6-11-p13

Dowen, R. H., Pelizzola, M., Schmitz, R. J., Lister, R., Dowen, J. M., et al. (2012). Widespread dynamic DNA methylation in response to biotic stress. *Proceedings of the National Academy of Sciences of the United States of America, 109*, E2183–E2191. https://doi.org/10.1073/pnas.1209329109

Fahlgren, N., Jogdeo, S., Kasschau, K. D., Sullivan, C. M., Chapman, E. J., et al. (2010). MicroRNA gene evolution in *Arabidopsis lyrata* and *Arabidopsis thaliana*. *The Plant Cell, 22*(4), 1074–1089. https://doi.org/10.1105/tpc.110.073999

Fei, Q., Zhang, Y., Xia, R., & Meyers, B. C. (2016). Small RNAs add Zing to the Zig-Zag-Zig model of plant defenses. *Molecular Plant-Microbe Interactions®, 29*(3), Article 3. https://doi.org/10.1094/MPMI-09-15-0212-FI

Gao, Z., Nie, J., & Wang, H. (2021). MicroRNA biogenesis in plant. *Plant Growth Regulation, 93*(1), 1–12. https://doi.org/10.1007/s10725-020-00654-9

Guria, A., Sharma, P., Natesan, S., & Pandi, G. (2020). Circular RNAs – The road less traveled. *Frontiers in Molecular Biosciences, 6*, 146.

Hao, K., Wang, Y., Zhu, Z., Wu, Y., Chen, R., & Zhang, L. (2022). miR160: An indispensable regulator in plant. *Frontiers in Plant Science, 13*. https://www.frontiersin.org/articles/10.3389/fpls.2022.833322

Howell, M. D., Fahlgren, N., Chapman, E. J., Cumbie, J. S., Sullivan, C. M., et al. (2007). Genome-wide analysis of the RNA-dependent RNA Polymerase6/DICER-LIKE4 pathway in *Arabidopsis* reveals dependency on miRNA- and tasiRNA-directed targeting. *The Plant Cell, 19*(3), 926–942. https://doi.org/10.1105/tpc.107.050062

Huang, J., Yang, M., & Zhang, X. (2016). The function of small RNAs in plant biotic stress response. *Journal of Integrative Plant Biology, 58*(4), Article 4–Articl327. https://doi.org/10.1111/jipb.12463

Hung, Y.-H., & Slotkin, R. K. (2021). The initiation of RNA interference (RNAi) in plants. *Current Opinion in Plant Biology, 61*, 102014. https://doi.org/10.1016/j.pbi.2021.102014

Jagadeeswaran, G., Saini, A., & Sunkar, R. (2009). Biotic and abiotic stress down-regulate miR398 expression in Arabidopsis. *Planta, 229*(4), 1009–1014.

Jin, H. (2008). Endogenous small RNAs and antibacterial immunity in plants. *FEBS Letters, 582*(18), Article 18. https://doi.org/10.1016/j.febslet.2008.06.053

Jones-Rhoades, M. W., & Bartel, D. P. (2004). Computational identification of plant microRNAs and their targets, including a stress-induced miRNA. *Molecular Cell, 14*(6), 787–799. https://doi.org/10.1016/j.molcel.2004.05.027

Katiyar-Agarwal, S., & Jin, H. (2010). Role of small RNAs in host-microbe interactions. *Annual Review of Phytopathology, 48*, 225–246. https://doi.org/10.1146/annurev-phyto-073009-114457

Katiyar-Agarwal, S., Morgan, R., Dahlbeck, D., Borsani, O., Villegas, A., Jr., et al. (2006). A pathogen inducible endogenous siRNA in plant immunity. *Proceedings of the National Academy of Sciences of the United States of America, 103*, 18002–18007.

Katiyar-Agarwal, S., Gao, S., Vivian-Smith, A., & Jin, H. (2007). A novel class of bacteria-induced small RNAs in *Arabidopsis*. *Genes & Development, 21*(23), 3123–3134. https://doi.org/10.1101/gad.1595107

Knip, M., Constantin, M. E., & Thordal-Christensen, H. (2014). Trans-kingdom cross-talk: Small RNAs on the move. *PLoS Genetics, 10*(9), e1004602.

Lamb, C., & Dixon, R. A. (1997). The oxidative burst in plant disease resistance. *Annual Review of Plant Physiology and Plant Molecular Biology, 48*(1), 251–275. https://doi.org/10.1146/annurev.arplant.48.1.251

Li, H., Deng, Y., Wu, T., Subramanian, S., & Yu, O. (2010a). Misexpression of miR482, miR1512, and miR1515 increases soybean nodulation. *Plant Physiology, 153*, 1759–1770. https://doi.org/10.1104/pp.110.156950

Li, Y., Zhang, Q., Zhang, J., Wu, L., Qi, Y., & Zhou, J.-M. (2010b). Identification of MicroRNAs involved in pathogen-associated molecular pattern-triggered plant innate immunity1. *Plant Physiology, 152*(4), 2222–2231. https://doi.org/10.1104/pp.109.151803

Liu, W., Liu, J., Triplett, L., Leach, J. E., & Wang, G.-L. (2014). Novel insights into rice innate immunity against bacterial and fungal pathogens. *Annual Review of Phytopathology, 52*(1), 213–241. https://doi.org/10.1146/annurev-phyto-102313-045926

Navarro, L., Dunoyer, P., Jay, F., Arnold, B., Dharmasiri, N., et al. (2006). A plant miRNA contributes to antibacterial resistance by repressing auxin signaling. *Science, 312*(5772), 436–439. https://doi.org/10.1126/science.1126088

Noshi, M., Mori, D., Tanabe, N., Maruta, T., & Shigeoka, S. (2016). *Arabidopsis* clade IV TGA transcription factors, TGA10 and TGA9, are involved in ROS-mediated responses to bacterial PAMP flg22. *Plant Science, 252*, 12–21. https://doi.org/10.1016/j.plantsci.2016.06.019

Palauqui, J. C., Elmayan, T., Pollien, J. M., & Vaucheret, H. (1997). Systemic acquired silencing: Transgene specific post-transcriptional silencing is transmitted by grafting from silenced stocks to non-silenced scions. *The EMBO Journal, 16*(15), 4738–4745.

Park, M. Y., Wu, G., Gonzalez-Sulser, A., Vaucheret, H., & Poethig, R. S. (2005). Nuclear processing and export of microRNAs in *Arabidopsis*. *Proceedings of the National Academy of Sciences of the United States of America, 102*(10), 3691–3696. https://doi.org/10.1073/pnas.0405570102

Park, Y. J., Lee, H. J., Kwak, K. J., Lee, K., Hong, S. W., & Kang, H. (2014). MicroRNA400-guided cleavage of pentatricopeptide repeat protein mRNAs renders *Arabidopsis thaliana* more susceptible to pathogenic bacteria and fungi. *Plant and Cell Physiology, 55*(9), 1660–1668. https://doi.org/10.1093/pcp/pcu096

Pattanayak, D., Solanke, A. U., & Kumar, P. A. (2013). Plant RNA interference pathways: Diversity in function, similarity in action. *Plant Molecular Biology Reporter, 31*(3), 493–506.

Pelaez, P., & Sanchez, F. (2013). Small RNAs in plant defense responses during viral and bacterial interactions: Similarities and differences. *Frontiers in Plant Science, 4*, 343.

Pratt, A. J., & MacRae, I. J. (2009). The RNA-induced silencing complex: A versatile gene-silencing machine. *Journal of Biological Chemistry, 284*(27), 17897–17901.

Raja, I., & Tennyson, J. (2023). Small RNA–Regulator of biotic stress and pathogenesis in food crops. In P. Guleria, V. Kumar, & B. Mo (Eds.), *Plant small RNA in food crops* (pp. 233–269). Academic.

Reinhart, B. J., Weinstein, E. G., Rhoades, M. W., Bartel, B., & Bartel, D. P. (2002). MicroRNAs in plants. *Genes & Development, 16*, 1616–1626.

Ren, B., Wang, X., Duan, J., & Ma, J. (2019). Rhizobial tRNA-derived small RNAs are signal molecules regulating plant nodulation. *Science, 365*, 919–922.

Rose, L. E., Overdijk, E. J. R., & Van Damme, M. (2019). Small RNA molecules and their role in plant disease. *European Journal of Plant Pathology, 154*, 1.

Shivaprasad, P. V., Chen, H.-M., Patel, K., Bond, D. M., Santos, B. A. C. M., & Baulcombe, D. C. (2012). A MicroRNA superfamily regulates nucleotide binding site–leucine-rich repeats and other mRNAs. *The Plant Cell, 24*(3), 859–874. https://doi.org/10.1105/tpc.111.095380

Stork, J., Harris, D., Griffiths, J., Williams, B., Beisson, F., Li-Beisson, Y., et al. (2010). CELLULOSE SYNTHASE9 serves a nonredundant role in secondary cell wall synthesis in Arabidopsis epidermal testa cells. *Plant Physiology, 153*, 580–589. https://doi.org/10.1104/pp.110.154062

Subramanian, S., Fu, Y., Sunkar, R., Barbazuk, W. B., Zhu, J.-K., & Yu, O. (2008). Novel and nodulation-regulated microRNAs in soybean roots. *BMC Genomics, 9*(1), 160. https://doi.org/10.1186/1471-2164-9-160

Sun, X., Wang, L., Ding, J., Wang, Y., Wang, J., et al. (2016). Integrative analysis of *Arabidopsis thaliana* transcriptomics reveals intuitive splicing mechanism for circular RNA. *FEBS Letters, 590*(20), 3510–3516. https://doi.org/10.1002/1873-3468.12440

Sunkar, R., & Zhu, J.-K. (2004). Novel and stress-regulated microRNAs and other small rnas from *Arabidopsis*. *The Plant Cell, 16*(8), 2001–2019. https://doi.org/10.1105/tpc.104.022830

Vaucheret, H. (2006). Post-transcriptional small RNA pathways in plants: Mechanisms and regulations. *Genes & Development, 20*(7), Article 7. https://doi.org/10.1101/gad.1410506

Vazquez, F., Blevins, T., Ailhas, J., Boller, T., & Meins, F., Jr. (2008). Evolution of *Arabidopsis* MIR genes generates novel microRNA classes. *Nucleic Acids Research, 36*(20), 6429–6438. https://doi.org/10.1093/nar/gkn670

Wang, Z., Liu, Y., Li, D., Li, L., Zhang, Q., Wang, S., & Huang, H. (2017). Identification of circular RNAs in kiwifruit and their species-specific response to bacterial canker pathogen invasion. *Frontiers in Plant Science, 8*. https://doi.org/10.3389/fpls.2017.00413

Weiberg, A., Wang, M., Lin, F.-M., Zhao, H., Zhang, Z., et al. (2013). Fungal small RNAs suppress plant immunity by hijacking host RNA interference pathways. *Science, 342*(6154), 118–123. https://doi.org/10.1126/science.1239705

Yamasaki, H., Abdel-Ghany, S. E., Cohu, C. M., Kobayashi, Y., Shikanai, T., & Pilon, M. (2007). Regulation of copper homeostasis by micro-RNA in Arabidopsis. *Journal of Biological Chemistry, 282*(22), 16369–16378. https://doi.org/10.1074/jbc.M700138200

Ye, C.-Y., Chen, L., Liu, C., Zhu, Q.-H., & Fan, L. (2015). Widespread non-coding circular RNAs in plants. *New Phytologist, 208*(1), 88–95. https://doi.org/10.1111/nph.13585

Zhang, Y., Zhang, X.-O., Chen, T., Xiang, J.-F., Yin, Q.-F., et al. (2013). Circular Intronic long non-coding RNAs. *Molecular Cell, 51*(6), 792–806. https://doi.org/10.1016/j.molcel.2013.08.017

Zhang, Y., Waseem, M., Zeng, Z., Xu, J., Chen, C., et al. (2022). MicroRNA482/2118, a miRNA superfamily essential for both disease resistance and plant development. *New Phytologist, 233*(5), 2047–2057. https://doi.org/10.1111/nph.17853

Zhao, H., Sun, R., Albrecht, U., Padmanabhan, C., Wang, A., et al. (2013). Small RNA profiling reveals phosphorus deficiency as a contributing factor in symptom expression for citrus Huanglongbing disease. *Molecular Plant, 6*(2), 301–310. https://doi.org/10.1093/mp/sst002

Zhou, R., Zhu, Y., Zhao, J., Fang, Z., Wang, S., Yin, J., Chu, Z., & Ma, D. (2017). Transcriptome-wide identification and characterization of potato circular RNAs in response to *Pectobacterium carotovorum* subspecies *brasiliense* infection. *International Journal of Molecular Sciences, 19*, 71.

Zhu, C., Ding, Y., & Liu, H. (2011). MiR398 and plant stress responses. *Physiologia Plantarum, 143*, 1–9.

Chapter 23
Deciphering the Rhizosphere Microbiome Through Metagenomics

Ayon Pal, Madhumita G. Chaki, and Samarpita Adhikary

Abstract The rhizosphere is the region of soil that is in immediate contact with the plant roots. The chemical composition of the root exudates defines the entire microbial community composition. A thorough metagenomic study is of immense significance, enabling a bird's eye view into the whole rhizospheric micro-domain. Diversified populations of plant growth-promoting rhizobacteria exhibiting an indispensable role in mitigating stress responses to both biotic and abiotic stress have been deciphered by rhizosphere microbiome study. This chapter deals with the noteworthy role of metagenomics in unraveling the rhizosphere community composition, dynamic diversity, and its potential role in plant stress mitigation and bioprospecting. As a consequence of priming, the plant growth-promoting rhizobacterial population releases antagonistic agents such as VOCs, ACC deaminase, antibiotics, siderophores, LOCs, AHLs, and lipopolypeptides, which serve as defense priming agents against phytopathogenic fungi and bacteria thus exhibiting resistance to biotic agents. Metagenomics has helped elucidate the rhizosphere microbiome's remarkable role in mitigating abiotic stress responses specifically heavy metals, salinity, chilling, iron deficiency chlorosis, and nitrogen deficiency. A deep insight into the genetic profile of the rhizosphere of stressed and non-stressed plants has revealed significant differences relating to the predominance of a few plant growth-promoting rhizobacterial phyla in the rhizosphere community, such as Alphaproteobacteria, Betaproteobacteria, Gammaproteobacteria, and Acidobacteria. Mitigation of the stress responses is also exhibited due to the dynamic augmentation of genes such as *czcA*, *arsH*, *copA*, *merC*, *nifHDK*, *norB*, *nosZ*, *nirSK*, *nrfA*, *napA*, *narG*, and *hzo*.

Keywords Abiotic stress · Microbial diversity · Omics · Priming · Rhizobacteria

A. Pal (✉) · M. G. Chaki · S. Adhikary
Microbiology and Computational Biology Laboratory, Department of Botany, Raiganj University, Raiganj, West Bengal, India

P. Mathur, S. Roy (eds.), *Plant Microbiome and Biological Control*, Sustainability in Plant and Crop Protection 20,
https://doi.org/10.1007/978-3-031-75845-4_23

Abbreviations

ACCD	Aminocyclopropane-1-carboxylate deaminase
AHLs	N-Acyl homoserine lactones
DCD	Dicyandiamide
ISR	Induced systemic resistance
NBPT	N-(n-butyl) thiophosphoric triamide
PGPR	Plant growth promoting rhizobacteria
VOCs	Volatile organic compounds

23.1 Introduction

The rhizosphere is the domain of top soil, which is proximate to the plant roots (Jamil et al., 2022). This region represents a "miniature microbial ecosystem" or "microbial storehouse." The microbial community in the soil rhizosphere can also be called a "microbial consortia," as many species symbiotically associate with each other, forming an infinitesimal ecosystem (Palmieri et al., 2017; Santoyo et al., 2021; Sarma et al., 2015). This region hosts a varied population of microbial species, including everything from beneficial to harmful to opportunistic organisms (Lareen et al., 2016). All of these species participate in a certain kind of species interaction with the host plant, which sustains their survival in the soil microenvironment. The most dominant ones are mutualism and commensalism (Chagas et al., 2018; Martin & Schwab, 2012). Also, the microbial communities benefit themselves by getting the organic and inorganic materials and the root exudates, an essential component of the "rhizodeposition" that encompasses the total carbon content transferred from the root to the soil, primarily serving as the food for the persisting species in that micro-zone (Bukhat et al., 2020; Venturi & Fuqua, 2013; Venturi & Keel, 2016; Zipfel & Oldroyd, 2017). The rhizosphereic micro-domain is subdivided into three distinct zones: the endorhizosphere, rhizoplane, and ectorhizosphere (Wang et al., 2022). The region of the bulk soil, which lies away from the plant root and demonstrates a relatively poor microbiological activity than the rhizospheric soil, is referred to as non-rhizospheric soil or edaphosphere (Herreño et al., 2023). The rhizosphere effect, represented by the R/S ratio, is a critical quantitative measure of the number of microbial species present in the rhizosphere to that of the edaphosphere (Pandey & Palni, 2007). The R/S ratio ranges from 0.2 to 0.35 for bacteria, 0.3 to 0.29 for actinomycetes, and 0.3 to 0.34 for fungi (Mira et al., 2022). According to some sources, the R/S ratio for bacteria can be as high as 20:1 (Garbeva et al., 2008).

Rhizosphere microbiome research has identified several communities of plant growth-promoting rhizobacteria that play an essential role in reducing stress responses to both biotic and abiotic stressors. This chapter primarily focuses on the importance of metagenomics in understanding rhizosphere community composition, dynamic diversity, and its potential function in plant stress reduction and bioprospecting. As a result of priming, the plant-growth-promoting rhizobacterial population produces various antagonistic agents which act as defence priming agents against phytopathogenic fungi and bacteria, resulting in biotic resistance. Metagenomics has shed light on the rhizosphere microbiome's unique function in alleviating abiotic stress responses. This chapter also discusses the effect of the rhizosphere microbiome in the mitigation of abiotic stress responses, notably salinity, cold, iron deficiency chlorosis, nitrogen deficiency, and heavy metal detoxification, by comparing the genetic profiles of stressed and unstressed plants. Furthermore, it discusses how quorum sensing works to boost beneficial bacterial populations in the rhizosphere microbiome via communication cascades.

23.2 Significance of Rhizosphere Microbiome

The rhizosphere encompasses a greater microbial diversity than the bulk soil revealed by the R/S ratio. It contains root exudates that serve as the sole source of nutrition for the available microflora (Fan et al., 2018; Ling et al., 2022). A mutual symbiotic relation is established between the plants and plant growth-promoting rhizobacteria (PGPR) by the interchange of about 20 to 50% of photosynthate, out of which only 17% is unleashed into the soil, which adequately serves as the nourishment source for the microbiome (Olanrewaju et al., 2019). In turn, the microorganisms help alleviate abiotic stresses like heat, heavy metals, drought, cold, etc. (Khan et al., 2020; Mellidou & Karamanoli, 2022). They also endow plants with plasticity and durability to combat phytopathogens (Jiao et al., 2021) and boost plant growth by providing nutrients and phytohormones (Kudoyarova et al., 2019).

Having an insight into the functional attributes of the rhizosphere microbiome or rhizomicrobiome renders a broad new perspective to unravel novel industrially significant membrane proteins, antibiotics, and antibiotic-resistance enzymes. With the advent of rapid urbanization and industrialization, considerable pressure is being exerted upon agricultural land to escalate the crop yield by unit area since an increment in the available area is not feasible as it is a non-renewable natural resource (Saeed et al., 2021). Several alternative strategies have been imposed to reduce the detrimental effects of chemical pesticides and fertilizers by switching to biological control methods to balance present and future food requirements judiciously (Saeed et al., 2021). In this context, PGPR plays an invincible part. It provides an impetus boost in up-regulating almost every aspect of plant augmentation (Mendes et al., 2013).

Broadly, the mechanism of action of PGPR can be highlighted in depth with an overview of its direct and indirect action (Backer et al., 2018). Hence, the comprehensive aspect of the action of PGPR is inferred with respect to its action as a significant biocontrol agent exhibiting antibiosis and out competition of pathogenic microbes by synthesizing hydrogen cyanide (HCN) or other lytic enzymes across the vicinity of phytopathogens. Besides, induced immunity, i.e., induced systemic resistance (ISR), is coupled with PGPR elicitation (Mekonnen et al., 2022; Siddiqui, 2006). In addition, they play a significant role in bioremediation by strengthening siderophores production, biosurfactant biosorption, and ACC deaminase activity (Gupta et al., 2023). Another significant function is manifested in modern agriculture to assist in plant growth promotion directly by phosphate and potassium solubilization, and phytohormone synthesis (Mohanty et al., 2021). PGPR are an alternative natural remedy in the restoration and mitigation of the hindrance faced by the ecosphere (Chandran et al., 2021). They subject the plant into a number of metabolic, genetic and physiological adjustments by effectuating resistance to biotic and abiotic stress and heavy metal contamination (Khan et al., 2019; Lucas et al., 2023; Reddy et al., 2023).

23.3 Microbiome Hotspots and Microbial Dark Matter

Microbiome hotspots are vibrant, diversified areas dominated by prevalent microbial populations representing the entire consortia of culturable, accessible, and exploitable communities. The human gut, thermal vent of oceans, hot springs, and rhizospheric soil are some of the most critical microbiomes with a huge microbiota population (Berg et al., 2020). Nevertheless, only analyzing the so-called "hotspot domain" will not give an entire blueprint since out of the whole population of microorganisms that can reproduce in predetermined laboratory conditions, only 0.1% are culturable. Hence, it is essential to bridge the gap of the negative feedback provided by conventional cultivation techniques (Connon & Giovannoni, 2002). However, a significant proportion of these uncultivable bacteria also comprises novel phylotypes, families, and divisions, which should be equally explored in depth (Handelsman, 2004; Sharma et al., 2005).

The field that deals with the study of any microbiome is known as metagenomics or microbial community analysis (Cloete & Muyima, 1997; Huson & Weber, 2013; Myrold et al., 2014; Nannipieri et al., 2020; Wooley et al., 2010). Metagenomics is the optimum toolkit to divulge deep into the rhizosphere to unravel the unprecedented enigma. In fact, it is the zenith to explore the enormous diversity of uncharacterized microflora of the soil microbiome (Breitwieser et al., 2019). The rigorous evaluation and interpretation of the rhizosphere microbiome is attainable through metagenomics (Alves et al., 2018; Breitwieser et al., 2019). Hence, it is necessary and sufficient to gain insight into how the soil system functions and how it may be manipulated to promote a healthy ecosystem and sustainable agriculture. Both variables are, therefore, inextricable. Thus, contouring the rhizosphere microbiome would remain abridged rather than conceivable without the metagenomics aspect (Alves et al., 2018). Since "the sky is the limit," likewise metagenomics has enabled us to outstretch the utmost boundary of any microbiome. Nature has provided us with an enormous microbial consortium; how deep we can plunge into it is a million-dollar question. There exists a fine line of difference between the terms microbiota and microbiome the former of which relates to assemblage of microbial species or more precisely only the biotic component portion of the microbial ecosystem (Dubey et al., 2019; Jiao et al., 2022; van der Heijden & Schlaeppi, 2015). Besides, the microbiome embraces both the biotic (microbial species) and the abiotic factors associated with the particular habitat (Abd-Elgawad, 2021; Dubey et al., 2019).

Metagenomics can help unravel different vistas of the rhizosphere like host-pathogen interaction, delimitation of nitrogen deficiency, mitigating the adverse effects of iron deficiency chlorosis, scaling down unfavourable responses of salinity stress, ameliorating toxic effects of heavy metals and also minimizing the responses of chilling stress. The role of metagenomics in elucidating a few of these is briefly dealt with in this chapter.

23.4 Generalized Outline for Rhizosphere Soil Metagenomics

The metagenomic analysis consists of a series of steps (Fig. 23.1). Initially total genome is extracted from the rhizosphere region, followed by digestion, ligation, and transformation in an appropriate host (Sabale et al., 2019). This is followed by

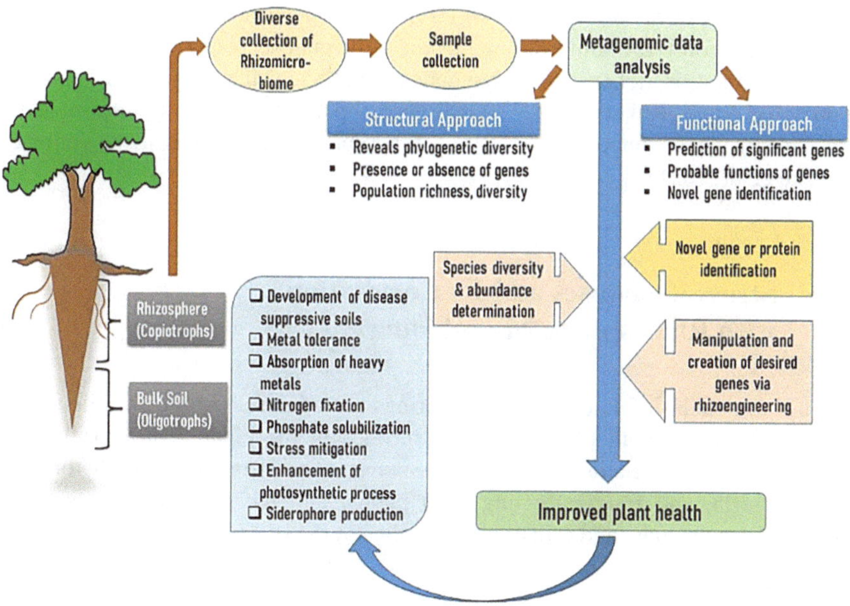

Fig. 23.1 A generalized depiction of the metagenomic analysis procedure from the rhizospheric soil microbiome and its role in plant growth enhancement.

library construction, which is then subjected to functional or phylogenetic screening (Pudake et al., 2021). Shearing of DNA is a common constraint faced during DNA extraction, overcoming this using density gradient or centrifugation using cation exchange resins is important (Unno & Shinano, 2013). Humic acid is a common soil adulterant causing soil contamination, which, when filtered, frequently results in the shearing of DNA (Prayogo et al., 2020; Roslan et al., 2017; Sharma et al., 2007). Preciseness should be maintained while isolating DNA from the rhizospheric zone, proper host vector selection, and appropriate screening methods must be adopted (Katzke et al., 2017). High-quality pure rhizospheric soil DNA extraction kits (such as Power Soil®, Ultra Clean®, RNA Fast DNA®, and Soil Master®) are frequently used for this purpose (Jia et al., 2022).

Metagenomics library construction, metagenomics data sequencing, and analysis using a combinatorial approach of amplicon sequencing followed by shotgun metagenomics based on the results of the 16S sequencing data would be much more beneficial. For culturable species, marker-based sequencing is the most appropriate (those species available in databases), and whole genome shotgun sequencing (for novel gene and species identification) can be done employing high throughput NGS sequencing methods (Pudake et al., 2021). The obtained metagenomic dataset can be subjected to either structural metagenomic, functional metagenomic, or both approaches. The former provides an overview of the metagenome in the microbiome, while the latter offers an insight into deciphering the functional aspects of the genes (Fig. 23.1). Although both approaches can be thought of as a sequential cascade, the purpose of the research will sometimes determine the approach to be undertaken. If only revealing the phylogenetic diversity, presence or absence of a species, population richness and abundance is sufficient then structural genomics approach is opted (Kui et al., 2021). Otherwise, if prediction of the significant genes and their probable function is essential, functional approach has to be followed (Pudake et al., 2021).

23.5 Stress Responses and Its Correlation with Rhizosphere Microbiome

Stress is any undesirable condition that brings out adverse detrimental effects and alteration in the overall physiological responses of the plant. It is broadly subcategorized into biotic and abiotic stress responses. The former include phytopathogens, bacteria, and fungi affecting crop plants, and the latter include salinity, drought, heavy metals, and mineral deficiencies in the soil.

23.5.1 *Defense Priming Mechanisms Contrasting Biotic Stress*

Compared to humans, priming in plants is analogous to human vaccination (Tiwari et al., 2022). The active or passive priming correlates to vaccination with respect to application strategy. As the former develops upon pathogen attack, it is more specific and induced, which can be local or systemic. However, the passive resistance develops upon defenses constitutively expressed in the plants (Hönig et al., 2023). Priming is a mechanism that renders the plant better options when multiple kinds of stress or other contradictory forces act upon it. It is a physiological and genetic alteration that ensures better chances of survival, similar to adaptive immunity or, more specifically, the B lymphocyte's memory cells. It provides the plant signals about the stress or pathogen type and the explicit or appropriate response needed to delimit it (Mauch-Mani et al., 2017). Therefore, defense priming is about exaggerating the innate immune response so that plants are built with an adaptive response resulting in transcriptional, physiological, epigenetic, and behavioural changes (Espinas et al., 2016). In fact, plant defense priming is also perpetual and transferrable (Conrath et al., 2015). Priming plants against secondary stresses is only achieved after innate resistance is developed in response to abiotic stresses, pathogens, chemicals, and arthropods, which serve as signature indicator molecules (Pare et al., 2005). Primed plants outcompete the non-primed ones by being more competent and exerting better versatility over phytopathogens and abiotic stresses (Pastor et al., 2013). As a consequence of priming the PGPR population releases VOCs, ACC deaminase, antibiotics, siderophores, LOCs, AHLs, and lipopolypeptides, which serve as defense priming agents (Kegge et al., 2013).

Microbiome-wide association studies in Cassava (*Manihot esculenta* ssp. *flabellifolia)* have unveiled that the structural composition of the rhizosphere microbiome plays a remarkable role in developing disease resistance against *Xanthomonas axonopodis* pv. *manihotis* (*Xam*) by triggering the release of lantibiotic, nisin (lanthionine containing cyclic peptide bactericidal antibiotic). The rhizosphere has been found to be predominated by genera such as *Lactococcus* sp., *Pantoea dispersa* and *Saccharomyces cerevisiae*. Also, it contains all the genes necessary for nisin biosynthesis, like *nisC*, *nisB*, *nisR*, *nisT*, *nisK* and *nisP* (Zhang et al., 2021). Another in-depth metagenomic analysis deciphered the functional role of the bacteria *Pantoea dispersa* in exhibiting antifungal effects against pathogen *Fusarium oxysporum* and *Ceratocytis fimbriata* in rice and sweet potato, respectively (Jiang et al., 2019). A noteworthy mechanism of disease resistance has been observed in *Ralstonia solanacearum*, a soil-borne phytopathogenic bacterium causing bacterial wilting disease in tomato plants. *Flavobacterium* is being assigned as the predominant genera of resistant varieties as unveiled by an in-depth metagenomic analysis. The rhizosphere microbiome of resistant varieties upholds a clearly distinct variable diversity as compared to the susceptible varieties. The former is dominated by members of the genera *Pseudomonas, Achromobacter, Brachymonas, Sphingopyxis,*

Acidobacteria, Solimonas, Pedomicrobium, Geomonas, Desulfuromonas belonging to phylum Pseudomonadota, while the latter being dominated by Bacteroidota and Verrucomicrobiota (Gu et al., 2022). Further evidence of disease resistance is provided by developing disease-suppressive soils, which are much more tolerant than conductive or susceptible soils. Metagenomic studies of wheat field soils from the Netherlands and Germany have revealed antifungal effects against root rot of wheat fungus, *Fusarium culmorum,* by the production of biosynthetic gene clusters (BGCs) most prominently nonribosomal peptide synthetases (NRPSs) and polyketide synthases (PKSs).

Metagenomic analysis has revealed an indispensable role of siderophore-producing bacteria, most notably pyoverdine from *Pseudomonas*, scabichelin from *Streptomyces*, and delftibactin from *Delftia* exhibiting antifungal activity in the rhizosphere of suppressive soils. Apart from these genera, siderophore biosynthesis is also increased by a few other predominant ones residing in the rhizosphere including *Variovorax*, *Burkholderia*, and *Collimonas* (Tracanna et al., 2021). There have been reports of *Janthinobacterium* displaying antagonistic properties against soil-borne pathogens like *Pythium ultimum, Rhizoctonia solani* AG8, and *R. oryzae* (Li et al., 2021; Yin et al., 2021). Bacteria belonging to Comamonadaceae, *Pseudomonas, and Stenotrophomonas* have been found to be abundant in the rhizosphere of cucumber plants, thereby exhibiting antifungal antagonism against *Fusarium* wilt (Li et al., 2021; Wen et al., 2020). Metagenomic insights into the rhizosphere of wheat plants demonstrate the storehouse of an immensely varied collection of bacteria belonging to the genera *Chitinophaga, Pseudomonas, Chryseobacterium, Flavobacterium, Pedobacter, Variovorax,* and *Rhizobium* exerting antagonistic activities against the soil-borne pathogen *Rhizoctonia solani* AG8 in soil (Yin et al., 2021). Metagenomics has revealed the presence of species of *Microbacterium, Stenotrophomonas,* and *Xanthomonas* in the rhizosphere of *Arabidopsis* to impede the downy mildew pathogen *Hyaloperonospora arabidopsidis* causing foliar attack (Berendsen et al., 2018). Metagenomic investigation of the *Arabidopsis* rhizosphere showed taxa belonging to *Bacillus* and *Pseudomonas* to colonize in response to root exudates, limiting *Pseudomonas syringae* pv. *tomato* invasion in *Arabidopsis*, thus conferring resistance to foliar pathogens (Yuan et al., 2018). The PGPR upsurge in response to corresponding phytopathogens is depicted in Table 23.1.

Table 23.1 PGPR in response to phytopathogens

Phytopathogen	Rhizosphere PGPR impeding pathogens
Ralstonia solanacearum	*Flavobacterium* sp.
Fusarium oxysporum	*Pantoea dispersa*
Rhizoctonia solani AG8	*Pseudomonas, Variovorax, Chitinophaga, Rhizobium* spp.
Pythium ultimum	*Janthinobacterium* sp.

23.5.2 *Amelioration of Abiotic Stress Response*

Like biotic stresses, plants are also subjected to several abiotic stresses to which they develop tolerance and resistance. Abiotic stress responses may occur due to nitrogen delimitation, iron deficiency chlorosis, salinity stress, and toxic effects of heavy metals and chilling stress. The role of metagenomics in understanding these has been discussed in this chapter's sub-sections.

23.5.2.1 Delimitation of Nitrogen Deficiency

Nitrogen is an essential macronutrient and is of utmost importance for a number of significant physiological and biological processes in plants; hence, its deficiency leads to detrimental effects on plant growth and metabolism. Nitrogen is available in the soil, and plants take it as ammonium (NH^{4+}) or nitrate (NO^{3-}) in the presence of several nitrifying bacteria. However, sometimes, the inadequacy of nitrogen emphasizes the need for nitrogenous fertilizers. The global nitrogen cycle consists of a conversion of ammonia or ammonium (NH^{3+}/NH^{4+}) ion to nitrite (NO^{2-}) catalyzed by ammonia-oxidizing bacteria (AOB) like *Nitrosomonas* and *Nitrospira* and further nitrite to nitrate (NO^{3-}) by nitrite-oxidizing bacteria (NOB) like *Nitrospira* and *Nitrobacteria*. *Nitrospira* is also a commamox that can directly convert ammonia to nitrate.

Typical inhibitors of the nitrogen cycle, dicyandiamide (DCD) and N-(n-butyl) thiophosphoric triamide (NBPT), which are potential nitrification and urease inhibitors, respectively are chief constituents of nitrogen fertilizers. Both DCD and NBPT are supposed to lessen the ammonium monoxygenase (*amo*) gene copy number. Metagenomic analysis has revealed that the switchgrass (*Panicum virgatum* L.) microbiome is predominated by members of Proteobacteria (~75%), mostly *Janinthobacterium,* a potent associative nitrogen fixer. The bulk metagenome unveiled an enriched entire nitrogen fixation (*nif/anf/vnf*) cluster of the *nif* gene found in Betaproteobacteria members like *Azonexus*, *Hydrophilus*, *Herbaspirillum*, *Dechloromonas*, *Rhodocyclaceae*, and *Sulfuriferula*. A deep insight into functional metagenomic screen revealed that in the rhizospheric metagenome of switchgrass, an exaggerated and reckless amount of nitrogen fertilizer application leads to a severe reduction in the number of genes involved in the nitrogen cycle, such as those for nitrogen fixation (*nifHDK*), denitrification (*norB, nosZ, nirSK*), ammonification (*nrfA, napA, narG, nasA*), urea catabolism (*ureABC*), and anammox (*hzo*). Hence, the maintenance of a proper trade-off is of utmost importance (White III et al., 2023).

Metagenomic study on the Miyake-jima volcanic deposits of Japan provided evidence of a primary succession where the pioneer species *Mischantus condensatus* dominated the area and framed the rhizosphere microbiota. The concentration of NH^{3+} was greater than NO^{3-} which suggests that a population of dissimilatory nitrate reduction to ammonium (DNRA) and denitrifiers bacteria belonging to phyla Betaproteobacteria (*Cupriavidus* spp.), Acidobacteria, Actinobacteria, and

Chloroflexi were the most prevalent ones. A deep insight into the genetic profile of the dominant genera revealed that there was an overabundance of *nirK, norB, nrfA, nirB, nirD* genes unlike *nifH, nifD, amoA, amoB, amoC* and *hao* which thereby affirmed the occurrence of denitrifying genera and not nitrogen fixing or nitrifying bacteria in the rhizosphere microbiome (Arsyadi et al., 2023). Metagenomic analysis on different alfalfa (*Medicago sativa*) cultivars at an experimental station located in Yangling Shaanxi, China, revealed the predominance of Proteobacteria and Acidobacteria (predominantly Vicinamibacterales and Vicinamibacteraceae). This study further claimed that the total soil nitrogen content positively increased with an increase in the microbial community composition (An et al., 2022).

23.5.2.2 Mitigating the Adverse Effects of Iron Deficiency Chlorosis

The basic needs for plant growth include optimal proportions of micro and macronutrients, water, sunlight, fertile soil, and rainfall for better production of good quality and quantity of crop. In this context, those plants grown in places with a poor drainage system often face the problem of water logging. This directly affects the aeration capacity of the root due to the low availability of oxygen content in the soil, which is obligatory for the growth of aerobic microbes in the root zone in the immediate vicinity of the soil.

Apart from delimiting oxygen, the water-logged soils also directly hurt the iron uptake. Since iron is an essential immobile micronutrient required for chlorophyll synthesis and the photophosphorylation process, so even a tiny deficiency causes significant physiological and metabolic imbalances in plant growth and development. Interveinal leaf chlorosis in the younger leaves is the most common symptom of iron deficiency in the leaves. Besides iron deficiency, chlorosis also occurs due to other elemental deficiencies like manganese, zinc, and calcium, all delimited in the high soil pH. Many factors account for the availability and solubility of iron in the soil, including soil pH, aerobic conditions, and soil redox potential.

A thorough metagenomic analysis of the Asian pear plants (*Pyrus pyrifolia* Nakai) of Xiao Xian County of Anhui province in China revealed a comparative difference between rhizospheric soils of normal and chlorotic plants. Metagenomic data insights of the chlorotic plants revealed *feoAB* (high affinity ferrous [Fe^{2+}] iron transporter), *FTR1* (high-affinity iron permease), and *fet3p* (multicopper oxidase) gene curtailment. The chlorotic plants displayed significant changes in the leaf morphology and other physicochemical properties, including a smaller and lesser number of chloroplasts, decreased size of guard cells, enlarged epidermal cells, and large vacuoles. Hence, many changes also occur in the microbial community composition to fight iron deficiency chlorosis. Since the overall oxygen content of the soil is impeded due to the iron delimitation, a direct consequence would be the switch in microbial diversity from aerobic species prevalent in the regular soil to anaerobic species in the chlorotic soil. A detailed metagenomic microbiome analysis provided a discernment regarding the species diversity of the rhizospheric zone of the normal and chlorotic plants. Of several phyla, the most prevalent genera in decreasing abundance in both zones included *Nitrospira defluvii*, *Gemmatirosa kalamazoonesis*, and *Sulfuricella denitrificans*. In addition, chlorotic soil also unveiled the occurrence of two most predominant anaerobic genera, *Candidatus* Koribacter versatilis and *Candidatus* Nitrosoarchaeum koreensis, whereas *Candidatus* Solibacter usitatus was prevalent in the rhizosphere of normal soils.

To facilitate Fe^{2+} availability, an adequate number of genes responsible for iron uptake, storage, and transport is necessary and is sufficiently present in the genomes of plants locally grown in normal soils. However, the same is non-existent in the genomes of chlorotic plants as deciphered through metagenomics of the rhizospheric soil of both zones. Another important metagenomic finding revealed the presence of both ferrireductase and ferrioxidase genes in the genome of normal plants, which is essential for maintaining iron homeostasis in plants. The most absorbable, non-toxic form of iron is ferric iron (Fe^{3+}), which is reduced into ferrous (Fe^{2+}) form in the presence of ferrireductase enzyme. However, the ferrous (Fe^{2+}) form, being toxic to the plants, needs to be oxidized by ferritin to a ferric (Fe^{3+}) state to prevent disruption of redox misbalance (Jia et al., 2022).

FER-like iron deficiency-induced transcription factor (FIT) is expressed to counter rhizosphere acidification in iron-deficient conditions, regulating the enzymes necessary to minimize such iron deficiencies (Molnár et al., 2023). A remarkable report of attenuating iron deficiency chlorosis through significant production of a chelating agent desferrioxamine by the PGPR *Rahnella aquatilis* JZ-GX1 has been demonstrated (Kong et al., 2022) in *Cinnamomum camphora*, which significantly helped in easing the deficit by an increased diversified collection of fungal genera such as *Glomus*, *Mortierella*, *Trichoderma*, and *Penicillium* (Kong et al., 2022).

23.5.2.3 Scaling Down Unfavourable Responses to Salinity Stress

The rhizosphere microbiomes significantly influence the growth and performance of crops. One of the most critical challenges impeding agricultural production and food security is soil salinization (Wang et al., 2023). Excessive salt in the soil can cause oxidative stress, nutritional imbalances, organ senescence, and other undesirable effects that might hinder plant development and productivity. Thus, an appropriate methodology is required to identify the precise functions of rhizospheric microorganisms in mitigating plant salt stress. PGPR has been shown to enhance plant growth and production when exposed to salt stress (Wang et al., 2023). For instance, tomatoes' resistance to a 10% salt stress was increased when the maize rhizosphere bacterium *Bacillus sp.* MT7 was introduced into the tomato rhizosphere (Wang et al., 2023). Furthermore, liquorice grew more successfully when rhizobia and *Pseudomonas* were applied together under salt stress (Wang et al., 2023).

A few recent studies have found that salt-resistant peanut cultivars had higher relative abundances of beneficial bacteria, including Cyanobacteria and Proteobacteria. In contrast, salt-sensitive cultivars had lower abundances of these bacteria. In saline soil, Alphaproteobacteria are the major contributors to nitrogen fixation (Yousuf et al., 2014), and Gammaproteobacteria are also considered denitrifiers and nitrogen fixers (Baskaran & Prabavathy, 2022). These key species are the dominant nitrogen-fixing phyla in saline environments. Gammaproteobacteria have been demonstrated to participate in sulfur oxidation and inorganic carbon fixation in saline soil. Soil salinization has been reported to decrease the relative abundance of Solirubrobacterales significantly.

Furthermore, by generating exopolysaccharides or aminocyclopropane-1-carboxylate deaminase (ACCD) (Mathew et al., 2020), *Actinobacteria* in the rhizosphere allowed plants to adapt to salt marsh conditions. By breaking down the precursor molecule ACC into α-ketobutyrate and ammonia, ACCD can shield plants from salt stress by reducing the synthesis of the stress hormone, ethylene (Wang et al., 2023). Plants stressed by salt can undoubtedly benefit from the assistance of rhizosphere-associated microorganisms. Rhizospheric microbiomes significantly influence the growth and productivity of crops. Therefore, exploring the processes behind species coexistence in these impacted microbiomes, and the interactions between plants and microbiomes at the root-soil interface, may open up new possibilities for utilizing native rhizobiomes to support crop productivity and soil health.

23.5.2.4 Minimizing the Responses to Chilling Stress

Chilling is the exposure of plants to temperatures above freezing (0 °C) but below the optimum temperature, which causes significant morphological and physiological disturbances in the plant. In addition, the dissimilarity is based on the plasma membrane lipid composition of saturated fatty acids for chilling sensitive plants (CSP) and unsaturated fatty acids in chilling resistant plants (CRP). Besides, there also exists a discrepancy based upon inoculation of PGPR *Burkholderia phytofirmans* strain PsJN into *Vitis vinifera* L. cv. which enabled a heightened tolerance to cold stress, mostly in chilling-resistant plants (Ait Barka et al., 2006). Metagenomic study on vascular plants *Deschampsia antarctica* and *Colobanthus quitensis* from Antarctica demonstrated Pseudomonas, Serratia, and Staphylococcus genera as predominant ones occurring in the rhizosphere thus providing pronounced tolerance to cold stress in CSP as well as CRP (Araya et al., 2020).

23.5.2.5 Ameliorating Toxic Effects of Heavy Metals

A class of metallic and metalloid elements with comparatively high densities (> 5 g/cm^3), atomic weights, and atomic numbers are known as heavy metals (Prieto et al., 2018). They represent a long-term hazard to the ecosystem. Therefore, remediation is required to minimize heavy metal pollution (Yan et al., 2020). Plants are used in phytoremediation to remove and reduce metal contaminants in the environment, which could decrease their bioavailability in the soil. Plants may take up ions from the soil through the root system, even in tiny amounts, allowing them to recover contaminated soil and stabilize soil fertility (Bhat et al., 2022).

Table 23.2 Mitigation of abiotic stresses, and corresponding up-regulated genes

N delimitation	Fe deficiency, chlorosis	Salinity	Heavy metal	Chilling
nif HDK, norB, hzo, nrfA, amoA, amoB, amoC	*feoAB, FTR1*	*ACCD* (Aminocyclopropane-1 carboxylate deaminase)	*czcA, arsH, copA, merC, merT, merE*	Cold tolerant hyper ACC degrading bacteria

Heavy metals tend to accumulate in the food chain at the topmost consumer level, resulting in bioaccumulation and biomagnification. A metagenomic and genomic approach-based study on *Avicennia germinans* from a semi-arid mangrove of La Guajira, Colombia, unveiled several heavy metal tolerance (HMT) and heavy metal resistance (HMR) genes from the rhizosphere of mangroves. These ecosystems are areas near marine coastal regions, thus dominated by halophiles. In general, HMT genes reported included *cusA*, *silA* (copper/silver efflux system protein belonging to RND family of H+ antiporters and helps in Cu transport from the cytoplasm to the periplasm), and efflux pumps ATP Binding cassette (ABC transporter). Yet, a bulk amount of genes were related to HMR, such as *copA*, *ctpA* (P-type Cu^{+} transporter), *czcA* (Co, Zn, Cd efflux system protein), *arsH* (arsenical resistance protein), *narJ*, *narW* (nitrate reductase molybdenum cofactor assembly chaperone associated with molybdenum reduction). Functional metagenomic screening demonstrated the presence of HMT and HMR genes in the bacterium *Vibrio fluvialis* MG 1.1 isolated from the mangrove rhizosphere (Muñoz-García et al., 2022).

Metagenomic analysis of rhizosphereic soil of plants such as *Rumex induratus*, *Rumex bucephalophorus*, *Avena sativa*, *Medicago sativa* and *Vicia benghalensis* from a mining district in Spain, and of bulk soil as well, demonstrated a distinction in the relative abundance of the population between the two soils. The rhizospheric soil was predominated by genera belonging to the class Actinobacteria (*Kribbella* sp., *Propionibacterium acnes*), Alphaproteobacteria (*Mesorhizobium loti*), Betaproteobacteria (*Variovorax* sp.) and Gammaproteobacteria (*Pseudomonas* sp.). Acidobacteria and Cyanobacteria were dominant in the bulk soil (González et al., 2022). Mitigation of different forms of abiotic stresses and the corresponding genetic upregulation and significant responses have been summarized in Table 23.2.

23.6 Bidirectional Rhizosphere Communication System

The discreteness of microbial consortia within the rhizosphere microbiome depends on the chemical profile of root exudate (Saleem et al., 2018). The beneficial or pathogenic makeup of the microbial communities in the vicinity is contingent on the intermolecular cascade circuit that alluringly operates within. The divergent heterogeneity of the discharge includes several low and high-molecular weight compounds like fatty acids, proteins, mucilage, flavonoids, etc. (More et al., 2020). All these compounds serve as chemical attractants, establishing symbiotic, parasitic, or antagonistic relationships between the host plant and the microbiota. The gene expression in bacteria is regulated by the signal molecules that serve as ligands for bacterial receptor proteins. Volatiles like VOCs, including phenylpropanoids, terpenoids, fatty acids, and strigolactones, are primarily released from the phyllosphere (above ground parts of plants) (Nakayasu et al., 2023). Cucumber secretes citric acid, and bananas secrete fumaric acid, initiating biofilm formation accompanied by *Bacillus amyloliquefaciens* and *B. subtilis* N11, respectively (Jain et al., 2020).

Another antagonistic association has been noticed in broccoli against the rhizospheric microbes *Pseudomonas syringae, Sphingomonas suberifaciens*, and *Fusarium oxysporum* aided by isothiocyanates (ITC) and glucosinolates (GLS), a significant secretion of root exudate (Jamil et al., 2022). However, the big picture involves further interspecies communication between microbe-microbe and an interkingdom association between the plant and the microbe exerted by several signaling intermediates like lipo-chitooligosaccharides, volatile organic compounds, quorum sensing molecules, autoinducers, N-acyl homoserine lactone (Jamil et al., 2022). Hence, such complex interassociation between the interacting partners reveals the importance of metagenomics in opening up all the covet gates and fruitfully establishing such a relationship that can be agronomically useful (Jamil et al., 2022).

Root exudates such as flavones have been found to enable nitrogen acquisition, improve the growth of maize plants, and render protection against root microbes in association with bacteria from Oxalobacteriaceae, particularly *Janinthobacterium* sp. (Yin et al., 2021). There also have been reports of ROS production by FERONIA mediated receptor kinase that gradually increases the population of *Pseudomonas* in the rhizosphere microbiome (Song et al., 2021). Barley plants also control their rhizosphere community by antagonizing *Fusarium graminearum,* and recruiting antifungal microbes by reconstructing their root exudates (Li et al., 2021).

Studies have revealed significant volatiles (hexenal isomers and 2, 3 -butanediol) to be released due to *Erwinia amylovora* infection of apple trees, which subsequently leads to salicylic acid synthesis from healthy plants (Cellini et al., 2018). Another finding reported that β-caryophyllene, from *Bacillus amyloliquefaciens* GB03, released sufficient amounts of salicylic acid from tomato seedlings' root exudate, thereby eliciting salicylic acid responses against pathogens (Kong et al., 2021). A detailed metagenomic study has demonstrated protection against sudden-wilt fungal disease through inoculating *Nicotiana attenuata* seeds with a rhizosphere bacterial community that elicits antifungal effects (Santhanam et al., 2015).

The stable root exudate breakdown product, 6-methoxy-benzoxazolin-2-one (MBOA), of benzoxazinoids (BX) in maize and several other members of Poaceae has been known to serve as defensive signals against *Spodoptera frugiperda* caterpillar. A high concentration of salicylic acid and jasmonic acid has been reported in BX+ soils compared to BX- soils. High throughput metagenomic study has shown that the BX+ soils are abundant in Proteobacteria species, most importantly *Pseudomonas putida*, which triggers ISR against maize anthracnose fungus *Colletotrichum graminicola* (Hu et al., 2018).

23.7 Reciprocal Exchanges Amongst Interacting Partners

The microbial community exhibits three major interactions: antagonism, amensalism, and parasitism (Begum et al., 2023; Mishra et al., 2022). Arbuscular mycorrhizal fungi (AMF) and PGPR are the most relevant components of soil microbiota (Afridi et al., 2022). Since the environment is pretty much different in the rhizosphere as compared to the bulk soil, therefore the microbiome, which is entirely composed of bacteria, fungi, oomycetes, nematodes, viruses, archaea, and arthropods is also distinct (Kim et al., 2021; Zhou et al., 2020). However, the abundance of bacteria is most noteworthy, as revealed through metagenomics.

Regarding the niche differences, the phylogenetic diversity prevails by oligotrophs (organisms dominating in nutrient-poor environments), which are K-strategists (organisms with low growth rate of increase) relying on minimal nutrients in the environment and occur in the bulk soil (the sporulating bacteria being most common) (Ling et al., 2022; López et al., 2022). In contrast, copiotrophs (organisms with an utmost nutritional requirement) tend to depend on the nutrients (the root exudate) significantly and exhibit high growth and metabolic rates (i.e., r-strategists that reside in the rhizospheric soil) (Ling et al., 2022; López et al., 2022). They include Bacteriodetes, Proteobacteria, Actinobacteria, Nitrospirae, Acidobacteria, and Chloroflexi (Kim et al., 2021; Ling et al., 2022).

Quorum sensing (QS) is carried out to strengthen the microbial community signaling, and quorum quenching (QQ) is done to restrain the pathogen communication attained by molecules like (Kamath et al., 2022; Zhang et al., 2016) auto-inducer synthase, acyl homoserine lactones (AHLs) (in Gram-negative bacteria) (Zhang & Li, 2016) and auto-inducing peptides (AIPs) (in Gram-positive bacteria) (Bhatt, 2018). A schematic representation of QS and QQ in growth promotion and restraining plant pathogens is depicted in Fig. 23.2.

With the advent of a constant co-evolutionary arms race between the host and the pathogen, both the interacting species seem to show a dynamic transitional alteration in their interaction, with the host trying to develop new ways to minimize the pathogens' fitness cost (Baltrus, 2017). Four sets of genes carry out this kind of reciprocal relationship; resistance genes (*R*), consist of a nucleotide binding domain (NB) and a leucine rich repeat (LRR), susceptible gene (*r*) and also the avirulent gene (*A*) and virulent gene (*a*) occurring in the plant and pathogen, respectively (Cook et al., 2015). As per the gene-for-gene hypothesis, a disease is said to occur only when at least one gene, either from the plant or the pathogen is carrying the recessive allele (Cook et al., 2015). The plant builds up immunity against the phytopathogens through a number of mechanisms or signaling pathways. Susceptibility or virulence relationship among the pathogen and plant and the probability of disease development in presence of at least one recessive gene is depicted in Table 23.3.

Genes coding for *A* (avirulent, dominant) or *a* (virulent, recessive) factors are present in the pathogen, whereas *R* (resistant, dominant) or *r* (susceptible, recessive) genes occur in the host plant. For successful disease development the pathogen must be virulent, the host must be susceptible and the environment should be favorable for the disease.

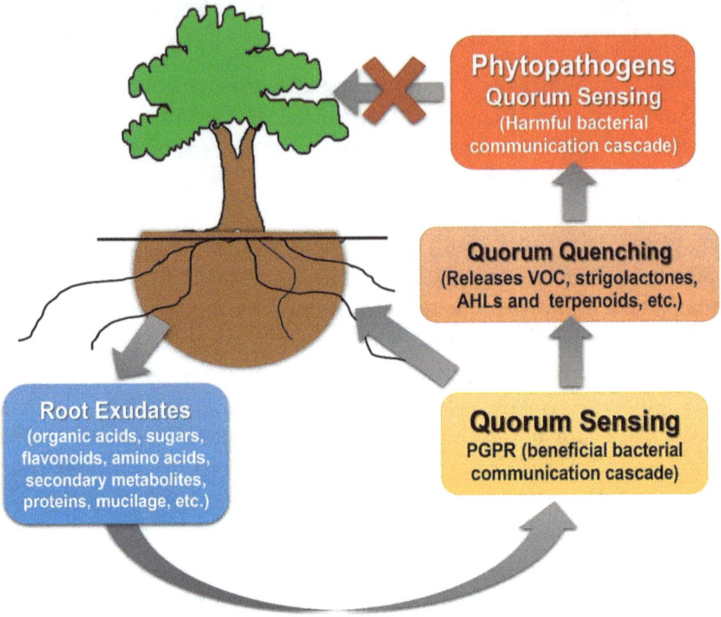

Fig. 23.2 A two-way rhizosphere communication system mediated by the chemical attractants of root exudates leading to quorum sensing and subsequent colonization of PGPR which thereby results in release of certain compounds that assists in quorum quenching in harmful phytopathogens.

Table 23.3 Interaction among host plant gene and pathogen gene

Pathogen gene[a]	Host plant gene[b]	Combined effect	Disease
A	R	*AR* (resistant)	No
a	r	*Ar* (susceptible)	Yes
A	r	*Ar* (susceptible)	Yes
a	R	*aR* (susceptible)	Yes

[a] *A* = avirulent, dominant; *a* = virulent, recessive
[b] *R* = resistant, dominant; *r* = susceptible, recessive

23.8 Challenges

As everything has pros and cons, the molecular techniques and the omics approach are also nothing exceptional. Representing the entire rhizospheric ecosystem through a small sample is challenging and tedious. The dispute between screening culturable and non-culturable organisms is yet another issue. Translating, interpreting, and decoding the functions of novel genes is very obscure. There are many pitfalls in the screening procedure as such studies demand huge preciseness, sophistication, and a robust economic backup. Despite all odds, we have the omics branch in our hands for assistance. When we want to derive a complete picture of the analyzed microbiome, we often tend to restrict ourselves to deciphering the information concerning the most prominent and the most abundant genera; thereby, the least abundant species are frequently left unexploited (Ekschmitt & Griffiths, 1998).

Comprehensive bioprospecting of the rhizosphere microbiome flora, aided in alliance with meta transcriptomics and proteomics, depicts a wholesome picture of the area, which needs to be analyzed since they also offer functional attributes and all these perspectives unless considered, the entire screening process would remain abridged (Priya et al., 2021). Since the bioinformatics aspect is also wrapped around the metagenomic data analysis, the technical and economic feasibility and labour intensiveness should be considered, which can sometimes become impractical considering the enormous data generated. Moreover, unsheared quality DNA extraction demands advanced commercial kits and the removal of humic acid, the most prevalent soil contaminant (Barra Caracciolo & Terenzi, 2021). To the greatest extent, in vitro application and the actual aim of the metagenome analysis is to procure maximum infield application in agricultural soils, which often poses many complications and needs to be sorted out.

References

Abd-Elgawad, M. M. (2021). Optimizing safe approaches to manage plant-parasitic nematodes. *Plants, 10*, 1911.

Afridi, M. S., Fakhar, A., Kumar, A., Ali, S., Medeiros, F. H., Muneer, M. A., Ali, H., & Saleem, M. (2022). Harnessing microbial multitrophic interactions for rhizosphere microbiome engineering. *Microbiological Research, 265*, 127199.

Ait Barka, E., Nowak, J., & Clément, C. (2006). Enhancement of chilling resistance of inoculated grapevine plantlets with a plant growth-promoting rhizobacterium, *Burkholderia phytofirmans* strain PsJN. *Applied and Environmental Microbiology, 72*, 7246–7252. https://doi.org/10.1128/AEM.01047-06

Alves, L. F., Westmann, C. A., Lovate, G. L., de Siqueira, G. M. V., Borelli, T. C., & Guazzaroni, M.-E. (2018). Metagenomic approaches for understanding new concepts in microbial science. *International Journal of Genomics, 2018*, 2312987.

An, Y., Sun, H., Zhang, W., Sun, Y., Li, S., et al. (2022). Distinct rhizosphere soil responses to nitrogen in relation to microbial biomass and community composition at initial flowering stages of alfalfa cultivars. *Frontiers in Plant Science, 13*, 938865.

Araya, M. A., Valenzuela, T., Inostroza, N. G., Maruyama, F., Jorquera, M. A., & Acuña, J. J. (2020). Isolation and characterization of cold-tolerant hyper-ACC-degrading bacteria from the rhizosphere, endosphere, and phyllosphere of Antarctic vascular plants. *Microorganisms, 8*, 1788.

Arsyadi, A., Guo, Y., Ebihara, A., Sakagami, N., Sakoda, M., Tago, K., et al. (2023). A nitrate-transforming bacterial community dominates in the miscanthus rhizosphere on nitrogen-deficient volcanic deposits of Miyake-jima. *Microorganisms, 11*, 260.

Backer, R., Rokem, J. S., Ilangumaran, G., Lamont, J., Praslickova, D., Ricci, E., et al. (2018). Plant growth-promoting Rhizobacteria: Context, mechanisms of action, and roadmap to commercialization of biostimulants for sustainable agriculture. *Frontiers in Plant Science, 9*, 1473.

Baltrus, D. A. (2017). Adaptation, specialization, and coevolution within phytobiomes. *Current Opinion in Plant Biology, 38*, 109–116.

Barra Caracciolo, A., & Terenzi, V. (2021). Rhizosphere microbial communities and heavy metals. *Microorganisms, 9*, 1462.

Baskaran, V., & Prabavathy, V. (2022). Diverse key nitrogen cycling genes *nifH*, *nirS* and *nosZ* associated with Pichavaram mangrove rhizospheres as revealed by culture-dependent and culture-independent analyses. *Archives of Microbiology, 204*, 109.

Begum, N., Zhao, T., Antwi-Boasiako, A., Zhang, L., & Ashraf, M. (2023). Role of strigolactones in rhizobiology: Plant-microbe interactions. In R. A. Bhat, M. Ashraf, F. M. Policarpo Tonelli, B. T. Unal, & G. H. Dar (Eds.), *Ozturk M* (pp. 13–25). Elsevier.

Berendsen, R. L., Vismans, G., Yu, K., Song, Y., de Jonge, R., Burgman, W. P., et al. (2018). Disease-induced assemblage of a plant-beneficial bacterial consortium. *The ISME Journal, 12*, 1496–1507.

Berg, G., Rybakova, D., Fischer, D., Cernava, T., Vergès, M.-C. C., et al. (2020). Microbiome definition re-visited: Old concepts and new challenges. *Microbiome, 8*, 103.

Bhat, S. A., Bashir, O., Haq, S. A. U., Amin, T., Rafiq, A., Ali, M., Américo-Pinheiro, J. H. P., & Sher, F. (2022). Phytoremediation of heavy metals in soil and water: An eco-friendly, sustainable and multidisciplinary approach. *Chemosphere, 303*, 134788.

Bhatt, V. S. (2018). Quorum sensing mechanisms in gram positive bacteria. In P. V. Bramhachari (Ed.), *Implication of quorum sensing system in biofilm formation and virulence* (pp. 297–311). Springer.

Breitwieser, F. P., Lu, J., & Salzberg, S. L. (2019). A review of methods and databases for metagenomic classification and assembly. *Briefings in Bioinformatics, 20*(4), 1125–1136.

Bukhat, S., Imran, A., Javaid, S., Shahid, M., Majeed, A., & Naqqash, T. (2020). Communication of plants with microbial world: Exploring the regulatory networks for PGPR mediated defense signaling. *Microbiological Research, 238*, 126486.

Cellini, A., Buriani, G., Rocchi, L., Rondelli, E., Savioli, S., Rodriguez Estrada, M. T., Cristescu, S. M., Costa, G., & Spinelli, F. (2018). Biological relevance of volatile organic compounds emitted during the pathogenic interactions between apple plants and *Erwinia amylovora*. *Molecular Plant Pathology, 19*(1), 158–168.

Chagas, F. O., de Cassia, P. R., Caraballo-Rodriguez, A. M., & Pupo, M. T. (2018). Chemical signaling involved in plant–microbe interactions. *Chemical Society Reviews, 47*(5), 1652–1704.

Chandran, H., Meena, M., & Swapnil, P. (2021). Plant growth-promoting Rhizobacteria as a green alternative for sustainable agriculture. *Sustainability, 13*(19), 10986.

Cloete, T. E., & Muyima, N. Y. O. (1997). *Microbial community analysis*. IWA Publishing.

Connon, S. A., & Giovannoni, S. J. (2002). High-throughput methods for culturing microorganisms in very-low-nutrient media yield diverse new marine isolates. *Applied and Environmental Microbiology, 68*(8), 3878–3885.

Conrath, U., Beckers, G. J., Langenbach, C. J., & Jaskiewicz, M. R. (2015). Priming for enhanced defense. *Annual Review of Phytopathology, 53*, 97–119.

Cook, D. E., Mesarich, C. H., & Thomma, B. P. (2015). Understanding plant immunity as a surveillance system to detect invasion. *Annual Review of Phytopathology, 53*, 541–563.

Dubey, A., Malla, M. A., Khan, F., Chowdhary, K., Yadav, S., Kumar, A., Sharma, S., Khare, P. K., & Khan, M. L. (2019). Soil microbiome: A key player for conservation of soil health under changing climate. *Biodiversity and Conservation, 28*, 2405–2429.

Ekschmitt, K., & Griffiths, B. (1998). Soil biodiversity and its implications for ecosystem functioning in a heterogeneous and variable environment. *Applied Soil Ecology, 10*(3), 201–215.

Espinas, N. A., Saze, H., & Saijo, Y. (2016). Epigenetic control of defense signaling and priming in plants. *Frontiers in Plant Science, 7*, 1201.

Fan, K., Weisenhorn, P., Gilbert, J. A., & Chu, H. (2018). Wheat rhizosphere harbors a less complex and more stable microbial co-occurrence pattern than bulk soil. *Soil Biology and Biochemistry, 125*, 251–260.

Garbeva, P., Van Elsas, J., & Van Veen, J. (2008). Rhizosphere microbial community and its response to plant species and soil history. *Plant and Soil, 302*, 19–32.

González, D., Robas, M., Fernández, V., Bárcena, M., Probanza, A., & Jiménez, P. A. (2022). Comparative metagenomic study of rhizospheric and bulk mercury-contaminated soils in the mining district of Almadén. *Frontiers in Microbiology, 13*, 797444.

Gu, Y., Banerjee, S., Dini-Andreote, F., Xu, Y., Shen, Q., Jousset, A., & Wei, Z. (2022). Small changes in rhizosphere microbiome composition predict disease outcomes earlier than pathogen density variations. *The ISME Journal, 16*(10), 2448–2456.

Gupta, R., Khan, F., Alqahtani, F. M., Hashem, M., & Ahmad, F. (2023). Plant growth-promoting Rhizobacteria (PGPR) assisted bioremediation of heavy metal toxicity. *Applied Biochemistry and Biotechnology*. https://doi.org/10.1007/s12010-023-04545-3

Handelsman, J. (2004). Metagenomics: Application of genomics to uncultured microorganisms. *Microbiology and Molecular Biology Reviews, 68*(4), 669–685.

Herreño, B., De la Colina, F., & Delgado-Iniesta, M. J. (2023). Edaphosphere: A perspective of soil inside the biosphere. *Earth, 4*(3), 691–697.

Hönig, M., Roeber, V. M., Schmülling, T., & Cortleven, A. (2023). Chemical priming of plant defense responses to pathogen attacks. *Frontiers in Plant Science, 14*. https://doi.org/10.3389/fpls.2023.1146577

Hu, L., Robert, C. A., Cadot, S., Zhang, X., Ye, M., Li, B., Manzo, D., Chervet, N., Steinger, T., & Van Der Heijden, M. G. (2018). Root exudate metabolites drive plant-soil feedbacks on growth and defense by shaping the rhizosphere microbiota. *Nature Communications, 9*(1), 2738.

Huson, D. H., & Weber, N. (2013). *Microbial community analysis using MEGAN* (pp. 465–485). Elsevier.

Jain, A., Chakraborty, J., & Das, S. (2020). Underlying mechanism of plant–microbe crosstalk in shaping microbial ecology of the rhizosphere. *Acta Physiologiae Plantarum, 42*, 1–13.

Jamil, F., Mukhtar, H., Fouillaud, M., & Dufossé, L. (2022). Rhizosphere signaling: Insights into plant–rhizomicrobiome interactions for sustainable agronomy. *Microorganisms, 10*(5), 899.

Jia, B., Chang, X., Fu, Y., Heng, W., Ye, Z., Liu, P., Liu, L., Al Shoffe, Y., Watkins, C. B., & Zhu, L. (2022). Metagenomic analysis of rhizosphere microbiome provides insights into occurrence of iron deficiency chlorosis in field of Asian pears. *BMC Microbiology, 22*(1), 1–16.

Jiang, L., Jeong, J. C., Lee, J.-S., Park, J. M., Yang, J.-W., Lee, M. H., Choi, S. H., Kim, C. Y., Kim, D.-H., & Kim, S. W. (2019). Potential of *Pantoea dispersa* as an effective biocontrol agent for black rot in sweet potato. *Scientific Reports, 9*(1), 16354.

Jiao, X., Takishita, Y., Zhou, G., & Smith, D. L. (2021). Plant associated Rhizobacteria for biocontrol and plant growth enhancement. *Frontiers in Plant Science, 12*. https://doi.org/10.3389/fpls.2021.634796

Jiao, S., Chen, W., & Wei, G. (2022). Core microbiota drive functional stability of soil microbiome in reforestation ecosystems. *Global Change Biology, 28*(3), 1038–1047.

Kamath, A., Shukla, A., & Patel, D. (2022). Quorum sensing and quorum quenching: Two sides of the same coin. *Physiological and Molecular Plant Pathology, 101927*.

Katzke, N., Knapp, A., Loeschcke, A., Drepper, T., & Jaeger, K.-E. (2017). Novel tools for the functional expression of metagenomic DNA. *Metagenomics: Methods and Protocols, 1539*, 159–196.

Kegge, W., Gankema, P., & Pierik, R. (2013). *Plant-produced volatile organic compounds*. Access Science.

Khan, N., Bano, A., Rahman, M. A., Guo, J., Kang, Z., & Babar, M. A. (2019). Comparative physiological and metabolic analysis reveals a complex mechanism involved in drought tolerance in chickpea (Cicer arietinum L.) induced by PGPR and PGRs. *Scientific Reports, 9*(1), 2097. https://doi.org/10.1038/s41598-019-38702-8

Khan, N., Bano, A., Ali, S., & Babar, M. A. (2020). Crosstalk amongst phytohormones from planta and PGPR under biotic and abiotic stresses. *Plant Growth Regulation, 90*, 189–203.

Kim, H., Jeon, J., Lee, K. K., & Lee, Y.-H. (2021). Compositional shift of bacterial, archaeal, and fungal communities is dependent on trophic lifestyles in Rice Paddy soil. *Frontiers in Microbiology, 12*, 719486.

Kong, H. G., Song, G. C., Sim, H.-J., & Ryu, C.-M. (2021). Achieving similar root microbiota composition in neighbouring plants through airborne signalling. *The ISME Journal, 15*(2), 397–408.

Kong, W.-L., Wang, Y.-H., Lu, L.-X., Li, P.-S., Zhang, Y., & Wu, X.-Q. (2022). *Rahnella aquatilis* JZ-GX1 alleviates iron deficiency chlorosis in *Cinnamomum camphora* by secreting desferrioxamine and reshaping the soil fungal community. *Frontiers in Plant Science, 13*. https://doi.org/10.3389/fpls.2022.960750

Kudoyarova, G., Arkhipova, T., Korshunova, T., Bakaeva, M., Loginov, O., & Dodd, I. C. (2019). Phytohormone mediation of interactions between plants and non-symbiotic growth promoting bacteria under edaphic stresses. *Frontiers in Plant Science, 10*, 1368. https://doi.org/10.3389/fpls.2019.01368

Kui, L., Chen, B., Chen, J., Sharifi, R., Dong, Y., Zhang, Z., & Miao, J. (2021). A comparative analysis on the structure and function of the *Panax notoginseng* rhizosphere microbiome. *Frontiers in Microbiology, 12*. https://doi.org/10.3389/fmicb.2021.673512

Lareen, A., Burton, F., & Schäfer, P. (2016). Plant root-microbe communication in shaping root microbiomes. *Plant Molecular Biology, 90*, 575–587.

Li, J., Wang, C., Liang, W., & Liu, S. (2021). Rhizosphere microbiome: The emerging barrier in plant-pathogen interactions. *Frontiers in Microbiology, 12*, 772420.

Ling, N., Wang, T., & Kuzyakov, Y. (2022). Rhizosphere bacteriome structure and functions. *Nature Communications, 13*(1), 836.

López, J. L., Pappas, N., Poppeliers, S. W., Sanchez-Gil, J. J., Fourie-Fouche, A., de Jonge, R., & Dutilh, B. E. (2022). Copiotrophs dominate rhizosphere microbiomes and growth rate potential is a major factor explaining the rhizosphere effect. bioRxiv: 2022.2011.2024.517860.

Lucas, J. A., Garcia-Villaraco, A., Montero-Palmero, M. B., Montalban, B., Ramos Solano, B., & Gutierrez-Mañero, F. J. (2023). Physiological and genetic modifications induced by plant-growth-promoting rhizobacteria (PGPR) in tomato plants under moderate water stress. *Biology, 12*(7), 901.

Martin, B. D., & Schwab, E. (2012). Current usage of symbiosis and associated terminology. *International Journal of Biology, 5*(1), 32–45.

Mathew, B. T., Torky, Y., Amin, A., Mourad, A.-H. I., Ayyash, M. M., et al. (2020). Halotolerant marine rhizosphere-competent actinobacteria promote *Salicornia bigelovii* growth and seed production using seawater irrigation. *Frontiers in Microbiology, 11*, 552.

Mauch-Mani, B., Baccelli, I., Luna, E., & Flors, V. (2017). Defense priming: An adaptive part of induced resistance. *Annual Review of Plant Biology, 68*(1), 485–512. https://doi.org/10.1146/annurev-arplant-042916-041132

Mekonnen, H., Kibret, M., & Assefa, F. (2022). Plant growth promoting Rhizobacteria for biocontrol of tomato bacterial wilt caused by *Ralstonia solanacearum*. *International Journal of Agronomy, 2022*, 1489637. https://doi.org/10.1155/2022/1489637

Mellidou, I., & Karamanoli, K. (2022). Unlocking PGPR-mediated abiotic stress tolerance: What lies beneath. *Frontiers in Sustainable Food Systems, 6*. https://doi.org/10.3389/fsufs.2022.832896

Mendes, R., Garbeva, P., & Raaijmakers, J. M. (2013). The rhizosphere microbiome: Significance of plant beneficial, plant pathogenic, and human pathogenic microorganisms. *FEMS Microbiology Reviews, 37*(5), 634–663. https://doi.org/10.1111/1574-6976.12028

Mira, S., Emily, M., Mougel, C., Ourry, M., & Le Cadre, E. (2022). A field indicator for rhizosphere effect monitoring in arable soils. *Plant and Soil*, 1–22.

Mishra, R., Rathore, U. S., Pandey, S., Mishra, M., Sharma, N., Kumar, S., & Tripathi, K. M. (2022). *Plant–rhizospheric microbe interactions: Enhancing plant growth and improving soil biota* (pp. 445–459). Springer.

Mohanty, P., Singh, P. K., Chakraborty, D., Mishra, S., & Pattnaik, R. (2021). Insight into the role of PGPR in sustainable agriculture and environment. *Frontiers in Sustainable Food Systems, 5*. https://doi.org/10.3389/fsufs.2021.667150

Molnár, Z., Solomon, W., Mutum, L., & Janda, T. (2023, 1945). Understanding the mechanisms of Fe deficiency in the rhizosphere to promote plant resilience. *Plants, 12*(10).

More, S., Shinde, S., & Kasture, M. (2020). Root exudates a key factor for soil and plant: An overview. *The Pharma Innovation Journal, 8*, 449–459.

Muñoz-García, A., Arbeli, Z., Boyacá-Vásquez, V., & Vanegas, J. (2022). Metagenomic and genomic characterization of heavy metal tolerance and resistance genes in the rhizosphere microbiome of *Avicennia germinans* in a semi-arid mangrove forest in the tropics. *Marine Pollution Bulletin, 184*, 114204.

Myrold, D. D., Zeglin, L. H., & Jansson, J. K. (2014). The potential of metagenomic approaches for understanding soil microbial processes. *Soil Science Society of America Journal, 78*(1), 3–10.

Nakayasu, M., Takamatsu, K., Yazaki, K., & Sugiyama, A. (2023). Plant specialized metabolites in the rhizosphere of tomatoes: Secretion and effects on microorganisms. *Bioscience, Biotechnology, and Biochemistry, 87*(1), 13–20.

Nannipieri, P., Ascher-Jenull, J., Ceccherini, M. T., Pietramellara, G., Renella, G., & Schloter, M. (2020). Beyond microbial diversity for predicting soil functions: A mini review. *Pedosphere, 30*(1), 5–17.

Olanrewaju, O. S., Ayangbenro, A. S., Glick, B. R., & Babalola, O. O. (2019). Plant health: Feedback effect of root exudates-rhizobiome interactions. *Applied Microbiology and Biotechnology, 103*, 1155–1166.

Palmieri, D., Vitullo, D., De Curtis, F., & Lima, G. (2017). A microbial consortium in the rhizosphere as a new biocontrol approach against *Fusarium* decline of chickpea. *Plant and Soil, 412*, 425–439.

Pandey, A., & Palni, L. (2007). The rhizosphere effect in trees of the Indian central Himalaya with special reference to altitude. *Applied Ecology and Environmental Research, 5*(1), 93–102.

Pare, P. W., Farag, M. A., Krishnamachari, V., Zhang, H., Ryu, C.-M., & Kloepper, J. W. (2005). Elicitors and priming agents initiate plant defense responses. *Photosynthesis Research, 85*, 149–159.

Pastor, V., Luna, E., Mauch-Mani, B., Ton, J., & Flors, V. (2013). Primed plants do not forget. *Environmental and Experimental Botany, 94*, 46–56.

Prayogo, F. A., Budiharjo, A., Kusumaningrum, H. P., Wijanarka, W., Suprihadi, A., & Nurhayati, N. (2020). Metagenomic applications in exploration and development of novel enzymes from nature: A review. *Journal of Genetic Engineering and Biotechnology, 18*(1), 1–10.

Prieto, M., Acevedo, S., Prieto, G., & González, N. T. (2018). Phytoremediation of soils contaminated with heavy metals. *Biodiversity International Journal, 2*(4), 362–376.

Priya, P., Aneesh, B., & Harikrishnan, K. (2021). Genomics as a potential tool to unravel the rhizosphere microbiome interactions on plant health. *Journal of Microbiological Methods, 185*, 106215.

Pudake, R. N., Sahu, B. B., Kumari, M., & Sharma, A. K. (2021). *Omics science for rhizosphere biology*. Springer.

Reddy, C. S., Cho, M., Kaul, T., Joeng, J. T., & Kim, K. M. (2023). *Pseudomonas fluorescens* imparts cadmium stress tolerance in *Arabidopsis thaliana* via induction of AtPCR2 gene expression. *Journal of Genetic Engineering and Biotechnology, 21*(1), 8. https://doi.org/10.1186/s43141-022-00457-7

Roslan, M. A. M., Mohamad, M. A. N., & Omar, S. M. (2017). High quality DNA from peat soil for metagenomic studies a minireview on dna extraction methods. *Science, 1*(2), 01–06.

Sabale, S. N., Suryawanshi, P. P., & Krishnaraj, P. (2019). Soil metagenomics: Concepts and applications. In *Metagenomics-basics, methods and applications*. IntechOpen.

Saeed, Q., Xiukang, W., Haider, F. U., Kučerik, J., Mumtaz, M. Z., et al. (2021). Rhizosphere bacteria in plant growth promotion, biocontrol, and bioremediation of contaminated sites: A comprehensive review of effects and mechanisms. *International Journal of Molecular Sciences, 22*(19), 10529.

Saleem, M., Law, A. D., Sahib, M. R., Pervaiz, Z. H., & Zhang, Q. (2018). Impact of root system architecture on rhizosphere and root microbiome. *Rhizosphere, 6*, 47–51.

Santhanam, R., Luu, V. T., Weinhold, A., Goldberg, J., Oh, Y., & Baldwin, I. T. (2015). Native root-associated bacteria rescue a plant from a sudden-wilt disease that emerged during continuous cropping. *Proceedings of the National Academy of Sciences, 112*(36), E5013–E5020.

Santoyo, G., Guzmán-Guzmán, P., Parra-Cota, F. I., de los Santos-Villalobos, S., Orozco-Mosqueda, M. C., & Glick, B. R. (2021). Plant growth stimulation by microbial consortia. *Agronomy, 11*(2), 219.

Sarma, B. K., Yadav, S. K., Singh, S., & Singh, H. B. (2015). Microbial consortium-mediated plant defense against phytopathogens: Readdressing for enhancing efficacy. *Soil Biology and Biochemistry, 87*, 25–33.

Sharma, R., Ranjan, R., Kapardar, R. K., & Grover, A. (2005). 'Unculturable' bacterial diversity: An untapped resource. *Current Science*, 72–77.

Sharma, P. K., Capalash, N., & Kaur, J. (2007). An improved method for single step purification of metagenomic DNA. *Molecular Biotechnology, 36*, 61–63.

Siddiqui, Z. A. (2006). PGPR: Prospective biocontrol agents of plant pathogens. In Z. A. Siddiqui (Ed.), *PGPR: Biocontrol and biofertilization* (pp. 111–142). Springer Netherlands.

Song, Y., Wilson, A. J., Zhang, X.-C., Thoms, D., Sohrabi, R., Song, S., Geissmann, Q., Liu, Y., Walgren, L., & He, S. Y. (2021). FERONIA restricts *Pseudomonas* in the rhizosphere microbiome via regulation of reactive oxygen species. *Nature Plants, 7*(5), 644–654.

Tiwari, M., Pati, D., Mohapatra, R., Sahu, B. B., & Singh, P. (2022). The impact of microbes in plant immunity and priming induced inheritance: A sustainable approach for crop protection. *Plant Stress, 4*, 100072. https://doi.org/10.1016/j.stress.2022.100072

Tracanna, V., Ossowicki, A., Petrus, M. L., Overduin, S., Terlouw, B. R., Lund, G., Robinson, S. L., Warris, S., Schijlen, E. G., & Van Wezel, G. P. (2021). Dissecting disease-suppressive rhizosphere microbiomes by functional amplicon sequencing and 10× metagenomics. *MSystems, 6*(3), e01116–e01120.

Unno, Y., & Shinano, T. (2013). Metagenomic analysis of the rhizosphere soil microbial community. *Molecular Microbial Ecology of the Rhizosphere, 1*, 1099–1103.

van der Heijden, M. G., & Schlaeppi, K. (2015). Root surface as a frontier for plant microbiome research. *Proceedings of the National Academy of Sciences of the United States of America, 112*(8), 2299–2300.

Venturi, V., & Fuqua, C. (2013). Chemical signaling between plants and plant-pathogenic bacteria. *Annual Review of Phytopathology, 51*, 17–37.

Venturi, V., & Keel, C. (2016). Signaling in the rhizosphere. *Trends in Plant Science, 21*(3), 187–198.

Wang, J., Man, Y., Ruan, W., Tam, N. F.-Y., Tao, R., Yin, L., Yang, Y., Dai, Y., & Tai, Y. (2022). The effect of rhizosphere and the plant species on the degradation of sulfonamides in model constructed wetlands treating synthetic domestic wastewater. *Chemosphere, 288*, 132487.

Wang, B., Wang, X., Wang, Z., Zhu, K., & Wu, W. (2023). Comparative metagenomic analysis reveals rhizosphere microbial community composition and functions help protect grapevines against salt stress. *Frontiers in Microbiology, 14*, 1102547.

Wen, T., Yuan, J., He, X., Lin, Y., Huang, Q., & Shen, Q. (2020). Enrichment of beneficial cucumber rhizosphere microbes mediated by organic acid secretion. *Horticulture Research*, 7.

White, R. A., III, Garoutte, A., Mclachlan, E. E., Tiemann, L. K., Evans, S., & Friesen, M. L. (2023). Genome-resolved metagenomics of nitrogen transformations in the switchgrass rhizosphere microbiome on marginal lands. *Agronomy, 13*(5), 1294.

Wooley, J. C., Godzik, A., & Friedberg, I. (2010). A primer on metagenomics. *PLoS Computational Biology, 6*(2), e1000667.

Yan, A., Wang, Y., Tan, S. N., Mohd Yusof, M. L., Ghosh, S., & Chen, Z. (2020). Phytoremediation: A promising approach for revegetation of heavy metal-polluted land. *Frontiers in Plant Science, 11*, 359.

Yin, C., Casa Vargas, J. M., Schlatter, D. C., Hagerty, C. H., Hulbert, S. H., & Paulitz, T. C. (2021). Rhizosphere community selection reveals bacteria associated with reduced root disease. *Microbiome, 9*, 1–18.

Yousuf, B., Kumar, R., Mishra, A., & Jha, B. (2014). Differential distribution and abundance of diazotrophic bacterial communities across different soil niches using a gene-targeted clone library approach. *FEMS Microbiology Letters, 360*(2), 117–125.

Yuan, J., Zhao, J., Wen, T., Zhao, M., Li, R., et al. (2018). Root exudates drive the soil-borne legacy of aboveground pathogen infection. *Microbiome, 6*(1), 1–12.

Zhang, W., & Li, C. (2016). Exploiting quorum sensing interfering strategies in gram-negative bacteria for the enhancement of environmental applications. *Frontiers in Microbiology, 6*, 1535.

Zhang, L., Guo, Z., Gao, H., Peng, X., Li, Y., Sun, S., Lee, J.-K., & Lin, W. (2016). Interaction of *Pseudostellaria heterophylla* with quorum sensing and quorum quenching bacteria mediated by root exudates in a consecutive monoculture system. *Journal of Microbiology and Biotechnology, 26*, 2159–2170. https://doi.org/10.4014/jmb.1607.07073

Zhang, L., Zhang, J., Wei, Y., Hu, W., Liu, G., Zeng, H., & Shi, H. (2021). Microbiome-wide association studies reveal correlations between the structure and metabolism of the rhizosphere microbiome and disease resistance in cassava. *Plant Biotechnology Journal, 19*(4), 689–701.

Zhou, Y., Coventry, D. R., Gupta, V. V., Fuentes, D., Merchant, A., et al. (2020). The preceding root system drives the composition and function of the rhizosphere microbiome. *Genome Biology, 21*(1), 1–19.

Zipfel, C., & Oldroyd, G. E. (2017). Plant signalling in symbiosis and immunity. *Nature, 543*(7645), 328–336.

Chapter 24
Harnessing the Power of Phytomicrobiome Interaction Involving Multi-Omics Techniques to Control Biotic Stresses

Kakan Ball, Kuntal Bera, Sanjoy Sadhukhan, and Puspendu Dutta

Abstract In agriculture, crop productivity and food security are continuously being threatened by various biotic stresses as imposed by pests and pathogens. Agricultural lands are getting polluted due to the injudicious application of agricultural chemicals, including herbicides, fungicides, insecticides, fertilisers, etc. Concerns about environmental sustainability and food safety have been raised with the conventional methods used for managing pests and pathogens, which frequently rely on chemical interventions. There are many ways to boost crop yields, nutrient availability, and plant's biotic and abiotic stress tolerance, but one of the more recent emerging approaches is to comprehend phytomicrobiome structure and associated chemical communications. The phytomicrobiome emphasises its function as a dynamic and intricate ecosystem of microorganisms living in plant tissues, including the endosphere, phyllosphere, and rhizosphere. These microorganisms are crucial in determining the health and resilience of plant biotic stress. The use of multi-omics approaches, including genomics, transcriptomics, proteomics, and/or metabolomics, as practical tools for decoding the molecular pathways behind phytomicrobiome interactions is at the core of this study. These methods provide a comprehensive understanding of the complex network of genetic, transcriptomic,

K. Ball · S. Sadhukhan
Plant Molecular Biology Laboratory, Department of Botany, Raiganj University, Raiganj, Uttar Dinajpur, West Bengal, India

K. Bera
Plant Molecular Biology Laboratory, Department of Botany, Raiganj University, Raiganj, West Bengal, India

Department of Seed Science and Technology, Uttar Banga Krishi Viswavidyalaya, Cooch Behar, West Bengal, India

P. Dutta (✉)
Department of Seed Science and Technology, Uttar Banga Krishi Viswavidyalaya, Cooch Behar, West Bengal, India
e-mail: puspendu@ubkv.ac.in

P. Mathur, S. Roy (eds.), *Plant Microbiome and Biological Control*, Sustainability in Plant and Crop Protection 20,
https://doi.org/10.1007/978-3-031-75845-4_24

and metabolic processes that plants use in response to biotic stresses and simultaneously make it easier to understand the ecological dynamics and functional roles of the phytomicrobiome.

Keywords Biotic stress · Multi-omics · Phytomicrobiome · Rhizosphere

Abbreviations

Ecosis	Ecological Spectral Information System
Ecosml	Ecological Spectral Model Library
FTIR	Fourier-transform infrared spectroscopy
GCMS	Gas chromatography mass spectrometry
MIR	Mid infrared
NDVI	Normalized difference vegetation index
NGS	Next generation sequencing
NIR	Near infrared
PGPR	Plant growth-promoting rhizobacteria
PM	Phytomicrobiome
PTMs	Post-translational modifications
RDP	Ribosomal database project
SWIR	Short wave infrared
UV	Ultraviolet
VIS	Visible
VOCs	Volatile organic compounds
VSWIR	Visible to short wavelength infrared

24.1 Introduction

Environmental stressors under changing climatic scenario have made crop productivity and food security a worldwide concern. By 2050, the world population could reach approx. 9 billion, while environmental disruptions lower crop productivity and soil fertility, making agriculture less sustainable (Muller et al., 2017; Chouhan et al., 2021). In order to feed the enormous population, it will therefore probably increase more than 60% of the food demand, particularly for the cereal crops, mainly wheat, maize, and rice, leguminous crops, pulses (peas, lentils, and soybeans), etc. (Majeed et al., 2017). Due to poor soil quality and biotic and abiotic stressors, the yields of these crops are frequently hampered (Suzuki et al., 2014; Kumar et al., 2021). Pesticides and fertilizers are most frequently utilized in traditional agricultural practices, which are solely used to supply the continued demand for food, but they result in a contaminated ecosystem. Such stresses end up with a wide range of harmful effects on plants in terms of their morphology, biochemistry, and reactivity at the molecular, physiological, and metabolic levels. Henceforth, implementing new and environmentally friendly approaches is necessary to protect agricultural sustainability and increase food yields for future generations (Majeed et al., 2017; Chouhan et al., 2021).

Numerous studies have shown for years that crop plants and the sky-scraping diversity of the microbiome are interrelated (Lemanceau et al., 2017). The phytomicrobiome (PM) is the microbial population connected to a plant and includes the entire root and the shoot sections (Bhatt et al., 2020). As the plant is the primary entry source of energy, all living organisms depend on plants for their need of energy. Microorganisms develop a complicated relationship with the plant parts. This phytocommunity system, which consists of plants and microbes, comprises regulated molecular signals that modulate each other's activity and serve mutual benefits (Bákonyi et al., 2020). The pool of microbial strains that function as plant growth-promoting rhizobacteria (PGPR) is found in the rhizospheric region of the plant.

Further investigation into the rhizosphere's biochemical substances and molecular signals has shown a sophisticated and complicated signalling process that enables plants to flourish in their environment. The phyllosphere, as well as the rhizospheric zone of plants, is inhabited by various living organisms like bacteria, fungi, yeasts, algae, nematodes, etc. It impacts on overall phytophysiology and development (Nuccio et al., 2020). With the change in biotic factors, the PM also varies (Blaustein et al., 2017). The biotic and abiotic conditions regulate the microbiome composition of the phyllosphere (Copeland et al., 2015). PM analysis requires a comprehensive study involving the various techniques of genomics, transcriptomics, proteomics, and metabolomics to ascertain the relationship between the plant and its associated microbial community.

Since many plant pathogens and their hosts coevolved in antagonistic interactions in natural systems, plant disease is a significant factor in controlling plant populations (Burdon, 1991; Jeger et al., 2014; Bever et al., 2015). Plant health is essential for maintaining yield in managed systems, such as forest plantations or agro-ecosystems. However, understanding and managing the ecology has traditionally relied on a reductionist approach to research individual components (e.g., just one or few microbe-plant interactions) related to specific disease pathologies rather than the complex ecological interactions between communities of microbes, hosts, and the environment. The changing environmental conditions are crucial to adopting a comprehensive approach to understanding how and why pathogenesis arises to cause diseases. Using a multi-omics method, one can get a clear picture of plant-microbial interactions (Crandall et al., 2020). A new era of scientific investigation has begun with the development of multi-omics approaches, which provide a comprehensive understanding of plants' genetic, transcriptomic, and metabolic responses. Understanding how plants respond to the intricate problems posed by pests and pathogens has been made possible due to this depth of awareness that has never before been attained. These methods also provide insight into the ecological dynamics and functional roles the microbial populations play in the phytomicrobiome.

24.2 The Phytomicrobiome: An Overview

Different microbial populations surrounding plants might have either a beneficial or adverse impact. All plant parts, including the roots and shoots, are surrounded by these microbial communities, among which the underground or root-associated microbial communities are the most well-studied (Turner et al., 2013). Microbes associated with roots comprise the rhizomicrobiome, while those associated with shoots comprise the phyllomicrobiome and those connected with the interior plant tissue system known as the endomicrobiome. Almost every plant tissue is inhabited by a particular microbiome, which promotes its hosts' development, overall health, and fitness. Thus, plants and microbes develop a close mutualistic interaction (Sharma et al., 2022).

Phytomicrobiome members are broadly categorised into specialists or generalists. Specialists, as the name suggests, are species-specific and only impact plants after receiving particular signals (most likely created during stressful situations during nodulation). However, generalists have a greater impact on a wider range of plant species (Lyu et al., 2020). Additionally, they help plants fix nitrogen and increase the uptake of several immobile minerals, including zinc and phosphorus. Recent research has gathered several advantageous characteristics that the phytomicrobiome community conferred upon the host plant (Kaul et al., 2021). Numerous research studies have examined that the various PM components show a great diversity of microorganisms, varying from crop to crop (Brisson et al., 2019). It changes depending on the crop developmental stage and the specific developmental stage when stress conditions are present (da Silva et al., 2014).

Phyllosphere community structure changes with seasonal variations, and these community variations are very active, according to a study with common bean, soybean, and canola related to the effects of biotic and abiotic stress on the composition at three locations in Ontario, Canada (Nuccio et al., 2020). Studies on PM compositions show that plants and the microbial populations around them interact dynamically. These interactions vary depending on the stage of plant development and the ambient conditions. It has been shown that endophytic microorganisms and the rhizosphere and phyllosphere help plants establish themselves in their surroundings. The same plant growing in various soils exhibits the recruitment of the same microbial communities for their establishment, which is quite intriguing to note and indicates how plants may alter the soil microbiota. This supports a complex signalling chemical pathway for community assembly, opening a new line of inquiry. Actinobacteria, Proteobacteria, and Bacteriodetes are present as conserved phyla in the rhizomicrobiome of these plant species, and the community structure of the roots of various types of plants varies (Redford et al., 2010). Numerous members in the rhizosphere micro-community can combat soil-borne phytopathogens in the root tissue at both the primary and secondary stages of infection (Mendes et al., 2011).

Fungi and bacteria in the rhizosphere produce metabolic substances that can stop phytopathogenic microorganisms (Brakhage & Schroeckh, 2011). The rhizosphere's microbes depend on the nutrients that the plant releases and are a member of a complicated food chain. By promoting the growth of specific beneficial bacteria, plants may be able to modify the microbiome of the rhizosphere for their benefit (Cook et al., 1995). Plants can control their productivity by limiting their accumulation of rhizo-deposits. The rhizo-deposits support intricate physiological processes or activities including cell division, cell development, and plant pathogenic microbe suppression.

Plant growth-promoting rhizobacteria (PGPRs), also known as beneficial soil bacteria, can be recruited by plants from a variety of genera, including *Azospirillum*, *Bacillus*, *Pseudomonas*, and *Rhizobium*. These bacteria promote plant development, create biofertilizers, solubilise phosphate, kill pathogens, and induce tolerance to biotic and abiotic conditions (Bhattacharyya & Jha, 2012; De-la-Peña & Loyola-Vargas, 2014; Singh et al., 2015; Shameer & Prasad, 2018; Shefali et al., 2020). Numerous researchers in phyllosphere studies have noted that similar plant species, even when planted in different geographic areas, exhibit little variation in their phyllomicrobiomes (Lebeis, 2014; Maignien et al., 2014).

Although the rhizosphere communities have a well-established role, researchers have recently discovered that other phyllospheric communities also have a part in reducing plant stress. Previous studies have suggested that the phyllospheric bacterial genus (*Bacillus*) can reduce rice's drought stress (Kumar, 2017). By increasing proline-like osmolytes, phenolic content, and the activity of antioxidative enzymes, including superoxide dismutase, catalase, peroxidase, etc., it was discovered that the phyllosphere isolates were involved in the activation of the plant antioxidant system (He et al., 2012). The manipulation of the plant microbiome can assist in lowering biotic stresses in plants and boosting productivity. Reduction in pathogenic disease also results in a low amount of chemical inputs into the fields, encouraging sustainable agriculture (Turner et al., 2013).

24.3 Biotic Stress in Agriculture and Phytomicrobiome

The host plant and the microcolonies that dwell on it develop distinct and stable microbiomes as they coevolve. So, the plants are regarded as meta-organisms. All eukaryotic organisms, not just plants, can be considered metaorganisms since they consist of a macroscopic host that coexists near a diverse community of bacteria, archaea, fungi, and protists, supporting the holobiont concept. A holobiont is an assemblage of microorganisms living in or around a host plant and constituting a separate ecological unit. Holobionts are also a unit of selection in evolution, making their study an exciting field for plant breeders (Sharma et al., 2022). It is suggested that the plants and the related bacteria should no longer be viewed as an "individual" but as a holobiont (Dessaux et al., 2016).

According to the "symbiosis cascade effect" that was further advanced by (Uroz et al., 2019) the effects of the host plant and its symbionts combined affect the composition of the plant microbiome. The need for sustainable crop management necessitates a positive intervention in this interaction. Metaorganisms are assemblages of coevolved species. These fall into three categories: mycorrhiza, endophytic fungi, and bacteria, which promote plant growth. Arbuscular mycorrhiza is the oldest and most significant symbiotic interaction between plants and microorganisms (dating back about 400 million years). It is also crucial in developing terrestrial plants (Selosse & Tacon, 1998; Smith et al., 2015; Vandenkoornhuyse et al., 2015). Numerous studies demonstrated the ability of phytomicrobes to combat many biotic stressors (Table 24.1).

Table 24.1 A list of phytomicrobiomes effective against various host plant specific pathogens

Host plant	Phytomicrobiome	Causal organism	References
Cucumis sativus	*Azotobacter* sp., *Pseudomonas* sp.	Cucumber mosaic virus	El-Borollosy and Oraby (2012)
Arabidopsis thaliana	*Bacilluscereus AR156*	*Pseudomonas syringae* pv. tomato	Niu et al. (2016)
Vigna radiata	*Pseudomonas putida* CRN-09, *Bacillus subtilis* CRN-16	*Macrophomina phaseolina*	Sharma et al. (2018)
Brassica oleracea var. *capitata*	*Paenibacillus* sp. P16	*Xanthomonas campestris* pv. *campestris*	Ghazalibiglar et al. (2016)
Solanum lycopersicum	*Bacillus amyloliquefaciens*	*Ralstonia solanacearum*	Li et al. (2017)
Oryza sativa	*Pseudomonas* sp. (BaC1–38)	*Xanthomonas campestris*	Lucas et al. (2014)
Zea mays	*Pseudomonas putida* KT2440	*Colletotrichum graminicola*	Planchamp et al. (2015)
Solanum tuberosum	*Pseudomonas sp.* R41805 in association with *Rhizophagus irregularis* MUCL 41833	*Rhizoctonia solani*	Velivelli et al. (2015)
Panax ginseng	*Bacillus amyloliquifaciens* strain HK34	*Phytophthora cactorum*	Song et al. (2015)
Solanum lycopersicum	*Funneliformis mosseae*	*Alternaria solani sorauer*	Lai et al. (2016)
Zea mays	*Bacillus cereus*	*Cochliobolus heterostrophus*	Tonelli et al. (2017)
Glycine max	*Bacillus sp.* CHEP5 in association with *Bradyrhizobium japonicum* E109	*Cercospora sojina*	Tonelli et al. (2017)
Vegetables	*Pseudomonas sp.* S2 and S4	*Salmonella enterica*	Hsu and Micallef (2017)
Cucumis sativus	*Trichoderma atroviride (TRS25)*	*Rhizoctonia solani*	Nawrocka et al. (2018)
Oryza sativa	*Bacillus sp.*	*Pyricularia oryzae*	Rais et al. (2017)
Arachis hypogea	*Bacillus* sp. (CHEP5 specie) and *Bradyrhizobium* sp. (SEMIA6144)	*Sclerotium rolfsii* (causal organism of plant stem wilt disease)	Figueredo et al. (2017)
S. lycopersicum, *Piper nigrum*, *Cucumis sativus*	*Bacillus* sp.	*Psedomonas* sp., *Xanthomonas, Pythium* sp.	Liu et al. (2018)

24.4 Omics Techniques: Tools for Studying Phytomicrobiome Interaction

The emergence of "-omics" techniques makes it possible to study plant-microbe interactions and microbial diversity in various ecological groups. One of the major hurdles is finding complex microbial communities in varied habitats, as only a small subset of microorganisms has been characterized so far. The "Omics" branch of study, which arose from the idea of integrative analysis, aids in elucidating complicated plant-microbe interactions at the molecular level (Diwan et al., 2022). Multi-omics technique can play a role in exploring the molecular mechanisms and interpreting the functionality underlying these interactions, which impact the growth and development of plants, to completely comprehend the complexity between plant(s) and the associated microorganisms (Doni et al., 2022). Multi-omics techniques in plant-microbe interactions need concurrently analysing different omics data sets, such as the epigenome, metabolome, proteome, transcriptome, and genomes, to fully comprehend the complex interactions between the plant(s) and the associated microorganisms. This integrated multi-omics approach makes the identification of key molecular markers, regulatory networks, and metabolic pathways involved in host-microbe interactions possible (López-Mondéjar et al., 2017).

24.4.1 Genomics

The study of the entire genetic makeup of an organism is called genomics. Johannsen discovered the gene concept in the early 1900s, and Hans Winkler coined the term "genome" in 1920. Microbes are the organisms where genomes were sequenced first, which introduced the beginning of a period of tool development and exponential growth in the number of whole genome sequences. The bacterium *Haemophilus influenzae* was the first free-living organism to have its entire genome sequenced by shotgun sequencing (Fleischmann et al., 1995). The nematode *Caenorhabditis elegans*, the first multicellular eukaryote (Consortium, 1998), and *Saccharomyces cerevisiae* (yeast), the first eukaryote to have its entire genome sequenced (Goffeau et al., 1996). The cost of sequencing has been significantly decreased with advancements in Next Generation Sequencing (NGS) technology, which also speeds up the availability of whole genome sequences, *de novo* sequence assemblies, and the resequencing of several strains of a single species.

The structure, function, evolution, mapping, and editing of the genomes of bacteria, fungi, archaea, viruses, and other microscopic organisms are the main topics of the interdisciplinary area of microbial genomics. Ecological questions, particularly those that deal with environmental change, can be helped by integrating genomics questions and technologies. Merging genomes and transcriptomics has improved the understanding of several elements relating to plant-pathogen interactions, plant-pathogen biology, and plant healing (Lindeberg, 2012; Sundin et al., 2016; Ramlal et al., 2023). Two significant types of microbial genomes can be helpful in studying plant disease ecology and plant health. A physical map of the entire genome is generated by structural genomics, which focuses on allocating and mapping genes and markers to individual chromosomes. To characterize gene functions and connections, functional genomics combines genome sequences with an organism's transcriptome (the transcripts it produces) and proteome (the encoded proteins) (Hieter & Boguski, 1997; Barone et al., 2008).

Gene annotation and gene prediction have been greatly enhanced by combining genomic data and NGS-based RNA sequencing (Yandell & Ence, 2012). Nevertheless, there are numerous additional "types" of genomics to consider depending on the scope and environment of the current plant disease ecology study. Comparative genomics aims to identify structural and functional genomic components that are conserved either inter-specifically or intra-specifically (Nobrega & Pennacchio, 2004). Using this method, we have learned that the genomic architecture of many fungal plant diseases is exceptionally varied and dynamic, consisting of areas that have undergone fast development. These regions range from transposon-rich chromosome compartments to complete auxiliary chromosomes (Möller & Stukenbrock, 2017). In addition to understanding genomic architecture, phylogenomics and population genomics approach related issues at various biological dimensions, such as links between species and individuals and populations within species, respectively. It can address the questions including those pertaining to population structure, speciation, and the divergence of closely related taxa, as well as locus-specific effects acting on particular genes or chromosomes that affect adaptation or specific phenotypes (Stukenbrock et al., 2011; Grunwald, 2016; de Vries et al., 2020).

24.4.2 *Metagenomics*

The term metagenomics is derived from the Greek prefix meta-, which means to "transcend," and it refers to the way that this field goes beyond standard genomic approaches to identify the genomic diversity and gene function of microbes from what some researchers define as uncultivable microbial "dark matter" (Solden et al., 2016). Investigating DNA that is derived directly from the environment requires the use of metagenomics, which relies on the techniques and theories of molecular genomics. The desire to comprehend the variety and functional importance of unculturable bacteria gave rise to the field of metagenomics. In the late 1990s, approaches and instruments were first invented by microbiologists and plant pathologists (Handelsman et al., 1998; Chen & Pachter, 2005). In this discipline, microbial DNA targets are extracted from soil, water, air, plants, and animals and amplified.

Although microbes are thought to be widespread and present in every ecosystem on Earth, only about 1% of all microbes can be cultured and identified using conventional first-generation Sanger sequencing technologies (Solden et al., 2016). It is now possible to sequence a microorganism's entire genome or just a portion of it due to recent developments in high-throughput DNA sequencing techniques. This allows researchers to provide answers to questions about the structure, function, and phylogeny of microorganisms on the level of entire microbial communities (Caporaso et al., 2012). A significant challenge in metagenomics is defining the microbial species concept itself. The biological species concept works well with more derived taxa (such as ourselves, *Homo sapiens*) where reproductive boundaries and barriers are strongly delineated. Bacteria, on the other hand, are currently assigned to a common species if their reciprocal, pairwise DNA re-association values are less than equal to 70% during DNA hybridization experiments (Staley, 2006). Traditionally, strains within a species must have a certain degree of phenotypic consistency, and species descriptions should be based on more than one type strain, such as a genetic variant or subtype. A microbial species name should technically only be assigned if its members can be distinguished from other species by at least one diagnostic phenotypic trait (Staley, 2006). Metagenomic studies on plant health have been concentrated on two main categories to date.

Research work to incorporate populations of endophytic bacteria in the shoots or roots needs to be broadened to include other crops and more natural systems, because most studies continue to use model plant species such as maize. Other studies have explored the rhizosphere and rhizoplane beyond the root, using metagenomics to investigate the relevance of functional genes for plant and soil health (Zhang et al., 2020; Kumar & Dubey, 2020). Understanding how soil microbes inhibit plant diseases is the focus of the second category of plant health research (Larkin, 2015). There is a need to research the phenomenon of suppressive soils in different habitats, such as managed and unmanaged grasslands and forests, since there have only been a few studies on it in crop pathosystems.

24.4.3 Metabolomics

Metabolomics, which is the thorough profiling of all small molecules within an organism, records the outcomes of the information cascade, starting with the genome and moving via the transcriptome and proteome, regarded as the phenotypic end of the -omics spectrum (Liu & Locasale, 2017). Metabolomics started in the 1970s with the expansion of medical analysis of relevant substances utilizing Gas Chromatography-Mass Spectrometry (GC-MS) (Sumner et al., 2003). Beginning in the late 1990s, the integration of functional genomics and metabolomics has grown in step with the development of chromatography, mass spectrometry (MS), and imaging technologies (Sauter et al., 1991; Oliver et al., 2002; Sumner et al., 2003). Though the discipline of metabolomics has advanced significantly over the past 20 years, it is still firmly based on the chemical examination of individual constituent chemicals. Metabolites may be quickly and thoroughly assessed in many physical and temporal dimensions (Bartels & Svatoš, 2015; Rinschen et al., 2019).

With modern metabolomics technology, which typically combines ultra-high-pressure liquid chromatography with high-resolution MS or nuclear magnetic resonance spectroscopy, the organisms chemical phenotypes can be displayed with high details, allowing the profiling of thousands of compounds (Ward et al., 2007; De Vos et al., 2007; Allwood et al., 2008). However, these developments are not without drawbacks. The enormous range of possible metabolites present in any given sample is a significant issue in metabolomics. A single plant species can produce up to 5000 distinct chemicals, in contrast to the 1500 and 2500 unique metabolites produced by microbes or mammals, respectively. The plant kingdom contains an estimated 200,000 different metabolites (Hall et al., 2002; Oksman-Caldentey & Saito, 2005). Targeted metabolomics methods overcome these issues through constraining profiling to a pre-determined list of annotated molecules (Griffiths et al., 2010). By comparing these compounds to well-known libraries of analytical standards, precise quantitative measurement of these compounds is made possible (Lu et al., 2008; Roberts et al., 2012). Targeted metabolomics techniques provide methodological optimization and higher resolution into biological processes with recognized relevance. They are beneficial for analyzing known processes and responses.

Conversely, untargeted metabolomics methods shed light on the unknowns. Untargeted metabolomics aims to characterize a significant amount of substances present in biological samples by comparing relative intensities, coming close to the ideal of true global metabolomics (Schrimpe-Rutledge et al., 2016). Constraints-free comparisons between and within experimental units enable the identification of the substances and pathways that regulate biological processes. Untargeted metabolomics, in particular, is crucial for multi-omics techniques because it makes connections easier, discovery combining spectromic, volatilomic, metagenomic, and genomic methods (Oksman-Caldentey & Saito, 2005; Johnson et al., 2016).

24.4.4 Volatilomics

One category of metabolomics that might be included is volatilomics (Cumeras & Correig, 2018). Volatilomics is more than just a subset of metabolomics. It is a complete profiling of high vapour pressure substances emitted from organisms (Bicchi & Maffei, 2012). Volatilomics evaluates molecules emitted by an organism, the essential elements of chemically mediated inter-organismal communication, in contrast to metabolomics, which aims to profile mostly intra-organism compounds (Crandall et al., 2020; Midzi et al., 2022). Volatile organic compounds (VOCs) serve as plants produce signalling molecules. These VOCs are critical to plant interaction with biotic and abiotic factors in the environment as well as to plant defence against herbivores and insects (Kasote et al., 2023).

Volatilomics primarily focuses on communication and offers exceptional prospects for comprehending community dynamics when combined with a multi-omics approach. These volatile compounds play numerous roles: influencing plant pollination, dispersal, and herbivory and providing defences against abiotic and biotic stress, such as plant pathogens (Rosenkranz & Schnitzler, 2016). Plants alone are thought to release on the order of 30,000 different volatiles into the atmosphere and rhizosphere (Majchrzak et al., 2020). Indeed, volatile communication is potentially even more critical belowground, where plant volatiles mediate everything from plant-pathogen interactions to plant herbivore-natural enemy interactions, in environments where visual information processing fails (Insam & Seewald, 2010; van Dam & Bouwmeester, 2016).

Advances in volatile organic compound collection techniques have been essential to the growth of volatilomics as a field. Headspace collection or sorptive materials are frequently utilized to store volatile chemicals used in plant communication in anticipation of their introduction to analysis by GC/MS (Bicchi & Maffei, 2012). These more contemporary methods have started to be applied in high-throughput phenotyping, where a quick evaluation of plant volatilomic profiles is associated with gene-by-environment interactions in a non-destructive multi-omics method (Jud et al., 2018). By extending this strategy to the plant microbiome, new options for locating and eliminating pathogens and pests may be discovered (Bailly & Weisskopf, 2017; Seethapathy et al., 2018; Sarkar & Sadhukhan, 2023).

24.4.5 Spectranomics

Spectranomics is an approach which combines spectroscopy with chemistry, taxonomy, and community ecology (Asner & Martin, 2009, 2016). The spectranomics technique is based on two fundamental ideas: (1) plants have chemical fingerprints that become increasingly distinctive as more constituents are added (Ustin et al., 2004), and (2) spectroscopic signatures identify a variety of compounds present in plants (Jacquemoud et al., 1995).

Plant pathologists have only just started to benefit from the groundwork set in place by terrestrial ecologists in the field of spectranomics. Recent research has helped to develop *in situ* (or foliar) and imaging spectroscopy (also known as "hyperspectral imaging") as useful methods for early, nondestructive, and scalable biotic stress detection in natural and agroecosystems (Mahlein, 2016; Mahlein et al., 2018; Zarco et al., 2018; Fallon et al., 2020). Plant-microbe interactions that are both beneficial (Fisher et al., 2016) and parasitic (Mahlein et al., 2019) have an impact on several plant properties that may be non-destructively detected. By producing systemic effectors or secondary metabolites or by physically present pathogen structures like hyphae and spores, microbes harm, impede, and/or modify foliar function both directly and indirectly, changing the chemical composition. With proximal and remote spectroscopy, the overall changes microorganisms contribute to plant health can be assessed using a statistical approach (Mahlein, 2016; Mahlein et al., 2018).

This strategy has been effective in improving accuracy and efficiency of agronomic crop breeding lines (Ge et al., 2019; Meacham-Hensold et al., 2019; Meacham-Hensold et al., 2020). As a result, we can identify, visualize, and model the biochemical and physiological pathosystem processes that underlie our ability to use spectroscopy for plant-microbe interaction, sensing through the spectrum quantification of foliar functional characteristics (Gold et al., 2020; Fallon et al., 2020).

Since the 1980s, late-stage plant disease has been detected using broad-band and multispectral techniques that largely rely on visible (VIS) and near-infrared (NIR) reflectance indices, such as normalized difference vegetation index (NDVI) (Nagarajan et al., 1984; Jackson, 1986; Hatfield & Pinter, 1993; Nilsson, 1995). Due to SWIR sensitivity to a variety of foliar traits, including nutrient content (Gillon et al., 1999; Serbin et al., 2012; Zhai et al., 2013; Singh et al., 2015), water (Gao, 1996), photosynthetic capacity (Oren et al., 1986), physiology (Serbin et al., 2019), phenolics and secondary metabolites (Kokaly & Skidmore, 2015; Couture et al., 2016) can be measured. SWIR changes have been proven to be helpful for plant-microbe interaction sensing, all of which are influenced by plant-microbe interactions. The field of plant-microbe interaction sensing has been revitalized by the development of VSWIR (400–2500 nm) sensing, due to its newly established potential for reliable, pre-visual disease detection (Wu et al., 2008; Rumpf et al., 2010; Xie et al., 2017; Bienkowski et al., 2019).

24.4.6 Proteomics

Plants have evolved a series of molecular programs to combat the adversities of a constantly changing environment. Biotic stress is one of the major problems faced by plants. In such cases, proteins play a crucial role. There are two main reasons which bring proteins into focus: (1) new phenotypic expression takes place by the direct participation of proteins, which is regulated by physiological characteristics and helps the plants to adopt environmental changes, and (2) proteins help in monitoring cellular mechanisms and act as a key player for the maintenance of cellular equilibrium. It is not possible to understand the complex interlinked signals and continuous cellular processes which are involved in plant biotic responses. The plant immune system is thought to respond to biotic stress as a complex system with connections and crosstalk between various signals and with a wide range of stress tolerance-related proteins. Proteomics is now a comprehensive tool for biological research that may be used to explain biological functions, evolving from a stand-alone field of study. Proteomics has been effectively employed in several research as a method for discovering plant resistance pathways. The spatial proteome and post-translational modifications (PTMs) of proteins appear to play a direct role in the plant immune response, according to accumulating data (Liu et al., 2019).

The proteins in each component are then enriched by selective fractionation, immunoprecipitation, chromatography, electrophoresis or centrifugal technologies. As a result of recent advances, spatial and PTMs proteomics have become tools capable of delivering answers to vital biological questions, and the roles of these tools in basic and applied science will likely expand in the coming decades. Spatial proteomics has developed into a significant research area in recent years (Pankow et al., 2019; Lundberg & Borner, 2019). However, because the affinity purification-based technique depends on the specificity of antibodies, it can be challenging to find the proper antibodies. A subcellular separation method based on various separation principles has been devised to address the drawbacks of the methods mentioned above. Crops are increasingly at risk of having their yields reduced by biotic stress. Thus, it is crucial to understand how plants react to it.

The fundamental knowledge of how plants react to biotic stress needs to be further developed. Precise control targets for plant immunity will be revealed by an understanding of proteomics at the cellular and subcellular levels. Considering the quantitative and qualitative characterisation of proteins and their associations at the gene level, proteomics is a rapidly developing field of biological science (Sarkar & Sadhukhan, 2022).

24.4.7 Multi-omics for the Detection of Plant Stress

Pigments, nitrogen, protein, structural components (cellulose and lignin), and mineral compounds were the main focus of early vegetative spectroscopic investigations addressing the restoration of foliar biochemicals (Kokaly et al., 2009; Thulin et al., 2014). This capacity was quickly increased to include phenolics and other groups of secondary metabolites connected to defence and the salicylic acid route (Kokaly & Skidmore, 2015; Couture et al., 2016; Couture et al., 2018). The boundaries of traits specificity have dramatically enlarged over the past 5 years to include particular photosynthesis-related characteristics such as maximal rates of carboxylation and photosynthetic activity, sucrose, starch, fructose, and free amino acids in leaves (Yendrek et al., 2017; Serbin et al., 2019; Meacham-Hensold et al., 2019). To better evaluate and comprehend plant-microbe interactions, spectroscopic techniques such as fluorescence spectroscopy, Fourier-transform infrared (FTIR) spectroscopy, and chemometrics are used; however, these techniques require sampling and processing (Barth, 2000; Ni et al., 2011).

A new generation of plant and plant-microbe phenotyping techniques would be possible if it were feasible to quantify and forecast plant metabolites via passive monitoring with vegetative spectroscopy, particularly in field settings. Many subgroups in the plant sciences, most lately, have proposed *in vivo* metabolomics as having relevance for precision agriculture. Its specificity is considerably lower than traditional metabolomics, and it uses hyperspectral detection (Martins et al., 2019) (Fig. 24.1). The first-ever evidence of the capability of hyperspectral field sensors to non-destructively estimate foliar metabolite profiles was provided by groundbreaking work (Vergara Díaz, 2019). According to this study, hyperspectral data may successfully predict roughly 25% of the metabolites discovered in wheat leaves and ear bracts by GC-MS profiling with a validation accuracy above 50%.

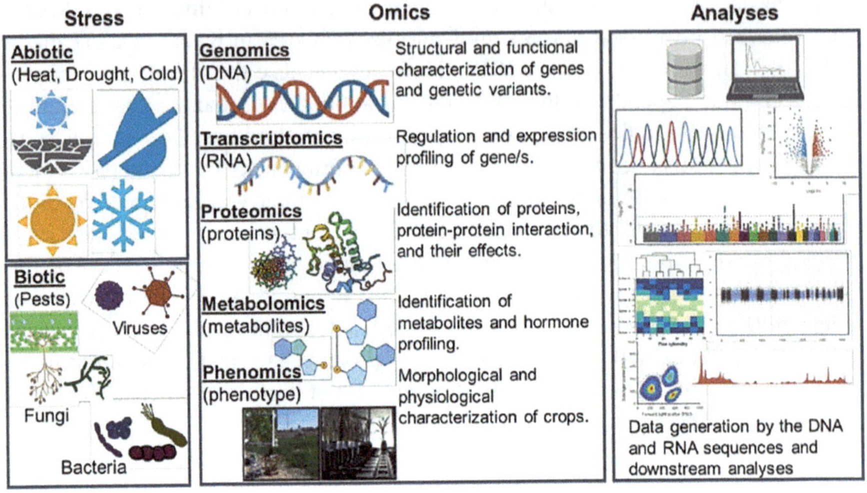

Fig. 24.1 Comprehensive use of multi-omics in the detection of plant stress

They were able to anticipate many metabolites, including organic acids, sugars, and amino acids, that are essential for both primary and secondary metabolism. Regardless of the metabolite investigated in either of the two plant parts under study (leaf or ear), Vergara-Diaz et al. (2020) discovered the blue region to be the most pertinent waveband for metabolite prediction. The 1300–1400 nm and 2200–2400 nm zones were consistently the best predictors for metabolite prediction, and they coincided with absorption bands known to be related with sugars and nitrogen compounds. These authors also discovered consistency in their canopy models. Combining spectranomics and metabolomics would not be without its difficulties: the measurement of plant biochemical features at the canopy scale still confronts significant obstacles due to difficulties with canopy structure/density and the resulting effects on light reflection. The fact that other methods like UV, MIR, X-ray, Raman, or FTIR spectroscopy have traditionally been utilized limits previous understanding of the VIS-NIR-SWIR spectral characteristics (spectral signals in plant reflectance), linked with plant metabolites (Barth, 2000). These more sensitive methods can measure isolated metabolite concentrations much more precisely, but they do not take into consideration how plant macro- and micro-physical characteristics affect, how light interacts with these bonds.

24.5 Challenges and Future Directions

We can better comprehend the microbiological participants in the system, their distribution through time and space, and their individual and collective functions by integrating multi-omic techniques. The mechanisms underlying core concepts in plant disease ecology, such as plant defence, stress response, and the potential for disease suppression, can be better understood with the help of -omics. It is particularly true when contrasting the use of a single-omic. These methods have the potential to provide a multifaceted understanding of plant disease when combined adequately at the appropriate scale and for the appropriate research problem. However, researchers must be aware of numerous challenges and restrictions when integrating and employing these methodologies before developing a study proposal and project. Moving away from purely observational research questions and toward hypothesis-driven studies is necessary when developing projects that could benefit from multi-omic techniques.

Although the number of research articles published has increased significantly over the past decades, reflecting advances in high throughput sequencing, only a few have been applied to experimental problems in the disciplines of environmental and conservation genomics (Galla et al., 2016). Combining and analysing data from several -omics platforms (including genomics, transcriptomics, proteomics, and metabolomics) might be difficult. It is a continuous struggle to develop reliable bioinformatics workflows and tools to manage and analyze multi-omics data. It might be challenging to combine and analyze data from these omics platforms. Developing reliable bioinformatics workflows and tools to manage and analyze multi-omics data is a continuous effort.

The field of plant disease ecology might benefit from a transition away from hypotheses that determine correlation to those that determine an underlying cause for an ecological phenomenon. For instance, it is critical to developing hypothesis-based research that verifies ecological theory to comprehend how microbial diversity and dynamics change in response to environmental change. There is value in observational studies that outline the role of microorganisms in a system because they enable researchers to focus their ideas and inquiries after examining the taxa present and specific environmental variables. This early research frequently employs amplicon-based sequencing techniques. Although employing amplicon-based sequencing to identify taxa is the first step in defining the taxonomic and correlative limits of a system, descriptive studies are unable to shed light on the mechanisms or forces that underlie the patterns that are seen.

As more microorganisms are sequenced and discovered globally, taxonomy or chemical compound databases, which are still in the early stages, constitute a significant barrier to doing multi-omics. However, a few carefully managed databases do exist and are the norm for carrying out -omics research. Nevertheless, there are a few carefully controlled databases that are the norm for carrying out -omics research. Open-source databases for amplicon-based and metagenomic research include RDP (Cole et al., 2014), Silva (Quast et al., 2013), Greengenes (McDonald et al., 2012), and UNITE (Nilsson et al., 2019). However, with many researchers keeping private instrument-specific libraries, there are still limitations in their scope and capacity to be used as reference databases for compound matching. Ecosml.org (Ecological Spectral Model Library) and ecosis.org (Ecological Spectral Information System) are two spectroscopy (spectranomics) resources that are helpful for discovering spectral models and data on leaf nutrition, cellulose, and other physiological factors.

Cost-per-sample (particularly when considering different evaluation techniques) and challenges combining diverse data sets are additional barriers to integrating multi-omics approaches in plant disease ecology. Omics data naturally come in a variety of formats and can be pre-processed, analyzed, and interpreted in a variety of ways. However, it can be challenging to integrate those findings throughout the genotype-phenotype continuum.

24.6 Conclusions

Multi-omics is a valuable tool that can be applied to improve our comprehension of the fundamentals of interactions between plants and microbiomes. The interaction between plants and microbes requires relatively little energy. With the growth and development of plants, the microorganisms get the same benefits that are associated with the plant root system. The study of PM signals (hormones or exo-hormones) can help to improve this synergy between microbes and plants. The management of agro-ecosystems can benefit from this advanced omics research. The development of alternative sequencing-data integration methodologies with potential uses beyond plant biology will be necessary to support the implementation of this method. Adopting this strategy would encourage ecologists, statisticians, computational scientists, and biologists of plants and microbes to work together to develop integrated experimental frameworks, that incorporate various points of view.

It is feasible to assume that PMs can help people produce beneficial and sustainable food materials to meet the growing demand for food worldwide, while also minimizing the adverse effects on the ecosystem and its stakeholders. Cost-per-sample is a barrier to integrating multi-omics methodologies in plant disease ecology, particularly when different evaluation techniques are taken into account. Another barrier is the challenge of merging diverse data sets. Omics data are inherently available in various formats, requiring varying methods for pre-processing, analyzing, and interpreting its results. Collaboration is necessary to address these issues and fully utilize the potential of multi-omics research. A single individual cannot be an expert in all multi-omics disciplines, so integrating expertise is just as crucial for performing multiomics research as integrating datasets.

This collaborative element of multi-omics research may be its most promising part. An integration of perspectives results from the merger of a multiplicity of disciplines. New insights from another field will likely help address the issues that are evident in that discipline. Understanding plant defence, spotting plant stress, and treating disease with suppressive soils have all been made possible by the multi-omics approach to solve challenges in plant disease ecology. Additionally, it appears that this strategy will also advance our knowledge on endophytes. Beyond these fields, however, multi-omics has great promise for revolutionizing our knowledge of how microbial populations adapt to changing environmental conditions in the upcoming years.

References

Allwood, J. W., Ellis, D. I., & Goodacre, R. (2008). Biomarker metabolites capturing the metabolite variance present in a rice plant developmental period. *Physiologia Plantarum, 132*, 117–135.

Asner, G. P., & Martin, R. E. (2009). Airborne spectranomics: Mapping canopy chemical and taxonomic diversity in tropical forests. *Frontiers in Ecology and the Environment, 7*, 269–276. https://doi.org/10.1890/070152

Asner, G. P., & Martin, R. E. (2016). Spectranomics: Emerging science and conservation opportunities at the interface of biodiversity and remote sensing. *Global Ecology and Conservation, 8*, 212–219.

Bailly, A., & Weisskopf, L. (2017). Mining the volatilomes of plant-associated microbiota for new biocontrol solutions. *Frontiers in Microbiology, 8*, 1638.

Bákonyi, N., Kisvarga, S., Barna, D., Tóth, I. O., El-Ramady, H., Abdalla, N., et al. (2020). Chemical traits of fermented alfalfa brown juice: Its implications on physiological, biochemical, anatomical, and growth parameters of *celosia. Agronomy, 10*, 247.

Barone, A., Chiusano, M. L., Ercolano, M. R., Giuliano, G., Grandillo, S., & Frusciante, L. (2008). Structural and functional genomics of tomato. *International Journal of Plant Genomics, 2008*, 820274.

Bartels, B., & Svatoš, A. (2015). Spatially resolved *in vivo* plant metabolomics by laser ablation-based mass spectrometry imaging (MSI) techniques: LDI-MSI and LAESI. *Frontiers in Plant Science, 6*, 471.

Barth, A. (2000). The infrared absorption of amino acid side chains. *Progress in Biophysics and Molecular Biology, 74*, 141–173.

Bever, J. D., Mangan, S. A., & Alexander, H. M. (2015). Maintenance of plant species diversity by pathogens. *Annual Review of Ecology, Evolution, and Systematics, 46*, 305–325.

Bhatt, P., Verma, A., Verma, S., Anwar, M. S., Prasher, P., Mudila, H., & Chen, S. (2020). Understanding phytomicrobiome: A potential reservoir for better crop management. *Sustainability, 12*, 5446.

Bhattacharyya, P. N., & Jha, D. K. (2012). Plant growth-promoting rhizobacteria (PGPR): Emergence in agriculture. *World Journal of Microbiology and Biotechnology, 28*, 1327–1350.

Bicchi, C., & Maffei, M. (2012). The plant volatilome: Methods of analysis. In J. Normanly (Ed.), *High-throughput phenotyping in plants: Methods and protocols* (pp. 289–310). Humana Press.

Bienkowski, D., Aitkenhead, M. J., Lees, A. K., Gallagher, C., & Neilson, R. (2019). Detection and differentiation between potato (*Solanum tuberosum*) diseases using calibration models trained with non-imaging spectrometry data. *Computers and Electronics in Agriculture, 167*, 105056.

Blaustein, R. A., Lorca, G. L., Meyer, J. L., Gonzalez, C. F., & Teplitski, M. (2017). Defining the core citrus leaf- and root-associated microbiota: Factors associated with community structure and implications for managing Huanglongbing (citrus greening) disease. *Applied and Environmental Microbiology, 83*, e00210–e00217.

Brakhage, A. A., & Schroeckh, V. (2011). Fungal secondary metabolites—Strategies to activate silent gene clusters. *Fungal Genetics and Biology, 48*, 15–22.

Brisson, V. L., Schmidt, J. E., Northen, T. R., Vogel, J. P., & Gaudin, A. C. M. (2019). Impacts of maize domestication and breeding on rhizosphere microbial community recruitment from a nutrient depleted agricultural soil. *Scientific Reports, 9*, 15611.

Burdon, J. J. (1991). Fungal pathogens as selective forces in plant populations and communities*. *Australian Journal of Ecology, 16*, 423–432. https://doi.org/10.1111/j.1442-9993.1991.tb01072.x

Caporaso, J. G., Lauber, C. L., Walters, W. A., Berg-Lyons, D., Huntley, J., Fierer, N., et al. (2012). Ultra-high-throughput microbial community analysis on the Illumina HiSeq and MiSeq platforms. *The ISME Journal, 6*, 1621–1624.

Chen, K., & Pachter, L. (2005). Bioinformatics for whole-genome shotgun sequencing of microbial communities. *PLoS Computational Biology, 1*, e24. https://doi.org/10.1371/journal.pcbi.0010024

Chouhan, G. K., Verma, J. P., Jaiswal, D. K., Mukherjee, A., Singh, S., de Araujo Pereira, A. P., Liu, H., Abd Allah, E. F., & Singh, B. K. (2021). Phytomicrobiome for promoting sustainable agriculture and food security: Opportunities, challenges, and solutions. *Microbiological Research, 248*, 126763. https://doi.org/10.1016/j.micres.2021.126763

Cole, J. R., Wang, Q., Fish, J. A., Chai, B., McGarrell, D. M., Sun, Y., Brown, C. T., Porras-Alfaro, A., Kuske, C. R., & Tiedje, J. M. (2014). Ribosomal database project: Data and tools for high throughput rRNA analysis. *Nucleic Acids Research, 42*, D633–D642. https://doi.org/10.1093/nar/gkt1244

Consortium. (1998). Genome sequence of the nematode *C. elegans*: A platform for investigating biology. *Science, 282*, 2012–2018. https://doi.org/10.1126/science.282.5396.2012

Cook, R. J., Thomashow, L. S., Weller, D. M., Fujimoto, D., Mazzola, M., Bangera, G., & Kim, D. S. (1995). Molecular mechanisms of defense by rhizobacteria against root disease. *Proceedings of the National Academy of Sciences, 92*, 4197–4201. https://doi.org/10.1073/pnas.92.10.4197

Copeland, J. K., Yuan, L., Layeghifard, M., Wang, P. W., & Guttman, D. S. (2015). Seasonal community succession of the Phyllosphere microbiome. *MPMI, 28*, 274–285. https://doi.org/10.1094/MPMI-10-14-0331-FI

Couture, J. J., Singh, A., Rubert-Nason, K. F., Serbin, S. P., Lindroth, R. L., & Townsend, P. A. (2016). Spectroscopic determination of ecologically relevant plant secondary metabolites. *Methods in Ecology and Evolution, 7*, 1402–1412. https://doi.org/10.1111/2041-210X.12596

Couture, J. J., Singh, A., Charkowski, A. O., Groves, R. L., Gray, S. M., Bethke, P. C., & Townsend, P. A. (2018). Integrating spectroscopy with potato disease management. *Plant Disease, 102*, 2233–2240. https://doi.org/10.1094/PDIS-01-18-0054-RE

Crandall, S. G., Gold, K. M., del Jiménez-Gasco, M., Filgueiras, C. C., & Willett, D. S. (2020). A multi-omics approach to solving problems in plant disease ecology. *PLoS One, 15*, e0237975. https://doi.org/10.1371/journal.pone.0237975

Cumeras, R., & Correig, X. (2018). *Volatile organic compound analysis in biomedical diagnosis applications*. CRC Press.

da Silva, D. A. F., Cotta, S. R., Vollú, R. E., de Azevedo Jurelevicius, D., Marques, J. M., Marriel, I. E., & Seldin, L. (2014). Endophytic microbial community in two transgenic maize genotypes and in their near-isogenic non-transgenic maize genotype. *BMC Microbiology, 14*, 332. https://doi.org/10.1186/s12866-014-0332-1

De Vos, R. C., Moco, S., Lommen, A., Keurentjes, J. J., Bino, R. J., & Hall, R. D. (2007). Untargeted large-scale plant metabolomics using liquid chromatography coupled to mass spectrometry. *Nature Protocols, 2*, 778–791. https://doi.org/10.1038/nprot.2007.95

de Vries, S., Stukenbrock, E. H., & Rose, L. E. (2020). Rapid evolution in plant-microbe interactions – An evolutionary genomics perspective. *The New Phytologist, 226*, 1256–1262. https://doi.org/10.1111/nph.16458

De-la-Peña, C., & Loyola-Vargas, V. M. (2014). Biotic interactions in the rhizosphere: A diverse cooperative Enterprise for Plant Productivity. *Plant Physiology, 166*, 701–719. https://doi.org/10.1104/pp.114.241810

Dessaux, Y., Grandclément, C., & Faure, D. (2016). Engineering the rhizosphere. *Trends in Plant Science, 21*, 266–278. https://doi.org/10.1016/j.tplants.2016.01.002

Diwan, D., Rashid, M. M., & Vaishnav, A. (2022). Current understanding of plant-microbe interaction through the lenses of multi-omics approaches and their benefits in sustainable agriculture. *Microbiological Research, 265*, 127180. https://doi.org/10.1016/j.micres.2022.127180

Doni, F., Miranti, M., Mispan, M. S., Mohamed, Z., & Uphoff, N. (2022). Multi-omics approaches for deciphering the microbial modulation of plants' genetic potentials: What's known and what's next? *Rhizosphere, 24*, 100613. https://doi.org/10.1016/j.rhisph.2022.100613

El-Borollosy, A. M., & Oraby, M. M. (2012). Induced systemic resistance against *cucumber mosaic cucumovirus* and promotion of cucumber growth by some plant growth-promoting rhizobacteria. *Annals of Agricultural Sciences, 57*, 91–97. https://doi.org/10.1016/j.aoas.2012.08.001

Fallon, B., Yang, A., Lapadat, C., Armour, I., Juzwik, J., Montgomery, R. A., & Cavender-Bares, J. (2020). Spectral differentiation of oak wilt from foliar fungal disease and drought is correlated with physiological changes. *Tree Physiology, 40*, 377–390. https://doi.org/10.1093/treephys/tpaa005

Figueredo, M. S., Tonelli, M. L., Ibáñez, F., Morla, F., Cerioni, G., Del Carmen, T. M., & Fabra, A. (2017). Induced systemic resistance and symbiotic performance of peanut plants challenged with fungal pathogens and co-inoculated with the biocontrol agent bacillus sp. CHEP5 and Bradyrhizobium sp. SEMIA6144. *Microbiological Research, 197*, 65–73. https://doi.org/10.1016/j.micres.2017.01.002

Fisher, J. B., Sweeney, S., Brzostek, E. R., Evans, T. P., Johnson, D. J., Myers, J. A., Bourg, N. A., Wolf, A. T., Howe, R. W., & Phillips, R. P. (2016). Tree-mycorrhizal associations detected remotely from canopy spectral properties. *Global Change Biology, 22*, 2596–2607. https://doi.org/10.1111/gcb.13264

Fleischmann, R. D., Adams, M. D., White, O., Clayton, R. A., Kirkness, E. F., Kerlavage, A. R., Bult, C. J., Tomb, J.-F., Dougherty, B. A., Merrick, J. M., McKenney, K., Sutton, G., FitzHugh, W., Fields, C., Gocayne, J. D., Scott, J., Shirley, R., Liu, L., Glodek, A., Kelley, J. M., Weidman, J. F., Phillips, C. A., Spriggs, T., Hedblom, E., Cotton, M. D., Utterback, T. R., Hanna, M. C., Nguyen, D. T., Saudek, D. M., Brandon, R. C., Fine, L. D., Fritchman, J. L., Fuhrmann, J. L., Geoghagen, N. S. M., Gnehm, C. L., McDonald, L. A., Small, K. V., Fraser, C. M., Smith, H. O., & Venter, J. C. (1995). Whole-genome random sequencing and assembly of Haemophilus influenzae Rd. *Science, 269*, 496–512. https://doi.org/10.1126/science.7542800

Galla, S. J., Buckley, T. R., Elshire, R., Hale, M. L., Knapp, M., McCallum, J., Moraga, R., Santure, A. W., Wilcox, P., & Steeves, T. E. (2016). Building strong relationships between conservation genetics and primary industry leads to mutually beneficial genomic advances. *Molecular Ecology, 25*, 5267–5281. https://doi.org/10.1111/mec.13837

Gao, B. (1996). NDWI—A normalized difference water index for remote sensing of vegetation liquid water from space. *Remote Sensing of Environment, 58*, 257–266. https://doi.org/10.1016/S0034-4257(96)00067-3

Ge, Y., Atefi, A., Zhang, H., Miao, C., Ramamurthy, R. K., Sigmon, B., Yang, J., & Schnable, J. C. (2019). High-throughput analysis of leaf physiological and chemical traits with VIS–NIR–SWIR spectroscopy: A case study with a maize diversity panel. *Plant Methods, 15*, 66. https://doi.org/10.1186/s13007-019-0450-8

Ghazalibiglar, H., Hampton, J. G., van Zijll de Jong, E., & Holyoake, A. (2016). Is induced systemic resistance the mechanism for control of black rot in *Brassica oleracea* by a *Paenibacillus* sp.? *Biological Control, 92*, 195–201. https://doi.org/10.1016/j.biocontrol.2015.10.014

Gillon, D., Houssard, C., & Joffre, R. (1999). Using near-infrared reflectance spectroscopy to predict carbon, nitrogen and phosphorus content in heterogeneous plant material. *Oecologia, 118*, 173–182. https://doi.org/10.1007/s004420050716

Goffeau, A., Barrell, B. G., Bussey, H., Davis, R. W., Dujon, B., Feldmann, H., Galibert, F., Hoheisel, J. D., Jacq, C., Johnston, M., Louis, E. J., Mewes, H. W., Murakami, Y., Philippsen, P., Tettelin, H., & Oliver, S. G. (1996). Life with 6000 genes. *Science, 274*, 546–567. https://doi.org/10.1126/science.274.5287.546

Gold, K. M., Townsend, P. A., Chlus, A., Herrmann, I., Couture, J. J., Larson, E. R., & Gevens, A. J. (2020). Hyperspectral measurements enable pre-symptomatic detection and differentiation of contrasting physiological effects of late blight and early blight in potato. *Remote Sensing, 12*, 286. https://doi.org/10.3390/rs12020286

Griffiths, W. J., Koal, T., Wang, Y., Kohl, M., Enot, D. P., & Deigner, H.-P. (2010). Targeted metabolomics for biomarker discovery. *Angewandte Chemie International Edition, 49*, 5426–5445. https://doi.org/10.1002/anie.200905579

Grunwald, S. (2016). *Environmental soil-landscape modeling: Geographic information technologies and pedometrics*. CRC Press.

Hall, R., Beale, M., Fiehn, O., Hardy, N., Sumner, L., & Bino, R. (2002). Plant metabolomics: The missing link in functional genomics strategies. *The Plant Cell, 14*, 1437–1440. https://doi.org/10.1105/tpc.140720

Handelsman, J., Rondon, M. R., Brady, S. F., Clardy, J., & Goodman, R. M. (1998). Molecular biological access to the chemistry of unknown soil microbes: A new frontier for natural products. *Chemistry & Biology, 5*, R245–R249. https://doi.org/10.1016/s1074-5521(98)90108-9

Hatfield, P. L., & Pinter, P. J. (1993). Remote sensing for crop protection. *Crop Protection, 12*, 403–413. https://doi.org/10.1016/0261-2194(93)90001-Y

He, Z., Piceno, Y., Deng, Y., Xu, M., Lu, Z., DeSantis, T., Andersen, G., Hobbie, S. E., Reich, P. B., & Zhou, J. (2012). The phylogenetic composition and structure of soil microbial communities shifts in response to elevated carbon dioxide. *The ISME Journal, 6*, 259–272. https://doi.org/10.1038/ismej.2011.99

Hieter, P., & Boguski, M. (1997). Functional genomics: It's all how you read it. *Science, 278*, 601–602. https://doi.org/10.1126/science.278.5338.601

Hsu, C.-K., & Micallef, S. A. (2017). Plant-mediated restriction of *Salmonella enterica* on tomato and spinach leaves colonized with *Pseudomonas* plant growth-promoting rhizobacteria. *International Journal of Food Microbiology, 259*, 1–6. https://doi.org/10.1016/j.ijfoodmicro.2017.07.012

Insam, H., & Seewald, M. S. A. (2010). Volatile organic compounds (VOCs) in soils. *Biology and Fertility of Soils, 46*, 199–213. https://doi.org/10.1007/s00374-010-0442-3

Jackson, R. D. (1986). Remote sensing of biotic and abiotic plant stress. *Annual Review of Phytopathology, 24*, 265–287. https://doi.org/10.1146/annurev.py.24.090186.001405

Jacquemoud, S., Verdebout, J., Schmuck, G., Andreoli, G., & Hosgood, B. (1995). Investigation of leaf biochemistry by statistics. *Remote Sensing of Environment, 54*, 180–188. https://doi.org/10.1016/0034-4257(95)00170-0

Jeger, M. J., Salama, N. K. G., Shaw, M. W., van den Berg, F., & van den Bosch, F. (2014). Effects of plant pathogens on population dynamics and community composition in grassland ecosystems: Two case studies. *European Journal of Plant Pathology, 138*, 513–527. https://doi.org/10.1007/s10658-013-0325-1

Johnson, C. H., Ivanisevic, J., & Siuzdak, G. (2016). Metabolomics: Beyond biomarkers and towards mechanisms. *Nature Reviews. Molecular Cell Biology, 17*, 451–459. https://doi.org/10.1038/nrm.2016.25

Jud, W., Winkler, J. B., Niederbacher, B., Niederbacher, S., & Schnitzler, J.-P. (2018). Volatilomics: A non-invasive technique for screening plant phenotypic traits. *Plant Methods, 14*, 109. https://doi.org/10.1186/s13007-018-0378-4

Kasote, D., Lee, J., & Sreenivasulu, N. (2023). Editorial: Volatilomics in plant and agricultural research: Recent trends. *Frontiers in Plant Science, 14*.

Kaul, S., Choudhary, M., Gupta, S., & Dhar, M. K. (2021). Engineering host microbiome for crop improvement and sustainable agriculture. *Frontiers in Microbiology, 12*.

Kokaly, R. F., & Skidmore, A. K. (2015). Plant phenolics and absorption features in vegetation reflectance spectra near 1.66 μm. *International Journal of Applied Earth Observation and Geoinformation, 43*, 55–83. https://doi.org/10.1016/j.jag.2015.01.010

Kokaly, R. F., Asner, G. P., Ollinger, S. V., Martin, M. E., & Wessman, C. A. (2009). Characterizing canopy biochemistry from imaging spectroscopy and its application to ecosystem studies. *Remote Sensing of Environment, 113*, S78–S91. https://doi.org/10.1016/j.rse.2008.10.018

Kumar, A. S., Sridar, R., & Uthandi, S. (2017). Mitigation of drought in rice by a phyllosphere bacterium Bacillus altitudinis FD48. *African Journal of Microbiology Research, 11*, 1614–1625. https://doi.org/10.5897/AJMR2017.8610

Kumar, A., & Dubey, A. (2020). Rhizosphere microbiome: Engineering bacterial competitiveness for enhancing crop production. *Journal of Advanced Research, 24*, 337–352. https://doi.org/10.1016/j.jare.2020.04.014

Kumar, A., Singh, S., Mukherjee, A., Rastogi, R. P., & Verma, J. P. (2021). Salt-tolerant plant growth-promoting *Bacillus pumilus* strain JPVS11 to enhance plant growth attributes of rice and improve soil health under salinity stress. *Microbiological Research, 242*, 126616. https://doi.org/10.1016/j.micres.2020.126616

Lai, Y. R., Lin, P. Y., Chen, C. Y., & Huang, C. J. (2016). Feasible Management of Southern Corn Leaf Blight via induction of systemic resistance by Bacillus cereus C1L in combination with reduced use of Dithiocarbamate fungicides. *Plant Pathology Journal, 32*, 481–488. https://doi.org/10.5423/PPJ.OA.02.2016.0044

Larkin, R. P. (2015). Soil health paradigms and implications for disease management. *Annual Review of Phytopathology, 53*, 199–221. https://doi.org/10.1146/annurev-phyto-080614-120357

Lebeis, S. L. (2014). The potential for give and take in plant–microbiome relationships. *Frontiers in Plant Science, 5*.

Lemanceau, P., Blouin, M., Muller, D., & Moënne-Loccoz, Y. (2017). Let the Core microbiota be functional. *Trends in Plant Science, 22*, 583–595. https://doi.org/10.1016/j.tplants.2017.04.008

Li, C., Hu, W., Pan, B., Liu, Y., Yuan, S., Ding, Y., Li, R., Zheng, X., Shen, B., & Shen, Q. (2017). Rhizobacterium *Bacillus amyloliquefaciens* strain SQRT3-mediated induced systemic resistance controls bacterial wilt of tomato. *Pedosphere, 27*, 1135–1146. https://doi.org/10.1016/S1002-0160(17)60406-5

Lindeberg, M. (2012). Genome-enabled perspectives on the composition, evolution, and expression of virulence determinants in bacterial plant pathogens. *Annual Review of Phytopathology, 50*, 111–132. https://doi.org/10.1146/annurev-phyto-081211-173022

Liu, X., & Locasale, J. W. (2017). Metabolomics: A primer. *Trends in Biochemical Sciences, 42*, 274–284. https://doi.org/10.1016/j.tibs.2017.01.004

Liu, K., McInroy, J. A., Hu, C.-H., & Kloepper, J. W. (2018). Mixtures of plant-growth-promoting Rhizobacteria enhance biological control of multiple plant diseases and plant-growth promotion in the presence of pathogens. *Plant Disease, 102*, 67–72. https://doi.org/10.1094/PDIS-04-17-0478-RE

Liu, Y., Lu, S., Liu, K., Wang, S., Huang, L., & Guo, L. (2019). Proteomics: A powerful tool to study plant responses to biotic stress. *Plant Methods, 15*, 135. https://doi.org/10.1186/s13007-019-0515-8

López-Mondéjar, R., Kostovčík, M., Lladó, S., Carro, L., & García-Fraile, P. (2017). Exploring the plant microbiome through multi-omics approaches. In V. Kumar, M. Kumar, S. Sharma, & R. Prasad (Eds.), *Probiotics in agroecosystem* (pp. 233–268). Springer.

Lu, W., Bennett, B. D., & Rabinowitz, J. D. (2008). Analytical strategies for LC–MS-based targeted metabolomics. *Journal of Chromatography B, 871*, 236–242. https://doi.org/10.1016/j.jchromb.2008.04.031

Lucas, J. A., García-Cristobal, J., Bonilla, A., Ramos, B., & Gutierrez-Mañero, J. (2014). Beneficial rhizobacteria from rice rhizosphere confers high protection against biotic and abiotic stress inducing systemic resistance in rice seedlings. *Plant Physiology and Biochemistry, 82*, 44–53. https://doi.org/10.1016/j.plaphy.2014.05.007

Lundberg, E., & Borner, G. H. H. (2019). Spatial proteomics: A powerful discovery tool for cell biology. *Nature Reviews. Molecular Cell Biology, 20*, 285–302. https://doi.org/10.1038/s41580-018-0094-y

Lyu, D., Backer, R., Subramanian, S., & Smith, D. L. (2020). Phytomicrobiome coordination signals hold potential for climate change-resilient agriculture. *Frontiers in Plant Science, 11*.

Mahlein, A.-K. (2016). Plant disease detection by imaging sensors–parallels and specific demands for precision agriculture and plant phenotyping. *Plant Disease, 100*, 241–251. https://doi.org/10.1094/PDIS-03-15-0340-FE

Mahlein, A.-K., Kuska, M. T., Behmann, J., Polder, G., & Walter, A. (2018). Hyperspectral sensors and imaging Technologies in Phytopathology: State of the art. *Annual Review of Phytopathology, 56*, 535–558. https://doi.org/10.1146/annurev-phyto-080417-050100

Mahlein, A.-K., Kuska, M. T., Thomas, S., Wahabzada, M., Behmann, J., Rascher, U., & Kersting, K. (2019). Quantitative and qualitative phenotyping of disease resistance of crops by hyperspectral sensors: Seamless interlocking of phytopathology, sensors, and machine learning is needed! *Current Opinion in Plant Biology, 50*, 156–162. https://doi.org/10.1016/j.pbi.2019.06.007

Maignien, L., DeForce, E. A., Chafee, M. E., Eren, A. M., & Simmons, S. L. (2014). Ecological succession and stochastic variation in the assembly of Arabidopsis thaliana Phyllosphere communities. *MBio, 5*. https://doi.org/10.1128/mbio.00682-13

Majchrzak, T., Wojnowski, W., Rutkowska, M., & Wasik, A. (2020). Real-time volatilomics: A novel approach for analyzing biological samples. *Trends in Plant Science, 25*, 302–312. https://doi.org/10.1016/j.tplants.2019.12.005

Majeed, A., Muhammad, Z., Islam, S., Ullah, Z., & Ullah, R. (2017). Cyanobacterial application as bio-fertilizers in rice fields: Role in growth promotion and crop productivity. *PSM Microbiology, 2*, 47–50.

Martins, R. C., Magalhães, S., Jorge, P., Barroso, T., & Santos, F. (2019). Metbots: Metabolomics robots for precision viticulture. In P. Moura Oliveira, P. Novais, & L. P. Reis (Eds.), *Progress in artificial intelligence* (pp. 156–166). Springer International Publishing.

McDonald, D., Price, M. N., Goodrich, J., Nawrocki, E. P., DeSantis, T. Z., Probst, A., Andersen, G. L., Knight, R., & Hugenholtz, P. (2012). An improved Greengenes taxonomy with explicit ranks for ecological and evolutionary analyses of bacteria and archaea. *The ISME Journal, 6*, 610–618. https://doi.org/10.1038/ismej.2011.139

Meacham-Hensold, K., Montes, C. M., Wu, J., Guan, K., Fu, P., Ainsworth, E. A., Pederson, T., Moore, C. E., Brown, K. L., Raines, C., & Bernacchi, C. J. (2019). High-throughput field phenotyping using hyperspectral reflectance and partial least squares regression (PLSR) reveals genetic modifications to photosynthetic capacity. *Remote Sensing of Environment, 231*, 111176. https://doi.org/10.1016/j.rse.2019.04.029

Meacham-Hensold, K., Fu, P., Wu, J., Serbin, S., Montes, C. M., Ainsworth, E., Guan, K., Dracup, E., Pederson, T., Driever, S., & Bernacchi, C. (2020). Plot-level rapid screening for photosynthetic parameters using proximal hyperspectral imaging. *Journal of Experimental Botany, 71*, 2312–2328. https://doi.org/10.1093/jxb/eraa068

Mendes, R., Kruijt, M., de Bruijn, I., Dekkers, E., van der Voort, M., Schneider, J. H. M., Piceno, Y. M., DeSantis, T. Z., Andersen, G. L., Bakker, P. A. H. M., & Raaijmakers, J. M. (2011). Deciphering the rhizosphere microbiome for disease-suppressive bacteria. *Science, 332*, 1097–1100. https://doi.org/10.1126/science.1203980

Midzi, J., Jeffery, D. W., Baumann, U., Rogiers, S., Tyerman, S. D., & Pagay, V. (2022). Stress-induced volatile emissions and Signalling in inter-plant communication. *Plants, 11*, 2566. https://doi.org/10.3390/plants11192566

Möller, M., & Stukenbrock, E. H. (2017). Evolution and genome architecture in fungal plant pathogens. *Nature Reviews. Microbiology, 15*, 756–771. https://doi.org/10.1038/nrmicro.2017.76

Muller, A., Schader, C., El-Hage Scialabba, N., Brüggemann, J., Isensee, A., Erb, K.-H., Smith, P., Klocke, P., Leiber, F., Stolze, M., & Niggli, U. (2017). Strategies for feeding the world more sustainably with organic agriculture. *Nature Communications, 8*, 1290. https://doi.org/10.1038/s41467-017-01410-w

Nagarajan, S., Seibold, G., Kranz, J., Saari, E. E., & Joshi, L. M. (1984). Monitoring wheat rust epidemics with the Landsat-2 satellite. *Phytopathology, 74*, 585–587.

Nawrocka, J., Małolepsza, U., Szymczak, K., & Szczech, M. (2018). Involvement of metabolic components, volatile compounds, PR proteins, and mechanical strengthening in multilayer protection of cucumber plants against Rhizoctonia solani activated by Trichoderma atroviride TRS25. *Protoplasma, 255*, 359–373. https://doi.org/10.1007/s00709-017-1157-1

Ni, J., Tian, Y., Yao, X., Zhu, Y., & Cao, W. (2011). Application of monitoring system about plant growth information based on spectroscopy technique. In *PIAGENG 2010: Photonics and imaging for agricultural engineering* (pp. 321–326). SPIE.

Nilsson, H. (1995). Remote sensing and image analysis in plant pathology. *Annual Review of Phytopathology, 33*, 489–528. https://doi.org/10.1146/annurev.py.33.090195.002421

Nilsson, R. H., Larsson, K.-H., Taylor, A. F. S., Bengtsson-Palme, J., Jeppesen, T. S., Schigel, D., Kennedy, P., Picard, K., Glöckner, F. O., Tedersoo, L., Saar, I., Kõljalg, U., & Abarenkov, K. (2019). The UNITE database for molecular identification of fungi: Handling dark taxa and parallel taxonomic classifications. *Nucleic Acids Research, 47*, D259–D264. https://doi.org/10.1093/nar/gky1022

Niu, D., Wang, X., Wang, Y., Song, X., Wang, J., Guo, J., & Zhao, H. (2016). *Bacillus cereus* AR156 activates PAMP-triggered immunity and induces a systemic acquired resistance through a *NPR1*-and SA-dependent signaling pathway. *Biochemical and Biophysical Research Communications, 469*, 120–125. https://doi.org/10.1016/j.bbrc.2015.11.081

Nobrega, M. A., & Pennacchio, L. A. (2004). Comparative genomic analysis as a tool for biological discovery. *The Journal of Physiology, 554*, 31–39. https://doi.org/10.1113/jphysiol.2003.050948

Nuccio, E. E., Starr, E., Karaoz, U., Brodie, E. L., Zhou, J., Tringe, S. G., et al. (2020). Niche differentiation is spatially and temporally regulated in the rhizosphere. *The ISME Journal, 14*, 999–1014.

Oksman-Caldentey, K.-M., & Saito, K. (2005). Integrating genomics and metabolomics for engineering plant metabolic pathways. *Current Opinion in Biotechnology, 16*, 174–179. https://doi.org/10.1016/j.copbio.2005.02.007

Oliver, D. J., Nikolau, B., & Wurtele, E. S. (2002). Functional genomics: High-throughput mRNA, protein, and metabolite analyses. *Metabolic Engineering, 4*, 98–106. https://doi.org/10.1006/mben.2001.0212

Oren, R., Schulze, E.-D., Matyssek, R., & Zimmermann, R. (1986). Estimating photosynthetic rate and annual carbon gain in conifers from specific leaf weight and leaf biomass. *Oecologia, 70*, 187–193. https://doi.org/10.1007/BF00379238

Pankow, S., Martínez-Bartolomé, S., Bamberger, C., & Yates, J. R. (2019). Understanding molecular mechanisms of disease through spatial proteomics. *Current Opinion in Chemical Biology, 48*, 19–25. https://doi.org/10.1016/j.cbpa.2018.09.016

Planchamp, C., Glauser, G., & Mauch-Mani, B. (2015). Root inoculation with pseudomonas putida KT2440 induces transcriptional and metabolic changes and systemic resistance in maize plants. *Frontiers in Plant Science, 5*.

Quast, C., Pruesse, E., Yilmaz, P., Gerken, J., Schweer, T., Yarza, P., Peplies, J., & Glöckner, F. O. (2013). The SILVA ribosomal RNA gene database project: Improved data processing and web-based tools. *Nucleic Acids Research, 41*, D590–D596. https://doi.org/10.1093/nar/gks1219

Rais, A., Jabeen, Z., Shair, F., Hafeez, F. Y., & Hassan, M. N. (2017). Bacillus spp., a bio-control agent enhances the activity of antioxidant defense enzymes in rice against Pyricularia oryzae. *PLoS One, 12*, e0187412. https://doi.org/10.1371/journal.pone.0187412

Ramlal, A., Rani, A., Nautiyal, A., Kalra, C., Kumari, R., Kumar, J., Veeranna, S., & Mishra, V. (2023). Importance of omics approaches in plant-microbe interaction for plant disease control. *Physiological and Molecular Plant Pathology, 128*, 102153. https://doi.org/10.1016/j.pmpp.2023.102153

Redford, A. J., Bowers, R. M., Knight, R., Linhart, Y., & Fierer, N. (2010). The ecology of the phyllosphere: Geographic and phylogenetic variability in the distribution of bacteria on tree leaves. *Environmental Microbiology, 12*, 2885–2893. https://doi.org/10.1111/j.1462-2920.2010.02258.x

Rinschen, M. M., Ivanisevic, J., Giera, M., & Siuzdak, G. (2019). Identification of bioactive metabolites using activity metabolomics. *Nature Reviews. Molecular Cell Biology, 20*, 353–367. https://doi.org/10.1038/s41580-019-0108-4

Roberts, L. D., Souza, A. L., Gerszten, R. E., & Clish, C. B. (2012). Targeted metabolomics. *Current Protocols in Molecular Biology, 98*, 30.2.1–30. 2.24. https://doi.org/10.1002/0471142727.mb3002s98

Rosenkranz, M., & Schnitzler, J.-P. (2016). Plant volatiles. In *Encyclopedia of life sciences* (pp. 1–9). Wiley.

Rumpf, T., Mahlein, A.-K., Steiner, U., Oerke, E.-C., Dehne, H.-W., & Plümer, L. (2010). Early detection and classification of plant diseases with support vector machines based on hyperspectral reflectance. *Computers and Electronics in Agriculture, 74*, 91–99. https://doi.org/10.1016/j.compag.2010.06.009

Sarkar, A. K., & Sadhukhan, S. (2022). Proteomics—A powerful tool for understanding saline stress response in germinating seed. In S. Roy, P. Mathur, A. P. Chakraborty, & S. P. Saha (Eds.), *Plant stress: Challenges and Management in the new Decade* (pp. 375–399). Springer International Publishing.

Sarkar, A. K., & Sadhukhan, S. (2023). Unearthing the alteration in plant volatiles induced by mycorrhizal fungi: A shield against plant pathogens. *Physiologia Plantarum, 175*, e13845. https://doi.org/10.1111/ppl.13845

Sauter, H., Lauer, M., & Fritsch, H. (1991). Metabolic profiling of plants. In *Synthesis and chemistry of agrochemicals II* (pp. 288–299). American Chemical Society.

Schrimpe-Rutledge, A. C., Codreanu, S. G., Sherrod, S. D., & McLean, J. A. (2016). Untargeted metabolomics strategies—Challenges and emerging directions. *Journal of the American Society for Mass Spectrometry, 27*, 1897–1905. https://doi.org/10.1007/s13361-016-1469-y

Seethapathy, P., Gurudevan, T., Subramanian, K., Paliyath, G., Jayasankar, S., & Kuppusamy, P. (2018). Volatile metabolites fingerprinting to discriminate the major post harvest diseases of mango caused by Colletotrichum gloeosporioides Penz. and Lasiodiplodia theobromae Pat. Annals of Phytomedicine: An. *International Journal, 6*. https://doi.org/10.21276/ap.2017.6.2.4

Selosse, M.-A., & Tacon, F. L. (1998). The land flora: A phototroph-fungus partnership? *Trends in Ecology & Evolution, 13*, 15–20. https://doi.org/10.1016/S0169-5347(97)01230-5

Serbin, S. P., Dillaway, D. N., Kruger, E. L., & Townsend, P. A. (2012). Leaf optical properties reflect variation in photosynthetic metabolism and its sensitivity to temperature. *Journal of Experimental Botany, 63*, 489–502. https://doi.org/10.1093/jxb/err294

Serbin, S. P., Wu, J., Ely, K. S., Kruger, E. L., Townsend, P. A., Meng, R., Wolfe, B. T., Chlus, A., Wang, Z., & Rogers, A. (2019). From the Arctic to the tropics: Multibiome prediction of leaf mass per area using leaf reflectance. *New Phytologist, 224*, 1557–1568. https://doi.org/10.1111/nph.16123

Shameer, S., & Prasad, T. N. V. K. V. (2018). Plant growth promoting rhizobacteria for sustainable agricultural practices with special reference to biotic and abiotic stresses. *Plant Growth Regulation, 84*, 603–615. https://doi.org/10.1007/s10725-017-0365-1

Sharma, C. K., Vishnoi, V. K., Dubey, R. C., & Maheshwari, D. K. (2018). A twin rhizospheric bacterial consortium induces systemic resistance to a phytopathogen *Macrophomina phaseolina* in mung bean. *Rhizosphere, 5*, 71–75. https://doi.org/10.1016/j.rhisph.2018.01.001

Sharma, A., Raina, M., Kumar, D., Singh, A., Chugh, S., Jain, S., Kumar, M., & Rustagi, A. (2022). Harnessing phytomicrobiome signals for phytopathogenic stress management. *Journal of Biosciences, 47*, 6. https://doi.org/10.1007/s12038-021-00240-9

Shefali, Sankhla, M. S., Kumar, R., Sonone, S.S. (2020) Phytomicrobiome Studies for Combating the Abiotic Stress.

Singh, M., Awasthi, A., Soni, S. K., Singh, R., Verma, R. K., & Kalra, A. (2015). Complementarity among plant growth promoting traits in rhizospheric bacterial communities promotes plant growth. *Scientific Reports, 5*, 15500. https://doi.org/10.1038/srep15500

Smith, D. L., Subramanian, S., Lamont, J. R., & Bywater-Ekegärd, M. (2015). Signaling in the phytomicrobiome: Breadth and potential. *Frontiers in Plant Science, 6*.

Solden, L., Lloyd, K., & Wrighton, K. (2016). The bright side of microbial dark matter: Lessons learned from the uncultivated majority. *Current Opinion in Microbiology, 31*, 217–226. https://doi.org/10.1016/j.mib.2016.04.020

Song, Y., Chen, D., Lu, K., Sun, Z., & Zeng, R. (2015). Enhanced tomato disease resistance primed by arbuscular mycorrhizal fungus. *Frontiers in Plant Science, 6*.

Staley, J. T. (2006). The bacterial species dilemma and the genomic–phylogenetic species concept. *Philosophical Transactions of the Royal Society B: Biological Sciences, 361*, 1899–1909. https://doi.org/10.1098/rstb.2006.1914

Stukenbrock, E. H., Bataillon, T., Dutheil, J. Y., Hansen, T. T., Li, R., Zala, M., McDonald, B. A., Wang, J., & Schierup, M. H. (2011). The making of a new pathogen: Insights from comparative population genomics of the domesticated wheat pathogen Mycosphaerella graminicola and its wild sister species. *Genome Research, 21*, 2157–2166. https://doi.org/10.1101/gr.118851.110

Sumner, L. W., Mendes, P., & Dixon, R. A. (2003). Plant metabolomics: Large-scale phytochemistry in the functional genomics era. *Phytochemistry, 62*, 817–836. https://doi.org/10.1016/S0031-9422(02)00708-2

Sundin, G. W., Wang, N., Charkowski, A. O., Castiblanco, L. F., Jia, H., & Zhao, Y. (2016). Perspectives on the transition from bacterial Phytopathogen genomics studies to applications enhancing disease management: From promise to practice. *Phytopathology, 106*, 1071–1082. https://doi.org/10.1094/PHYTO-03-16-0117-FI

Suzuki, N., Rivero, R. M., Shulaev, V., Blumwald, E., & Mittler, R. (2014). Abiotic and biotic stress combinations. *New Phytologist, 203*, 32–43. https://doi.org/10.1111/nph.12797

Thulin, S., Hill, M. J., Held, A., Jones, S., & Woodgate, P. (2014). Predicting levels of crude protein, digestibility, lignin and cellulose in temperate pastures using hyperspectral image data. *American Journal of Plant Sciences, 2014*, 997–1019. https://doi.org/10.4236/ajps.2014.57113

Tonelli, M. L., Magallanes-Noguera, C., & Fabra, A. (2017). Symbiotic performance and induction of systemic resistance against Cercospora sojina in soybean plants co-inoculated with bacillus sp. CHEP5 and Bradyrhizobium japonicum E109. *Archives of Microbiology, 199*, 1283–1291. https://doi.org/10.1007/s00203-017-1401-2

Turner, T. R., Ramakrishnan, K., Walshaw, J., Heavens, D., Alston, M., Swarbreck, D., Osbourn, A., Grant, A., & Poole, P. S. (2013). Comparative metatranscriptomics reveals kingdom level changes in the rhizosphere microbiome of plants. *The ISME Journal, 7*, 2248–2258. https://doi.org/10.1038/ismej.2013.119

Uroz, S., Courty, P. E., & Oger, P. (2019). Plant symbionts are engineers of the plant-associated microbiome. *Trends in Plant Science, 24*, 905–916. https://doi.org/10.1016/j.tplants.2019.06.008

Ustin, S. L., Roberts, D. A., Gamon, J. A., Asner, G. P., & Green, R. O. (2004). Using imaging spectroscopy to study ecosystem processes and properties. *Bioscience, 54*, 523–534. https://doi.org/10.1641/0006-3568(2004)054[0523:UISTSE]2.0.CO;2

van Dam, N. M., & Bouwmeester, H. J. (2016). Metabolomics in the rhizosphere: Tapping into belowground chemical communication. *Trends in Plant Science, 21*, 256–265. https://doi.org/10.1016/j.tplants.2016.01.008

Vandenkoornhuyse, P., Quaiser, A., Duhamel, M., Le Van, A., & Dufresne, A. (2015). The importance of the microbiome of the plant holobiont. *New Phytologist, 206*, 1196–1206. https://doi.org/10.1111/nph.13312

Velivelli, S. L., Lojan, P., Cranenbrouck, S., de Boulois, H. D., Suarez, J. P., Declerck, S., Franco, J., & Prestwich, B. D. (2015). The induction of ethylene response factor 3 (ERF3) in potato as a result of co-inoculation with pseudomonas sp. R41805 and Rhizophagus irregularis MUCL 41833–A possible role in plant defense. *Plant Signaling & Behavior, 10*, e988076. https://doi.org/10.4161/15592324.2014.988076

Vergara Díaz, O. (2019). *High-throughput field phenotyping in cereals and implications in plant ecophysiology*. Ph.D. Thesis,. Universitat de Barcelona.

Vergara-Diaz, O., Vatter, T., Kefauver, S. C., Obata, T., Fernie, A. R., & Araus, J. L. (2020). Assessing durum wheat ear and leaf metabolomes in the field through hyperspectral data. *The Plant Journal, 102*, 615–630. https://doi.org/10.1111/tpj.14636

Ward, J. L., Baker, J. M., & Beale, M. H. (2007). Recent applications of NMR spectroscopy in plant metabolomics. *The FEBS Journal, 274*, 1126–1131. https://doi.org/10.1111/j.1742-4658.2007.05675.x

Wu, D., Feng, L., Zhang, C., & He, Y. (2008). Early detection of *B*otrytis *C*inerea on eggplant leaves based on visible and near-infrared spectroscopy. *Early detection of botrytis cinerea on eggplant leaves based on visible and near-infrared spectroscopy, 51*, 1133–1139.

Xie, C., Yang, C., & He, Y. (2017). Hyperspectral imaging for classification of healthy and gray mold diseased tomato leaves with different infection severities. *Computers and Electronics in Agriculture, 135*, 154–162. https://doi.org/10.1016/j.compag.2016.12.015

Yandell, M., & Ence, D. (2012). A beginner's guide to eukaryotic genome annotation. *Nature Reviews. Genetics, 13*, 329–342. https://doi.org/10.1038/nrg3174

Yendrek, C. R., Tomaz, T., Montes, C. M., Cao, Y., Morse, A. M., Brown, P. J., McIntyre, L. M., Leakey, A. D. B., & Ainsworth, E. A. (2017). High-throughput phenotyping of maize leaf physiological and biochemical traits using hyperspectral reflectance. *Plant Physiology, 173*, 614–626. https://doi.org/10.1104/pp.16.01447

Zarco. T. P. J., Camino, C., Beck, P., Calderon, R., Hornero, A., Hernandez-Clemente, R., Kattenborn, T., Montes-Borrego, M., Susca, L., Morelli, M., Gonzalez-Dugo, V., North, P. R. J., Landa, B., Boscia, D., Saponari, M., Navas-Cortes, J. A. (2018) Previsual symptoms of Xylella fastidiosa infection revealed in spectral plant-trait alterations. JRC Publications Repository. https://publications.jrc.ec.europa.eu/repository/handle/JRC107247. Accessed 18 Feb 2024.

Zhai, S., Chen, H., Ding, C., & Zhao, X. (2013). Double-negative acoustic metamaterial based on meta-molecule. *Journal of Physics D, Applied Physics, 46*, 475105. https://doi.org/10.1088/0022-3727/46/47/475105

Zhang, Y., Xu, J., Wang, E., & Wang, N. (2020). Mechanisms underlying the rhizosphere-to-rhizoplane enrichment of Cellvibrio unveiled by genome-centric metagenomics and Metatranscriptomics. *Microorganisms, 8*, 583. https://doi.org/10.3390/microorganisms8040583

FSC
www.fsc.org
MIX
Papier | Fördert gute Waldnutzung
FSC® C083411

Kolibri 360 GmbH
Ferdinand-Jühlke-Straße 7
99095 Erfurt, Deutschland
produktsicherheit@kolibri360.de